Smart Power Distribution Systems

Smart Power Distribution Systems

Control, Communication, and Optimization

Edited by

Qiang Yang

Ting Yang

Wei Li

Academic Press is an imprint of Elsevier
125 London Wall, London EC2Y 5AS, United Kingdom
525 B Street, Suite 1650, San Diego, CA 92101, United States
50 Hampshire Street, 5th Floor, Cambridge, MA 02139, United States
The Boulevard, Langford Lane, Kidlington, Oxford OX5 1GB, United Kingdom

Library of Congress Cataloging-in-Publication Data
A catalog record for this book is available from the Library of Congress

British Library Cataloguing-in-Publication Data
A catalogue record for this book is available from the British Library

ISBN: 978-0-12-812154-2

For information on all Academic Press publications
visit our website at https://www.elsevier.com/books-and-journals

Publisher: Joe Hayton
Acquisition Editor: Maria Convey
Editorial Project Manager: Mariana L. Kuhl
Production Project Manager: R.Vijay Bharath
Cover Designer: Christian J. Bilbow

Typeset by SPi Global, India

Contents

List of contributors

Sandhya Armoogum Department of Industrial Systems Engineering, School of Innovative Technologies and Engineering, University of Technology Mauritius, La Tour Koenig, Pointe-aux-Sables, Mauritius

Birgitte Bak-Jensen Department of Energy Technology, Aalborg University, Aalborg, Denmark

Vandana Bassoo Department of Electrical and Electronic Engineering, Faculty of Engineering, University of Mauritius, Reduit, Mauritius

Bishnu P. Bhattarai Department of Power and Energy System, Idaho National Laboratory, Idaho Falls, ID, United States

Rajeev Kumar Chauhan School of Computing and Electrical Engineering, Indian Institute of Technology, Mandi, India

Kalpana Chauhan Department of Electrical and Electronics Engineering, Galgotias College of Engineering and Technology, Greater Noida, India

Weirong Chen School of Electrical Engineering, Southwest Jiaotong University, Chengdu, China

Bo Chen Department of Electrical Engineering, Guangxi University, Nanning, People's Republic of China

B. Chitti Babu University of Nottingham Malaysia Campus, Semenyih, Malaysia

S.S. Choi School of Electrical, Electronic and Computer Engineering, University of Western Australia, Perth, WA, Australia

Sanchari Deb Centre for Energy, Indian Institute of Technology, Guwahati, India

Sanjoy Debbarma Electrical Engineering Department, NIT Meghalaya, Shillong, India

Ali Ehsan College of Electrical Engineering, Zhejiang University, Hangzhou, People's Republic of China

Xinli Fang PowerChina Huadong Engineering Corporation Limited; Hangzhou Huachen Electric Power Control Co. LTD., Hangzhou, People's Republic of China

Deqiang Gan College of Electrical Engineering, Zhejiang University, Hangzhou, China

Yajing Gao School of Electrical and Electronic Engineering, North China Electric Power University, Baoding, China

Ying Han School of Electrical Engineering, Southwest Jiaotong University, Chengdu, China

Xiaoqing Han Shanxi Key Laboratory of Power System Operation and Control, Taiyuan University of Technology, Taiyuan, China

Yujin Huang Department of Electrical Engineering, Guangxi University, Nanning, People's Republic of China

Le Jiang College of Electrical Engineering, Zhejiang University, Hangzhou, People's Republic of China

Karuna Kalita Department of Mechanical Engineering, Indian Institute of Technology, Guwahati, India

Mitja Kolenc ELES, d.o.o., Ljubljana, Slovenia

Qi Li School of Electrical Engineering, Southwest Jiaotong University, Chengdu, China

Jinghua Li Department of Electrical Engineering, Guangxi University, Nanning, People's Republic of China

Bo Lu Department of Electrical Engineering, Guangxi University, Nanning, People's Republic of China

Jin Ma School of Electrical and Information Engineering, The University of Sydney, Sydney, NSW, Australia

Pinakeswar Mahanta Department of Mechanical Engineering, Indian Institute of Technology, Guwahati, India

Yuhong Mo Department of Electrical Engineering, Guangxi University, Nanning, People's Republic of China

Qitian Mu School of Electrical and Electronic Engineering, North China Electric Power University, Baoding, China

Kurt S. Myers Department of Power and Energy System, Idaho National Laboratory, Idaho Falls, ID, United States

Bonu Ramesh Naidu Department of Electrical Engineering, Indian Institute of Technology Kharagpur, Kharagpur, India

Gayadhar Panda Department of Electrical Engineering, National Institute of Technology Meghalaya, Shillong, India

Feng Qiao School of Electrical and Information Engineering, The University of Sydney, Sydney, NSW, Australia

M.M. Rana College of Science and Engineering, James Cook University, Townsville, QLD, Australia

Mohammad Seydali Seyf Abad School of Electrical and Information Engineering, The University of Sydney, Sydney, NSW, Australia

Rituraj Shrivastwa Department of Electrical Power Engineering (IEE), Grenoble Institute of Technology, Grenoble, France

Nermin Suljanović Milan Vidmar Electric Power Research Institute, Ljubljana, Slovenia

Robert J. Turk Department of Power and Energy System, Idaho National Laboratory, Idaho Falls, ID, United States

Zhen Wang College of Electrical Engineering, Zhejiang University, Hangzhou, China

Eric Wang College of Science and Engineering, James Cook University, Townsville, QLD, Australia

Zhibang Wang Department of Electrical Engineering, Guangxi University, Nanning, People's Republic of China

Shanyang Wei Department of Electrical Engineering, Guangxi University, Nanning, People's Republic of China

Kit Po Wong School of Electrical, Electronic and Computer Engineering, University of Western Australia, Perth, WA, Australia

Wei Xiang College of Science and Engineering, James Cook University, Townsville, QLD, Australia

Zhengqing Yang College of Electrical Engineering, Zhejiang University, Hangzhou, China

Qiang Yang College of Electrical Engineering, Zhejiang University, Hangzhou, People's Republic of China

Hanqing Yang School of Electrical Engineering, Southwest Jiaotong University, Chengdu, China

Ting Yang Key Laboratory of Smart Grid of Ministry of Education, School of Electrical and Information Engineering, Tianjin University, Tianjin, China

Matej Zajc Faculty of Electrical Engineering, University of Ljubljana, Ljubljana, Slovenia

Jiasheng Zhou Department of Electrical Engineering, Guangxi University, Nanning, People's Republic of China

Xiaojie Zhou School of Electrical and Electronic Engineering, North China Electric Power University, Baoding, China

Jing Zhu Maintenance Branch Company of State Grid Fujian Electric Power Co., Ltd, Xiamen, China

Yi Zong Center for Electric Power and Energy (CEE), Department of Electrical Engineering, Technical University of Denmark, Roskilde, Denmark

About the editors

Dr. Qiang Yang holds a BS degree (first class honors) in electrical engineering and received an MSc (with distinction) and a PhD degree both in electronic engineering and computer science from Queen Mary College, University of London, London, UK, in 2003 and 2007, respectively. He has worked as a postdoctoral research associate at the Department of Electrical and Electronic Engineering, Imperial College London, UK, from 2007 to 2010 and involved in a number of high-profile UK EPSRC and European IST research projects. He had visited University of British Columbia and University of Victoria Canada as a visiting scholar in 2015 and 2016. Currently, he is an associate professor at College of Electrical Engineering, Zhejiang University, China, and has published more than 150 technical papers, applied for more than 50 national patents, and coauthored 2 books and 10 book chapters. He has received more than 20 research grants in the past 5 years, including the National Key Research and Development Program of China, the National Natural Science Foundation of China, and the National High-Tech Research and Development Program of China (863 Program). His research interests over the years include communication networks, smart energy systems, and large-scale complex network modeling, control, and optimization. He is the senior member of IEEE, member of IET, and IEICE as well as the senior member of China Computer Federation (CCF).

Prof. Ting Yang is currently a chair professor of Electrical Theory and Advanced Technology, at the School of Electrical Engineering and Automation, Tianjin University, China. He was the cooperative research staff of Imperial College London (2008); visiting professor of University of Sydney, Australia (2015). Prof. Yang is the winner of the "New Century Excellent Talents in University Award" from Chinese Ministry of Education. He is the leader of tens of research grant projects, including the International S&T Cooperation Program of China, the National High-Tech Research and Development Program of China (863 Program), the National Natural Science Foundation of China, and so on. Prof. Yang is the chairman of two workshops of IEEE International Conference, and the editor in chief of one of the special issues of the International Journal of Distributed Sensor Network (DSN). He has authored/ coauthored four books, more than 100 publications in internationally refereed journals and conferences. Prof. Yang is a senior member of the Chinese Institute of Electronic, the fellow of Circuit and System Committee, the fellow of Theory and Advanced Technology of Electrical Engineering, and the member of International Society for Industry and Applied Mathematics. His research fields include smart energy systems, artificial intelligence, and internet of things.

Dr. Wei Li has received his PhD from School of Information Technologies at The University of Sydney. He is currently a research fellow of Centre for Distributed and High Performance Computing, and School of Information Technologies at The University of Sydney. His research is supported by Early Career Researcher (ECR) funding scheme and Clean Energy and Intelligent Networks Cluster funding scheme, The University of Sydney. He is the recipient of four IEEE or ACM conference best paper awards. His research interests include Internet of things, edge computing, energy efficient, task scheduling, and optimization. He is a senior member of IEEE and a member of ACM.

Preface

In recent decades, the electric power grid is experiencing a fundamental transformation. Smart grids are considered as one of the biggest technological revolutions since the advent of Internet, as they have the potential to reduce carbon dioxide emissions, increase the reliability of electricity supply, and increase the efficiency of our energy infrastructures. In particular, the electrical network is integrated with the information and communication network in order to improve the stability, efficiency, and robustness. The control, communication, and optimization technologies explain how diverse technologies play hand in hand in building and maintaining smart grids around the globe. This book aims to highlight the recent advances in the control, communication, and optimization aspects of smart distribution grids, provides incredible insight into power system control, sensing and communication, and optimization technologies, and points out the potentials for new technologies and markets.

First, the current resurgence of interest in the use of renewable energy is driven by the need to reduce the high environmental impact of fossil-based energy systems. Smart distribution grids promise to facilitate the integration of distributed renewable sources and provide other benefits as well. Advanced control technology is a key enabling technology for the deployment of renewable energy systems including solar, wind, and other small-scale renewable sources in a reliable and friendly fashion. Second, the underlying communication network of smart distribution grids is a critical enabler of new functions, such as demand response, dynamic pricing, robust distributed generation system, and so forth. It is expected to support the functionalities of enhancing energy savings, cost reductions, and increased reliability and security. In fact, the smart grid sets a novel context for addressing communication problems and devising innovative solutions. Finally, the optimization-based techniques have received a lot of attention in overcoming the outstanding challenges of optimal planning and operation of the smart power distribution systems.

This book is an excellent reference for researchers and postgraduate students working in the area of smart power distribution networks. It also targets professionals interested in gaining deeper knowledge and technical challenges of the smart distribution grids. This book is mainly for readers who have a good knowledge of electrical engineering. We also tried to include sufficient details and provide the necessary background information in each chapter to help the readers to easily understand the content. We hope the readers will enjoy reading this book.

Qiang Yang
College of Electrical Engineering, Zhejiang University,
Hangzhou, China

Acknowledgments

We would like to express our gratitude to everyone who participated in this project and made this book a reality. In particular, we would like to acknowledge the hard work of the authors and their patience during the revisions of their chapters.

We would also like to acknowledge the outstanding comments of the reviewers, which enabled us to select these chapters out of the many we received and improved their quality. Some of the authors also served as referees and hence their double task is highly appreciated.

We also thank our family members for realizing the importance of this project and their consistent support throughout the project. Special thanks to Angelia for her continuous love, support, and encouragement over the years.

The editors hope to acknowledge the following grants for supporting their research work: the National Key Research and Development Program of China (Basic Research Class 2017YFB0903000)-Basic Theories and Methods of Analysis and Control of the Cyber Physical Systems for Power Grid, the Natural Science Foundation of China (51777183, 61571324, 51407146, 61473238, 51607068, 51377027), the Natural Science Foundation of Zhejiang Province (LZ15E070001) and the Natural Science Foundation of Tianjin (16JCZDJC30900).

Lastly, we are very grateful to the editorial team at Elsevier Press for their support through the stages of this project. We enjoyed working with Mariana Kühl Leme, who was involved in all the phases, from the time this project was just an idea through the writing and editing of the chapters, and then during the production process. We would also like to thank R.Vijay bharath, from Elsevier, for managing the production process of this book.

Qiang Yang
Ting Yang
Wei Li

Organization of this book

There have been quite a few books out related to the smart grid. However, most of the books attempt to cover the concepts and technologies in the scope of very broad area: smart grid, covering the power generation, transmission, and distribution networks. This book aims to restrict the view to the control, communication, and optimization of smart active power distribution networks with integration of renewable distributed generations. Also, some of them lack comprehensive analysis of the control and optimization-related models and algorithms, besides the communication and data traffic models and analysis are not discussed in details.

This book aims to present the latest technology advances in electric power distribution networks, with a particular focus on the scientific innovations of the methodologies, approaches, and algorithms in enabling efficient and secure operation of smart distribution networks.

This book is divided into three parts, each of which is devoted to a distinctive area.

Part One: Modeling and control of smart power distribution network

Part I of this book aims to focus on theoretical, experimental, and proof-of-concept results of the modeling and control methodologies in the context of active distribution networks.

Chapter 1 presents an overview of grid-integration codes on wind turbine generator (WTG's) frequency regulation in several typical countries. The typical WTG frequency control strategies, such as inertial emulation, de-loading control, overproduction, and droop control are further investigated in detail to reflect the development of related WTG technologies.

Chapter 2 investigates one of the active power control technologies, WTG de-loading control, which is briefly reviewed first. A bi-level optimization model of two-stage reserve scheduling problem considering WTG integration is then proposed to evaluate the benefits of WTG de-loading control on system reserve scheduling.

Chapter 3 presents the control and energy management in a grid-connected DC microgrid comprising solar PV array, battery, and supercapacitor banks in order to maximize the synchronized grid integration of the renewable energy along with the objective of stabilizing the DC microgrid.

Chapter 4 studies the system structure and the math model of each DC MG subsystem. Also, a DC microgrid experiment platform is set up to verify its control ability and energy management performance.

Chapter 5 incorporates the VSC-based MG into the voltage control scheme of the distribution network by implementing a two-layer control structure. With the distribution network central controller at the upper layer, the tap position of on-load tap changer (OLTC), the position of shunt capacitors (SCs), and reactive power injection from MGs can be efficiently coordinated.

Chapter 6 proposes a discrete-time linear-quadratic Gaussian (LQG) controller to stabilize the microgrid states under fading channel condition. Numerical studies have confirmed the effectiveness of using smaller number of step sizes and fading parameters to stabilize the distributed energy resource (DER) states.

Chapter 7 investigates the optimal coordinated operation of multiple autonomous MGs and shows the potential technical benefits. The proposed solution identifies the optimal network topologies and allocates the critical loads (CLs) to appropriate DGs based on the minimum spanning tree (MST) algorithm with power loss and reliability considerations.

Part Two: ICT technologies for smart power distribution networks

Part II of this book focuses on the advanced information and communication technologies, high-performance computing, and cyber security issues in the smart distribution grids.

Chapter 8 identifies different security vulnerabilities in smart grids. The impact of consumer data privacy and confidentiality breach are discussed and existing techniques as proposed in the literature to protect the privacy of customer information in a smart grid are presented.

Chapter 9 studies the structure and categories of microgrid communication system, and overviews the application of microgrid communication system integrating consensus algorithm. In addition, a distributed hierarchical control method is established based on the communication system to balance the battery state of charge in the decentralized battery energy storage system.

Chapter 10 begins with a brief introduction to the basic concepts of smart grids and active distribution networks, followed by an introduction to the areas of active grid that require information and communication technology (ICT) technologies, the power system communication standard, and the supervisory control and data acquisition (SCADA) systems. In addition, the cyber-security issues are also discussed.

Chapter 11 investigates the communication system architecture of VPPs, giving an overview of current communication technologies and communication protocols. The study focused on the downstream communication between the virtual power plant and distributed energy resources, and upstream communication between the virtual power plant, transmission system operator, distribution system operator, electricity market, and retailers.

Chapter 12 presents virtual synchronous power concept (SPC)-based high-voltage DC transmission (HVDC) systems for emulating the virtual inertia in order to regulate grid frequency. This work also considers mobile electric vehicles (EVs) network as energy storage with smart V2G algorithm taking into consideration future driving demand of EV owners.

Chapter 13 discusses the current challenges and opportunities of internet of things (IoT)-enabled smart energy systems from a number of aspects. Existing approaches and recent solutions with respect to domestic demand response, IoT cyber security, and modeling and simulation challenges faced by current smart grid are provided.

Chapter 14 proposes an H-infinity-based microgrid state estimation algorithm. First of all, the renewable microgrid is represented by the state-space framework. The IoT-based smart sensors are used to obtain the system measurements. The energy management system adopts the H-infinity-based state estimation algorithm where it is not required to know the exact noise statistics.

Part Three: Optimization models and methods in smart power distribution networks

Part III of this book aims to focus on the optimization models, methodologies, and techniques of smart distribution network planning and operation management.

Chapter 15 presents an intelligent control strategy for photovoltaic (PV) and multi-battery bank for a DC microgrid in grid connected as well as isolated mode.

Chapter 16 introduces scenario-based methods for tackling the uncertainties of renewable generation in the electricity distribution network planning problem and two scenario-generation methods are adopted.

Chapter 17 proposes two approaches for network operation in consideration of renewable generation uncertainties using probabilistic-based and scenarios-based approaches.

Chapter 18 proposes a method to access the wind power accommodation by considering adequacy indexes aiming at the statistical characteristics of wind power. The optimal planning of wind power capacity and energy storage capacity is addressed based on the bilinear interpolation theory.

Chapter 19 explores the optimal energy dispatch problem in the scope of residential community with penetration of renewable DGs and energy storage in the presence of real-time pricing. An efficient algorithmic solution is presented and implemented at two levels: optimal control within individual households and energy trading among neighboring households.

Chapter 20 carries out the evaluation on the short-term power supply capacity of active distribution system based on multiple scenarios in order to fully consider the operational uncertainties.

Chapter 21 proposes an integrated multi-time-scale energy management approach for active distribution networks. The suggested solution can not only maximize the deployment of flexibility from spatially distributed resources, but also enable single flexible resource to provide multiple grid support functionalities.

Chapter 22 investigates the impact of EV charging station load on different operational parameters of the distribution network. The method of optimal placement of the EV charging stations in the distribution network is also studied considering different operating network parameters.

Chapter 23 studies the relaxation and linearization methods for the hosting capacity assessment in power distribution systems.

Qiang Yang
College of Electrical Engineering, Zhejiang University, Hangzhou, China

Part One

Modeling and control of smart power distribution network (control aspect)

An overview of codes and control strategies for frequency regulation in wind power generation

Zhen Wang, Kit Po Wong†, S.S. Choi†, Deqiang Gan*, Yi Zong‡*
*College of Electrical Engineering, Zhejiang University, Hangzhou, China, †School of Electrical, Electronic and Computer Engineering, University of Western Australia, Perth, WA, Australia, ‡Center for Electric Power and Energy (CEE), Department of Electrical Engineering, Technical University of Denmark, Roskilde, Denmark

1.1 Introduction

In recent years, significant increase of wind power generation has emerged for the purpose of releasing the pressure of energy shortage and low-carbon power supply. However, with a large amount of wind power integrated, the daily operation of conventional power systems can be inevitably affected (Ummels et al., 2007; Banakar et al., 2008). For example, more synchronous reserve capacities are desired for frequency regulation because of the intermittent, volatile, and antipeaking characteristics of wind power (Ernst et al., 2007).

Those variable-speed wind turbine generators (VSWTGs) such as double-fed induction generators (DFIGs) and permanent magnet synchronous generators (PMSGs), have gained high market share during the past few years due to their high efficiency and easy-controllability performance over a wide range of wind speeds. Generally, typical VSWTG is connected with a grid through power electronic converters. Therefore, its rotational speed is decoupled with system frequency and it cannot provide any rotational inertia to the main power grid (Mullane and O'Malley, 2005; Holdsworth et al., 2004). As a result, the system equivalent inertia will greatly decrease when large numbers of VSWTGs are introduced to replace synchronous generators (SGs). On the one hand, a low inertia system becomes more prone to large frequency oscillation when any serious disturbance occurs (Lalor et al., 2005). On the other hand, as most VSWTGs are designed to operate at the maximum power point tracking (MPPT) mode and they usually have no frequency regulation ability, in consequence the frequency regulation burden of conventional SGs will be aggravated; even the risk of system frequency collapse is thus exposed (Lalor et al., 2005; Conroy and Watson, 2008). Therefore, when large-scale wind power is integrated into a power grid, more challenges regarding frequency regulation and system reserve schedule and dispatch will emerge for the independent system operator (ISO).

Compared with conventional SG, WTG has its own characteristics regarding frequency control: (1) WTG usually has a faster frequency response than conventional

Smart Power Distribution Systems. https://doi.org/10.1016/B978-0-12-812154-2.00001-8

units, but its response capability is greatly limited by real wind condition; (2) WTG designed as other control scheme may exhibit quite different characteristics. Therefore, it's very important to investigate WTGs' frequency regulation ability according to their control schemes.

Only years ago, most WTGs were still regarded as unscheduled units due to their intermitted power output characteristics and were exempted from system frequency regulation. With the development of control technologies and increasing pressure of frequency regulation, WTGs now can participate in different levels of frequency control events. In fact, in many European countries such as Denmark and Germany, there exist corresponding operation codes (Energinet, 2004; Tennet TSO GmbH, 2012)· As a result, the potentials of frequency support function by WTGs are gradually attracting attention from the power industry as well as academia (Sun et al., 2010).

The purpose of this chapter is to investigate in depth and summarize the development of frequency regulation issues in wind power generation. The remainder of the chapter is organized as follows: in Section 1.2, existing grid codes of frequency regulation for WTG in several countries are investigated and compared. A three-level hierarchical dispatch framework is introduced in Section 1.3. System-level and plant-level control schemes are presented in Sections 1.4 and 1.5, respectively. Section 1.6 introduces four types of control schemes for WTGs to provide frequency support and their control mechanisms and characteristics are elaborated. In addition, future trends on a three-level frequency regulation framework are discussed in Section 1.7. Section 1.8 gives the conclusions.

1.2 Grid codes on frequency regulation

After dozens of years' development, several representative countries have established frequency regulation rules on wind farms in various grid codes, which are summarized in Fig. 1.1, including representative countries such as Germany (Tennet TSO GmbH, 2012), South Africa (Eskom System Operation and Planning, 2012), England and Scotland (UK National Grid, 2009), Sweden (Nordel, 2007), Ireland (Irish EirGrid, 2015), Denmark (Energinet, 2004), USA ERCOT (USA ERCOT, 2013) and Canada (Hydro-Québec TransÉnergie, 2009), China (GB Institute, 2011). It can be seen that in most grid codes the regulated system frequency can be divided into five bands (based on a 50 Hz rate frequency):

- serious supersynchronous band (SSUP, >52.0 Hz)
- medium supersynchronous band (MSUP, [51.0,52.0]Hz)
- synchronous band (SYN, [49.0,51.0]Hz)
- medium subsynchronous band (MSUB, [47.0,49.0]Hz)
- severe subsynchronous band (SSUB, <47.0 Hz)

It is very clear that in the SYN range, in most countries above WTGs will maintain MPPT operation (full production). When the system frequency increases and enters into the MSUP range, WTGs need to actively reduce their output in order to prevent the system from overfrequency. On the other hand, when the system frequency is in

f(Hz)	Germany	South Africa	England and Scotland	Sweden	Ireland
(SSUP)	Disconnection (> 52.0 Hz) or absorb active power				Disconnection (> 52.5 Hz)
52.0 – (MSUP) – 51.0	Reduce power with a ramp rate 10% P_{MPPT}/min (51.0–52.0) Hz	Reduce power at least with a ramp rate of 1% P_{rate}/s (51.0–52.0) Hz	(1) Reduce power 2% P_{MPPT}/0.1 Hz (50.5–52.0) Hz (2) Online>90min in (51–51.5) Hz	(1) Reduce power with a ramp rate of 10% P_{rate}/min (2) Reduce from 100% to 20% of P_{rate} in less than 5s (51.0–52.0) Hz	Reduce power till 30% P_{MPPT} according to P-f curves (51.0–52.5) Hz
(SYN) – 50.0 – (SYN)	100% P_{MPPT} (49.0–51.0) Hz	95% P_{MPPT} (49.5–51.0) Hz	100% P_{MPPT} (49.5–50.5) Hz	100% P_{MPPT} (49.0–51.0) Hz	100% P_{MPPT} (49.5–51.0) Hz
49.0 – (MSUB) – 47.0	Online>10–30 min (47.0–49.0) Hz	(1) Ramp rate< 5% P_{MPPT} (2) Actual operating points must be agreed by the ISO (47.0–49.5) Hz	(1) Ramp rate < 5% P_{MPPT}/2.5 Hz (2) ΔP<60% P_{MPPT} (49.5–50.0) Hz	(1) Ramp rate <5% P_{MPPT} (2) Online >30 min (47.0–49.0) Hz	Increase power till 100% P_{MPPT} according to P-f curves (48.0–49.5) Hz; 100% P_{MPPT} (46.5–48.0) Hz
(SSB)	Disconnection (< 47.0 Hz)		Disconnection (< 47.0 Hz)	Disconnection (< 47.0 Hz)	Disconnection (< 46.5 Hz)

f(Hz)	Denmark (1)	Denmark (2)	China Old (2005)	China New (2011)	USA ERCOT	Canada Hydro–Quebec	f(Hz)
(SSUP) – 52.0	Disconnection (> 51.5 Hz)		Disconnection (> 52.0 Hz)		All WTGs should be equipped with 2%–5% droop control	Disconnection (> 61.7 Hz)	(SSUP) – 61.7
(MSUP) – 50.5	Reduce power according to P-f curves (50.5–51.5) Hz		100% P_{MPPT} (48.0–52.0) Hz	(1) Reduce power according to ISO command (2) Online > 5 min. (50.2–52.0) Hz		P-f curves (60.6–61.7) Hz	(MSUP) – 60.6
(SYN) – 49.0	100% P_{MPPT} (47.5–50.5) Hz	50% P_{MPPT} (49.8–50.5) Hz; Reduce power according to P-f curves (49.0–49.8) Hz		100% P_{MPPT} (47.0–50.5) Hz	100% P_{MPPT} (59.96–60.04) Hz	100% P_{MPPT} (59.4–60.6) Hz	(SYN) – 59.4
(MSUB)		100% P_{MPPT} (47.0–49.0) Hz	Ensure WTG safety under emergency (<48.0 Hz)	Disconnection (<48.0 Hz)	All WTGs should be equipped with 2%–5% droop control	P-f curves (55.5–59.4) Hz	(MSUB) – 55.5
47.0 – (SSUB)	Disconnection (< 47.5 Hz)					Disconnection (<55.5 Hz)	(SSUB)

* P_{MPPT}, the MPPT power output; P_{rate}, the rated power output.

Fig. 1.1 A comparison of grid codes according to the frequency regulation requirement.

the MSUB range, power reduction is strictly controlled to prevent the underfrequency situation. And if system frequency is in the SSUP or SSUB range, WTGs are mostly allowed to be tripped off entirely.

The representative five-band frequency regulation can be clearly seen in the Denmark *P-f* curve of Fig. 1.2A, which can be outlined by the five *P-f* points A-B-C-D-E. In Fig. 1.2, there particularly exist two separate operation codes: (1) under MPPT operation or unreduced production, WTG farms can only make downward regulation; and (2) under active power control (APC) or reduced production, WTG farms will

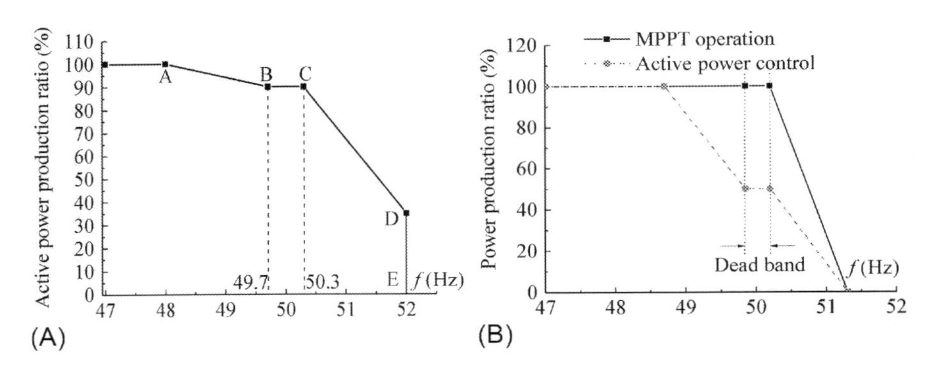

Fig. 1.2 Denmark and Ireland *P-f* curves. (A) Ireland *P-f* curve (Irish EirGrid, 2015) (B) Denmark *P-f* curve (Energinet, 2004).

operate at 50% MPPT output when the system frequency is in the 49.5–50.5 Hz range, in which the opportunity cost incurred would be compensated accordingly; as a result the WTG can make an up- or down-regulation. The corresponding *P-f* curves are given in Fig. 1.2B, which can reflect the relationship between the system frequency and WTG power output. In these curves, there usually exists a dead band for WTG control to prevent frequency action near synchronous frequency.

According to different thresholds and regulation requirements in Fig. 1.1, European countries such as Germany and Sweden have relatively high requirements regarding ramp rate and lasting time than Canada and China. As a matter of fact, the grid codes continuously evolved as WTG's control technologies advance or administration manner changes. For example, from April 1, 2005 the UK Grid Code has taken effect in England, Scotland, and Wales, replacing the Scottish Grid Code (Scottish TSO, 2010). In China, WTGs had no duty to participate in frequency regulation before 2011, but the new national standards drafted in 2011 clearly raise a similar frequency regulation requirement for European countries. In addition, there is evidence that the grid code on a wind farm's power ramp rate very specifically caters for the frequency regulation requirement (GB Institute, 2011):

- If $P_R < 30$ MW, then $R_{10} \leq 10$ MW/10 min, and $R_1 \leq 3$ MW/10 min
- If 30 MW $\leq P_R \leq 150$ MW, then $R_{10} \leq P_R/3$, and $R_1 \leq P_R/10$ within 10 min
- If $P_R \geq 150$ MW, then $R_{10} \leq 50$ MW/10 min, and $R_1 \leq 15$ MW/10 min

where P_R, R_{10}, and R_1 denote the installed capacity and the 10- and 1-min power ramp rate, respectively.

Despite nonspecific requirement regarding WTGs' frequency regulation in the FERC's grid code (USA FERC, 2005), some regional power systems in United States such as Texas ERCOT also issue rules that wind farms/plants should provide a proper frequency response to severe system disturbances (USA ERCOT, 2013; Ackermann, 2012). In addition, the frequency ancillary service can be captured by those mature frequency regulation markets in United States, for example, PJM and ERCOT.

It should be noted that the *P-f* curve described in Fig. 1.2 actually is the WTG frequency response resulting from APC technologies, which have enabled WTG to adjust its actual operating point according to the operator's command or system grid codes.

1.3 Frequency regulation framework

As wind power's penetration increases in a power system, the coordination of WTGs, wind farms, and conventional power system equipment would attract considerable attention for the purpose of secure and economic operation of the power grid. When there are wind farms participating in system frequency regulation, a hierarchical three-level dispatch framework can be utilized as there exist different time-scale characteristics among all participants involved (Wang et al., 2015a): the WTG level, the plant level, and the system level as illustrated in Fig. 1.3. On the system level, ISO will concentrate on co-dispatching wind farms with conventional generators or energy storage system for frequency regulation; on the plant level, wind farms are committed to organize and dispatch the WTGs, and determine which control mode the WTG will rest on, MPPT or APC; on the WTG level, some network-friendly and closed-loop control strategies, such as pitch control and de-loading control, can be adopted for WTG to participate in the system ancillary service.

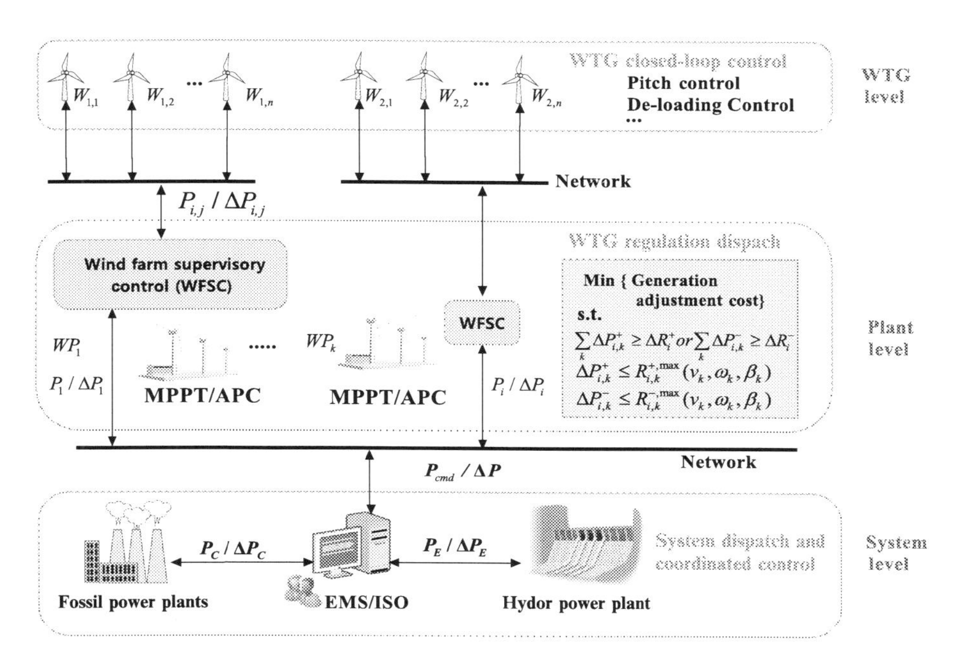

Fig. 1.3 Three-level frequency regulation framework considering a wind farm's active power control ability.

In the three-level dispatch framework, a bi-directional data and information exchange is established among WTGs, wind farms, ISO, and other conventional plants, including: (a) WTG operating condition, including the instant wind speed v_k and the WTG's state such as rotor speed ω_k and pitch angle β_k related to the kth WTG in a wind farm; and (b) the active power adjustment of other fossil plants $P_C/\Delta P_C$ and hydropower plants, $P_E/\Delta P_E$;. In this framework, there exist two implementation schemes: passive frequency regulation (Wang et al., 2015a) and active frequency regulation (Kroposki et al., 2017).

1.3.1 Passive frequency regulation

Normally, all WTGs in a wind farm will operate at MPPT in the SSF band ($|\Delta f| \leq 0.1\,\text{Hz}$). All aggregated WTG information is sent to the ISO and a specific frequency regulation scheme is therefore formulated according to the grid-code requirement (Energinet, 2004). When there is frequency deviation beyond the SSF band, the ISO will send a frequency regulation command in the form of power adjustment to the ith wind farm, either the upward ΔP_i^+ (MSF or SSBF) or the downward ΔP_i^-, in which the WTGs will receive the command of power output adjustment from the wind farm-level supervisory control (WFSC).

1.3.2 Active frequency regulation

The main difference compared with the above scheme is that wind farms will actively reschedule their operating curves when they participate in competitive ancillary services markets and provide a frequency regulation service (Singarao and Rao, 2016), which will also prespecify the wind farm's operation mode such as MPPT or APC (Badihi et al., 2015). In this way, some WTGs will actively keep reserve capacities for frequency regulation. Therefore, the active regulation scheme can be more credible. However, the establishment of a prescheduling mechanism in the active regulation scheme is quite complicated and will be discussed below.

1.4 System-level control

On the system level, the required power regulation, ΔP_{total}, can be captured by the ISO in two ways: (1) the ISO can purchase the required capacity via some optimal bidding in a joint energy and regulation market (He et al., 2017); and (2) there is preallocation among conventional thermal/hydro/gas power plants as well as wind farms (if available) according to some negotiated rate and allocation coefficients (Wang et al., 2017), similar to conventional AGC procurement. For example, the corresponding adjustment on the power reference for the ith power plant, ΔP_i, can be proportionally allocated:

$$\Delta P_i = \Delta P_{total} \times pf_i \tag{1.1}$$

where the allocation coefficient pf_i meets $\sum pf_i = 1$ and can be specified according to the power plant's available reserve capacity, ramp ratio, rated capacity, etc. (Zhang et al., 2014; Li et al., 2015).

To fulfil the system control purpose, some two-level hierarchical control schemes can be an alternative to achieve satisfactory performance of dynamic frequency and damping response (Leon et al., 2012) without taking WTG's APC ability into account. The upper level includes a centralized controller based on synchronized wide-area signals, and the decentralized controller in the lower level can coordinate all conventional SGs, wind farm converters, energy storage system, FACTS, etc. (He et al., 2017; Leon et al., 2012; Kothari, 2005).

On the system level, the ISO needs to consider an overall dispatch among wind farms and conventional power plants to maintain system reserve capacity adequacy.

1.5 Plant/farm-level coordinated control

The purpose of plant/farm-level control is to cooperate all aggregated WTGs in a wind farm. Usually, the wind farm will arrange the WTG sequence to participate in frequency regulation according to the WTG status (on or off service) in the WFSC system and thus the wind farm can effectively respond to the ISO's command to adjust its power outputs. The operating points of the active and reactive power for these WTGs can be scheduled according to some dynamic distribution factor (Chang-Chien et al., 2008), or systematically determined by some advanced optimization algorithms, for example, in (de Almeida et al., 2006) a primal-dual predictor-corrector interior point method is developed to carry out ISO requests.

A general optimal model related to this reserve scheduling problem can be formulated as follows:

$$\text{Min} \sum_k a_k |\Delta P_{i,k}| \tag{1.2}$$

$$s.t. \begin{cases} \sum_k \Delta P_{i,k}^+ \geq \Delta R_i^+ \text{ or } \sum_k \Delta P_{i,k}^- \geq \Delta R_i^- \\ 0 \leq \Delta P_{i,k}^+ \leq R_k^{+,\max}(v_k, \omega_k, \beta_k) \\ 0 \leq \Delta P_{i,k}^- \leq R_k^{-,\max}(v_k, \omega_k, \beta_k) \end{cases} \tag{1.3}$$

The object is to pursue a minimum regulation cost, where a_k and $\Delta P_{i,k}$ are the cost coefficiency and power adjustment of the kth WTG in the ith wind farm, respectively. The related constraints in Eq. (1.3) include: (1) the total frequency regulation capacity ($\Delta R_i^+/\Delta R_i^-$) required should be respected; and (2) each WTG has its power adjustment bounds ($R_k^{-,\max}/R_k^{+,\max}$) that are dependent on its operating conditions. In addition, there exist other plant-level control targets, such as minimum power reduction of WTG for maintaining a frequency regulation reserve (Diaz et al., 2013) and minimum number of acted WTGs, for example, a WTG with more reserve has priority to respond to the dispatch command.

The challenge on this level lies in the fact that the operating condition for each WTG in a farm actually differs due to environmental and spatial effects, and thus a predictive evaluation of the WTG's adjustment bounds requires those advanced real-time measurement technologies.

1.6 WTG-level control strategy

WTG-level control is usually implemented by an auxiliary closed-loop control such as inertial emulation (Gautam et al., 2011), overproduction control with step input (Keung et al., 2009), de-loading control (Chang-Chien et al., 2011), and droop control (Vidyanandan and Senroy, 2013), which will be elaborated below.

1.6.1 Inertial emulation control

By inertial emulation control, WTG can emulate the inertial response and frequency regulation similar to that of SG, which is mainly implemented by adding some PI loop and introducing the frequency signal into the torque or power control (Anaya-Lara et al., 2006). Typical control strategies are given in Fig. 1.4A. In mathematics, the control law in Fig. 1.4A can be formulated in Eq. (1.4):

$$T_{gen} = T_m - K_D \frac{df}{dt} \tag{1.4}$$

where f is the system frequency; T_{gen} and T_m are the electromagnetic torque and mechanical torque, respectively; and K_D is the derivative time constant. Further, Eq. (1.4) can be rewritten as Eq. (1.5) by introducing the rotor speed ω:

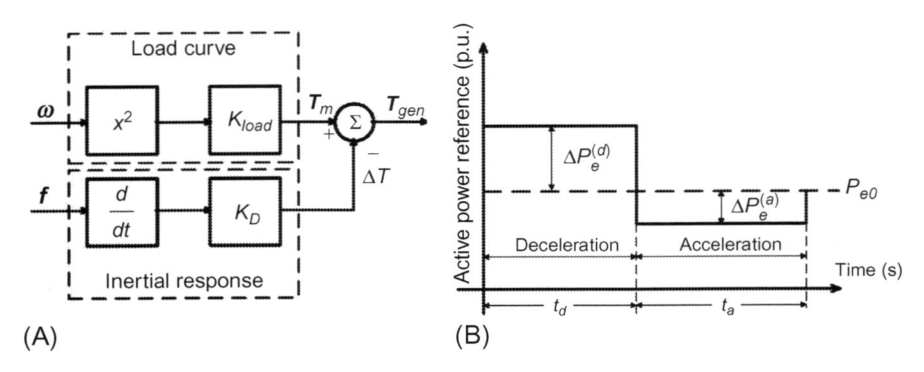

Fig. 1.4 (A) Inertial emulation control schemes; (B) overproduction power reference signal.

$$T_{gen} = T_m - \frac{K_D}{2\pi} \frac{d\omega}{dt} \tag{1.5}$$

On the other hand, the torque-rotor speed relationship of any SG has similar expression in Eq. (1.6), where H denotes the inertial constant and other variables are SG counterparts:

$$2H \frac{d\omega}{dt} = T_m - T_{gen} \tag{1.6}$$

By comparing Eq. (1.5) with Eq. (1.6) above, it is observed that WTG can emulate a conventional generator with an inertia time constant $K_D/4\pi$. In short, WTGs with inertia emulation control are capable of increasing system rotational inertia (Kayikci and Milanovic, 2009). The emulated inertia will be released or absorbed when the system frequency varies. The corresponding power injection or absorption is completely uncontrollable and only depends on df/dt or final Δf once the PI parameters are set. That means, WTGs with inertia emulation control only cannot maintain long-term APC ability.

Despite the underlying common philosophy, the inertial emulation control can be classified into several categories (Wu et al., 2017): (1) natural inertial control, which can emulate the inertial response of a conventional SG by introducing a df/dt loop control (Hwang et al., 2016; Van de Vyver et al., 2016; Zhao et al., 2016); and (2) virtual synchronous control, which enables a DFIG to deliver the inertial response to enhance the frequency stability without the traditional PLL loop (Wang et al., 2015b,c; Huang et al., 2017) and is particularly helpful when WTG is integrated into a weak AC power grid with low short-circuit ratio.

1.6.2 Overproduction

By adding a step input signal, WTG with overproduction control could increase the power output when the system frequency drops. A typical step input signal of overproduction is shown in Fig. 1.4B, where P_{e0} represents the WTG steady power. When the system frequency drops, there will be a power reference increment $\Delta P_e^{(d)}$ and this process will last for t_d. Kinetic energy stored in rotational mass is extracted to increase electrical power output and the rotor decelerates during the deceleration period. In order to avoid stalling, there exists an acceleration process $(\Delta P_e^{(a)}, t_a)$ (Rawn et al., 2010).

Another overproduction scheme can be implemented by overloading a converter for a short time when the wind speed is above a rated value (Rawn and Lehn, 2008). It should be noted that, the over-loading capability of a wind generator is determined by the maximal excess power that the drive train, the generator, and the converter can withstand without adding damaging fatigue loads on wind turbine structure.

Overproduction is similar to inertia emulation in making use of kinetic energy stored in rotational mass to serve extra output. The characteristic of these kinetic energy releasing-based control strategies (inertia emulation and overproduction) is that there exists an acceleration process to recover rotor speed for preventing rotor stalling. In the acceleration process, the power output will be less than in the normal state. Therefore, there exists a secondary frequency drop event (Wang et al., 2015a). The difference is that the output power profile of the overproduction scheme can be set manually. Hence, overproduction has APC capability and can be used as an active frequency regulation strategy.

1.6.3 De-loading operation

The de-loading control can make WTG maintain a higher rotate speed and less power output compared with MPPT that has the potential to keep reserve capacity for frequency regulation (de Almeida and Lopes, 2007). The de-loading operation principle is illustrated in Fig. 1.5, in which a group of bell curves reflects the relationship between mechanical power and rotor speed; the MPPT curve and de-loading curve reflect the relationship between electrical power output and rotor speed. In Fig. 1.5, P_b represents de-loading power while P_a represents MPPT power. There are two ways to implement the de-loading operation of WTG: (1) the rotational speed control, which means the P-ω curve changes from the MPPT curve to the de-loading curve (Fig. 1.5A); and (2) the pitch angle control, which means the P_m curve changes from a solid to a dashed bell curve (Fig. 1.5B). Under rotational speed control (Fig. 1.5A), the power reserve for frequency regulation can be calculated by Eq. (1.7):

$$P_r = (P_a - P_b) + \frac{H}{T}\left(\omega_b^2 - \omega_a^2\right) \tag{1.7}$$

where $(P_a - P_b)$ means power deviation between the MPPT mode and the de-loading mode; $H(\omega_b^2 - \omega_a^2)$ means kinetic energy stored in rotational masses, which will release when the operation point moves from point b to point a; H is the inertia

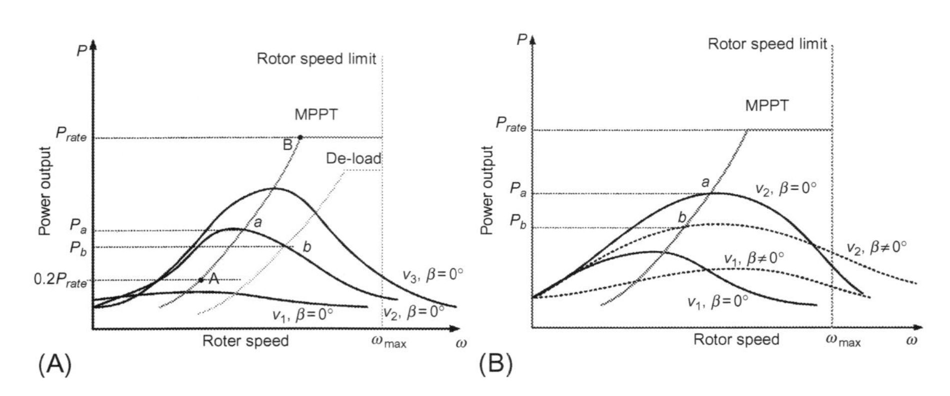

Fig. 1.5 Principle illustration of two de-loading schemes: (A) rotational speed control and (B) pitch angle control.

constant and T is the conversion time from point b to point a. Under pitch angle control (Fig. 1.5B), the power reserve of WTG can be calculated as follows:

$$P_r = P_a - P_b \tag{1.8}$$

Differing from formula (1.7), there is no additional item caused by kinetic energy transformation, because acceleration energy for a rotor to move from point b to point a is extracted from wind energy. Learning from Eqs. (1.7), (1.8), rotational speed control usually has more reserve capability than the pitch angle control under the same wind condition. But limited by the maximum rotor speed, rotational speed control can only be applied under middle wind speed.

A typical rotational speed control scheme is proposed in Fig. 1.6A, in which the WTG is de-loaded in advance and there will not be an additional power signal $\triangle P$ until the system frequency drops. And, a typical pitch angle control scheme is proposed in Fig. 1.6B.

Under de-loading control, WTG's output power is not fully rated in advance and the power reserve can be released faster and last for a more long-term timescale than the inertial emulation control. What's more, the pitch angle control usually acts more slowly than rotational speed control, because the pitch angle control relies more on mechanical action while rotational speed control basically relies on a converter's fast response (Attya and Hartkopf, 2014).

1.6.4 Droop control

A typical droop control scheme is given in Fig. 1.7, where the additional power signal ΔP is calculated through Eq. (1.9):

$$\Delta P = \frac{1}{R}\Delta f \tag{1.9}$$

The droop control is to make a WTG turbine emulate SG's droop characteristic in frequency regulation, which is usually combined with other strategies such as the

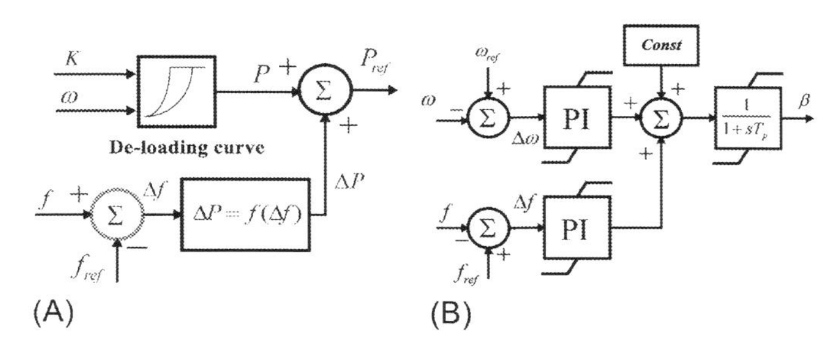

Fig. 1.6 Two types of de-loading control strategies: (A) rotational speed control and (B) pitch angle control.

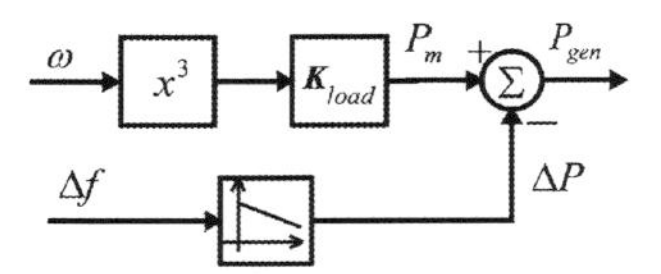

Fig. 1.7 Droop control scheme.

de-loading control to ensure that the output of WTG increases with the drop of system frequency gradually. However, droop control also has its own disadvantage: the droop control with fixed droop coefficient may lead to WTG instability. A variable droop control strategy can not only solve this problem but also improve the efficiency of frequency regulation (Vidyanandan and Senroy, 2013).

1.6.5 Hybrid control schemes

To make full use of the above WTG-level control strategies, several of the above control strategies can be combined. For example, a joint inertia emulation and droop control can achieve remarkable frequency dynamics (Hwang et al., 2016; Lee et al., 2016); a hybrid strategy scheme with the droop control and the de-loading control combined has long-term frequency regulation performance similar to SG (Huang et al., 2017; Lu et al., 2014).

Another hybrid scheme is to switch over among control strategies according to the real wind condition (Attya and Hartkopf, 2014). As listed in Table 1.1, the wind speed can be classified into four continuous intervals: (1) the cut-in speed v_{in}; (2) the cut-out speed v_{out}; (3) the low speed v_{low} and (4) the high speed v_{high}. Here, v_{low} and v_{high} are the wind speed at which the WTG power is equal to 0.2 and 1.0 p.u., respectively, as denoted at points A and B of Fig. 1.5A, respectively.

Normally, WTG operates in the MPPT mode. When there is any frequency support requirement ordered from the ISO, WTG can be mutually switched to frequency control modes, for example, when there is $v_{low} \leq v \leq v_{high}$, WTG will be switched to the de-loading mode to hold a frequency regulation reserve.

Table 1.1 WTG's control strategies based on the real wind speed

Wind speed	Hybrid strategy	
	Normal operation	Frequency support
$v < v_{in}$, $v > v_{out}$	–	–
$v_{in} \leq v \leq v_{low}$	MPPT	Overproduction (kinetic energy extraction)
$v_{low} \leq v \leq v_{high}$	MPPT	De-loading
$v_{high} \leq v \leq v_{out}$	MPPT	Overproduction (converter overloading)

1.6.6 Performance comparison

A performance comparison of the various control strategies above is given in Table 1.2, in which C1-C4 represent the implementation methods these control schemes will adopt; F1-F2 are two typical types of frequency controls with different timescales. It is obvious that only in the middle or high wind speed band can WTGs have large frequency regulation capacity. And there is no single control strategy that can provide steady frequency support under full-scope wind conditions.

1.7 Discussion

In this section, future work under the three-level frequency regulation framework is discussed.

1.7.1 WTG level

1.7.1.1 Advanced control strategies

During the process of inertial emulation control, a secondary frequency drop phenomenon is likely to occur during the rotor speed recovery due to the reduced output power; thus, it is desirable to design advanced control schemes to provide inertial response support while preventing the rotor speed from overdeceleration and mitigating the impact of rotor speed recovery on the overall frequency performance (Wu et al., 2017).

1.7.1.2 WTG regulation margin assessment

At some given wind speed, the regulation margin, that is, the maximum de-loading ratio (Lu et al., 2014) or the additional power reference signal $\Delta P_e^{(d)}$ in Fig. 1.4B, needs to be accurately assessed for each type of WTG in order to establish the link between the wind speed and the frequency regulation capacity and fully exploit frequency regulation potential.

1.7.2 Wind plant/farm level

1.7.2.1 WTG inner coordination

Due to the impact of local-scale terrain, WTGs' site-specific information should be collected for decision-making, so that the regulation reference can be determined to achieve a smooth frequency regulation (Piwko et al., 2012).

1.7.2.2 Frequency regulation capacity assessment

As previously mentioned, diverse control strategies will be triggered as the wind speed changes, as reported by Lei and Infield (2013). As terrain wind difference exists, assessment of the aggregated frequency regulation capacity considering site-specific information is desired for system planning and scheduling (Yan and Saha, 2015).

Table 1.2 Comparison of control strategies on WTG level

Control strategy		Control type	Response speed	Reserve duration	APC capability	Optimal wind speed
Inertia emulation		F1	Fast kinetic energy release	Short term	×	Middle/high
Overproduction	C1	F1	Fast kinetic energy release	Short term	√	Middle
	C2	F1	Fast kinetic energy release	Short term	√	High
De-loading	C3	F1 + F2	Fast reserve release	Long term	√	Middle
	C4	F1 + F2	Slower than converter control	Long term	√	High
Droop control		F1 + F2	Coexisting with other control	Long term	√	Not related

C1, step reference command; C2, converter overloading; C3, rotational speed control; C4, pitch angle control; F1, the primary control; F2, the secondary control.

1.7.3 System level

1.7.3.1 Dynamic allocation scheduling

An ordinary allocation mechanism among various aggregated devices has been introduced in Section 1.4, in which there is a fixed allocation coefficient. Actually, the allocation scheme will be affected by the system operation status and wind condition. That means the allocation coefficient pf_i in Eq. (1.1) should be dynamic and time varying. The solution process of this problem can be built on the actual system scheduling scheme (Wang et al., 2015d).

1.7.3.2 Economics of frequency service

As active power output and generation benefits will decrease when wind power participates in system frequency regulation, a market mechanism needs to be introduced to price the ancillary service provided by related wind farms (Saiz-Marin et al., 2012).

1.8 Conclusion

In this chapter, the frequency regulation issues in wind power integration are reviewed and discussed, including gird codes, frequency regulation framework, and advanced control strategies. Existing grid codes also witness the technical trend of WTGs participating in frequency regulation. With the increase of the wind power penetration ratio, grid codes gradually require wind power to be equipped with frequency regulation capability and they become more and more detailed. On the other hand, though various control strategies have been developed for integrating WTG, especially VSWTG, with frequency regulation ability, there is not any specific control scheme that can cover the whole operation scope of WTG. Hence, those hybrid control strategies that can make full use of each control scheme's advantages would become promising. The control strategy on the wind-plant level forms a connecting link between WTG and ISO, playing the role of power allocation for each WTG by using some optimization methods. Regarding the system level, ISO must act as an administrator to coordinate all participants.

Acknowledgment

The National Natural Science Foundation of China (No. 51677165) and the Danish Agency for Science, Technology and Innovation (No. 4070-00023B) are to be acknowledged for their funding support.

References

Ackermann, T., 2012. Wind Power in Power System, second ed. John Wiley & Sons, Chichester, West Sussex, United Kingdom.

Anaya-Lara, O., Hughes, F.M., Jenkins, N., et al., 2006. Contribution of DFIG-based wind farms to power system short-term frequency regulation. IEE Proc. Gener. Transm. Distrib. 153 (2), 164–170.

Attya, A.B., Hartkopf, T., 2014. Wind turbine contribution in frequency drop mitigation—modified operation and estimating released supportive energy. IET Gener. Transm. Distrib. 8 (5), 862–872.

Badihi, H., Zhang, Y., Hong, H., 2015. Active power control design for supporting grid frequency regulation in wind farms. Annu. Rev. Control. 40 (Supplement C), 70–81.

Banakar, H., Changling, L., Boon Teck, O., 2008. Impacts of wind power minute-to-minute variations on power system operation. IEEE Trans. Power Syst. 23 (1), 150–160.

Chang-Chien, L.-R., Hung, C.-M., Yin, Y.-C., 2008. Dynamic reserve allocation for system contingency by DFIG wind farms. IEEE Trans. Power Syst. 23 (2), 729–736.

Chang-Chien, L.-R., Lin, W.-T., Yin, Y.-C., 2011. Enhancing frequency response control by DFIGS in the high wind penetrated power systems. IEEE Trans. Power Syst. 26 (2), 710–718.

Conroy, J.F., Watson, R., 2008. Frequency response capability of full converter wind turbine generators in comparison to conventional generation. IEEE Trans. Power Syst. 23 (2), 649–656.

de Almeida, R.G., Lopes, J.A.P., 2007. Participation of doubly fed induction wind generators in system frequency regulation. IEEE Trans. Power Syst. 22 (3), 944–950.

de Almeida, R.G., Castronuovo, E.D., Lopes, J.A.P., 2006. Optimum generation control in wind parks when carrying out system operator requests. IEEE Trans. Power Syst. 21 (2), 718–725.

Diaz, G., Casielles, P.G., Viescas, C., 2013. Proposal for optimising the provision of inertial response reserve of variable-speed wind generators. IET Renew. Power Gener. 7 (3), 225–234.

Energinet, 2004. Regulation TF 3.2.6: Grid Connection of Wind Turbines to Networks With Voltages Below 100kV. Available from: http://www.wt-certification.dk/Common/WindTurbinesConnectedtoGridswithVoltageabove100kV.pdf.

Ernst, B., Oakleaf, B., Ahlstrom, M.L., et al., 2007. Predicting the wind. IEEE Power Energy Mag 5 (6), 78–89.

Eskom System Operation & Planning, 2012. Grid Code Requirements for Wind Energy Facilities Connected to Distribution or Transmission System in South Africa (Version 5.4). Available from: http://www.nersa.org.za/Admin/Document/Editor/file/Electricity/TechnicalStandards.

Gautam, D., Goel, L., Ayyanar, R., et al., 2011. Control strategy to mitigate the impact of reduced inertia due to doubly fed induction generators on large power systems. IEEE Trans. Power Syst. 26 (1), 214–224.

GB Institute, 2011. China Standard: GB/T 19963–2011 Technical Rule for Connecting Wind Farm to Power System. China Standards Publication, Beijing.

He, G., Chen, Q., Kang, C., et al., 2017. Cooperation of wind power and battery storage to provide frequency regulation in power markets. IEEE Trans. Power Syst. 32 (5), 3559–3568.

Holdsworth, L., Ekanayake, J.B., Jenkins, N., 2004. Power system frequency response from fixed speed and doubly fed induction generator-based wind turbines. Wind Energy 7 (1), 21–35.

Huang, L., Xin, H., Zhang, L., et al., 2017. Synchronization and frequency regulation of DFIG-based wind turbine generators with synchronized control. IEEE Trans. Energy Convers. 32 (3), 1251–1262.

Hwang, M., Muljadi, E., Park, J.-W., et al., 2016. Dynamic droop–based inertial control of a doubly-fed induction generator. IEEE Trans. Sustain. Energy 7 (3), 924–933.

Hydro-Québec TransÉnergie, 2009. Technical Requirements for the Connection of Generation Facilities to the Hydro-Quebec Transmission System: Supply Requirements for Wind Generation. Available from: http://www.hydroquebec.com/transenergie/fr/commerce/pdf/eolienne_transport_en.pdf.

Irish EirGrid, 2015. EirGrid Grid Code (Version 6). Available from: www.eirgridgroup.com/site-files/library/EirGrid/GridCodeVersion6.pdf.

Kayikci, M., Milanovic, J.V., 2009. Dynamic contribution of DFIG-based wind plants to system frequency disturbances. IEEE Trans. Power Syst. 24 (2), 859–867.

Keung, P.-K., Li, P., Banakar, H., et al., 2009. Kinetic energy of wind-turbine generators for system frequency support. IEEE Trans. Power Syst. 24 (1), 279–287.

Kothari, I.P.K.D.P., 2005. Recent philosophies of automatic generation control strategies in power systems. IEEE Trans. Power Syst. 20 (1), 346–357.

Kroposki, B., Johnson, B., Zhang, Y., et al., 2017. Achieving a 100% renewable grid: operating electric power systems with extremely high levels of variable renewable energy. IEEE Power Energy Mag 15 (2), 61–73.

Lalor, G., Mullane, A., O'Malley, M., 2005. Frequency control and wind turbine technologies. IEEE Trans. Power Syst. 20 (4), 1905–1913.

Lee, J., Muljadi, E., Srensen, P., et al., 2016. Releasable kinetic energy-based inertial control of a DFIG wind power plant. IEEE Trans. Sustain. Energy 7 (1), 279–288.

Lei, W., Infield, D.G., 2013. Towards an assessment of power system frequency support from wind plant-modeling aggregate inertial response. IEEE Trans. Power Syst. 28 (3), 2283–2291.

Leon, A.E., Mauricio, J.M., Gomez-Exposito, A., et al., 2012. Hierarchical wide-area control of power systems including wind farms and FACTS for short-term frequency regulation. IEEE Trans. Power Syst. 27 (4), 2084–2092.

Li, Z., Wu, W., Zhang, B., et al., 2015. Adjustable robust real-time power dispatch with large-scale wind power integration. IEEE Trans. Sustain. Energy 6 (2), 357–368.

Lu, Z., Wang, Z., Xin, H., et al., 2014. Small signal stability analysis of a synchronized control-based microgrid under multiple operating conditions. J. Mod. Power Syst. Clean Energy 2 (3), 244–255.

Mullane, A., O'Malley, M., 2005. The inertial response of induction-machine-based wind turbines. IEEE Trans. Power Syst. 20 (3), 1496–1503.

Nordel, 2007. Nordic Grid Code 2007 (Nordic Collection of Rules). Available from: https://www.entsoe.eu/fileadmin/user_upload/_library/publications/nordic/planning/070115_entsoe_nordic_NordicGridCode.pdf.

Piwko, R., Meibom, P., Holttinen, H., et al., 2012. Penetrating insights: lessons learned from large-scale wind power integration. IEEE Power Energy Mag 10 (2), 44–52.

Rawn, B.G., Lehn, P., 2008. In: Wind rotor inertia and variable efficiency: fundamental limits on their exploitation for inertial response and power system damping.Proc. of Euro Wind Energy Conf (EWEC2008), Brussels, Belgium, 31 March-3 April.

Rawn, B.G., Gibescu, M., Kling, W.L., 2010. In: A static analysis method to determine the availability of kinetic energy from wind turbines.Proc. of IEEE PES Gen Meet (PES2010), Minneapolis, USA, 25–29 July 2010.

Saiz-Marin, E., Garcia-Gonzalez, J., Barquin, J., et al., 2012. Economic assessment of the participation of wind generation in the secondary regulation market. IEEE Trans. Power Syst. 27 (2), 866–874.

Scottish TSO, 2010. Scottish Grid Code. Available from: http://www.hi-energy.org.uk/Downloads/General%20Documents/Xero-Energy-Grid-Guide-May%202010.pdf.

Singarao, V.Y., Rao, V.S., 2016. Frequency responsive services by wind generation resources in United States. Renew. Sust. Energ. Rev. 55 (Supplement C), 1097–1108.

Sun, Y., Zhang, Z., Li, G., et al., 2010. In: Review on frequency control of power systems with wind power penetration.Proc. of the 2010 International Conference on Power System Technology (POWERCON2010), Hangzhou, China, 24–28 October.

Tennet TSO GmbH, 2012. Grid Code: High and Extra High Voltage. Available from: http://www.tennettso.de/site/binaries/content/assets/transparency/publications/grid-connection/tennet-nar2012eng.pdf.

UK National Grid, 2009. Grid Code Documents: Connection Conditions. Available from: http://www2.nationalgrid.com/UK/Industry-information/Electricity-codes/Grid-code/The-Grid-code/.

Ummels, B.C., Gibescu, M., Pelgrum, E., et al., 2007. Impacts of wind power on thermal generation unit commitment and dispatch. IEEE Trans. Energy Convers. 22 (1), 44–51.

USA ERCOT, 2013. Nodal Protocols. Available from: http://www.ercot.com/mktrules/nprotocols/current.

USA FERC, 2005. Order 661-Interconnection With Wind Energy. Available from: https://www.ferc.gov/whats-new/comm-meet/052505/E-1.pdf.

Van de Vyver, J., Kooning, J.D.M.D., Meersman, B., et al., 2016. Droop control as an alternative inertial response strategy for the synthetic inertia on wind turbines. IEEE Trans. Power Syst. 31 (2), 1129–1138.

Vidyanandan, K.V., Senroy, N., 2013. Primary frequency regulation by deloaded wind turbines using variable droop. IEEE Trans. Power Syst. 28 (2), 837–846.

Wang, H., Chen, Z., Jiang, Q., 2015a. Optimal control method for wind farm to support temporary primary frequency control with minimised wind energy cost. IET Renew. Power Gener. 9 (4), 350–359.

Wang, S., Hu, J., Yuan, X., 2015b. Virtual synchronous control for grid-connected DFIG-based wind turbines. IEEE J. Emerg. Sel. Top. Power Electron. 3 (4), 932–944.

Wang, S., Hu, J., Yuan, X., et al., 2015c. On inertial dynamics of virtual-synchronous-controlled DFIG-based wind turbines. IEEE Trans. Energy Convers. 30 (4), 1691–1702.

Wang, Y., Bayem, H., Giralt-Devant, M., et al., 2015d. Methods for assessing available wind primary power reserve. IEEE Trans. Sustain. Energy 6 (1), 272–280.

Wang, G.Z., Bian, Q.Y., Xin, H.H., et al., 2017. A robust reserve scheduling method considering asymmetrical wind power distribution. IEEE/CAA J. Automat. Sin., 1–7. https://doi.org/10.1109/JAS.2017.7510652.

Wu, Z., Gao, W., Gao, T., et al., 2017. State-of-the-art review on frequency response of wind power plants in power systems. J. Mod. Power Syst. Clean Energy, 1–16. https://doi.org/10.1007/s40565-40017-40315-y.

Yan, R., Saha, T.K., 2015. Frequency response estimation method for high wind penetration considering wind turbine frequency support functions. IET Renew. Power Gener. 9 (7), 775–782.

Zhang, X., Xu, Z.Y., Iqbal, M.J., et al., 2014. In: Dispatch strategy of large-scale wind farm automatic generation control system.Proc. of IEEE PES Gen Meet (PES2014), Washington, USA, 27–31 July.

Zhao, J., Lyu, X., Fu, Y., et al., 2016. Coordinated microgrid frequency regulation based on DFIG variable coefficient using virtual inertia and primary frequency control. IEEE Trans. Energy Convers. 31 (3), 833–845.

A two-stage reserve scheduling considering wind turbine generator's de-loading control

2

Zhen Wang, Zhengqing Yang*, Kit Po Wong†, S.S. Choi†*
*College of Electrical Engineering, Zhejiang University, Hangzhou, China, †School of Electrical, Electronic and Computer Engineering, University of Western Australia, Perth, WA, Australia

2.1 Introduction

The past 10 years have witnessed the rapid wind power development for the purpose of solving the worldwide energy and environmental crises (Global Wind Energy Council (GWEC), 2017). The new challenges come from the wind power's intermittent and volatility or even nondispatchable characteristic, accordingly additional operating reserves should be scheduled to maintain the system's frequency stability during sudden imbalances (Piwko et al., 2012). The variable nature of wind power increases the importance of the reserves due to the uncertain nature of these renewable resources, and even various reserves over different time horizons need to be scheduled (Holttinen et al., 2012; Wang et al., 2016). However, extra operating reserve scheduling will result in increasing overall operation cost and emissions from the fossil power plants, which will impede sustainable development of renewable energy (Piwko et al., 2012).

Given this background, the problem of reserve scheduling with wind power integration has attracted intense attention to system economic operation. The operational and economic impact of wind power uncertainty on the reserve scheduling has been investigated in many literatures, which can be generally classified into two categories (Holttinen et al., 2012): (1) deterministic methods; and (2) probability or uncertainty-based methods. As for the former, although deterministic approaches may be based in part on statistical variations of the wind and load, and therefore possess a probabilistic component, for example, a Lévy α-s Table 2 distribution is proposed by Bruninx and Delarue (2014) as an improved description of the WPFE based on statistical analysis, which can quantify a probabilistic reserve sizing. For the latter, there exist advanced probabilistic methods, such as calculating dynamic LOLP values to derive a reserve level that holds the risk constant (Holttinen et al., 2012). Other approaches include those systematic optimization methods, for example, a reserve optimization model considering the wind power forecast error (WPFE) is proposed by Ortega-Vazquez and Kirschen (2009). The net demand error is assumed to be normal distribution and divided into several intervals and a cost/benefit analysis is carried out in each

Smart Power Distribution Systems. https://doi.org/10.1016/B978-0-12-812154-2.00002-X

interval to calculate the optimal reserve requirements. A stochastic programming market-clearing model considering network constraints and costs of wind spillage is developed to find the optimal reserve levels and costs (Morales et al., 2009). A two-stage stochastic programming model is proposed to procure the required load-following reserves from both generation and demand side resources under high wind power penetration, in which load shedding and wind spillage are also taken into account (Paterakis et al., 2015). A distributionally robust coordinated reserve scheduling model is proposed by Wang et al. (2016), aiming to minimize the total procurement cost of conventional generation and reserve, while satisfying the security requirement over all possible probability distribution of WPFE.

Meanwhile, the control technologies of wind turbine generators (WTGs) are evolved rapidly with the widespread use of advanced power electronic converters. With the advent of advanced control technologies, WTGs now can participate in different levels of frequency control activities to make contribute to system ancillary service similar to conventional generation (Kroposki et al., 2017). The active power control (APC) technologies such as de-loading control (Wang-Hansen et al., 2013), inertial emulation control (Attya and Hartkopf, 2013), and overproduction control (Rawn and Lehn, 2008) make WTG's frequency regulation response possible. For example, the double-fed induction generator (DFIG) equipped with frequency modulation function can exhibit outstanding system dynamics and frequency recovery (Ramtharan et al., 2007).

In this chapter, the de-loading control is firstly introduced to fulfil WTG reserve capability. Then, a bi-level reserve scheduling model considering the reserve requirement for the prescheduling stage and the real-time stage is proposed. By applying the Karush-Kuhn-Tucker (KKT) condition, the proposed bi-level model can be transformed into a single-level optimization model. Finally, the effectiveness of the proposed bi-level model is validated using a modified IEEE 6-geneartor and 30-bus system.

2.2 WTG-integrated dispatch mode and DFIG de-loading operation

2.2.1 WTG-integrated dispatch mode

To make full use of WTG's APC capability and to accommodate as much wind power as possible, a three-level hierarchical dispatch framework (Wang et al., 2015) to realize a dispatchable power operation is recommended to replace a conventional dispatch mode with the maximum power point tracking (MPPT)-based WTG operation (Wu et al., 2012), as illustrated in Fig. 2.1. On the system level, independent system operator (ISO) will concentrate on co-dispatching wind farms with conventional generators for frequency regulation; on the plant level wind farms are committed to organize and dispatch the WTGs, and determine which control mode the WTG will rest on, MPPT or APC; on the WTG level, network friendly closed-loop control

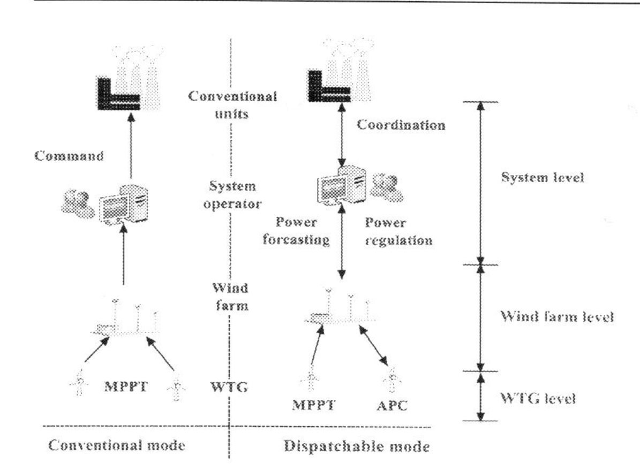

Fig. 2.1 Diagram of system dispatch mode with wind power integration.

strategies, such as pitch control and de-loading control, can be adopted for WTG to participate in system ancillary service. Compared with the conventional mode, there exists bidirectional information flow in dispatchable mode and WTG operation status can be adjusted according to power regulation commands from system operator.

2.2.2 DFIG de-loading operation

The de-loading operation of DFIG means it will operate at some operating points deviated from MPPT points by a combination of pitch control and converter control according to actual wind condition. Under de-loading operation mode, there is reserve capacity for WTG to regulate its output power. Compared with other APC strategies, de-loading control can maintain more long-term power reserve (Attya and Hartkopf, 2014). The control principle of de-loading control is illustrated in Fig. 2.2 and a typical implementation of the control scheme can be found by Lu et al. (2014). Under the de-loading operation, the DFIG is not operated along MPPT curve, but actively along some tracks below the MPPT curve (called de-loading curve) so that part of available wind energy can be stored in the form of rotating kinetic energy. As explained by Lu et al. (2014), this energy can be released when needed, that is, the de-loaded power serves as providing additional reserve capability.

A simple linear relationship between MPPT curve and de-loading curve is supposed. Given certain wind velocity, the corresponding power on MPPT point (the intersection between the MPPT curve and the power efficient curve) is denoted as P_{MPPT}, then the corresponding active power reference P_{ref} and the upward/downward reserve capacity R^{+}/R^{-} derived in the de-loading curve are as follows:

$$P_{ref} = (1-K)P_{MPPT}, \quad 0 \leq K < K^{\max} \tag{2.1}$$

$$R^{+} = K \cdot P_{MPPT}, \quad R^{-} = (K^{\max} - K) \cdot P^{MPPT} \tag{2.2}$$

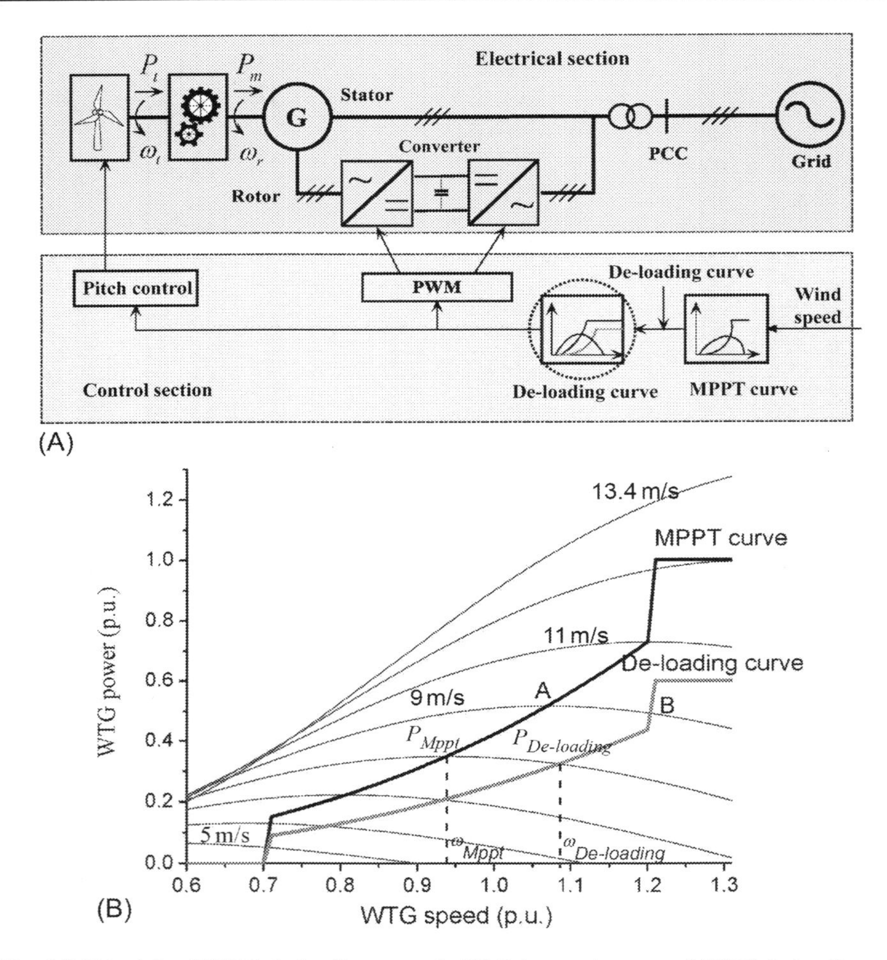

Fig. 2.2 Principle of WTG de-loading control. (A) Primary structure of WTG de-loading control. (B) Typical WTG de-loading operation curve.

where K is the DFIG reserve capability level coefficient. Larger K means more frequency regulating capability is retained in DFIG, thus better frequency support can be provided and larger reserve kept. As far as DFIG is concerned, $K^{\max}$ is about 0.1–0.2 (Wang-Hansen et al., 2013).

2.3 A bi-level optimization model for the two-stage reserve scheduling problem

In this section, a bi-level optimization model of two-stage reserve scheduling model considering WTG integration is proposed to evaluate the benefits of WTG de-loading control on system reserve scheduling. The idea behind lies in that, in the prescheduling stage, the total system scheduling cost is to be optimized and in the real-time stage, the change or adjustment to the scheduling scheme made in the first stage is to be minimized. In essence, if WTG de-loading capability is considered in the real-time stage, then the above two successive problems are coupled and can be formulated into a bi-level optimization model. For ease of explanation, two problems corresponding to these two stages are introduced respectively and then the bi-level optimization model is proposed.

2.3.1 The prescheduling stage

In this stage, the overall system cost for joint operation as well as reserve requirement is minimized:

$$\min Q_1 = \sum_{m=1}^{N_g} \left[f_m\left(P_m^{(0)}\right) + c_m^{R^+} R_m^+ + c_m^{R} R_m^- \right] + \sum_{n=1}^{N_w} \left(c_n^P P_n^{(0)} + c_n^{R^+} R_n^+ + c_n^{R} R_n^- \right) \tag{2.3}$$

$$P_n^{(0)} = (1 - K_n)\widetilde{P}_n \tag{2.4}$$

$$R_n^+ = K_n \widetilde{P}_n \tag{2.5}$$

$$R_n^- = \left(K_n^{\max} - K_n\right) \cdot \widetilde{P}_n \tag{2.6}$$

where $P_m^{(0)}$ and R_m^+/R_m^- are the mth conventional unit's power output and upward/downward reserve capacity to be decided, respectively ($m=1, 2, \ldots, N_g$). $P_n^{(0)}$ is the forecasted wind power of the nth wind farm. R_n^+/R_n^- is upward/downward reserve capacity held by the nth wind farm ($n=1, 2, \ldots, N_w$); $f_m(\cdot)$ is a quadratic cost function of conventional units; $c_m^{R^+}/c_m^{R^-}$ are the reserve prices for conventional units; c_n^P and $c_n^{R^+}/c_n^{R^-}$ are WTG power price and reserve prices, respectively. K_n is the de-loading ratio to be decided in the prescheduling stage; $\widetilde{P}_n$ is the predicted power in MPPT mode.

The objective function Eq. (2.3) consists of power and reserve costs of conventional units and wind farms. The wind power energy/reserve prices can be regarded as some negotiated rates between ISO and wind farms before mature market mechanism exists (Conejo et al., 2010).

In prescheduling stage, the de-loading ratio of wind power is determined in a conservative way, by choosing the minimum value of all possible de-loading ratios under credible scenarios $K_n^{(s)}$, as given in formula (2.7), which means that more reserve capacities served by conventional units is assumed.

$$K_n = \min\left(K_n^{(s)}\right) \tag{2.7}$$

where $K_n^{(s)}$ represents the de-loading ratio under the sth scenario in the next real-time stage.

The constraints in the prescheduling stage include

(1) *Active power limits*: The limits of the active power outputs of conventional units are considered:

$$P_m^{\min} \leq P_m^{(0)} \leq P_m^{\max}, \ \forall m \tag{2.8}$$

where $P_m^{\min}$ and $P_m^{\max}$ are the minimum and the maximum power limits of the mth conventional unit.

(2) *Ramp rate limits*:

$$\begin{cases} P_m^{(0)}(t) - P_m^{(0)}(t-1) \leq r_m^+ D \\ P_m^{(0)}(t-1) - P_m^{(0)}(t) \leq r_m^- D \end{cases}, \ \forall m \tag{2.9}$$

where r_m^+/r_m^- are the up/down ramp rate; D is the scheduling time interval and set to be 1 h in the study.

(3) *Reserve constrains:* the up and down reserve capacities of conventional units are limited by regulation margin or ramp capacity.

$$0 \leq R_m^+ \leq \min\left(P_m^{\max} - P_m^{(0)}, r_m^+ D\right) \tag{2.10}$$

$$0 \leq R_m^- \leq \min\left(P_m^{(0)} - P_m^{\min}, r_m^- D\right) \tag{2.11}$$

(4) *System reserve requirement*: sufficient reserve is to be ensured for secure system operation.

$$\begin{cases} \sum_{m=1}^{N_g} R_m^+ + \sum_{n=1}^{N_w} R_n^+ \geq R_D^+ \\ \sum_{m=1}^{N_g} R_m^- + \sum_{n=1}^{N_w} R_n^- \geq R_D^- \end{cases} \tag{2.12}$$

$$R_D^+ = 3\% \widetilde{P}_L + 5\% P_n^{(0)}, \ R_D^- = R_D^+ \tag{2.13}$$

where R_D^+/R_D^- is the upward/downward regulation requirement and is set to be the sum of 3% forecasted load $\widetilde{P}_L$ and 5% forecasted wind power$P_n^{(0)}$ (USA NREL, 2010).

(5) *DC power flow:*

$$P_m^{(0)} + P_n^{(0)} - P_{load,i} - \sum_j B_{ij}\theta_{ij} = 0, \ \forall i \tag{2.14}$$

where B_{ij} and θ_{ij} are the susceptance and the voltage angle difference between node i and node j, respectively.

(6) *Maximum power flow:*

$$-P_{ij}^{\max} \leq P_{ij} \leq P_{ij}^{\max} \tag{2.15}$$

where P_{ij} and $P_{ij}^{\max}$ are the power flow and its limit value through line between node i and node j, respectively.

2.3.2 The real-time stage

In the real-time stage, the objective is to minimize the change cost Q2 of two stages, considering N_s possible scenarios which can reflect the uncertainty of wind power:

$$\min\ Q_2 = \sum_{s=1}^{N_s} \sigma^{(s)} \left\{ \sum_{m=1}^{N_g} \left(P_m^{(s)} - P_m^{(0)} \right)^2 + \sum_{n=1}^{N_w} \left(P_n^{(s)} - P_n^{(0)} \right)^2 \right\} \tag{2.16}$$

$$P_n^{(s)} = \left(1 - K_n^{(s)}\right) \widetilde{P}_n^{(s)} \tag{2.17}$$

where $\sigma^{(s)}$ is the occurring possibility of the sth scenario ($s=1.2, \ldots, N_s$); $P_n^{(s)}$ and $\widetilde{P}_n^{(s)}$ are the real-time power and the MPPT power in the sth scenario, respectively; $K_n^{(s)}$ is the de-loading ratio of the nth farm in the same scenario. From Eqs. (2.16) to (2.17), it can be observed that by properly adjusting $K_n^{(s)}$ the deviation of the real-time power $P_n^{(s)}$ and the scheduled power $P_n^{(0)}$ can be reduced.

The constraints in the real-time stage include

(1) *Active power limits*:

$$P_m^{\min} \leq P_m^{(s)} \leq P_m^{\max}, \quad \forall m, \forall s \tag{2.18}$$

(2) *De-loading ratio limits*:

$$0 \leq K_n^{(s)} \leq K_n^{\max}, \quad \forall n, \forall s \tag{2.19}$$

(3) *Power balance*:

$$\sum_{m=1}^{N_g} P_m^{(s)} + \sum_{n=1}^{N_w} P_n^{(s)} = P_{load}, \quad \forall s \tag{2.20}$$

2.3.3 The two-stage and bi-level optimization model

From the models in these two stages above, the reserve scheduling problem can be formulated into a bi-level optimization, as summarized in Table 2.1. In the lower level, the WTG de-loading ratio $K_n^{(s)}$ and the conventional unit power $P_m^{(s)}$ ($s=1,2, \ldots, N_s$) are

Table 2.1 The proposed bi-level optimization model

Level	Model	Coupled condition
The prescheduling stage (upper level)	$\min_{P_m^{(0)}, R_m^+, R_m^-} Q_1\left(K_n, \widetilde{P}_n\right)$ s.t. (2.4)–(2.15)	$K_n = \min\left(K_n^{(s)}\right)$
The real-time stage (lower level)	$\min_{K_n^{(s)}, P_m^{(s)}} Q_2\left(P_m^{(0)}, P_n^{(0)}\right)$ s.t. (2.17)–(2.20)	

to be decided to pursue the scheduling change minimization; in the upper level, $P_m^{(0)}$, R_m^+/R_m^- ($m=1, 2, \ldots, N_g$) are to be decided to pursue the overall system cost minimization.

The scheduled power in the lower level is derived from the upper level and the de-loading ratio in the upper level is the minimum de-loading ratio in all possible scenarios. In other words, the model in the prescheduling stage and the model in the real-time control stage are coupled.

2.3.4 The equivalent single-level model

By applying the KKT condition (Bard, 1987; Bazaraa et al., 2006), the bi-level model can be decoupled and transformed into a single-level optimization model with the derivation process presented in the Appendix and the resultant single-level model is given as follows:

The objective function: (2.3)

s.t. (1) The upper level constraints: Eqs. (2.4)–(2.15).

(2) The lower level constraints ($\forall m, \forall n, \forall s$):

$$\begin{cases} P_m^{\min} - P_m^{(s)} \leq 0, \quad P_m^{(s)} - P_m^{\max} \leq 0 \\ K_n^{(s)} \geq 0, \quad K_n^{(s)} - K_n^{\max} \leq 0 \end{cases} \tag{2.21}$$

$$\sum_{m=1}^{N_g} P_m^{(s)} + \sum_{n=1}^{N_w} P_n^{(s)} - P_{load}^{(s)} = 0 \tag{2.22}$$

$$\begin{cases} \displaystyle\sum_{s=1}^{N_s} \sigma^{(s)} \left[2\left(P_m^{(s)} - P_m^{(0)}\right) - \alpha_m^{(s)} + \beta_m^{(s)} + \lambda^{(s)}\right] = 0 \\ \displaystyle\sum_{s=1}^{N_s} \sigma^{(s)} \left\{2\left[P_n^{(s)} - P_n^{(0)}\right]\left(-\widetilde{P}_n^{(s)}\right) - \mu_n^{(s)} + \nu_n^{(s)} + \lambda^{(s)}\left(-\widetilde{P}_n^{(s)}\right)\right\} = 0 \end{cases} \tag{2.23}$$

$$\begin{cases} \alpha_m^{(s)}\left(P_m^{\min} - P_m^{(s)}\right) = 0 \\ \beta_m^{(s)}\left(P_m^{(s)} - P_m^{\max}\right) = 0 \\ \mu_n^{(s)}\left(-K_n^{(s)}\right) = 0 \\ \nu_n^{(s)}\left(K_n^{(s)} - K_n^{\max}\right) = 0 \end{cases} \tag{2.24}$$

$$\alpha_m^{(s)}, \beta_m^{(s)}, \mu_m^{(s)}, \nu_m^{(s)} \geq 0 \tag{2.25}$$

where $\alpha_m^{(s)}$, $\beta_m^{(s)}$, $\mu_n^{(s)}$, $\nu_n^{(s)}$ are the Lagrange multipliers corresponding to inequalities Eqs. (2.18), (2.19), $\lambda^{(s)}$ is the one corresponding to equality Eqs. (2.20); Eqs. (2.21)–(2.25) are the equivalent constraints derived from the lower level model by applying KKT condition, which include: (a) the original inequality and equality Eqs. (2.21), (2.22); (b) as for Eq. (2.23), the first partial derivatives of the Lagrange function for the single-level model is zero; (c) the linear complementarity condition Eq. (2.24); and (d) the bounds for the Lagrange multipliers in Eq. (2.25). In consideration of the complementarity condition in Eq. (2.24), the AMPL/MINOS solver is adopted to solve this single-level optimization (Fourer et al., 2003).

2.4 Case studies

A modified IEEE 30-bus system illustrated in Fig. 2.3 is adopted to validate the proposed bi-level model. There are four conventional units G1-G4 and two wind farms W1-W2 in the system. Parameters of conventional units and wind farms are given in

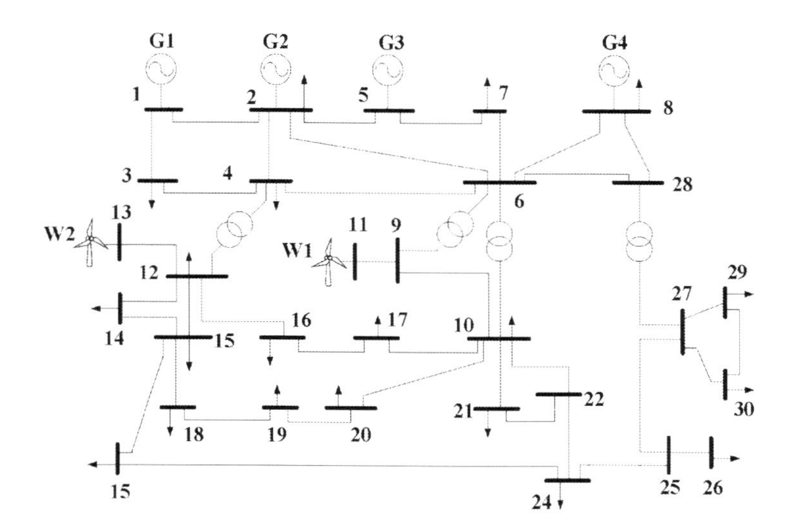

Fig. 2.3 The IEEE 30-bus system with two wind farms.

Table 2.2 Parameters of conventional units

No.	Quadratic cost coefficients			P_{min} (MW)	P_{max} (MW)	r_m^+, r_m^- (MW/h)	c_m^R ($/MW)
	a	b	c				
G1	0.00398	19.70	450	50	120	20	13
G2	0.00413	25.92	660	10	60	10	21
G3	0.00222	27.27	665	10	60	10	23
G4	0.00173	27.97	700	10	60	10	19

Table 2.3 Parameters of wind farms

No.	Rated capacity (MW)	c_n^P ($/MW)	c_n^R ($/MW)	K^{max}
W1	60	22	17	0.2
W2	40	19	15	0.2

Table 2.4 Load profiles in the system

Bus	Bus load ratio (%)	Bus	Bus load ratio (%)
6	33	10,30	4
8	11	4,15,17,19,24	3
2,7	8	10,14	2
21	6	3,16,18,20,23,26,29	1

Tables 2.2 and 2.3. All load profiles are listed in Table 2.4, with the ratio of each load to the system total load given. The network parameters can be found in Power Systems Test Case Archive (1999).

In the prescheduling stage, two kinds of operating conditions are investigated according to 24-h curves of the forecasted wind power and the total system load, as illustrated in Fig. 2.4: (1) C1: wind power plays a negative role in increasing peak reserve requirement; (2) C2: wind power plays a positive role in peak-load shaving, which can reflect a representative situation of wind power and system load fluctuation.

In the real-time stage, the scenarios of wind power are generated as follows: (1) wind power $\widetilde{P}_n^{(s)}$ ($s=1.2, \ldots, N_s$) satisfy normal distribution; (2) their mean values and variances are set as the forecasted value $\widetilde{P}_n$ and $5\%\widetilde{P}_n$, respectively; (3) the prediction error is set as $\pm 15\%$. The scenarios of bus load are generated as follows: (1) system loads have uniform distribution; (2) each load mean value is equal to the forecasted value $\widetilde{P}_L$, and the prediction error is $\pm 3\%$. To solve the scenario scale

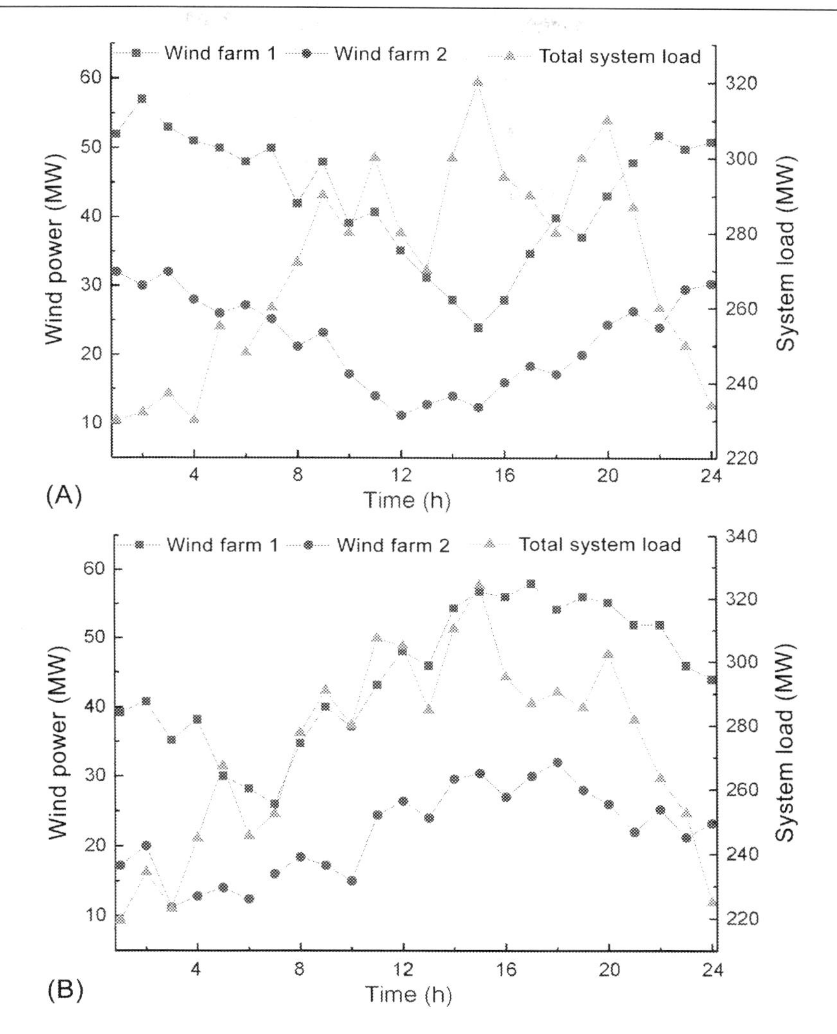

Fig. 2.4 The forecasted wind power and system load: (A) C1: WTG increasing reserve requirement and (B) C2: peak-load shaving by WTG.

problem, scenario reduction technique such as that by Growe-Kuska et al. (2003) can be adopted. A resultant group of wind power scenarios for W1 and W2 derived from C1 are given in Fig. 2.5A and B and will be used in the following studies.

2.4.1 Bi-level optimization results

The resultant de-loading ratios of the bi-level optimization model corresponding to two conditions C1 and C2 are given in Fig. 2.6A and B, respectively. Meanwhile, a penetration ratio is defined in Eq. (2.26) to measure the integration level of wind power in the system. The time-series penetration ratio is also given in Fig. 2.6A and B.

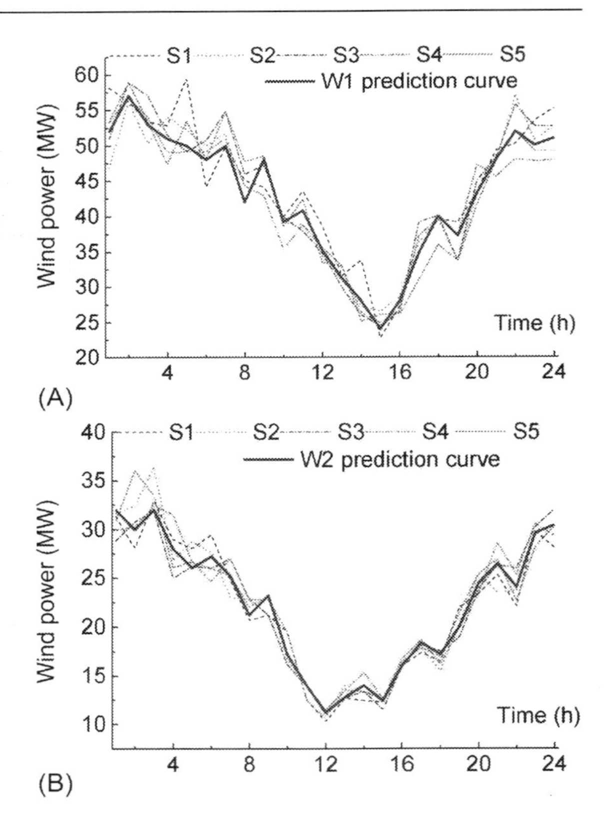

Fig. 2.5 The real-time scenarios for wind farms in the system: (A) W1 scenarios and (B) W2 scenarios.

$$\rho(t) = \frac{\sum_{n} \widetilde{P}_n(t)}{\sum_{i} \widetilde{P}_{L,i}(t)} \tag{2.26}$$

It can be easily learned from Fig. 2.6 that the WTG de-loading status is closely related to wind power penetration ratio. In the WTG increasing reserve period (C1), when penetration ratio is high (larger than 20%), WTG de-loading control is preferred to provide reserve capacity via de-loading control, as illustrated in Fig. 2.6A; in the peak-load shaving period (C2), since the reserve requirement is relatively low and the overall conventional unit reserve capacities are more economical than the WTG counterparts, and as a result the WTGs will stay at MPPT operation in most of the time.

2.4.2 Comparison between MPPT and de-loading control

The two optimization results under MPPT ($K=0$) and de-loading control ($K>0$) independently are compared in this subsection and the results are presented in Fig. 2.7. According to two operating conditions, the operation cost Q1, the scheduled

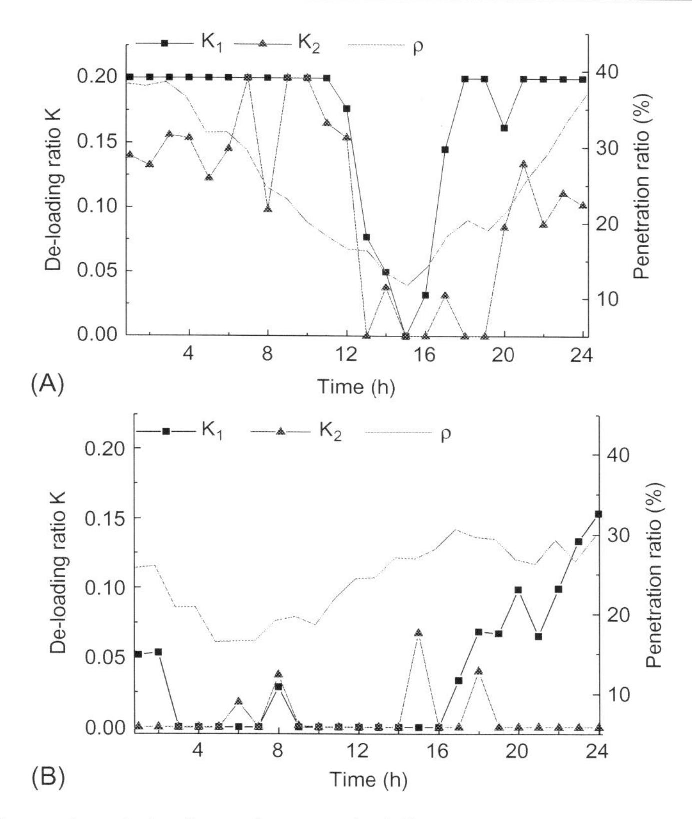

Fig. 2.6 The resultant de-loading ratio at prescheduling stage: (A) operating condition C1 and (B) operating condition C2.

wind power at the prescheduling stage, and the scheduling change cost Q2 at the real-time stage are given in Fig. 2.7A and B, respectively. Some interesting observations can be found and concluded as follows:

(1) Compared with MPPT control, the scheduling scheme using de-loading control can achieve relatively lower overall cost Q1, and particularly in the WTG increasing reserve period more benefits can be earned such as 3–13 and 20–24 h.

(2) In the real-time stage, de-loading control has less scheduling change than MPPT (the middle diagram) and the scheduled total wind power under de-loading control is less than that of MPPT (on the right), but the former is more smooth than the latter. Accordingly, the reserve requirement from conventional units under de-loading control is less.

2.4.3 The real-time control effect

In the real-time stage, if the *s*th de-loading scenario happens, that is, $K_n^{(s)}$ will be executed for wind farms control to realize the regulation capability mentioned before.

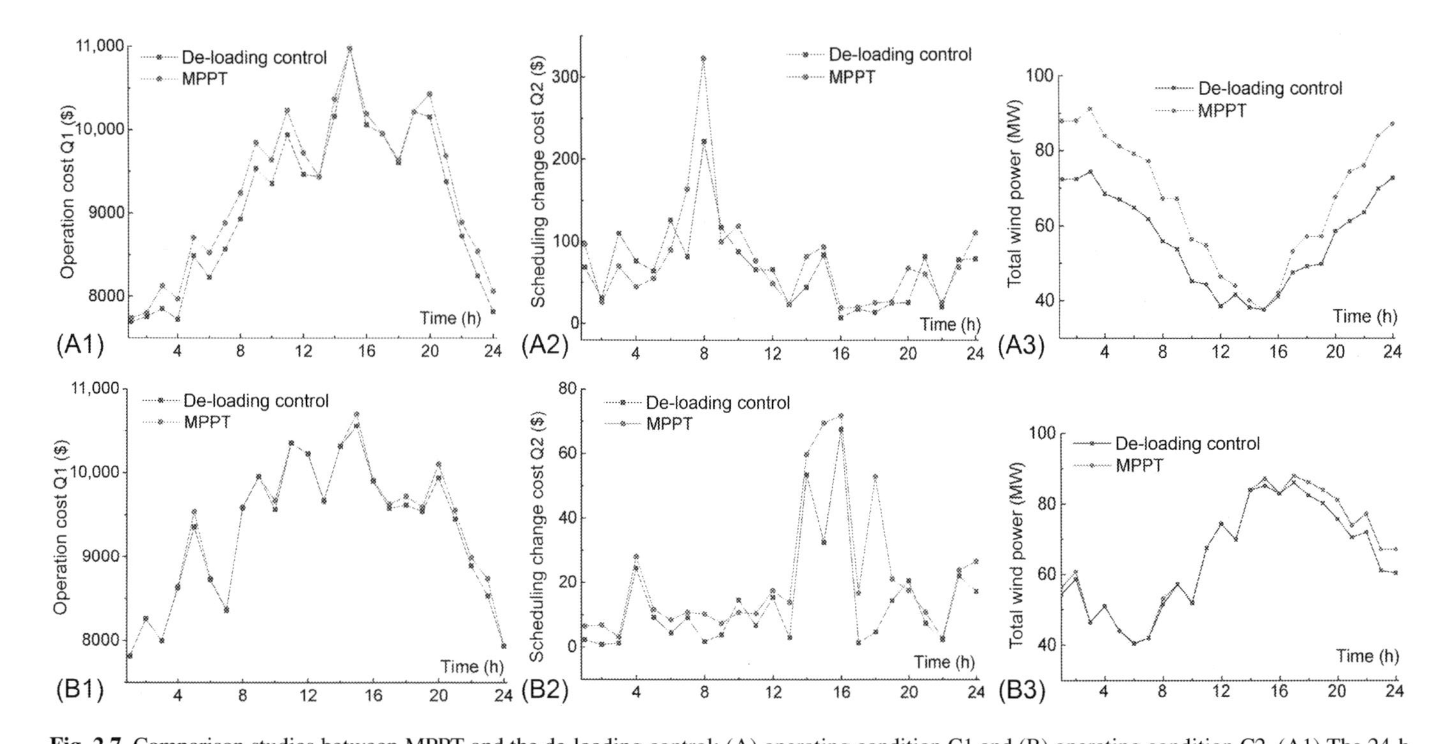

Fig. 2.7 Comparison studies between MPPT and the de-loading control: (A) operating condition C1 and (B) operating condition C2. (A1) The 24-h operation cost Q1 under operating condition C1. (A2) The 24-h scheduling change cost Q2 under operating condition C1. (A3) The 24-h total wind power under operating condition C1. (B1) The 24-h operation cost Q1 under operating condition C2. (B2) The 24-h scheduling change cost Q2 under operating condition C2. (B3) The 24-h total wind power under operating condition C2.

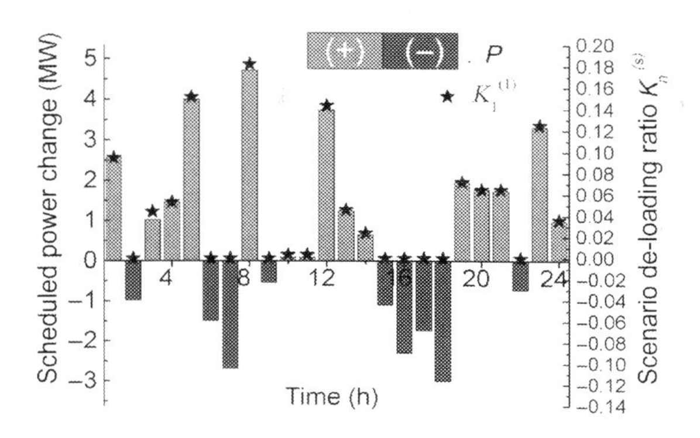

Fig. 2.8 Illustration of real-time WTG control solution.

Take the wind farm W1 under the operating condition C1 and scenario S1 for an example, the real-time de-loading ratio $K_1^{(1)}$ is presented in Fig. 2.8 and the scheduled power change between the prescheduled wind power $P_n^{(0)}$ and the real-time power $\widetilde{P}_n^{(s)}$, that is, $\Delta P = \widetilde{P}_n^{(s)} - P_n^{(0)}$ is simultaneously given. Evidently, WTG will stay at MPPT ($K_n^{(s)} = 0$) if $\Delta P < 0$; WTG will stay at the de-loading operation ($K_n^{(s)} > 0$) when $\Delta P > 0$ and larger ΔP means larger $K_n^{(s)}$ under these circumstances. It is easy to understand that the scheduling change cost Q2 will decrease by executing de-loading control scheme mentioned above to make the real-time power $P_n^{(s)}$ close to the prescheduled power $P_n^{(0)}$.

2.4.4 Results of reserve scheduling

The comparison of reserve scheduling between two control modes is listed in Table 2.5. It can be seen that de-loading control can achieve less reserve cost and less reserve capacities served by conventional units.

Table 2.5 Reserve scheduling comparison between MPPT and de-loading control

Operating conditions		Conventional units		WTGs		Total reserve cost ($)
		Reserve capacity (MW)	**Reserve cost ($)**	**Reserve capacity (MW)**	**Reserve cost ($)**	
C1	MPPT	275.85	4037.87	0	0	4037.87
	De-loading	136.56	1956.28	104.67	1664.13	3620.41
C2	MPPT	275.75	4006.95	0	0	4006.95
	De-loading	230.95	3386.34	31.84	509.44	3895.78

2.5 Conclusion

APC of WTG is a promising technology which can make WTG provide system reserve and frequency support. In this chapter, a two-stage reserve scheduling problem, or an equivalent bi-level reserve scheduling optimization model that can consider WTG's de-loading control has been developed: in the prescheduling stage, the total system scheduling cost is to be optimized; in the real-time stage, the change or adjustment to the scheduling scheme made in the first stage is to be minimized. Potential benefits under WTG de-loading control are evaluated by this model. Case studies validate the effectiveness of the proposed model and the following conclusions can be drawn:

(1) De-loading control can suppress wind power fluctuation, reduce system reserve, and enhance the overall system economy.
(2) The de-loading ratio can be selected as an effective information for active and flexible wind power integration.

Appendix. The single-level model formulation

The formulation procedure to transform the bi-level optimization into an equivalent single-level model is explained further, in which the KKT condition is applied to the lower-level optimization model.

The original bi-level optimization model is listed as follows:

$$\min_{P_m^{(0)}, R_m^+, R_m^-} \quad Q_1\left(K_n, \widetilde{P}_n\right)$$

s.t. (2.4)–(2.15)

$$K_n = \min\left(K_n^{(s)}\right)$$

$$\min_{K_n^{(s)}, P_m^{(s)}} Q_2\left(P_m^{(0)}, P_n^{(0)}\right)$$

s.t. (2.17)–(2.20).

The procedure of the derivation process can be summarized as follow:

Step 1. Establishing the Lagrange function for the lower-level in Eq. (A.1):

$$\begin{aligned} f = &\sum_{s=1}^{N_s} \sigma^{(s)} \left\{ \sum_{m=1}^{N_g} \left(P_m - P_m^{(0)}\right)^2 + \sum_{n=1}^{N_w} \left[P_n^{(s)} - P_n^{(0)}\right]^2 \right\} \\ &+ \sum_{s=1}^{N_s} \left\{ \sum_{m=1}^{N_g} \alpha_m^{(s)} \left(P_m^{\min} - P_m^{(s)}\right) + \sum_{m=1}^{N_g} \beta_m^{(s)} \left(P_m^{(s)} - P_m^{\max}\right) + \sum_{n=1}^{N_w} \mu_n^{(s)} \left(-K_n^{(s)}\right) \right. \\ &\left. + \sum_{n=1}^{N_w} \nu_n^{(s)} \left(K_n^{(s)} - K_n^{\max}\right) + \lambda^{(s)} \left[\sum_{m=1}^{N_g} P_m^{(s)} + \sum_{n=1}^{N_w} P_n^{(s)} - P_{load}^{(s)} \right] \right\} \end{aligned} \quad (A.1)$$

where $\alpha_m^{(s)}$, $\beta_m^{(s)}$, $\mu_n^{(s)}$, $\nu_n^{(s)}$ are the Lagrange multipliers corresponding to inequalities Eqs. (2.18), (2.19), $\lambda^{(s)}$ is the one corresponding to equality Eq. (2.20).

Step 2: Taking the derivatives of the Lagrange function w.r.t. decision variables and forming the following equalities:

$$\frac{df}{dP_m} = \sum_{s=1}^{N_s} \sigma_s \left[2\left(P_m^{(s)} - P_m^{(0)}\right) - \alpha_m^{(s)} + \beta_m^{(s)} + \lambda^{(s)} \right] = 0 \quad \forall m \tag{A.2}$$

$$\frac{df}{dK_n} = \sum_{s=1}^{N_s} \sigma_s \left\{ 2\left(P_n^{(s)} - P_n^{(0)}\right)\left(-\widetilde{P}_n^{(s)}\right) - \mu_n^{(s)} + \nu_n^{(s)} + \lambda^{(s)}\left(-\widetilde{P}_n^{(s)}\right) \right\} = 0 \quad \forall n \tag{A.3}$$

Step 3: Listing other inequalities and equalities related:

$$\begin{cases} P_m^{\min} - P_m^{(s)} \le 0, \quad P_m^{(s)} - P_m^{\max} \le 0 \quad \forall m, \forall s \\ K_n^{(s)} \ge 0, \quad K_n^{(s)} - K_n^{\max} \le 0 \quad \forall n, \forall s \end{cases} \tag{A.4}$$

$$\sum_{m=1}^{N_g} P_m^{(s)} + \sum_{n=1}^{N_w} P_n^{(s)} - P_{load}^{(s)} = 0 \quad \forall s \tag{A.5}$$

$$\alpha_m^{(s)}, \beta_m^{(s)}, \mu_m^{(s)}, \nu_m^{(s)} \ge 0 \quad \forall m, \forall s \tag{A.6}$$

Step 4: Forming the following linear complementarity condition:

$$\begin{cases} \alpha_m^{(s)}\left(P_m^{\min} - P_m^{(s)}\right) = 0 \\ \beta_m^{(s)}\left(P_m^{(s)} - P_m^{\max}\right) = 0 \end{cases} \quad \forall m, \forall s \tag{A.7}$$

$$\begin{cases} \mu_n^{(s)}\left(-K_n^{(s)}\right) = 0 \\ \nu_n^{(s)}\left(K_n^{(s)} - K_n^{\max}\right) = 0 \end{cases} \quad \forall n, \forall s \tag{A.8}$$

Then the optimal solution in the lower-level optimization model will satisfy Eqs. (A.2), (A.3), (A.7), (A.8). Thus the final single-level model can be formulated as follows:

The objective function:

$$\min Q_1 = \sum_{m=1}^{N_g} \left[f_m\left(P_m^{(0)}\right) + c_m^R R_m \right] + \sum_{n=1}^{N_w} \left[c_n^P P_n^{(0)} + c_n^R R_n \right]$$

s.t. (1) The original constraints in the upper-level model:

$$\begin{cases} P_m^{\min} \le P_m^{(0)} \le P_m^{\max} \quad \forall m \\ \begin{cases} P_m^{(0)}(t) - P_m^{(0)}(t-1) \le r_m^+ D \\ P_m^{(0)}(t-1) - P_m^{(0)}(t) \le r_m^- D \end{cases} \quad \forall m \\ 0 \le R_m \le \min\left(P_m^{\max} - P_m^{(0)}, r_m D\right) \\ \sum_{m=1}^{N_g} R_m + \sum_{n=1}^{N_w} K_n \widetilde{P}_n \ge R_D \\ P_m^{(0)} + P_n^{(0)} - P_{load,i} - \sum B_{ij}\theta_{ij} = 0, \forall i \\ -P_{ij}^{\max} \le P_{ij} \le P_{ij}^{\max} \\ 0 \le K_n \le K_n^{\max} \quad \forall n \\ K_n = \min\left(K_n^{(s)}\right) \end{cases} \tag{A.9}$$

(2) *The equivalent constraints* in the lower-level model after applying KKT:

$$\begin{cases} \dfrac{df}{dP_m} = \sum_{s=1}^{N_s} \sigma_s \left[2\left(P_m^{(s)} - P_m^{(0)}\right) - \alpha_m^{(s)} + \beta_m^{(s)} + \lambda^{(s)}\right] = 0 \quad \forall m \\ \dfrac{df}{dK_n} = \sum_{s=1}^{N_s} \sigma_s \left\{2\left(P_n^{(s)} - P_n^{(0)}\right)\left(-\widetilde{P}_n^{(s)}\right) - \mu_n^{(s)} + \nu_n^{(s)} + \lambda^{(s)}\left(-\widetilde{P}_n^{(s)}\right)\right\} = 0 \quad \forall n \\ \begin{cases} P_m^{\min} - P_m^{(s)} \le 0, \;\; P_m^{(s)} - P_m^{\max} \le 0 \quad \forall m, \forall s \\ K_n^{(s)} \ge 0, \;\; K_n^{(s)} - K_n^{\max} \le 0 \quad \forall n, \forall s \end{cases} \\ \sum_{m=1}^{N_g} P_m^{(s)} + \sum_{n=1}^{N_w} P_n^{(s)} - P_{load}^{(s)} = 0 \quad \forall s \\ \alpha_m^{(s)}, \beta_m^{(s)}, \mu_m^{(s)}, \nu_m^{(s)} \ge 0 \quad \forall m, \forall s \\ \begin{cases} \alpha_m^{(s)}\left(P_m^{\min} - P_m^{(s)}\right) = 0 \\ \beta_m^{(s)}\left(P_m^{(s)} - P_m^{\max}\right) = 0 \end{cases} \quad \forall m, \forall s \\ \begin{cases} \mu_n^{(s)}\left(-K_n^{(s)}\right) = 0 \\ \nu_n^{(s)}\left(K_n^{(s)} - K_n^{\max}\right) = 0 \end{cases} \quad \forall n, \forall s \end{cases} \tag{A.10}$$

Acknowledgment

The National Natural Science Foundation of China (No. 51677165) is to be acknowledged for the funding support.

References

Attya, A.B.T., Hartkopf, T., 2013. Control and quantification of kinetic energy released by wind farms during power system frequency drops. IET Renew. Power Gener. 7 (3), 210–224.

Attya, A.B., Hartkopf, T., 2014. Wind turbine contribution in frequency drop mitigation - modified operation and estimating released supportive energy. IET Gener. Transm. Distrib. 8 (5), 862–872.

Bard, J.F., 1987. Convex two-level optimazation. Math. Program. 40, 15–27.

Bazaraa, M.S., Sherali, H.D., Shetty, C.M., 2006. Nonlinear Programming: Theory and Algorithms. John Wiley & Sons, Inc., Hoboken, New Jersey.

Bruninx, K., Delarue, E., 2014. A statistical description of the error on wind power forecasts for probabilistic reserve sizing. IEEE Trans. Sustain. Energy 5 (3), 995–1002.

Conejo, A.J., Carrión, M., Morales, J.M., 2010. Decision Making Under Uncertainty in Electricity Markets. Springer, New York.

Fourer, R., Gay, D.M., Kernighan, B.W., 2003. AMPL-A Modeling Language for Mathematical Programming, second ed. Thomson, Duxbury, MA.

Global Wind Energy Council (GWEC), 2017. Global Wind Report 2009–2016. Available from: http://gwec.net/publications/global-wind-report-2/.

Growe-Kuska, N., Heitsch, H., Romisch, W., 2003. Scenario reduction and scenario tree construction for power management problems. Proc. of 2003 IEEE Bologna Power Tech Conference, Bologna, Italy, 23–26 June.

Holttinen, H., Milligan, M., Ela, E., et al., 2012. Methodologies to determine operating reserves due to increased wind power. IEEE Trans. Sustain. Energy 3 (4), 713–723.

Kroposki, B., Johnson, B., Zhang, Y., et al., 2017. Achieving a 100% renewable grid: operating electric power systems with extremely high levels of variable renewable energy. IEEE Power Energy Mag. 15 (2), 61–73.

Lu, Z., Wang, Z., Xin, H., et al., 2014. Small signal stability analysis of a synchronized control-based microgrid under multiple operating conditions. J Mod Power Syst Clean Energy 2 (3), 244–255.

Morales, J.M., Conejo, A.J., Perez-Ruiz, J., 2009. Economic valuation of reserves in power systems with high penetration of wind power. IEEE Trans. Power Syst. 24 (2), 900–910.

Ortega-Vazquez, M.A., Kirschen, D.S., 2009. Estimating the spinning reserve requirements in systems with significant wind power generation penetration. IEEE Trans. Power Syst. 24 (1), 114–124.

Paterakis, N.G., Erdinc, O., Bakirtzis, A.G., et al., 2015. Load-following reserves procurement considering flexible demand-side resources under high wind power penetration. IEEE Trans. Power Syst. 30 (3), 1337–1350.

Piwko, R., Meibom, P., Holttinen, H., et al., 2012. Penetrating insights: lessons learned from large-scale wind power integration. IEEE Power Energy Mag. 10 (2), 44–52.

Power Systems Test Case Archive, 1999. IEEE 30-Bus Test System. Available from: https://www2.ee.washington.edu/research/pstca/. (Accessed 10 September 2017).

Ramtharan, G., Ekanayake, J.B., Jenkins, N., 2007. Frequency support from doubly fed induction generator wind turbines. IET Renew. Power Gener. 1 (1), 3–9.

Rawn, B.G., Lehn, P., 2008. Wind rotor inertia and variable efficiency: fundamental limits on their exploitation for inertial response and power system damping.Proc. of Euro Wind Energy Conf (EWEC2008), Brussels, Belgium, 31 March-3 April.

USA NREL, 2010. Western Wind and Solar Integration Study. Available from: https://www.nrel.gov/docs/fy10osti/47781.pdf.

Wang, H., Chen, Z., Jiang, Q., 2015. Optimal control method for wind farm to support temporary primary frequency control with minimised wind energy cost. IET Renew. Power Gener. 9 (4), 350–359.

Wang, Z., Bian, Q., Xin, H., et al., 2016. A distributionally robust co-ordinated reserve scheduling model considering CVaR-based wind power reserve requirements. IEEE Trans. Sustain. Energy 7 (2), 625–636.

Wang-Hansen, M., Josefsson, R., Mehmedovic, H., 2013. Frequency controlling wind power modeling of control strategies. IEEE Trans. Sustain. Energy 4 (4), 954–959.

Wu, W., Zhang, B., Chen, J., et al., 2012. Multiple time-scale coordinated power control system to accommodate significant wind power penetration and its real application.Proc. of IEEE PES Gen Meet (PES2012), San Diego, USA, 22-26 July 2012.

Dynamic energy management and control of a grid-interactive DC microgrid system

Bonu Ramesh Naidu, Gayadhar Panda†, B. Chitti Babu‡*
*Department of Electrical Engineering, Indian Institute of Technology Kharagpur, Kharagpur, India, †Department of Electrical Engineering, National Institute of Technology Meghalaya, Shillong, India, ‡University of Nottingham Malaysia Campus, Semenyih, Malaysia

3.1 Introduction

The significant improvements in the solar photovoltaic (PV) generation technologies indicate the future with a noninertial generation dominated utility grids in the near future. A growing number of electric industry leaders agree that it is only a matter of time before renewable energy resources dominate their grid systems. In the context of increasing interest for smart grids, grid-connected renewable energy systems play a major role. However, the increasing renewable energy penetration poses typical challenges for the existing power system like stability, protection, and power quality issues. This indicates the necessity of a proper channel for the efficient and stable inclusion of such sporadic generation. DC microgrid acts as an effective platform for such purposes due to its renowned characteristics like increased efficiency, reduced complexity in control, negligible power quality issues, etc.

3.2 System description

The system under study comprises of a grid-connected DC microgrid as shown in Fig. 3.1. The DC microgrid employs solar PV array-based energy source assisted by battery and supercapacitor bank-based hybrid energy storage devices (HESD). The components of the HESD are so chosen to equip significant specific power density and energy density to the system and are controlled to address the indeterminacy associated with the solar PV generation. DC-DC boost converter interfaces the PV array with the DC-link whereas bidirectional converters are used to interface supercapacitor and battery banks with the DC-link. The DC microgrid is connected to the utility grid through a two-level three-phase voltage source inverter with a facilitation for bidirectional power flow. The sole purpose of the interfacing converters is to facilitate a controlled power flow between the respective devices and the DC-link. The inductor filter smoothens the switching frequency harmonics at the voltage source inverter (VSI) output. The power output from the VSI is either fed to the local loads

Smart Power Distribution Systems. https://doi.org/10.1016/B978-0-12-812154-2.00003-1

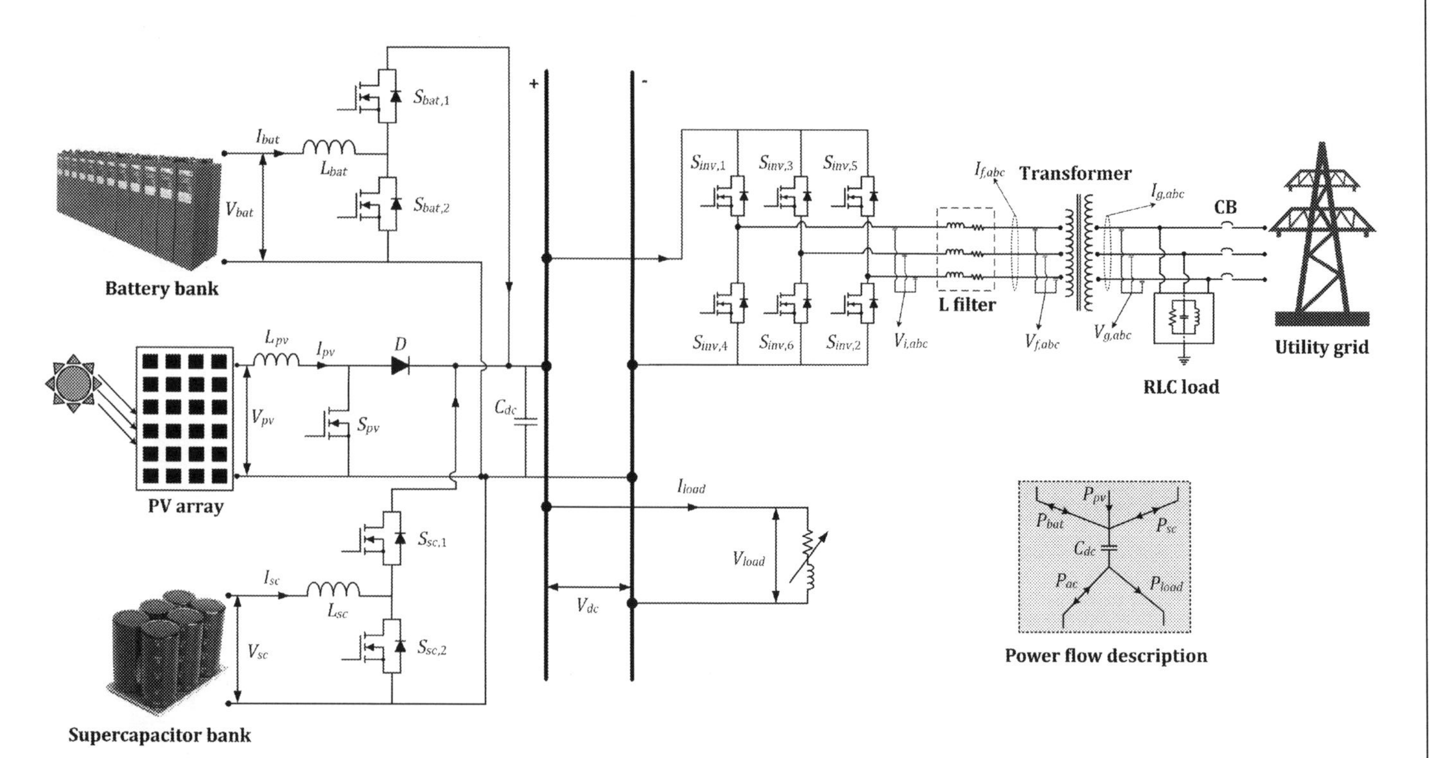

Fig. 3.1 Schematic representation of the grid-connected DC microgrid under study.

or is fed into the utility grid (slack terminal) using a proper control and energy management strategy. Throughout this work, the utility grid is assumed to be balanced and connected to the DC microgrid. The complete block diagram of the system along with the assumed power flow is presented in Fig. 3.1. Before moving on to the operating constraints associated with the chosen system, different components, their operating principle and modeling are briefed in the following section.

3.2.1 Modeling of the solar PV generation

A PV cell is the most basic generation part in PV system. There are many kinds of solar cells with respect to the type of materials used to fabricate the cell. The single-diode mathematical model is applicable to simulate silicon PV cells, which consist of a photocurrent source I_L, a nonlinear diode, internal resistances R_s and R_{sh} as shown in Fig. 3.2 (Wang and Nehrir, 2008).

The mathematical relationship between the terminal current (I) and the voltage (V) in the single-diode equivalent circuit is described by

$$I = I_L - I_D\left(e^{\frac{q(V+IR_{se})}{AkT}}\right) - \frac{V+IR_{se}}{R_{sh}} \tag{3.1}$$

where I_D is the diode saturation current; T is the cell temperature in kelvin; A is the ideality factor of the diode; R_{se} and R_{sh} are the series and shunt resistance of the photo cell. Photocurrent (I_L) is the function of solar radiation and cell temperature, described as follows:

$$I_L = \frac{S}{S_{ref}}\left[I_{L,ref} + C_T\left(T - T_{ref}\right)\right] \tag{3.2}$$

where S is the incident solar radiation; S_{ref}, $I_{L,ref}$, and T_{ref} are the solar irradiance, cell absolute temperature, and photocurrent in standard test conditions, respectively; and C_T is the temperature coefficient.

3.2.2 Modeling of supercapacitor bank

The supercapacitor is an electrostatic device that comes with a huge capacitance due to its constructional feature, that is, the electrodynamic double layer. Since its operation is like any other capacitors, it can be discharged/charged endlessly unlike the batteries

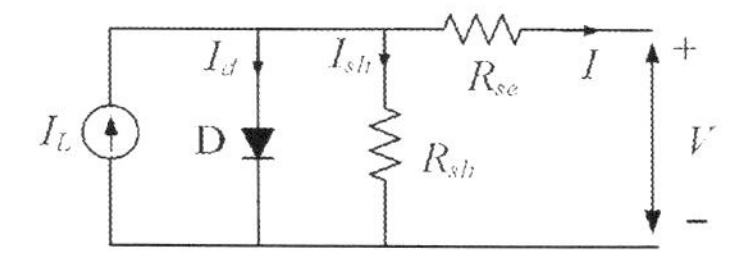

Fig. 3.2 Equivalent circuit of a solar cell.

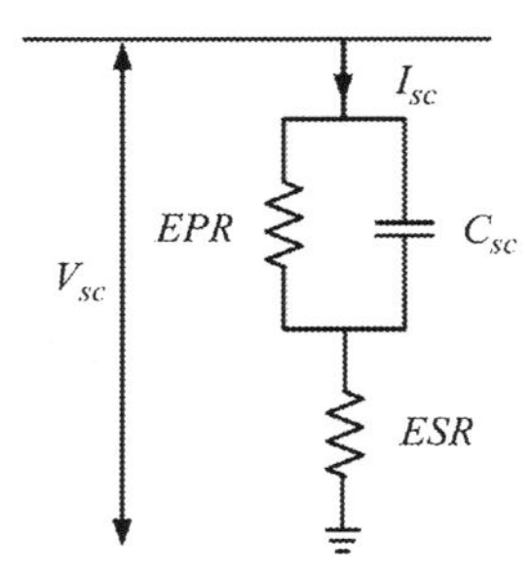

Fig. 3.3 Classical RC equivalent circuit of a supercapacitor.

that show degradation with the increasing cycles of operation. Hardly the supercapacitor degrades to 80% of its rated capacity over a duration of 10 years. The supercapacitor is comparable to batteries in terms of higher energy density and to capacitors in terms of higher power density and thus finds its place in energy storage applications. The electrical equivalent circuit of the supercapacitor is shown in Fig. 3.3. *ESR* and *EPR* represent the equivalent series and parallel resistance offered by the supercapacitor, respectively. C_{eq} is the equivalent capacitance of the supercapacitor (Cheng et al., 2010; Thounthong et al., 2009).

3.2.3 Modeling of battery bank

A battery consists of one or more electrochemical cells, connected in series or parallel. In these cells, chemically stored energy is converted into electrical energy through an electrochemical reaction. The battery is modeled using a simple controlled voltage source in series with a constant resistance (Zhou and Sun, 2014). The battery is modeled as a nonlinear voltage source using the following equation:

$$V_{bat} = E_o - k_{bat}\frac{Q_{max}}{Q_{max} - \int i_{bat}dt} + A_{bat}e^{B_{bat}\int i_{bat}dt} - i_{bat}R_i \tag{3.3}$$

where V_{bat} and E_o are the terminal voltage and battery constant voltage, respectively; K_{bat} is the polarization constant; $Q_{\max}$ is maximum capacity of the battery; A_{bat} is the exponential zone voltage in volts; B_{bat} is the exponential zone capacity Ah^{-1}; R_i is the internal resistance of the battery; i_{bat} is the battery current in amperes; and dt is the time step in seconds.

3.2.4 Modeling of the DC-DC boost converter

DC-DC converters usually interface two different DC voltage levels to facilitate required controlled power flow. DC-DC boost converter presents a higher output voltage compared to its input, keeping the total power constant. In doing so, it has two

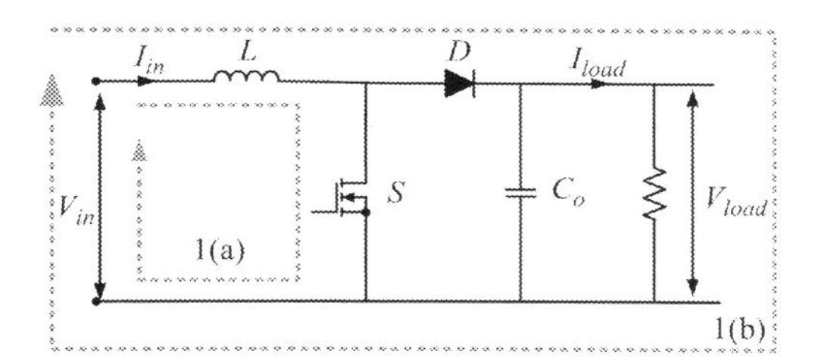

Fig. 3.4 Schematic representation of a DC-DC boost converter with its modes of operation.

modes of operation: continuous conduction mode or discontinuous conduction mode. The continuous current mode has many inherent advantages like feasibility of output voltage to duty cycle control, the absence of ringing phenomenon and reduction in power losses for the same amount of power transfer. Schematic representation of the boost converter is shown in Fig. 3.4 and typical modes of operation are indicated as 1(a) and (b). Generally, a switch belonging to transistor family is used based on the application (MOSFET for low-voltage high-current application, IGBT for high-voltage application). The inductor (L) in series with the supply voltage helps reduce the ripple in input current whereas the output capacitor (C) helps in reducing the voltage ripple at the output terminals.

The average voltage across the inductor over one switching time (T_s) is zero. Consequently, the average output voltage V_{load} given by

$$V_0 = V_{in} \times \frac{1}{1-D} \tag{3.4}$$

where $D = \frac{T_{on}}{T_s}$ is termed as duty cycle of the switch in the boost converter and T_{on} is the total ON period of the switch. Always $0 \leq D \leq 1$. The output voltage is always greater than the input voltage.

3.2.5 Modeling of the DC-DC bidirectional converter

Bidirectional converter incorporates both the buck and boost modes of operation. Generally they are used to interface low-voltage energy storage devices with the high-voltage DC bus. The energy storage device voltage can be kept lower than the reference DC-link voltage (V_{dc}) and hence less number of series combinations are sufficient to obtain the required voltage. This could reduce the overall cost and maintenance of the energy storage device. The DC-DC bidirectional converter shown in Fig. 3.5 is operated in continuous conduction mode. The switches S_1 and S_2 are switched in such a way that the converter operates in a steady state with four subintervals. All the four subintervals are marked on the figure.

The boost mode of operation of the bidirectional converter is represented by 1 (a) and 1(b) that facilitates the transfer of power from the low-voltage side to the high voltage. Similarly 2(a) and (b) represents the buck mode of operation where the power transfer is reversed.

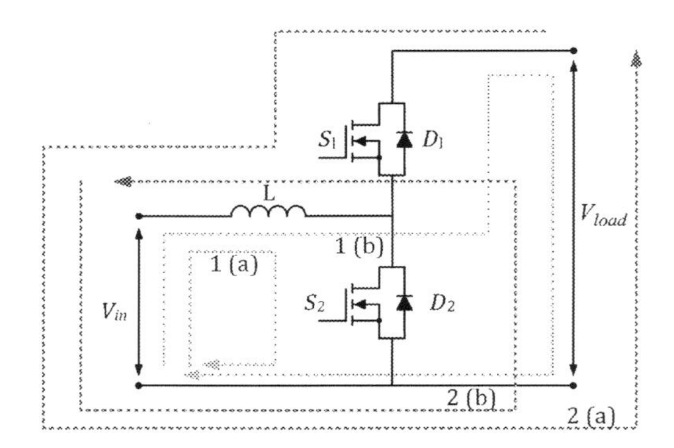

Fig. 3.5 Schematic representation of the DC-DC bidirectional converter with its modes of operation.

3.2.6 Modeling of the voltage source inverter

VSI is mainly used to convert a constant DC voltage into an AC voltage with variable magnitude and frequency. Fig. 3.6 shows a schematic diagram of a three-phase VSI. The inverter is composed of six switches $S_{inv,1}$ through $S_{inv,6}$ with each phase output connected to the middle of each inverter leg. In the simplest form, three reference signals are compared to a high-frequency carrier waveform to control the output AC voltage of the inverter. The result of that comparison in each leg is used to turn the switches ON or OFF. It should be noted that the switches in each leg should be operated interchangeably, in order to avoid a dead short circuit of the DC supply.

The dynamic equations of the output side of the VSI are given by (Chittibabu and Mohanty, 2017)

$$V_{i,abc} = V_{f,abc} + L_f \frac{dI_{f,abc}}{dt} + R_f I_{f,abc} \tag{3.5}$$

where L_f and R_f represents the per phase filter inductance and resistance, respectively; $I_{f,abc}$ and $V_{i,abc}$ are the three-phase inverter output current and voltage, respectively, whereas $V_{f,abc}$ is the filter output voltage. The conventional method to interface

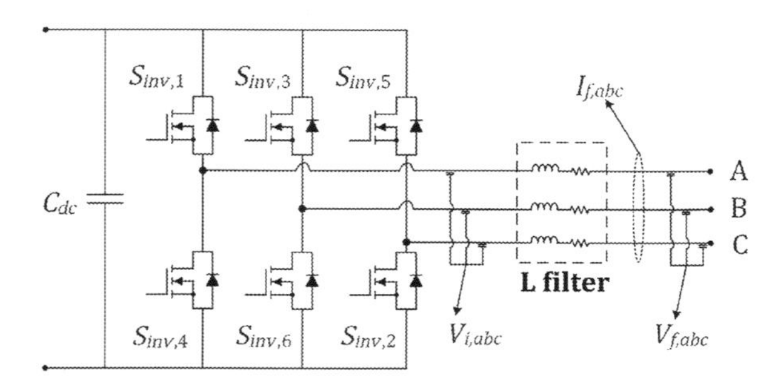

Fig. 3.6 Schematic representation of the voltage source inverter along with the output filter.

VSI of distributed generation to PCC is through first-order low pass filter (L_f). This interface filter inductor is required for proper shaping of current being injected at PCC. L_f is determined by the following equation (Geddada et al., 2012).

$$L_f = \frac{V_{dc} \times D(1-D)}{f_{sw} \times \Delta_{i,p-p}} \tag{3.6}$$

where V_{dc} is the total DC-link voltage and $\Delta_{i,\,p-p}$ is the allowable peak-to-peak current ripple expressed as the percentage of maximum current in inductor. f_{sw} is the switching frequency of triangular carrier signal. D is the duty cycle of the switch, and the worst-case current ripple occurs at 50% duty cycle.

3.3 Dynamic energy management and control

The system under study primarily deals with the grid integration of the sporadic PV generation and demands for the coexistence of different HESD along with the utility grid. Under such a scenario, there exists a necessity for dynamic energy management (DEM) among the system participants in order to effectively schedule discharging/charging commands for the microgrid participants and track the same for varying system conditions. The main challenges that are to be addressed by the DEM employed to this system are

- To employ maximum power point tracking (MPPT) scheme to control solar PV converter such that it will always equip maximum power into the system. Such a source can be treated as a nondispatchable unit.
- To regulate the DC-link voltage irrespective of the operational scenarios of the system.
- To establish proper power sharing among the utility grid, battery bank, and supercapacitor bank irrespective to the varying system scenarios.
- To establish proper synchronization and required bidirectional power transfer between the utility grid and the DC microgrid.
- To employ suitable current controllers in AC and DC side to track the respective reference current commands.

To deal with these operational challenges a DEM has been presented. The DEM mainly comprises of different functions such as generation of reference currents using a dedicated power management strategy (PMS), tracking of the reference current commands and switching pulse generation for the interfacing converters. The complete overview of the energy management strategy along with the control is shown in Fig. 3.7. The overall energy management can be divided into three important parts.

(1) DC-link voltage control.
(2) PMS for the HESD and the utility grid.
(3) Grid synchronization and control.

3.3.1 DC-link voltage control

The PV array is controlled to operate at its maximum power point by the incremental conductance-based MPPT control (Harini and Syama, 2015; Faraji et al., 2011) giving

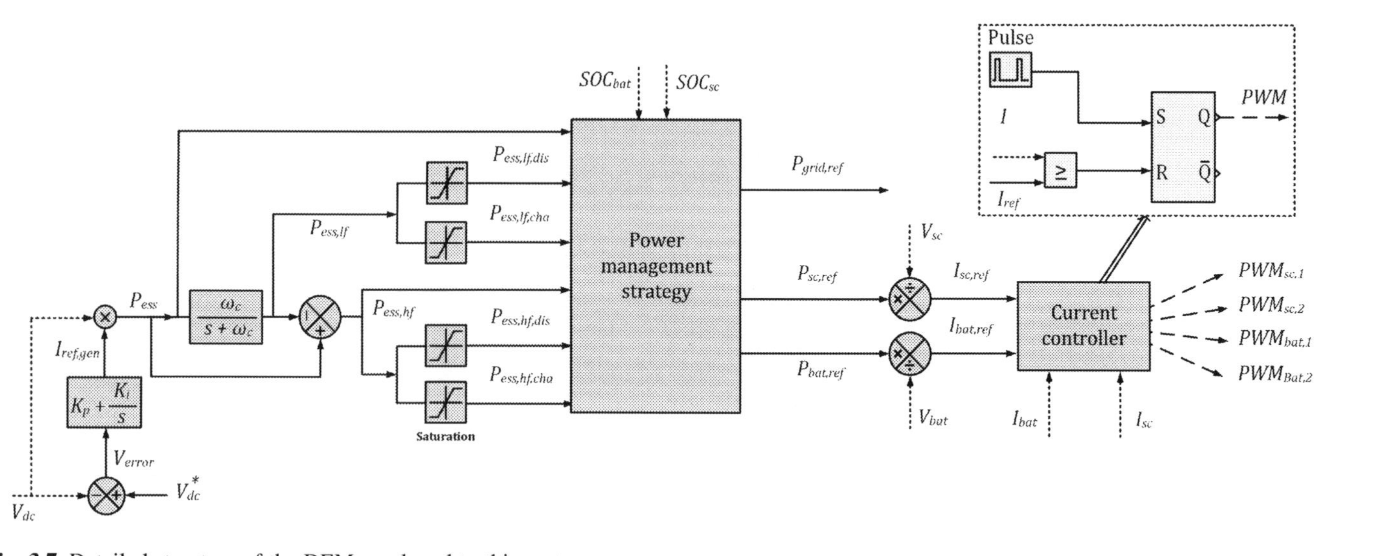

Fig. 3.7 Detailed structure of the DEM employed to this system.

rise to a possibility of excess or deficient power at the DC-link (as a proper load-generation match is rarely achieved) giving rise to variations in DC-link voltage. The DC side controller is responsible for the control of the DC-link voltage.

In a DC microgrid, the variation of the DC-link voltage occurs majorly due to the variation in power flow from/to the DC-link. The primary task of the DC side controller is to estimate near accurately the amount of power imbalance arising at the DC-link and thereby reduce the same by dispatching the HESD accordingly. To this effect a voltage controller utilizing PI control is utilized to generate the equivalent current that is to be injected into or taken away from the DC-link (Tummuru et al., 2015; Nehrir et al., 2011). The expression for reference current generated by the voltage controller is given by

$$I_{ref,gen} = K_p V_{error} + K_i \int V_{error} dt \tag{3.7}$$

where $I_{ref,\ gen}$ is the DC-link reference current generated by the voltage controller based on the voltage error given by $V_{error} = V_{dc}^{*} - V_{dc}$; V_{dc}^{*} and V_{dc} are the reference and actual voltages of the DC-link, respectively; and K_p and K_i are the proportional and integral gains of the voltage controller, respectively.

The reference current generated by Eq. (3.7) is converted into the equivalent power command by Eq. (3.8). The power command (P_{ess}) dispatched by the voltage controller is then passed on to the PMS to liberate the equivalent current command for the operation of the DC microgrid participants using Eqs. (3.12), (3.13). The SR flip-flop-based current controller then tracks the respective current commands to make the equipment operate as a dispatchable unit (Ramesh Naidu et al., 2017). Switching frequency of the SR flip-flop utilized in this analysis is 20 kHz. The detailed view of the DC side controller is shown in Fig. 3.7.

$$P_{ess} = V_{dc} \times I_{ref,gen} \tag{3.8}$$

3.3.2 PMS for the HESD and utility grid

In general, the charging or discharging of the battery bank is limited to low-frequency variations due to its electrochemical reactions whereas supercapacitor bank is good at absorbing high-frequency power variations due to its electrostatic operation (Tummuru et al., 2015; Nehrir et al., 2011). Also, the state of charge (SOC) of the energy storage devices needs to be limited to their extreme limits. Taking all these facts into consideration, it is a best practice to segregate the total power command as high-frequency and low-frequency commands and supply it separately to the battery bank and supercapacitor bank. Hence,

$$P_{ess} = P_{ess,\,hf} + P_{ess,\,lf} \tag{3.9}$$

where $P_{ess,\ hf}$ and $P_{ess,\ lf}$ constitute the high-frequency and low-frequency variations in power command. $P_{ess,\ lf}$ and $P_{ess,\ hf}$ can individually be segregated into discharging and charging power commands based on their sign.

$$P_{ess,\,hf} = P_{ess,\,hf,dis} + P_{ess,\,hf,cha} \tag{3.10}$$

$$P_{ess,\,lf} = P_{ess,\,lf,dis} + P_{ess,\,lf,cha} \tag{3.11}$$

where $P_{ess,\,hf,\,dis}$, $P_{ess,\,hf,\,cha}$, $P_{ess,\,lf,\,dis}$, and $P_{ess,\,lf,\,cha}$ are the high-frequency discharging, charging, low-frequency discharging, and charging power commands, respectively. A low-pass filter with a cut-off frequency "ω_c" is used to filter out the low-frequency power command from P_{ess}.

The SOC of the supercapacitor should be limited between $SOC_{sc,\min}$ and $SOC_{sc,\max}$. Similarly, the SOC of the battery should be limited between its extremes $SOC_{bat,\min}$ and $SOC_{bat,\max}$. It is best advised to stop the storage device from absorbing further energy when its SOC limit is at its upper limit and vice versa. To do so, the utility grid should replicate the operation of the storage device during their operation at extreme SOC limits. Irrespective of the P_{ess} being positive or negative or zero, the following combinations may exist during different system conditions. They are

- Battery + Supercapacitor
  This combination prevails as long as the SOC limits of both the devices are within their extreme limits, or when the SOC of the HESD is at/below its lower limit and there exists a charging reference power on the ESD or when the SOC of the HESD is at its upper limit and the HESD is insisted to discharge at that instant. During such combination of devices, the utility grid stays connected to the DC microgrid under proper synchronization but does not transfer power, that is, utility grid stays in standby mode.
- Battery + Utility grid
  This combination of devices comes into operation only when the SOC of the supercapacitor is at its lower limit and there exists a discharging reference power to the device and/or when the SOC of the supercapacitor is at its upper limit and there is the availability of excess power at the DC-link. During this combination of devices, the utility grid stays connected to the DC microgrid under proper synchronization and facilitates the required bidirectional high-frequency power flow, that is, it substitutes the supercapacitor.
- Supercapacitor + Utility grid
  The utility grid performs the task of the battery when the SOC of the battery is at its lower limit and there is demand for low-frequency power at the DC-link or when the SOC of the battery is at its upper limit and there is the availability of low-frequency excess power at the DC-link. During this combination of devices, the utility grid stays connected to the DC microgrid under proper synchronization and facilitates the required bidirectional power flow.
- Only utility grid
  The utility grid supplements the function of both battery bank and supercapacitor bank only when the SOC limits of both the devices are at their maximum/minimum limits and there is a requirement for absorbing/releasing power from/to the DC-link. Here, utility grid facilitates both high-frequency and low-frequency bidirectional power flow.

Based on the above discussions, the reference power commands during different SOC conditions are given as follows:

(1) If $SOC_{sc,\min} < SOC_{sc} < SOC_{sc,\max}$ and $SOC_{bat,\min} < SOC_{bat} < SOC_{bat,\max}$
  * During discharging power command ($P_{ess} \geq 0$)
    - If $P_{ess,hf} \geq 0$

$$P_{bat,ref} = P_{ess,lf,dis}$$

$$P_{sc,ref} = P_{ess,hf,dis}$$

$$P_{grid,ref} = 0$$

- If $P_{ess,hf} < 0$

$$P_{bat,ref} = P_{ess,lf,dis}$$

$$P_{sc,ref} = P_{ess,hf,cha}$$

$$P_{grid,ref} = 0$$

* During charging power command ($P_{ess} < 0$)
  - If $P_{ess,hf} \geq 0$

$$P_{bat,ref} = P_{ess,lf,cha}$$

$$P_{sc,ref} = P_{ess,hf,dis}$$

$$P_{grid,ref} = 0$$

  - If $P_{ess,hf} < 0$

$$P_{bat,ref} = P_{ess,lf,cha}$$

$$P_{sc,ref} = P_{ess,hf,cha}$$

$$P_{grid,ref} = 0$$

(2) If $SOC_{sc} \leq SOC_{sc,\min}$ and $SOC_{bat} \geq SOC_{bat,\max}$
* During discharging power command ($P_{ess} \geq 0$)
  - If $P_{ess,hf} \geq 0$

$$P_{bat,ref} = P_{ess,lf,dis}$$

$$P_{sc,ref} = 0$$

$$P_{grid,ref} = P_{ess,hf,dis}$$

  - If $P_{ess,hf} < 0$

$$P_{bat,ref} = P_{ess,lf,dis}$$

$$P_{sc,ref} = P_{ess,hf,cha}$$

$$P_{grid,ref} = 0$$

* During charging power command ($P_{ess} < 0$)
  - If $P_{ess,hf} \geq 0$

$$P_{bat,ref} = 0$$

$$P_{sc,ref} = 0$$

$$P_{grid,ref} = P_{ess,lf,cha} + P_{ess,hf,dis}$$

- If $P_{ess,hf} < 0$

$$P_{bat,ref} = 0$$

$$P_{sc,ref} = P_{ess,hf,cha}$$

$$P_{grid,ref} = P_{ess,lf,cha}$$

Similarly, all other combinations are addressed in this PMS. The complete PMS applied to the HESD is shown in Figs. 3.8 and 3.9. After deciding on the reference powers for each equipment, the reference currents for the battery and supercapacitor bank are generated using the following equations:

$$I_{bat,ref} = \frac{P_{bat,ref}}{V_{bat}} \tag{3.12}$$

$$I_{sc,ref} = \frac{P_{sc,ref}}{V_{sc}} \tag{3.13}$$

The liberated reference currents are given to the respective SR flip-flop-based current controllers for proper tracking.

3.3.3 Grid synchronization and control

After finalizing the respective power command from the PMS, the VSI coupled between the DC microgrid and the utility grid is controlled to facilitate the required bidirectional power flow. Generally, the VSI control is divided into two main components, that is, voltage control and current control. The three-phase variables (voltage or current) with a phase sequence of *abc* are transformed from natural *abc* frame to the synchronously rotating reference (*dq0*) frame to get the advantage of reduced complexity, reduced controller resources, and independent active and reactive power control. The primary control objectives of the interfacing VSI in context of this work are

- To facilitate proper synchronization between the DC microgrid and the utility grid.
- To transfer the required active power as insisted by the PMS.

To establish this objective, the three-phase voltages ($V_{f,abc}$) and currents ($I_{f,abc}$) at the filter output are measured. The three-phase measured variables at unity power factor are represented as

$$\begin{bmatrix} V_{f,a} \\ V_{f,b} \\ V_{f,c} \end{bmatrix} = \begin{bmatrix} V_{fm}\sin(\omega t) \\ V_{fm}\sin\left(\omega t - \frac{2\pi}{3}\right) \\ V_{fm}\sin\left(\omega t + \frac{2\pi}{3}\right) \end{bmatrix} \tag{3.14}$$

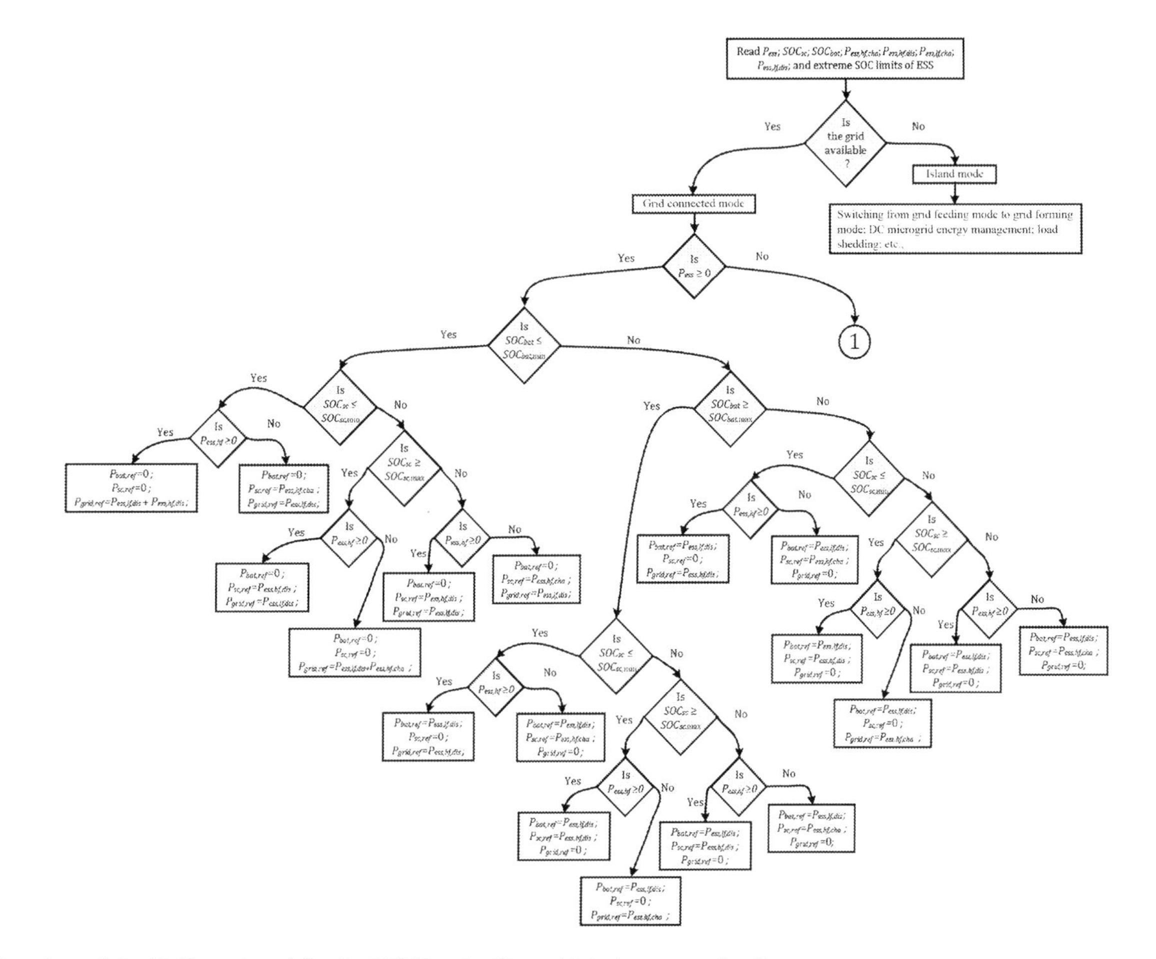

Fig. 3.8 Flowchart of the PMS employed for the HESD and utility grid during power dearth.

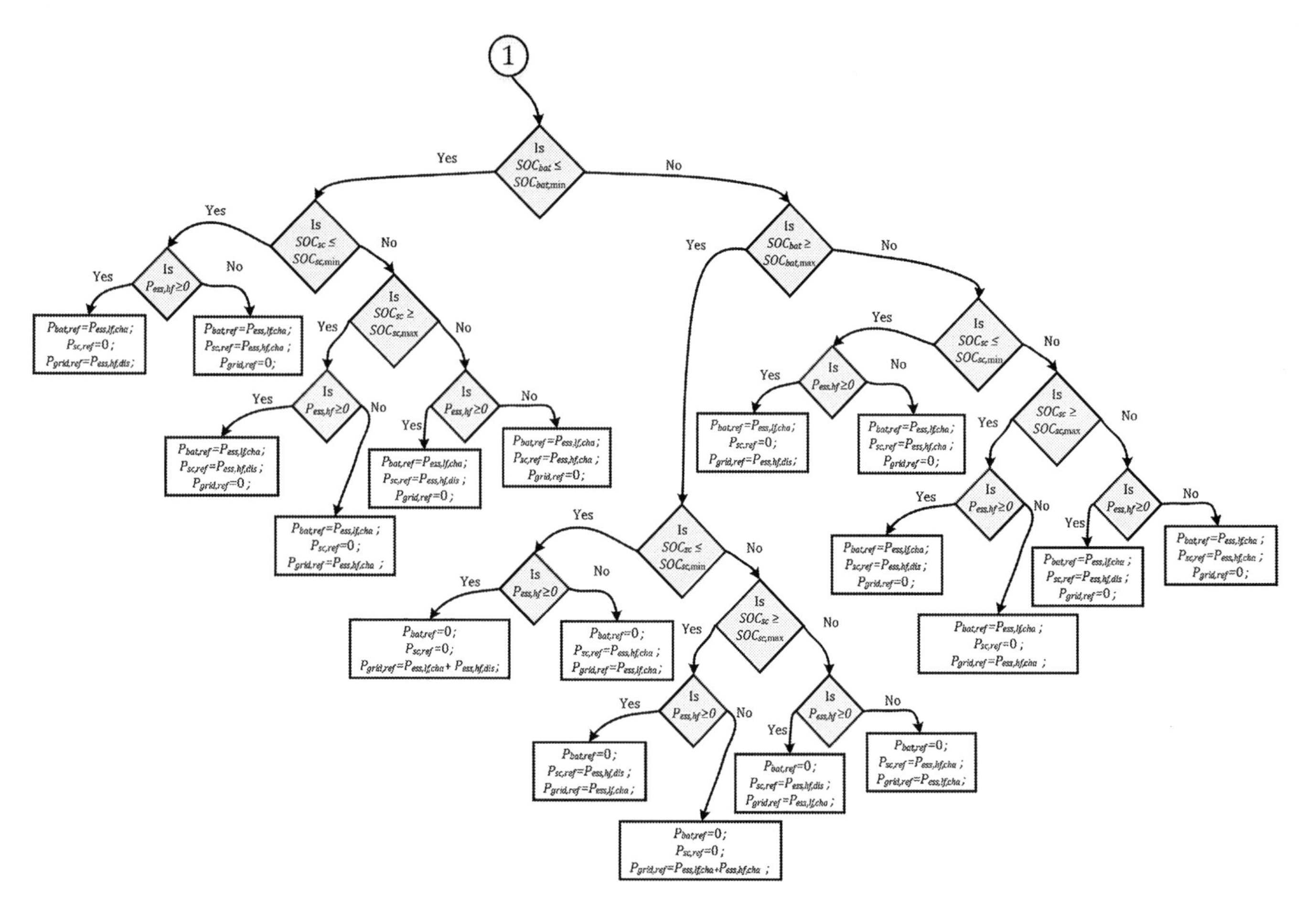

Fig. 3.9 Flowchart of the PMS employed for the HESD and utility grid during excess power.

$$\begin{bmatrix} I_{f,a} \\ I_{f,b} \\ I_{f,c} \end{bmatrix} = \begin{bmatrix} I_{fm}\sin(\omega t) \\ I_{fm}\sin\left(\omega t - \frac{2\pi}{3}\right) \\ I_{fm}\sin\left(\omega t + \frac{2\pi}{3}\right) \end{bmatrix} \tag{3.15}$$

Since the voltage at the PCC should remain constant during all modes of operation of the system, the V_{fm} is fixed and is equal to the peak magnitude of the grid voltage. The magnitude of the $I_{f,abc}$ varies w.r.t to the power command received from the PMS. The measured variables ($V_{f,abc}$ and $I_{f,abc}$) are transformed into *dq0* components ($V_{f,dq0}$ and $I_{f,dq0}$) using the transformation T_{dq0} as shown below (Geddada et al., 2012):

$$\begin{bmatrix} V_{f,d} \\ V_{f,q} \\ V_{f,0} \end{bmatrix} = T_{dq0} \begin{bmatrix} V_{f,a} \\ V_{f,b} \\ V_{f,c} \end{bmatrix} \tag{3.16}$$

$$T_{dq0} = \sqrt{\frac{2}{3}} \begin{bmatrix} \cos(\theta) & \cos\left(\theta - \frac{2\pi}{3}\right) & \cos\left(\theta + \frac{2\pi}{3}\right) \\ \sin(\theta) & \sin\left(\theta - \frac{2\pi}{3}\right) & \sin\left(\theta + \frac{2\pi}{3}\right) \\ \frac{1}{\sqrt{2}} & \frac{1}{\sqrt{2}} & \frac{1}{\sqrt{2}} \end{bmatrix} \tag{3.17}$$

where

$$\theta = \omega t + \alpha \tag{3.18}$$

where ω is the angular frequency of the utility grid and α is the phase angle of *d*-axis with respect to $V_{f,a}$ at time zero. Now, when $\theta = \omega t$ the *dq0* components rotates at system (line) frequency. Generally, phase locked loop (PLL) is applied to obtain the value of θ.

Blaabjerg et al. (2006) give the dynamic equation of the VSI presented in Eq. (3.5) when transformed into dq0 frame as shown

$$V_{i,d} = V_{f,d} + L_f \frac{dI_{f,d}}{dt} + R_f I_{f,d} - \omega L_f I_{f,q} \tag{3.19}$$

$$V_{i,q} = V_{f,q} + L_f \frac{dI_{f,q}}{dt} + R_f I_{f,q} - \omega L_f I_{f,d} \tag{3.20}$$

whereas the V_o term in V_{dq0} tends to zero under balanced grid condition. It can be inferred from Eqs. (3.19), (3.20) that there exists cross-coupling terms between the *d*- and *q*-axis voltages of the inverter. Therefore, for the inverter voltage to have a straightforward control on the grid-injected currents, the current controller should have cross-coupling terms, which will compensate the system coupling terms. The inverter voltages should be controlled according to the following relations such that direct control of the filter currents is achieved.

$$V_{i,d}^{*}=V_{f,d}+U_{f,d}-\omega L_f I_{f,q} \tag{3.21}$$

$$V_{i,q}^{*}=V_{f,q}+U_{f,q}+\omega L_f I_{f,d} \tag{3.22}$$

where $V_{i,\,d}^{*}$ and $V_{i,\,q}^{*}$ are the d- and q-axis reference voltage of the inverter output voltage; $-\omega L_f I_{f,\,q}$ and $+\omega L_f I_{f,\,d}$ are the cross-coupling terms; $V_{f,\,d}$ and $V_{f,\,q}$ are the feed forward filter output d and q-axis voltages, respectively; $U_{f,\,d}$ and $U_{f,\,q}$ in Eqs. (3.21), (3.22) can be obtained as shown in Fig. 3.10 (Geddada et al., 2012).

$$U_{f,d}=\left(K_{id,p}+\frac{K_{id,i}}{s}\right)\times \Delta I_{f,d} \tag{3.23}$$

$$U_{f,q}=\left(K_{iq,p}+\frac{K_{iq,i}}{s}\right)\times \Delta \mathrm{I}_{f,q} \tag{3.24}$$

where $K_{id,\,p}$ and $K_{id,\,i}$ are the proportional and integral gains of the d-axis current controller, respectively; $K_{iq,\,p}$ and $K_{iq,\,i}$ are the proportional and integral gains of the q-axis current controller respectively; $\Delta I_{f,\,d}$ and $\Delta I_{f,\,q}$ are the d-axis and d-axis current errors, respectively, as shown in Fig. 3.10.

$$\Delta I_{f,d}=I_{f,d}^{*}-I_{f,d} \tag{3.25}$$

$$\Delta I_{f,q}=I_{f,q}^{*}-I_{f,q} \tag{3.26}$$

The reference active power command received from the PMS is converted into the equivalent d-axis current command using the following relation:

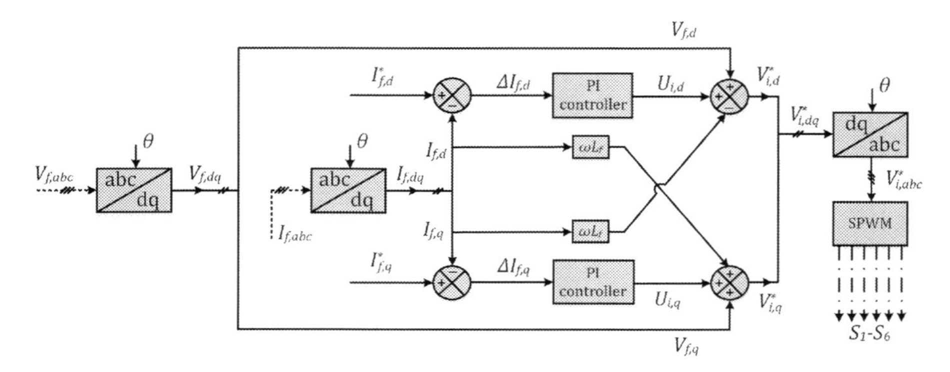

Fig. 3.10 Block diagram of the VSI controller.

$$I_{f,d}^{*}=\frac{P_{grid,ref}}{V_{f,d}} \tag{3.27}$$

Since the converter is intended to operate at unity power factor, reactive power current command is zero, that is, $I_{f,q}^{*}=0$.

After obtaining the reference voltage $V_{i,dq}^{*}$ from Eqs. (3.21), (3.22), *dq0* to *abc* transformation is performed to generate the three-phase reference voltage or modulating signals for sinusoidal pulse width modulator operating at a switching frequency of f_{sw} using the following transformation.

$$\begin{bmatrix} V_{i,a}^{*} \\ V_{i,b}^{*} \\ V_{i,c}^{*} \end{bmatrix} = T_{dq0}^{-1} \begin{bmatrix} V_{i,d}^{*} \\ V_{i,q}^{*} \\ V_{i,0}^{*} \end{bmatrix} \tag{3.28}$$

where

$$T_{dq0}^{-1}=\sqrt{\frac{2}{3}}\begin{bmatrix} \cos(\theta) & \sin(\theta) & \frac{1}{\sqrt{2}} \\ \cos\left(\theta-\frac{2\pi}{3}\right) & \sin\left(\theta-\frac{2\pi}{3}\right) & \frac{1}{\sqrt{2}} \\ \cos\left(\theta+\frac{2\pi}{3}\right) & \sin\left(\theta+\frac{2\pi}{3}\right) & \frac{1}{\sqrt{2}} \end{bmatrix} \tag{3.29}$$

The three-phase modulating signal ($V_{i,abc}^{*}$) acts as the reference waveform for the SPWM block shown in Fig. 3.10, which generates the switching pulses ($S_{inv,1}$ to $S_{inv,6}$) for the VSI.

3.4 Results and discussion

The grid-connected DC microgrid along with the DEM and control is analyzed for perturbations in solar PV array output power, DC load power, and SOC of HESD in MATLAB/SIMULINK platform. The parameters considered for simulation studies are listed in Tables 3.1–3.4. In all operating modes of the system, the grid is assumed to be balanced and connected throughout.

The proposed system is simulated for the following sequence of actions:

- A dynamic variation in solar irradiance has been programmed to study the performance of the system under varying renewable power generation. Firstly, the irradiance of the PV array is simulated to be 1000 W/m^2 until 2 s, reduces dynamically till 4 s and stays constant at a value of 500 W/m^2. At 6 s, the irradiance is programmed to rise to a value of 1000 W/m^2 till 8 s and stays constant thereon.
- Transient R, R-L load changes have been applied on to the system with the sequence of actions presented in Table 3.3.

Table 3.1 Component rating of the system components

Device	Specification
PV array	95 V, 0–52A
Supercapacitor bank	63 F, 120 V, ±200 A (peak)
Battery bank	110 Ah, 125 V
VSI	400 V (DC)/208 V filter output (L-L), 10 kVA
Filter inductor (l_f)	10 mH

Table 3.2 Control parameters used in the analysis

Parameter	Specification
$V_{dc,ref}$	400 V
$V_{g,abc}$	400 V
$V_{i,abc}$	208 V
ω_c	2 × 3.14 × 50
$SOC_{sc,\min}$	75%
$SOC_{sc,\max}$	95%
$SOC_{bat,\min}$	50%
$SOC_{bat,\max}$	95%
f_{sw}	2000 Hz
Sampling period	5 μs
$K_{v,p}$ and $K_{v,i}$	1.6 and 20
$K_{id,p}$ and $K_{id,i}$	1.5 and 20
$K_{iq,p}$ and $K_{iq,i}$	1.5 and 20

Table 3.3 Programmed DC load variations

Load type	Value	Actions and time
R-L	50 Ω and 0.5 mH	Present throughout
Step R-L	50 Ω and 0.5 mH	Removed at 2 s
Pulse R-L	90 Ω and 1 mH	Applied at 3 s and removed at 3.2 s
Step R-L	200 Ω and 1 mH	Applied at 5 s
Step R-L	200 Ω and 1 mH	Removed at 9 s

Table 3.4 SOC of the HESD assumed at the starting of each case

Mode	SOC_{sc} (%)	SOC_{bat} (%)
Case I	85.98	80.33
Case II	74.98	50.01
Case III	95.02	95.01

- SOC of HESD is simulated to be at their (upper and lower) extreme limits and within their extreme limits during different cases in order to study the expeditious support extended by the utility grid under the supervision of the PMS. The set of SOC of the HESD applied at the start of each simulation is listed in Table 3.4. Case I deals with the operation of the system with HESD operating in their normal SOC range whereas case II analyses the system performance during the HESD operation near to or below minimum SOC limits. Case III tests the system operation with the HESD at their upper limit of SOC.

The microgrid undergoing the above-stated operating conditions is observed to stabilize its DC-link voltage and is able to obtain proper power sharing among the participants due to the timely action of the PMS and the current controllers. In addition, the microgrid is able to dispatch the required amount of power to/from the utility grid under proper synchronization in order to stabilize the DC-link voltage, assure continuity of supply to the local loads, and protect the HESD from under and overcharging. The variation in PV and load power being same for all the cases (programmed so for proper evaluation of the system performance), the power-sharing among microgrid participants, DC-link voltage, filter output voltage and current, terminal parameters of all the microgrid participants for different cases of SOC are presented in Figs. 3.11–3.25.

3.4.1 Case I

The power shared by each device in the DC microgrid during case I, case II, and case III is presented in Figs. 3.11, 3.16, and 3.21, respectively. It is imperative from the figures that there is a power mismatch between the PV generation and the load demand throughout the analysis in all the cases except between 5 and 6 s (where they are approximately equal). During 1–2 s and 3–3.2 s, the load demand is greater than the PV generation resulting in a power dearth at the DC-link whereas rest of the time load power is less than the PV power. In Fig. 3.11, that is, during case I, since HESD is operating well within its extreme SOC limits, both supercapacitor and the battery bank share the available power dearth or surplus without any assistance from the utility grid. The utility grid stays connected to the DC microgrid in standby mode, that is, it does not transfer any power. The DC-link voltage presented in Fig. 3.12 suggests that there is a maximum deviation of around 4 V throughout the analysis resulting in a tight voltage regulation by the EMS employed to this system. Similarly, though the utility grid

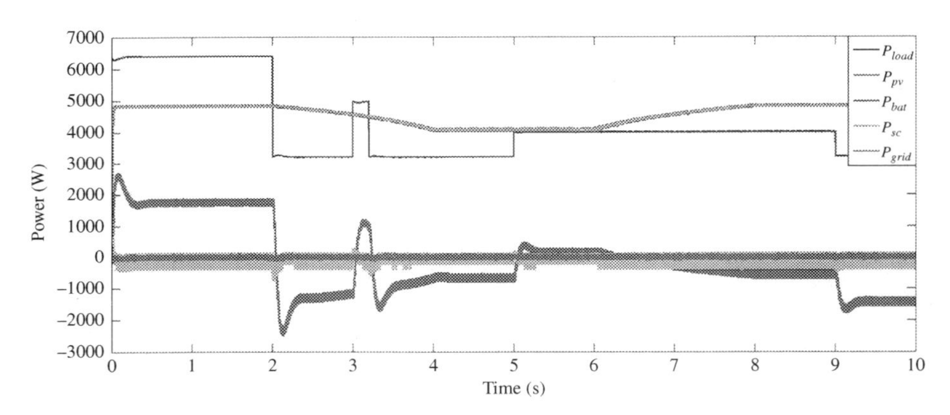

Fig. 3.11 Power flow from/to the microgrid participants during case I.

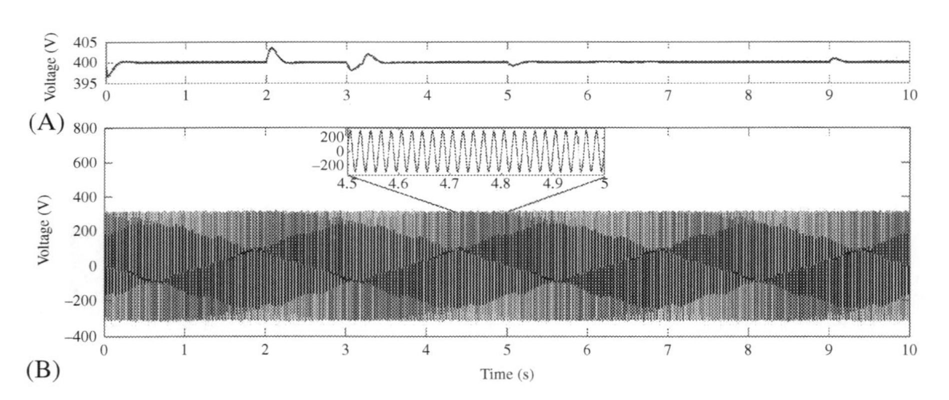

Fig. 3.12 Voltage variations during case I (A) DC-link voltage and (B) line voltage (V_{AB}) at filter output.

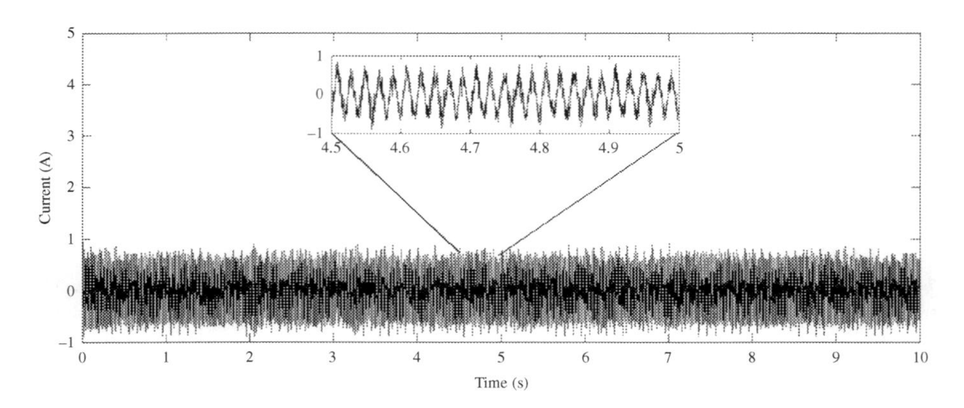

Fig. 3.13 Filter output current of phase A in case I.

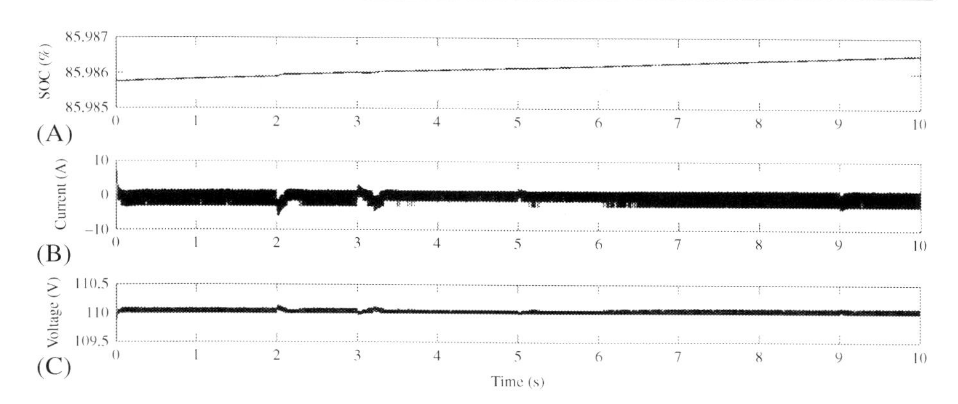

Fig. 3.14 Supercapacitor bank terminal parameters during case I (A) SOC, (B) current, and (C) voltage.

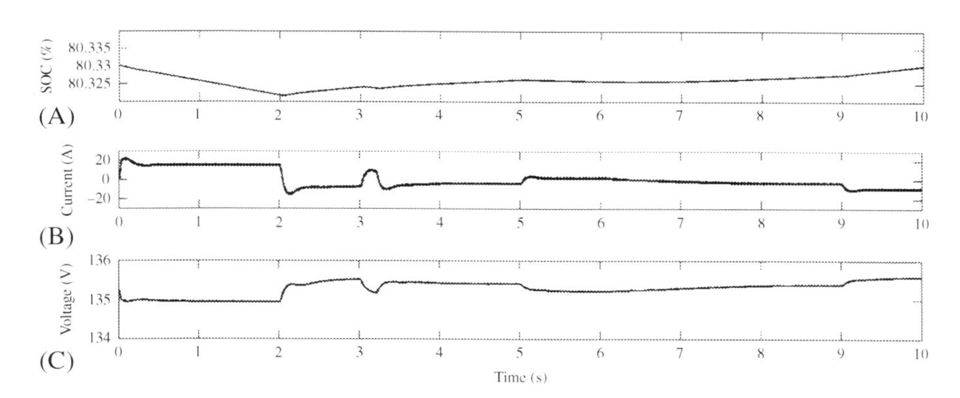

Fig. 3.15 Battery bank terminal parameters during case I (A) SOC, (B) current, and (C) voltage.

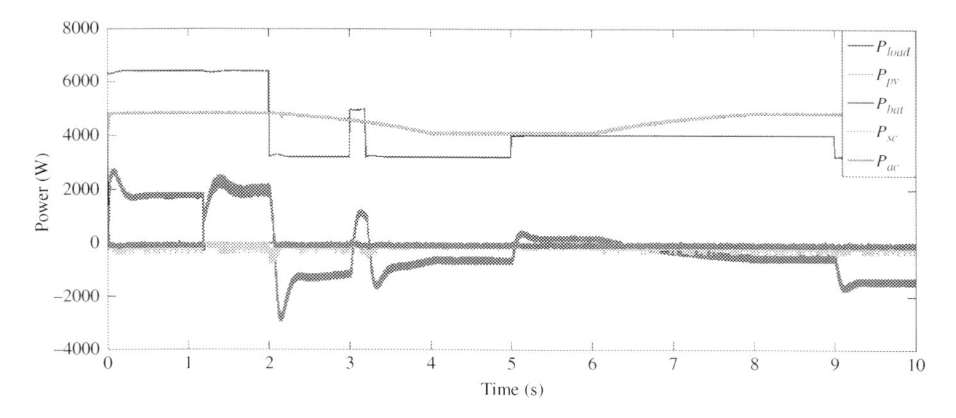

Fig. 3.16 Power flow from/to the microgrid participants during case II.

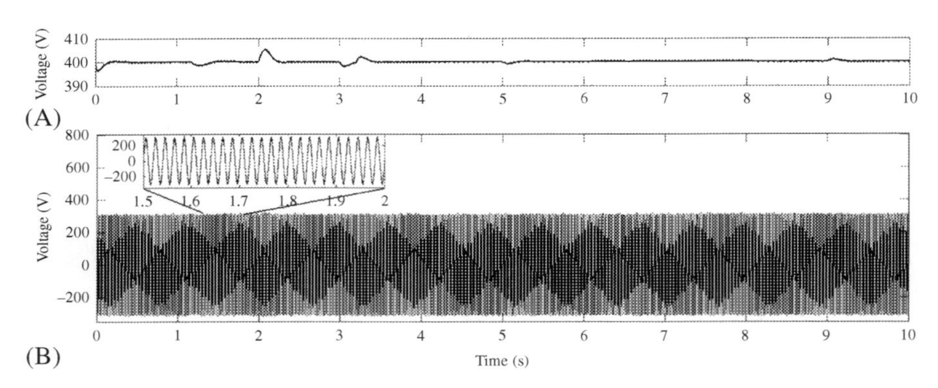

Fig. 3.17 Voltage variations during case II (A) DC-link voltage and (B) line voltage (V_{AB}) at filter output.

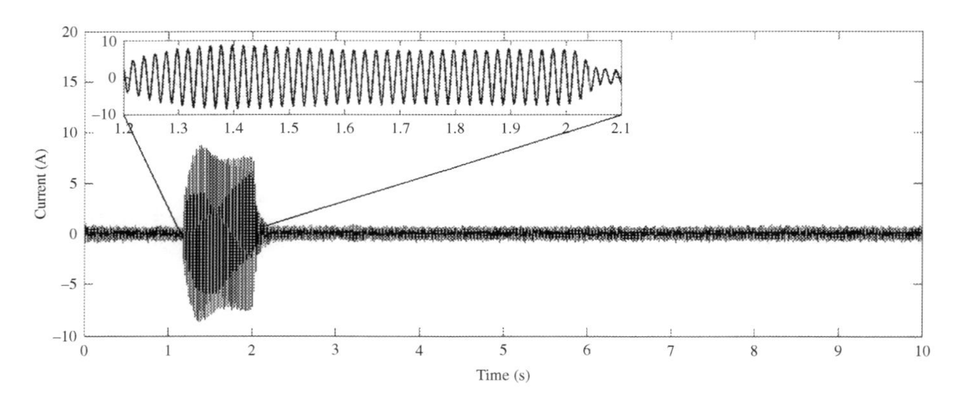

Fig. 3.18 Filter output current of phase A in case II.

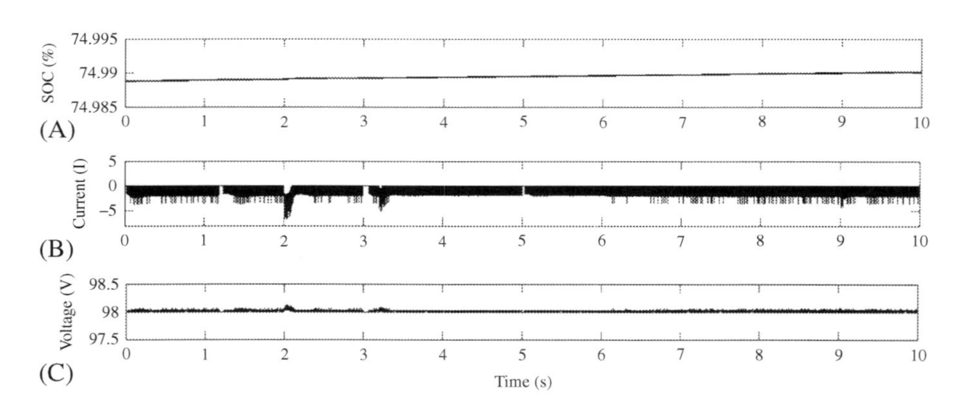

Fig. 3.19 Supercapacitor bank terminal parameters during case II (A) SOC, (B) current, and (C) voltage.

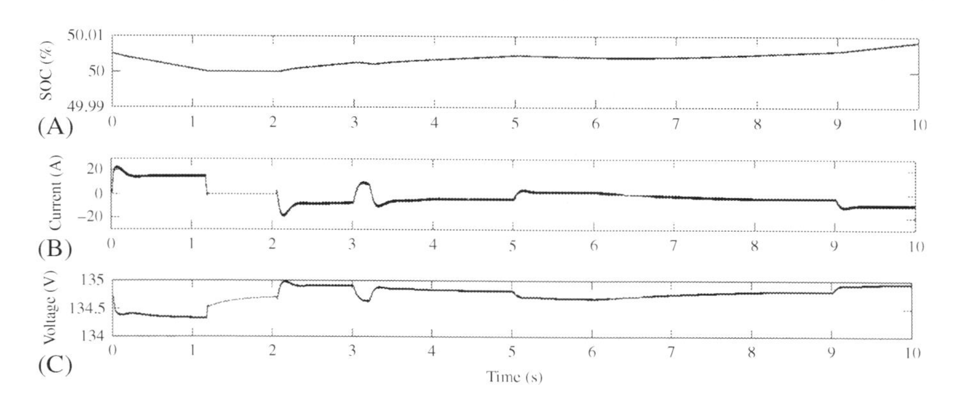

Fig. 3.20 Battery bank terminal parameters during case II (A) SOC, (B) current, and (C) voltage.

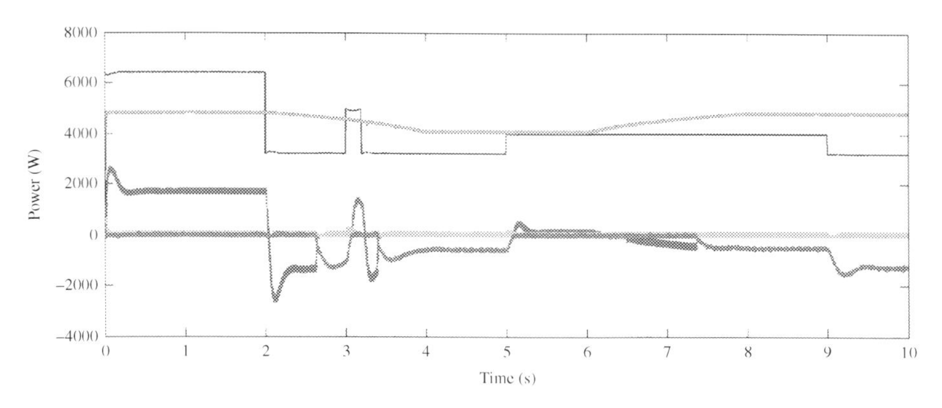

Fig. 3.21 Power flow from/to the microgrid participants during case III.

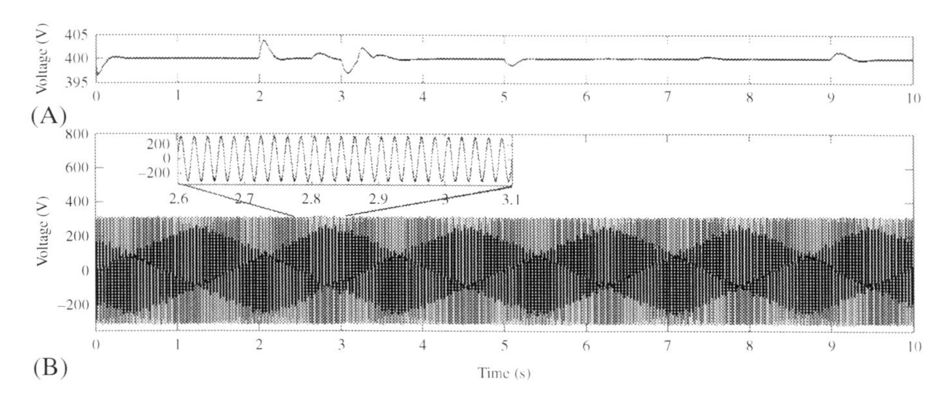

Fig. 3.22 Voltage variations during case III (A) DC-link voltage and (B) line voltage (V_{AB}) at filter output.

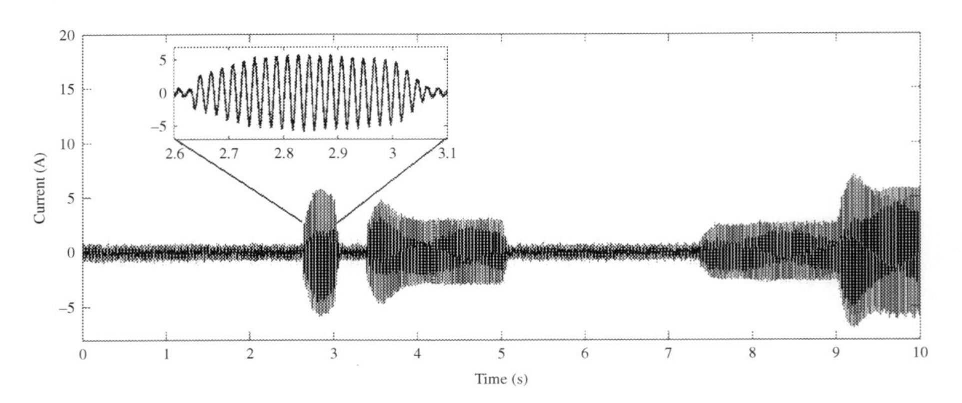

Fig. 3.23 Filter output current of phase A in case III.

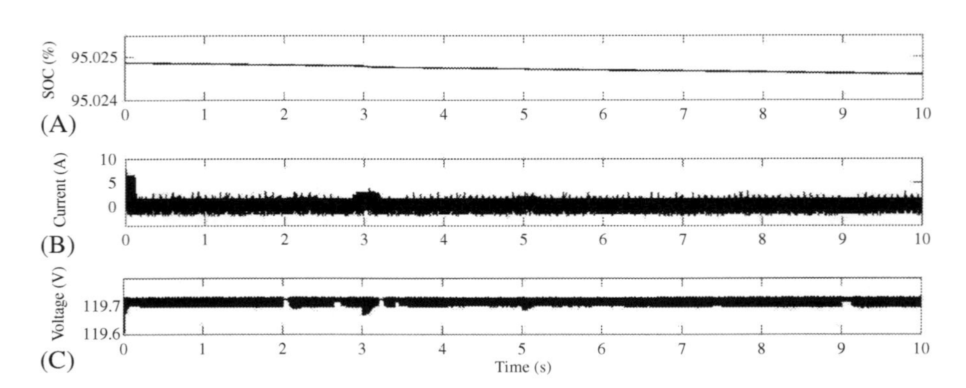

Fig. 3.24 Supercapacitor bank terminal parameters during case III (A) SOC, (B) current, and (C) voltage.

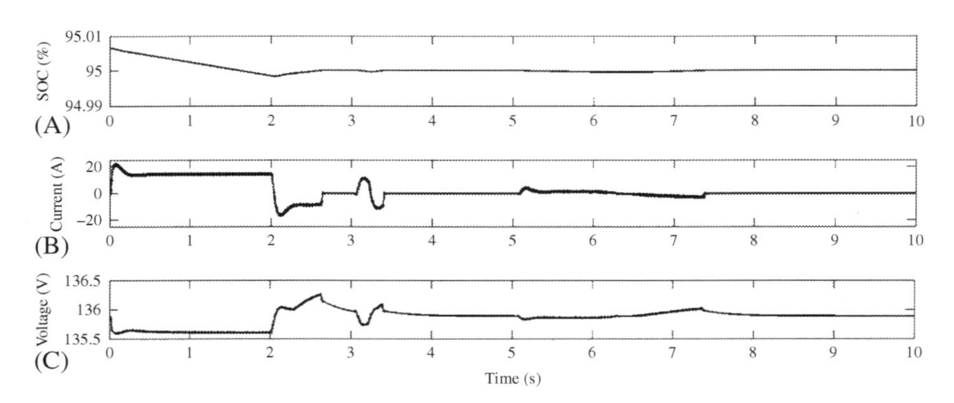

Fig. 3.25 Battery bank terminal parameters during case III (A) SOC, (B) current, and (C) voltage.

is connected and is operating in standby mode, the filter output voltage is sinusoidal with a negligible filter current throughout as shown in Figs. 3.12 and 3.13. The supercapacitor bank and battery bank parameters are shown in Figs. 3.14 and 3.15. It can be inferred from Figs. 3.14B and 3.15B that the supercapacitor bank and battery bank are controlled to output high-frequency and low-frequency currents as insisted by the PMS.

3.4.2 Case II

Figs. 3.16–3.20 present the variation in the system parameters for the above-said load and PV power variations during case II, that is, minimum SOC condition of HESD. During minimum SOC mode, the HESD should not react to any power dearth at the DC-link, and the utility grid takes that duty. The active power absorbed/released by each microgrid participant is shown in Fig. 3.16. The variation in SOC, terminal current and voltage of the supercapacitor bank and battery bank are shown in Figs. 3.19 and 3.20, respectively. It can be observed from these figures that the HESD reduce their discharge current to zero during any power dearth at the DC-link due to their minimum SOC condition. During 0–2 s, the DC microgrid sees a power dearth, which is supplied by the battery and utility grid because the supercapacitor bank SOC is already below its minimum level. At around 1.2 s, the battery SOC has reached its lower limit (50%) due to which the battery reduced its generation to zero and the utility grid takes over the duty of battery from now onward. Except at the pulse load and in between 5 and 6 s, there is an excess power available at the DC-link. The supercapacitor and battery bank share the excess power among them to strengthen their SOC. Throughout this operating scenarios, the variations in the DC-link voltage, filter output line voltage and filter output line current are shown in Figs. 3.19 and 3.20, respectively. The grid injected current is seen only when a power command is given to the VSI and rest of the time it is below 0.5 A.

3.4.3 Case III

Figs. 3.21–3.25 present the controlled response of the system components during case III, that is, maximum SOC mode of HESD. During maximum SOC mode, the HESD is prohibited from absorbing any excess power from the DC-link and transfers this duty to the utility grid. The active power absorbed/released by each microgrid participant is shown in Fig. 3.21. The variation in SOC, terminal current and voltage of the supercapacitor bank and battery bank are shown in Figs. 3.24 and 3.25, respectively. It can be observed from Figs. 3.24 and 3.25 that in spite of the availability of the excess power at the DC-link, the HESD is restrained from charging beyond their maximum SOC limit. During 0–2 s, the DC microgrid faces a power dearth which is supplied by the battery and supercapacitor bank as insisted by the PMS. From 2 to 2.65 s, the excess power available at the DC-link is observed to be absorbed by the battery bank and utility grid as the SOC of the supercapacitor is above 95% and that of the battery is below 95%. At around 2.65 s, the battery ceases to absorb further excess power due to maximum SOC condition. At 3 s, the battery and supercapacitor bank combine their generation to suffice the pulse load power requirement. Similarly, the further power variations at the DC-link are absorbed/released by suitable dispatchable units to regulate the DC-link voltage throughout. The variation in the DC-link voltage, filter output line voltage and filter output line current are shown in Figs. 3.22 and 3.23, respectively. The grid injected current is appearing whenever a power command is given to the VSI and rest of the time it is below 0.5A. Also, the mode changes between the HESD and utility grid are done seamlessly without stressing the battery with high-frequency current commands.

It can be inferred from the results presented in Figs. 3.11–3.25 that the control and energy management applied to the DC microgrid with the chosen system components is proven to address its objectives and has provided a stable subsystem for integrating the renewable energy into the utility grid.

3.5 Conclusion

In this chapter, a utility grid-integrated DC microgrid comprising of PV array, HESD is analyzed. Firstly, the system under study is introduced and the system components are modeled. Important research objectives concerning the microgrid are presented and are addressed using a DEM strategy. The DEM comprising of DC side controllers, PMS and AC side controllers is explained in detail. A flowchart of the PMS with all possible combinations of the HESD and utility grid and their associated power commands is presented. The complete system is modeled and tested in MATLAB/SIMULINK for different perturbations in PV power, load power, and SOC of the HESD. The combined action of the voltage controller, PMS, and the respective current controllers lead to a proper integration of the renewable energy into the utility grid.

References

Blaabjerg, F., Teodorescu, R., Liserre, M., Timbus, A., 2006. Overview of control and grid synchronization for distributed power generation systems. IEEE Trans. Ind. Electron. 53 (5), 1398–1409.

Cheng, Z.I., Chen, W.R., Li, Q., Jiang, Z.I., Yang, Z.H., 2010. Modelling and dynamic simulation of an efficient energy storage component-supercapacitor.Asia-Pacific Power and Energy Engineering Conference, pp. 1–4.

Chittibabu, B., Mohanty, K.B., 2017. Experimental validation of improved control strategy of grid-interactive power converter for wind power system. Technol. Econ. Smart Grid. Sust. Energ. 2 (5), 1–12.

Faraji, R., Rouholamini, A., Naji, H.R., Fadaeinedjad, R., Chavoshian, M.R., 2011. FPGA-based real-time incremental conductance maximum power point tracking controller for photovoltaic systems. IET Power Electron. 7 (5), 1294–1304.

Geddada, N., Karanki, S.B., Mishra, M.K., 2012. Synchronous reference frame based current controller with SPWM switching strategy for DSTATCOM applications.IEEE International Conference on Power Electronics, Drives and Energy Systems, pp. 1–6.

Harini, K., Syama, S., 2015. Simulation and analysis of incremental conductance and perturb and observe MPPT with DC-DC converter topology for PV array.IEEE International Conference on Electrical, Computer and Communication Technologies (ICECCT), pp. 1–5.

Nehrir, M.H., Wang, C., Strunz, K., 2011. A review of hybrid renewable/alternative energy systems for electric power generation: configurations, control, and applications. IEEE Trans. Sustainable Energy 2 (4), 392–403.
Ramesh Naidu, B., Panda, B., Siano, P., 2017. A self-reliant DC microgrid: sizing, control, adaptive dynamic power management and experimental analysis. IEEE Trans. Ind. Inf. Early Access Article.
Thounthong, P., Rael, S., Davat, B., 2009. Analysis of supercapacitor as second source based on fuel cell power generation. IEEE Trans. Energy Convers. 24 (1), 247–255.
Tummuru, N.R., Mishra, M.K., Srinivas, S., 2015. Dynamic energy management of hybrid energy storage system with high-gain PV converter. IEEE Trans. Energy Convers. 30 (1), 150–160.
Wang, C., Nehrir, M.H., 2008. Power management of a stand-alone wind/photovoltaic/fuel cell energy system. IEEE Trans. Energy Convers. 23 (3), 957–967.
Zhou, T., Sun, W., 2014. Optimization of battery–supercapacitor hybrid energy storage station in wind/solar generation system. IEEE Trans. Sustainable Energy 5 (2), 408–415.

Further reading

Jain, S., Agarwal, V., 2008. An integrated hybrid power supply for distributed generation applications fed by nonconventional energy sources. IEEE Trans. Energy Convers. 23 (2), 622–631.

Modeling, control, and energy management for DC microgrid

Ying Han, Weirong Chen, Qi Li
School of Electrical Engineering, Southwest Jiaotong University, Chengdu, China

4.1 Introduction

With the increasing demand for electrical power and energy and more severe environmental problems caused by fossil energy, renewable power generations have turned to be promising in recent years (Nejabatkhah and Li, 2015; Katiraei and Iravani, 2006; Li et al., 2016; Chen et al., 2014). Microgrids (MGs), the controllable small-scale power networks, have been widely studied and play an important role in the future smart power system by facilitating an effective and smooth integration of distributed energy resources, loads, and energy-storage devices in the local vicinity (Thounthong et al., 2011; Chen et al., 2013; Kakigano et al., 2010; Kwasinski, 2011).

Although the studies in literatures are mainly focused on AC MGs, DC MGs are continuously demonstrating the advantages compared with AC MGs (Fig. 4.1). (i) The distributed generations [e.g. photovoltaic (PV) systems, fuel cells (FCs)], energy-storage systems (e.g., batteries, supercapacitors), and loads (e.g. electrolyzes, charging piles, LEDs) can be connected without DC/AC conversions, and in this way system efficiency and stability are enhanced because of less number of power conversion stages; (ii) the control and management of DC MGs are much simpler than those in AC MGs; (iii) DC MGs are not afflicted with frequency synchronization, reactive power flow, harmonic current, and other power quality issues that are common in AC MGs (Liu et al., 2007; Schulz, 2007). Because of the aforementioned factors, DC MGs are receiving increased attention as small-scale power systems in a close physical vicinity, which can reduce energy loss losses from power transmission lines over long distances (Du et al., 2010). DC MGs will be widely used in building electrical systems, data centers, remote communication stations, and plug-in hybrid electric vehicles in the future (Gu et al., 2014).

After a long period of development, the number of academic research works and demonstration projects on PV/battery DC MGs (Sechilariu et al., 2014a,b) and PV/wind/battery DC MGs (Chen et al., 2013) are increasing. However, the output power of PV and wind generators fluctuate considerably depending on weather conditions. Because of the limitation of battery capacity and scale, this kind of DC MG is difficult to keep steady operate for a long time (Yoo et al., 2013; Sechilariu et al., 2014a,b). In order to overcome the challenge, the PV/fuel cell/battery DC MGs have been found in

Smart Power Distribution Systems. https://doi.org/10.1016/B978-0-12-812154-2.00004-3

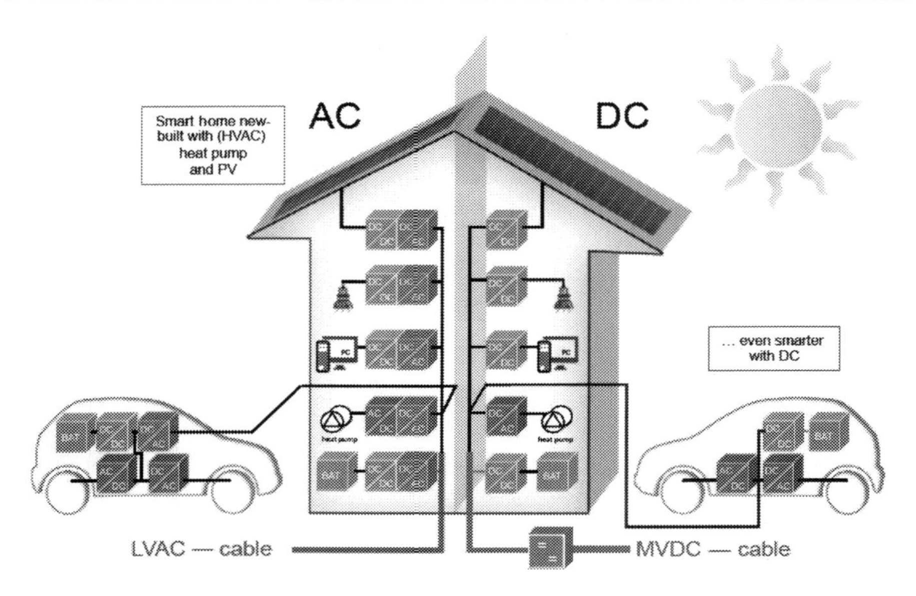

Fig. 4.1 Comparison of DC microgrid and AC microgrid for building.

several literatures, which rely on fuel cell for long-term generation system while PV array works as a primary source and battery bank works as a short-term energy storage. Fuel cells are proved reliable, but are still expensive and not specifically designed to cope with renewable energy and fast demand response. The development of the energy management strategy of hydrogen DC MG has been addressed by many researchers in recent years.

Different energy management strategies for fuel cell MGs and hybrid power systems have been reported in literatures. Valverde et al. (2013) proposes a classical PI control method depending on the battery state of charge (SoC) to manage the operation of the fuel cell, and the PI controller can be easily tuned online for better tracking. The state machine control is a simple and well-known rule-based control strategy, and Han et al. (2017) presents a DC MG energy management control strategy based on state machine with 13 operating states to justly distribute the power of different distributed generations. Brka et al. (2015) presents a predictive energy management strategy, which predicts the load demand and the power of renewable generations to control the output of fuel cell and batteries. In order to maintain the DC bus voltage and limit the current slope of battery and fuel cell, Bambang et al. (2014) and Valverde et al. (2016) develop energy management systems based on model predictive control approach for a grid-tied hydrogen MG.

In this chapter, we will focus on the modeling, control, and energy management for the PV/fuel cell/energy-storage DC MG. First, the math model of each subsystem, the system structure and the experimental platform for the DC MG are studied. Moreover, on this basis, the distributed generations control methods based on their operating characteristic are introduced. Two energy management strategies based on classical PI control and state machine control are evaluated in the PV/fuel cell/energy-storage DC MG under different operating conditions.

4.2 DC MG structure and modeling

In this chapter, the structure of the PV/fuel cell/energy-storage DC MG is shown in Fig. 4.2. The DC MG system consists of a kW PV array, a fuel-cell generation system, a battery bank, DC loads (building load, electric vehicles, etc.), and the distributed generation subsystems are connected to the dc bus by the necessary power electronics. In the proposed DC MG, PV array and FC system work as the primary and auxiliary sources, respectively, while battery bank is the energy-storage system to ensure the power balance and stability of the system. Furthermore, the generation output powers are adequately distributed by an energy management system.

4.2.1 PV system modeling

PV generation is a kind of renewable energy generation, which converts light energy directly into electrical energy by PV effect and its key component is PV cells. In this chapter, the PV cell model uses the engineering analytical model, which calculated four electrical parameters ($I_{sc}^*, U_{oc}^*, I_m^*, U_m^*$) in standard test condition ($S^* = 1000$ W/m^2, $T^* = 25°$C) supplied by manufacturer. The engineering analytical model of PV system is expressed as

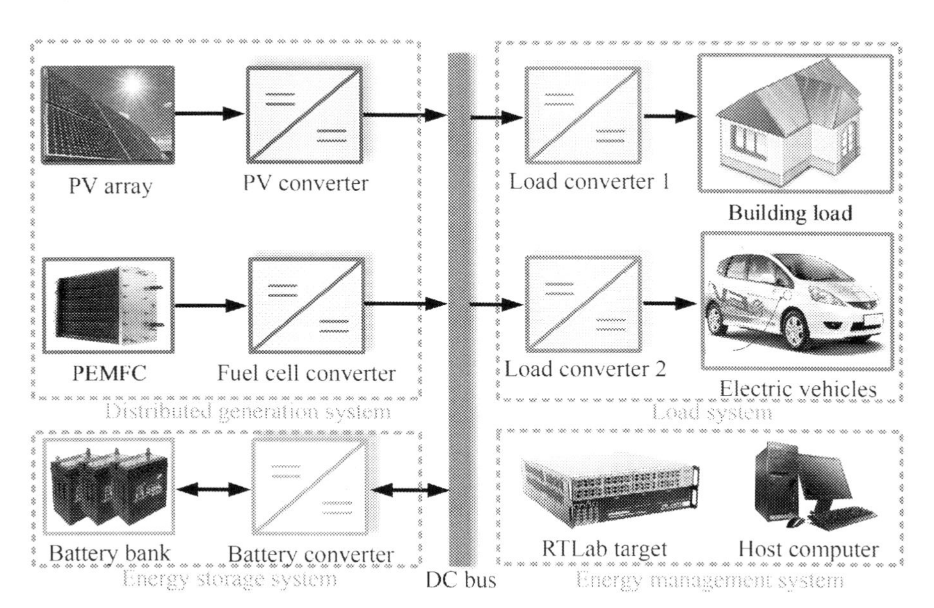

Fig. 4.2 The structure of the proposed DC microgrid.

$$\begin{cases} I = I_{sc}\left[1 - C_1\left(e^{U/C_2U_{oc}} - 1\right)\right] \\ I_{sc} = I_{sc}^*[1 + a(T - T^*)] \cdot S/S^* \\ U_{oc} = U_{oc}^*[1 - c(T - T^*)] \cdot \ln[1 + b \cdot (S/S^* - 1)] \\ I_m = I_m^*[1 + a(T - T^*)] \cdot S/S_{ref} \\ U_m = U_m^*[1 - c(T - T^*)] \cdot \ln[1 + b \cdot (S/S^* - 1)] \\ C_1 = (1 - I_m/I_{sc})e^{-U_m/(C_2U_{oc})} \\ C_2 = (U_m/U_{oc} - 1)/\ln(1 - I_m/I_{sc}) \end{cases} \tag{4.1}$$

where I is the output current of PV cell; T is ambient temperature under nonstandard condition; S is actual illumination strength; I_{sc} and U_{oc} are the short-circuit current and open-circuit voltage (OCV) of PV cell, respectively; I_m and U_m are the current and voltage when PV cell works at the maximum power point, respectively; a, b, and c are the compensation factors; and the superscript $*$ represents the reference value of the variable.

4.2.2 Fuel-cell generation system modeling

Proton exchange membrane fuel cell (PEMFC) has many excellent characteristics such as high energy density, well reliability, and so on. Therefore, it is considered as a reliable power source, and individuals are trying to use PEMFC to different application. In this chapter, the fuel-cell generation system uses PEMFC. The fuel-cell stack output voltage is obtained from the sum of its equilibrium potential and three irreversible losses, namely activation overvoltage, ohmic overvoltage, and concentration overvoltage. Fig. 4.3 shows the equivalent circuit model of PEMFC and the output voltage equation is expressed as

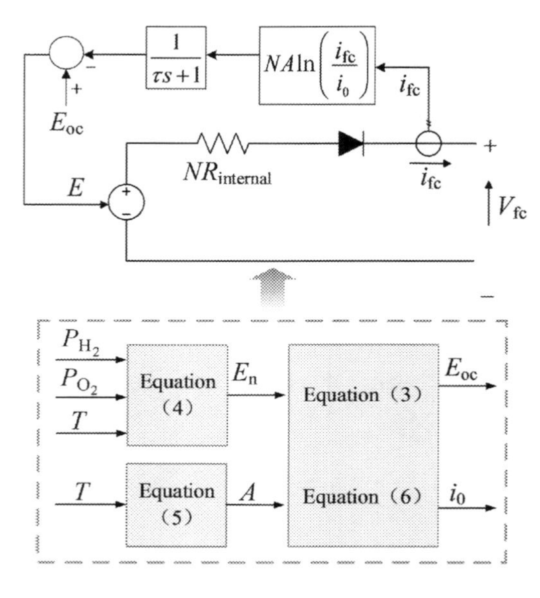

Fig. 4.3 The equivalent circuit model of fuel cell.

$$V_{fc} = E_{oc} - V_{act} - V_{ohmic} - V_{conc} \quad (4.2)$$

where E_{oc} is the OCV, V_{act} is the activation voltage drop which is mainly responsible for the voltage drop at low current density, V_{ohmic} is the ohmic voltage drop, and V_{conc} is the concentration voltage drop which is often neglected in practical working condition because of the reason that it becomes more significant at high current density that is not allowed in practical condition. These voltages are computed theoretically as

$$\begin{cases} E_{oc} = N(E_n - A\ \ln(i_0)) \\ V_{act} = \dfrac{1}{\tau s + 1} \cdot NA\ \ln\left(\dfrac{i_{fc}}{i_0}\right) \\ V_{ohmic} = R_{internal} \cdot i_{fc} \end{cases} \quad (4.3)$$

where N is the number of cells, E_n is the Nernst's instantaneous voltage, A is Tafel slope, i_o is the exchange current, τ is the dynamic response time which will make a time delay when occurring a sudden change in order to ensure a more practical reaction, i_{fc} is the cell output current, and $R_{internal}$ is the equivalent inner resistance of the stack (Ohms).

Besides, E_n, A, and i_o can be calculated as Eqs. (4.4)–(4.6).

$$E_n = 1.229 + (T - 298)\frac{-44.43}{zF} + \frac{RT}{zF}\ \ln\left(P_{H_2} + P_{O_2}^{1/2}\right) \quad (4.4)$$

where T is the environment temperature, z is transfer electron number, R is gas constant, P_{H_2} and P_{O_2} are, respectively, the hydrogen and oxygen gas pressure, and F is the Faraday constant.

$$A = \frac{RT}{z\alpha F} \quad (4.5)$$

where α is the electron transfer coefficient.

$$i_0 = \frac{zFk(P_{H_2} + P_{O_2})}{Rh} e^{\frac{-\Delta G}{RT}} \quad (4.6)$$

where k is Boltzmann's constant, h is Planck's constant, and ΔG is the change value of Gibbs free energy.

4.2.3 Battery bank modeling

The battery bank based on lead-acid, nickel cadmium (NiCd), nickel metal hydride (NiMH), and lithium ion (Li-ion) have been used widely. Compared with them, lead-acid battery remains the most used for MGs because of its low cost and high rate of recycling of 97%. In this chapter, the battery bank composes of three Pb-acid batteries in parallel. The behavior of the model is represented by a modified Shepherd

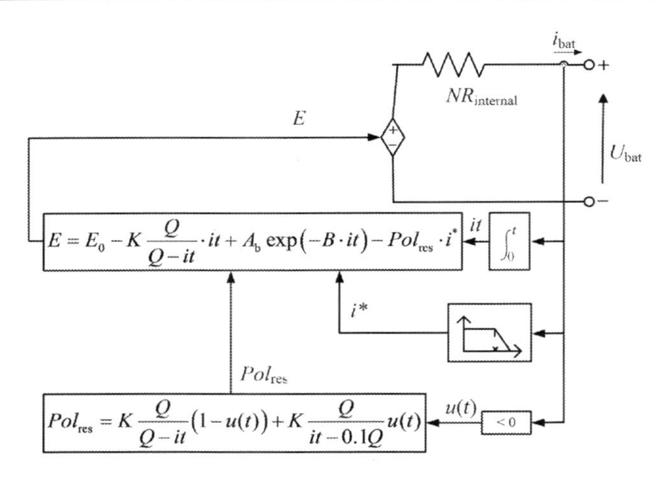

Fig. 4.4 The equivalent circuit model of battery.

curve-fitting model, in which a polarization voltage is introduced into the battery charge expression to better reflect influence of SoC on battery output performance. The equivalent circuit is shown in Fig. 4.4.

The battery output voltage can be obtained through the OCV and voltage drop caused by the battery equivalent internal impedance, as

$$\begin{aligned} U_{\text{bat}} &= E - R_{\text{b}} i_{\text{bat}} \\ &= E_0 - K\frac{Q}{Q - it}\cdot it + A_{\text{b}}\exp\left(-B\cdot it\right) - Pol_{\text{res}}\cdot i^* - R_{\text{b}} i_{\text{bat}} \end{aligned} \tag{4.7}$$

where U_{bat} is the battery output voltage, E is OCV, E_0 is the battery internal equivalent voltage, K is polarization constant, Q is the maximum battery capacity, it is the extracted capacity, R_{b} is internal impedance, i_{bat} is the battery current, and i^* is the filtered battery current. Pol_{res} is polarization resistance which is different when charging or discharging, as is written in the following equation.

$$Pol_{\text{res}} = \begin{cases} K\dfrac{Q}{Q - it} & (i_{\text{bat}} > 0) \\ K\dfrac{Q}{it - 0.1Q} & (i_{\text{bat}} < 0) \end{cases} \tag{4.8}$$

When battery is used for public transport application, its SOC must be kept between suitable range to achieve higher charge and discharge efficiency. As an important parameter for the battery bank, SoC can be calculated as the following Eq. (4.9) shows, in which SOC_0 is the initial SOC, $Q_{\max}$ is the battery bank's normal capacity (Ah), and $i(t)$ is the battery current.

$$\text{SOC}(t) = \text{SOC}_0 + (1/Q_{\max})^* \int i(t)dt\cdot 100\% \tag{4.9}$$

4.3 DC MG experimental set-up

4.3.1 DC MG configuration

Fig. 4.5 shows the structure of the PV/fuel cell/energy-storage DC MG and the main specifications of the MG entities are listed in Table 4.1. The DC MG system comprises of a 1-kW PV generation system, a 1-kW PEMFC generation system, a 36 V/36 Ah battery bank, a DC electronic load emulates the demand power (building load, electric vehicles, etc.), an energy management system and the necessary power electronics. The proposed MG is formed a standard 60 V bus, which is a kind of low-voltage DC power distribution system that is widely used. The PV array, the fuel-cell system and battery bank are controlled by the real-time energy management system based on RT-Lab target computer.

4.3.2 PV array simulator

Because the output characteristic of PV array changes under various environmental conditions, PV array should access to DC bus by DC/DC converter. In this work, the PV array has three working modes under different operating condition: maximum power point tracking (MPPT) mode, power-voltage droop mode, and non-working mode. The switch of the three working modes is implemented by controlling the DC/DC converter. Because of the limitations on PV array availability in a laboratory environment, a PV simulator is used for emulating the PV array of the proposed MG system test bench, and the output characteristic of PV simulator is shown in Fig. 4.6.

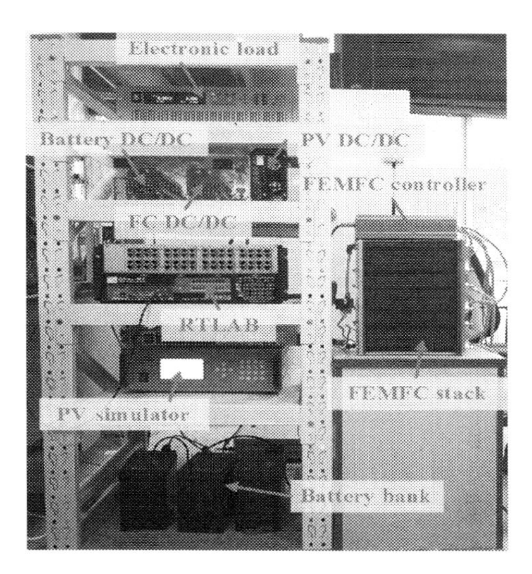

Fig. 4.5 The experiment platform of the proposed DC microgrid.

Table 4.1 Parameters of the proposed DC microgrid

Subsystems	Descriptions	Values
PV array simulator	Open-circuit voltage	50 V
	Voltage of the maximum power	36 V
	Short-circuit current	30 A
	Current of the maximum power	28 A
	Rated power	1000 W
Fuel cell	Rated power	1000 W
	Voltage range	32–34 V
	Max current	32 A
	Max temperature	65°C
Battery bank	Rated voltage	12*3 V
	Rated capacity	36 Ah
	Set numbers	3 series
Load	DC bus voltage	60 V
	Rated power	1500 W

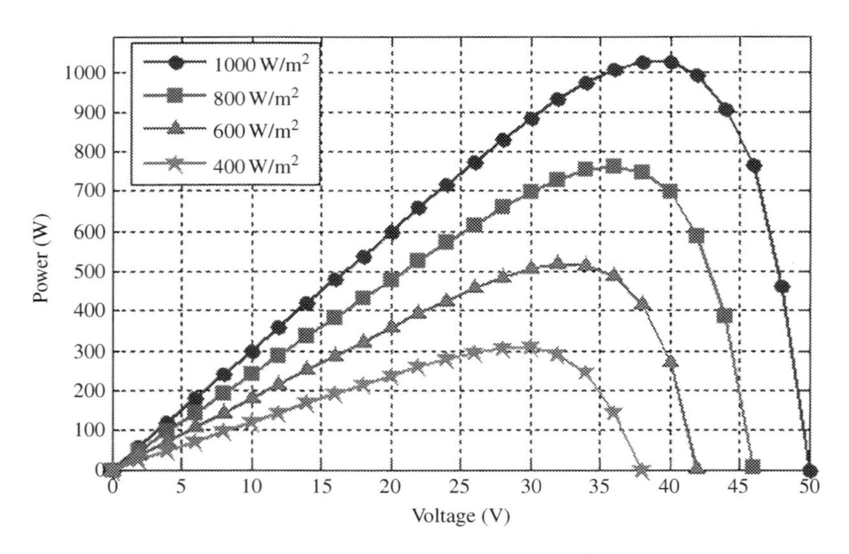

Fig. 4.6 The comparison of photovoltaic array simulator and theoretical model.

4.3.3 PEMFC generation system

In this study, the PEMFC generation system has three components: PEMFC stack, fuel-cell subsystem, and the unidirectional DC/DC converter, as shown in Fig. 4.7. The rated power of the stack is 1000 W and the output voltage varies between 32 and 45 V with maximum current 35 A. The output characteristic curve of fuel-cell stack is displayed in Fig. 4.8, and the hydrogen usage rate curve of stack is shown in

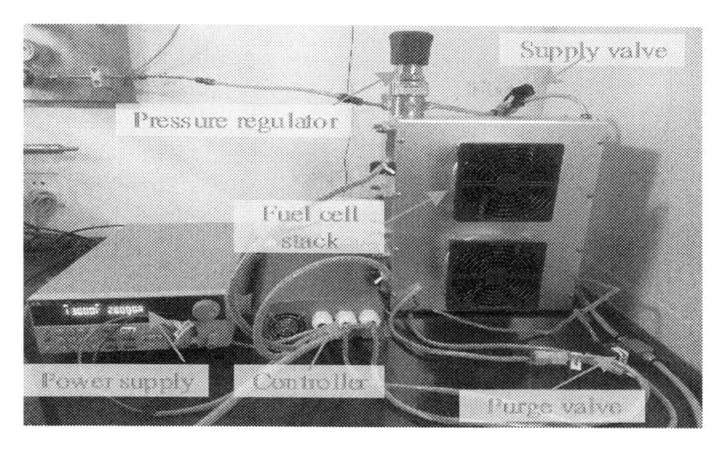

Fig. 4.7 PEMFC generation system.

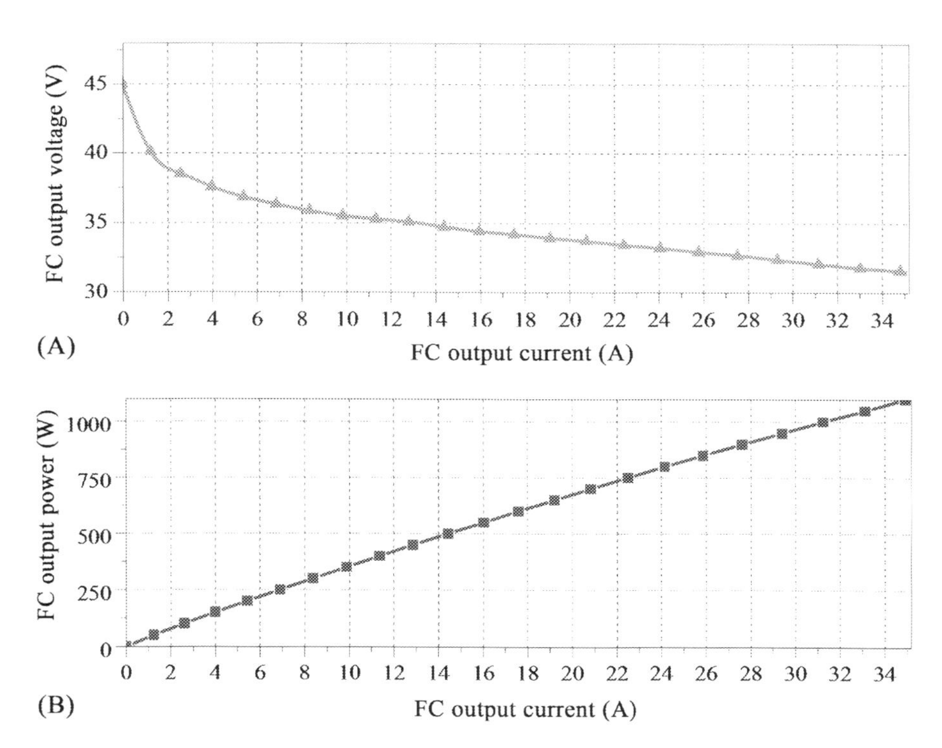

Fig. 4.8 PEMFC voltage and power polarization curve. (A) Relationship between current and voltage of FC. (B) Relationship between current and power of FC.

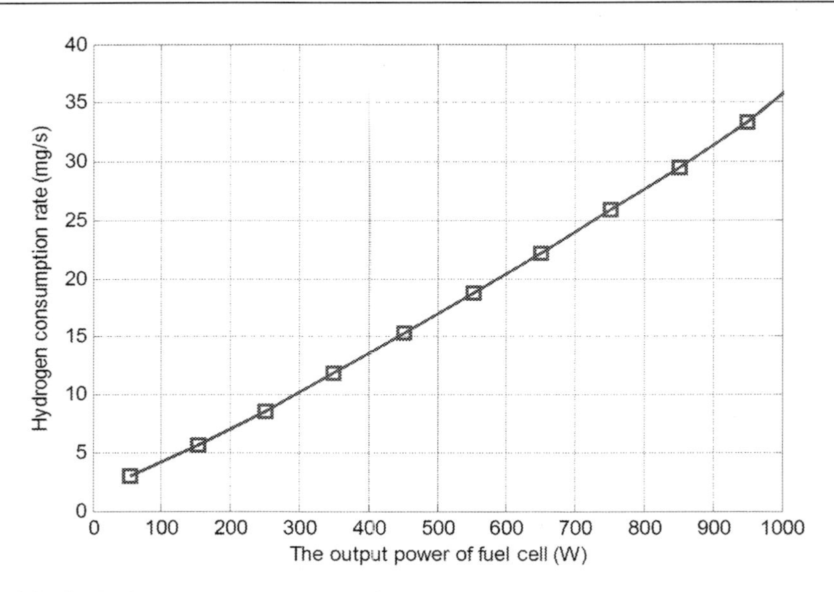

Fig. 4.9 The hydrogen usage rate curve of PEMFC stack.

Fig. 4.9. The subsystem consists of an air-cooled stack, blower, H_2 pressure regulator, and valves, controller for interfacing with the main controller and protections. Because the fuel-cell output is unregulated, meaning the stack voltage drops as the current it provides increases, and a unidirectional DC/DC converter is required to connect the PEMFC system with the DC bus to maintain the PEMFC system stable in spite of the variations from load. Furthermore, the PEMFC system can also be protected from high frequency and surge load power demand so that the lifetime could be extended.

The fuel-cell efficiency is inversely proportional to the output power, while the fuel-cell converter has a higher output efficiency at higher output power. Therefore, the combined efficiency curve of the PEMFC generation unit is shown in Fig. 4.10, and the peak of the curve represents the optimal operating point.

4.3.4 Battery energy-storage system

In this paper, the energy-storage system of the DC MG consists of a battery bank and a bidirectional DC/DC converter. The battery bank is composed of three lead-acid batteries in parallel rated 36 V, 36 Ah, and the battery parameters (*OCV*, internal charge resistance R_{chg}, and internal charge resistance R_{dis}) vary with the SoC, as shown in Fig. 4.11.

A bidirectional DC/DC converter (Zahn DC6350F-SU) is used to regulate the power delivered by the battery bank to the bus. The acceptable output current range is 36 A, and the maximum output power is 2880 W. The converter works in buck mode during charging process, similarly the converter is in boost mode in the course of discharge process.

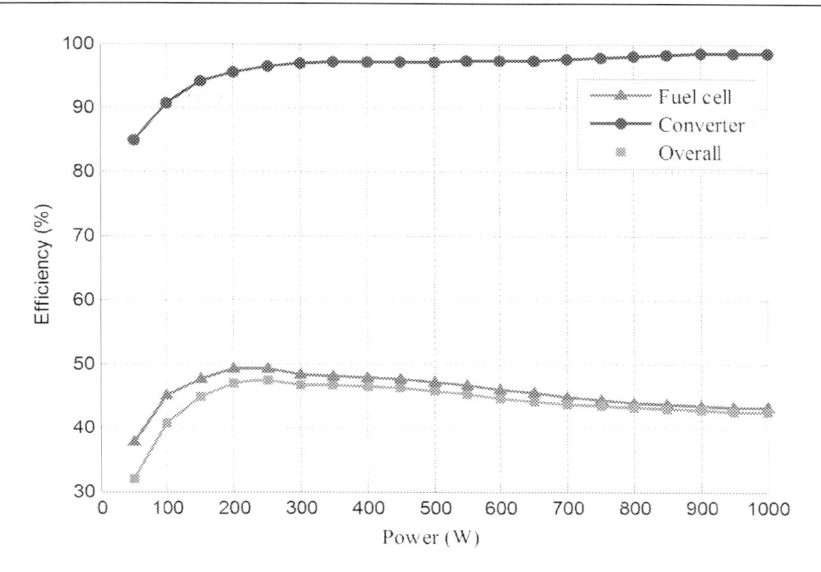

Fig. 4.10 PEMFC generation system combined efficiency.

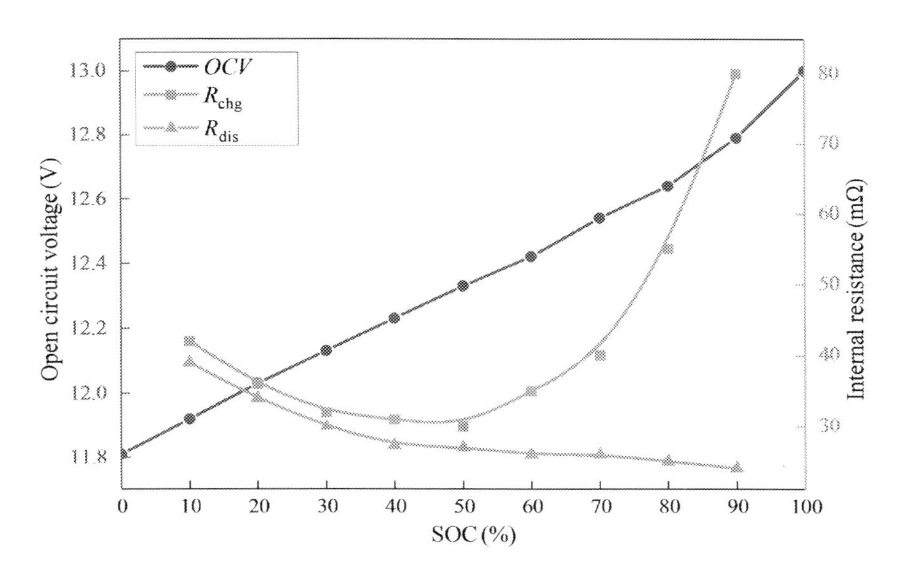

Fig. 4.11 Battery parameters as a function of SoC.

4.4 Optimal control and energy management for DC MG

4.4.1 Control objectives

Because of the variety of types and characteristics of the distributed generations, the network structure, the effective and steady optimal control, and energy management of DC MG have become key issues that need urgent solutions during the process of practical application. The energy management strategy has a great effect on the fuel economy, dynamic performance, and service life of distributed generations (Han et al., 2014). In the DC MG, the primary objective of the energy management strategy is to keep DC bus voltage stable. In addition to this, extending the life of the distributed generations and reducing the system cost are also important control objectives. All in all, the energy management of DC MG should fulfill the following objectives:

(1) maintain the stability of the dc bus voltage
(2) fully utilize the renewable resources (PV, wind power generation) under varying environment conditions
(3) minimize the fuel consumption and improve generation efficiency
(4) protect the distributed generations from damage

4.4.2 The optimal control of distributed generations

In the DC MG, the distributed generations are controlled to maintain the DC bus voltage in a reasonable range and realize the system power balance and stable operation. Based on the different types and characteristics, the distributed generations are controlled by different optimal control methods to realize the reasonable scheduling of power and the stable operation of the power system. The control structure of the DC MG is shown in Fig. 4.12. In addition, it can be seen that: the PV generation system works at three operating modes under different operating conditions; the battery bank is controlled to maintain the DC bus voltage; and the fuel-cell generation system output power is controlled by the energy management system of the DC MG.

4.4.2.1 The control of PV generation

In this DC MG, PV generation provides power to the DC bus by DC/DC converter. Due to the different operating conditions of the DC MG, the PV generation system has three working modes according to the control the DC/DC converter: MPPT mode, droop mode, and nonworking mode. Because the output characteristic of PV array changes under various environmental conditions, PV generation system generally works at MPPT mode to make an efficient utilization of renewable energy when the battery is not fully charged. MPPT strategies are out of the scope of this chapter, interested readers may also refer to Salas et al. (2006). When the battery bank is approaching to be fully charged or the charge power is higher than the limit, battery bank should limit its charge power according to SoC conditions. Coordinately, in order to balance the energy of generated and demanded, PV generation system should decrease power based on the droop relationship between the bus voltage and the duty

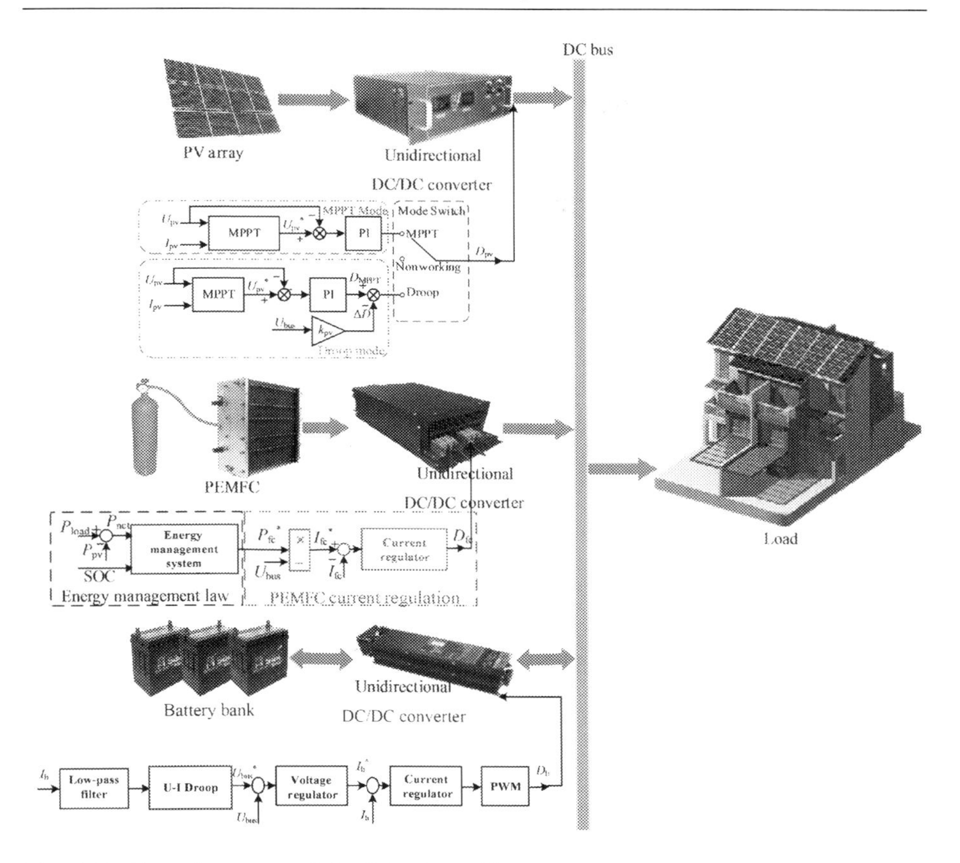

Fig. 4.12 Control structure of the DC microgrid system.

ratio of the DC/DC converter. However, when the system works in cloudy or at night, the PV generation system cannot continuously provide power, and in this way the PV generation system works at nonworking mode. In this operating mode, the fuel-cell generation system and battery bank provide power to load collectively. The control structure of the PV array DC/DC converter is shown in Fig. 4.13.

4.4.2.2 The control of PEMFC generation

In the PEMFC generation system, the output of the EMS is the fuel-cell generation system reference power P_{fc}^*, and then the P_{fc}^* is divided by the bus voltage to obtain the fuel-cell generation system reference current I_{fc}^*. Furthermore, a current regulator adjusts output current of DC/DC converter to this value. The control structure of the PEMFC generation DC/DC converter is shown in Fig. 4.14.

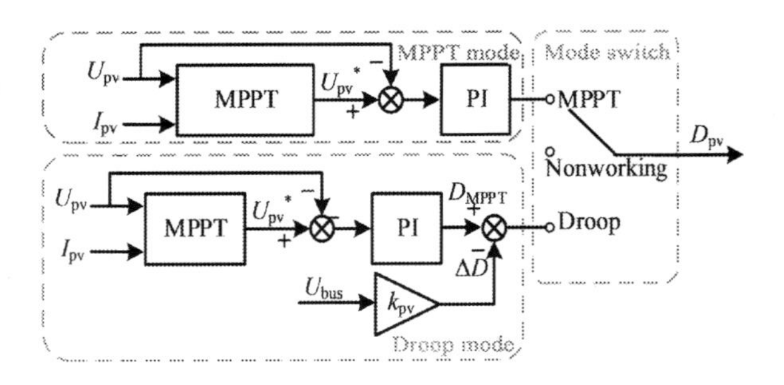

Fig. 4.13 Control diagram of the PV array DC/DC converter.

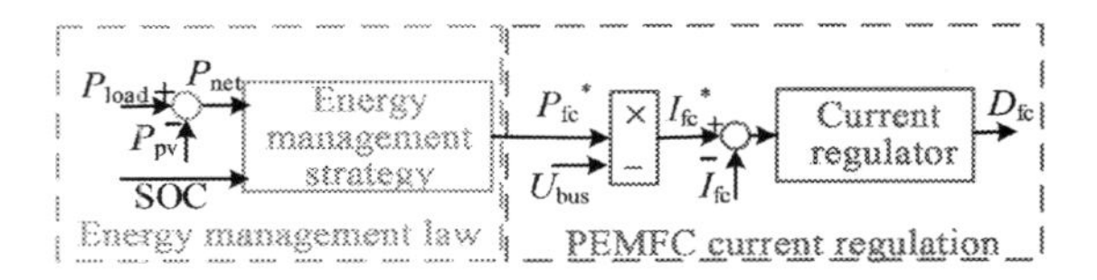

Fig. 4.14 Control diagram of the PEMFC generation DC/DC converter.

4.4.2.3 The control of battery bank

In the DC MG, the battery bank is responsible for balancing the power of DC MG and adjusting the voltage of the DC bus. In this chapter, the droop control method is used to maintain the DC bus voltage stability and automatic charge and discharge of the battery bank, and to achieve the power balance and stable operation of the DC MG system. The droop control diagram is shown in Fig. 4.15. In addition, the expression of droop control in DC MG can be shown as

$$U_{\mathrm{dc}}^{*}=\begin{cases} U_{\mathrm{H}}-mI_{\mathrm{b}},\text{charging},I_{\mathrm{b}}<0 \\ U_{\mathrm{L}}-mI_{\mathrm{b}},\text{discharging},I_{\mathrm{b}}>0 \end{cases} \tag{4.10}$$

where U_{dc}^{*} is the reference of DC bus voltage, U_{H} and U_{L} are the critical voltage value of charging and discharging, I_{b} is the filtered output current of battery DC/DC by the low-pass filter, and m is the adaptive droop coefficient.

The adaptive droop coefficient m is determined by SoC of the battery bank, which is calculated by

$$m=\begin{cases} m^{*}\text{SoC},\text{charging},I_{\mathrm{b}}<0 \\ m^{*}/\text{SoC},\text{discharging},I_{\mathrm{b}}>0 \end{cases} \tag{4.11}$$

where m^{*} is the initial droop coefficient, which is calculated by

$$m^{*}=\frac{\Delta U}{I_{\max}} \tag{4.12}$$

where ΔU is the maximum voltage fluctuation range, and $I_{\max}$ is the maximum charge current.

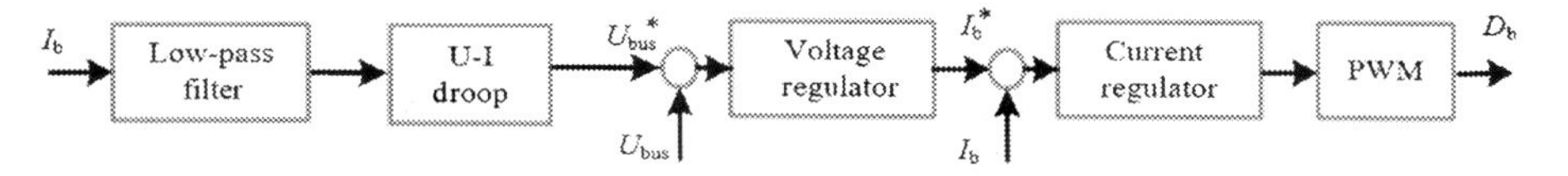

Fig. 4.15 Control diagram of the battery bank DC/DC converter.

In the discharge process, the droop coefficient increases with the decrease of SoC, and the discharge current gradually increases accordingly. Similarly, in the charge process, the droop coefficient decreases with the increase of SoC, and the discharge current gradually decreases accordingly (Fig. 4.15).

4.4.3 The energy management strategy of DC MG

4.4.3.1 Classical PI control strategy

The energy management strategy diagram based on classical PI control is shown in Fig. 4.16. The battery SoC is controlled by a PI regulator, and the output of the PI regulator is the battery reference output power, which is afterward removed from the load power to obtain the fuel-cell reference power. When the battery SoC is above the reference, the fuel-cell output power is low, and the battery discharges power to decrease the battery SoC. When the SoC is below the reference, the fuel cell provides power to the load and the battery bank, and the battery SoC will increase to a suitable range.

4.4.3.2 State machine control strategy

The state machine control strategy is one of the classical rule-based control strategies for the energy management system of MG. In this DC MG, the implemented state machine control strategy consists of 13 states under three SoC states, as shown in Table 4.2. The SoC is divided into three states based on hysteresis cycle: high state, normal state, and low state, which is shown in Fig. 4.17. It can be seen that the fuel-cell reference power is determined based on the battery SOC range and system net demand power P_{net}, and the following limits have been considered at the same time: P_{load} and P_{pv} are the real-time measured load demand power and PV generation system output power, respectively; P_{fcmin}, P_{fcopt}, and P_{fcmax} are the minimum, optimum, and maximum output power of the fuel-cell system; P_{batmax} and P_{batopt} are the maximum and optimum efficiency power of the battery bank.

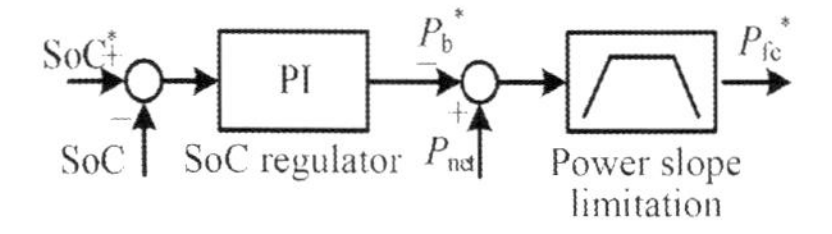

Fig. 4.16 Energy management strategy diagram based on PI control.

Table 4.2 Control decisions based on the state machine control strategy

SOC level	State	Net demand power characteristics	$P_{fc}*(t)$
Low SoC	1	$P_{net}(t) < P_{fcopt}$	P_{fcopt}
	2	$(P_{fc}*(t-\Delta t) = P_{fcopt})\&(P_{net}(t) < P_{batopt})$	P_{fcopt}
	3	$P_{net}(t) > P_{fcmax}$	P_{fcmax}
	4	$(P_{fc}*(t-\Delta t) = P_{fcmax})\&(P_{net}(t) > P_{batopt} + P_{fcopt})$	P_{fcmax}
	5	Otherwise	P_{net}
Normal SoC	6	$P_{net}(t) < P_{fcopt}$	0
	7	$(P_{fc}*(t-\Delta t) = P_{fcopt})\&(P_{net} < P_{batopt})$	0
	8	$P_{net}(t) > P_{fcmax}$	P_{fcmax}
	9	Otherwise	$P_{net}(t)$
High SoC	10	$P_{net}(t) < P_{fcmin}$	0
	11	$(P_{fc}*(t-\Delta t) = P_{fcmin})\&(P_{net}(t) < P_{fcopt})$	0
	12	$P_{net}(t) > P_{fcmax}$	P_{fcmax}
	13	Otherwise	$P_{net}(t)$

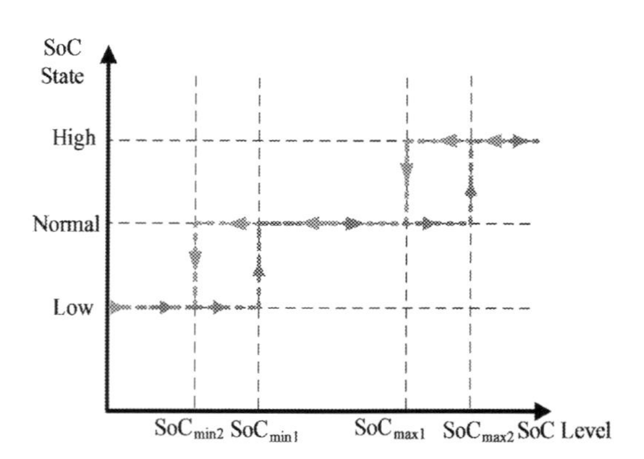

Fig. 4.17 Hysteresis cycles for SoC levels of battery bank.

The P_{net} is real-time calculated based on SOC state as

$$P_{net}(t) = \begin{cases} P_{load}(t) - P_{pv}(t) + P_{batopt} & \text{Low SoC} \\ P_{load}(t) - P_{pv}(t) & \text{Normal SoC} \\ P_{load}(t) - P_{pv}(t) - P_{batopt} & \text{High SoC} \end{cases} \tag{4.13}$$

4.5 Results and discussions

In order to authenticate the energy management strategies under different operating conditions, the energy management systems based on classical PI control strategy

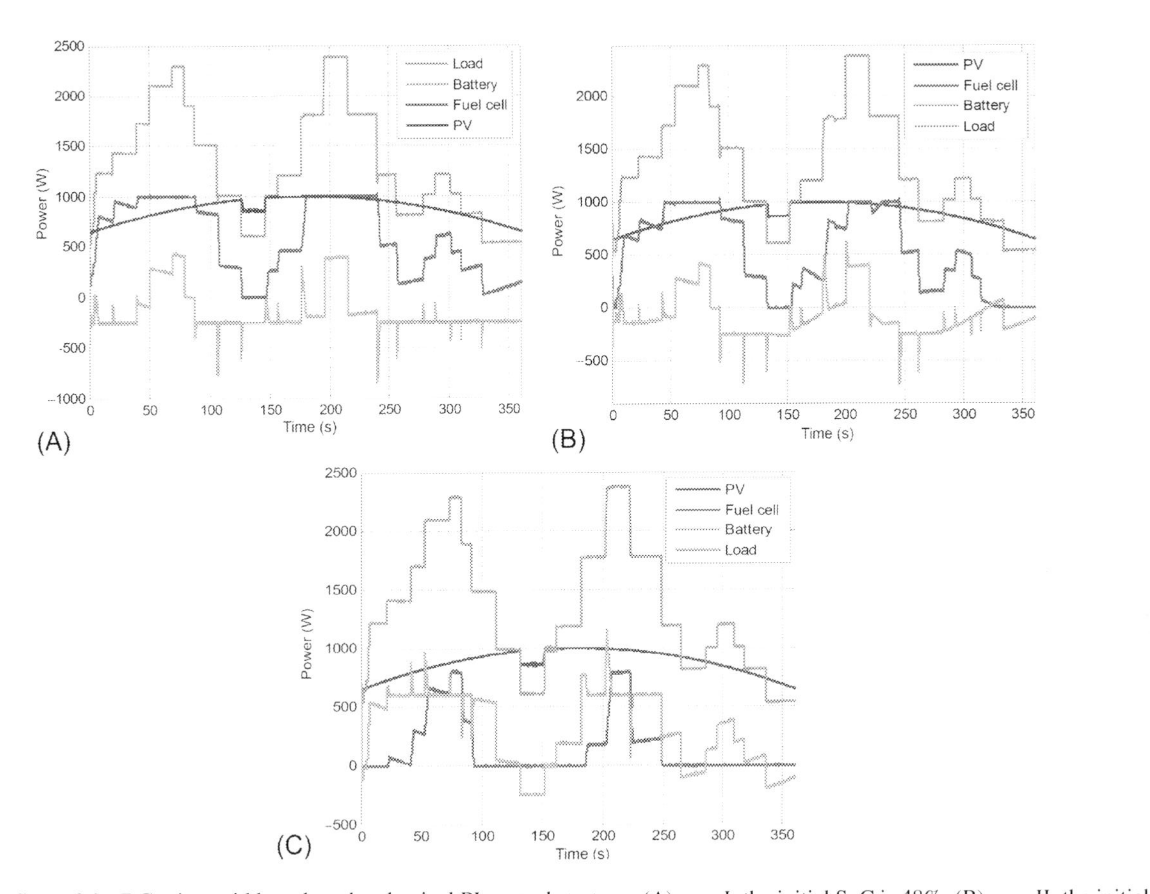

Fig. 4.18 Power flow of the DC microgrid based on the classical PI control strategy: (A) case I: the initial SoC is 48%, (B) case II: the initial SoC is 60%, and (C) case III: the initial SoC is 85%.

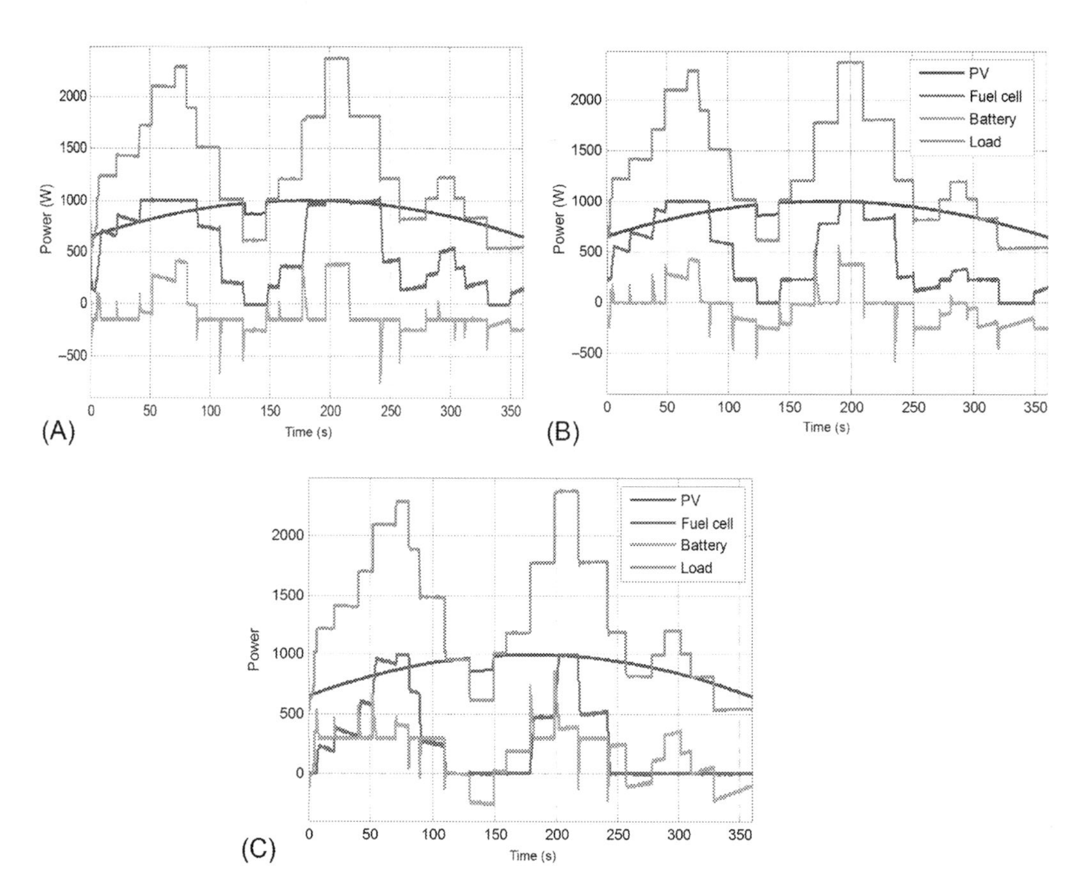

Fig. 4.19 Power flow of the DC microgrid based on the state machine control strategy: (A) case I: the initial SoC is 48%, (B) case II: the initial SoC is 60%, and (C) case III: the initial SoC is 85%.

Table 4.3 Performance comparison based on classical PI control strategy and state machine control strategy

	Classical PI control strategy			**State machine control strategy**		
	Case I	**Case II**	**Case III**	**Case I**	**Case II**	**Case III**
Average efficiency of FC generation system (%)	43.85	43.99	44.87	44.13	44.68	45.19
Hydrogen consumption of FC generation system (g)	7.14	6.42	1.42	6.54	5.56	3.01
SOC of battery bank (%)	48 → 48.8	60 → 60.2	85 → 81.1	48 → 48.4	60 → 59.8	85 → 82.7

and state machine control strategy are developed by the RT-Lab. In addition, the PV/fuel cell/energy-storage DC MG experiment platform in Section 4.3 operates under three cases: the initial SoC is 48%, 60%, and 85%. The experimental results are shown in Figs. 4.18 and 4.19.

The experimental results show that the energy management strategies can realize the reasonable distribution of the power flow and the stable operation of the DC MG under different cases. In case I, the battery bank with a low SoC is charged as much as possible to increase the SoC to a suitable range; the PV system and fuel-cell system provide the power to the battery bank and the DC load. In case II, the battery bank tries to keep the SoC in a suitable range; the PV system and fuel-cell system are controlled by the energy management system to balance the energy of that generated and demanded. In case III, the battery bank with a high SoC is discharged as much as possible to decrease the SoC to a suitable range, and the fuel-cell system works at a low power to reduce the hydrogen consumption. Meanwhile, PV generation system generally works at MPPT mode to make efficient utilization of renewable energy under different environment conditions, and decreases the output power based on droop control when the charge power is higher than the limit.

The comparison of control performance based on classical PI control strategy and state machine control strategy for DC MG is summarized in Table 4.3. The results show that the classical PI control strategy has a stronger control ability of SoC, the SoC of battery bank can controlled to a suitable range faster; the state machine control strategy performs a better economy, and the fuel-cell system works at a higher operating efficiency.

4.6 Conclusions

In this chapter, the DC MG system structure and the math model of each subsystem are studied. On this basis, the distributed generations control methods based on their output characteristic are introduced. In addition, two energy management strategies based on classical PI control and state machine control are employed to realize the optimal power distribution of the multisource hybrid DC MG. In order to verify the energy management strategies under different operating conditions, a lab-scale DC MG platform is developed in this chapter. The results show that the classical PI control strategy has a stronger control ability of SoC, and the SoC of battery bank can controlled to a suitable range faster; the state machine control strategy performs a better operating efficiency and a better operating economy in case I and case II.

Acknowledgments

This work was supported by National Natural Science Foundation of China (61473238), NEEC Open-end Fund of China (NEEC-2017-B01), and Cultivation Program for the Excellent Doctoral Dissertation of Southwest Jiaotong University (D-YB201704).

References

Bambang, R.T., Rohman, A.S., Dronkers, C.J., Ortega, R., Sasongko, A., 2014. Energy management of fuel cell/battery/supercapacitor hybrid power sources using model predictive control. IEEE Trans. Ind. Inform. 10 (4), 1992–2002.

Brka, A., Kothapalli, G., Al-Abdeli, Y.M., 2015. Predictive power management strategies for stand-alone hydrogen systems: lab-scale validation. Int. J. Hydrog. Energy 40 (32), 9907–9916.

Chen, Y.K., Wu, Y.C., Song, C.C., Chen, Y.S., 2013. Design and implementation of energy management system with fuzzy control for DC microgrid systems. IEEE Trans. Power Electron. 28 (4), 1563–1570.

Chen, W., Han, Y., Li, Q., Liu, Z., Peng, F., 2014. Design of proton exchange membrane fuel cell grid-connected system based on resonant current controller. Int. J. Hydrog. Energy 39 (26), 14402–14410.

Du, Y., Zhou, X., Bai, S., Lukic, S., Huang, A., 2010. Review of non-isolated bi-directional DC-DC converters for plug-in hybrid electric vehicle charge station application at municipal parking decks.Applied Power Electronics Conference and Exposition (APEC), 2010 Twenty-Fifth Annual IEEE, California, February, pp. 1145–1151.

Gu, Y., Xiang, X., Li, W., He, X., 2014. Mode-adaptive decentralized control for renewable DC microgrid with enhanced reliability and flexibility. IEEE Trans. Power Electron. 29 (9), 5072–5080.

Han, J., Charpentier, J.F., Tang, T., 2014. An energy management system of a fuel cell/battery hybrid boat. Energies 7 (5), 2799–2820.

Han, Y., Chen, W., Li, Q., 2017. Energy management strategy based on multiple operating states for a photovoltaic/fuel cell/energy storage DC microgrid. Energies 10 (1), 136.

Kakigano, H., Miura, Y., Ise, T., 2010. Low-voltage bipolar-type DC microgrid for super high quality distribution. IEEE Trans. Power Electron. 25 (12), 3066–3075.

Katiraei, F., Iravani, M.R., 2006. Power management strategies for a microgrid with multiple distributed generation units. IEEE Trans. Power Syst. 21 (4), 1821–1831.

Kwasinski, A., 2011. Quantitative evaluation of DC microgrids availability: effects of system architecture and converter topology design choices. IEEE Trans. Power Electron. 26 (3), 835–851.

Li, C., Vasquez, J.C., Guerrero, J.M., 2016. Convergence analysis of distributed control for operation cost minimization of droop controlled DC microgrid based on multiagent. Applied Power Electronics Conference and Exposition (APEC), 2016 IEEE, California, March, pp. 3459–3464.

Liu, Y., Pratt, A., Kumar, P., Xu, M., Lee, F.C., 2007. 390V input VRM for high efficiency server power architecture.Applied Power Electronics Conference, APEC 2007-Twenty Second Annual IEEE, California, February, pp. 1619–1624.

Nejabatkhah, F., Li, Y.W., 2015. Overview of power management strategies of hybrid AC/DC microgrid. IEEE Trans. Power Electron. 30 (12), 7072–7089.

Salas, V., Olias, E., Barrado, A., Lazaro, A., 2006. Review of the maximum power point tracking algorithms for stand-alone photovoltaic systems. Sol. Energy Mater. Sol. Cells 90 (11), 1555–1578.

Schulz, W., 2007. ETSI standards and guides for efficient powering of telecommunication and datacom.Telecommunications Energy Conference, 2007. INTELEC 2007. 29th International, Rome, September, pp. 168–173.

Sechilariu, M., Wang, B.C., Locment, F., 2014a. Supervision control for optimal energy cost management in DC microgrid: design and simulation. Int. J. Electr. Power Energy Syst. 58, 140–149.

Sechilariu, M., Wang, B.C., Locment, F., Jouglet, A., 2014b. DC microgrid power flow optimization by multi-layer supervision control. Design and experimental validation. Energy Convers. Manag. 82, 1–10.

Thounthong, P., Chunkag, V., Sethakul, P., Sikkabut, S., Pierfederici, S., Davat, B., 2011. Energy management of fuel cell/solar cell/supercapacitor hybrid power source. J. Power Sources 196 (1), 313–324.

Valverde, L., Rosa, F., Del Real, A.J., Arce, A., Bordons, C., 2013. Modeling, simulation and experimental set-up of a renewable hydrogen-based domestic microgrid. Int. J. Hydrog. Energy 38 (27), 11672–11684.

Valverde, L., Bordons, C., Rosa, F., 2016. Integration of fuel cell technologies in renewable-energy-based microgrids optimizing operational costs and durability. IEEE Trans. Ind. Electron. 63 (1), 167–177.

Yoo, C.H., Chung, I.Y., Lee, H.J., Hong, S.S., 2013. Intelligent control of battery energy storage for multi-agent based microgrid energy management. Energies 6 (10), 4956–4979.

Hybrid AC/DC distribution network voltage control

Feng Qiao, Jin Ma*, Xiaoqing Han[†]*
*School of Electrical and Information Engineering, The University of Sydney, Sydney, NSW, Australia, [†]Shanxi Key Laboratory of Power System Operation and Control, Taiyuan University of Technology, Taiyuan, China

5.1 Introduction

The voltage control in distribution network becomes more challenging after the integration of large-scale distributed generations (DGs). Extensive research was done to regulate the voltage along the distribution feeders by incorporating DGs' output into the voltage control scheme and utilizing traditional voltage regulators such as the transformer tap changer and shunt capacitor (SC) to decrease the impact from DGs' reversed power flow. However, due to the intermittent nature of the renewable generation, it is not practical to treat all DGs as controllable units contributing to the voltage in the distribution network, and their fickle power generation would result in extra operation of traditional voltage regulators. Moreover, the increasing number of the integrated DG in the distribution network would cause massive investment on communication channels to regulate them in a two-way manner.

Apart from coordinating individual DGs with traditional voltage regulators, compacting local DGs in a microgrid (MG) and interacting with the distribution network as a controllable aggregation is an alternative way to resolve the voltage issue caused by DGs' integration. The initial AC MG is connected to the distribution network via a static switch, so the interaction between MG and distribution network should be restricted to avoid deterioration on the local power quality. However, due to the advancement in electronic technic such as the voltage-source converter (VSC), MG has evolved to be a hybrid AC/DC community dominated by electronic interfaced DGs. The up-to-date VSC-based hybrid AC/DC MG has shown a great potential to actively participate in the voltage control scheme.

As shown in Fig. 5.1, the VSC-based hybrid AC/DC MG includes an AC subsystem, a DC subsystem, a VSC (namely VSC1) interfaces the distribution network, and another VSC (namely VSC2) located between two subsystems. The details of this MG will be introduced in Section 5.2. By forming multiple DGs into AC and DC subsystems and operating them by a proper control scheme, the MG in Fig. 5.1 is capable to manage the varying output from local DGs (Loh et al., 2013a,b). Moreover, it is also capable to provide a dispatchable power injection to the distribution network irrespective of the node voltage since the interfaced VSC enables the decomposed control of active and reactive power (Yazdani and Iravani, 2010). Therefore, this

Smart Power Distribution Systems. https://doi.org/10.1016/B978-0-12-812154-2.00005-5

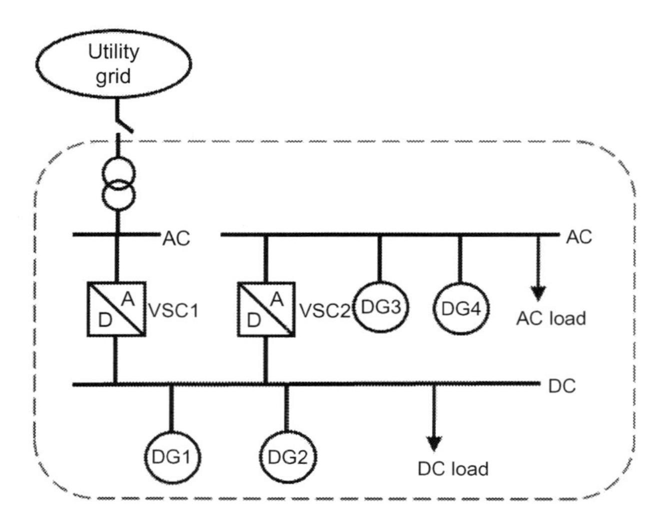

Fig. 5.1 Voltage-source converter (VSC)-based hybrid AC/DC microgrid (MG).

up-to-date VSC-based hybrid AC/DC MG is promising to provide the auxiliary service of voltage control in the distribution network when it is geographically close to the distribution feeder and electrically coupled to the distribution network.

In this chapter, we propose a two-layer voltage control scheme to incorporate the reactive power support from the MG in Fig. 5.1 into the distribution network voltage control scheme, aiming to satisfy the voltage quality in both distribution network and each grid-tied MG. At the upper layer, system parameters like node voltage, reactive power reserve of MGs are collected by the central controller, and optimized set points for the on load tap changer (OLTC), SC, and MGs are determined by calculating a multiobjective voltage optimization problem. At the lower layer, OLTC, SC, and MGs adjust their control parameters in response to the optimized set points from the upper layer. Specifically, each VSC-based hybrid AC/DC MG with the corresponding controllers is employed at the lower layer to enable the fast reactive power support to the distribution network and maintain its local power quality. The proposed voltage control scheme was implemented in Matlab to evaluate its performance, and a distribution network coupled with multiple hybrid AC/DC MGs was modeled in PowerFactory to validate the advantage and effectiveness of the control scheme. This chapter is organized as follows: Section 5.1 presents an introduction for the voltage control problem after integration of DGs and concludes the potential of MG to support the voltage in the distribution network. Section 5.2 introduces the VSC-based hybrid AC/DC MG and its corresponding controllers. Section 5.3 proposes the multiobjective voltage optimization problem and introduces an evolutionary algorithm called multiobjective evolutionary algorithm based on decomposition (MOEA/D) to solve it. Section 5.4 performs several case studies to validate the advantages and effectiveness of the proposed voltage control scheme. Section 5.5 concludes the whole chapter.

5.1.1 Voltage control after DGs' integration

As shown in Fig. 5.2, when a DG is connected to the downstream distribution feeder, its power injection will influence the voltage of the nearest bus. The magnitude of $\dot{U}_2$ and active power loss P_{loss} can be described as

$$
\begin{aligned}
U_2 &= U_1 - \Delta U = U_1 - \frac{(P_L - P_G)R + (Q_L - Q_G)X}{U_2} \\
P_{loss} &= \frac{(P_L - P_G)^2 + (Q_L - Q_G)^2}{U_2^2} R
\end{aligned}
\tag{5.1}
$$

It can be seen from Eq. (5.1) that the reversed power flow from DG increases U_2 if P_G and Q_G take positive values or decreases U_2 if P_G and Q_G take negative values. This voltage variation makes traditional voltage regulators failed to control the voltage of downstream buses. For example, the tap changing transformer at substation usually adjusts its tap position according to the downstream load profile and system impedance, but it would no longer maintain the voltage within the rated range after DG's integration because it is blinded to the extra power injection from downstream DGs. In addition, the voltage deviation ΔU in Eq. (5.1) will be minimized when $Q_G = Q_L - (P_L - P_G)$ while the active power loss P_{loss} can be minimized when $Q_G = Q_L$. This indicates a conflict between the optimized voltage and minimized active power in the system, a trade-off between these two aims are always existed during the system operation.

Extensive methods have been proposed to address the voltage control issue caused by DGs. One approach is to utilize a centralized energy management system (EMS) to enable DGs to collaborate with traditional voltage regulators. In Ref. (Cagnano and De Tuglie, 2015), a centralized control scheme was employed to optimize the nodal voltages by allocating reactive power set points for photovoltaics (PVs). The set points were calculated in real time by solving a constrained dynamic optimization problem aimed at minimizing the voltage deviation from a reference value. The economical drawback from the limitation of active power output caused by reactive power injection by PV was also discussed in this paper. In Ref. (Ranamuka et al., 2014), an online voltage control was proposed to mitigate operational conflicts from multiple voltage regulation devices. This approach utilizes the measurements like voltage, power, and tap position from distributed management system (DMS), then work out an optimal solution to maximize the voltage support from DGs while prioritizes the operation of other voltage regulation devices.

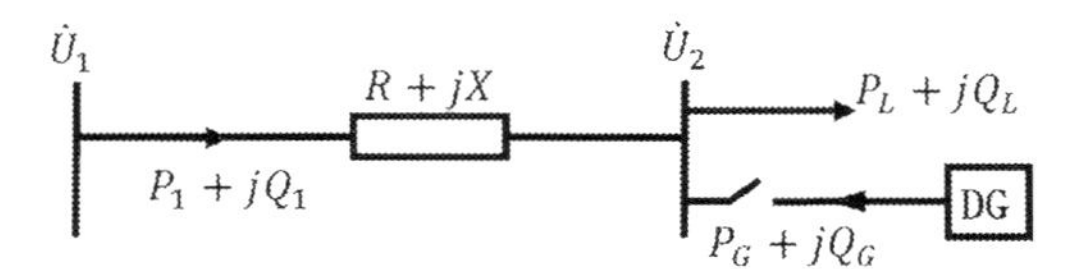

Fig. 5.2 Single-line diagram of a section of distribution feeder.

Another approach is the distributed voltage control, which mainly relies on the local measurements to regulate voltage. In Ref. (Viawan et al., 2007), a variety of voltage regulation methods by using OLTC, line drop compensation (LDC), and DG were discussed. The influences of DG on the operational frequency of OLTC and LDC were studied. The results indicated that using OLTC and LDC can significantly increases the hosting capability of DGs without sacrificing the performance of the voltage regulation. In Ref. (Casavola et al., 2011), DGs were treated as the voltage regulators in coordination with the traditional voltage control devices. The control method was developed based on the concept of control zone in which each voltage regulator is responsible for its local voltage.

Multiagent-based technic controls the voltage by using smart agents located at traditional voltage regulators, distributed controllable generators and loads. Energy surplus or shortage at these controllable nodes are sensed and broadcasted to an upper-layer controller, then optimal solutions can be carried into execution to maintain the voltage stable. In Ref. (Baran and El-Markabi, 2007), a multiagent-based technic is proposed to regulate the voltage in the distribution network. With the aid of smart agents integrated into DGs' controller and a two-way communication system, the reactive power deficiency is measured and the calculated reactive power set points are sent to each controller to regulate voltage efficiently. The detailed numerical deduction of reactive power dispatch is also performed, which makes this method more tangible. The multiagent-based voltage regulation scheme in Ref. (Baran and El-Markabi, 2007) was further extended to accommodate more voltage control devices in Ref. (Farag et al., 2012), where not only the DGs but also the LTCs (load-tap-changers), SVRs (static voltage regulator), and SCs were equipped with a smart agent contributing to a coordinated voltage control.

Although extensive research has been done regarding the voltage regulation in a distribution network with high penetration of DGs, there are many shortcomings existed which call for further study:

- Due to the small size of DG compares with other generators in the grid and the intermittent output from the renewable energy generation, it is difficult to consider each DG as a dispatched unit for the task of voltage control. Extra investment on voltage regulators is inevitable.
- Power exchange between DG and grid is unidirectional, which may lead to the restriction of DG's active power generation when the reactive support is needed. For example, when a DG has reached its maximum capacity, it may reduce its active power to meet the reactive power demand.
- With more and more DGs integrated to the distribution system, the number of the controllable unit increases dramatically. The computation burden is not affordable if the centralized control is used, and the investment on communications could be soaring.

5.1.2 Evolution of MG and its control method

The above shortcomings come from DG's inherent features are expected to be eliminated by gathering multiple DGs into an MG. The MG regulates multiple DGs to energize the local loads while exchanging power with the distribution network in case of need. The superior controllability of MG compared with dispersed DGs endow it a great potential to be an auxiliary of voltage control scheme in the distribution network.

It can be found in literature that there are various layouts of MG, and they can be categorized into AC MG, DC MG, and hybrid AC/DC MG. Meanwhile, corresponding power control and energy management strategies were proposed to make these MGs highly efficient and cost-effective building blocks in the future distribution network.

The concept of MG was proposed by the Consortium for Electric Reliability Technology Solutions (CERTS) in the early 21st century. Fig. 5.3 shows an example of AC MG proposed in Ref. (Lasseter and Paigi, 2004). This initial example of MG consists of four radial feeders. The feeder D has sensitive loads, while the rest three feeders have nonsensitive loads which can be energized by DGs. This AC MG connects to the distribution network via a single point of common coupling (PCC), and a static switch is implemented to isolate nonsensitive loads in case of need. AC MG is popular because it can fit in existing AC-dominated power system with a minor transformation. Many construction projects have been built around the world (Lidula and Rajapakse, 2011), proving that AC MG is capable to achieve a reliable operation in the distribution network. For example, it is reported in Ref. (Shahidehpour and Khodayar, 2013) that a reliable, energy-efficient, and sustainable AC MG equipped with the wind, solar, gas generators were built on the campus of Illinois Institute of Technology (IIT). It provides a high quality of electricity for each building in IIT main campus with a lower cost. A hierarchical control scheme is behind the IIT MG to support its operation, which includes droop-based primary control allocating load for distributed generators, secondary control for restoring voltage and frequency deviation, and tertiary control achieving economical and reliable criterions.

However, with the increasing number of DC elements such as computer, printer, TV, light-emitting diode (LED), and electric vehicle, the demand of AC/DC converters is significantly soared, which in turn increases the financial cost to build an AC MG. Hence, the DC MG becomes attractive as it accommodates DC devices without augment of electronic converters. A typical DC MG modeled in Ref. (Chen and Xu, 2012) is shown in Fig. 5.4 where a wind generator, an energy storage system (ESS), and DC loads are connected to a common DC bus with corresponding electronic converters. An interlinking converter (IC) is employed to exchange power between DC MG and the utility grid. The DC MG avoids extra AC/DC conversion,

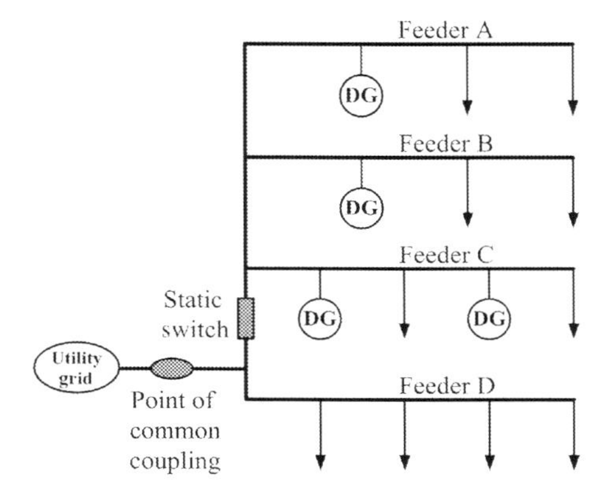

Fig. 5.3 AC MG.

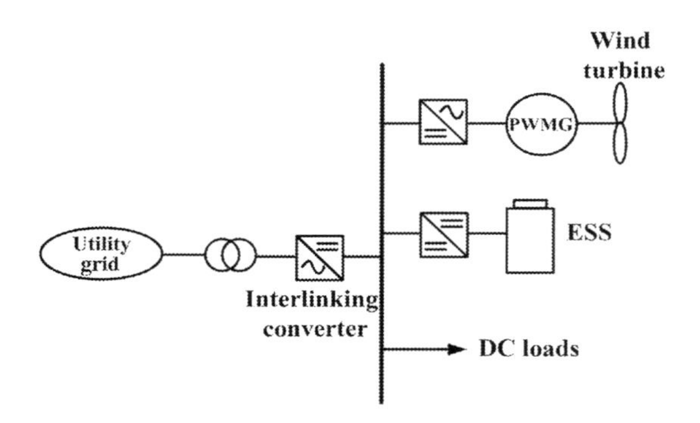

Fig. 5.4 DC MG.

which contributes to a better economical and efficient operation. For a DC MG, a proper control strategy is essential to distribute the load current to each converter interfaced generator to eliminate the circulating current caused by the mismatch between converters' terminal voltage. It is suggested in Ref. (Guerrero et al., 2011) that the circulating current between different DC/DC power converters can be avoided by implementing the droop control method to each converter without extra communication facilities. The droop control method for electronic converter shares load current by emulating the behavior of the traditional synchronous generator by adding a virtual output impedance in the controller. The inherent trade-off between voltage regulation and load current sharing introduced by the droop control method is mitigated by the secondary control by changing the virtual resistance. The tertiary control is utilized to enhance the power quality at the PCC.

The hybrid AC/DC MG combines advantages of AC MG and DC MG. As shown in Fig. 5.5, this up-to-date layout deploys AC and DC components into AC and DC subsystems respectively and utilizes a main converter to exchange power between two subsystems. Since the control method of main converter is of crucial for the efficiency and reliability of the hybrid AC/DC MG, it becomes a hot topic in recent years. In Ref. (Liu et al., 2011), a coordinated control scheme for the hybrid AC/DC MG is proposed. On the grid-connected mode, the AC subsystem is synchronized to the utility grid, and the power exchange between AC and DC subsystems is enabled by the IC to maintain the DC voltage. Meanwhile, the battery in DC subsystem acts as an energy buffer to smooth the power transfer between two subsystems. The renewable energy from the wind and solar generators is fully utilized on this mode. On the islanding mode, different operational cases are distinguished by EMS, and corresponding coordinated control scheme is activated to balance the power flow within the MG. The energy from the wind and solar generation may be curtailed if the total power demand is less than the total power generation. In Ref. (Loh et al., 2013b), the IC in hybrid AC/DC MG is utilized to enable a bidirectional power exchange between AC and DC subsystems. The normalized AC frequency is compared with the normalized DC voltage under a unified coordinate, then an error is produced to indicate the power surplus or shortage in AC or DC subsystem. Base on this principle, either a

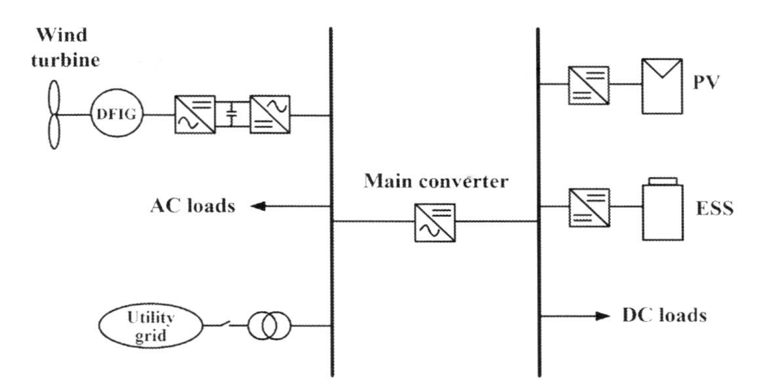

Fig. 5.5 Hybrid AC/DC MG.

proportional and integral (PI) controller or a droop controller can be implemented for IC to enable a bidirectional power exchange.

Compared with the AC and DC MG, the advantages of the hybrid AC/DC MG are concluded as:

- The power loss is reduced by avoiding the extra energy conversion between AC and DC energy.
- The DC elements can be integrated without extra investment on the electronic converters.
- With the main converter in between two subsystems, the bidirectional power transfer can be well regulated without harmonics.

5.1.3 Potential of voltage support from MG

When an MG is coupled to the distribution network, the interactions between these two entities could be varied depending on the size and location of the MG. For example, if a VSC-based MG with a redundant power is connected to a bus with low voltage, the reactive power support can be easily achieved by setting the controller of interfaced VSC in the MG. However, this should be taken with carefulness for not interfering the power quality inside the MG.

To coordinate MG with the distribution network, a hierarchical control with multiple control layers is usually employed. In Ref. (Tian et al., 2016), a hierarchical energy management structure is utilized to minimize the daily operational cost for the distribution network where an MG community is coupled in parallel with individual DGs and energy storages. The optimal exchanged power between MGs is determined at the lower level first, then this exchanged power is sent to the upper level as the constraints to perform a global optimization. In Ref. (Wang et al., 2015), a coordinated control strategy for networked MGs is proposed where the distribution network and each MG have their own objective to minimize the operational cost. The problem is formulated as a stochastic bi-level optimization problem with the distribution network in the upper level and MGs in the lower level. Each level has two stages where the first stage optimizes the objective according to the forecast and the second stage adjusts the calculation based on the uncertainty of the renewable generation. In Ref. (Che et al., 2015), the control method for a single MG in Refs. (Liu et al., 2011; Loh et al., 2013b) was extended to control a community of MG, where the droop-based primary control enables MGs to share the total load in the community, the secondary control restores the deviation of voltage and frequency caused by the primary droop control, and the tertiary control enhances the economic operation and the reliability by dispatching each MG in the community.

Above works have shown that it is viable to operate MG in a large community, but merely pay attention to the reactive power interaction between MG and the distribution network. It has been proposed in many works that the electronic converters are used in various scenarios to provide the reactive power injection for retaining the operation parameters in power system. In Ref. (Turitsyn et al., 2011), the PV's interfaced inverter is controlled to regulate the local voltage at its PCC, and various local control methods are tested to demonstrate the advantage of incorporating distributed reactive power support into the voltage control scheme. In Ref. (Renedo et al., 2017), the reactive power injections of VSCs in a high-voltage direct current multiterminal system are controlled with respect to the frequency of their corresponding AC bus to enhance the transient stability of the power system. The reactive power injection from interfaced converter of wind generators is coordinated with their active power injection in Ref. (Zhao et al., 2017) to improve the voltage control capability of a wind farm. Two control strategies are designed to face the normal and violated voltage conditions. Above technics can be transplanted to a VSC-interfaced MG in the distribution network to endow the MG the ability to participate in the distribution network voltage control scheme, and the reactive power from VSC-interfaced MGs should be dispatched in a centralized manner since the coordination between different regulators and MGs is necessary to achieve an efficient operation and reduced power losses.

For incorporating MGs' reactive power support into the centralized voltage control scheme, it should be stressed that an MG is distinct from the traditional voltage regulator as it needs to supply the local loads and maintain the local power quality before being able to support the voltage of the distribution network, and the reactive power support from an MG should not conflict its active power injection. Thus, several requirements should be met beforehand:

- The MG should have the required capacity when being considered to supply the reactive power to the distribution network, and the power content it follows should not be surplus to the capacity of the IC.
- The specific control scheme should be employed for the MG to meet the active and reactive power demand from the distribution network and maintain the local power quality at the same time.
- The available power reserve of MG should be updated regularly for maximizing the utilization of the reactive power from it.

It should be noted that the capacity of the MG and its location can be determined by solving a planning problem, which is not the scope of this chapter. In this chapter, it is assumed that all VSC-based MGs are equipped with proper controllers for the interfaced VSC and local DGs, and MG's location and capacity are planned in advance. Moreover, due to the advantages of hybrid AC/DC MG introduced in Section 5.1.2, the VSC-based hybrid AC/DC MG is employed to respond to the reactive power set point from the central controller while regulating the local power quality. The details of this MG and its control method will be introduced in the following section.

5.2 VSC-based hybrid AC/DC MG (lower layer)

5.2.1 Layout of proposed MG

The VSC-based hybrid AC/DC MG applied in this study is shown in Fig. 5.6. The DGs in this MG are categorized into the grid-forming DG and grid-following RDG (renewable distributed generator). The DG1 and DG2 are grid-forming DGs which have the adequate capacity to regulate voltage and frequency within MG while the RDG1 and RDG2 are grid-following DGs inject power under the rated voltage and frequency with respect to their own benefit (Katiraei et al., 2008). IC1 and IC2 are VSC-based converters. IC1 either electrically isolates the MG with distribution network if no power injection is required or injects the dispatched power to the distribution network in case of need. IC2 is in charge of smoothing power transfer between AC and DC subsystems. In addition, there is a local controller implemented in this MG to monitor the local power flow and communicate with the central controller in the distribution

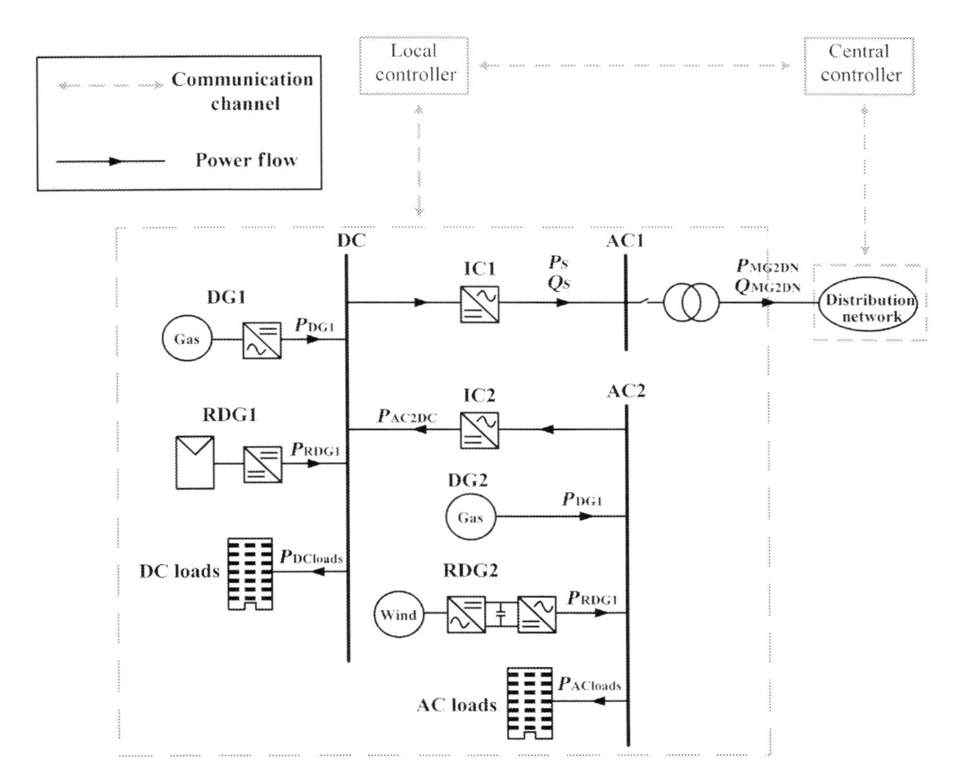

Fig. 5.6 Proposed hybrid AC/DC MG.

network. The active power injection from MG to DN (distribution network) can be described as

$$P_{\text{MG2DN}} = P_{\text{DG1}} + P_{\text{RDG1}} + P_{\text{AC2DC}} - P_{\text{DCloads}} \tag{5.2}$$

where P_{DG1} and P_{RDG1} are power injections from DG1 and RDG1, respectively, P_{DCloads} is active power consumption of DC subsystem, and P_{AC2DC} is the amount of active power transferred from the AC subsystem to the DC subsystem.

The reactive power injection Q_{MG2DN} is limited by the capacity of IC1 and its active power injection P_{MG2DN}

$$Q_{\text{MG2DN}} \le \sqrt{S_{nom}^2 - P_{\text{MG2DN}}^2} \tag{5.3}$$

For simplicity, the transformer in Fig. 5.6 is assumed to be the ideal model and its power loss is neglected, so the power injection to the distribution system $P_{\text{MG2DN}} + jQ_{\text{MG2DN}}$ equals to the power output from IC1's AC terminal $P_S + jQ_S$.

5.2.2 Control method of IC1

Fig. 5.7 shows the schematic diagram of IC1. The current mode is selected here as its immunity to overcurrent problem and the better dynamic performance (Yazdani and Iravani, 2010). Using the system voltage V_S-oriented control scheme, the real and reactive power delivered to the distribution system can be obtained in the dq frame

$$\begin{aligned} P_S &= \frac{3}{2} V_{sd} i_d \\ Q_S &= -\frac{3}{2} V_{sd} i_q \end{aligned} \tag{5.4}$$

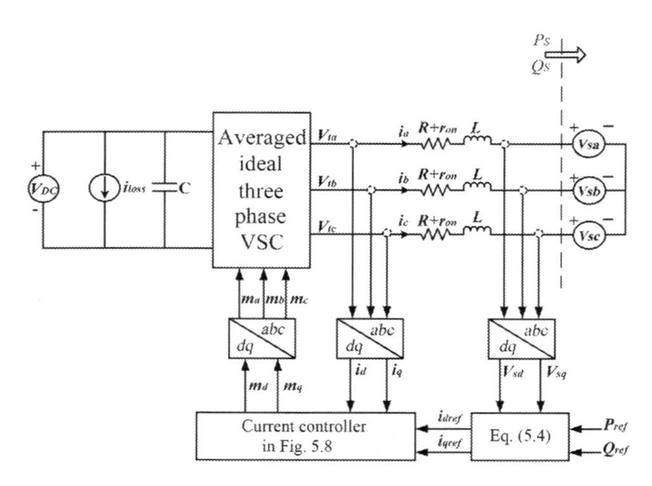

Fig. 5.7 Schematic diagram of IC1.

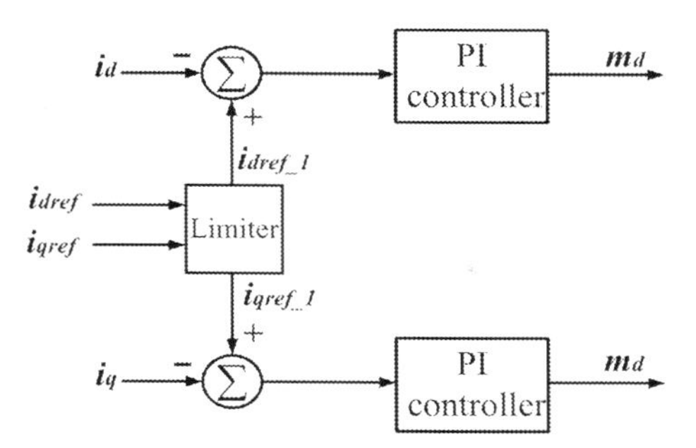

Fig. 5.8 Current controller in *dq*-frame.

Based on Eq. (5.4), the power injection P_S and Q_S of VSC can be controlled by adjusting i_d and i_q, respectively. With the power demand P_{ref} and Q_{ref}, the current reference value i_{dref} and i_{qref} can be obtained by a PI controller as shown in Fig. 5.8.

The limiter in Fig. 5.8 prevents the VSC from saturation by limiting its current reference in *dq* frame. Let S_{nom} represents the nominal capacity of the VSC, U_{nom} and I_{nom} represent the nominal voltage and current at AC side, respectively, then the constraint for i_d and i_q can be obtained

$$\sqrt{i_d^2 + i_q^2} = I \leq I_{nom} = \frac{S_{nom}}{\sqrt{3}U_{nom}} \tag{5.5}$$

5.2.3 Control method of IC2

Fig. 5.9 shows a schematic diagram of the IC2. Different to IC1 that P_{ref} and Q_{ref} are dispatched by the distribution network central controller, a normalization-based droop method proposed in Ref. (Loh et al., 2013b) is employed to generate the P_{ref} and Q_{ref} for IC2. This method is illustrated in Fig. 5.10 and its control parameters can be calculated as

$$\begin{aligned} P_{ref} &= K_p{}^*e_p = K_p\left(f_{pu} - V_{pu}\right) = K_p\left(\frac{f_{ac} - f_{rated}}{\Delta f} - \frac{V_{dc} - V_{dc\_rated}}{\Delta V}\right) \\ Q_{ref} &= K_q{}^*e_q = K_q\left(V_{ac} - V_{ac_{rated}}\right) \end{aligned} \tag{5.6}$$

where K_p and K_q are droop coefficients and e_p and e_q are errors produced by control parameters as shown in Eq. (5.6). The subscript *pu* represents the normalized value of the variable and the subscript *ac* and *dc* indicate the AC and DC subsystem, respectively, while the subscript *rated* represents the rated value of variable. The term $\Delta f = 0.5(f_{ac}^{\max} - f_{ac}^{\min})$ represents the acceptable range of frequency deviation, and the $\Delta V = 0.5(V_{dc}^{\max} - V_{dc}^{\min})$ represents the acceptable range of DC voltage deviation.

It is assumed that the distance between the maximum voltage and rated voltage is assumed to be equal to the distance between a minimum voltage and rated voltage,

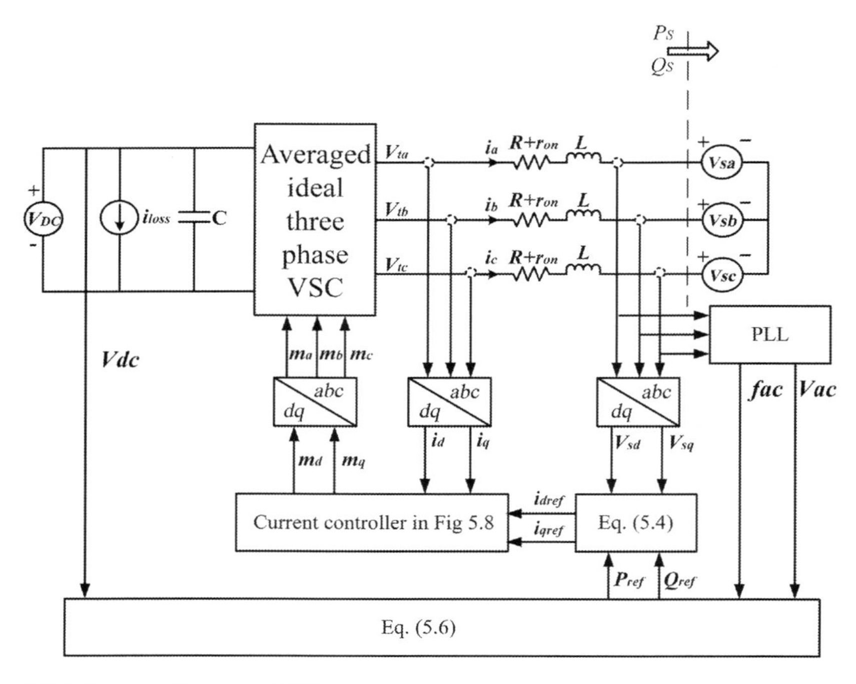

Fig. 5.9 Schematic diagram of IC2.

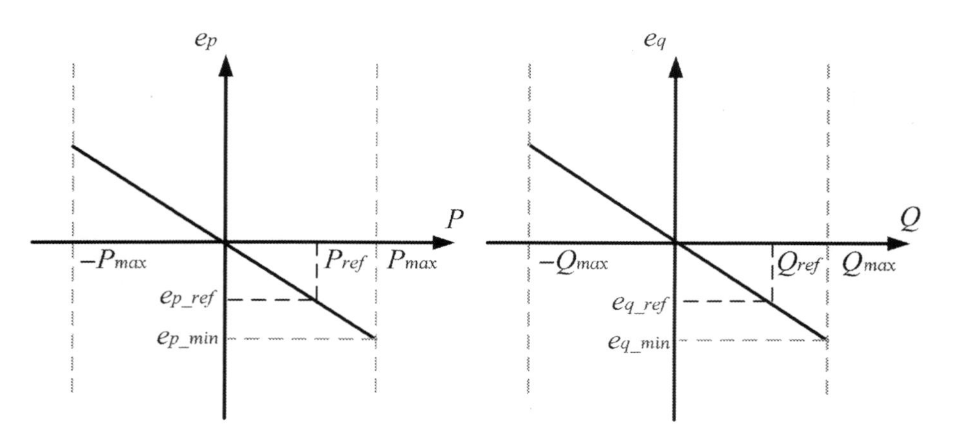

Fig. 5.10 Normalization-based droop control used by IC2.

similarly, the distance between the maximum frequency and rated frequency is equal to the distance between a minimum frequency and rated frequency. If f_{rated} is 50 Hz, V_{dc_rated} is set to be 1 kV, and Δf and ΔV are 1% and 5% of their rated value, respectively, then the f_{pu} and V_{pu} will be ranged from -1 to 1.

The values of K_p and K_q can be determined by system parameters and the rated capacity of IC2 by Eq. (5.7).

$$K_p = \frac{P_{\max}}{e_{p\_min}} = \frac{P_{\max}}{f_{pu\_min} - V_{pu\_max}} = \frac{P_{\max}}{\frac{f_{ac\_min} - f_{rated}}{\Delta f} - \frac{V_{dc\_max} - V_{dc\_rated}}{\Delta V}}$$
$$K_q = \frac{Q_{\max}}{e_{q\_min}} = \frac{Q_{\max}}{V_{ac\_min} - V_{ac\_rated}} \tag{5.7}$$

In Fig. 5.10, it can be observed that P_{ref} will be negative when f_{pu} is greater than V_{pu} or positive when f_{pu} is less than V_{pu}. With the well-known relationship between active power and frequency in AC system, Fig. 5.10 and Eq. (5.6) indicate that IC2 will transfer active power between AC and DC subsystems according to the power surplus or deficiency in them. Meanwhile, IC2 will also provide reactive power to AC subsystem in response to the gap between actual AC voltage and its rated value as shown in Fig. 5.10.

The normalization-based method drives IC2 to exchange active power from AC to DC to support the voltage of DC subsystem or from DC to AC to maintain the frequency of AC subsystem. Moreover, it enables IC2 to inject reactive power to the AC subsystem for retaining the voltage of AC bus. With this control method implemented in IC2, all energy sources in proposed hybrid AC/DC MG share the total load in proportion to their power ratings rather than their physical placements in two subsystems. Thus, P_{AC2DC} in Eq. (5.2) can be described as

$$\mathrm{P_{AC2DC} = P_{DG2} + P_{RDG2} - P_{ACloads}} \tag{5.8}$$

where P_{DG2} and P_{RDG1} are power output from DG2 and RDG2, respectively, and P_{ACloads} is the active power consumption in AC subsystem.

5.3 Proposed voltage control scheme (upper layer)

5.3.1 Multiobjective voltage optimization problem formulation

As stated in Section 5.1.1, the optimal voltage regulation and the minimized active power loss cannot be achieved simultaneously. Therefore, the voltage optimization problem is modeled with two objectives where the first is to minimize the total power loss and the second is to minimize the voltage deviation of each bus. The formulation of the two-objective optimization problem will provide several optimal solutions to the system operator and each of the solutions represents a certain trade-off between two objectives, so the operator is able to choose a solution corresponding to different needs. In this chapter, the active power injection from MGs is assumed to be determined in advance, while the reactive power injections from MGs are treated as the additional control variables. The control variables during the optimization are position of the OLTC t_T, position of the SC k_C, and reactive power injection from MG Q_{MG}.

Multiple sets of optimal solution can be obtained by solving the multiobjective optimization problem described as follows:

$$\begin{aligned} &\text{minimize}\, F = (f_1, f_2) \\ &f_1 = \sum_{i,j\epsilon N} G_{ij}\left(U_i^2 + U_j^2 - 2U_iU_j\cos\theta_{ij}\right) \\ &f_2 = \sum_{i=1}^{N} \frac{|U_i - U_{rated}|}{U_{rated}} \end{aligned} \tag{5.9}$$

where f_1 represents the total active power loss of the system branches and f_2 represents the voltage deviation on all the buses. The second objective f_2 prevents the voltage of certain buses from being equal to the highest or lowest boundary after optimization. The constraints of above objective functions are defined as

- Equality constraints

$$\begin{cases} P_i - U_i \sum_{j=1}^{N} U_j\left(G_{ij}\cos\theta_{ij} + B_{ij}\sin\theta_{ij}\right) = 0,\ i, j\epsilon N \\ Q_i - U_i \sum_{j=1}^{N} U_j\left(G_{ij}\sin\theta_{ij} - B_{ij}\cos\theta_{ij}\right) = 0,\ i, j\epsilon N \end{cases} \tag{5.10}$$

- Inequality constraint for control variables:

$$\begin{cases} t_T^{\min} \le t_T \le t_T^{\max} \\ k_{C,i}^{\min} \le k_{C,i} \le k_{C,i}^{\max}, i\epsilon N_C \\ -\sqrt{1 - P_{MG,i}^2} \le Q_{MG,i} \le \sqrt{1 - P_{MG,i}^2}, i\epsilon N_{MG} \end{cases} \tag{5.11}$$

- Inequality constraint for state variables:

$$U_i^{\min} \le U_i \le U_i^{\max}, i\epsilon N_{PQ} \tag{5.12}$$

The nomenclature of variables in above objective function and constraints is defined in Table 5.1.

5.3.2 Proposed solution (MOEA/D and decision making)

The two-objective optimization problem modeled in Section 5.3.1 contains continuous variables like $Q_{MG,\,i}$ and discontinuous variables like t_T and $k_{C,\,i}$, so it is a mixed integer nonlinear programming problem (MINLP). It can be solved by either the classical approaches such as the ε-constraint method or the heuristic methods such as the evolutionary algorithm. Since the evolutionary algorithm has a better performance in

Table 5.1 Nomenclature of variables in Eqs. (5.10), (5.11), (5.12)

Variable	Nomenclature
N	Set of buses in the network
N_C	Set of buses where a capacitor is connected
N_{MG}	Set of buses where an MG is connected
U_i, U_j	Voltage of the bus i and j
θ_{ij}	Voltage angle difference between the bus i and j
U_{rated}	Rated voltage of the network
G_{ij}	Conductance of the line section between bus i and j
B_{ij}	Susceptance of the line section between bus i and j
P_i	Active power generation from bus i
Q_i	Reactive power generation from bus i
$k_{C,\,i}$	Position of the shunt capacitor at bus i
t_T	Tap position of the OLTC
$P_{MG,\,i}$	Active power injection from the MG at bus i
$Q_{MG,\,i}$	Reactive power injection from the MG at bus i

parallel searches and it does not need the mathematic characteristic of the problem such as the slope of the objective function, it is applied here to solve the objective function. Among many evolutionary algorithms, the MOEA/D (Zhang and Li, 2007) is a relatively new one which has the lower computational complexity at each generation compared with MOGLS (Ishibuchi and Murata, 1998) and NSGA-II (Deb et al., 2002). Therefore, it is selected to solve the multiobjective optimization problem. In addition, a decision-making process is taken to select a solution for the final execution after multiple Pareto optimal solutions are obtained. The decision making is based on the principle of prioritizing the second objective f_2 (minimization of the voltage deviation) if the voltage of any bus violate the acceptable boundary, while minimizing the active power loss if the voltage of all buses stays within a normal range.

5.3.2.1 MOEA/D

MOEA/D requires a decomposed form of the multiobjective optimization problem. The Tchebycheff approach (Zhang and Li, 2007) as described in Eq. (5.13) is used to decompose the two-objective voltage optimization problem in Eq. (5.8). Let λ^1, …, λ^j, …, λ^N be a set of evenly spread weight vectors, where $\lambda^j = (\lambda_1^j, \lambda_2^j)$ for the two-objective optimization problem here. The problem in Eq. (5.9) can be decomposed into N scalar optimization subproblems and the objective function of the jth subproblem is

$$\min\left[g^{te}\left(x|\,\lambda^j,z^*\right) = \max\left\{\lambda_1^j\left|f_1(x)-z_1^*\right|,\lambda_2^j\left|f_2(x)-z_2^*\right|\right\}\right] \tag{5.13}$$

where z_1^* and z_2^* are the minimum values of f_1 and f_2, respectively. MOEA/D optimizes all these N subproblems simultaneously in a single run.

Proposed in Ref. (Zhang and Li, 2007), the optimal solution of $g^{te}(x|\lambda^i, z^*)$ is close to that of $g^{te}(x|\lambda^j, z^*)$ if the Euclidean distance between $\lambda^i = (\lambda_1^i, \lambda_2^i)$ and $\lambda^j = (\lambda_1^j, \lambda_2^j)$ is close. Therefore, each subproblem is optimized by using the distance information only from its neighboring subproblems, and the best solution found so far is kept for each subproblem to form the current population.

In this chapter, the number of subproblems N is 50, and the number of the weight vectors in the neighborhood of each weight vector T is 8. The maximum iteration is set as 300. The number of Pareto solutions will be kept during the calculation for no more than 200.

The major steps of MOEA/D are introduced below, and more information can be found in Ref. (Zhang and Li, 2007).

Step 1. Initialization

Step 1.1. Input data which contains N, T, objective function in Eq. (5.9), and constraints in Eqs. (5.10), (5.11), (5.12).

Step 1.2. Generate N evenly spread weight vectors: $\lambda^1, \ldots, \lambda^j, \ldots, \lambda^N$.

Step 1.3. Compute the Euclidean distance between any two weight vectors and then find the T closest weight vectors to each vector. For each $i = 1, \ldots, N$, set $B(i) = \{i_1, \ldots, i_T\}$, where $i_1, \ldots, i_T$ are indexes of T closest weight vectors to λ^i.

Step 1.4. Generate N initial solutions $x^1, \ldots, x^N$ with respect to the constraints defined in Eqs. (5.10)–(5.12). Calculate the objective function value $FV^i = F(x^i) = [f_1(x^i), f_2(x^i)]$ by Eq. (5.8).

Step 1.5. Initialize $z = (z_1, z_2)^T$, where z is the minimum value found in $FV^1, \ldots, FV^N$.

Step 2. Update

For $i = 1, 2, \ldots, N$, do

Step 2.1 Reproduction. Randomly select two indexes i_k, i_l from $B(i)$, and then generate a new solution y from x^k and x^l by using problem-specific generic operators.

Step 2.2. Constraints checking. Check solution y by using constraints defined in Eqs. (5.10)–(5.12), and replace any element with its maximum or minimum value if it violates the upper or lower bound.

Step 2.3 Update of z. If $F(y) < z$, then set $z = F(y)$.

Step 2.4 Update of neighboring solutions. For each index $j \in B(i)$, if $g^{te}(y|\lambda^j, z^*) \leq g^{te}(x^j|\lambda^j, z^*)$, then set $x^j = y$ and $FV^j = F(y)$.

Step 2.5 Update of EP(external population). Remove from EP all the vectors dominated by $F(y)$, and add $F(y)$ to EP if no vectors in EP dominate $F(y)$.

Step 3 Stopping criteria checking. If the maximum iteration number is reached or no update is found in EP in 30 successive iterations, stop and output EP. Otherwise, go to Step 2.

5.3.2.2 *Decision-making process*

The Pareto solutions obtained from above procedures represent the potential candidates for the multiobjective optimization problem. None of them can bring an optimal solution for both objectives but show a trade-off between them. The two objectives are minimizing total power loss and voltage deviation, respectively, so it can be expected that the minimum power loss is obtained only if the voltage of the system is of a high quality. Therefore, the decision-making process is performed based on following principle, and a flow chart based on this principle is shown in Fig. 5.11.

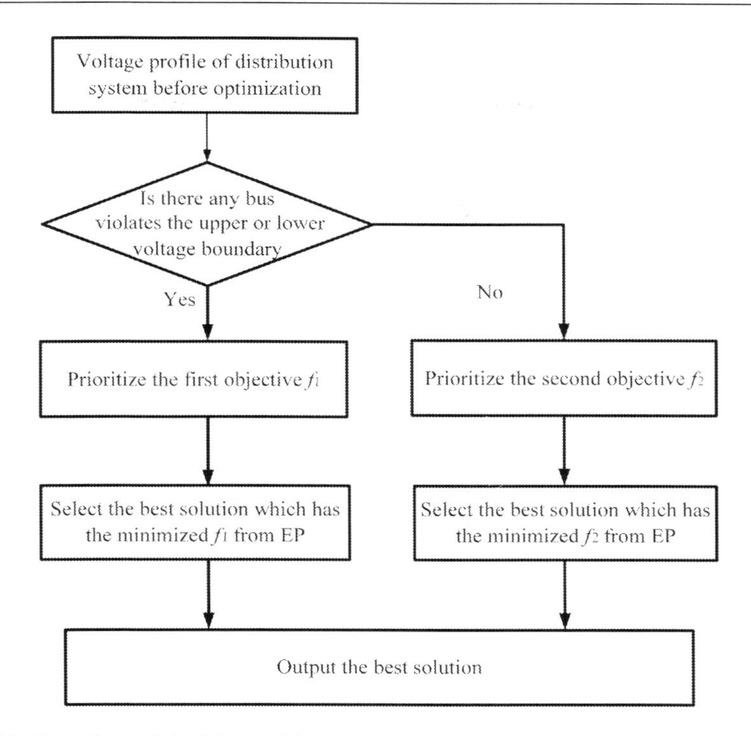

Fig. 5.11 Flow chart of decision making.

- If the voltage deviation of all nodes staying within the rated range, prioritize the first objective: to minimize the total power lossf_1.
- If the voltage deviation of any node violates the rated range, prioritize the second objective: to minimize the voltage deviationf_2.

5.4 Case study

The modified IEEE 33 nodes distribution system in Fig. 5.12 is used to evaluate the effectiveness of the proposed voltage control scheme (Baran and Wu, 1989). The power base and voltage base of the system is set to be 100 MVA and 12.66 kV, respectively. Loads of the system are increased by 1.5 times. The allowed deviation of the system voltage is set from 0.95 to 1.05 p.u. Case studies were first performed in Matlab to validate the voltage control scheme in the upper layer. Then, PowerFactory was used to perform the time-series simulation for examining the performance of the voltage control scheme at the lower layer. The simulation was practiced on a desktop which is equipped with 8 GB memory and Intel Core i5-6600 processor.

As shown in Fig. 5.12, this test system is modified by adding an OLTC for the the transformer and an SC at bus 2. Four MGs are connected to bus 30, 11, 18, and 22, respectively, and they take the same layout as shown in Fig. 5.6. The capability of

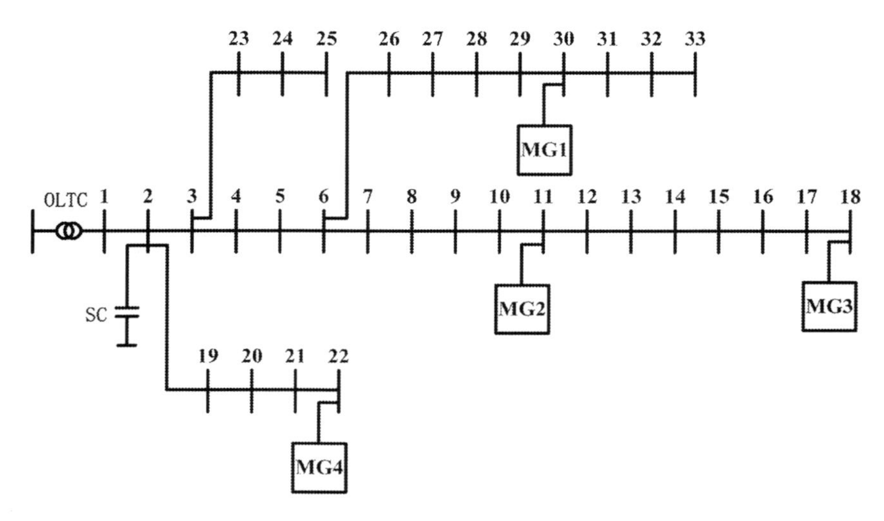

Fig. 5.12 Modified IEEE 33 nodes distribution network.

Table 5.2 Capability of OLTC and SC

Voltage regulator	Bus no.	Specifications
OLTC	1	Having 20 tap positions, 0.005 p.u. for each step, tap position T ranges from -10 to 10, so the secondary bus voltage ranges from 0.95 to 1.05 p.u.
SC	2	Contains 5 unit capacitors, and 0.1 MVar for each unit, position K ranges from 0 to 5

OLTC and SC are shown in Table 5.2 and the specifications of four MGs are set to be identical as shown in Table 5.3.

As stated in Section 5.2, the DGs in each MG are grid-forming units controlling the voltage in AC and DC subsystems while the RDGs are grid-following units changing their outputs with time. In this chapter, it is assumed that the RDGs only generate active power. RDGs' generation and loads in AC and DC subsystems are changing with time according to their power bases in Table 5.3 and the multipliers in Fig. 5.13 (Wang et al., 2016). The IC1 and IC2 are equipped with controllers as introduced in Section 5.2. Therefore, the IC1 responds to the power set point from the central controller, and the IC2 is in charge of smoothing the power exchange between AC and DC subsystems.

The initial voltage profile of the test system is shown in Fig. 5.14. Since there is no voltage regulator in the system, the test system is initially poorly compensated and the lower voltage boundary is violated at some buses. The total power loss f_1 is 0.2590 MW, and the voltage deviation f_2 is 1.7599 p.u.

Table 5.3 Specifications of the VSC-based hybrid AC/DC MG

System parameters		**Rated voltage of AC1 and AC2**	**0.4 kV**
		Rated AC frequency	**50 Hz**
		Rated DC voltage	**1 kV**
IC1		Capacity	0.5 MVA
		Initial active power injection	0.4 MW
		Initial range of reactive power injection	[−0.3,0.3] MVar
IC2		Capacity	0.5 MVA
Elements in DC subsystem	DG1	Maximum generation	1 MW
	RDG1	Maximum generation	0.2 MW
	DC loads	Maximum consumption	0.6 MW
Elements in AC subsystem	DG2	Maximum generation	0.8 MW, 0.6 MVar
	RDG2	Maximum generation	0.2 MW
	AC loads	Maximum consumption	0.8 MW, 0.6 MVar

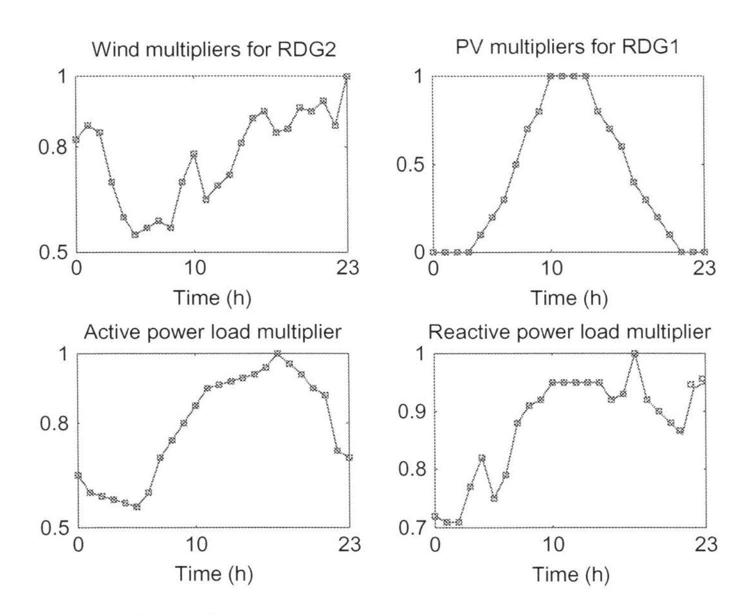

Fig. 5.13 Multiplier for RDGs and loads.

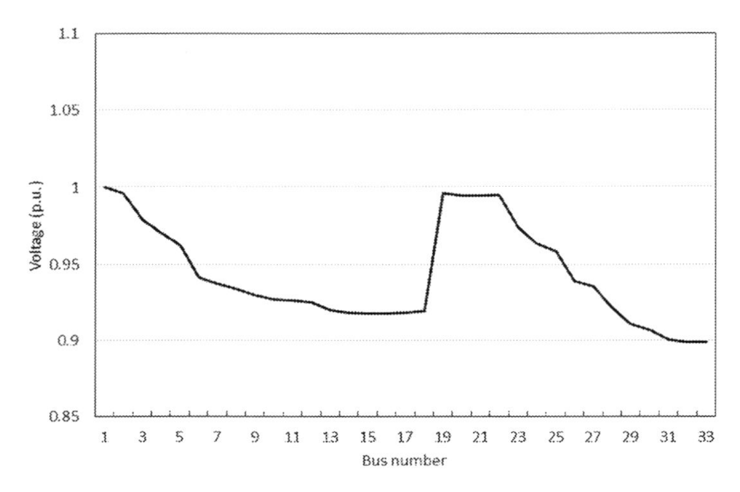

Fig. 5.14 Initial voltage profile before optimization.

5.4.1 Voltage control in the upper layer

5.4.1.1 Case A: Voltage control without reactive power injection from MGs

The optimization without reactive support from MGs is performed first for comparison. The active power injection from each MG is set as 0.4 MW, while the reactive power injection from them is not considered.

Due to the initially poor voltage profile as shown in Fig. 5.14 and the fact that there is no reactive power compensator at the downstream area of the system, it is expected that OLTC and SC will be working at their maximum position to maintain a high voltage at the upstream area for retaining the downstream voltage within an acceptable range. This is validated by the calculation result that only one solution was obtained from solving the multiobjective optimization problem. The optimal solution is shown in Table 5.4, and the voltage profile after optimization is shown in Fig. 5.15 (solid line in the figure). Compared with the initial case, the values of f_1 and f_2 are decreased to 0.2291 and 0.8601, respectively.

5.4.1.2 Case B: Voltage control with reactive power injection from MGs

In this case, MGs are incorporated into the voltage control scheme. The active power injections from four MGs are all 0.4 MW as shown in Table 5.3. The power ratings of their interfaced VSC IC1 are all set to be 0.5 MVA. Thus, the range of reactive power injection from each MG can be determined by Eq. (5.3), which is ranged from −0.3 to 0.3 MVar as shown in Table 5.3.

Fig. 5.16 shows the Pareto optimal solutions from solving the two-objective voltage optimization problem modeled in Section 5.3.1 by using the MOEA/D introduced in Section 5.3.2. The 182 solutions as shown in Fig. 5.16 were obtained after 300

Table 5.4 Optimized set-points for OLTC and SC

Voltage regulator	Bus no.	Set-point
OLTC	1	$T = 10$
SC	2	$K = 5$

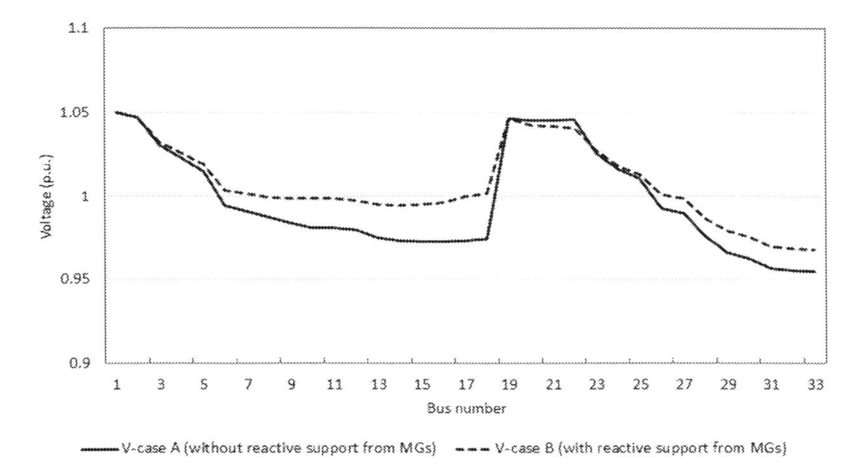

Fig. 5.15 Voltage profile after optimization in cases A and B.

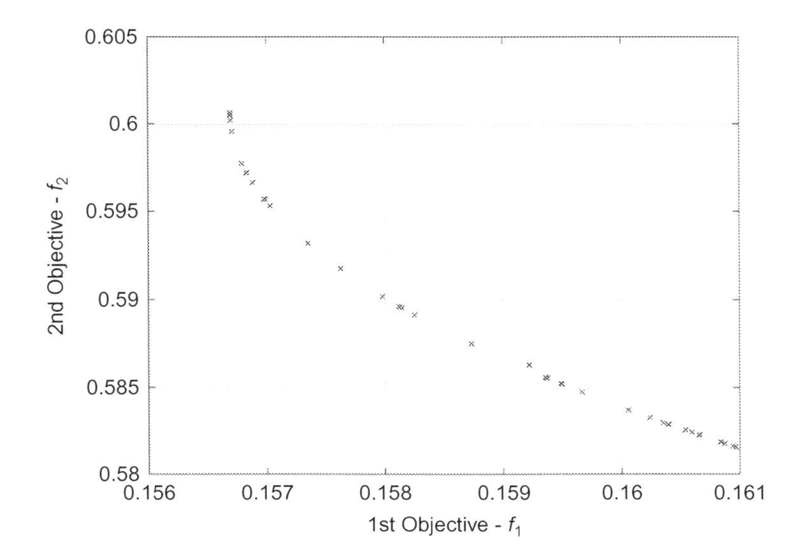

Fig. 5.16 Pareto optimum solutions.

Table 5.5 Optimized set-points for OLTC, SC, and MGs

Voltage regulator	Bus no.	Set-point
OLTC	1	$T = 9$
SC	2	$K = 5$
MG1	30	0.3 MVar
MG2	11	0.3 MVar
MG3	18	0.3 MVar
MG4	22	−0.3 MVar

iterations in 67 s. This calculation speed could be further reduced by optimizing the MOEA/D procedures, but it is already very fast for practical usage.

After applying the decision-making process proposed in Section 5.3.2.2, one set of optimized set points was obtained in Table 5.5. It can be observed that the OLTC takes a lower position compared with the Case A in Section 5.4.1.1 due to the reactive power support from four MGs located at downstream buses. It can be seen that MG1, MG2, and MG3 maximize their capacitive reactive power injection because the initial voltage of their coupled buses is low, while the MG4 injects its maximum inductive reactive power in response to the initially high voltage of its coupled bus 22. The reactive power injection from the downstream MGs flatten the voltage along the feeders, thus the burden of the OLTC is released. Moreover, a lower position of OLTC brings the distribution system more redundancy to face the upcoming voltage violation. This is the advantage of incorporating MGs into the distribution network voltage control scheme.

The optimized voltage profile is shown in Fig. 5.15 with the dotted line. Compared with Case A in Section 5.4.1.1, the voltage of all buses become closer to the rated value due to the participation of MGs. This superior voltage quality can be also reflected by the lower value of voltage deviation f_2 which is 0.5816 p.u. In addition, the total power loss f_1 is also reduced to 0.1610 MW because the power demand tends to be local fulfilled and power loss from the power flow through the feeders is decreased.

5.4.2 Voltage control in the lower layer

In order to validate the MG's local controllers introduced in Section 5.2, the modified IEEE 33 nodes system in Fig. 5.17 was modeled in PowerFactory. The time-series simulation was starting at 0 s and ended at 24 s with the step size 0.0001 s, and the generation and load profiles are changing with time to simulate an operational period of 24 h in real time. The set points in Table 5.5 were applied at 1 h after the initialization.

The simulation result of MG1 is selected to be the representative. It can be seen from Fig. 5.18 that IC1 in MG1 follows the reactive set point after 1 h. The reactive power injection from IC1 in MG1 increases from 0 to 300 kVar very quickly due to the fast response of the IC1 controller in Fig. 5.7, while the active power injection stays at 400 kW during the period.

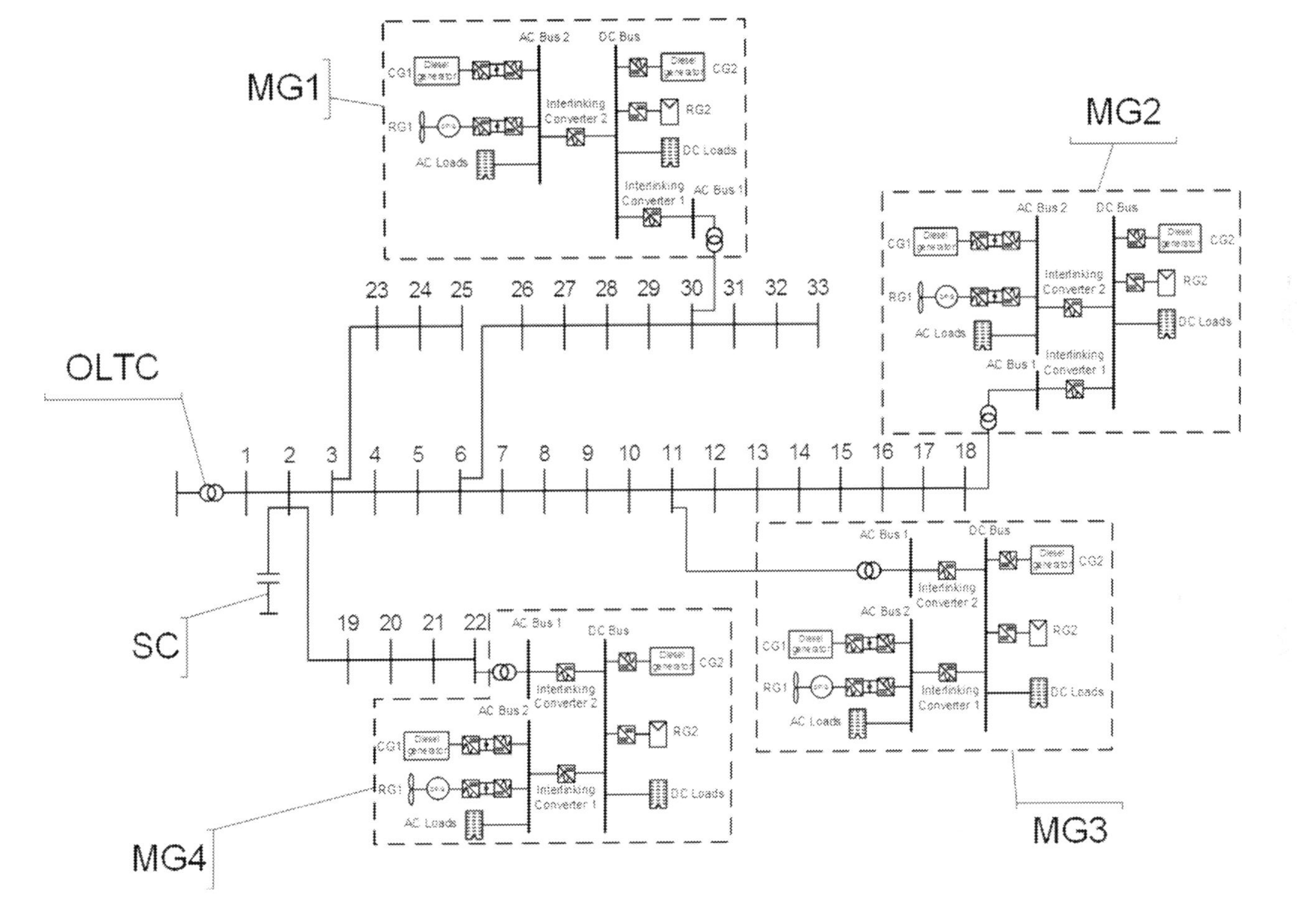

Fig. 5.17 Modified IEEE 33 nodes system in PowerFactory.

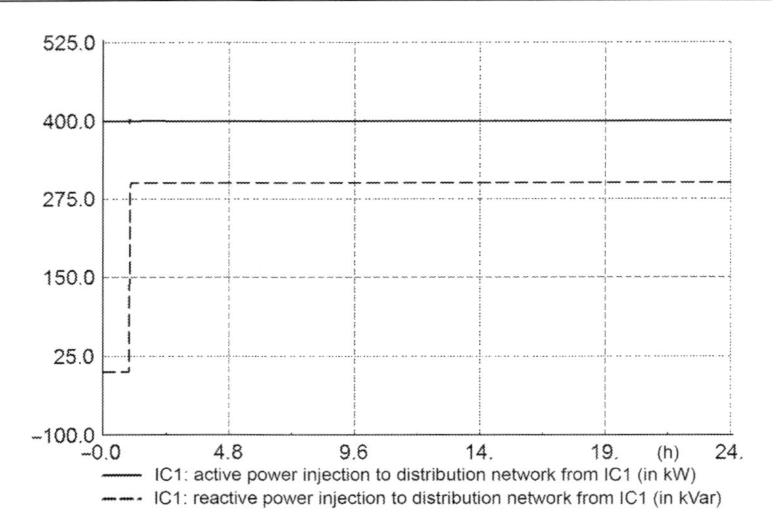

Fig. 5.18 Active and reactive power injection of IC1 in MG1.

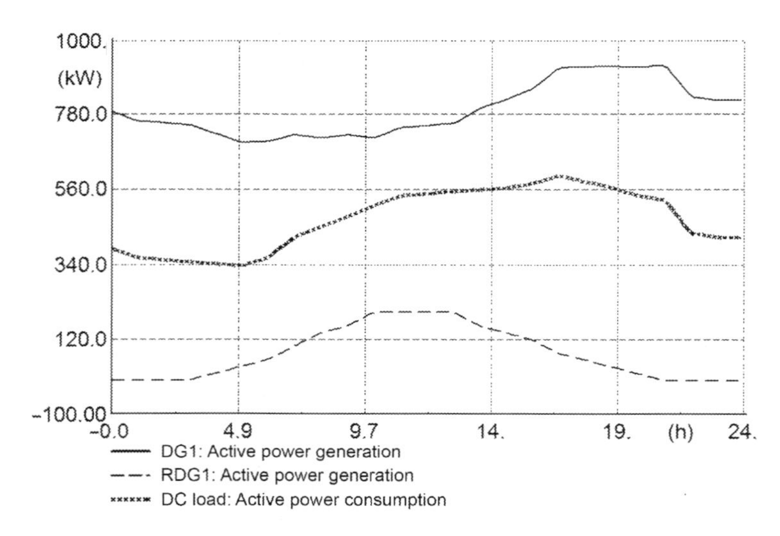

Fig. 5.19 Active power generation and consumption in DC subsystem.

Figs. 5.19–5.21 show the power generation and consumption in DC and AC subsystem. The generation from RDGs and power consumption of loads are changing with time as stated earlier. During the period, it can be seen that grid-forming units DG1 and DG2 adjust their generation in response to the changing loads and intermittent outputs of RDG1 and RDG2. Meanwhile, it is shown in Fig. 5.22 that the IC2 exchanges power between AC and DC subsystems in response to the power transients for smoothing the voltage and frequency in AC and DC subsystems.

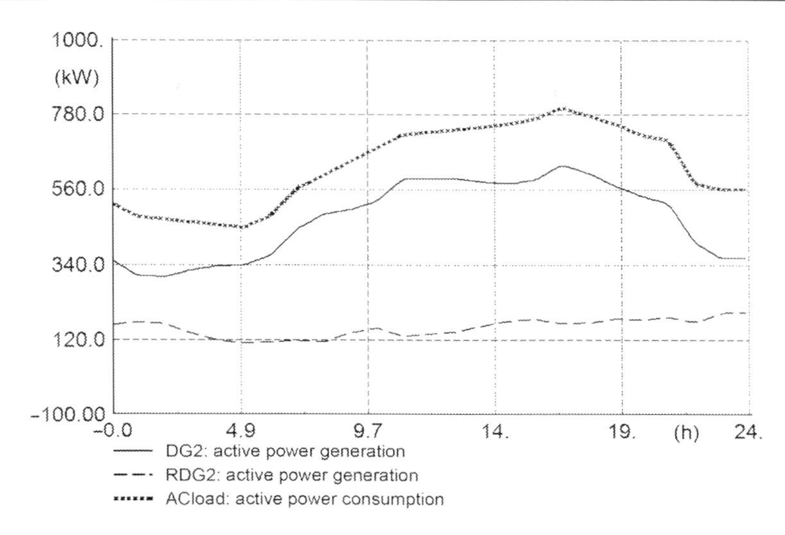

Fig. 5.20 Active power generation and consumption in AC subsystem.

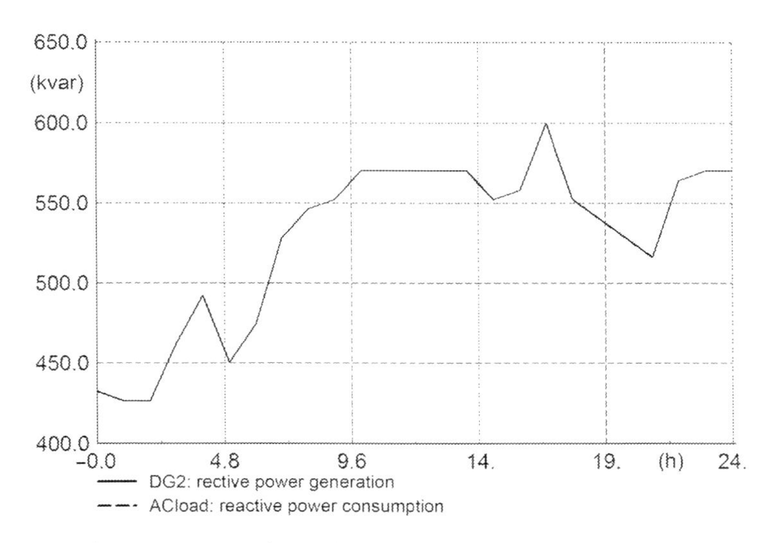

Fig. 5.21 Reactive power generation and consumption in AC subsystem.

With the coordination between DGs and ICs at the lower layer, the voltage and frequency in MG1 are maintained at their rated values as shown in Fig. 5.23.

The time-series simulation demonstrates that the voltage scheme at the lower layer is capable to control the voltage within the MG, and it quickly responses to the reactive power dispatch from the central controller at the upper layer. However, it should be pointed out that the reactive power set point for IC1 in this case study does not violate the capacity of IC1. When the distribution network is operated with a very poor

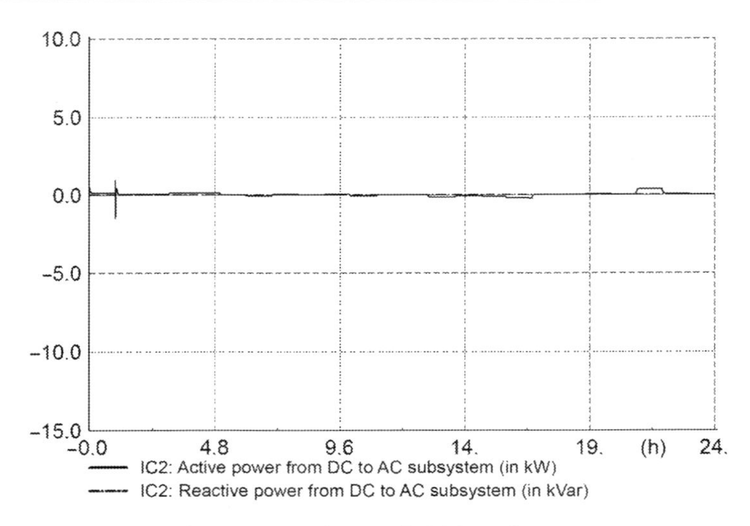

Fig. 5.22 Active and reactive power exchange of IC2 in MG1.

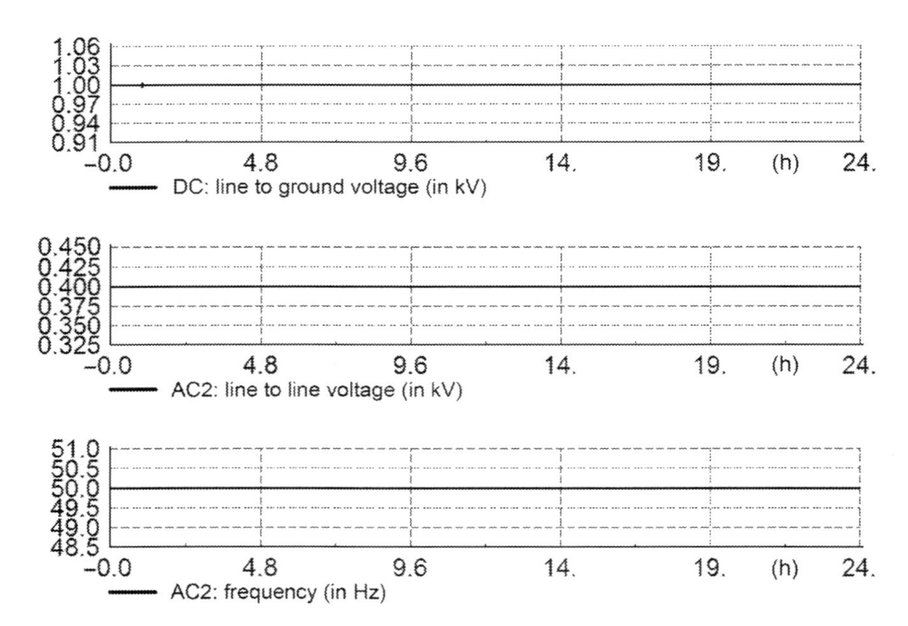

Fig. 5.23 Voltage and frequency in DC and AC subsystems of MG1.

voltage, it is expected that the active power injection from IC1 can be reduced for providing a higher reactive power injection. In this case, the reduced amount of active power injection to the distribution network can be controlled to charge the local battery system in MG, so the energy utilization in MG is not sacrificed. The further study should focus on coordinating MG's active power injection with its reactive power injection without deterioration of its local energy usage.

5.5 Conclusions

The VSC-based hybrid AC/DC MG has emerged to be a smart building block in future power system. This chapter incorporates the VSC-based hybrid AC/DC MG into the distribution voltage control scheme by using a two-layer control structure. The multiobjective voltage optimization problem is modeled at the upper layer to minimize the power loss and voltage deviation in the distribution network, and the reactive power injections from MGs are treated as the extra control variables. The MOEA/D is utilized to solve the multiobjective optimization problem, and a decision-making process is performed to select a proper set of solutions for the final execution. In the lower layer, the VSC-based hybrid AC/DC MG with corresponding controllers is modeled to meet the dispatched reactive power set point from the upper layer while maintaining the MG's local voltage and frequency.

Case studies validated the effectiveness of proposed two-layer voltage control scheme. It is shown that proposed voltage control scheme fully utilizes the reactive power from VSC-based MGs without deteriorating their local power quality under the changing power generation within MGs. The proposed voltage control scheme is promising to alleviate the burden of traditional voltage regulators and reduce the investment on extra reactive power compensators.

Acknowledgment

This research is supported by the University of Sydney Bridging Funding and Faculty of EIT Mid-Career Researcher Development Scheme.

References

Baran, M.E., El-Markabi, I.M., 2007. A multiagent-based dispatching scheme for distributed generators for voltage support on distribution feeders. IEEE Trans. Power Syst. 22 (1), 52–59.

Baran, M.E., Wu, F.F., 1989. Network reconfiguration in distribution systems for loss reduction and load balancing. IEEE Trans. Power Delivery 4 (2), 1401–1407.

Cagnano, A., De Tuglie, E., 2015. Centralized voltage control for distribution networks with embedded PV systems. Renew. Energy 76, 173–185.

Casavola, A., Franze, G., Menniti, D., Sorrentino, N., 2011. Voltage regulation in distribution networks in the presence of distributed generation: a voltage set-point reconfiguration approach. Electr. Power Syst. Res. 81 (1), 25–34.

Che, L., Shahidehpour, M., Alabdulwahab, A., Al-Turki, Y., 2015. Hierarchical coordination of a community microgrid with AC and DC microgrids. IEEE Trans. Smart Grid 6 (6), 3042–3051.

Chen, D., Xu, L., 2012. Autonomous DC voltage control of a DC microgrid with multiple slack terminals. IEEE Trans. Power Syst. 27 (4), 1897–1905.

Deb, K., Pratap, A., Agarwal, S., Meyarivan, T., 2002. A fast and elitist multiobjective genetic algorithm: NSGA-II. IEEE Trans. Evol. Comput. 6 (2), 182–197.

Farag, H.E., El-Saadany, E.F., Seethapathy, R., 2012. A two ways communication-based distributed control for voltage regulation in smart distribution feeders. IEEE Trans. Smart Grid 3 (1), 271–281.

Guerrero, J.M., Vasquez, J.C., Matas, J., de Vicuna, L.G., Castilla, M., 2011. Hierarchical control of droop-controlled AC and DC microgrids—a general approach toward standardization. IEEE Trans. Ind. Electron. 58 (1), 158–172.

Ishibuchi, H., Murata, T., 1998. A multi-objective genetic local search algorithm and its application to flowshop scheduling. IEEE Trans. Syst. Man Cybern. Part C Appl. Rev. 28 (3), 392–403.

Katiraei, F., Iravani, R., Hatziargyriou, N., Dimeas, A., 2008. Microgrids management. IEEE Power Energy Mag. 6 (3), 54–65.

Lasseter, R.H., Paigi, P., 2004. Microgrid: a conceptual solution. IEEE Power Electron 4285–4290.

Lidula, N.W.A., Rajapakse, A.D., 2011. Microgrids research: a review of experimental microgrids and test systems. Renew. Sust. Energ. Rev. 15 (1), 186–202.

Liu, X., Wang, P., Loh, P.C., 2011. A hybrid AC/DC microgrid and its coordination control. IEEE Trans. Smart Grid 2 (2), 278–286.

Loh, P.C., Li, D., Chai, Y.K., Blaabjerg, F., 2013a. Autonomous control of interlinking converter with energy storage in hybrid AC-DC microgrid. IEEE Trans. Ind. Appl. 49 (3), 1374–1382.

Loh, P.C., Li, D., Chai, Y.K., Blaabjerg, F., 2013b. Autonomous operation of hybrid microgrid with AC and DC subgrids. IEEE Trans. Power Electron. 28 (5), 2214–2223.

Ranamuka, D., Agalgaonkar, A.P., Muttaqi, K.M., 2014. Online voltage control in distribution systems with multiple voltage regulating devices. IEEE Trans. Sustain. Energy 5 (2), 617–628.

Renedo, J., Garcia-Cerrada, A., Rouco, L., 2017. Reactive-power coordination in VSC-HVDC multi-terminal systems for transient stability improvement. IEEE Trans. Power Syst. 32 (5), 3758–3767.

Shahidehpour, M., Khodayar, M., 2013. Cutting campus energy costs with hierarchical control: the economical and reliable operation of a microgrid. IEEE Electrif. Mag. 1 (1), 40–56.

Tian, P.G., Xiao, X., Wang, K., Ding, R.X., 2016. A hierarchical energy management system based on hierarchical optimization for microgrid community economic operation. IEEE Trans. Smart Grid 7 (5), 2230–2241.

Turitsyn, K., Sulc, P., Backhaus, S., Chertkov, M., 2011. Options for control of reactive power by distributed photovoltaic generators. Proc. IEEE 99 (6), 1063–1073.

Viawan, F.A., Sannino, A., Daalder, J., 2007. Voltage control with on-load tap changers in medium voltage feeders in presence of distributed generation. Electr. Power Syst. Res. 77 (10), 1314–1322.

Wang, Z.Y., Chen, B.K., Wang, J.H., Begovic, M.M., Chen, C., 2015. Coordinated energy management of networked microgrids in distribution systems. IEEE Trans. Smart Grid 6 (1), 45–53.

Wang, Z.Y., Chen, B.K., Wang, J.H., Chen, C., 2016. Networked microgrids for self-healing power systems. IEEE Trans. Smart Grid 7 (1), 310–319.

Yazdani, A., Iravani, R., 2010. Voltage-Sourced Converters in Power Systems: Modeling, Control, and Applications. IEEE Press/John Wiley, Hoboken, NJ.

Zhang, Q., Li, H., 2007. MOEA/D: a multiobjective evolutionary algorithm based on decomposition. IEEE Trans. Evol. Comput. 11 (6), 712–731.

Zhao, H.R., Wu, Q.W., Wang, J.H., Liu, Z.X., Shahidehpour, M., Xue, Y.S., 2017. Combined active and reactive power control of wind farms based on model predictive control. IEEE Trans. Energy Convers. 32 (3), 1177–1187.

Controlling the distributed energy resources under fading channel

M.M. Rana, Wei Xiang, Eric Wang
College of Science and Engineering, James Cook University, Townsville, QLD, Australia

6.1 Introduction

Green energy technologies with distributed energy resources (DERs) such as wind turbines and solar panels have recently huge attention in smart grid due to their harmonious relationships with nature (Yan et al., 2013; Rana and Li, 2015a). However, their power generation patterns are generally intermittent in nature, so they create big problems for control and reliability of the smart grid. Therefore, smart grid has a strong requisite of an efficient communication infrastructure to monitor and control the system states. Generally speaking, the smart grid can spread the intelligence of the energy distribution and control from the central system to the far away remote areas, thus enabling accurate state estimation and real-time monitoring of these intermittent energy resources (Li et al., 2012). Consequently, the reliability tools and power system operations are totally depending on the results obtained by the state estimation and stabilization (Huang et al., 2012).

Generally speaking, the point common coupling (PCC) voltage deviations are increased dramatically at the interfacing points, so it is necessary to apply a proper control method. After applying suitable control strategy, the PCC voltage deviations are driven to zero, otherwise it is very dangerous in terms of network stability and operation of the DERs. Driven by this motivation, this chapter proposes a discrete-time linear quadratic Gaussian (LQG) controller to regulate the system states under the condition of fading channel. Finally, extensive simulation results depict that the proposed LQG controller can stabilize the system states with few iterations even if there are large disturbances and fading channel between controller and microgrid.

Organization: The chapter is organized as follows. Section 6.2 describes the microgrid state-space model. Furthermore, the linear quadratic regulator (LQR) controller under fading channel is presented in Section 6.3. Moreover, the effectiveness of the proposed algorithm is demonstrated in Section 6.4. The paper is wrapped up in Section 6.5.

Notation: Upper case letters and bold face lower are used to represent matrices and vectors, respectively; superscripts $x^{'}$ denotes the transpose of x and I is the identity matrix.

Smart Power Distribution Systems. https://doi.org/10.1016/B978-0-12-812154-2.00006-7

6.2 Microgrid state-space model

Fig. 6.1 shows the schematic diagram of the microgrid incorporating multiple DERs.

It can be seen that each DER is connected to the local load through a converter (Rana and Li, 2015b). Moreover, each DER is represented by a DC voltage source in series with a voltage source converter (VSC) and an resistor-inductor (RL) filter. After applying Kirchhoff's laws in the islanded microgrid incorporating three DERs with discretization parameter μ, the system dynamic can be written as follows (Karimi et al., 2008; Rana et al., 2017):

$$\mathbf{x}(k+1) = \mathbf{A}\mathbf{x}(k) + \mathbf{B}\mathbf{u}(k) + \mathbf{n}(k)$$

where $\mathbf{A}$ is system state matrix, $\mathbf{x} = (\mathrm{i}_{l1}\ \mathrm{i}_{d1}\ \mathrm{i}_{t1} v_1\ \mathrm{i}_{l2}\ \mathrm{i}_{d2}\ \mathrm{i}_{t2}\ v_2\ \mathrm{i}_{l3}\ \mathrm{i}_{d3}\ v_3)'$ is the system state, $\mathbf{B}$ is the input matrix, $\mathbf{u} = (v_{d1}\ v_{d2}\ v_{d3})'$ is the system control effort, and $\mathbf{n}$ is the system noise whose covariance matrix is $\mathbf{N}$. Here, v_i is the ith PCC voltage and v_{di} is the ith DER input voltage. The system state and input matrices are given in Karimi et al. (2008) and Rana et al. (2017). In order to control the microgrid, the proposed controller presents in the following section.

6.3 LQG controller under fading channel

Generally speaking, the controller and the microgrid are located in a far way. If so, the control action is lost due to fading channel. Fig. 6.2 shows the controller under the condition of fading channel.

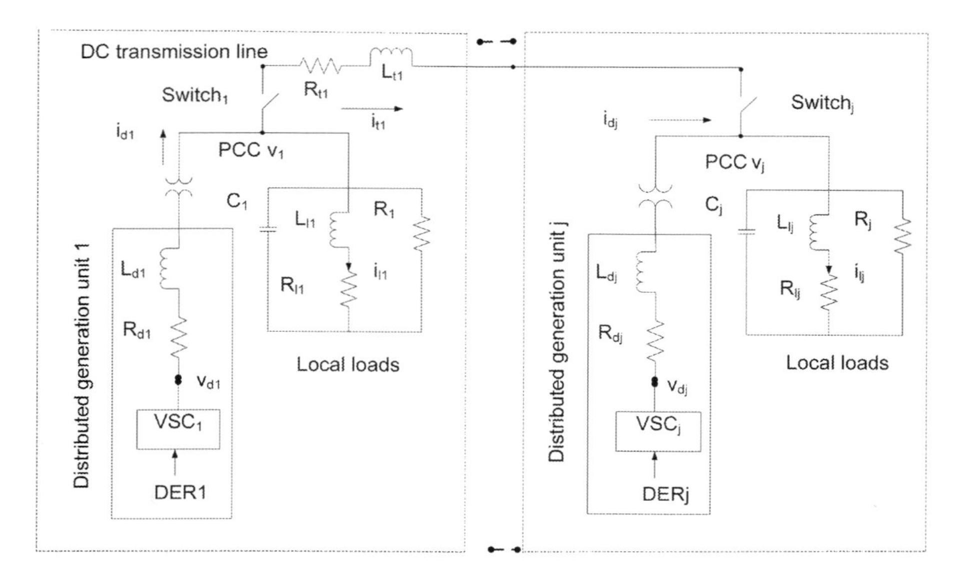

Fig. 6.1 Block diagram of the microgrid incorporating multiple distributed energy resources (DERs) (Karimi et al., 2008; Rana et al., 2017).

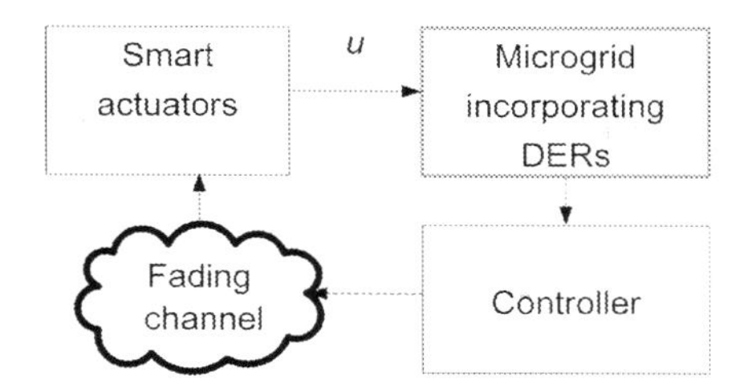

Fig. 6.2 Proposed controller under fading channel.

In order to stabilize the system states, we considering the following feedback control law (Su and Chesi, 2015; Hu and Yan, 2009; Xiao et al., 2012):

$$u(k) = -\beta(k)Fx(k)$$

by minimizing the following objective function:

$$L = \sum_{i=1}^{M} [x(i)\mathbf{Q}x(i) + u'(i)\mathbf{R}u(i)],$$

where **Q** and **R** are positive definite weighting matrices. Here, F is the feedback gain, and $\beta(k)$ is the packet loss parameter with

$$\beta(k) = \begin{cases} 1, & \text{no loss,} \\ \alpha(k), 0<\alpha(k)<1 & \text{losyy condition,} \end{cases}$$

where $\alpha(k)$ is the fading parameter at the actuator due to the unreliable communication network. Considering the fixed fading coefficient, the feedback gain is given by

$$\mathrm{F} = (\mathrm{B}\mathbf{P}\mathrm{B}' + \mathrm{R})^{-1}\mathrm{B}'\mathbf{P}\mathrm{A}$$

where **P** is the unique stabilizing solution of the discrete-time algebraic Riccati equation:

$$\mathbf{P} = \mathrm{A}'\mathbf{P}\mathrm{A} - \beta\mathrm{A}'\mathbf{P}\mathrm{B}(\mathrm{B}\mathbf{P}\mathrm{B}' + \mathrm{R})^{-1}\mathrm{B}'\mathbf{P}\mathrm{A} + \mathbf{Q}$$

Considering the controller, the simulation results are presented in the following section.

6.4 Simulation results and discussions

The performance of the LQG controller is illustrated considering different simulation environments.

6.4.1 Simulation Case 1

In this simulation environment, we used weighted matrices, $\mathbf{Q} = [1\ 1\ 1\ 10^2\ 1\ 1\ 1\ 10^2\ 1\ 1\ 10^4]'$, $R = [0.15^* eye(3)]'$, fading parameter $\alpha = 0.9$, and discretization step size parameter is 0.2.

From the simulation result in Fig. 6.3, it can be seen that the LQG controller is able to stabilize the system state within 800 iterations. The corresponding control effort is illustrated in Fig. 6.4. It is observed that the proposed controller is required minimum control effort.

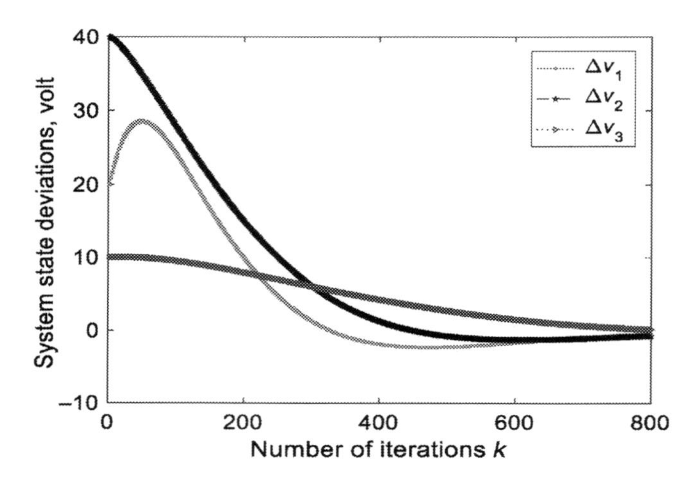

Fig. 6.3 Point common coupling (PCC) voltage deviations when $\mu = 0.2$ and $\alpha = 0.8$.

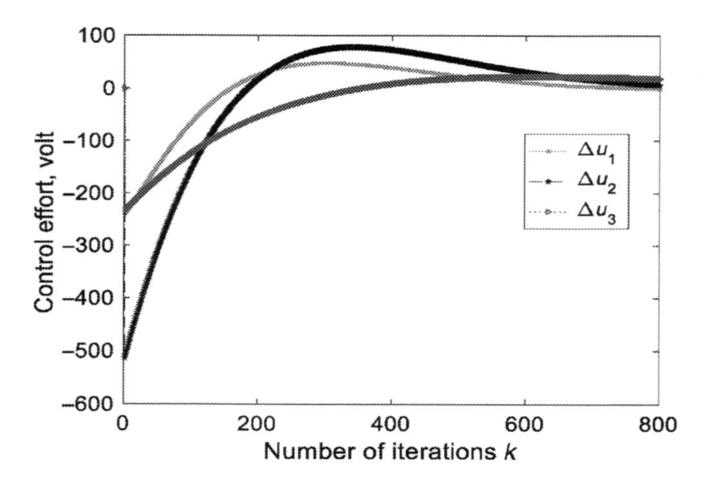

Fig. 6.4 Input voltage deviations when $\mu = 0.2$ and $\alpha = 0.8$.

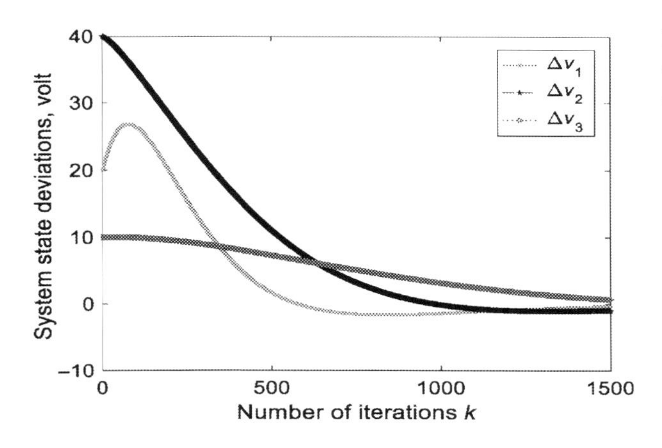

Fig. 6.5 PCC voltage deviations when $\mu = 0.1$ and $\alpha = 0.7$.

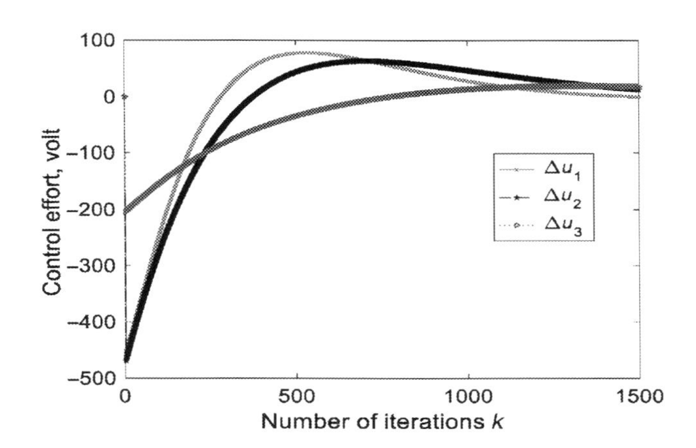

Fig. 6.6 Input voltage deviations when $\mu = 0.1$ and $\alpha = 0.7$.

When the step size is decreased from 0.2 to 0.1 and fading parameter $\alpha = 0.7$, the simulation results are presented in Figs. 6.5 and 6.6. It can be observed that the LQG controller is required more time to stabilize the system states.

6.4.2 Simulation Case 2

In this case, we used weighted matrices $\mathbf{Q} = [1\ \ 1\ \ 1\ \ 10^5\ \ 1\ \ 1\ \ 1\ \ 10^6\ \ 1\ \ 1\ \ 10^7]'$, $\mathrm{R} = [0.5^* eye(3)]'$, fading parameter $\alpha = 0.9$, and discretization step size parameter is 0.2. From the simulation result in Fig. 6.7, it can be seen that the LQG controller is able to stabilize the system state within 100 iterations. The corresponding control effort is illustrated in Fig. 6.8. It is noticed that more control effort is required.

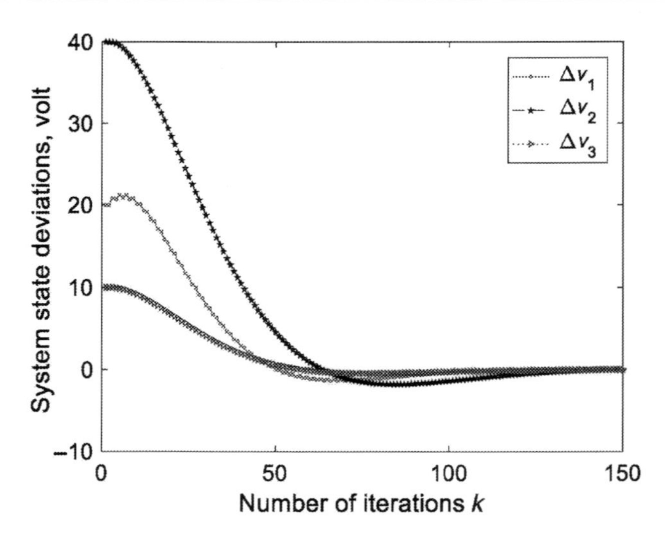

Fig. 6.7 PCC voltage deviations when $\mu = 0.2$ and $\alpha = 0.9$.

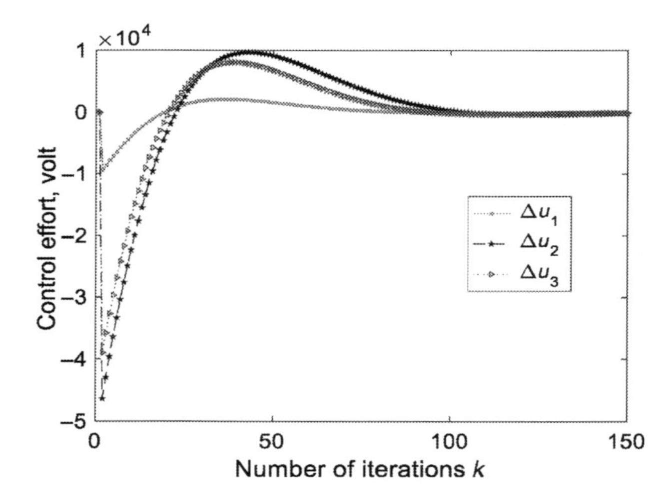

Fig. 6.8 Input voltage deviations when $\mu = 0.2$ and $\alpha = 0.9$.

When the step size is decreased from 0.2 to 0.1 and fading parameter $\alpha = 0.7$, the simulation results are presented in Figs. 6.9 and 6.10. It can be observed that the LQG controller is required more time to stabilize the system states. In order to design a LQR, the weight matrices, step size, and fading parameters should be properly chosen in smart grid communications.

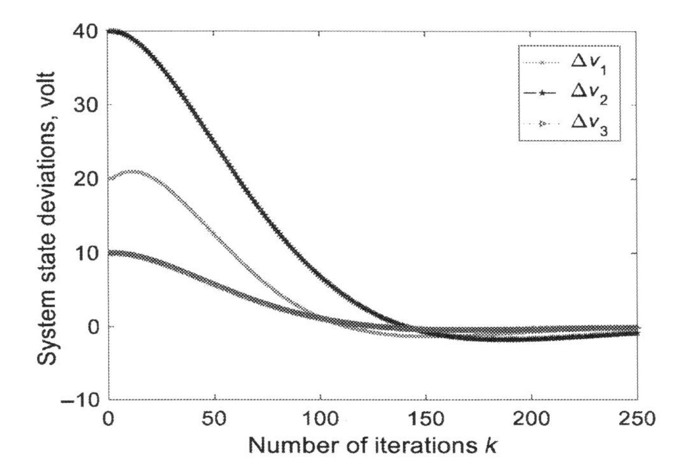

Fig. 6.9 PCC voltage deviations when $\mu = 0.1$ and $\alpha = 0.7$.

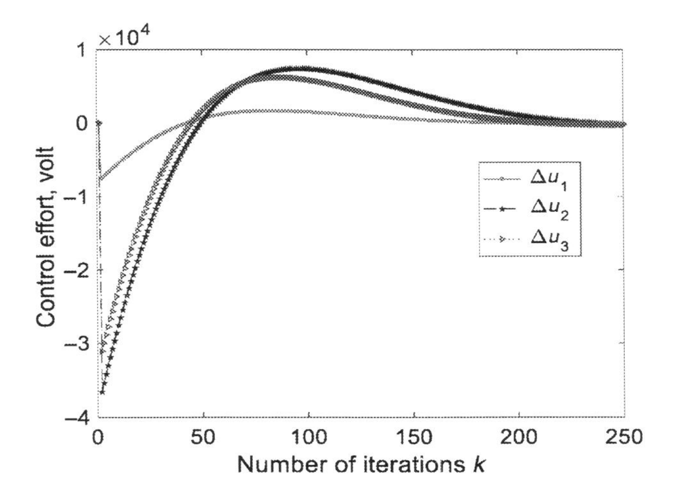

Fig. 6.10 Input voltage deviations when $\mu = 0.1$ and $\alpha = 0.7$.

6.5 Conclusions and future work

This chapter proposes a LQR-based state feedback controller considering fading channel. The extensive simulation results illustrate that the proposed controller can stabilize the system states within a few iterations even if there are large disturbances and fading channel. Results indicate that it is better to use the smaller number of step size to stabilize the DER states. In the future, we will use dynamic fading communication channel with a distributed controller.

References

Hu, S., Yan, W.-Y., 2009. Stability of networked control systems subject to input and output packet loss.Proceedings of the Joint 48th IEEE Conference on Decision and Control and 28th Chinese Control Conference.

Huang, Y.F., Werner, S., Huang, J., Kashyap, N., Gupta, V., 2012. State estimation in electric power grids: meeting new challenges presented by the requirements of the future grid. IEEE Signal Process. Mag. 29 (5), 33–43.

Karimi, H., Nikkhajoei, H., Iravani, R., 2008. Control of an electronically coupled distributed resource unit subsequent to an islanding event. IEEE Trans. Power Delivery 23 (1), 493–501.

Li, H., Lai, L., Poor, H.V., 2012. Multicast routing for decentralized control of cyber physical systems with an application in smart grid. IEEE J. Sel. Areas Commun. 30 (6), 1097–1107.

Rana, M.M., Li, L., 2015a. In: Controlling the distributed energy resources using smart grid communications. Proceedings of the International Conference on Information Technology-New Generations, pp. 490–495.

Rana, M.M., Li, L., 2015b. Microgrid state estimation and control for smart grid and the internet of things communication network. Electron. Lett. 51 (2), 149–151.

Rana, M.M., Li, L., Su, S.W., 2017. Controlling the renewable microgrid using semidefinite programming technique. Int. J. Electr. Power Energy Syst. 84, 225–231.

Su, L., Chesi, G., 2015. In: On the robust stability of uncertain discrete time networked control systems over fading channels. Proceedings of the American Control Conference, pp. 6010–6015.

Xiao, N., Xie, L., Qiu, L., 2012. Feedback stabilization of discrete time networked systems over fading channels. IEEE Trans. Autom. Control 57 (9), 2176–2189.

Yan, Y., Qian, Y., Sharif, H., Tipper, D., 2013. A survey on smart grid communication infrastructures: motivations, requirements and challenges. IEEE Commun. Surv. Tutorials 15 (1), 5–20.

Further reading

Rana, M.M., 2017. Modelling the microgrid and its parameter estimations considering fading channels. IEEE Access 5, 10953–10958.

Cooperative energy dispatch for multiple autonomous microgrids with distributed renewable sources and storages

7

Xinli Fang[*,†], *Qiang Yang*[‡]
[*]PowerChina Huadong Engineering Corporation Limited, Hangzhou, People's Republic of China, [†]Hangzhou Huachen Electric Power Control Co. LTD., Hangzhou, People's Republic of China, [‡]College of Electrical Engineering, Zhejiang University, Hangzhou, People's Republic of China

Nomenclature

P_{CLs}	the supplied CLs
P_{NLs}	the noncritical loads (NLs), which can be supplied or shed flexibly at any time during MG operations
P_{DG_min}, P_{DG_max}	the upper and lower bounds of DG capacity
U_{min}, U_{max}	the upper and lower voltage limits of node i and U_i is node voltage
p_i, q_i	the injected active and reactive power of node i
$p_{DG\to i}$, $q_{DG\to i}$	the injected active and reactive power from DG to node i
P_j, Q_j	the active and reactive loads of node i
$\dot{Y}$	the admittance matrix
$\sum \omega_{\forall(DG\to CL)}(t)$	the weight sums from any DG to any CL at time t

7.1 Introduction

With the penetration of a massive number of small-scale distributed generators (DGs) (e.g., wind turbines and photovoltaic equipments) into the current medium/low voltage power distribution networks, a new form of power supply paradigm, that is, microgrid (MG), emerges. MGs can be flexibly operated in conjunction with the power utility or in an autonomous mode where the power demand is expected to be met by the DGs (Bhandaria et al., 2014). MGs have been adopted with renewable DGs to promote the power supply reliability with reduced cost. However, the power output of wind and solar-based generators exhibits nondeterministic and intermittent characteristics due to the local meteorological factors (Obara, 2015). It also has been recognized that the current-operated MGs may not have adequate energy storage system (ESS) with appropriate location and capacity (Schroeder, 2011). In addition, MGs

Smart Power Distribution Systems. https://doi.org/10.1016/B978-0-12-812154-2.00007-9

need to operate in an autonomous mode under certain circumstances, for example, no power supply from the power utilities upon network outage due to cascading failures or nature disasters, and the places without power utility infrastructures (e.g., offshore islands). Not surprisingly, this may directly undermine the performance of power supply in autonomous MGs (AMGs) in many aspects: the critical loads (CLs) which need to operate at their power ratings may have to be curtailed without stable power supply; it may also happen that within the individual MGs, the DGs generate more power than their own demands and the excess energy cannot be utilized by neighboring MGs. This imposes an urgent technical challenge that the cooperative operation across multiple AMGs needs to be studied such that the global resource utilization efficiency and energy security of the power distribution network encompassing multiple MGs can be optimized at the network level.

In AMGs, it is important to keep generation-demand balance while guarantee the power supply security to CLs with economical benefits maximization (e.g., Javadian et al., 2013). In the literature, much research effort has been made to address this technical challenge. Zhao et al. (2015) presented a multi-agent system (MAS)-based solution to coordinate the operation of islanded MGs in the power market context. However, the operational complexity as well as the network reinforcement cost imposed by this control structure can be prohibitive for large-scale networks as a large number of control devices need to be embedded to fulfil the management functionalities. Hawkes and Leach (2009) investigated the economical and environmental benefits and presented a cost minimization-based optimization approach to appropriately coordinate a large number of MG components, but without including the crucial network operational metrics, for example, system reliability, power loss, into the optimization process. A "token ring" algorithm was proposed to acquire the DG operational details so as to carry out the optimization tasks for MG operations. In fact, the operational information can be inaccurate or even unavailable due to the lack of a sufficient number of high-performance electronic power processors (EPPs), which can significantly deteriorate the effectiveness of this solution (Tenti et al., 2012). Inspired by the tree knapsack problem (TKP), Jikeng et al. (2012) followed the similar line of research and proposed a strategy based on the islanded division model to optimize the design of MG control structure. These attempts have been carried out merely based on the knowledge of physical network topologies without considering the operational characteristics of the power network. In parallel, an autonomous division strategy of power distribution network with DGs was proposed for enhancing system operation and self-healing capability (Caldon et al., 2008), but again the operational aspects have not been assessed. In addition, Wang et al. (2012) proposed an optimal cooperative design solution for MGs with distributed wind and PV (photovoltaic) generators. The studied network scenario considers the fuel-based energy storage devices, and thus the potential benefits of cooperative operation cannot be clearly identified. Most recently, Arefifar et al. (2013a,b) presented the smart grid design solutions with the focus on improving the self-healing capability and power security in network planning based on the heuristic Tabu search (TS) algorithm. Such design solutions are carried out offline due to the computational complexity and searching uncertainties, and hence they are suitable for network planning tasks, rather than optimizing the network performance during operation. These previous results have confirmed our intuition that the power supply reliability and DG energy utilization in multiple AMGs can be significantly promoted by appropriately coordinating their operations in a cooperative fashion, even without sufficient energy storage facilities and hardware devices running complex control applications. An energy management approach for a residential grid-tied MG was proposed to control energy exchange through a battery system, but its performance firmly relied on the storage capacity and prediction of state-of-charge (SOC) (Pascual et al., 2015). The issue of simultaneous energy supply and demand planning in MGs was addressed through a rolling horizon optimization-based approach (Silvente et al., 2015). In addition, Comodia et al. (2015) carried out the study in realistic residential MGs considering both electrical and thermal storage devices, and Fathima and Palanisamy (2015) reviewed the optimization techniques and considerations in MGs with hybrid energy systems. Velik and Nicolay (2015) developed a modified simulated annealing triple-optimizer for energy management considering the electricity prices, which provided an optimal or suboptimal solution with limited computation efforts. However, the aforementioned studies of energy dispatch and operational optimization have restricted the view to either individual or small-scale MGs, which have not explored the potential benefits of coordination across multiple operational MGs.

This chapter follows this line of research and presents a collective energy dispatch solution for multiple AMGs embedded with small-scale distributed renewable and storage units (SUs) based on a "tree stem-leaves" approach. It is assumed that the MG operational states can be made available, for example, based on MASs (Hernandez et al., 2013; Karavas et al., 2015; Dou and Liu, 2013), and the network topology can be reconfigured through manipulating network connecting switches, for example, Xie et al. (2012). The key technical contributions made in this research can be summarized as follows: (1) we exploit the coordinated energy dispatch across multiple AMGs embedded with different forms of intermittent renewable DGs, and present a cost-effective algorithmic solution which can be adopted in practice with minimal deployment hurdles and (2) a set of key operational metrics, for example, power loss and system reliability statistics, are explicitly incorporated into the MG operation optimization to promote the level of DG utilization efficiency and demand supply security. In this chapter, aiming to directly quantify the potential benefits of cooperative dispatch across multiple MGs, distributed energy SUs, for example, ESS, are not explicitly included into consideration. The basic idea behind this proposed approach is to optimally match the DG generation to the demands through adaptive mapping between DGs and critical demands with optimized power loss and supply reliability. The approach can consistently lead to optimal energy management with significantly improved CL supply security, global DG utilization efficiency, and generation-demand matching performance across multiple AMGs. The performance of the suggested solution is verified and assessed through carrying out a set of simulation experiments for a range of network scenarios based on the IEEE 33-bus network model through a comparative study.

7.2 Autonomous microgrid optimization model

When a MG is operated in the autonomous mode, the demand can only be supplied by DGs and other onsite generation (e.g., diesel generators). However, the dynamic fluctuation of the power output from DGs and loads can lead to a significant mismatch between the generation and demand, for example, the power generation cannot meet the total MG power loads when the power output of the embedded DGs are at the low level during certain periods. In a MG, the dynamics of DG generation and load profiles (both CLs and NLs) can lead to a significant mismatch overtime, for example, the DGs cannot meet the demand requirement at certain periods in certain AMGs. Appropriate mappings of DGs and demands are required across multiple connected AMGs to maintain an optimal generation-demand match dynamically. In a distribution network embedded with many AMGs, such problem can be addressed through coordination of generation and SUs across multiple connected AMGs, with the additional benefits of improving DG utilization efficiency and power security. It can be formulated as an optimization problem subject to a collection of constraints as follows:

$$\varepsilon = \min\left[\sum P_{DGs} - \left(\sum P_{CLs} + \sum P_{NLs}\right)\right] \tag{7.1}$$

$s.t.:$

$$\varepsilon \geq 0 \tag{7.2}$$

$$P_{DG\_i}^{\min} \leq P_{DG\_i} \leq P_{DG\_i}^{\max} \tag{7.3}$$

$$U_i^{\min} \leq U_i \leq U_i^{\max} \tag{7.4}$$

$$p_i + p_{DG\rightarrow i} = P_i + U_i \sum U_i \dot{Y} \tag{7.5}$$

$$q_i + q_{DG\rightarrow i} = Q_i + U_i \sum U_i \dot{Y} \tag{7.6}$$

$$\delta = \max \sum P_{CLs} \tag{7.7}$$

$$\eta = \min \sum \omega_{\forall(DG\rightarrow CL)}(t) \tag{7.8}$$

In Eq. (7.1), ε is the optimization objective, minimizing the mismatching between the generation of DGs and the network demand (including CLs and NLs); Eqs. (7.2)–(7.6) give the power flow constraints: Eq. (7.2) ensures the DG generation can meet the loads demands in an AMG; Eqs. (7.3) and (7.4) are the DG capacity (within the range of $[P_{DG\_i}^{\min}, P_{DG\_i}^{\max}]$) and node voltage constraints (within the range of $[U_i^{\min}, U_i^{\max}]$); the active and reactive power flow of node i is given in Eqs. (7.5) and (7.6), respectively; Eq. (7.7) means the CLs should be maximum supplied; and Eq. (7.8) means the power loss and unreliability between the power supply DG and power consumption CL need to be minimized.

To address such problem mentioned in Eq. (7.8), an AMG is first mathematically modeled in the form of a weighted matrix by incorporating two operational metrics, that is, power loss and reliability statistics. For each feeder of the distribution network, these two metrics are normalized and expressed as follows:

$$\omega_{ij} = \beta \cdot P^{ij}_{norm\_loss} + (1-\beta) \cdot K^{ij}_{norm\_risk} \tag{7.9}$$

In Eq. (7.9), ω_{ij} is the weight of the network feeder between any two network nodes i and j; the coefficient $\beta \in [0,1]$ can be determined based on the operational requirements. $P^{ij}_{norm\_loss} \in [0,1)$ and $K^{ij}_{norm\_loss} \in [0,1)$ are the normalized power loss and unavailability of the network feeder from any node i to node j, respectively, that is:

$$\begin{cases} P^{ij}_{norm\_loss} = \dfrac{P^{ij}_{loss}}{\sum\limits_{i,j=1;i\neq j}^{N} P^{ij}_{loss}} \\[2ex] K^{ij}_{norm\_risk} = \dfrac{K^{ij}_{risk}}{\sum\limits_{i,j=1;i\neq j}^{N} K^{ij}_{risk}} \end{cases} \tag{7.10}$$

In Eq. (7.10), P^{ij}_{loss} refers to the power loss from node i to node j, which can be obtained from Eq. (7.11). K^{ij}_{risk} is the failure rate of the power line from node i to node j, which can be quantified as Eq. (7.12) based on the expert experiences and historical statistics (Comodia et al., 2015), that is:

$$P^{ij}_{loss} = \frac{P_j^2 + Q_j^2}{U_j^2} \cdot R_{ij} \tag{7.11}$$

$$K^{ij}_{risk} = \eta \cdot E_{ij} + (1-\eta) \cdot K_{ij} \tag{7.12}$$

where η is the configurable coefficient to make the trade-off between E_{ij} and K_{ij}. E_{ij} refers to the failure rate from the expert assessment, and K_{ij} is the unavailability between node i and node j, which can be obtained based on Eq. (7.13) (Fathima and Palanisamy, 2015), that is:

$$K = \frac{f \cdot r}{8760} \tag{7.13}$$

where f and r are the annual failure frequency and repair time, respectively.

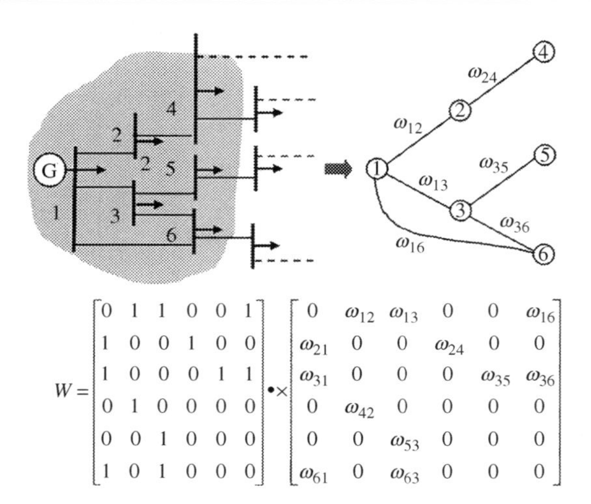

Fig. 7.1 The weighted matrix model of an autonomous MG (AMG).

Based on the derived normalized weights, the network topology matrix can be obtained. The weight of distribution line L_{ij} (between two connected nodes i and j) is denoted as ω_{ij} (i.e., $\omega_{ij} = \omega_{ji}$) based on Eqs. (7.9)–(7.13), as illustrated in Fig. 7.1.

In this chapter, the proposed cooperative control for coordinating multiple MGs is designed based on the extension of Kruskal's minimum spanning tree (MST) (Graham and Hell, 1985) and linear matrix inequality (LMI) algorithms. The MST algorithm is used to determine the mapping between DGs and CLs across different MGs while the LMI chooses NLs to be supplied by DGs. The mechanism of Kruskal's MST algorithm is simple and also scalable to cope with the computation task for coordinating a large number of MGs in large-scale distribution networks and the LMI can prove a fast convergence to determine how many NLs should be added or shed from the DG-CLs AMGs.

With the weighted power network incidence matrix, the MSTs for individual DGs in MGs can be derived by using each DG as a root node. The trees with the minimum weight can be identified from each DG to all the CL nodes across multiple MGs, that is, the mapping between DGs and CLs. Due to the fact that the adopted power network incidence matrix has included the power distribution line loss and supply reliability into its weight settings, the identified MSTs between DGs and CLs based on this algorithmic solution can represent the optimal mapping between generation and critical demands in terms of economical benefit and reliability consideration.

7.3 Cooperative operation control strategy (w/o storages)

This section presents the control strategy for optimally coordinating the operation of multiple AMGs. The cooperation of multiple MGs can be transformed into the problem of optimally remapping the DGs and the CLs across multiple MGs, if applicable and necessary, for example, some CLs are not sufficiently supplied in certain MGs while DGs in other MGs are underutilized. The proposed algorithmic solution can be described as follows in detail:

Step 1: Periodically monitoring the DG outputs and demands in MGs based on a reasonable sampling time interval (Δt).

Case 1a: If the DG power output is sufficient for meeting all demands, both CLs and NLs can be supplied by coordinating the MGs.

Case 1b: If the DG power output can supply all CLs but not all loads, then the network reconstruction is implemented.

Case 1c: If the DG power output cannot meet all CLs, some CLs need to be shed when DGs are fully utilized across multiple MGs.

Step 2: In Case 1b, multiple MGs are coordinated to meet all CLs while fully utilizing all DGs. In reality, frequent switching actions are not favored due to inevitable interruptions to the network operation. Thus, to avoid frequent switching operations and reduce the computational complexity, the changes in the matching between DG output and CLs are examined, which is defined as

$$\Psi = \frac{|P_{DG_i-CL_i}(t-\Delta t) - P_{DG_i-CL_i}(t)|}{P_{DG_i-CL_i}(t)} \tag{7.14}$$

In one or multiple MGs, if the predefined threshold θ (θ =30% is set in this study) is exceeded, that is, $\theta < \Psi$, then the multiple MGs are coordinated to update their structure by operating a set of switches. Otherwise, the MGs are not restructured and only NLs are added or shed in accordance to the DG output.

Step 3: Calculate the network link weights ω_{ij} based on Eqs. (7.9)–(7.13) and establish the topological relation matrixW^t.

Step 4: Find k MSTs with respect to k DGs as the root nodes based on the MST algorithm.

Step 5: Compute the total weight along the connection from a DG to individual CLs in MSTs, and determine the optimal DG-CL pair.

Step 6: Calculate the residual power output after all CLs are supplied.

Case 6a: If the power output exceeds the all CLs, certain NLs can be supplied to fully utilize the residual energy.

Case 6b: If a certain DG cannot meet all CLs, the CL will be mapped to other DGs.

Case 6c: If the DGs in a MG cannot supply any loads at certain time, they should be fused to another MG to collaborate with other DG or DGs.

Case 6d: If a DG can meet all its loads in a MG and the surplus energy is larger than a predefined threshold, Ω (e.g., it is set as $\Omega = 25\%$ of the DG power output in this study), it needs to cooperate with other DGs to supply as many loads as possible.

Case 6e: In all other cases, the MGs are not reconstructed and only NLs are added or shed.

Step 7: Verify the power flow constraint of the constructed AMGs based on Eqs. (7.2)–(7.6). If any constraint is violated, the DG-CL mapping needs to be updated with the subminimum weight sums.

Step 8: Count the numbers of different NLs between DG_i and CL_j and record them in variables n_M, n_T, and n_L.

Step 9: Construct LMI to determine the numbers of different NLs that are involved in the remapped MGs.

The adopted algorithm can carry out an exhaustive search to find all possible spanning trees for DG and CL/NL mappings and select the optimized ones with affordable complexity. The pseudo-code of the algorithmic solution is given as follows:

Require: $\theta, \Delta t$
Switch
1: **Case 1a**: $\sum P_{DGs} > \sum P_{Loads}$, **break**;
2: **Case 1b**: $\sum P_{Loads} > \sum P_{DGs} > \sum P_{CLs}$, **to** line 4 (**Step 2**);
3: **Case 1c**: $\sum P_{DGs} < \sum P_{CLs}$, delete $\forall CLs$ from the system with the **Target**: $\min(\sum P_{DGs} - \sum P_{CLs})$. // *Step 1*
4: **If** $\theta > \Psi$, $Loads \in \{NLs\}$ added to $\sum P_{Loads}$, **Target**: $\min(\sum P_{DGs} - \sum P_{loads})$;
5: **Else if** $\theta < \Psi$, **to** line 8 (**Step 3**).
6: **End**
7: **End If** // *Step 2*
8: calculate ω_{ij} with (9) to (13), get the topology matrix W^t. // *Step 3*
9: choose edge e_1 with minimum $\omega_{ij}(e_1)$;
10: **If** $e_1 \cdots e_i$ were chosen, **then** choose e_{i+1} from $E\{e_1 \cdots e_i, e_{i+1} \cdots e_n\}$, $\omega_{ij}(e_{i+1}) = \min(e_{i+1}, \cdots, e_n)$;
11: when the selection cannot go on, **break**. // *Step 4*
12: calculate every $\sum \omega_{ij}(DG_i \rightarrow CL_j)$;
13: **If** $\min \sum \omega_{ij}$, **then** $CL_j \in DG_i$
14: **End If** // *Step 5*
15: **If** $\sum P_{DGs} > \sum P_{CLs}$, **then** go to line 24 (**Step 7**);
16: **Else if** $P_{DG_i} < \sum P_{CLs}$, **then** choose some $CL_j \in DG_k$; % DG_k with subminimum $\sum \omega_{ik}$;
17: **Else if** $P_{DG_i} < \forall P_{CL_j}$, **then** $P_{DG_i} \cup \forall P_{DG_k}$;
18: **Else if** $P_{DG_i} >> \sum P_{Loads}$, then $\forall P_{DG_k} \cup P_{DG_i}$ and the scope is enlarged.
19: **Otherwise** go to line 24 (**Step 7**).
20: **End**
21: **End**
22: **End**
23: **End If** // *Step 6*
24: **Check** function (2)-(6).
25: **If** (2)-(6) all satisfied, go to 26 (**Step 8**);
Otherwise go to 13 and choose some $CL_j \in DG_k$ with subminimum $\sum \omega_{ik}$; // *Step 7*
26: record the numbers of different kind of NLs between DG_i and CL_j in n_M, n_T and n_L. // *Step 8*
27: construct LMI as:

$$\varepsilon = \min\{x \cdot M(t) + y \cdot T(t) + z \cdot L(t) - (DG_i - CL_j)\}$$
$$s.t.: \ 0 \leq \varepsilon$$
$$x \leq n_M; \quad y \leq n_T; \quad z \leq n_L$$

28: get x, y and z. // *Step 9*

Such load control can be implemented by operations of circuit breakers (CBs) through certain network management functionalities, for example, a MAS-based approach

(Kuznetsova et al., 2014). It can be seen from the algorithm that two mechanisms are introduced to guarantee the solutions are optimal: first, two optimization objects (power loss and reliability) are incorporated into a single metric in the optimization process and hence the problem is avoid to be NP (non-deterministic polynomial)-hard complexity, and can be solved without heuristics; second, by given the optimization metric, the identification of DG-CL mappings based on the MSTs is deterministic and optimal.

7.4 Numerical experiments and result

7.4.1 Simulation scenarios

In this work, we consider two network simulation scenarios as follows: (1) IEEE 33-bus network (Ref22, n.d.) which in total has four DGs, for example, photovoltaic (DG1) and wind turbines (DG2–DG4), connected to four network nodes (i.e., 4, 15, 20, and 30). The power output profiles range from 0 to 679.27 MW and from 13.765 to 340.0529 MW, respectively, obtained from the Belgian electricity transmission operator Elias (Ref23, n.d.) (May 30, 2013), as shown in Fig. 7.2A; (2) IEEE 300-bus network (Ref22, n.d.): the DGs (eight in total: two PV DGs and six wind DGs) and loads (i.e., CLs and NLs) are connected to the network nodes selected in a random fashion. The photovoltaic and wind DG power output profiles range from 0 to 1245.45 MW and from 24.25 to 270.04 MW, respectively, which are also adopted from Ref23 (n.d.). In addition, we consider four categories of network power loads, that is, municipal, tertiary, light industry, and heavy industry demand. For both simulation scenarios, the statistical profiles for different load categories are taken from Li et al. (2015), as shown in Fig. 7.2B. In particular, we look into the opertional scenarios that the generation of DGs is not sufficient (i.e., during the periods of 05:45–12:45 and 20:15–23:45, as shown in Fig. 7.2C). The details of network nodes connected with DGs and CLs can be found in Table 7.1.

7.4.2 Case study

Here, the proposed control strategy for coordinating multiple AMGs is first explained with the IEEE 33-bus network before presenting the simulation result. To explain the proposed cooperative solution, we take the network operation at 05:45 (May 30, 2013) as an example. Based on the algorithm presented in Section 7.3, the network matrix $W^t(t = 05:45)$ are first established and then four MSTs can be obtained by using four DGs as the root nodes, respectively, as shown in Fig. 7.3A–D. With the four MSTs, the weight sums from DGs to CLs can be obtained in Table 7.2.

From the algorithmic approach presented in Step 5 in Section 7.3 and Table 7.2, DGs supply a set of different CLs, that is, DG_1 (CL_7—the CL connected with node 7), DG_2 (CL_8 and CL_{18}), DG_3 (CL_{21}), and DG_4 (CL_{24} and CL_{32}) in the original DG-CL mapping. It should be noted that, as shown in Table 7.2, the minimum weight sum (0.0639) is abandoned and the suboptimal DG-CL mapping with the weight sum of 0.1195 is adopted. This is due to the fact that at the time of 05:45 (May 30, 2013),

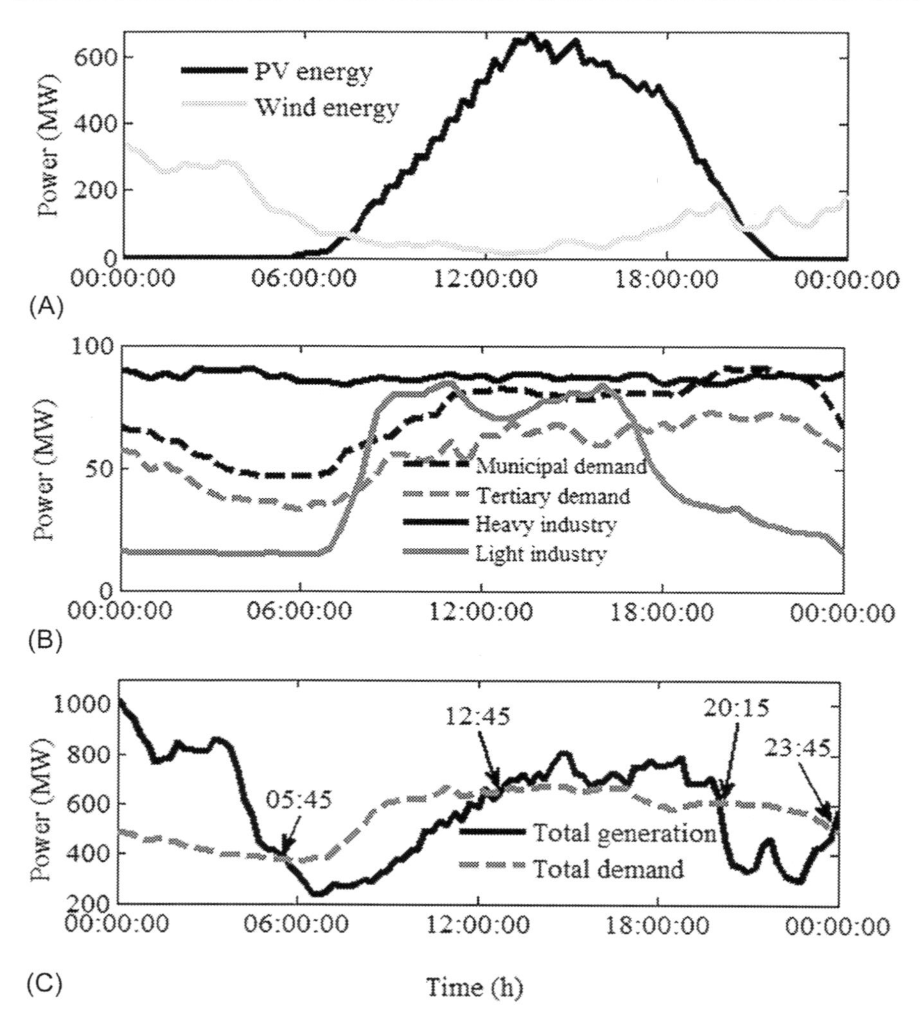

Fig. 7.2 (A) The distributed generator (DG) output (24 h, May 30, 2013); (B) typical demand profile; and (C) total demand and generation.

DG_1 (photovoltaic 6.15 MW) is not sufficient to supply CL_7 (25.905 MW) and CL_7 is switched to be supplied by the DG with the suboptimal mapping, that is, DG_4 (wind turbine 116.4632 MW) (Step 7). Simultaneously, the residual DG power can be used to supply NLs (Step 6: Case 6a). Thus, the original structure of multiple MGs (shown in Fig. 7.4A, where the dashed lines stand for the connecting switches) can be reshaped into the structure shown in Fig. 7.4B. In Fig. 7.4, the shadow areas represent individual AMGs, and the loads connected to the uncovered nodes (e.g., node 10 in Fig. 7.4B) cannot be supplied by DGs due to the generation inadequacy. Such remapping operations can be carried out instanteneously.

Table 7.1 Nodes connected with DGs and CLs in the IEEE 33-bus and 300-bus systems

Scenarios	Category of nodes	Node ID
IEEE 33-bus network	Municipal demand	3, 11, 14, 22, 27, 29, 33
	Tertiary demand	1, 2, 6, 12, 13, 16, 26, 28, 31
	Light industry	5, 9, 10, 17, 19, 23, 25
	Heavy industry (CLs)	7, 8, 18, 21, 24, 32
	Photovoltaic (DG_1)	4
	Wind turbines (DG_2–DG_4)	15, 20, 30
IEEE 300-bus network	Heavy industry (CLs)	77, 48, 77, 129, 143, 174, 210, 293, 526
	Photovoltaic	27, 169
	Wind turbines	70, 131, 213, 243, 293, 562

In the case that one DG can supply all loads (CLs and NLs) in its MG and the surplus energy is over the threshold ($\Omega = 25\%$), then it is fused with other DGs (Step 6: Case 6d). In Fig. 7.4C and D, we present the obtained remapping between the DGs and network demands across multiple MGs based on the proposed algorithmic solution at the time of 08:30 and 10:45, respectively.

7.4.3 Numerical result

Followed to the explanation, a set of numerical results are obtained to assess the suggested solution from different aspects. In all simulations, the sampling time interval is set to 15 min, to periodically evaluate and decide if the mapping between the DGs and loads needs to be revised. For the sake of clarity, the result is plotted at the time scale of 30 min (applied to Figs. 7.5–7.7).

7.4.3.1 Generation-demand match

Fig. 7.5 presents the result of the overall output of DGs against the CLs, NLs, and the total demand in a day (24 h). It can be observed from Fig. 7.5A that there exists a significant mismatch between the total generation and demand across multiple MGs. At certain time periods, for example, 00:00–05:30 and 13:00–20:00, the overall DG output power exceeds the total demand and hence all network demand should be met without load shedding if the appropriate mapping of DGs and loads is carried out. Also, during certain periods, for example, 05:45–12:45 and 20:15–23:45, the overall energy output is not sufficient to supply all the network loads. In this case, the DGs need to be coordinated to supply the CLs across multiple MGs through cooperative control and some loads have to be shed due to the inadequacy of energy. The optimal remapping between the DGs and loads ensures all CLs are supplied and DGs are optimally utilized to supply as many as possible of CLs and NLs across these MGs.

Different from the IEEE 33-bus scenario, in the IEEE 300-bus network, we assume that the total demand is larger than the total DG generation in multiple MGs. The total DG output is not sufficient to supply all CLs (period 00:00–01:30), when a set of CLs

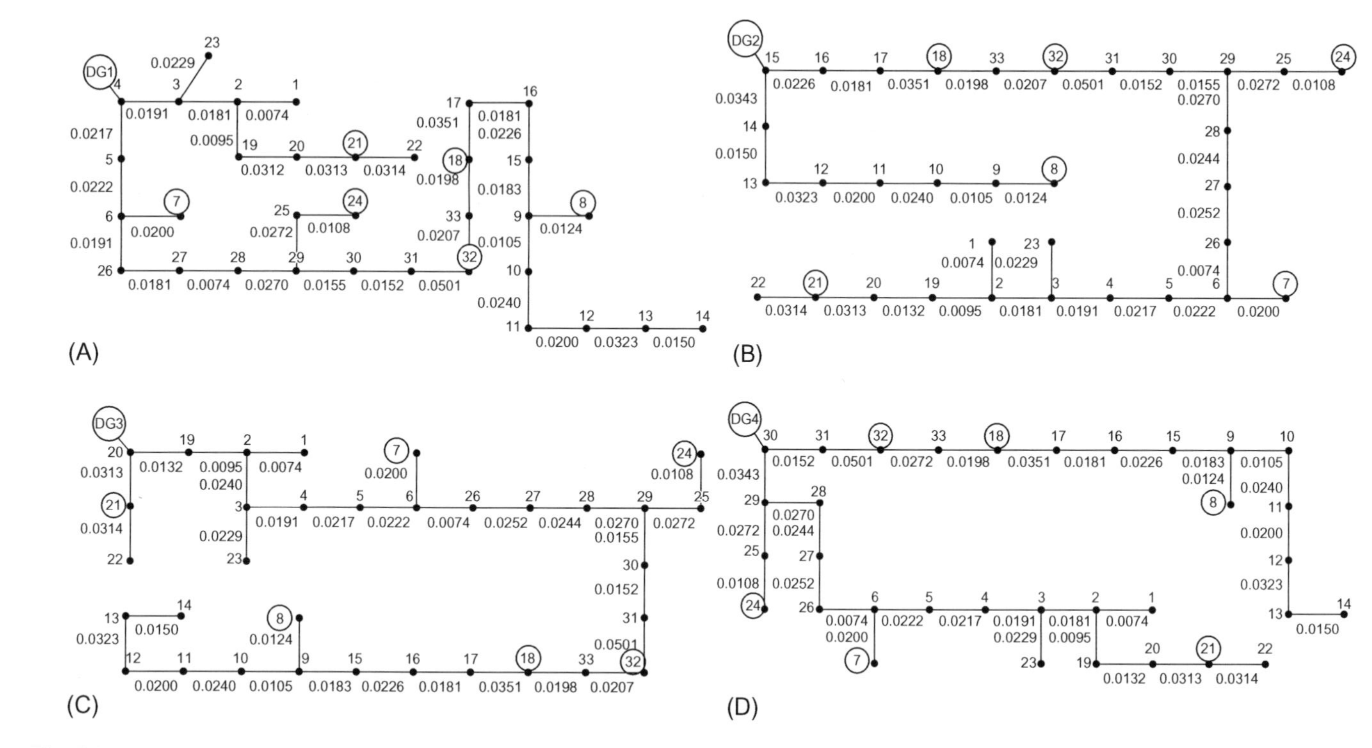

Fig. 7.3 The obtained four minimum spanning trees (MSTs) (A)–(D) with respect to individual DGs (DG1–DG4) as the root nodes (at 05:45).

Table 7.2 Weight sums from DGs to CLs

	DG_1	DG_2	DG_3	DG_4
CL_7	~~0.0639~~	0.3011	0.1238	**0.1195**
CL_8	0.3557	**0.1485**	0.4156	0.2123
CL_{18}	0.2492	**0.0758**	0.3091	0.1058
CL_{21}	0.0912	0.4162	**0.0313**	0.2346
CL_{24}	0.1659	0.2351	0.2258	**0.0535**
CL_{32}	0.2087	0.1163	0.2686	**0.0653**

The bold underline values given in this table means the minimum weight sums of every minimum spanning tree, i.e., the optimal solutions of every microgrid. The strike values in this table means the output of DG1 (solar PV) is 0, and hence the CL 7 should be rearranged to be supplied by another two DGs (as underlined in this table).

has to be shed, otherwise all the CLs can be supplied, as shown in Fig. 7.5B. The DG generation can be fully utilized over a day by adding and shedding NLs based on the proposed cooperative strategy. This result further confirms our expectation that the suggested solution can prioritize CLs in supply as well as maximize the DG facilities.

7.4.3.2 CLs power supply guarantee

To guarantee the power supply to the CLs with the highest priority is paramount to meet the promise to the premium customers. In Fig. 7.6, we present the performance of power supply to the CLs in all the AMGs based on the cooperative control strategy by using the scenario without cooperation as the comparison benchmark.

It clearly shows that, with the cooperative control solution, the power supply to the CLs in the networked MGs can be significantly enhanced due to the fact that the power output of DGs across multiple MGs can be shared and appropriately allocated to CLs. In comparison, the DG resources cannot be efficiently utilized or CLs cannot be supplied in individual MGs if different AMGs are not able to be coordinated. This indicates that through reallocating the resources, that is, DGs, to the network demands in an optimal and dynamic fashion, an obvious economical benefit—greatly enhanced power supply and global DG utilization efficiency, can be obtained.

7.4.3.3 Power supply reliability

Another important operational consideration of power distribution networks is the power supply reliability as the power equipments and lines of distribution networks are considered not as robust and reliable as those of the power transmission networks. In this study, we take the power supply reliability into the cooperative control strategy and the most reliable mapping between DGs and CLs will be selected. To quantify the reliability of power supply to CLs, we define the metric of "load safety (LS) (P_s)" to evaluate the power supply reliability of CLs across multiple MGs, which is expressed as

$$P_s = \sum_{i=1}^{n} \widetilde{L}_i \cdot \prod_{j=1}^{m} (1 - p_j) \tag{7.15}$$

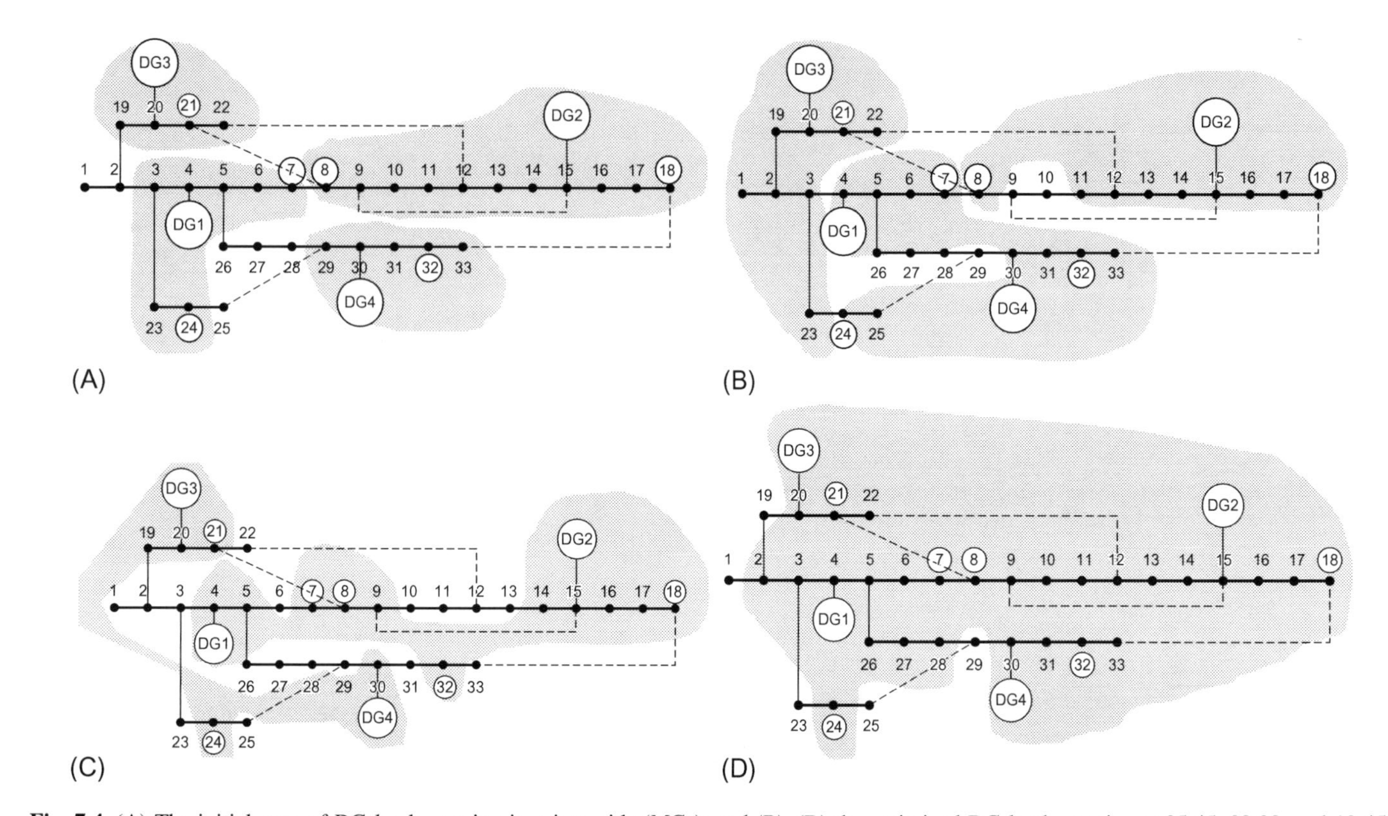

Fig. 7.4 (A) The initial state of DG-load mapping in microgrids (MGs); and (B)–(D) the optimized DG-load mapping at 05:45, 08:30, and 10:45, respectively. (The *dashed lines* stand for the connecting switches.)

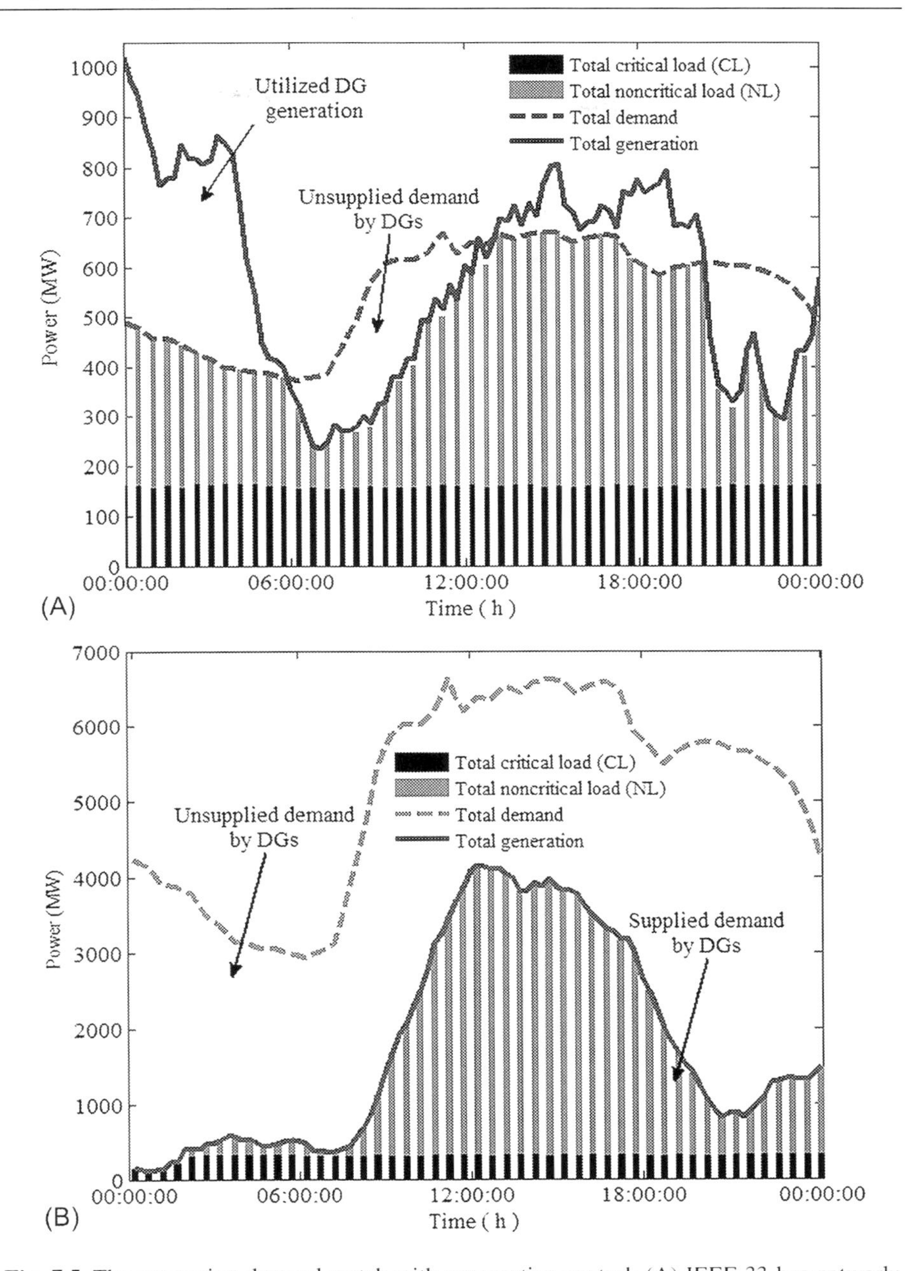

Fig. 7.5 The generation-demand match with cooperative control: (A) IEEE 33-bus network; and (B) IEEE 300-bus network.

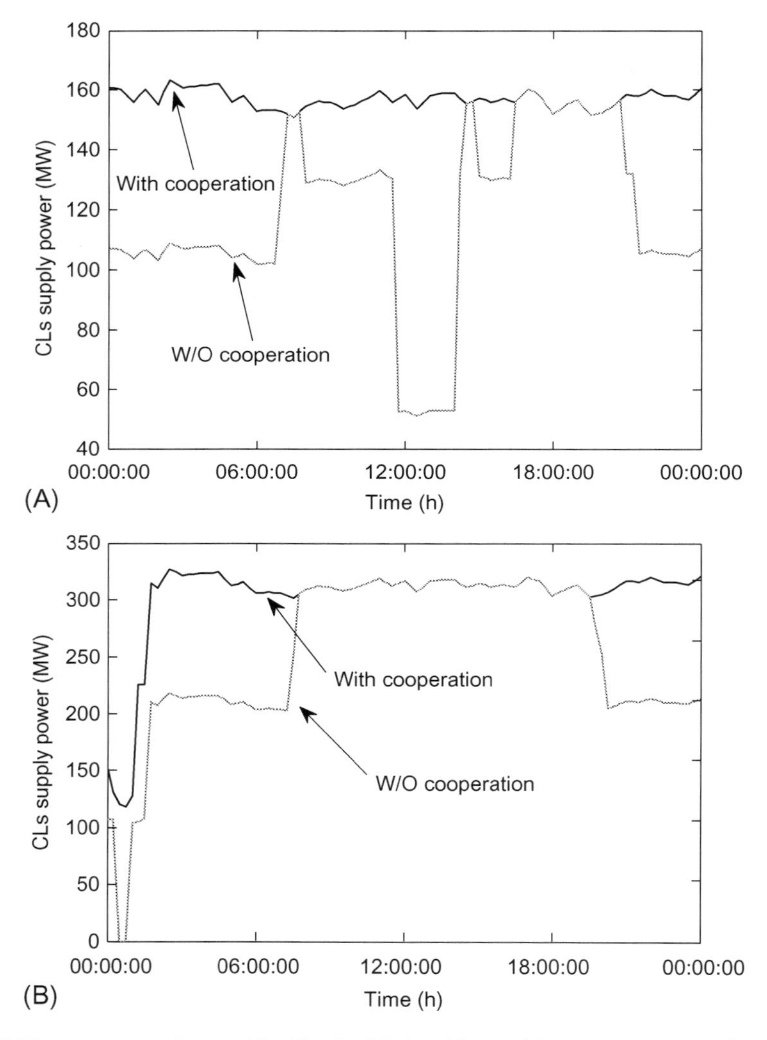

Fig. 7.6 The power supply to critical loads (CLs), with vs without cooperative control: (A) IEEE 33-bus network; and (B) IEEE 300-bus network.

where n is the total number of CLs in multiple MGs; $\widetilde{L_i}$ is the electric power of the *ith* CL; m denotes the number of power lines from DG to CL i; p_j is the failure rate of the *jth* power line; $i = 1, \cdots, n$ and $j = 1, \cdots, m$; the reliability statistics is derived from Li et al. (2008) as discussed in Section 7.2.

We consider two cases for optimizing the DG-CL mapping in the cooperative control: (1) only including the power loss; and (2) including both the power loss and the network reliability statistics. Fig. 7.7 shows the result of power supply provision to CLs for these two cases.

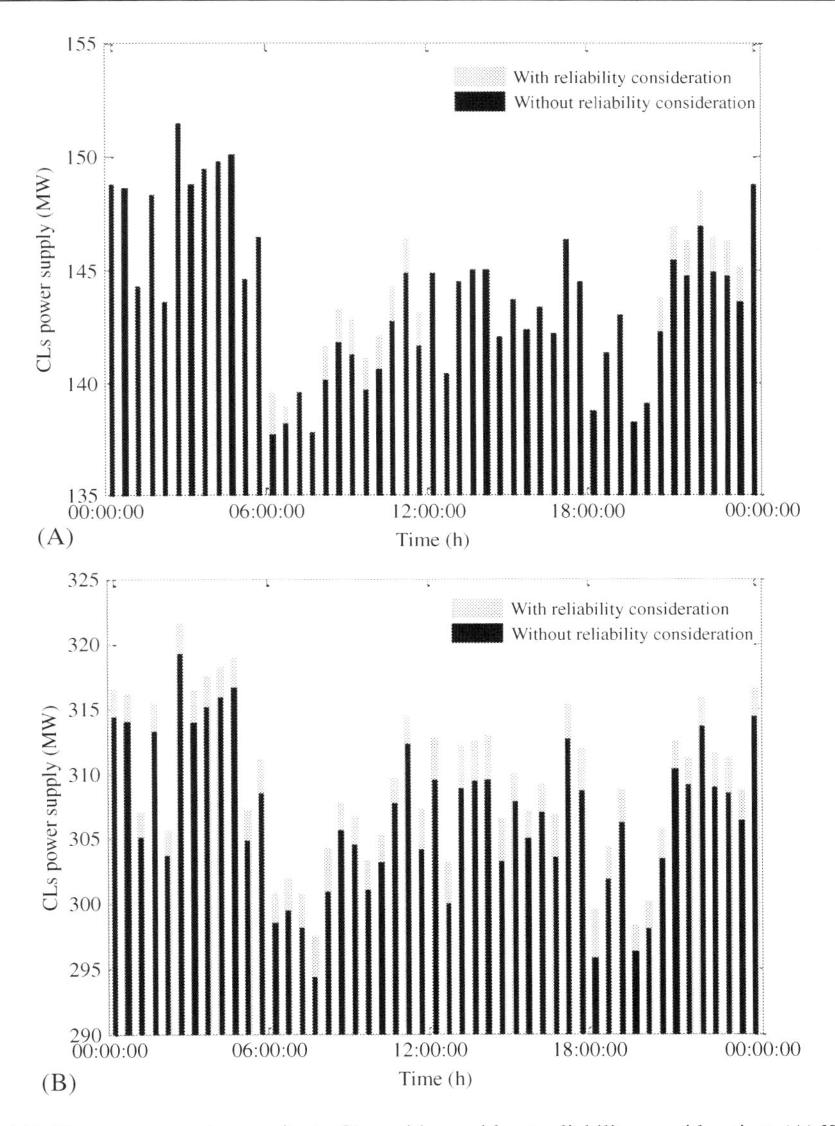

Fig. 7.7 The power supply security to CLs, with vs without reliability consideration: (A) IEEE 33-bus network; and (B) IEEE 300-bus network.

It can be seen that with the proposed cooperative control solution, during the periods that DG output is not sufficent [(a): 05:45–12:00 and 20:15–23:45 and (b): 00:00–24:00], the mappings between DGs and CLs across multiple MGs are optimized (no update on mappings are carried out in other operational periods) and including the consideration of power reliability can improve the security of power supply to CLs than the mappings only considering power loss. The result demonstrates that the proposed strategy can enhance the reliability of power supply to CLs across multiple AMGs without injection of power from the power distribution networks.

Finally, we observed that the number of remapping actions in the simulated IEEE 300-bus network is less than that of IEEE 33-bus network to implement the optimal cooperative operation across AMGs. One of the reasons can be that the large-scale power distribution network interconnected with a large number of loads (including CLs and NLs) can be more robust to the dynamics of DGs, particularly when the overall DG generation is less than the network demand.

7.5 Remarks

7.5.1 The energy storage deployment

This work studied the energy dispatch across multiple AMGs with the aim of directly identifying the technical benefits without the consideration of distributed energy storage. Our previous work confirmed the benefit in promoting demand response by exploiting three coordinated wind-PEV (pure electric vehicle) energy dispatching approaches in the vehicle-to-grid (V2G) context (Wu et al., 2013). Following this line of research, incorporating appropriate control of ESS (charging/discharging) into energy dispatch under certain operational considerations (e.g., real-time pricing) needs to be further exploited (e.g., Diaz et al., 2014). In addition, the optimal resource allocation (i.e., placement and sizing) of the energy storage facilities which can significantly affect the overall system performance needs further exploitation.

7.5.2 Network component reliability

The presented cooperative solution has included the reliability metric into the network model for optimal energy dispatching. As the network reliability assessment can be carried out in an offline manner, more *sophisticated* modeling techniques and tools can be adopted to obtain the system component outage parameters for large-scale power distribution networks, for example, a hybrid method of fuzzy set and Monte Carlo simulation for power system risk assessment was presented to capture both randomness and fuzziness of loads and component outage parameters (Xie et al., 2012). It can be envisaged that the large number of measurements devices along with the intelligent data mining techniques enable us to obtain solid reliability statistics. In addition, recent studies [e.g., Abdullah et al. (2014) and Gouveia et al. (2014)] highlighted that when DGs are integrated, the distribution networks need to be further assessed if some aspects, for example, power adequacy and security, are considered.

7.5.3 Accurate network model and interconnection with grid

The network model elaborated in our approach has adopted a combinatory weighted system which includes the power loss and network component reliability statistics into the network model. In realistic scenarios, a number of factors related to the physical (e.g., topological characteristics), electrical (e.g., real-time network dynamics, i.e., three-phase unbalance, shunt capacitors, zero sequence currents, transformer losses, and time-varying power factor), economical (e.g., demand response under real-time pricing), and operational (e.g., maintenance and operational complexity) constraints can also affect the performance of cooperative energy dispatch and can be included into consideration if applicable. In the case that the MGs are operated in the grid-connected mode, the network model needs to include DGs, demands, SUs, and power grid into a unified framework. In addition, the reasonable approximation and scalable modeling techniques are required to obtain the best trade-off between the accuracy and scalability to cope with the complexity induced from the network operational scenarios consisting of a large number of AMGs.

7.5.4 Technical feasibility in practical deployment

It can be envisaged that along with the penetration of various forms of renewable DGs and development of MGs in different scales in low/medium voltage distribution networks, the coordinated energy dispatch across multiple interconnected MGs are demanded to promote the DG utilization efficiency and power supply security. This work provided a preliminary cooperative energy management framework with an algorithmic solution which can be implemented in current distribution network without any network reinforcement and deployment hurdles. Pilot studies of energy dispatch spanning over multiple-islanded MGs are currently being carried out in Zhairuoshan Island, East China. However, a number of technical challenges still remain outstanding for future research effort, for example, the heterogeneity of interconnected MGs (in terms of power capacity, frequency, and voltage characteristics, and required level of power quality), and different energy conversions (AC bus or DC bus), which requires advanced planning and operation techniques to cope with the engineering complexities.

In this section, we exploited the potential benefit of coordinating multiple AMGs with distributed renewable generators in the power distribution work through a cooperative control strategy assuming that no energy SUs are available. An algorithmic solution based on the MST algorithm is proposed which includes the power loss in energy distribution and power network component reliability into the consideration through modeling the network as a weighted network matrix. Through optimally remapping the DGs and demand across multiple cooperative MGs, the match of DGs and demand in the overall network is approached and the resulted power loss and unavailability of power supply are minimized. The suggested solution has been assessed through the IEEE 33-bus network and large-scale IEEE 300-bus network scenarios, respectively. The numerical result confirms that the proposed cooperative control approach can effectively promote the capability of meeting the promise of secure power supply to CLs, and simultaneously significantly promotes the global utilization efficiency of DGs, even without any energy SUs. Next, we will present a collective energy dispatch solution to optimally coordinate DGs, distributed *SUs*, and critical demands across multiple *AMGs* based on a "tree stem-leaves" approach.

7.6 Cooperative scheduling strategies (with storages)

In the next two sections, we will further present a collective energy dispatch solution based on the MST to optimally coordinate DGs with distributed *SUs* and critical demands across multiple *AMGs* based on a "tree stem-leaves" approach.

The energy distribution network consisting of multiple AMGs is modeled mathematically as a weighted matrix simultaneously considering power loss and reliability statistics. Once the distribution network consisting of multiple AMGs is modeled as a weighted matrix (Fig. 7.1), the proposed "stem-leaves" approach can be proposed to collectively optimize the energy dispatch across multiple connected MGs embedded with DGs in different forms and SUs with different capacities.

In addition, the SUs can be operated in two modes: that is, charging and discharging, and the mode should be first determined according to the system supply-demand situation and SOC status at the sampling periods: (1) if total DGs power supply (P_{DGs}) is greater than the power demand (P_{Loads}) of system loads, that is, $P_{DGs} > P_{Loads}$, all storages inside the system are regarded as loads and at such time, if the SOC of one storage is less than 80%, that is, it is running in area "A" or "B" as Fig. 7.8 shows, then after all system CLs and NLs are supplied, the Ss will be charged, otherwise it without any actions. (2) If total DGs supply is less than the power demand of system loads, that is, $P_{DGs} < P_{Loads}$, all storages inside the system are regarded as generators and at such time, if the SOC of a storage is more than 20%, that is, it is running in area "B" or "C,"

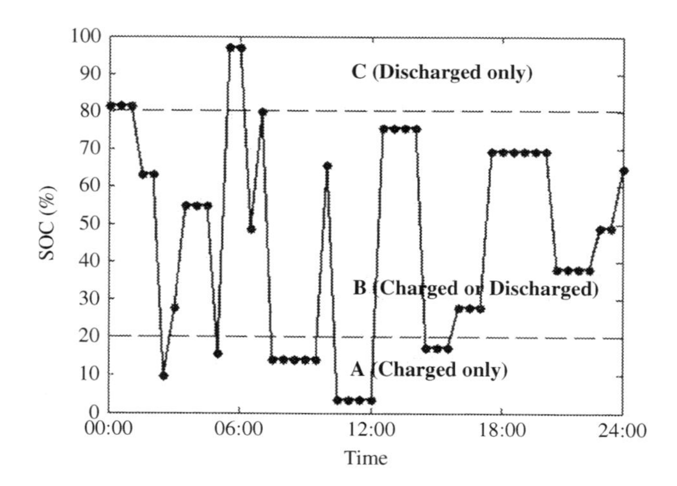

Fig. 7.8 The state-of-charge (SOC) characteristics of a storage unit (SU).

as shown in Fig. 7.8, then the SUs will be regarded as a root node and supply power to CLs and NLs, otherwise it without any actions.

It should be noted that: (1) to protect the lifetime of SUs, the charging/discharging operations need to be regulated in a certain range (20%–80% of SOC is adopted in this work); (2) the operational mode of SUs as "loads" (charging mode) or "generators" (discharging mode) at certain time slots depends on the operational conditions; and (3) the power supply to CLs and NLs needs to be met before SUs are charged (i.e., the priority of power supply is CLs > NLs > SUs).

The suggested approach can be described as follows: the improved MST algorithm is first adopted to determine the DG-CL power supply mapping with the minimum cost from DGs or SUs to CLs, that is, the "stems" are obtained, which are the bases for the new AM (autonomous microgrid) frameworks. After all system CLs are satisfied, to maximum use the DGs generated power energy, with the LMI algorithm, NLs are chosen as "leaves" and put on the "stems." The "stems" (including DGs and CLs) and their NLs together compose a new AM, and all AMs constitute the new structure of the multiple MGs. The cooperative energy dispatch approach can be implemented through the following steps:

Step 1: Initialization (lines 0–3).

The operational state of DGs, CLs, NLs, and SUs is collected periodically with the sampling time Δt. If the difference $(DG - CL)_i$ between the last sampling time and current sampling time is identified to exceed the predefined threshold, θ, in any AMG, the network can be reconfigured, otherwise additional NLs can be added or shed.

Step 2: Topology matrix model generation (line 4).

Based on Eqs. (7.9)–(7.13) (see Section 7.2), the network matrix model, $W(t)$, can be obtained.

Step 3: SU operation mode under different generation-demand matching scenarios (lines 5–13).

Through matching the total DG generation with the total demand in an AMG, it may have the following scenarios: (i) if the DG generation is sufficient to supply all CLs and NLs, the storages are operated in the charging mode (line 8); (ii) if the DG generation cannot meet all the loads, all storages are operated in the discharging mode (line 13); (iii) if the DG generation can only supply CLs, but not all of the demand (CLs and NLs), the operation of SU will be determined by their SOC information (line 12); and (iv) if the DG generation is insufficient to meet all CLs, all storages are operated in the discharging mode (line 13).

Step 4: SU operation mode with SOC constraints (lines 14–20).

The SOC status is periodically examined at each sampling time: if $SOC \geq 80\%$, the storage cannot be further charged in the time slot (lines 14–16); if $SOC \leq 20\%$, further discharging is not allowed (lines 17–19); otherwise (i.e., $20\ \% < SOC < 80\%$), the SUs may be operated in the charging (i.e., acting as load) or discharging (acting as DG) mode as required depending on the operational condition (line 20).

Step 5: Energy supply mapping generation between DG and CLs (lines 21–27).

All MSTs originated from any generators (DGs or SUs in discharging mode) to any CLs (line 21) are first identified through selecting the generator with the minimum MST weights sum (from G_i to CL_j) as the source node to CL_j (line 22). The aim is to identify the DG-CL mappings ("stem") with the minimum MST weights sum as well as meeting the constraints given in functions (7.2)–(7.6) (lines 24–27).

Step 6: Energy management of NLs in updated AGMs (lines 28–31).
Following to the determination of mappings between DGs/SUs and CLs (Step 5), the NLs (e.g., municipal, tertiary, and light industry loads) along the power supply path can be identified to be supplied (i.e., added as "leaves") using LMI based on the available energy generation.

The pseudo-code of the aforementioned algorithmic solution for optimal energy dispatch is given as follows:

---

0. **Requiring** states information of DGs, CLs, NLs and Ss, according to sampling time.
1. **Check** $\exists \forall \theta < \frac{|P_{(DG-CL)_i}(t-\Delta t) - P_{(DG-CL)_i}(t)|}{P_{(DG-CL)_i}(t)}, i = 1, \cdots, N_{AMG}(t-\Delta t)$:
2. **if** true, then go to line 4;
3. **otherwise**, goes to line 30.
4. **Calculate** the weights of topology matrix $W(t)$ with functions (1)-(5) (see Section 7.2).
5. **Check** $\sum_{i=1}^{N_{DG}} DG_i - \left(\sum_{i=1}^{N_{CL}} CL_i + \sum_{i=1}^{N_{NL}} NL_i\right) > 0$:
6. **if** true, then go to line 8;
7. **otherwise** go to line 9.
8. $S_i \in \{x|x \ is \ load\}$, $i = 1, \cdots, N_S$ and go to line 14.
9. **Check** $\sum_{i=1}^{N_{DG}} DG_i - \sum_{i=1}^{N_{CL}} CL_i > 0$:
10. **if** true, then go to line 12;
11. **otherwise** go to line 13.
12. $S_i \in \{x|x \ is \ load\} \cup \{y|y \ is \ generator\}$, $i = 1, \cdots, N_S$ and go to line 14.
13. $S_i \in \{y|y \ is \ generator\}$, $i = 1, \cdots, N_S$ and go to line 14.
14. **Check** $SOC_i \geq 80\%$, $i = 1, \cdots, N_S$ of every S_i:
15. **if** true, $S_j \notin \{z|z \ is \ G_i, \ i = 1, \cdots, N_G\}$, $j = 1, \cdots N_S^G$
16. **Otherwise**, go to line 20.
17. **Check** $SOC_i \leq 20\%$, $i = 1, \cdots, N_S$ of every S_i:
18. **if** true, $S_j \notin \{x|x \ is \ L_i, \ i = 1, \cdots, N_L\}$, $j = 1, \cdots N_S^L$
19. **otherwise**, go to line 20.
20. **If** $20\% < SOC_i < 80\%$, $i = 1, \cdots, N_S$
 $S_i \in \{z|z \ is \ G_i, \ i = 1, \cdots, N_G\} \cup \{x|x \ is \ L_i, \ i = 1, \cdots, N_L\}$.
21. **Calculate** $Min[sum(weight)]_{ij}$ from $\forall G_i \in \{z|z \ is \ G_i, \ i = 1, \cdots, N_G\}$ to $\forall CL_j$, $j = 1, \cdots, N_{CL}$ and N_G N_{CL} MSTs can be got.
22. **Choose** G_i with the minimum $Min[sum(weights)]_{ij}$ to supply CL_j.
23. **Determine** all N_G N_{CL} sets $\{G_i, CL_j\}$, $i \in (1, N_G)$, $j \in (1, N_{CL})$.
24. **Check** $G_i - CL_j > 0$ in every $\{G_i, CL_j\}$, $i \in (1, N_G)$, $j \in (1, N_{CL})$:
25. **if** it is true, then go to line 29;
26. **otherwise** go to line 27.
27. **Choose** G_i with the subminimum $Min[sum(weights)]_{ij}$ to supply CL_j.
28. **Determine** the number of municipal, tertiary and light industry loads from G_i to CL_j, and put the numbers in variables n_M, n_T and n_L.
29. **Construct** LMI as follows:

$$\varepsilon = \min\{(G_i - CL_j) - [x \cdot M(t) + y \cdot T(t) + z \cdot L(t)]\}$$
$$s.t.: 0 \leq \varepsilon$$
$$x \leq n_M; \ y \leq n_T; \ z \leq n_L$$

30. Determine the values of x, y and z.
31. Choose $\forall x$, y and z municipal, tertiary and light industry loads between G_i and CL_j to compose set $\{k|kth\ AMG,\ k \in (1,\ N_G\ N_{CL})\}$, i.e. the kth AM.

7.7 Case study with the IEEE 33-bus network scenario

In this work, the IEEE 33-bus network is used for a case study to further explain and verify the proposed cooperative energy dispatch solution, as shown in Fig. 7.9.

This simulated 33-bus network (the connecting switches are depicted as dashed lines) consists of three DGs: DG_1 is photovoltaic (0–624.205 MW), DG_2 and DG_3 (82.01–419.50 MW) are wind energy, where the DG generation profiles are obtained from Belgian electricity transmission operator Elias (Ref22, n.d.) (May 13, 2014), as shown in Fig. 7.2A. In this work, we consider the adoption of lead-acid-based SUs (S_1, S_2, and S_3), with the capacity of 900 MWh (Ref23, n.d.), resulting in the total capacity of 2700 MWh. In this scenario, 6 CLs and 21 NLs are connected in the system and the 24-h load profiles (Wu et al., 2013) are illustrated in Fig. 7.2B. The details of network nodes connected with DGs and loads can be found in Table 7.3. All simulations in this paper are carried out based on Matlab 2010 and the sampling time of 30 min is applied to all simulation experiments.

In this section, two network scenarios (scenario I and II) at time slots, 00:00 (SUs are in charging mode) and 05:00 (SUs are in discharging mode), are selected to further explain the proposed energy dispatch solution.

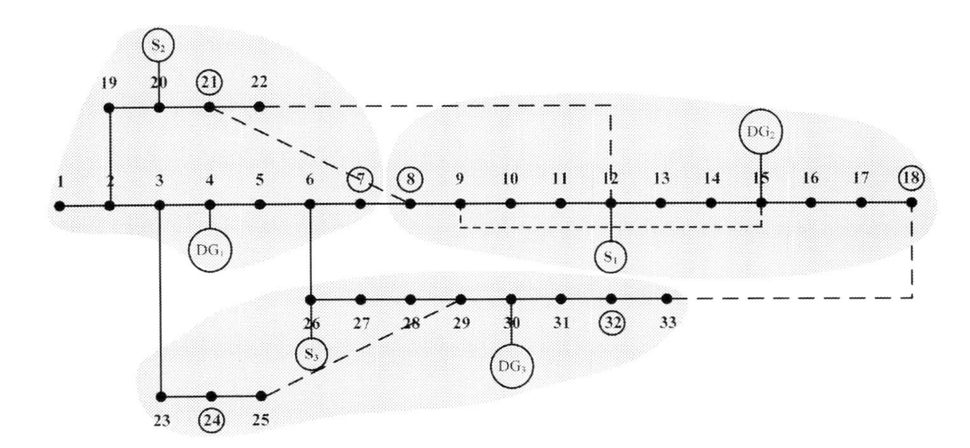

Fig. 7.9 IEEE 33-bus network and AMG configuration.

Table 7.3 Nodes connected with DGs and CLs in the IEEE 33-bus test systems

Scenarios	Category of nodes	Node ID
IEEE 33-bus network	Municipal demand	3, 11, 14, 22, 27, 29, 33
	Tertiary demand	1, 2, 6, 13, 16, 28, 31
	Light industry	5, 9, 10, 17, 19, 23, 25
	Heavy industry (CLs)	7, 8, 18, 21, 24, 32
	Photovoltaic (DG_1)	4
	Wind turbines (DG_2–DG_3)	15, 30
	Storages (S_1–S_3)	12, 20, 26

7.7.1 Scenario I: (time slot 00:00)

It can be seen from Fig. 7.2C, the total power generation can meet all the demands at time 00:00, and hence all SUs are operated in charging mode or without charging in accordance to their SOC. The system will only take the DGs as root nodes to search MSTs. At the time slot of 00:00, the obtained MSTs are given in Fig. 7.9 and the weight sums for all DG-CL mappings based on Fig. 7.10A–C are given in Table 7.4 (the minimum weights sum from every CL to three DGs is highlighted). It can be seen that the output of DG_1 (solar PV) is 0, and hence the CLs (7, 8, and 21) should be rearranged to be supplied by another two DGs (as underlined in Table 7.4). Therefore, two AMGs need to be reconfigured at time 00:00, as illustrated in Fig. 7.12A.

7.7.2 Scenario II (time slot 05:00)

Unlike the scenario at time slot 00:00, the total power generation can meet all the demands at 05:00. The SUs are operated in the discharging mode. With DGs and SUs as root nodes, the MSTs based on 05:00 time topology matrix $A(05:00)$ are obtained and shown in Fig. 7.11 and their weights sums are given in Table 7.5. In this case, in total five AMGs are formed, and based on DG_2, DG_3, S_1–S_3, and NLs, “leaves” to individual AMGs are selected and added to maximize the energy utilization, as shown in Fig. 7.12B.

7.8 Simulation experiment and numerical result

This section carries out a set of simulation experiments to assess the performance of the proposed optimal energy dispatch solution and analyzes the numerical result through a comparative study. Three aspects of the network performance, that is, power supply to CLs, DG generation and demand matching, as well as the DG utilization efficiency, are assessed and discussed.

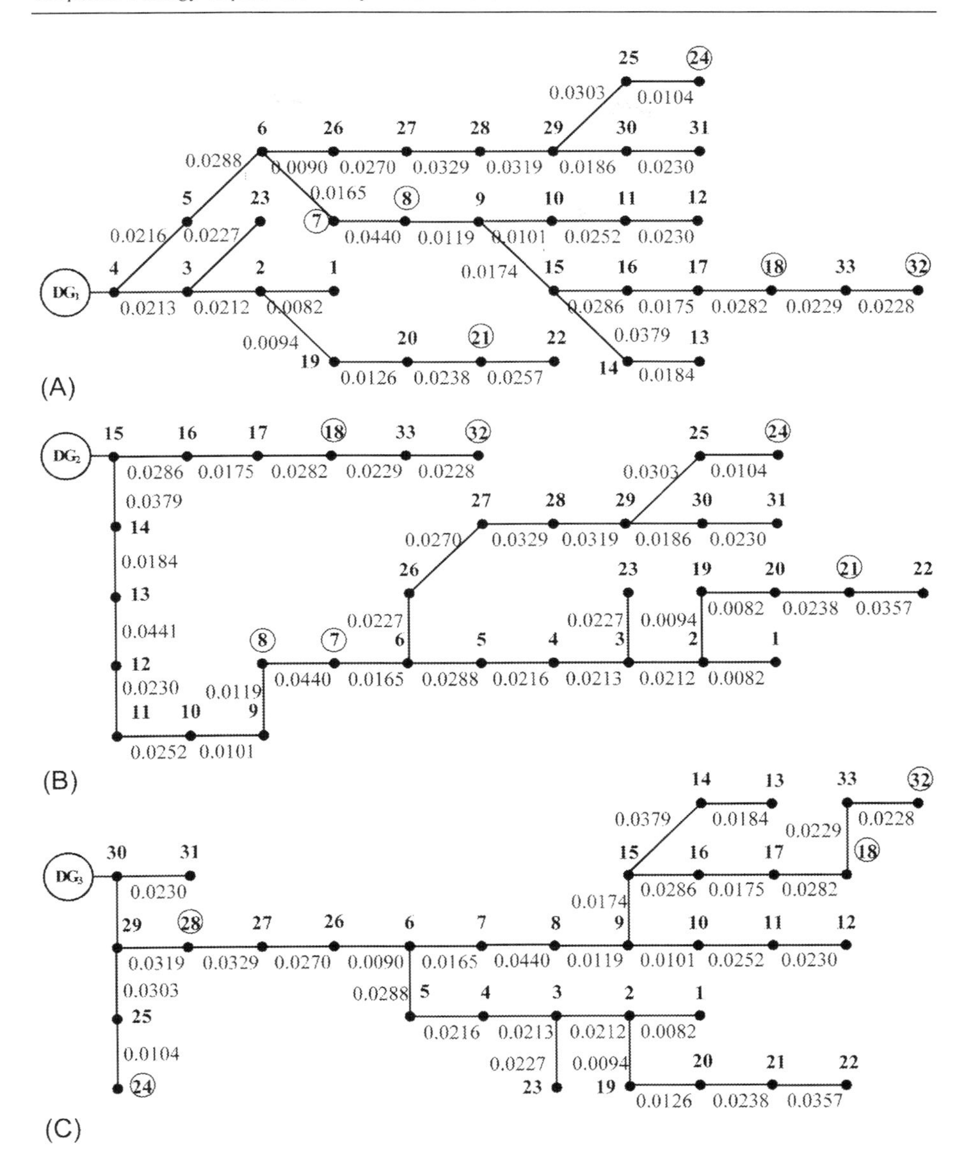

Fig. 7.10 The MSTs at 00:00 with DG1–DG3 (A–C) as the root nodes.

7.8.1 Power supply to CLs

We first look into the performance of power supply to CLs across multiple MGs, by using the IEEE 33-bus network scenario consisting of three AMGs (as shown in Fig. 7.9) as an example. In this case, there are in total one DG, one SU, and two CLs in each AMG. Fig. 7.13 shows that the power supply performance to CLs is

Table 7.4 Weight sums from DGs to CLs.

	DG_1	DG_2	DG_3
7	~~0.0669~~	0.2146	0.1359
8	~~0.1109~~	0.1706	0.1799
18	0.2146	**0.0743**	0.2837
21	~~0.0882~~	0.3697	0.2579
24	0.1918	0.3726	**0.0593**
32	0.2603	**0.1200**	0.3293

The bold underline values given in this table means the minimum weight sums of every minimum spanning tree, i.e., the optimal solutions of every microgrid. The strike values in this table means the output of DG1(solar PV) is 0, and hence the CLs (7, 8, and 21) should be rearranged to be supplied by another two DGs (as underlined in this table)

significantly improved through the adoption of proposed cooperative energy dispatch solution in comparison with the initial autonomous MGs.

7.8.2 Generation and demand match

Now, we evaluate the matching performance of the DG generation and demand (including CLs and NLs) in the simulated scenarios. The matching performance of power generation (DGs and SUs) and demand over a day (May 13, 2014) is shown in Fig. 7.14. It can be seen that even with storages in the electrical network system, for in the absence of reasonable scheduling strategy, the system cannot meet all load requirements in most of the time, as shown in Fig. 7.14A. Even without SUs, the strategy can promise most of the demand in mostly time over the day. In the case that total power generation is insufficient to meet all demand, part of the NLs cannot be supplied, as shown in Fig. 7.14B. Fig. 7.14C shows that the suggested solution can obtain the best matching between power generation and demand over the day. The result also indicates that if the SUs can be installed with appropriate capacity, all demand (CLs and NLs) is expected to be met.

7.8.3 DG utilization efficiency

Due to the fact that the DG utilization efficiency is one of the most important aspects in evaluation of MG planning and economical benefits, hence we assess the DG utilization efficiency based on the proposed optimal energy dispatch solution across multiple autonomous MGs. Fig. 7.15 shows the power utilization of DGs in the case of with and without the proposed energy dispatch solution which clearly demonstrates that with the proposed cooperative dispatch solution, the power energy generated by DGs is more effectively utilized. Through optimal reconfiguration of interconnected AMGs and regulation of charging/discharging operation of SUs, the energy utilization can be significantly promoted in terms of both space and temporal dimensions.

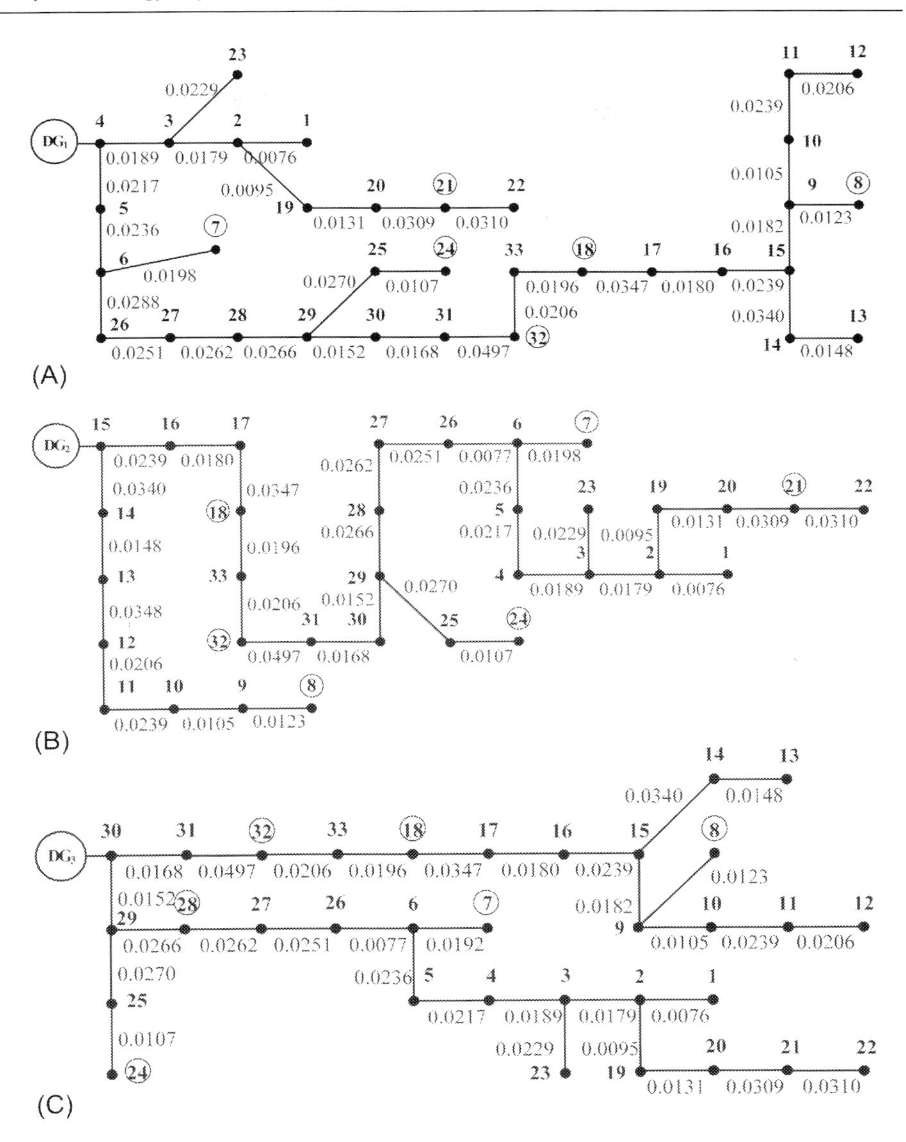

Fig. 7.11 The MSTs at 05:00 with DG1–DG3 (A–C) and S1–S3 (D–F) as the root nodes.

(Continued)

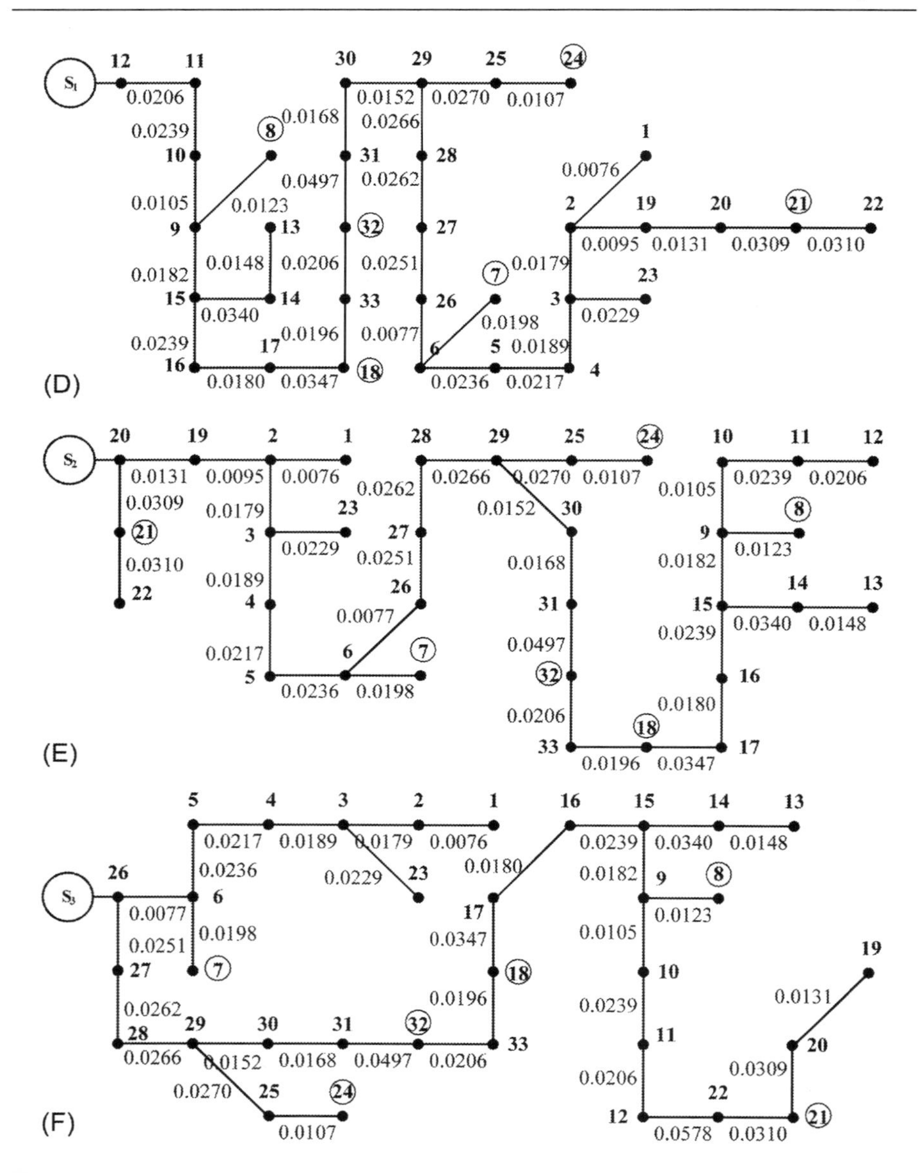

Fig. 7.11, Cont'd

Table 7.5 Weight sums from DGs and Ss to CLs

	DG_1	DG_2	DG_3	S_1	S_2	S_3
7	0.0651	0.3039	0.1206	0.3771	0.1245	**0.0275**
8	0.3599	0.1509	0.2138	**0.0674**	0.4193	0.3069
18	0.2528	**0.0766**	0.1067	0.1498	0.3122	0.1998
21	0.0903	0.4196	0.2364	0.4929	**0.0309**	0.4384
24	0.1686	0.2362	**0.0530**	0.3094	0.2280	0.0826
32	0.2127	0.1167	**0.0666**	0.1899	0.2721	0.1597

The bold values given in this table means the minimum weight sums of every minimum spanning tree, i.e., the optimal solutions of every microgrid.

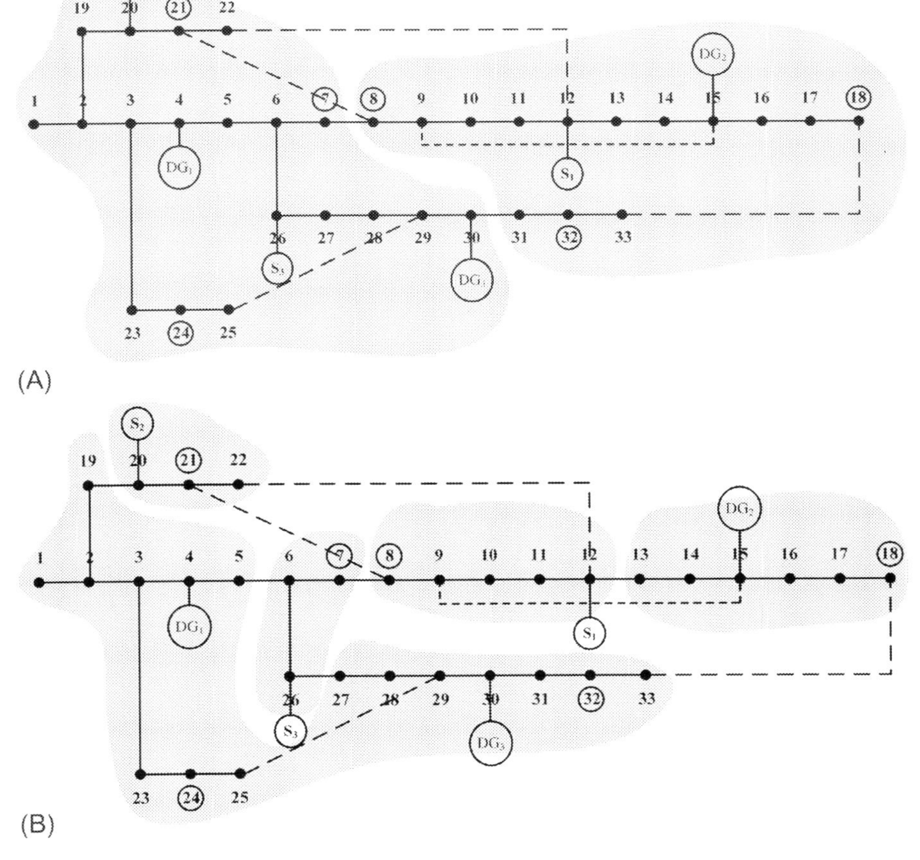

Fig. 7.12 The coordinated energy dispatch in AMGs at (A) 00:00; and (B) 05:00.

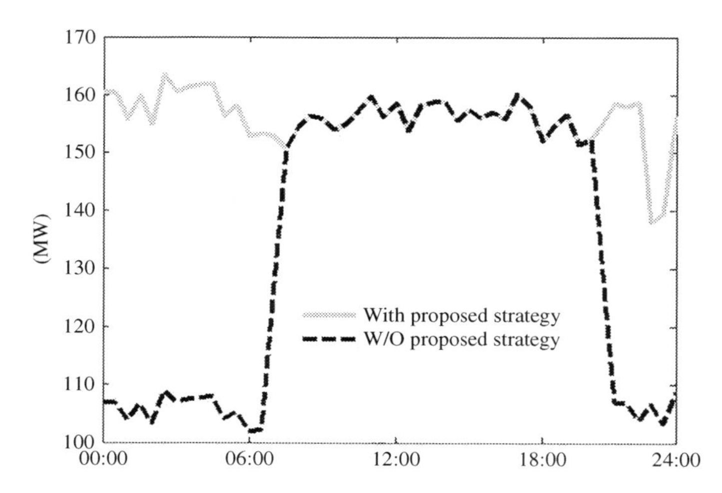

Fig. 7.13 The power supply to CLs (with vs W/O cooperative energy dispatch across AMGs).

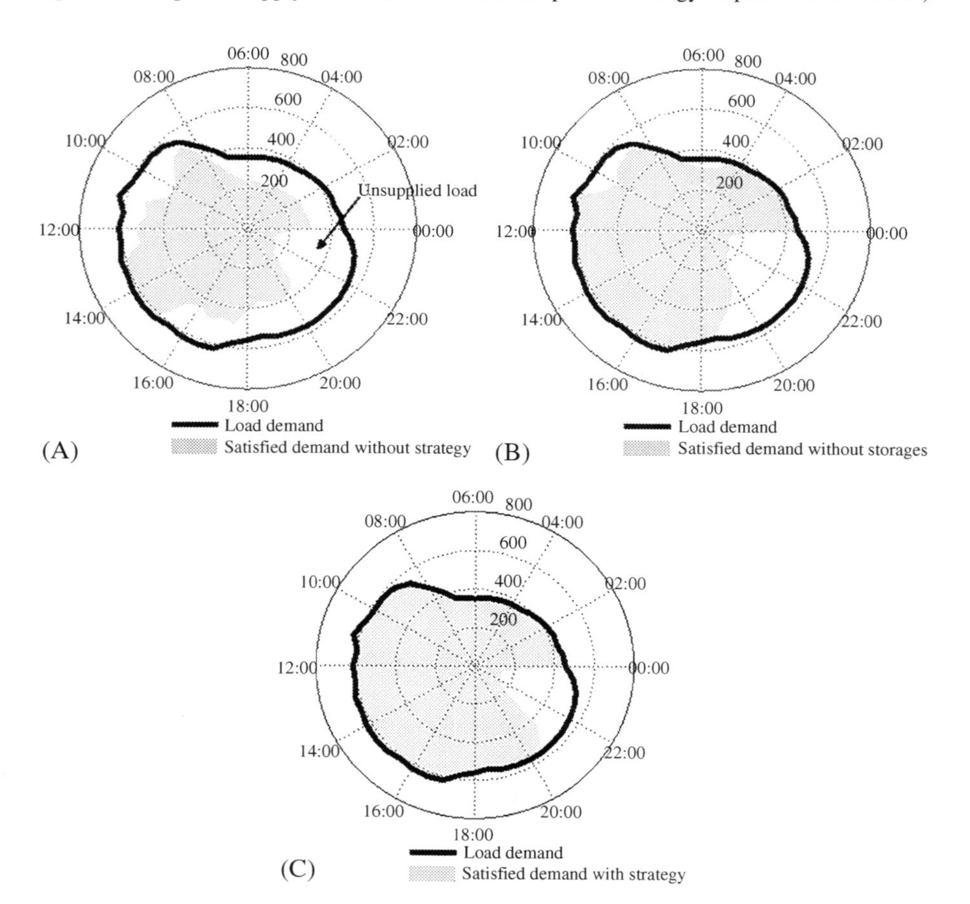

Fig. 7.14 The generation (DGs and SUs) and demand match performance: (A) without cooperative dispatch; (B) cooperative dispatch without SUs; and (C) cooperative dispatch with SUs (proposed solution).

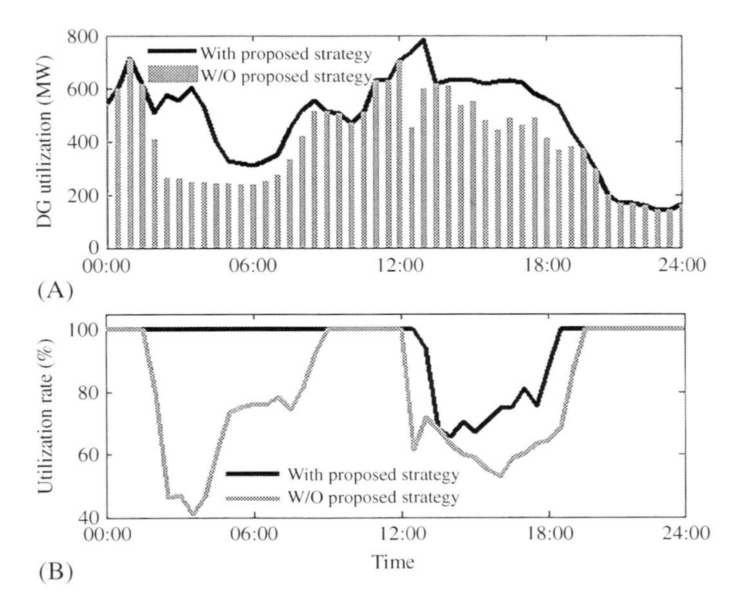

Fig. 7.15 DG utilization efficiency (with vs. W/O cooperative energy dispatch across AMGs).

7.9 Remarks

The network model elaborated in our approach has adopted a combinatory weighted system which includes the power loss and network component reliability statistics into the network model. A fuzzy algorithm to unify different metrics based on fuzzy mathematics theory was presented (Xie et al., 2012). In fact, more performance metrics can be further incorporated into the model covering multidimensional variables, including physical (e.g., topological characteristics), electrical (e.g., real-time network dynamics, i.e., three phase unbalance, shunt capacitors, zero sequence currents, transformer losses, and time-varying power factor), economical (e.g., demand response under real-time pricing), and operational (e.g., maintenance and operational complexity) constraints, if applicable.

In addition, the numerical result indicates that the capacity of distributed SUs can greatly affect the performance of the suggested energy dispatch solution. Sufficient capacity of storages can ensure the surplus energy to be fully stored and when the system energy is not enough, more loads can be supplied. However, excess storage capacity results in low utilization at the price of high purchase and maintenance cost. Therefore, reasonably planning of storage capacity considering the cooperative energy dispatch in multiple MGs can significantly reduce the cost in practical deployment. In addition, an energy dispatch considering the coordination of PEVs, renewable sources, and SUs have been exploited which can well improve the energy utilization efficiency in Wu et al. (2013). In the future work, this solution can be further improved by including different forms of mobile SUs, for example, PEVs, into the energy systems.

Finally, it should be noted that the lifetime of SUs should be protected properly during energy dispatch. In this work, the SOC-based limitations for charging/discharging are predefined to avoid over charging/discharging behaviors during energy dispatch. Future efforts can be made in the algorithmic design to avoid the frequent charging/discharging actions to minimize the deteriorate impact on the SUs.

7.10 Conclusions

This chapter proposed a "stem-leaves" algorithmic solution based on MST and LMI algorithms to implement the cooperative management in multiple autonomous MGs. A set of connected autonomous MGs can be modeled as a weighted matrix by incorporating a collection of operational-related metrics, for example, power loss and supply reliability. Through optimally identifying the mappings between the DGs and load demands in a dynamical fashion, the optimal coordination between DG, distributed storage and demands can be obtained, with significantly improved supply security of CLs, global utilization efficiency of DGs, as well as the generation-demand matching performance. The suggested solution has been evaluated through a set of simulation experiments for a range of network operational scenarios, and the result confirms its effectiveness and benefits.

References

Abdullah, M.A., Agalgaonkar, A.P., Muttaqi, K.M., 2014. Assessment of energy supply and continuity of service in distribution network with renewable distributed generation. Appl. Energy 113, 1015–1026.

Arefifar, S.A., Mohamed, Y.A.-R.I., EL-Fouly, T.H.M., 2013a. Optimum microgrid design for enhancing reliability and supply-security. IEEE Trans. Smart Grid 4, 1567–1575.

Arefifar, S.A., Mohamed, Y.A.-R.I., EL-Fouly, T.H.M., 2013b. Comprehensive operational planning framework for self-healing control actions. IEEE Trans. Power Syst. 28, 4192–4200.

Bhandaria, B., Leea, K., Leeb, C.S., Songc, C., Maskeyd, R.K., Ahna, S., 2014. A novel off-grid hybrid power system comprised of solar photovoltaic, wind, and hydro energy sources. Appl. Energy 133, 236–242.

Caldon, R., Stocco, A., Turri, R., 2008. Feasibility of adaptive intentional islanding operation of electric utility systems with distributed generation. Electr. Power Syst. Res. 78, 2017–2023.

Comodia, G., Giantomassib, A., Severinib, M., Squartinib, S., Ferracutib, F., Fontia, A., Cesarinic, D.N., Morodoc, M., Polonaraa, F., 2015. Multi-apartment residential microgrid with electrical and thermal storage devices: experimental analysis and simulation of energy management strategies. Appl. Energy 137, 854–866.

Diaz, N.L., Dragičević, T., Vasquez, J.C., Guerrero, J.M., 2014. Fuzzy-logic-based gain-scheduling control for state-of-charge balance of distributed energy storage systems for DC microgrids. Proceedings of the 29th Annual IEEE Applied Power Electronics Conference and Exposition. pp. 2171–2176.

Dou, C., Liu, B., 2013. Multi-agent based hierarchical hybrid control for smart microgrid. IEEE Trans. Smart Grid 4, 771–778.

Fathima, A.H., Palanisamy, K., 2015. Optimization in microgrids with hybrid energy systems: a review. Renew. Sust. Energ. Rev. 45, 431–446.

Gouveia, J.P., Dias, L., Martins, I., Seixa, J., 2014. Effects of renewables penetration on the security of Portuguese electricity supply. Appl. Energy 123, 438–447.

Graham, R.L., Hell, P., 1985. On the history of the minimum spanning tree problem. IEEE Ann. Hist. Comput. 7, 43–57.

Hawkes, A.D., Leach, M.A., 2009. Modelling high level system design and unit commitment for a microgrid. Appl. Energy 86, 1253–1265.

Hernandez, L., Baladron, C., Aguiar, J.M., Carro, B., Sanchez-Esguevillas, A., Lloret, J., Chinarro, D., Gomez-Sanz, J.J., Cook, D., 2013. A multi-agent system architecture for smart grid management and forecasting of energy demand in virtual power plants. IEEE Commun. Mag. 51, 106–113.

Javadian, S.A.M., Haghifam, M.R., Firoozabad, M.F., Bathaee, S.M.T., 2013. Analysis of protection system's risk in distribution networks with DG. Int. J. Electr. Power Energy Syst. 44, 688–695.

Jikeng, L., Xudong, W., Peng, W., Shengwen, L., Guanghui, S., Xin, M., Xingwei, X., Shanshan, L., 2012. Two-stage method for optimal island partition of distribution system with distributed generations. IET Gener. Transm. Distrib. 6, 218–225.

Karavas, C., Kyriakarakos, G., Arvanitis, K.G., Papadakis, G., 2015. A multi-agent decentralized energy management system based on distributed intelligence for the design and control of autonomous polygeneration microgrids. Energy Convers. Manag. 103, 166–179.

Kuznetsova, E., Li, Y., Ruiz, C., Zio, E., 2014. An integrated framework of agent-based modeling and robust optimization for microgrid energy management. Appl. Energy 129, 251–261.

Li, W.Y., Zhou, J.Q., Xie, K.G., Xiong, X.F., 2008. Power system risk assessment using a hybrid method of fuzzy set and Monte Carlo simulation. IEEE Trans. Power Syst. 23, 336–343.

Li, X.R., Jiang, X.J., Qian, J., Chen, H.H., Song, J.Y., Huang, L.G., 2015. A classifying and synthesizing method of power consumer industry based on the daily load profile. Autom. Electr. Power Syst. 34, 56–61.

Obara, S., 2015. Dynamic-characteristics analysis of an independent microgrid consisting of a SOFC triple combined cycle power generation system and large-scale photovoltaics. Appl. Energy 141, 19–31.

Pascual, J., Barricarte, J., Sanchis, P., Marroyo, L., 2015. Energy management strategy for a renewable-based residential microgrid with generation and demand forecasting. Appl. Energy 158, 12–25.

Ref22: n.d. Power systems test case achieve [online] http://www.ee.washington.edu/research/pstca/

Ref23: n.d. Belgian electricity transmission operator Elias [online] http://www.elia.be/en/about-elia

Schroeder, A., 2011. Modeling storage and demand management in power distribution grids. Appl. Energy 88, 4700–4712.

Silvente, J., Kopanos, G.M., Pistikopoulos, E.N., Espuña, A., 2015. A rolling horizon optimization framework for the simultaneous energy supply and demand planning in microgrids. Appl. Energy 155, 485–501.

Tenti, P., Costabeber, A., Mattavelli, P., Trombetti, D., 2012. Distribution loss minimization by token ring control of power electronic interfaces in residential microgrids. IEEE Trans. Ind. Electron. 59, 3817–3826.

Velik, R., Nicolay, P., 2015. Grid-price-dependent energy management in microgrids using a modified simulated annealing triple-optimizer. Appl. Energy 130, 384–395.

Wang, C.S., Liu, M.X., Guo, L., 2012. Cooperative operation and optimal design for islanded microgrid. Proceedings of IEEE International Conference on Innovative Smart Grid Technologies. pp. 1–8.

Wu, T., Yang, Q., Bao, Z., Yan, W., 2013. Coordinated energy dispatching in microgrid with wind power generation and plug-in electric vehicles. IEEE Trans. Smart Grid 4, 1453–1463.

Xie, C., Dong, D., Hua, S., Xu, X., Chen, Y., 2012. Safety evaluation of smart grid based on AHP-Entropy method. Syst Eng Procedia 2012. vol. 4. pp. 203–209.

Zhao, B., Xue, M., Zhang, X., Wang, C., Zhao, J., 2015. An MAS based energy management system for a stand-alone microgrid at high altitude. Appl. Energy 143, 251–261.

Part Two

ICT technologies for smart power distribution networks

Privacy of energy consumption data of a household in a smart grid

Sandhya Armoogum, Vandana Bassoo*[†]
*Department of Industrial Systems Engineering, School of Innovative Technologies and Engineering, University of Technology Mauritius, La Tour Koenig, Pointe-aux-Sables, Mauritius, [†]Department of Electrical and Electronic Engineering, Faculty of Engineering, University of Mauritius, Reduit, Mauritius

8.1 Introduction

The conventional electricity grid was developed over a century ago and consisted mainly of power generation plants and distribution systems (US Department of Energy, 2017). Customers were provided electricity and they were billed once a month. Such traditional electricity grid is characterized by one-way communication and the lack of interaction between the customer and the utility provider results in various problems such as energy loss and poor peak load management (Mahmood et al., 2015). Moreover, during the past decade, there has been an increase in the exploitation of renewable energy generation resources, which has proven to be challenging to manage because of the fluctuating and sporadic nature of those resources. Therefore, the complex energy demands of the modern world require a smart grid which can satisfy the capacity, provide reliable supply, and ensure the effective integration of renewable power generation sources.

8.2 Smart grid and its many benefits

Energy demand is not stable. At some point in time the demand can be too high, resulting in grid disruption and blackouts. Such power outages are very inconvenient for users and can also be costly to energy providers as they may lose clients to other companies. In view of satisfying the demand, mathematical algorithms are used to predict power consumption increases, so that corrective actions can be taken. However, it can also happen that more energy is generated than consumed, which results in energy wastes.

Smart grids involve the use of technologies that supports two-way communication between the utility supplier and its customers, for real-time monitoring and near-instantaneous balance of supply and demand of energy (Wang and Lu, 2013; Mahmood et al., 2015). It coordinates the needs and capabilities to operate the power grid as effectively as possible by lowering costs while boosting the reliability, stability, and resilience of the system (Ye et al., 2013). For instance, smart meters in the grid, enables the utility company to quickly and accurately track and record electricity

Smart Power Distribution Systems. https://doi.org/10.1016/B978-0-12-812154-2.00008-0

usage for better decision making which improves the reliability of power distribution. On the other hand, customers can collect information on their energy use to better adjust their habits to lower electricity bills, for example, a customer can choose to plug in an electric vehicle (EV) and program it to charge during off-peak hours.

Moreover, smart grids play a key role in minimizing environmental impacts. The current tendencies in energy supply are financially, ecologically, and socially unsustainable. Without crucial action, the increasing demand of fossil fuels will intensify concerns over the availability of supplies and will lead to the energy-related emissions of carbon dioxide (CO_2) to skyrocket by 2050 (Energy, 2011). Fortunately, smart grids enable the integration of renewable energies, and a more efficient distribution of energy may result in preventing the need for more power plants which would eventually add to pollution. A more efficient management of energy via smart grids can also help to reduce CO_2 emissions from the existing power plants.

Key technologies of a smart grid include integrated two-way communication between customers and the utility supplier, advanced components (e.g., excess electricity storage, smart devices, and diagnostics equipment), advanced control methods (which enable sophisticated data collection, diagnostics, and appropriate maintenance), sensing and measurement technologies (e.g., advanced metering infrastructure (AMI)), improved human and machine interfaces (HMI), and decision support (Komninos et al., 2014). Fig. 8.1 depicts the seven domains within the smart grid as identified by NIST (2014) and the communication within the smart grid.

From a network architecture perspective, the smart grid consists of several layers as follows (Kuzlu et al., 2014):

(1) a power system layer which refers to electricity generation, transmission, and distribution;
(2) a power control layer for monitoring and management of power in the smart grid;
(3) a communication layer which supports two-way communications between the customer and the utility supplier;
(4) a security layer for implementing security services such as authentication of customer, confidentiality and integrity of information transmitted, etc.; and
(5) an application layer which provides smart grid applications for customers to be able to view and control their energy usage online.

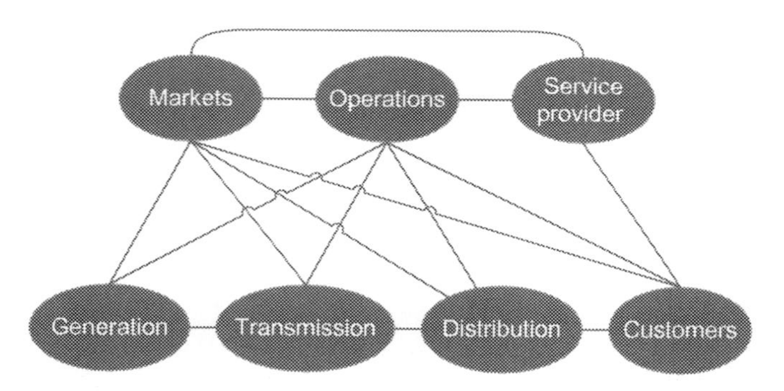

Fig. 8.1 Seven domains within the smart grid and their interaction.

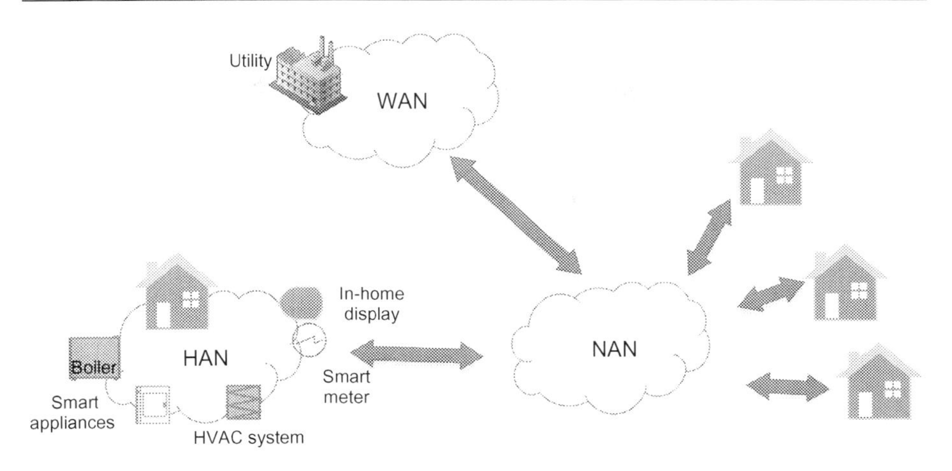

Fig. 8.2 Typical network architecture of a smart grid.

Applications for utility providers include pricing applications for sharing price information with customers, demand response applications for negotiating with customer for load reduction during peak demand times, and outage and restoration management (ORM) applications which allows an electric utility to detect power outages.

The communication layer is fundamental as it allows the transmission of data, which enables proper monitoring in the smart grid. The communication network in a smart grid consists of a hierarchical multilayer architecture, which uses a variety of communication protocols as shown in Fig. 8.2. The smart meter in the house communicates with the smart gas meter and other smart electric appliances such as plug-in hybrid electric vehicles (PHEVs) to collect readings, and send data to the in-home display (IHD), computer or smart phone. This small network constitutes the home area network (HAN) and it often uses short-range communication protocols such as ZigBee, Wi-Fi, ZWave, power-line carrier (PLC), Thread, and Bluetooth (Sabine et al., 2015). The smart meters are configured to send data collected to the closest concentrator node called the data aggregation point (DAP), which constitute the neighborhood area network (NAN), using such communication technologies as ZigBee mesh networks, Wi-Fi mesh networks, PLC, as well as long distance wired and wireless technologies, such as WiMAX, cellular M2M, digital subscriber line (DSL), and coaxial cable. Subsequently, the DAPs from the NAN communicate with the backend system of the utility data centers using a wireless area network (WAN). The WAN often uses WiMAX, 3G/GSM/LTE, or fiber optics for communication.

8.3 Security vulnerabilities of smart grid and its impact

Given the numerous components that make up the smart grid, the latter is very large and complex. Thus, the computing and communications technologies as well as the different smart grid components that are used to operate, control, and monitor smart grid is also vulnerable to a multitude of attacks. If the smart grid is under attack, consumers as well as the utility supplier may encounter inconvenience and even danger, for example, power outages during the depths of winter can cause loss of human lives. Such cyber-physical attack is illustrated by the recent blackout in Ukraine in 2015, 2016, and in January 2017 where hackers shutdown energy systems supplying heat and light to millions of homes (BBCNews, 2017; Polityuk, 2016). Similarly, in October 2016, a distributed denial of service (DDoS) attack disabled computers, which control heating systems of two apartment buildings in eastern Finland, thereby leaving residents in the cold for about a week (Kumar, 2016). Electricity shortages due to cyberattacks have also been reported in January 2016 in Israel (2016) and in December 2016 in Turkey (Ankara/Kocaeli, 2016).

Because smart grids allow to better balance electricity produced as well as integrate renewable resources, and it also allows consumers to monitor their energy consumption and thus reducing cost, many countries are adopting smart grid technologies. The European Union (EU) plans to have 200 million electricity smart meters deployed by 2020; the UK government wants to install 53 million smart meters by 2020 while the United States plans to roll 90 million smart meters (Cooper, 2016). Each of these smart meters can become an attack point and if hacked the consequences can be disastrous.

Smart meters are an essential component of smart grids. They support two-way communication where electricity and information can be exchanged between the utility company and the customers. Basically, a smart meter is an electrical meter, which registers usage at regular intervals and sends the information to the utility company for monitoring and invoicing purposes (Fang et al., 2012a,b). It also has the capability to connect, disconnect, and control smart home appliances remotely. The use of smart meters offers a number of potential benefits to both the consumers and the utility companies (Li et al., 2010). However, smart meters pose significant security threats to the grid. The widespread deployment of smart meters provides a high number of targets for malicious hackers and it is nearly impossible for a utility company to ensure that no smart meter is compromised. Moreover, most of these malicious attacks can be launched by remote computers. In the traditional grid, there were only a few critical assets and hence it was easier for the utility company to defend and protect the grid.

Hackers can compromise smart meters by fabricating false energy readings and therefore manipulating the electricity bill. Moreover, the injection of wrong information can mislead the utility company to take wrong decisions regarding the electricity usage and capacity in an area/region (Fang et al., 2012a,b). Hackers can also launch widespread cyberattacks on smart meters to cripple the electricity grid resulting in major disasters.

Historically, critical infrastructures such as power grid were physically isolated. However, due to the benefit of significant automation in operation and control of critical infrastructure systems, information and communication technologies have been integrated, resulting in smart grids becoming more accessible to attackers. According to the NIST Guidelines for Cybersecurity, denial of service (DoS) and DDoS attacks, which can disrupt the operation and reliability in the smart grid is an important concern due to the networking aspect of the smart grid (NIST, 2014). Such an attack can block access to the grid components, and information transmitted and collected, resulting in power failure and outages, which can have severe consequences in society. Smart grids are a typical example of cyber-physical system (CPS), which integrate computing and communication capabilities with monitoring and control of entities in the physical world. The emergence of malware such as Stuxnet (Vijayan, 2010), which operates on programmable logic controllers in CPS is of great concern for such critical infrastructure as the power grid. Similarly, if smart grid components such as the smart meter are not secured properly, it can be exploited to launch large-scale DDoS attacks as in the case of the Mirai bot where unsecured IoT devices such as CCTV cameras were used.

Security and privacy concerns must also be addressed such that citizens can embrace the smart grid benefits faster. Results from an Austrian survey (Döbelt et al., 2015), indicate that information transparency about the data capture, sharing and use in smart grid as well as privacy-preserving solutions are required for citizen's adoption of smart meters in their homes.

8.4 Security objectives of smart grid

Three main security objectives for the smart grid system as identified in Aloul et al. (2012) and NIST (2014) are to ensure:

(1) uninterrupted power supply to satisfy user demands—availability,
(2) the integrity of information, such as customer's usage data as well as account-related data, and
(3) confidentiality of consumer's information.

Although each node in the smart grid can be vulnerable to exploitation, the "last mile"—from the smart meter to the data collector, is considered the weakest link (Hayden, 2009) especially given that it supports two-way communication. If there are no proper security measures in place, it can be possible for an attacker to switch off a smart meter resulting in no power in the house (attack on availability). Similarly, an attacker can tamper with a customer's data in the grid (attack on integrity). For instance, illegal modifications of the smart meter data can result in service theft thereby reducing the bill of the customer. Furthermore, such tampering, in the case of customers selling energy (i.e., from solar panels) to the utility provider can result in a financial gain for the malicious customer. Sensor data manipulation can also give a fake view of the system and prevent proper monitoring of the entire system. Utility companies usually collect and store customer information such as name, address, national identity number, power consumption, credit cards details (for customers paying their bills online), etc. Such information should be secured such that unauthorized parties cannot access the user data resulting in a breach in confidentiality (Döbelt et al., 2015). Many existing techniques for smart grid are focused around these three security objectives.

Smart meters collect massive amount of sensitive information on households thereby raising justified privacy concerns (Sankar et al., 2012). Such information is transmitted to the DAP for onwards transmission to and storage by the utility provider. Malicious parties may be able to infer personal information such as habits and activities by analyzing the data gathered by the smart meters (NIST, 2014). For example, by accessing the fine granular information about power usage of a user, an attacker using nonintrusive load monitoring (NILM) algorithms can, among others: (1) estimate the number and types of electrical appliances based on the signatures of individual appliances; (2) estimate the composition of a home (number of inhabitants based on the energy usage pattern); (3) gather behavioral knowledge such as when the user is at home, when the user leaves for work and returns home or when the user is on vacation which can allow criminals activities; (4) infer what the user is doing when home (real-time surveillance); (5) perform location tracking based on EV usage patterns; and (6) gather personally identifiable information (PII) about users.

The privacy issue is exacerbated further since often utility providers contract out third-party cloud service providers for the storage, management, and analytics of consumer data (Fang et al., 2012a,b). Through the smart meters, utility companies can intervene by remotely controlling smart appliances within a home. This is also perceived as a threat to privacy, as it allows an external entity (which can also be malicious parties) to reach into the home. The increased risks to privacy in smart grids are likely to provoke opposition across age groups and other demographic characteristics and thus consumer privacy concerns must be addressed (Horne et al., 2015).

Privacy of data can be compromised at different places in the smart grid architecture as follows:

- Smart meters collect data in households. Usually, users can interact with their smart meters at home. However, if a malicious party is also able to interact with the smart meter then the attacker will be able to access private and confidential data of the user. Thus, strong and reliable authentication is often required to ensure that only the authorized user can interact with the smart meter. Similarly, authentication mechanisms can prevent attackers from interacting with other smart appliances in the house and in the smart grid.
- Data collected by the smart meters are transmitted to the DAP which then send the data to the control center of the utility provider. Communications between utility provider and DAPs are usually kept secure due to the implementation of the supervisory control and data acquisition (SCADA) system in the core network. On the other hand, communication between smart meters and DAPs needs to be secured (Vaidya et al., 2013). Strong mutual authentication, confidentiality, and privacy protection of data are inevitable between smart meter and the DAP.
- Breach in confidentiality and privacy of data can also occur in the control center where the utility company store collected customer information such as name, address, billing and payment information, and energy consumption data, if proper security measures are not taken.
- Smart grid technologies also allow consumers to get detailed information from web applications, mobile applications as well as social networking applications through the internet or a cloud service. Such applications aim to provide real-time usage statistics/profiling and dynamic pricing information together with recommendations for decreasing the utility bill, such as running the washing machine during off peak hours as opposed to in the evening when all residents are at home. A breach in security in such applications can compromise the confidentiality and privacy of consumer data.

In the remaining part of the chapter, we focus on the confidentiality and privacy of data collected and transmitted by the smart meters to the DAP. It is important that in a smart grid, no sensitive information about specific individuals is disclosed. When there is information disclosure, the disclosure can be of two types: (1) identity disclosure and (2) attribute disclosure. Identity disclosure occurs when an attacker, based on the information obtained, can associate the data with the individual for whom the data belongs. Attribute disclosure is mainly considered as a breach in confidentiality, whereby the attacker gets access to data, which is unauthorized and confidential. The two types of disclosure are independent. Thus, even if data confidentiality is implemented, identity disclosure, thus breach in privacy, may still be possible.

Smart meters are expected to send meter readings periodically after a short interval ranging from milliseconds to minutes. Data transmission from smart meters could have been sent in encrypted form using symmetric encryption with a key shared between the smart meter and the utility provider, for confidentiality and privacy. However, smart grids must capture, verify, transmit, and store large volume of high frequency metering data in near real time, which is no easy task. To address these challenges, in the smart grid, DAPs are used to aggregate data from a neighborhood before sending the data to the control center of the utility provider—smart meters may also aggregate readings from several smart appliances before sending data to the DAP in the NAN. Data aggregation decreases the packet traffic in the network, but data aggregation cannot be done normally on traditionally encrypted data. Data encrypted by the smart meter, must be decrypted by the DAP for aggregation with data from neighboring smart meters, which is not performance efficient and introduces security weaknesses in the system. Thus, traditional cryptographic techniques such as symmetric cryptography and public key cryptography are not suitable and other mechanisms are required to address the problem of confidentiality and privacy of data.

8.5 Privacy preserving techniques in smart grids

In this section, we present the different approaches that have been proposed for securing the privacy of consumer data in smart grids.

8.5.1 Data anonymization and perturbation

One general approach to provide privacy of data is through data anonymization, which aims to remove or hide/mask the PII from the data which may lead to user identification, that is, smart meters do not send the original data set X, but a modified version Y of it. Often the simple masking or removal of identifiers such as social security number (SSN) of a customer may not be enough to ensure privacy (Latanya, 2000). The combination of other identifiers/attributes of the data (e.g., customer's birth of date, gender) can also uniquely allow identification of individuals thereby compromising privacy (Ganti et al., 2011). Thus, when masking data for anonymization, the attributes to be masked should be chosen with caution.

The masking techniques can be either perturbative masking or non-perturbative masking. The non-perturbative masking method does not change the data; rather, it reduces the amount of information by suppressing some of the data or reducing the level of detail, thereby preserving the truthfulness of the data despite loss of data.

The perturbative masking method, also known as data perturbation, involves masking data by the addition of some random noise, including the uniform and Gaussian noise, which distorts the original data with the intent of hiding the customer's attributes. However, it has been found that it may be possible to separate the real data and the injected noise in the perturbed dataset by studying the spectral properties of the data and hence compromise privacy (Kargupta and Datta, 2003). Some of the popular models for data anonymization include k-anonymity (Latanya and Pierangela, 1998), l-diversity (Machanavajjhala et al., 2006), and t-closeness (Li et al., 2007).

The use of data anonymization for preserving privacy has been proposed for smart grids (Efthymiou and Kalogridis, 2010; Rottondi et al., 2012). Data anonymization is effective for large datasets and should not be used for small ones. Unfortunately, data anonymization such as random-data perturbation techniques does not fully preserve privacy (Kargupta et al., 2005). There exist data de-anonymization techniques such as linkage attack with auxiliary information which have proven to be successful by correctly identifying members of a Netflix anonymized dataset (Narayanan and Shmatikov, 2008). Similarly, de-pseudonymization can be done on metering data (Jawurek et al., 2011). In Ács and Castelluccia (2011), however, the authors, using the new distributed Laplacian perturbation algorithm (DLPA), adopt the data perturbation approach to add noise (random values) to the data before encrypting the data such that a utility company can collect data from smart meters, and compute aggregated values in such a way that the data does not divulge privacy-related information such as the individual households' activities.

8.5.2 Data aggregation and homomorphic encryption

Another approach for preserving privacy involves the spatial aggregation of data. To optimize the use of bandwidth, in wireless sensor networks, the nodes aggregate data before sending it to a data sink. For example, instead of sending temperature captured by individual sensors in an area, the average temperature in that area is computed and transmitted to the sink. This effectively eliminates redundant data, reduces the transmission bandwidth, and saves energy. A similar approach is adopted in smart grids regarding the collection and transmission of customer's data. In smart grids, data captured from different households' smart meters may be aggregated such that an attacker cannot determine the energy consumption of a specific household from the total. Similarly, the temporal aggregation approach can be used whereby, data collected over a period are aggregated by the smart meter or DAP before onwards transmission to the utility center. Data aggregation can provide some level of privacy; however, it may also result in information loss.

To further preserve customer usage and billing data, privacy-preserving data aggregations have been developed which relies on cryptographic algorithms like homomorphic encryption (Lagendijk et al., 2013; Tonyali et al., 2015, 2016; Rottondi et al., 2013a,b). A homomorphic encryption scheme is a cryptosystem, which allows computations to be performed on the encrypted data (no need to decrypt the data to process it). The result of such computations when decrypted produces the same value as when performing the mathematical operation on the unencrypted data. There are currently two types of homomorphic encryption: (1) partially homomorphic encryption (PHE) and (2) fully homomorphic encryption (FHE) (Gentry, 2009). With PHE, only a single mathematical operation can be performed on the ciphertext, either addition or multiplication. The two most popular PHE schemes that are usable today are the Paillier cryptosystem (Paillier, 1999) and the ElGamal cryptosystem (Morris, 2013). However, given that the ElGamal encryption scheme produces a 2:1 expansion in size from plaintext to ciphertext (Yi et al., 2014), the Paillier cryptosystem is mostly used for data aggregation in smart grids. Both the Paillier and ElGamal cryptosystem support addition and multiplication operations. FHE schemes can support arbitrary computation on ciphertext and can have a powerful impact for preserving privacy in many areas as then all computation can be done on encrypted data without revealing private information. However, FHE schemes are currently experimental. Existing FHE implementations are complex, and require immense computational power and relatively long time to perform tasks that can be done simply on the unencrypted data. Although, overtime, FHE schemes are being improved considerably.

Thus, data are encrypted before being transmitted to the DAP in the NAN, the DAP can aggregate the different captured data in the encrypted form using the Paillier additive homomorphic encryption scheme (González-Manzano et al., 2016; Baktir, 2014; Ozgur et al., 2016). The combination of aggregation and encryption enhances the privacy of the data. Likewise, Garcia and Jacobs devised a protocol based on homomorphic encryption (additive Paillier for aggregation) and additive secret sharing to securely handle the key management in such a system (Garcia and Jacobs, 2010).

A similar solution based on homomorphic encryption is proposed in Erkin and Tsudik (2012) that allow the aggregation of several smart meter readings from a single smart meter overtime (temporal aggregation) as well as the aggregation of smart meter readings for a group of neighboring customers (spatial aggregation) without the need for a trusted third party (TTP) (online aggregators). In Li et al. (2014), the authors propose a distributed and incremental data aggregation by defining an aggregation route (aggregation tree) across a neighborhood, where the smart meters aggregate the data sent to them with their own, using homomorphic encryption—namely the Paillier cryptosystem, without being able to read any intermediate or final result of the aggregation.

By exploiting the homomorphic properties of the Shamir's secret-sharing scheme, Rottondi et al. (2013a,b) proposed an infrastructure based on trusted third-party nodes called privacy preserving nodes (PPNs), which collect encrypted measurement data from smart meters and perform different spatial and temporal aggregation such that individual measurements cannot be obtained. The aggregated data are then sent to the utility provider. The secret sharing algorithm works by dividing a secret into shares that are distributed among a group of parties, in this case, meter readings (secrets) shared with different PPNs. One party cannot reconstruct the secret, unless it possesses at least a certain number of shares.

A few solutions based on FHE have also been proposed (Zhang and Chen, 2012; Tonyali et al., 2017). Moreover, the privacy-preserving aggregation protocol for the Internet of Things (PAgIoT) (González-Manzano et al., 2016), also uses the Paillier cryptosystem and can be adapted in smart grids.

Furthermore, in Kursawe et al. (2011), using data masking techniques together with Diffie-Hellman and a Bilinear-map based protocol instead of homomorphic encryption, it was shown that smart meter readings can be hidden in such a way that an attacker is not able to discover the correct meter measurements. Two protocols are defined to enable the data masking: an aggregation protocol and a comparison protocol. Both protocols are used to generate the aggregate smart meter readings. This mechanism also allows to detect fraud (i.e., if a customer is tampering with their meter measurements) and leakage without disclosing any additional data about the individual meter measurements.

8.5.3 Data aggregation and differential privacy

Differential privacy is another popular approach that provides strong privacy guarantees. Differential privacy was originally meant to provide privacy in an interactive setting, where queries are submitted to a database containing the original individual records. However, it can be applied to data in general. Differential privacy-preserving technique (Dwork, 2006) is used to give a guarantee to a data owner that the privacy of the owner will not be affected when they share their data, that is, a person is only affected by the conclusion of the analysis, not by the person's participation (Dwork, 2011). For instance, let us assume that the analysis of a dataset shows that drinking causes liver disorders. Jane, who is a drinker, is affected by the result of this analysis as her medical insurer can decide to increase her premiums. Jane will be affected by this analysis whether the dataset contains her data, given that she is impaired by the fact that "drinking causes liver disorder" and not necessarily by her sharing of the data. This can be demonstrated by considering if a given result change when a new data record is added or an existing data record is deleted in the dataset. In her research (Dwork, 2011), Cynthia Dwork has shown that the outcome will not significantly change due to one record being added or deleted provided that the dataset is big. This said, differential privacy provides inaccurate results for small populations and should be used when there already exists a big dataset of information.

In the context of smart grid, differential privacy implies that an attacker cannot learn anything regarding the appliances being used by the consumer from the aggregated power consumption data provided by the smart meter, which may be accessible to the attacker through eavesdropping. When the smart meter reading from one consumer is aggregated with others by the DAP, an attacker learns approximately the same information about any consumer, regardless of the presence or absence of the consumer's original data in the aggregated data. In Barbosa et al. (2016), the smart meter performs the data aggregation by adding the power consumption data from different smart appliances together with noise before sending data to the utility provider. Moreover, by generating the noise from a probability distribution such as Laplace distribution, the authors show that by varying the cumulative density function of the normal distribution, the level of privacy achieved can be controlled. The customer can still analyze his power consumption details inside his home since only data sent to the utility provider is masked.

8.5.4 Privacy-preserving using aggregation with a TTP

In Efthymiou and Kalogridis (2010), the authors propose a privacy-preserving technique, which relies on a TTP for aggregation of smart meter reading. Smart meter reading, which is not for billing purposes, does not have to be linked to a user. The utility provider only needs to be able to assess the location from which data have been aggregated for workload balancing and statistical analysis. Using a TTP, data from customers can be collected and aggregated such that the data are no longer linked to a customer.

8.5.5 Rechargeable batteries for masking and obfuscation of metering data

In Backes and Meiser (2014) and Kalogridis et al. (2010), the authors propose the use of the battery-based load hiding (BLH) approach for privacy instead of noise addition to guarantee differential privacy. The aim is to use a controllable and rechargeable battery in each home, which can be made to charge and discharge at strategic times to flatten the energy consumption of the household. BLH thus hide the real energy profile of the consumer. Zhao et al. (2014) further proposes an improved BLH algorithm to control the battery. This further obfuscates the actual consumption profile of the consumer. Instead of controlling the battery to flatten the power consumption, Egarter et al. propose a novel obfuscation technique, called load-based hiding (LLH), which fluctuate the power drawn by controlling energy-intensive household loads which are not driven by the user and which is used daily (e.g., water boiler) (Egarter et al., 2014). They show that the BLH technique works better for a home with mainly small appliances, while the LLH was more suitable at mitigating privacy leakage in household having appliances with high-energy consumption. Given that EVs are becoming more popular, it has also been proposed to use an EV-assisted battery load-hiding algorithm, which combines the use of an EV with a rechargeable battery, for obfuscation meter measurements as well as optimal charging (Sun et al., 2015). This solution is also more cost effective as compared with BLH, which relies only on the batteries. However, according to McLaughlin et al. (2011), the cost of equipping each home with the required amount of rechargeable batteries is not negligible. Moreover, the batteries should be replaced after every two to three years, and they can also have a consequential impact on the environment.

8.6 Conclusions

In this modern world, smart grid is the natural progression of the power grid, offering a promising future to the customers, utility companies, and the environment. The efficiency, reliability, and security of the smart grid are of utmost importance. The smart grid consists of a high number of devices spread over large geographical areas, which are recording, transmitting, or processing sensitive data. All these operations lead to complex security and privacy challenges.

In this chapter, the security vulnerabilities of the smart grid are examined. Focus is laid on the confidentiality and privacy of data collected and transmitted by the smart meters to the DAP. Five techniques to address or minimize confidentiality and privacy breaches are examined. Security in smart grids is an evolving topic and still has many open research challenges.

References

Ács, G., Castelluccia, C., 2011. I have a DREAM!: differentially private smart metering.Proceedings of the 13th International Conference on Information Hiding (IH'11), Prague, Czech Republic.

Aloul, F., et al., 2012. Smart grid security: threats, vulnerabilities and solutions. Int. J. Smart Grid Clean Energy 1 (1), 1–6.

Ankara/Kocaeli, 2016. Major cyber-attack on Turkish energy ministry claimed. [online]. Available from: http://www.hurriyetdailynews.com/major-cyber-attack-on-turkish-energy-ministry-claimed.aspx?PageID=238&NID=107981&NewsCatID=348.

Backes, M., Meiser, S., 2014. Differentially private smart metering with battery recharging. In: Data Privacy Management and Autonomous Spontaneous Security. Lecture Notes in Computer Science Book Series (LNCS), vol. 8247. Springer, Berlin, Heidelberg.

Baktir, S., 2014. Privacy preserving smart grid management in the cloud. Proceedings of the International Conference on IT Convergence and Security (ICITCS), Beijing, China.

Barbosa, P., Brito, A., Almeida, H., 2016. A technique to provide differential privacy for appliance usage in smart metering. Inf. Sci. 370–371 (2016), 355–367.

BBCNews, 2017. Ukraine power cut 'was cyber-attack'. [online]. Available from: http://www.bbc.com/news/technology-38573074. (Accessed 27 April 2017).

Cooper, A., 2016. Electric company smart meter deployments: Foundation for a smart grid. [online]. Available from: http://www.edisonfoundation.net/iei/publications/Documents/Final%20Electric%20Company%20Smart%20Meter%20Deployments-%20Foundation%20for%20A%20Smart%20Energy%20Grid.pdf.

Döbelt, S., Jung, M., Busch, M., Tscheligi, M., 2015. Consumers' privacy concerns and implications for a privacy preserving smart grid architecture—results of an Austrian study. J. Energy Res. Soc. Sci. 9, 137–145.

Dwork, C., 2006. Differential privacy. In: Automata, Languages and Programming. Lecture Notes in Computer Science, vol. 4052. Springer, Berlin/Heidelberg, pp. 1–12.

Dwork, C., 2011. The promise of differential privacy: a tutorial on algorithmic techniques. Proceedings of the 2011 IEEE 52nd Annual Symposium on Foundations of Computer Science.

Efthymiou, C., Kalogridis, G., 2010. Smart grid privacy via anonymization. First IEEE International Conference on Smart Grid Communications (SmartGridComm), Gaithersburg, MD, USA, USA.

Egarter, D., Prokop, C., Elmenreich, W., 2014. Load hiding of household's power demand.Proceedings of the IEEE International Conference on Smart Grid Communications (SmartGridComm), Venice, Italy.

Energy, A.O.I., 2011. Smart grids roadmap. [online]. Available from: https://www.iea.org/publications/.../smartgrids_roadmap.pdf. (Accessed 27 April 2017).

Erkin, Z., Tsudik, G., 2012. Private computation of spatial and temporal power consumption with smart meters. Proceeding ACNS'12 Proceedings of the 10th International Conference on Applied Cryptography and Network Security, Singapore.

Fang, X., Misra, S., Xue, G., Yang, D., 2012a. Managing smart grid information in the cloud: opportunities, models and applications. IEEE Netw. 26 (4), 32–38.
Fang, X., Misra, S., Xue, G., Yang, D., 2012b. Smart grid – the new and improved power grid: a survey. IEEE Commun. Surv. Tutorials 14 (4), 944–980.
Ganti, R.K., Ye, F., Lei, H., 2011. Mobile crowdsensing: current state and future challenges. IEEE Commun. Mag. 49 (11), 32–39.
Garcia, F., Jacobs, B., 2010. Privacy-friendly energy-metering via homomorphic encryption. 6th Workshop on Security and Trust Management (STM 2010), volume 6710 of Lecture Notes in Computer Science. Springer Verlag, pp. 226–238.
Gentry, C., 2009. A fully homomorphic encryption scheme. (Ph.D. thesis), Stanford University.
González-Manzano, L., de Fuentes, J.M., Pastrana, S., Peris-Lopez, P., Hernández-Encinas, L., 2016. PAgIoT–privacy-preserving aggregation protocol for internet of things. J. Netw. Comput. Appl. 71, 59–71.
Hayden, E., 2009. There is no SMART in smart grid without secure and reliable communications. [online] Available from: http://www.mclellancreative.com/files/Verizon_White_Paper_No%20SMART_in_Smart_Grid.pdf.
Horne, C., et al., 2015. Privacy, technology, and norms: the case of smart meters. Soc. Sci. Res. 51, 64–76.
Israel, T.o., 2016. Israel's electric authority hit by 'severe' cyber-attack. [online]. Available from: http://www.timesofisrael.com/steinitz-israels-electric-authority-hit-by-severe-cyber-attack/.
Jawurek, M., Johns, M., Rieck, K., 2011. Smart metering de-pseudonymization.Proceedings of the 27th Annual Computer Security Applications Conference (ACSA'11), pp. 11–13.
Kalogridis, G., et al., 2010. Privacy for smart meters: towards undetectable appliance load signatures. IEEE 1st International Conf. on Smart Grid Communications (SmartGridComm), Gaithersburg, USA.
Kargupta, H., Datta, S., 2003. Random data perturbation techniques and privacy preserving data mining. IEEE International Conference on Data Mining.
Kargupta, H., Datta, S., Wang, Q., Sivakumar, K., 2005. Random-data perturbation techniques and privacy-preserving data mining. Knowl. Inf. Syst. 7 (4), 387–414.
Komninos, N., Philippou, E., Pitsillides, A., 2014. Survey in smart grid and smart home security: issues, challenges and countermeasures. IEEE Commun. Surv. Tutorials 16 (4), 1933–1954.
Kumar, M., 2016. DDoS attack takes down central heating system amidst winter in Finland. [online]. Available from: http://thehackernews.com/2016/11/heating-system-hacked.html.
Kursawe, K., Danezis, G., Kohlweiss, M., 2011. Privacy-friendly aggregation for the smart-grid. In: Fischer, H.N., Hübner, S. (Eds.), Privacy Enhancing Technologies. PETS 2011. Lecture Notes in Computer Science, vol. 6794. Springer, Berlin, Heidelberg, Berlin, Heidelberg, pp. 175–191.
Kuzlu, M., Pipattanasomporn, M., Rahma, S., 2014. Communication network requirements for major smart grid applications in HAN, NAN and WAN. Comput. Netw. 67, 74–88.
Lagendijk, R., Erkin, Z., Barni, M., 2013. Encrypted signal processing for privacy protection: conveying the utility of homomorphic encryption and multiparty computation. IEEE Signal Process. Mag. 30, 82–105.
Latanya, S., 2000. Simple demographics often identify people uniquely. Data Privacy Working Paper 3.
Latanya, S., Pierangela, S., 1998. Protecting privacy when disclosing information: k-Anonymity and its enforcement through generalization and suppression. Technical Report SRI-CSL-98-04.

Li, F., Luo, B., Liu, P., 2010. Secure Information Aggregation for Smart Grids Using Homomorphic Encryption. IEEE, Gaithersburg, MD, USA.

Li, F., Luo, B., Liu, P., 2014. Secure information aggregation for smart grids using homomorphic encryption. Proceedings of the IEEE First IEEE International Conference on Smart Grid Communications (SmartGridComm), Gaithersburg, MD, USA, USA.

Li, N., Li, T., Venkatasubramanian, S., 2007. t-Closeness: privacy beyond k-anonymity and l-diversity. 23rd IEEE International Conference on Data Engineering.

Machanavajjhala, A., Gehrke, J., Kifer, D., 2006. L-diversity: privacy beyond k-anonymity. Proceedings of ICDE, p. 24.

Mahmood, A., Javaid, N., Razzaq, S., 2015. A review of wireless communications for smart grid. Renew. Sust. Energ. Rev. 41, 248–260.

McLaughlin, S., McDaniel, P., Aiello, W., 2011. Protecting consumer privacy from electric load monitoring. Proceedings of the 18th ACM Conference on Computer and Communications Security (CCS), Illinois, USA.

Morris, L., 2013. Analysis of partially and fully homomorphic encryption. [online]. Available from: http://www.liammorris.com/crypto2/Homomorphic%20Encryption%20Paper.pdf.

Narayanan, A., Shmatikov, V., 2008. Robust de-anonymization of large sparse datasets. Proceedings of the 2008 IEEE Symposium on Security and Privacy, pp. 111–125.

NIST, 2014. Guideline for smart grid cyber security. [Online]. Available from: http://nvlpubs.nist.gov/nistpubs/ir/2014/NIST.IR.7628r1.pdf.

Ozgur, U., Tonyali, S., Akkaya, K., Senel, F., 2016. Comparative evaluation of smart grid AMI networks: performance under privacy.IEEE Symposium on Computers and Communication (ISCC), Messina, Italy.

Paillier, P., 1999. Public-key cryptosystems based on composite degree residuosity classes. In: Stern, J. (Ed.), Advances in Cryptology–EuroCrypt99, Lecture Notes in Computer Science, vol. 1592. Springer, Berlin, Heidelberg, pp. 223–238.

Polityuk, P., 2016. Ukraine investigates suspected cyber attack on Kiev power grid. [online]. Available from: http://www.reuters.com/article/us-ukraine-crisis-cyber-attacks-idUSKBN1491ZF.

Rottondi, C., Mauri, G., Verticale, G., 2012. A Data Pseudonymization Protocol for SMART Grids. IEEE.

Rottondi, C., Verticale, G., Capone, A., 2013a. Privacy-preserving smart metering with multiple data consumers. Comput. Netw. 57, 1699–1713.

Rottondi, C., Verticale, G., Krauss, C., 2013b. Distributed privacy-preserving aggregation of metering data in smart grids. IEEE J. Sel. Areas Commun. 31 (7), 1342–1354.

Sabine, E., Bill, L., Markard, J., 2015. Smart meter communication standards in Europe – a comparison. Renew. Sust. Energ. Rev. 43, 1249–1262.

Sankar, L., Rajagopalan, S.R., Mohajer, S., Poor, V., 2012. Smart meter privacy: a theoretical framework. IEEE Trans. Smart Grid 4 (2), 837–846.

Sun, Y., Lampe, L., Wong, V.W.S., 2015. Combining electric vehicle and rechargeable battery for household load hiding. Proceedings of the IEEE International Conference on Smart Grid Communications (SmartGridComm), Miami, FL, USA.

Tonyali, S., Akkaya, K., Saputro, N., SelcukUlu, A., 2017. Privacy-preserving protocols for secure and reliable data aggregation in IoT-enabled smart metering systems. Futur. Gener. Comput. Syst. 78, 547–557.

Tonyali, S., Akkaya, K., Saputro, N., Uluagac, A., 2016. A reliable data aggregation mechanism with homomorphic encryption in smart grid AMI networks. The Proceedings of the IEEE, pp. 557–562.

Tonyali, S., Saputro, N., Akkaya, K., 2015. Assessing the feasibility of fully homomorphic encryption for smart grid AMI networks. Proceedings of IEEEpp. 591–596.
US Department of Energy, 2017. What is smart grid? [Online]. Available from: https://www.smartgrid.gov/the_smart_grid/smart_grid.html. (Accessed 19 July 2017).
Vaidya, B., Makrakis, D., Mouftah, H., 2013. Secure communication mechanism for ubiquitous smart grid infrastructure. J. Supercomput. 64 (2), 435–455.
Vijayan, J., 2010. Stuxnet renews power grid security concerns. [online]. Available at: http://www.computerworld.com/article/2519574/security0/stuxnet-renews-power-grid-security-concerns.html.
Wang, W., Lu, Z., 2013. Cyber security in the smart grid: survey and challenges. Comput. Netw. 1 (57), 1344–1371.
Ye, Y., Yi, Q., Hamid, S., David, T., 2013. A survey on smart grid communication infrastructures: motivations, requirements and challenges. IEEE Commun. Surv. Tutorials 15 (1), 5–20.
Yi, X., Paulet, R., Bertino, E., 2014. Homomorphic Encryption and Applications. Springer International Publishing, Cham, Switzerland. ISBN: 978-3-319-12229-8.
Zhang, Y., Chen, J., 2012. Wide-area SCADA system with distributed security framework. J. Commun. Netw. 14 (6), 597–605.
Zhao, J., Jung, T., Wang, Y., Li, X., 2014. Achieving differential privacy of data disclosure in the smart grid. Proceedings IEEE INFOCOM, 2014, Toronto, ON, Canada.

Further reading

Deepika, K., et al., 2014. A survey on approaches developed for data anonymization. Int. Adv. Res. J. Sci. Eng. Technol. 1 (3), 170–175.
Emam, K.E., Dankar, F.K., 2008. Protecting privacy using k-anonymity. J. Am. Med. Inform. Assoc. 15 (5), 627–637.
Rubner, Y., Tomasi, C., Guibas, L.J., 2000. The earth mover's distance as a metric for image retrieval. Int. J. Comput. Vis. 40, 99–121.
Solutions, S., 2013. Smart Meter Transmission Frequency Claims – "Misinformation" or "Missing Information"? [online]. Available from: https://smartgridawareness.org/2013/06/07/smart-meter-transmission-frequency-claims-misinformation-or-missing-information/.
Sweeney, L., 2002. k-Anonymity: a model for protecting privacy. Int. J. Uncertainty Fuzziness Knowledge Based Syst. 10 (5), 557–570.
Thoma, C., Cui, T., Franchetti, F., 2012. Secure multiparty computation based privacy preserving smart metering system. Proceedings of the North American Power Symposium (NAPS), 2012, Champaign, IL, USA.

Microgrid communication system and its application in hierarchical control

Hanqing Yang, Qi Li, Weirong Chen
School of Electrical Engineering, Southwest Jiaotong University, Chengdu, China

9.1 Introduction

In order to regulate the contradiction between distributed generation (DG) and the grid, the concept of microgrid is put forward to develop the value of DGs completely (Lassetter et al., 2002). However, the characteristics shown in all aspects of microgrid have some difference from the grid. For example, the inverter is mainly employed as the network interface for DG supplied in microgrid, while in the traditional grid, synchronous generators are often used, which need to use different control strategies (Shi et al., 2014). Some DGs, such as photovoltaic power generation, wind power, are always affected a lot by environmental factors. Furthermore, microgrid can operate in island mode or grid-connected mode, for which it is necessary to consider the microgrid island detection, grid-connected synchronization, and other issues, when switching between these two operation modes (Zhao et al., 2013). These characteristics lead to more complex information needed for decision making by microgrid control system compared with the traditional grid. Therefore, there are diversities of microgrid communication system and the traditional grid.

Due to the existence of DGs in the microgrid, it is possible to allow the energy to flow bidirectional, so that realizing the scheduling and protection must depend on bidirectional energy flow. To establish such a communication system, it is necessary to use a large number of measurement equipment in the network for accessing to the required information, which can also improve the efficiency and stability (Sood et al., 2009).

The microgrid communication system can realize the mutual communication among various intelligent electronic devices (IEDs) in the microgrid, and can be connected with other microgrids or grid to achieve the remote monitoring and remote control of the microgrid. The establishment of a mature microgrid system is not only conducive to improving the stability and intelligence of the system, but also can take advantage of resources more efficiently and improve the economy of microgrid (Xu and Li, 2010).

In this chapter, a general microgrid communication system, including the categories based on applications, the network technologies, and the functional requirements of communication system will be introduced, respectively.

Smart Power Distribution Systems. https://doi.org/10.1016/B978-0-12-812154-2.00009-2

9.1.1 Technologies

In microgrid, the communication system is always spread along the power line, so the structure of microgrid communication system is related with microgrid control methods closely. The typical communication system structure is basically a mixture of star, ring, and mesh structure. Ring structure, which can greatly improve the reliability of the system, is often regarded as a backbone network, due to its good self-healing.

9.1.1.1 Wired-line communication technology

(1) Optical fiber communication: optical fiber communication takes light wave as an information carrier and optical fiber as a transmission medium. The main features of optical fiber communication are excellent anti-electromagnetic interference ability, high transmission rate, large transmission capacity, good confidentiality, and so on. However, with employing optical fiber communication technology, the construction and maintenance costs are higher, which cause the large amount of data transmission and high reliability requirements application occasion. Ethernet-based passive optical network (EPON) is a new type of fiber access network technology. It adopts passive optical network (PON) technology in the physical layer, and uses Ethernet protocol at the link layer, which utilize the topology structure of PON to achieve the Ethernet access. Therefore, it is considered the first choice for the construction of digital distribution network communication system, which combines the advantages of PON technology and Ethernet (Shukla et al., 2014).

(2) Telephone network communication: telephone network communication mainly includes modem and ADSL (asymmetric digital subscriber line) way. Modem is used to convert digital signal into analog signal, so that the digital signal can be spread through the telephone line. However, this telephone network communication has the drawback of low transmission rate, usually about 56 kbps, which has been replaced by ADSL gradually. ADSL can provide up to 3.5 Mbps upstream speed and up to 24 Mbps downlink speed. The usage of local telephone network communication can reduce the cost of wiring, not needing to build a dedicated communication network.

(3) Power-line communication (PLC): PLC takes modern power lines as information transmission medium for voice or data transmission. There are some advantages of this communication technology, such as simple, low construction difficulty, low construction and operation costs, good security, and easy management. While the shortcomings is low transmission rate, vulnerable to interference, and so on. Using orthogonal frequency division multiple access (OFDMA), a communication system based on low voltage power line can be developed to meet the requirements of real-time monitoring system (Qinruo, 2003; Prasanna et al., 2009).

(4) Twisted pair/coaxial cable: these two kinds of media are mostly used for the construction of Ethernet. Ethernet defines the type of cable used in the local area network (LAN) and signal processing methods. The transmission rate of information packets between devices is usually at 10–100 Mbps. However, the Ethernet error control and real time are needed to be improved (Yigit et al., 2014).

9.1.1.2 Wireless communication technology

(1) GPRS (general packet radio service)/EDGE (enhanced data rate for GSM (global system for mobile communication) evolution)/CDMA (code division multiple access)/3G: GPRS and other technologies are regarded as a means of data transmission based on mobile communication network. This kind of technical signal can cover a wide range, adapt to the complex geographical conditions, and install conveniently. However, there are some limitation of an uncertain delay, low transmission rate, and low reliability. The suitable application of GPRS is the system with low real-time requirements.

(2) Wireless local area networks (WLANs): it is the use of radio frequency technology constituting the LAN. IEEE802.11 is the most commonly used transmission protocol in WLAN. With continuous development, transmission rate is improved from the initial 2 to 300 Mbps, even 600 Mbps. WLAN has the advantage of high transmission rate, flexible networking, and wide covering range, which can be applied to small-scale microgrid system.

(3) Worldwide interoperability for microware access system (WiMAX): WiMAX, namely IEEE802.16, is a new type of broadband wireless access technology. The data transmission distance is up to 50 km, with the advantage of large coverage, fast transmission, real time, low cost, and easy maintenance. While spectrum resources, standard, and security of WiMAX should be improved.

(4) Zigbee: Zigbee is synonymous with IEEE802.15.4 protocol, which determines a short-range, low-power wireless communication technology. Zigbee network is established mainly for data transmission in industrial field automation control, having characteristics of simple, low-power consumption, easy to use, reliable, low price, and short delay. It is suitable for a variety of sensor information collection and control, due to its low transmission rate (20–250 kbps). A variety of sensors and household appliances through the Zigbee module communicate with microgrid controller in the WINSmartGrid microgrid project, University of California, Los Angeles (UCLA). At the end of 2007, Zigbee Pro is introduced enhancing scalability and security of complex networks. Therefore, it can meet the requirements of the wireless network for supervisory control and data acquisition (SCADA) system (Franceschinis et al., 2013).

9.1.2 Categories

Microgrid communication system can be classified into the following categories according to the application scenarios (Kuzlu et al., 2014), as shown in Fig. 9.1. Fig. 9.2 shows the appropriate communication technologies in different categories:

- consumers' premises area networks (CPAN), including home area networks (HAN), building area networks (BAN), and industrial area networks (IAN)
- neighborhood area networks (NAN)
- wide area networks (WAN)

(1) CPAN: at consumers' premises, such as residential, commercial areas, and industrial areas, there are many IEDs sending signals to (receiving signals from) energy manage system, which not need to access data transmission system of very high frequency, due to the IEDs staying at the same premise. Therefore, the communication system of 100 kbs data rate and converging 100 m range is enough to satisfy the requirements of HAN, BAN, and IAN applications. There are many available communication technologies in this application, like PLCs, Bluetooth, Ethernet, ZigBee, and WiFi (Barmada et al., 2011).

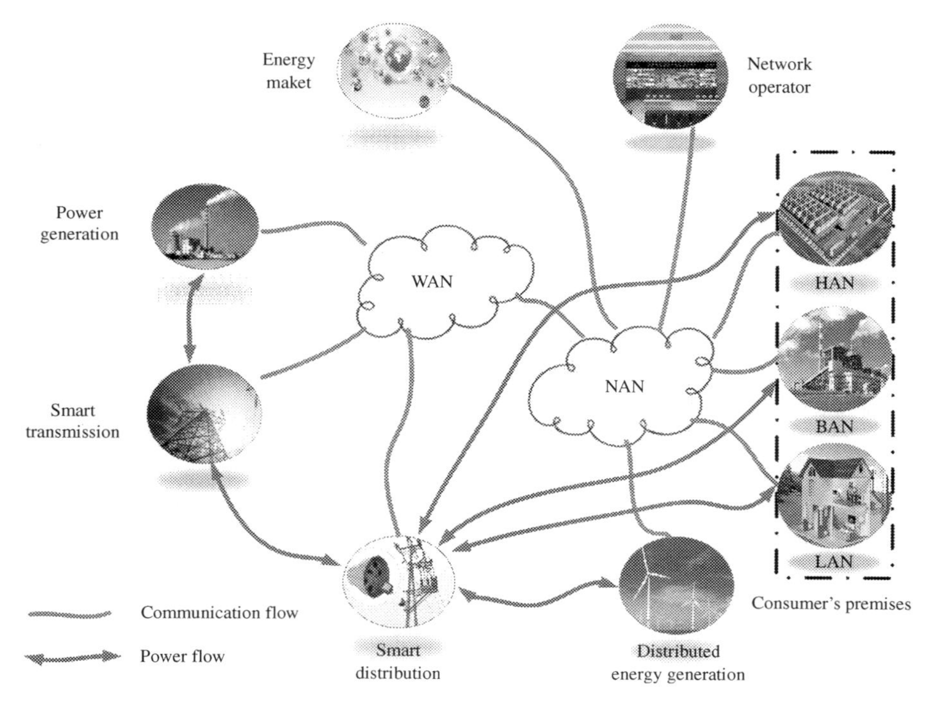

Fig. 9.1 The application scenarios of microgrid communication system.

Fig. 9.2 Communication technologies in different categories.

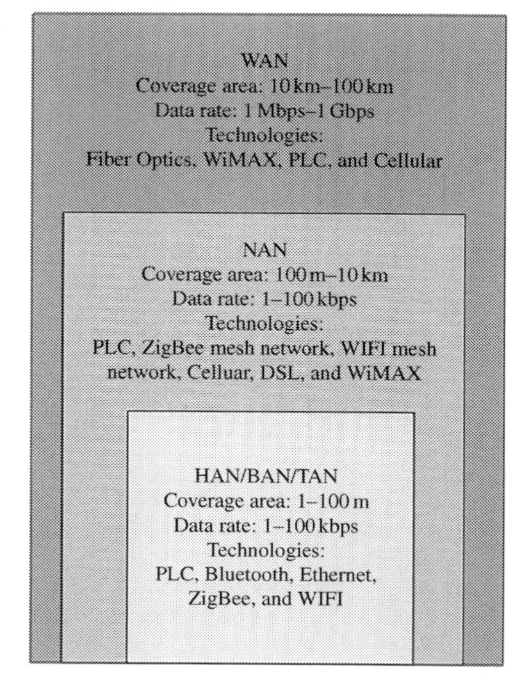

(2) NAN: the communication between the consumer premises and energy service provider requires communication technologies of 100 kbs to 10 Mbps data rate and converging 10 km area. Appropriate communication technologies for NAN applications include PLC, ZigBee, WiFi, cellular, digital subscriber line (DSL), and WiMAX (Barmada et al., 2011; Aalamifar et al., 2014).

(3) WAN: a wide range of monitoring and measurement equipment, such as sensors and power management controllers, are needed in the microgrid to exchange information with the grid. Compared with traditional SCADA system, the wide area monitoring and measurement requires higher data resolution and faster response time. The communication technologies needed in WAN application should be 100 Mbps to 1 Gbps data rate and converging 100 km area. The available communication technologies for WAN applications are fiber, WiMax, PLC, and cellular (Zimmermann and Dostert, 2002; Hochgraf et al., 2010; McGranaghan and Goodman, 2005).

9.1.3 Requirements

The microgrid communications infrastructure, which undertakes important responsibility of exchanging information, is the basis for coordinated operation of distributed power equipment. Poor communication performance not only causes the lower energy efficiency and service quality of microgrid, but also has brought potential danger to the grid. Protecting microgrid and realizing optimal operation depend on the communication infrastructure with many requirements (Adamiak et al., 1995).

9.1.3.1 Network latency

Network latency describes the maximum time that a message from origination to its destination in a communication network. There may be different requirements of network latency in communication between various IEDs, so that network architecture and communication medium must support the diverse necessities. Similarly, the data rates supported by the communication medium also determine the communication speed.

9.1.3.2 Reliability

The communicating devices in microgrid depend on the communication backbone sending/receiving messages to maintain the system stability. Therefore, it is exceedingly significant to exchange information timely and successfully as the communication backbone. There are several possible failures can influence the reliability of communication backbone, including timeout failures, network failures, and resource failures. A time-out failure occurs when the time spending exceeds the timing requiring. A network failure occurs when one of layers of the communication protocol fails. A resource failure means the fault of sending or receiving message. In addition, noise and interference in physical media can also affect communication, which make it necessary to improve the reliability of the system at the design stage.

9.1.3.3 Security

With the development of power system, the microgrid will spread over an extensively large range, like tens or hundreds of miles. So, the intruder's physical and cyber security is extremely important. In addition, if wireless communication technology, such as WiFi or ZigBee, is used as part of the communication network, the security problem increases due to the sharing and accessibility of these medium. Therefore, to provide security for the microgrid, it is essential to determine the different communication security solutions for each application.

9.1.3.4 Time synchronization

Some devices in microgrid need to be synchronized timely, which is decided on the criticality of the application. For example, involved in time synchronization tolerance and resolution requirements of IEDs are strict, due to the sensitive data processed. Time synchronization can be obtained in varieties ways, depending on resolution and jitter requirements. The precision time protocol (PTP) defined by the standard IEEE 1588 provides time synchronization for high-nanosecond precision over an ethernet network. Other ways, like global positioning system (GPS) and simple network time protocol (STNP) can also achieve time synchronization.

9.2 Communication construction based on hierarchical control

In this section, the communication configuration based on hierarchical control system is introduced, which can be divided into the three communication ways according to the primary control level, secondary control level, and tertiary control level, namely the communication in DGs, communication in microgrid, and communication between microgrid and the outside. The cooperation of these three communication methods can ensure the safe and stable operation of the system, by uploading data of primary control level and sending control signal of secondary/tertiary control level.0

There are some requirements of each communication methods to ensure safety and reliability of data transmission. First, the data of primary control level should be collected accurately and completely. Meanwhile, the quickly transmission of collected data from primary control level to secondary/tertiary control level should be achieved. Then, after verifying the integrity of received data, the calculation and energy management algorithm starts in secondary/tertiary control level, which can realize a more economical operation.

Due to the existence of different communications in the microgrid, it is inevitable to manage these communication ways effectively and uniformly. Therefore, a unified communication standard will be conducive to the microgrid management. What's more, with forming this standard, the communication of different control level can stay their own characteristics. The microgrid communication system should be designed according to the advantages and application of different communication technologies, including cost, application environment, and many other factors. A reasonable selection and design of communication system can utilize the advantage of different communication technologies to the utmost extent (Deng et al., 2008).

The typical network architecture of microgrid communication system is shown in Fig. 9.3, helping microgrid control center to manage the whole system by means of advanced communication technology (Ko et al., 2012). The management regulation mainly includes collecting electric information of microgrid key nodes, real-time monitoring of electric equipment, collecting environmental parameters of DGs, and monitoring electric information of relay protection.

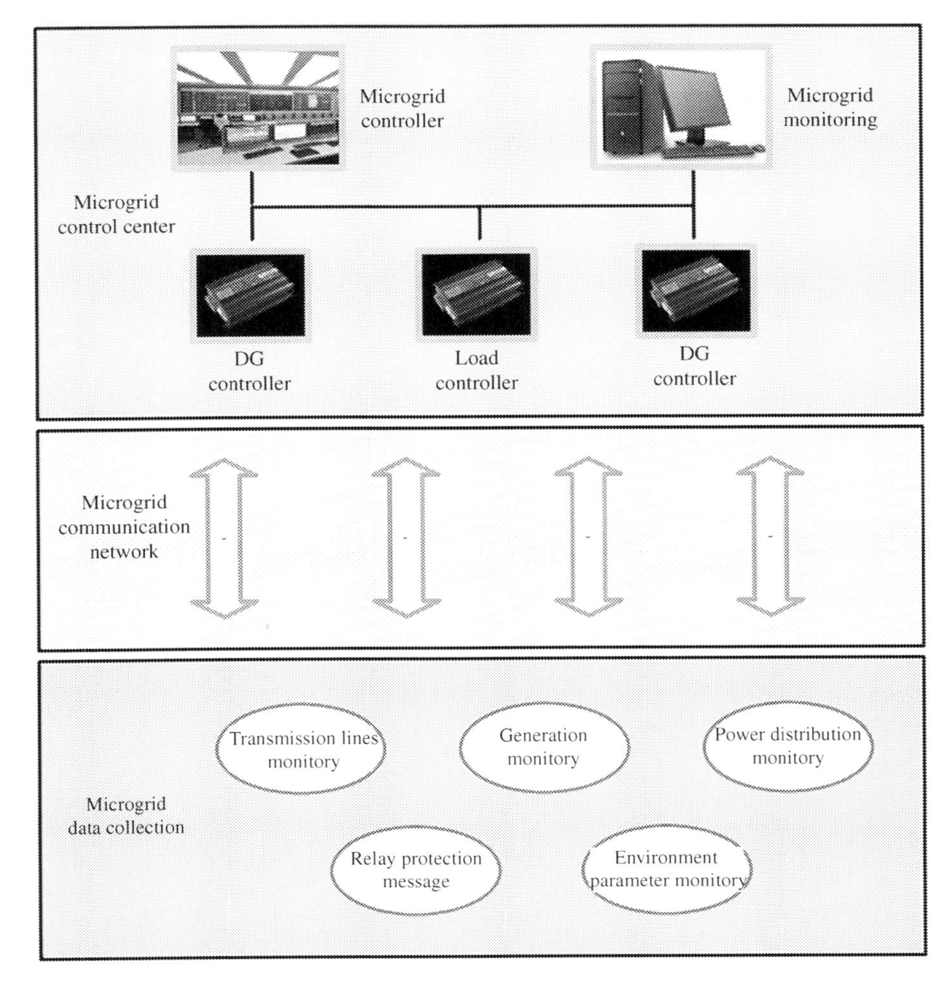

Fig. 9.3 Communication architecture of microgrid.

9.2.1 Communication in DGs

In the primary control level, there are varieties of DG control methods, which need to collect input and output electrical information to assist the corresponding DG control. For example, in droop control, the three-phase voltage and current on AC feeder need to be fed back to the controller to complete the active power control and reactive power control. Thus, in the process of communication among DGs, data collecting and data transmission will be involved at the input and output port of DGs.

The communication architecture of DG is shown in Fig. 9.4, which consists of sensor, DSP (digital signal processor) controller, and digital memory chip. The sensor, which obtains the electrical information of the main circuit by the means of mutual inductor, should be selected according to the system precision. Electrical signal is transmitted into the sensor in the way of analog signal, and then the conversion of analog to digital is executed in sensor. The filter calculation starts when conversed digital signal is transmitted into DSP controller, and some other electric information is obtained, which can be used in DG control program, such as output power and output reactive power. Some key electrical parameters can be stored in the digital memory chip for data transceiver using. Data transceiver serving as the only interface for maintaining communication between DGs and the outside, is an important IED for communication.

Communication in DGs is the basis of the whole communication system. Only depending on this communication, some control strategy can be achieved, such as the maximum power point tracking control of photovoltaic and wind power generation, switching strategy between PQ control and V/f control, and local control of lithium battery/super capacitor hybrid energy storage system.

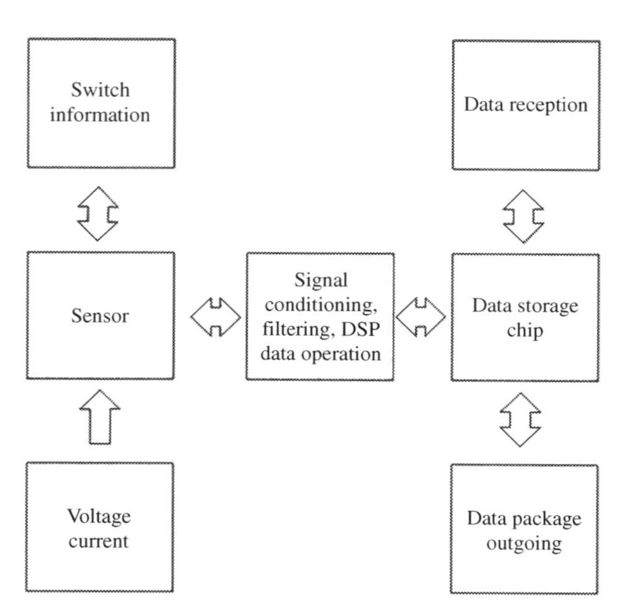

Fig. 9.4 Communication architecture of distributed generation (DG).

9.2.2 Communication in microgrid

Because of the different communication requirements of each microgrid parts and applications, there are some difference in the corresponding communication technology architecture.

In the small-scale microgrid, which has the characteristics of few communication nodes are few and simple network structure, the equipment supporting the simple communication protocol and the basic physical transmission media can basically meet the microgrid communication requirements to save the transmission cost, such as RS485 and industrial ethernet technology.

In the large-scale microgrid, the increase of communication nodes and the complexity of the network structure have caused the huge data traffic of communication system. In order to satisfy the communication real-time requirements, optical fiber transmission technology and international standard protocol (such as IEC61850 protocol) should be employed. In some special microgrid, wireless transmission communication is more flexible, which can cope with the wired network not configured.

The mainly task of communication in microgrid is to bring together the electrical information transmitted from the DG transceiver unit and send the control signal timely. DG will send the electric information collected locally to the communication medium continuously, and also receive the decision information from control center. Therefore, a large real-time bandwidth and good transmission quality are both required in microgrid communication.

The framework of communication in microgrid is shown in Fig. 9.5. Each DG has the corresponding measurement and control unit, and IED equipment. All the data should be transmitted to the microgrid coordination control unit through communication line, controlled by the combination of ARM (advanced RISC machines) and DSP. Grid-connected operation and control of DG can be completed through the communication between DG and the controller. Meanwhile, the data can be transferred to the microgrid monitor interface by the controller, for administrator analyzing and monitoring.

9.2.3 Communication between microgrid and the outside

Microgrid data can be transmitted to any place covered network to meet the further requirements of microgrid monitoring. In the communication between microgrid and the outside, data structure and interface of the entire database should be paid attention, due to large amount of microgrid data and complex data structure. Client-server way is always employed for the communication between microgrid and the outside, which can present all the data in web pages for people visiting. There is one more thing need to be noted that microgrid communication data is updated in real time, and data should to be kept refreshed for researchers analyzing.

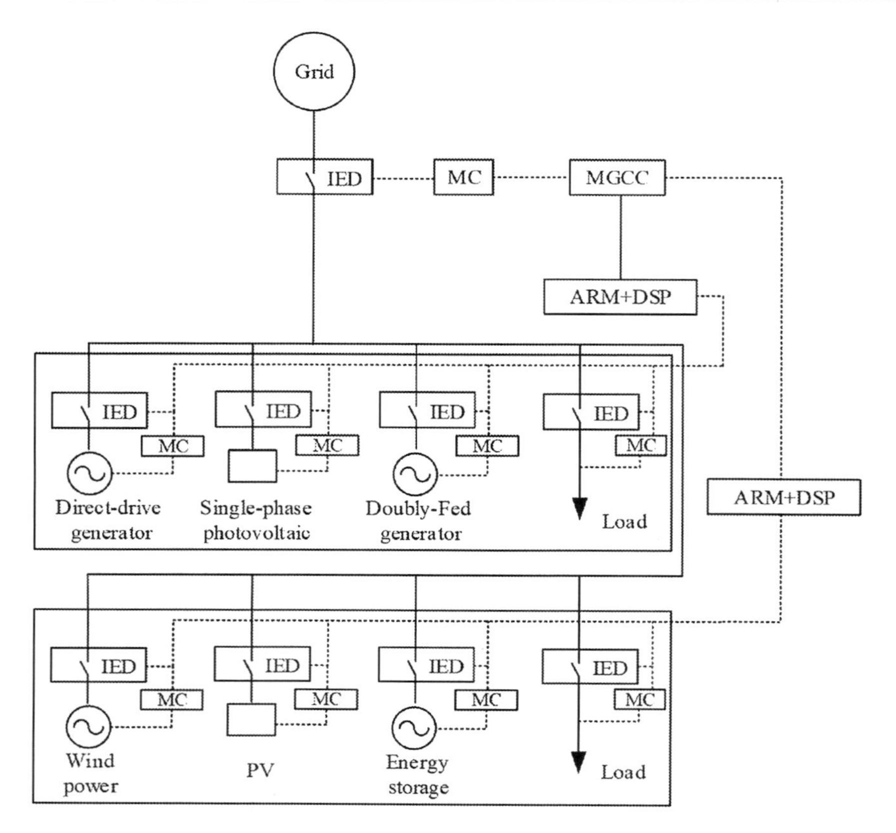

Fig. 9.5 Communications block diagram of multimicrogrid.

9.3 Consensus algorithm based on microgrid communication system

Consensus control can make some state of all individuals in the distributed system tend to be consistent with each other by exchanging information between individual and its adjacent individuals (Chen et al., 2015; Zhang and Chow, 2012). In recent years, the consensus algorithm has achieved much success in the group and formation control of space shuttle, and is gradually used for power system economic scheduling and hierarchical control research. In this section, the first-order discrete consensus algorithm is introduced.

9.3.1 Graph theory

For a microgrid system using a distributed hierarchical control structure, the communication network corresponds to a set of finite nodes (DG) with weightless undirected graphs $g(V,S)$. An undirected graph is a special case of a directed graph, where $S(i,j)$ represents the edge connecting nodes i and j.

The terms and definitions of the graphs of undirected graphs are as follows (Ren and Beard, 2008; Chen and Satyanarayana, 1978):

(1) Neighbor node: for a node i, all nodes directly connected to it are called neighbor nodes of the node i.
(2) Degree of node: the number of neighbor nodes of node i is the degree of the node i, donated by d_i.
(3) Connected: when each node of an undirected graph has a direct or indirect path connected to other nodes, this undirected graph is connected.
(4) Subgraph: for $g(V_p, S_p)$ and $g(V_{ps}, S_{ps})$, if (1) $V_{ps} \subseteq V_p$ and (2) $S_{ps} \subseteq S_p \bigcap V_{ps} \times V_{ps}$ are satisfied at the same time, $g(V_{ps}, S_{ps})$ is a subgraph of $g(V_p, S_p)$.
(5) Tree: for a connected undirected graph g, if a subgraph at the same time has the following three conditions, then the subgraph is a tree of the graph: (a) containing all nodes, (b) not containing loops, and (c) connected.
(6) Adjacency matrix: let $A = [A_{ij}] \in R^{n\times n}$ is the adjacency matrix of $g(V,S)$. When there is a direct connection between node i and node j, the value of A_{ij} is equal to the weight of the edge $S(i,j)$. In general, the weight is positive. When there is no direct connection between node i and node j, $A_{ij} = 0$. For any node i, $A_{ii} = 0$. For the undirected graph, the weight of $S(i,j)$ and $S(j,i)$ are equal, so the corresponding adjacency matrix A is a symmetric matrix.
(7) Laplacian matrix: let $L = [L_{ij}] \in R^{n\times n}$ is a Laplacian matrix of $g(V,S)$.

$$\begin{cases} L_{ij} = -A_{ij} & L \text{ is nondiagonal elements} \\ L_{ij} = \sum_{j=1}^{n} A_{ij} & L \text{ is diagonal elements} \end{cases} \tag{9.1}$$

It can be seen from Eq. (9.1), L is a symmetric matrix. And for any row of matrix L, the sum of this row elements is zero. Therefore, the matrix L has an eigenvalues of 0, and the corresponding eigenvector is 1.

(8) Stochastic matrix: for a matrix of all nonnegative elements P, if the sum of the elements of each row of the matrix is 1, then P is a row stochastic matrix; if each column element of row random matrix P is also 1, then P is a double stochastic matrix.

9.3.2 Consensus algorithm without communication delay

For a system like Eq. (9.2) shows, let x_i stands for the state variable of the ith node, such as position, velocity, power, voltage, and other physical information, and u represents the applied control. Communication network of this system is denoted by graph g. In the case where the communication network does not change with time, a common continuous consensus algorithm is shown in Eq. (9.3) (Ren and Cao, 2010).

$$\dot{x}_i = u_i \tag{9.2}$$

$$u_i = -\sum_{j=1}^{n} A_{ij}\left(x_i - x_j\right) \tag{9.3}$$

Combining Eqs. (9.2), (9.3) with matrix form, Eq. (9.4) is obtained.

$$\dot{x}_i = -Lx \tag{9.4}$$

where L is a Laplacian matrix of g, and $x \in R^{n\times 1}$ is the column vector consisting of the state variable for each node. For any x_i and x_j, and any initial state $x_{(0)}$, there is a relationship $|x_{i(k)}\text{-}x_{j(k)}| \to 0$ with the condition of $k \to \infty$, which is called convergence of consensus algorithm.

If different nodes communicate only at discrete time points, the consensus algorithm can be transformed into discrete form.

$$x_{(k+1)} = D \cdot x_{(k)} \tag{9.5}$$

where D is the weight matrix of the communication network, and the nondiagonal elements of D are the weight of $S(i,j)$. When $S(i,j) \notin g$ and $i \neq j$, $D_{ij} = 0$. When $S(i,j) \in g$, $D_{ij} \neq 0$. If A is the adjacency matrix of g, then $D_{ij} = A_{ij} = -L_{ij}$ $(i \neq j)$.

When the corresponding undirected graph of communication network is connected, D can be selected to be a row stochastic matrix, where 1 is the largest eigenvalue (Xin et al., 2013a,b) of matrix D and spectral radius $\rho(D) = 1$. With the condition of $k \to \infty$, the discrete consensus algorithm as Eq. (9.6) shows converge.

$$\lim_{k\to\infty} x_{(\mathrm{k})} = \lim_{k\to\infty} D^k x_{(0)} = 1_n v^T x_{(0)} \quad \left(v^T D = v^T \text{and } 1_n v = 1\right) \tag{9.6}$$

In the same communication network, there are many different ways of constructing matrix D, which can make Eq. (9.6) converge. The choice of D will affect the convergence speed of the consensus algorithm (Xin et al., 2013a,b).

9.3.3 Consensus algorithm with communication delay

In the last section, the consensus algorithm without communication delay is discussed. However, in a real system, it will take some time in the process of signal conversion and transmission, as shown in Fig. 9.6. The time required to transmit the signal from adjacent node j to node i is shown in Eq. (9.7).

$$\tau_{ij} = \tau_1 + \tau_2 + \tau_3 \tag{9.7}$$

where τ_1 is the total of signal sampling time and communication processing time, τ_2 is the time of transmitting over the communication line, and τ_3 is the time of received signal to be converted by node i.

The communication among all nodes is discrete in time when using the discrete consensus algorithm, so that there will be a time interval T between two communications. When T is greater than the communication delay τ_{ij}, the communication delay does not have an effect on system. In the case of τ_{ij} greater than T, the communication

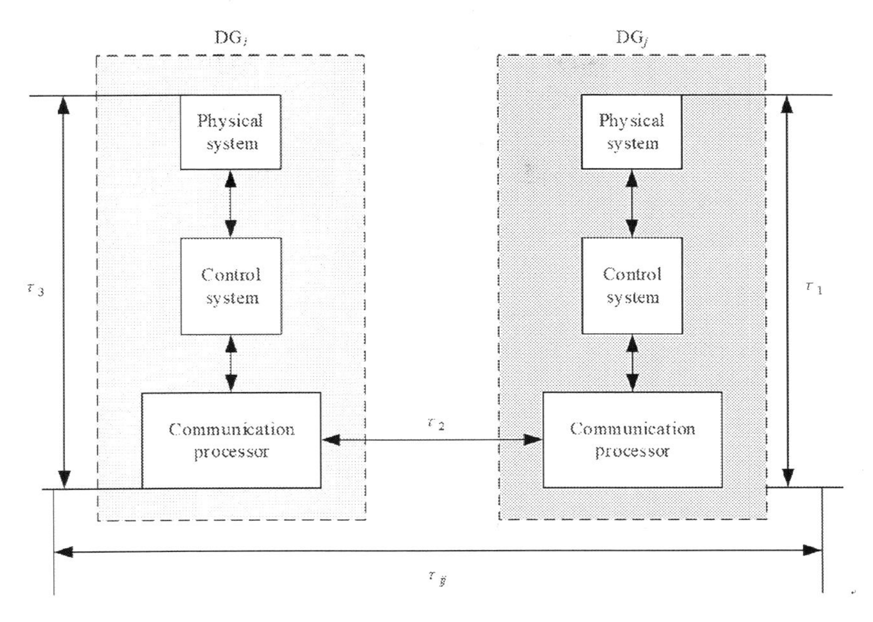

Fig. 9.6 The structure of the proposed DC microgrid.

delay will have an impact on the consensus algorithm, the corresponding algorithm expression should be rewritten as follows.

$$x_i[k+1] = x_i[k] + \sum_{j=1,\, j \neq i}^{m} d_{ij} x_j[k-1] \tag{9.8}$$

9.4 Case studies

A typical configuration of DC microgrid with decentralized battery energy storage system (DBESS) shown in Fig. 9.7 is regarded as the studied system in this chapter. And, the consensus control based on communication system is employed in DBESS to allocate output/input power among batteries.

The DBESS is used to stabilize the power fluctuation and maintain the DC bus voltage at a certain range (Huang and Qahouq, 2015). When the power generated by renewable energy generation is less than the power demanded by load, the DBESS works in the discharge mode, supplying power to the load together with the renewable energy generations. When the generated power is greater than the demanded power, the DBESS works in the charging mode to absorb the excess power in the system.

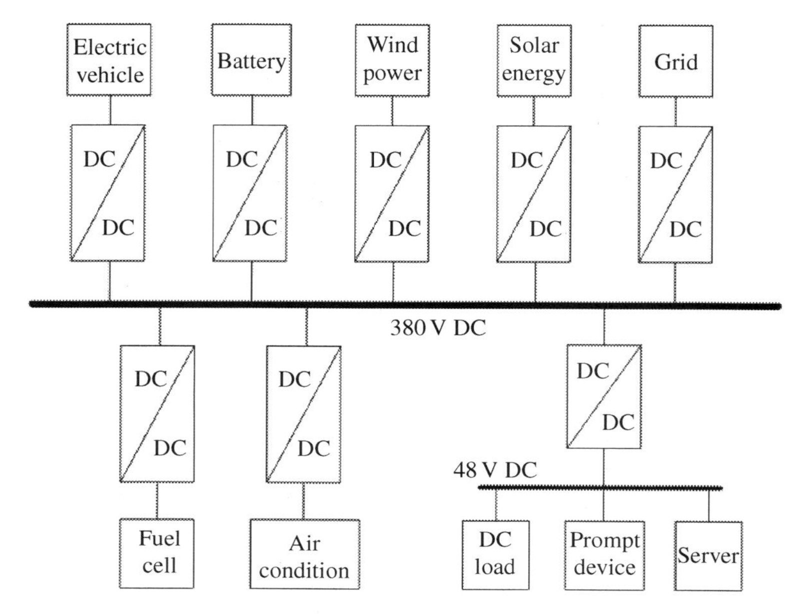

Fig. 9.7 A typical configuration of DC microgrid.

9.4.1 Hierarchical distributed control

In this section, the hierarchical distributed control method is applied to automatically adjust the charge/discharge power based on their own state of charge (SoC) condition. The hierarchical distributed control needs autonomous control units updating their own state by communicating with adjacent nodes, to complete complex tasks together. In Yang et al. (2017), a self-convergence droop control of no communication is proposed to realize SoC balance. Morstyn et al. (2017) proposes a sliding mode control strategy based on multiagent system to improve the converge speed of the existing SoC balancing strategies.

The structure of hierarchical distributed control is shown in Fig. 9.8, which can be divided into three levels, including the primary control, the secondary control, and tertiary control (Dou and Liu, 2013).

(1) Primary control. Control voltage and current of battery according to the control command sent by the secondary control. Meanwhile, transport the collected power, SoC, and other physical signals to the secondary control.

(2) Secondary control. Send the control command calculated by means of exchanging data with adjacent generation unit through preset communication network.

(3) Tertiary control. By applying the pinning control to the partial nodes, the dispatch control of the system can be realized by the coupling relationship between the following nodes and the pinning nodes.

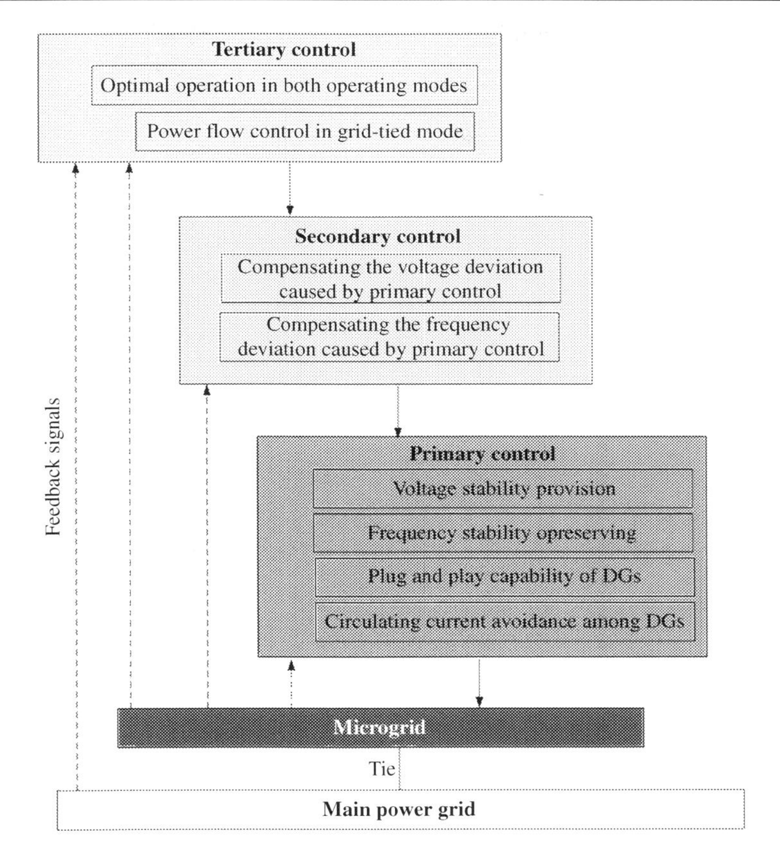

Fig. 9.8 Hierarchical control levels.

9.4.2 The consensus algorithm in hierarchical control

The main idea of the consensus control with pinning node is that regarding the DBESS shown in Fig. 9.7 as a network-connected distributed system. Each battery contains the primary control and the secondary control. The secondary control of battery, which is selected as the pinning node, can receive the power command from tertiary control. The principle of power allocation is that the charge/discharge power of each battery is the same as the ratio of the power to their respective SoC.

Let x_i denotes the state variable of ith battery. The battery can only communicate with its adjacent batteries. x_i can represent any physical information, such as voltage, current, and SoC. When x_i of all batteries are the same, the studied system achieves consensus convergence. The discrete consensus algorithm is shown in Eq. (9.9).

$$x_i[k+1] = \sum_{j=1}^{m} d_{ij} x_j[k] \tag{9.9}$$

where j is the adjacent node of ith battery, d_{ij} is the transfer coefficient between ith battery and its adjacent node j, and m is the number of batteries.

The transfer coefficient d_{ij} can affect the convergence and convergence speed of consensus algorithm. When $D = [d_{ij}]$ is a random matrix, the x_i can converge in some specific condition (Kriegleder et al., 2013). If D is a double random matrix, x_i can converge to the average of the system initial value. In order to improve the reliability of discrete consensus algorithm, the Metropolis method is employed to construct a double random matrix D, which diagonal elements are not equal to zero (Guerrero et al., 2011). The Metropolis method is described as follows.

$$d_{ij} = \begin{cases} \dfrac{1}{\max\left(n_i, n_j\right) + 1} & j \in N_i \\ 1 - \sum\limits_{j \in N_i} \dfrac{1}{\max\left(n_i, n_j\right) + 1} & i = j \\ 0 & \text{others} \end{cases} \tag{9.10}$$

where n_i and n_j are the adjacent number of node i and node j, $\max(n_i,n_j)$ stands for the maximum value of n_i and n_j.

First, the SoC of battery needs to be obtained. At present, the research on detecting battery SoC has made a lot of achievements. In this paper, the current integration method is employed to achieve SoC (Bhangu et al., 2005). The expression of current integration method is shown in Eq. (9.11).

$$SoC_i = SoC_i^* - \frac{1}{C_{ei}} \int I_{bi} dt \tag{9.11}$$

where I_{bi}, $SoC_i{}^*$, and C_{ei} are the output current, initial SoC, and rated capacity of battery of ith battery, respectively.

Ignoring the power loss of the DC/DC converter, there will be an expression in the following.

$$P_i = P_{\text{in}i} = U_{\text{in}i} I_{bi} \tag{9.12}$$

where $U_{\text{in}i}$, P_i, and $P_{\text{in}i}$ are output voltage, output power, and input power of DC/DC converter of ith BESU (battery energy storage unit), respectively.

The expression of detecting SoC can be rewritten combining Eqs. (9.11), (9.12).

$$SoC_i = SoC_i^* - \frac{1}{C_{ei} U_{\text{in}}} \int P_i dt \tag{9.13}$$

Next, the ratio is defined respectively in charging and discharging operation state, as shown in Eqs. (9.14), (9.15). The ratio of each battery will tend to be consistent after several iterations by using consensus algorithm. So that the purpose of power distribution is achieved.

$$r_{\mathrm{char}i}[k] = P_{\mathrm{BESU}i}[k] \cdot SoC_i[k] \tag{9.14}$$

$$r_{\mathrm{disc}i}[k] = \frac{P_{\mathrm{BESU}i}[k]}{SoC_i[k]} \tag{9.15}$$

where $r_{\mathrm{char}i}[k]$ and $r_{\mathrm{disc}i}[k]$ are the charging and discharging ratio of ith battery. $P_{\mathrm{BESU}i}[k]$ and $SoC_i[k]$ are the input power/output power and SoC of ith battery, which are used to calculating $r_{\mathrm{char}\_i}[k]$. $P_{\mathrm{BESU}i}[k] < 0$ stands for charging operation, while $P_{\mathrm{BESU}i}[k] > 0$ represents discharging operation.

It is assumed that the nearest battery to tertiary control center is battery 1#. In the DBESS, battery 1#, which is called pinning node, can get the reference power information from tertiary control center through network communication, while the remaining batteries in the DBESS which do not need to have a network connection to the tertiary control center, are called following node. Fig. 9.9 shows the flow chart of power allocating in a sampling period.

In the beginning of each sampling period, the initial power values of pinning node and following node are set as shown in the following.

$$\begin{cases} P_{\mathrm{BESU}1}[0] = P_{\mathrm{all}} \\ P_{\mathrm{BESU}i}[0] = 0 \end{cases} \tag{9.16}$$

$$\begin{cases} SoC_1[0] = SoC_1(0) \\ SoC_i[0] = SoC_i(0) \end{cases} \tag{9.17}$$

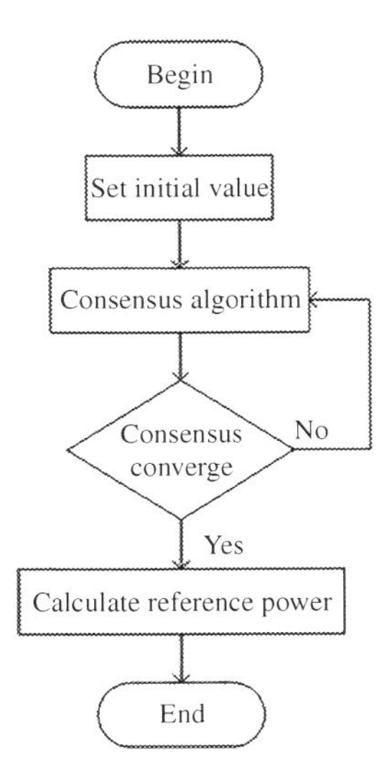

Fig. 9.9 The flowchart of consensus algorithm.

where P_{all} is the reference power value from dispatching center. $P_{\text{all}} < 0$ stands for charging operation, while $P_{\text{all}} > 0$ represents discharging operation.

Then, the discrete consensus algorithm is used to iterate the $P_{\text{BESU}i}$ and SoC_i.

$$P_{\text{BESU}i}[k+1] = P_{\text{BESU}i}[k] + \sum_{j=1, j\neq i}^{m} d_{ij} P_{\text{BESU}j}[k] \tag{9.18}$$

$$SoC_i[k+1] = SoC_i[k] + \sum_{j=1, j\neq i}^{m} d_{ij} SoC_j[k] \tag{9.19}$$

After several times iteration, Eqs. (9.18), (9.19) will converge to the average of initial value.

$$P^*_{\text{BESU}i} = \lim_{k\to+\infty} P_{\text{BESU}i}[k] = \frac{1}{n}\sum_{i=1}^{n} P_{\text{BESU}i}[0] = \frac{P_{\text{all}}}{n} \tag{9.20}$$

$$SoC^*_i = \lim_{k\to+\infty} SoC_i[k] = \frac{1}{n}\sum_{i=1}^{n} SoC_i(0) \tag{9.21}$$

Therefore, the charging and discharging ratio of ith battery will also converge to a constant.

$$r^*_{\text{char}i} = P^*_{\text{BESU}i} \cdot SoC^*_i = P_{\text{all}} \cdot \sum_{i=1}^{n} SoC_i(0)/n^2 \tag{9.22}$$

$$r^*_{\text{disc}i} = P^*_{\text{BESS}i}/SoC^*_i = P_{\text{all}}/\sum_{i=1}^{n} SoC_i(0) \tag{9.23}$$

Finally, the input power and output power can be calculated by the ratio of charging operation and discharging operation.

$$P^*_{\text{BESU}i} = \begin{cases} r^*_{\text{char}i}/SoC_i & \text{charging} \\ r^*_{\text{disc}i} \cdot SoC_i & \text{discharging} \end{cases} \tag{9.24}$$

9.4.3 The communication system

The above describe how to obtain the power reference value of each battery by the consensus control with pinning node. This method is a distributed algorithm, each battery gets its own charge/discharge power reference value, only according to their own information and its adjacent nodes information. The SoC of different BESUs can be

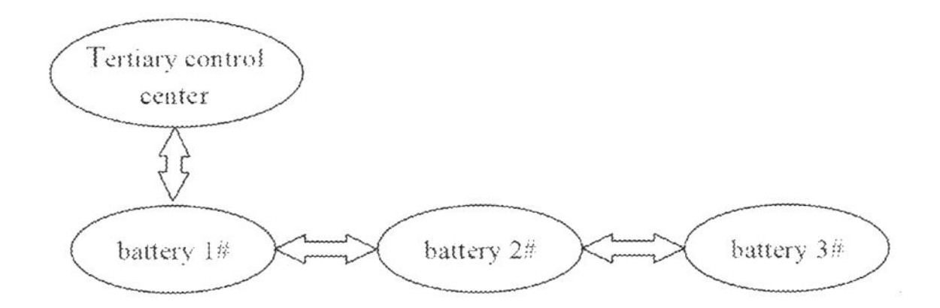

Fig. 9.10 The topology of communication network.

balanced and the total power of DBESS outputting according to the dispatching requirement is realized.

The communication network topology based on hierarchical distributed control is shown in Fig. 9.10. However, this communication network topology is not unique, as long as the network topology satisfies the requirement of connection. Different network topology can only affect the speed of consistency convergence.

In this control method, matrix D is constructed by Eq. (9.10), and the scheme of self-organization mechanism is designed. In this case, it is assumed that the pinning node is reliable and has been involved all the time. The self-organization mechanism is mainly used to constrain the switching of following nodes.

Taking battery 2# as an example, when battery 2# needs to be put in, battery 2# will send incoming packets to battery 1# and battery 3#, and then update its own matrix D. After receiving the incoming message, battery 1# and battery 3# broadcast the latest neighbor messages and update their own matrix D elements. Their own matrix D is all updated by Eq. (9.10), which can ensure matrix D is still a double random matrix with nonzero diagonal elements. According to the self-organization mechanism, the battery can connect or disconnect their neighbor nodes without affecting the system operation.

9.4.4 Results and discussion

The results of consensus control based on communication system in DBESS are discussed in this section, and the simulation model shown in Fig. 9.11 is established, which consists of three battery branches with line resistance. The model parameters are shown in Table 9.1.

9.4.4.1 Case 1: Charging operation

Figs. 9.12–9.14 show the simulation results of charging operation, and the total power of DBESS is given by dispatching center varying with time. The initial SoC of battery 1#, battery 2#, and battery 3# is respectively 40%, 45%, and 50%. Figs. 9.12 and 9.13 show the SoC and charging current of each battery, and Fig. 9.14 presents the average voltage of DBESS.

In 0–100 s, the total power of DBESS given by dispatching center is 800 W. At the beginning of operation, battery 1# with smallest SoC absorbs most of the total power.

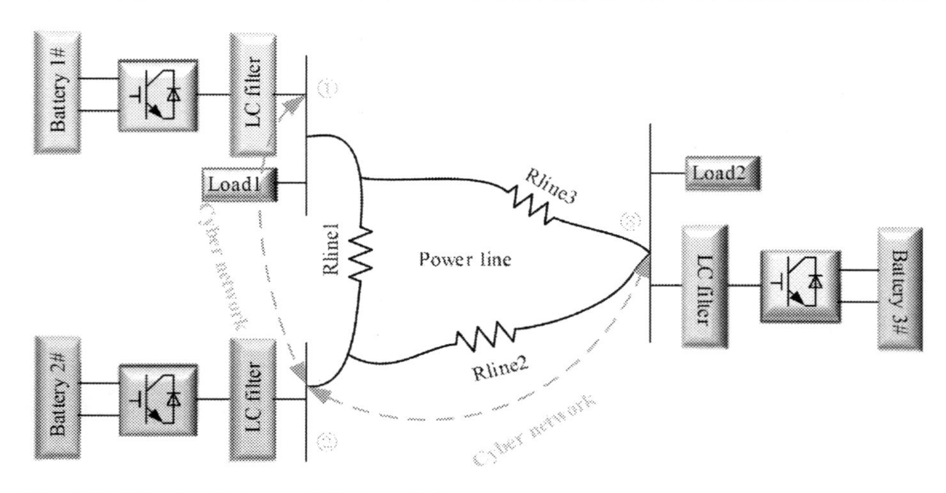

Fig. 9.11 The simulation model of studied system.

Table 9.1 System parameters

	Parameters	Value
Line resistance	R_{line1} (Ω)	0.5
	R_{line2} (Ω)	0.8
	R_{line3} (Ω)	0.9
Battery	Type	Lead-acid battery
	Rated voltage (V)	24
	Rated capacity (Ah)	5
DC bus	Rated voltage (V)	48

At 100 s, the SoC of battery 1# and battery 2# is almost consistent with each other, and the charging currents also tend to be equal. During the time of 100–200 s, the total power of DBESS given by dispatching center is 500 W. The charging currents of all batteries are reduced. At 200 s, the difference of SoC between battery 1#/battery 2# and battery 3# disappears gradually. In 200–300 s, the total power of DBESS given by dispatching center is 700 W. During this period, the SoC of these three batteries is balanced, also the charging currents are in the same value. Although the total power changes to 600 W at 300 s, the charging ratio of all batteries stays the same. According to Fig. 9.14, the voltage is maintained in the range of 48 ± 0.5 V during the operation.

Therefore, in the charging operation, it is desirable to give priority to the battery with the lowest SoC to obtain a larger charging current. The proposed control strategy can balance SoC and control output power varying with dispatching center requirements in DBESS.

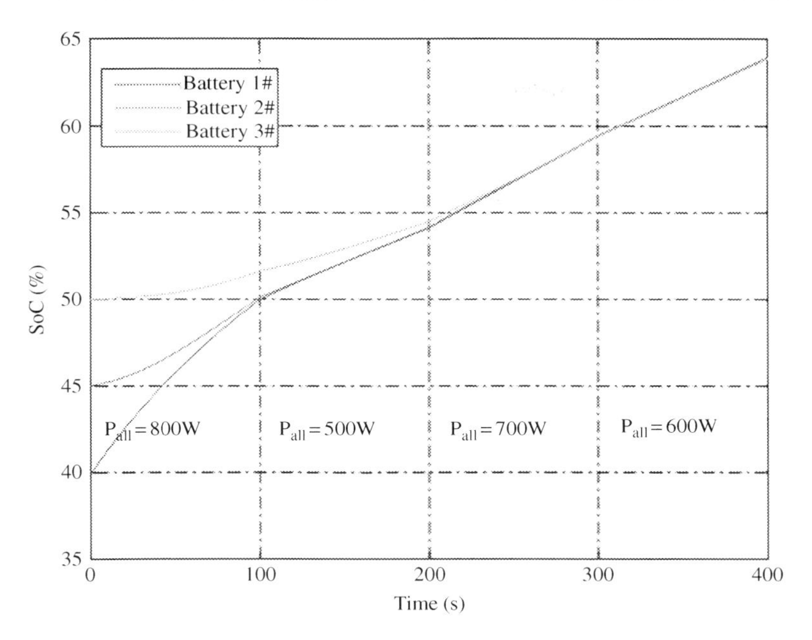

Fig. 9.12 The SOC results in charging operation mode.

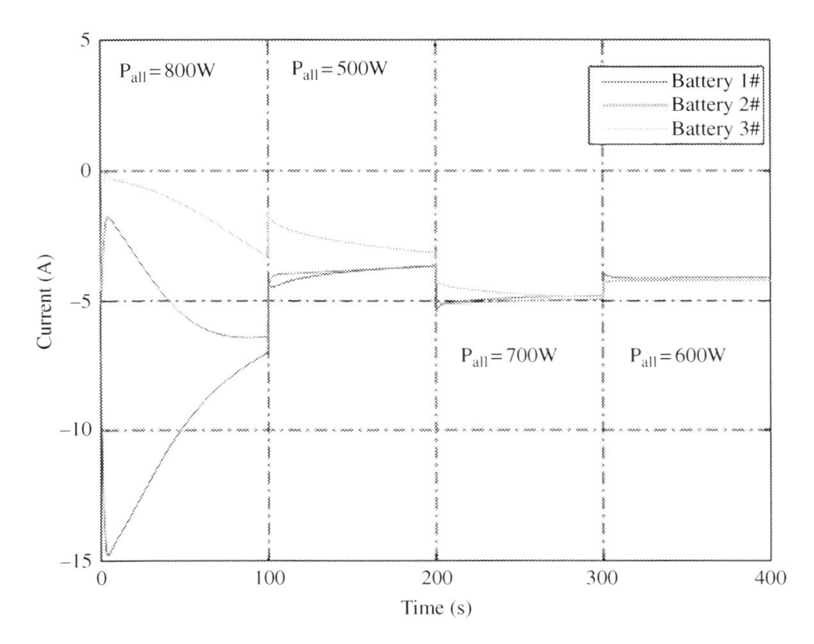

Fig. 9.13 The current results in charging operation mode.

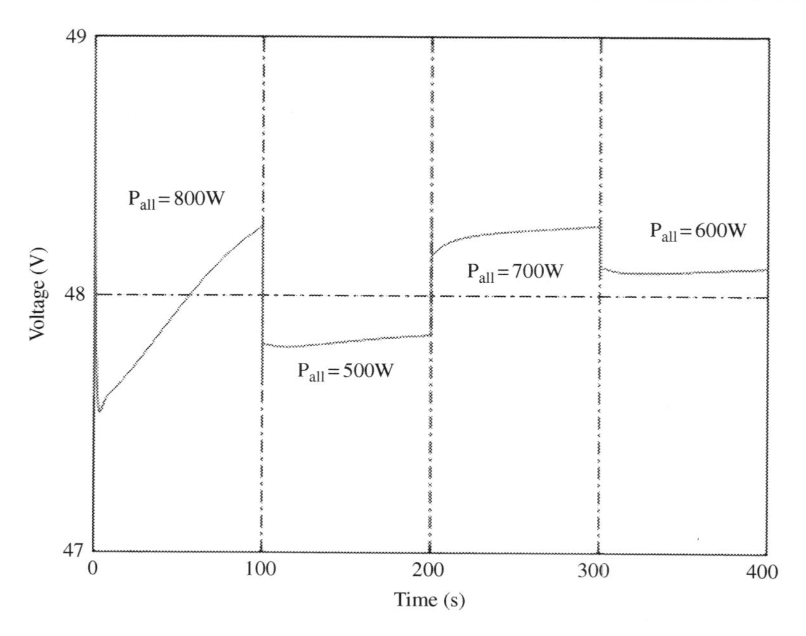

Fig. 9.14 The voltage results in charging operation mode.

9.4.4.2 Case 2: Discharging operation

Fig. 9.15 shows the simulation results of discharging operation, and the total power of DBESS is given by dispatching center varying with time. The initial SoC of battery 1#, battery 2#, and battery 3# is respectively 75%, 70%, and 65%. Figs. 9.15 and 9.16 show the SoC and discharging current of each battery, and Fig. 9.17 presents the average voltage of DBESS.

In 0–100 s, the total power of DBESS given by dispatching center is 500 W. At the beginning of operation, battery 1# with largest SoC outputs most of the total power. At 100 s, the difference of SoC among batteries is reduced <5%. During the time of 100–200 s, the total power of DBESS given by dispatching center are 700 W. The discharging currents of all batteries are increased. At 200 s, the SoC of the three batteries is almost consistent with each other, and the discharging currents also tend to be equal. In 200–300 s, the total power of DBESS given by dispatching center is 600 W, while the total power changes to 500 W at 300 s. During this period, the battery with the highest SoC always provides more power than the battery with smallest SoC. Finally, the purpose of balancing SoC is achieved at 400 s. According to Fig. 9.17, the voltage is maintained in the range of 48 ± 1 V during the whole operation.

In the discharging operation, although the total power demanded is changeable, the battery with the lowest SoC can always to output less discharging current. The purpose of balancing SoC and controlling output power depending on dispatching center in discharging operation is realized by adopting proposed control strategy.

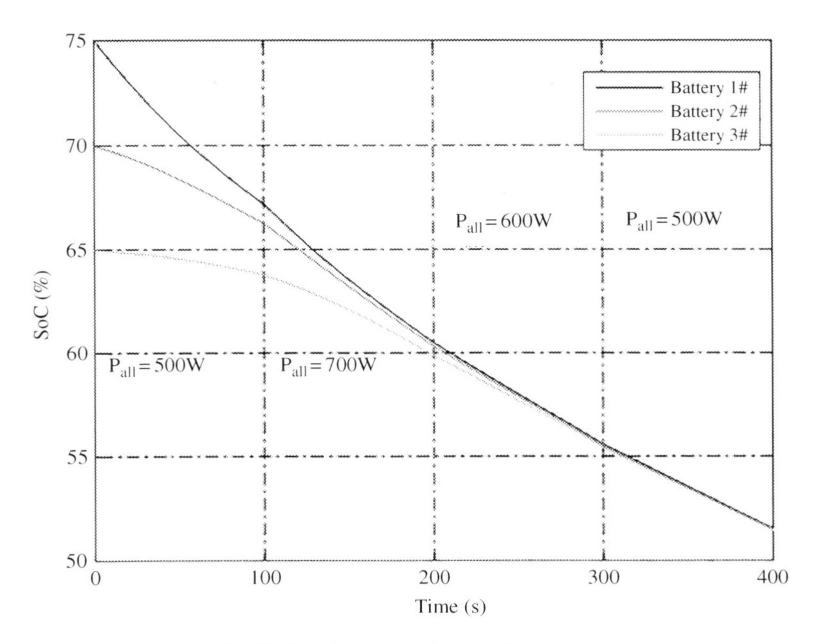

Fig. 9.15 The SOC results in discharging operation mode.

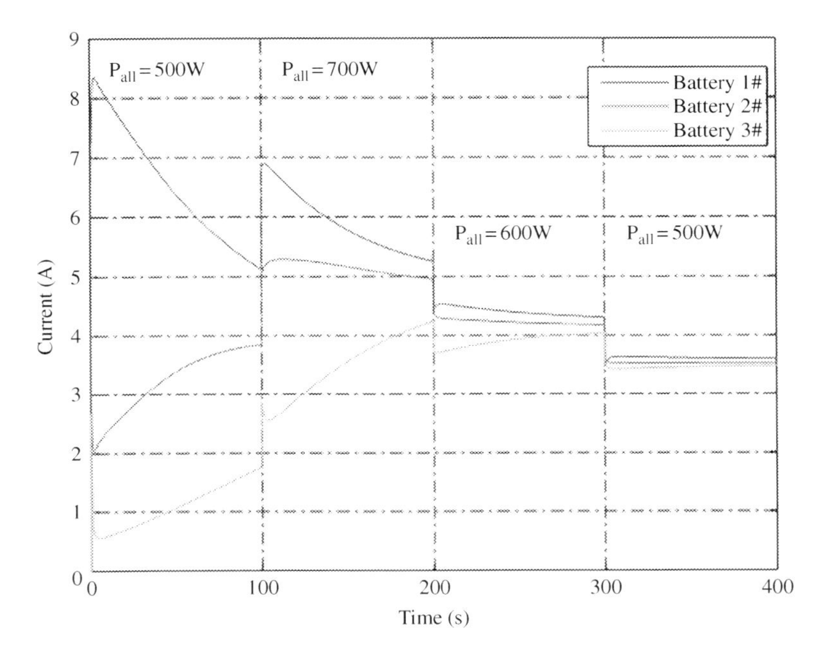

Fig. 9.16 The current results in discharging operation mode.

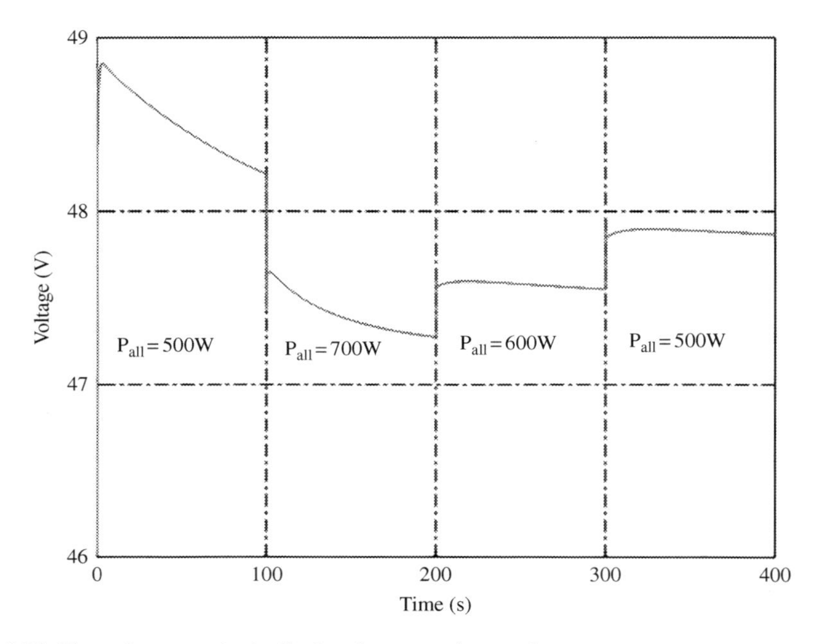

Fig. 9.17 The voltage results in discharging operation mode.

9.5 Conclusions

In this chapter, the microgrid communication and its application are studied. The communication technology and categories are introduced first. The consensus algorithm based on communication system is employed in distributed hierarchical control to maintain the microgrid operate stably and equalize the battery SoC. What's more, the structure of communication system is built according to the control goal. In order to verify this hierarchical control method, the simulation model of a DBESS with three batteries is developed. The results show that can also the total output power can be changed in the system according to the requirements, and the battery SoC is balanced by exchanging information through communication system both in charging operation and discharging operation.

Acknowledgments

This work was supported by National Natural Science Foundation of China (61473238, 51407146), Sichuan Provincial Youth Science and Technology Fund (2015JQ0016), and Doctoral Innovation Funds of Southwest Jiaotong University (D-CX201714).

References

Aalamifar, F., Lampe, L., Bavarian, S., Crozier, E., 2014. WiMAX technology in smart distribution networks: architecture, modeling, and applications. TandD Conference and Exposition, 2014 IEEE PES, pp. 1–5.

Adamiak, M., Patterson, R., Melcher, J., 1995. Inter and intra substation communications: requirements and solutions. American Power Conference, Chicago, IL, United States.

Barmada, S., Musolino, A., Raugi, M., Rizzo, R., Tucci, M., 2011. A wavelet based method for the analysis of impulsive noise due to switch commutations in power line communication (PLC) systems. IEEE Trans. Smart Grid 2 (1), 92–101.

Bhangu, B.S., Bentley, P., Stone, D.A., Bingham, C.M., 2005. Nonlinear observers for predicting state-of-charge and state-of-health of lead-acid batteries for hybrid-electric vehicles. IEEE Trans. Veh. Technol. 54 (3), 783–794.

Chen, C., Zhang, B., Lu, Q., Wu, Q., Wang, J., Liu, S., 2015. A consensus algorithm based on nearest second-order neighbors' information*. IFAC-PapersOnLine 48 (11), 545.

Chen, W.K., Satyanarayana, C., 1978. Applied graph theory: graphs and electrical networks. IEEE Trans. Syst. Man Cybern. 8 (5), 418.

Deng, W., Pei, W., Qi, Z., Kong, L., 2008. Design and implementation of data acquisition, communication and monitoring system for photovoltaic power station in microgrid. Proceedings of ISES World Congress 2007. Vol. I–V. Springer, Berlin, Heidelberg, pp. 1538–1542.

Dou, C.X., Liu, B., 2013. Multi-agent based hierarchical hybrid control for smart microgrid. IEEE Trans. Smart Grid 4 (2), 771–778.

Franceschinis, M., Pastrone, C., Spirito, M.A., Borean, C., 2013. On the performance of ZigBee pro and ZigBee IP in IEEE 802.15. 4 networks. Wireless and Mobile Computing, Networking and Communications (WiMob), 2013 IEEE 9th International Conference, pp. 83–88.

Guerrero, J.M., Vasquez, J.C., Matas, J., De Vicuña, L.G., Castilla, M., 2011. Hierarchical control of droop-controlled AC and DC microgrids—a general approach toward standardization. IEEE Trans. Ind. Electron. 58 (1), 158–172.

Hochgraf, C., Tripathi, R., Herzberg, S., 2010. Smart grid charger for electric vehicles using existing cellular networks and SMS text messages. Smart Grid Communications (SmartGridComm), 2010 First IEEE International Conference, pp. 167–172.

Huang, W., Qahouq, J.A.A., 2015. Energy sharing control scheme for state-of-charge balancing of distributed battery energy storage system. IEEE Trans. Ind. Electron. 62 (5), 2764–2776.

Ko, J., Seo, J., Kim, E.J., Shon, T., 2012. Monitoring agent for detecting malicious packet drops for wireless sensor networks in the microgrid and grid-enabled vehicles. Int. J. Adv. Robot. Syst. 9 (1), 31.

Kriegleder, M., Oung, R., D'Andrea, R., 2013. Asynchronous implementation of a distributed average consensus algorithm. Intelligent Robots and Systems (IROS), 2013 IEEE/RSJ International Conference, pp. 1836–1841.

Kuzlu, M., Pipattanasomporn, M., Rahman, S., 2014. Communication network requirements for major smart grid applications in HAN, NAN and WAN. Comput. Netw. 67, 74–88.

Lassetter, R., Akhil, A., Marnay, C., 2002. The CERTS microgrid concept.White paper on Integration of Distributed Energy Resources, CERTS.

McGranaghan, M., Goodman, F., 2005. Technical and system requirements for advanced distribution automation.18th International Conference and Exhibition on Electricity Distribution. vol. 5. p. 93.

Morstyn, T., Savkin, A., Hredzak, B., Agelidis, V., 2017. Multi-agent sliding mode control for state of charge balancing between battery energy storage systems distributed in a DC microgrid. IEEE Trans. Smart Grid. PP (99) 1–1.

Prasanna, G.S., Lakshmi, A., Sumanth, S., Simha, V., Bapat, J., Koomullil, G., 2009. Data communication over the smart grid. Power Line Communications and Its Applications, 2009. ISPLC 2009. IEEE International Symposium, pp. 273–279.

Qinruo, W., 2003, October. Communication reliability design of low-voltage power line communication system.Systems, Man and Cybernetics, 2003. IEEE International Conference-Vol. 4. pp. 4040–4044.

Ren, W., Beard, R.W., 2008. Distributed Consensus in Multi-Vehicle Cooperative Control. Springer London, London.

Ren, W., Cao, Y., 2010. Distributed Coordination of Multi-Agent Networks: Emergent Problems, Models, and Issues. Springer Science and Business Media, New York.

Shi, R., Zhang, X., Xu, H., Liu, F., Cao, W., 2014. Research on the synchronization control strategy for microgrid inverter. Power Electronics and Application Conference and Exposition (PEAC), 2014 International, pp. 210–213.

Shukla, S., Deng, Y., Shukla, S., Mili, L., 2014. Construction of a microgrid communication network. Innovative Smart Grid Technologies Conference (ISGT), 2014 IEEE PES, pp. 1–5.

Sood, V.K., Fischer, D., Eklund, J.M., Brown, T., 2009. Developing a communication infrastructure for the smart grid. Electrical Power and Energy Conference (EPEC), 2009 IEEE, pp. 1–7.

Xin, H., Gan, D., Li, N., Li, H., Dai, C., 2013b. Virtual power plant-based distributed control strategy for multiple distributed generators. IET Control Theory Appl. 7 (1), 90–98.

Xin, H., Zhang, M., Seuss, J., Wang, Z., Gan, D., 2013a. A real-time power allocation algorithm and its communication optimization for geographically dispersed energy storage systems. IEEE Trans. Power Syst. 28 (4), 4732–4741.

Xu, Z., Li, X., 2010, March. The construction of interconnected communication system among smart grid and a variety of networks. Power and Energy Engineering Conference (APPEEC), 2010 Asia-Pacific, pp. 1–5.

Yang, H., Qiu, Y., Li, Q., Chen, W., 2017. A self-convergence droop control of no communication based on double-quadrant state of charge in DC microgrid applications. J. Renewable Sustainable Energy. 9(3). 034102.

Yigit, M., Gungor, V.C., Tuna, G., Rangoussi, M., Fadel, E., 2014. Power line communication technologies for smart grid applications: a review of advances and challenges. Comput. Netw. 70, 366–383.

Zhang, Z., Chow, M.Y., 2012. Convergence analysis of the incremental cost consensus algorithm under different communication network topologies in a smart grid. IEEE Trans. Power Syst. 27 (4), 1761–1768.

Zhao, D., Zhang, N., Liu, Y., Zhang, X., 2013. Synthetical control strategy for smooth switching between grid-connected and islanded operation modes of microgrid based on energy storage system. Power Syst. Technol. 2 (4), 301–306.

Zimmermann, M., Dostert, K., 2002. Analysis and modeling of impulsive noise in broad-band powerline communications. IEEE Trans. Electromagn. Compat. 44 (1), 249–258.

ICT technologies standards and protocols for active distribution network

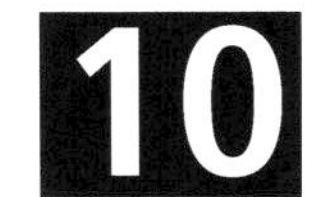

Ting Yang
Key Laboratory of Smart Grid of Ministry of Education, School of Electrical and Information Engineering, Tianjin University, Tianjin, China

10.1 Introduction to the concept of information and communication technology (ICT)

The concept of information and communication technology (ICT) was first born in the 1940s. At present, there is no uniform definition for ICT in the world. Generally, ICT can often be regarded as a synonym for information technology (IT) or the extension of IT concept. Compared with IT technology, the concept of ICT covers communication technology, computer technology, and assistive technology related to both. In the traditional sense, communication technology focuses on the sending and receiving technologies for message transmission, while IT focuses on encoding, decoding, as well as processing of information. As technology evolves, these two technologies are slowly becoming inextricably linked and are gradually merging into one category, namely, the concept of ICT.

On the basis of the above concept, the concept of ICT is often used to represent the entities converged by telecom networks, computer networks, and audio-video networks. Since the convergence of these networks can greatly reduce the network construction costs, people have been actively seeking technology solutions to realize the vision of an ICT network. In a broad sense, ICT includes all the media that can record information (such as hard disk, CDs, etc.), technologies that broadcast information (such as wireless technologies), and technologies that enable people to communicate via audio or video (such as video cameras, mobile phones, etc.). The advances in ICT have spawned everything from small home electronic data sheet to large enterprise software packages, online software services, and the hardware and software used to deliver information over the Internet.

In recent years, with the rise and rapid development of ICT technology, the demand for informatization in the power industry has also become stronger (Panajotovic et al., 2011). The informatization of power system follows the process of being from the component to the local and the system. At present, the information system of power enterprise can be divided into two parts: the automation of production process and the informationization of management process.

Smart Power Distribution Systems. https://doi.org/10.1016/B978-0-12-812154-2.00010-9

In the aspect of production process automation, supervisory control and data acquisition (SCADA) system is a typical representative of power system informationization. As the core of power system informationization and automation, SCADA system is responsible for collecting and processing all kinds of real-time and non-real-time data, which is the data source for a variety of applications; the monitoring and control capabilities of the SCADA system include not only direct control over the physical manufacturing process but also managerial controls. After the 1970s, the system of auto-FM (facility management) power economic allocation and automatic regulation system combines with SCADA system in the form of AGC/EDC (automatic generation control/economic dispatching control) software package. At the same time, software such as online power flow and breaking simulation was developed, SCADA has been gradually upgraded to an energy management system (EMS).

In the informationization of power system management process, with the popularization of enterprise resource planning (ERP) theory and systems, power companies will also introduce ERP system into the enterprise information management process. ERP expands the scope of management information integration. In addition to production management and quality management, it also covers functions such as finance, distribution, manpower, and decision support. The ERP system provides the following functions: dispatching automation, computer metering and accounting, electricity marketing management automation, customer service center, and so on. At present, the widely used information management systems include power transmission and transformation production management system, power market operation management system, power marketing management information system, and ERP and management system.

10.2 Introduction to active distribution network

The current power system basically consists of three parts, namely, power plants, power grids, and power consumers. The grid can be divided into transmission and distribution networks, thus forming the power generation—transmission—distribution areas. Since the grid has been adopting a centralized mode of operation, electricity is generated and output by geographically suitably located power plants, electric energy is delivered to load center by means of high-voltage cables, and central control strategies are taken at all levels to monitor the grid. The traditional distribution network is passive, and the high-voltage power on the transmission line is delivered to the users through the distribution network. Therefore, the medium- and low-voltage distribution network is regarded as the "passive" load of the power system. However, this mode of operation has a certain limitations:

1. The operating status of the grid is moving closer to the critical safety margin.
2. It is more and more difficult to build new transmission lines and large-scale power plants subject to various conditions.
3. There has been a sustained growth trend in electricity consumption in all countries.
4. Stimulated by a renewable economy, the government begins to encourage the development of low-power, environmentally friendly forms of power generation.

In the above background, more and more countries are beginning to develop vigorously distributed generation (DG). Geographically, distributed power stations are generally close to the load center; the capacity of which usually range from a few kilowatts to tens of megawatts. Such power supplies are generally designed to access the distribution network (low or medium voltage). However, to a certain extent, the above methods will affect the production, transmission, and distribution of the existing electric energy. The existing distribution network is not designed to be large-scale accessible from power generation units, distributed power supply will affect the power flow distribution, and thus affect the power grid control and protection system. Meanwhile, in the existing distribution network analysis and calculation, the loss, voltage, or reliability design are all based on the maximum load conditions or average load conditions. As a result, traditional power distribution system design rarely consider the access by large number of distributed resources. In addition, taking into account the intermittent, dispersive, and noncontrollability of most distributed power supplies and traditional structure of the power system, the network structure of the power distribution network needs to be further optimized.

To sum up, it has become the focus of future development of power grid that increasing the current reliability and flexibility of distribution network. With the access of new energy sources, electric vehicles and charging stations play an increasingly important role in improving the environment in the future. The existing distribution network has become even more complicated, and its control and protection has long been distinguished from the traditional distribution power grid, power distribution network is continuously developing to become even more "active."

10.2.1 Smart grid architecture

As one of the most important aspects of smart grid implementation, active distribution network is closely linked with the overall goal of smart grid and its architecture. The intellectualization of distribution network cannot be realized without the support of various subsystems in the smart grid (Fang et al., 2012). From a system architecture point of view, the smart grid encompasses advanced metering infrastructure (AMI), advanced distribution operation (ADO), advanced transmission operation (ATO), and advanced asset management (AAM), and the four systems work closely together to ensure the safe and stable operation of the smart grid (Yu and Luan, 2009). AMI is the network processing system responsible for measuring, collecting, storing, and analyzing power consumer information. This system is composed of smart meters, home area network (HAN), local communication network, and Back-Haul communication network which is connected to power company's data center. The main function of ADO is to realize the self-healing capability of the system. In order to achieve this goal, the power grid should have a flexible reconfigurable network topology and the ability to monitor and analyze the current status of the system in real time. The implementation of AAM requires the installation of a large number of sensors in the system that provide system parameters and "health status" of equipment and the integration of collected real-time information with equipment maintenance, engineering and construction, customer service, and other processes. ATO focuses on

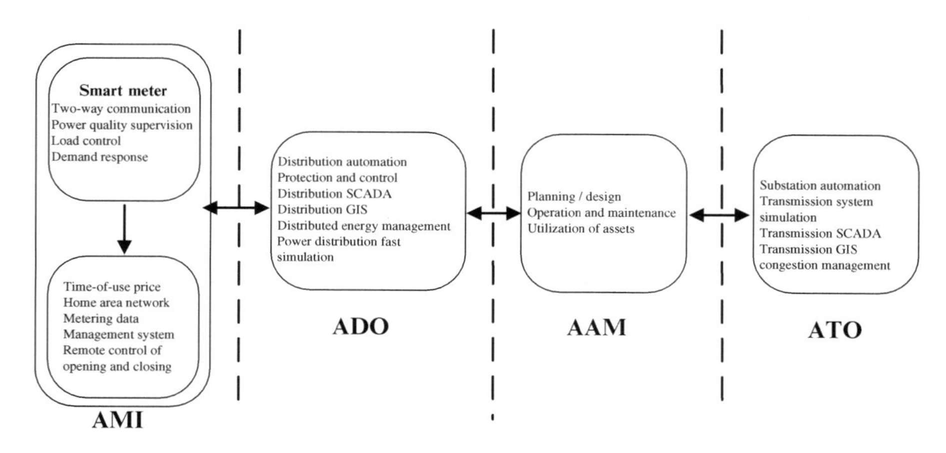

Fig. 10.1 Smart grid architecture.

blocking management and reducing the risk of mass outages and works closely with AMI, ADO, and AAM to optimize transmission systems. The relationship between the above subsystems in smart grid is shown in Fig. 10.1.

10.2.2 Active distribution network in the background of smart grid construction

In the background of smart grid construction, in order to cooperate with the access and regulation of distributed power sources, the concept of active distribution network has been proposed. The active distribution network is defined as the distribution network with operation and control capabilities and with distributed or decentralized energy within it (D'Adamo et al., 2009). Among them, the initiative is reflected in that the distribution network control center can monitor the distribution network and the user side of the load, as well as the status of distributed power operation, at the same time, the active distribution network could utilize a variety of sensing technology to predict its development trend and put forward coordinated optimization control strategy. The traditional transmission of electric energy depends on the transmission network; however, the transmission network is also the weakest link in the whole transmission system. After fully optimizing the access and management of DG, DG can reduce the load on the transmission network, thus reducing transmission losses and increasing the security margin of the system. The controllability of active distribution network is reflected in the effective control of distributed power, energy storage, and load. Through the dynamic perception and customization of the optimal control strategy, the control center can effectively implement the strategy. In addition, the active distribution network is also reflected in the initiative to be able to prejudice the possible factors that threaten the safety of the power grid, and to develop corresponding strategies in advance, and implementing such strategies by the control center, rather than the passive measures taken by the traditional distribution network only after the breakdown of the system.

The basic task of the power grid is still to provide users with normative power. However, the power cannot be stored in large quantities. The balance between the power generation and power consumption and loss must always be maintained. If power generation and electricity consumption suddenly lose balance, the whole area of power system may even crash. Therefore, after the introduction of distributed power supply, the distribution network needs to introduce new hardware devices and flexible intelligent control to enhance the automation, remote control, and informationization of the system so as to monitor the operation status of the power grid in real time and take the necessary measures in time ensure the supply of electricity.

In summary, the active distribution network needs to have the following main features: with distributed controllable resources; having a complete control capabilities; having the control center with a coordinated management abilities; and with flexible regulatory network topology. Since each device in the power system is likely to be equipped with an intelligent electronic device (IED), the development of IT has led to significant differences in the control and management of distribution networks from the traditional distribution network. In order to achieve the above objectives, the current research on the active distribution network mainly focuses on the following areas: wide area active control; adaptive protection and control; real-time distribution network simulation; and distributed communications.

10.3 ICT technologies in the active distribution networks

Active distribution network will be able to highly integrate and utilize a variety of distributed energy through real-time control system and two-way communication network to improve system reliability and operational efficiency. Distributed power supply will become an integral part of distribution management system (DMS) (voltage control, reactive power control, and load management) and integrated into the entire electricity market.

10.3.1 ICT in the active distribution networks

Active distribution networks require an effective wide area communication network to transfer data and control commands between the master and distribution terminals or onsite smart devices. Active distribution network communication system has the following characteristics:

1. A large number of communication terminals. Distribution network has a large number of substations, switching station, distribution transformers, load switches, and reactive power compensation devices. These substations and power distribution equipment need to be monitored, requiring a large amount of remote terminal units or onsite smart devices. Considering the large number of remote meter reading and control terminals that located at low-voltage users, the amount of end nodes in an active distribution system is quite large. In the

actual distribution network, even only some important distribution equipment and users are selectively monitored, the number of terminal nodes in the intelligent distribution network is an order of magnitude larger than that of the transmission dispatching automation system in the same area. In addition, remote terminals or on-site smart devices in active distribution networks are usually installed with distribution equipment. Due to the extensively geographical distribution of distribution equipment, communication nodes are not only large in number but also scattered in distribution.

2. Short communication range. The area covered by a distribution network is relatively small, and the distance between communication nodes in the smart distribution network is relatively short. Therefore, the distribution communication network generally adopts a combination of the communication main channel and the cell branch communication network, the communication distance between terminal data, and remote aggregation site is generally between 2 and 3 km.
3. Small data traffic. The monitoring objects of active distribution network RTU (remote terminal unit) are a large number of line switches, distribution transformers, etc., due to the limited amount of data communication, even including some distribution substations, substations or distribution substations, the data traffic is much less than the transmission substation.

10.3.1.1 Requirements for communication systems in the active distribution network

The requirements of the active distribution network for communication capability depend on the size and complexity of the network and the level of intelligence expected to be achieved (Yan et al., 2013). Meanwhile, in different power enterprises, the distribution network intelligence and the distribution network technology are developing at different speeds, which lead to the different requirements for an active distribution network. Generally speaking, the communication system applied to the active distribution network needs to meet the following requirements:

1. Capable of adapting to harsh operating conditions. Most devices in the distribution network are installed in outdoor harsh condition, these devices should have moisture proof and thunder prevention abilities. Communication equipment is usually installed in the power line column or modular distribution cabinet. In daily operation, these devices may experience high voltage, high current or strong electromagnetic interference, and noise, to ensure the normal operation of these devices, they need to have strong antiinterference ability.
2. High communication speed. The basic requirement for communication rates is to support high-speed, real-time communication required for power system protection and control. Distribution, substation monitoring, and feeder automation requirements for the communication rate should be no less than 300 bit/s; as for public transformer testing and load monitoring, about 10 bit/s will meet the requirement; while remote meter reading and billing systems require the minimum communication rate. When designing a communication system, it is crucial that not only the current active distribution network communication network needs to be satisfied, but also enough bandwidth needs to be reserved to support the power system to launch new services and new applications to meet the requirements for future system expansion.
3. High coverage. Ability to support real-time monitoring and control of all relevant locations.
4. The ability to communicate properly in the event of power line failure. Active distribution network is able to locate the fault location and implementing automatic isolation and

restoring power supply. In case of line faults or structural changes, the communication between the RTU and the master station cannot be affected, so as to ensure the dispatch automation function of the distribution network and the normal operation of the fault section isolation.

5. Two-way communication. Most functions in the active distribution network require two-way communication. Therefore, not only can the communication system be able to upload information, but also to achieve the downlink of control variable.
6. Economy. To meet the reliability requirements, we need to improve the system's cost performance, it is necessary to select the appropriate communication equipment and methods, but also to estimate the maintenance cost of long-term use.
7. Operational convenience. There are many communication terminals in the distribution network. The design should take full account of the convenience of installation, operation, and maintenance to reduce the installation man-hours and improve the construction efficiency.

10.3.1.2 ICT communication technologies in the active distribution network

A wide range of communication technologies can be used in active distribution networks, common types of such technologies include coaxial cable, power line carrier (PLC), microwave relay, and optical fiber communications. Through these new communication technologies, integration and connectivity of information and application can be realized in a wider area, it also realize the data flow between different main body and application of the distribution system, meeting the demand of the active distribution network for the communication system. The communication techniques employed in the power system are shown in Table 10.1.

Below is the introduction to some mature communication technology that can be used for active distribution network.

1. PLC technology. PLC communication uses the cables in the distribution lines as the information transmission medium. It has the advantages of low cost, wide coverage and convenient installation, and maintenance. As the dedicated power system communication network, PLC is mainly used for information transmission within the power system. With the continuous improvement of modulation, transmission technology, and signal processing technology, the broadband PLC technology has been put into use at present.
2. Fiber optic communications. The outstanding characteristic of optical fiber communication is high reliability, strong antiinterference ability (suitable for intense electromagnetic interference), good confidentiality, and high-transmission bandwidth, thus becoming the preferred communication method for backbone communication network in active distribution network. At present, optical fiber communication technologies used in distribution network automation include modem, Ethernet passive optical network (EPON), and industrial Ethernet. EPON technology has the advantages of saving fiber resources and high bandwidth. When adopting the star networking mode, EPON has certain advantages, however, its network management performance needs to be further strengthened. The advantage of industrial Ethernet mainly embodies in adapting to special environment, and the high reliability of ring networks, however, there is still a compatibility problem of communication protocols between different manufacturers' devices.
3. Wireless public network communication. The investment in wireless public network communication is relatively small, so as to avoid the risk of investment failure of communication

Table 10.1 Communication techniques employed in the power system

	Advantages	**Disadvantages**
Power line carrier	Communication function over existing power lines; small investment; wide coverage; Easy installation and maintenance	Narrow bandwidth; distribution transformers will block the power line carrier signal; three-phase power lines have greater signal loss; poor reliability;
Optical fiber communication	High reliability, antiinterference ability, long communication distance, high data bandwidth	High construction cost; shunt coupling inconvenience of de-multiplexing and multiplexing; incapable of transmit signals while powering the equipment
Wireless public network	High reliability; simplifying the communication network design and debugging process for power company	Lower real-time data transmission rate; the financial cost is closely related to the amount of data transmitted
Micro power wireless networks	Easy installation and wiring; does not depend on the network of telecommunications operator, large amount of network nodes, high redundancy	Short communication range; low data transmission rate, high failure rate of the single-point equipment
Field bus	High antiinterference ability, high reliability, providing real-time data transmission	Short communication range limited to a single LAN

equipment. At the same time, equipment construction, debugging, and maintenance based on communication network will be greatly simplified. At present, wireless public network is able to meet existing needs since the distribution network terminal data traffic is small and there is less demanding on the real-time data transmission. However, with the continuous development of active distribution network and smart grid, the data transmission rate and real-time requirements are becoming higher. Whether or not public network communication could meet the future needs of smart grid construction needs further verification.

4. Micro-power wireless technology. In recent years, with the development and integration of sensor technology, computer technology, and wireless communication technology, the development of wireless sensor network (WSN) technology has been promoted. Compared with the traditional wireless networks, WSNs have the advantages of low cost, wide detection coverage, quick deployment in some hard-to-wire areas in the power network, thus extends the wireless network coverage. Through the management node, the user could manage the WSNs by issuing monitoring tasks and collecting monitoring data to achieve data collection, data fusion, and task coordination control. In the power system, WSNs have been preliminarily applied in the fields of remote meter reading, load forecasting, substation automation, distribution network protection, etc. These applications effectively monitor the operation status of the power system and have improved the operation efficiency of the

power system, so that the WSNs have become a complement to the production, transmission, distribution, and consumption of electricity.

5. Fieldbus technology. Fieldbus technology mainly solves the problem of digital communication between field devices such as intelligent instruments and meters, controllers, and actuators in the industrial field and the transmission of information between control systems. Due to the reliability, economy, and practicability of the fieldbus technology, it has obtained great importance from many standard groups and computer manufacturers. The main current bus technology in use is RS485, Modbus, and LonWorks. RS485 serial bus is an improved serial bus standard, at present, the performance and security of RS485 can basically meet the needs of the active distribution network construction (input and output isolation, antistatic, lightning protection, etc.). Modbus was originally developed by Schneider Electric to control PLCs in 1979. Currently, Modbus has become the de facto standard protocol and is widely used in industrial electrical control equipment. In a SCADA system, Modbus bus is often used to connect the RTU to the control panel. Modbus protocol is supported by modems and gateways from many manufacturers. Meanwhile, Modbus protocol can also be used in wireless communications, such as GPRS (general packet radio service), and even used in micro-power wireless communications.

10.3.1.3 Communication protocol in the active distribution network

With the continuous development of IT, power enterprises have gradually deployed a large number of distributed communication systems and various application systems. These systems usually adopt different hardware platforms, different database technologies, and different communication protocols. Users need to switch between different systems during the operation of services, which frequently cause problems such as interoperability between network protocols and management information. In order to solve the above problems, a high degree of information sharing needs to be achieved between various constituent units and subsystems. Therefore, standardization is crucial for the construction of smart grid (Rohjans et al., 2010). First, it is necessary that the interface and product should be standardized while the standardization of applications and business cases should be avoided, otherwise the innovation and development of smart grid will be seriously hindered. Second, there should be the description of common requirements, however, the standardizing details should not be included. Based on the above requirements, the International Electrotechnical Commission (IEC) set up the "Smart Grid International Strategy Working Group (SG3)," which led the study on the technical standard system for smart grids and put forward a series of guidelines for smart grid communications standards (Gungor et al., 2011). Several commonly used communication protocols in power systems and distribution networks are introduced as follow.

IEC 60870

The power system communication protocol is the interface of information exchange between the station-side RTU and the dispatching system. With the development of software and hardware technology, the protocol is also constantly evolving. Remote protocol is the interface between RTU and dispatching system for information exchange. By use of standardized protocols, equipment from many different suppliers can be made to interoperate. IEC standard 60870 has six parts, defining general information related to the standard, operating conditions, electrical interfaces, performance requirements, and data transmission protocols. The 60870 standards are developed by IEC Technical Committee 57 (Working Group 03).

In order to improve the real-time communication capability, IEC60870-5-101 adopts the structure of three layers, including physical layer, the data link layer, and the application layer, the application layer is directly mapped to the data link layer to enhance the real-time information. In point-to-point and multiple point-to-point full-duplex configurations, this standard can employ balanced transmission to take advantage of the bandwidth of the transport network. IEC60870-5-101 provides a set of communication files for sending basic messages between the primary station and remote RTU. It is suitable for network topologies include point-to-point, multiple point-to-point, multipoint point-to-point, multipoint ring, and multipoint star network configuration, but it requires fixedly connected data circuits between the primary station and each remote mover station, which means that a fixed dedicated telecontrol channel must be used.

The IEC60870-5-104 telecontrol protocol employs a reference model derived from the ISO-OSI Reference Model for Open Systems Interconnection but uses only the physical, data link, network, transport, and application layers. IEC60870-5-104 is actually a combination of IEC60870-5-101 and network transmission functions provided by TCP/IP (transmission control protocol/internet protocol), making IEC60870-5-101 compatible with a variety of network types supported by TCP/IP. In the Internet, the transport layer can use the TCP protocol or UDP (user datagram protocol) protocol, in order to ensure reliable transmission of remote data, IEC60870-5-104 provides the transport layer using the TCP protocol, so the corresponding port number is TCP port, and the port number used by the IEC60870-5-104 standard is 2404, which has been approved by Internet Assigned Numbers Authority (IANA).

The structure of the application protocol data unit (APDU) of IEC60870-5-104 is shown in Fig. 10.2, which is composed of application protocol control information

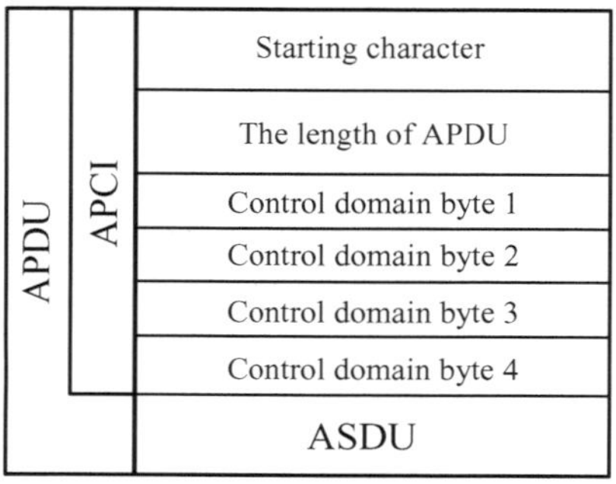

APDU	APCI	Starting character
		The length of APDU
		Control domain byte 1
		Control domain byte 2
		Control domain byte 3
		Control domain byte 4
		ASDU

Fig. 10.2 The structure of the application protocol data unit (APDU).

(APCI) and application service data unit (ASDU), compared with the frame structure of IEC60870-5-101. The application service data unit is the same, except that IEC60870-5-104 uses APCI, whereas IEC60870-5-101 uses the link protocol control information (LPCI). In APDU, the start character 68H defines the starting point within the data stream and the length of the application data unit defines the length of the APDU body. APDU control domain includes four 8-bit groups, according to its definition, APDU can also be divided into three kinds of message format.

IEC 61968

The power system production process is composed of dispatch centers, power plants, substations, distribution networks, and the power users, it is a hierarchical, distributed system, in order to achieve integration and information sharing between different systems, the concept of DMSs is introduced. With the continuous development and expansion of functions of DMS, it has become increasingly unrealistic that all the functions of DMS is provided by single specific supplier, and users are eagerly expecting DMS to be composed of different application subsystems, especially the application subsystems of different vendors to be concisely integrated in a standardized manner so as to realize data exchange and sharing among different applications and reduce data maintenance work, so as to reduce the total cost of construction of DMS.

In this background, the IEC TC57 WG14 proposed the IEC 61968 standard for standardizing the integration of DMS application functions. The IEC 61968 standard addresses DMS architecture, data modeling, functional design, and subsystem interface design approaches and tends to facilitate system integration of multiple distributed software applications that support grid management. IEC 61968 standard focuses on the interaction of different application modules rather than the specific implementation of the system, it does not define the specific implementation of a module and details and which execution modules constitute the DMS. The standard divides the DMS into several abstract application components and defines the interface specifications for each component from the overall business function point of view. The logical structure of the DMS is shown in Fig. 10.3. Power distribution management is the collaboration of many departments to complete the operation and management of distribution networks. The functions of various departments are logically referred to as transactional functions such as network operation (NO) function, operation plans and optimization (OP) function, maintenance and construction (MC) function, network expansion (NE) planning function, and so on.

IEC 61968 Common Information Model (CIM) is an important tool for application integration in information integration of power enterprises. It is based on the unified modeling language (UML), which include public classes, attributes, relationships, etc. Since its classes and objects are abstract, it can be applied to many power systems. CIM is the soul of logical data structures and extends the definition of data exchange models. The CIM defines the semantics of information exchange content and is the foundation and core of the entire protocol framework. The introduction of the IEC 61968-3 to 10 series of standards defines service abstraction, and Sections 3-10

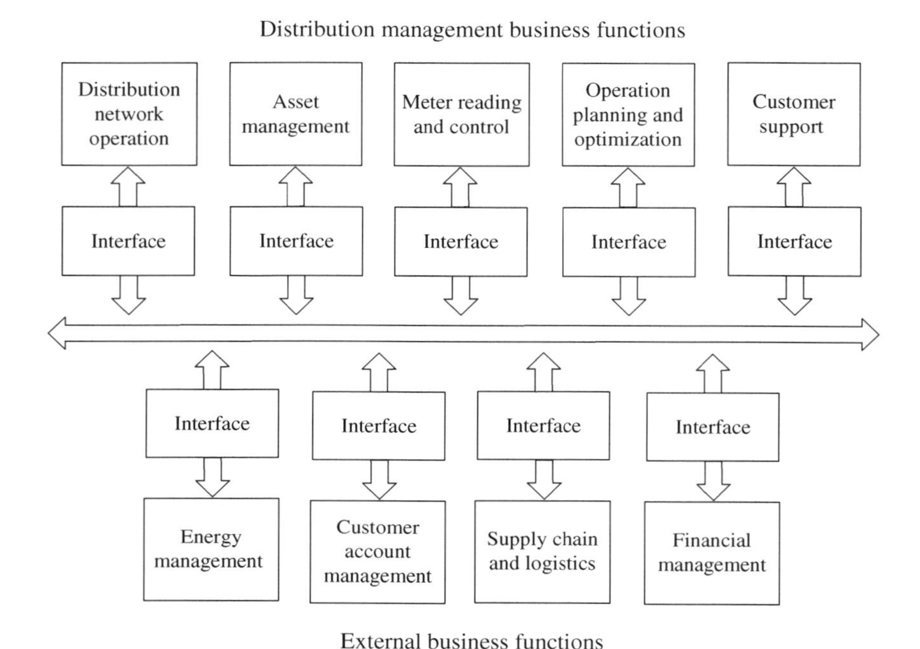

Fig. 10.3 Abstract business components of IEC 61968.

defines the exchange of messages between applications, covering the identified business functions, using cases and activity diagrams to illustrate the information exchange between interfaces. In the standard, message types are specifically defined for each information exchange requirement, the message types can be implemented by different protocols. One message type is a CIM-based canonical data model and the message type is defined using extensible markup language schema. The main work involved in the definition of service includes business process modeling, public information model extension, and definition of the message model.

IEC 61970

IEC 61970 is a series of international standards of "Energy Management System Application Programming Interface (EMS-API)" developed by the IEC. EMS is defined as: "A computer system consisting of a software platform that provides basic service support and a set of applications that provide efficient operation of power generation and transmission equipment to ensure sufficient safety when supplied with minimal power loss." An API is defined as: a set of public functions provided by an executable application component for use by another executable application. The IEC 61970 series of standards define the API of the EMS for the purpose of facilitating the integration of various applications within EMSs from different manufacturers, facilitating the interconnection of EMSs with other systems within the dispatch center, and the exchange of model between EMS at different scheduling center. The protocol consists of three parts: Interface Reference Model (IRM), CIM, and component interface specification (CIS). IRM describes the way of system integration, CIM defines the semantics of information exchange, and CIS defines the syntax of information exchange.

CIM and CIS are the core of IEC 61790. CIM is an abstract model representing the main objects of all power companies and is often included in an EMS information model. The model includes public class objects and its properties as well as the relationships between them. The overall CIM is split into several packs. Future IEC 61790-301 defines a set of basic software packages to provide both informational and physical views of the EMS. Including the core, topology, interruption, protection, measurement, load model, and so on. For a specific type of application and a particular type of exchange, the type and properties of the object being exchanged need to be defined. This means that the data are saved by defining the message structure of the interface. These structures can be CIM object classes or views of CIM object classes. The CIS actually defines the content and behavior of the data for each EMS application. For each exchange of information, CIS includes a set of specific interface categories expressed in UML. Currently, the CIS-defined application categories under preparation include: SCADA, alarm handling, load management, load forecasting, and generation control.

Although IEC 61970 is called "energy management system application program interface," in fact, the idea of IEC 61970 can be applied to power automation and information system integration. The complete IEC 61970 standard document includes the following five sections:

- 61970-1: Guideline.
- 61970-2: Terminology.
- 61970-3xx: CIM.
- 61970-4xx: CIS (component interface specifications).
- 61970-5xx: CIS technology mapping.

IEC 61850

Smart substation is the most basic component of smart grid. Therefore, the development of smart grid begins with the intelligent substation. Intelligent substation technology is developing with the extensive use of IEC 61850 standards, network communications technology, and smart devices in the substation. Computer network and optical fiber communication technology is the key technology that supports the construction of smart substation, however, in order to ensure that the communication system connects the computer and the terminal equipment to perform data transmission and exchange, the communication standard is an essential link. The communication standard defines the order of transmission of information, the format of information, thus ensuring the consistency of understanding of the meaning of the transmitted information between the sender and the receiver.

The purpose of IEC 61850 is to achieve interoperability between different electrical equipment manufacturers' products. This standard is not only the state-of-the-art communication standard, but also a comprehensive system-oriented standard from a system perspective. IEC 61850 not only achieved the standardization of system communications, but also the following aspects of the standardization:

(1) recommended solutions for system and project management;
(2) specific data models in the field of electricity;
(3) the services of the power system;
(4) substation configuration language; and
(5) consistency testing.

The IEC 61850 standard covers only interoperability, not interchangeability. Interchangeability refers to the ability of devices from different manufacturers to replace one another. Interchangeability requires that devices from different manufacturers not only have the same communication interface but also provide the same functionality without affecting the rest of the system. Interchangeability means a deeper standardization, however, this capability is not within the scope of the IEC 61850 standard. The contents of the IEC 61850 are shown in Table 10.2.

IEC 61850 is the first complete communication standard system about substation automation system. The core functions of the standard can be divided into three parts: information modeling, abstract service, and concrete mapping. Brief introduction of IEC 61850 standard specific features is as follows:

(1) Hierarchical structure. The IEC 61850 standard logically divides the required functions of a substation automation system into three levels: station level, bay level, and process level.
(2) Design of abstract communication service interfaces (ACSI) and specific communication service interfaces that is independent of the network. IEC 61850 adopts the mature international communication standard as its communication protocol stack. For different communication requirements, IEC 61850 defines four protocol stacks, including: ACSI service; generic object oriented substation event (GOOSE)/sample value service; GSSE (generic substation state event) service; and clock synchronization.
(3) The adoption of object-oriented, application-oriented self-description. The IEC 61850 standard uses an object-oriented description method that defines more than 80 logical nodes and more than 350 data objects and 23 public data classes covering all functions and data objects of the substation. The definition includes the following three aspects:
- Definition of a complete set of data objects and logical nodes, logical device code.
- Definition of how to use these codes to constitute a complete description of the data object.
- Definition of a set of object-oriented services.

(4) Interoperability. The major goal of IEC 61850 is to achieve interoperability among various smart electronic devices (IEDs) in substations. Interoperability will greatly protect the user's previous investment in assets and facilitate the direct integration of equipment from different manufacturers. By modeling with standard UML, the integration of heterogeneous systems would become simple and effective, the interoperability between devices will also be possible. In addition, the standard also defines interoperability testing methods, including conformance testing and application testing. Conformance testing is a test of whether IEDs meet specific criteria, and performance tests place IEDs in real-world applications to verify that the system meets operational performance requirements. The relationship between consistency and performance testing is shown in Fig. 10.4.

Table 10.2 Contents of IEC 61850 standard

Stand number	Name	Description of content
IEC 61850-1	Basic principle	Scope and purpose of the standard
IEC 61850-2	Terminology	Definition of specific terminology
IEC 61850-3	Overall requirements	Quality requirements, environmental conditions, ancillary services
IEC 61850-4	System and project management	Project requirements, project management requirements, special tools requirements
IEC 61850-5	Function and device model communication requirements	The path of the logical node, logical communication link
IEC 61850-6	Substation structure configuration language	Language description of device and system properties
IEC 61850-7-1	Communication structure of substation and feeder equipment: principles and models	
IEC 61850-7-2	Communication structure of substation and feeder equipment: abstract communication service interface	Interface description, communication service specification, service database model
IEC 61850-7-3	Communication structure of substation and feeder equipment: public data class	Definition of abstract public data levels and attributes
IEC 61850-7-4	Communication structure of substation and feeder equipment: compatible with logical node class and data object do addressing	The definition of logical nodes, data objects and their logical addressing
IEC 61850-8	Special communication service mapping: SCSM	Communication mapping within substation and bay and between substation and bay
IEC 61850-9	Special communication service mapping: SCSM	Mapping within bay and process layers and between bay and process layers
IEC 61850-10	Conformance testing	

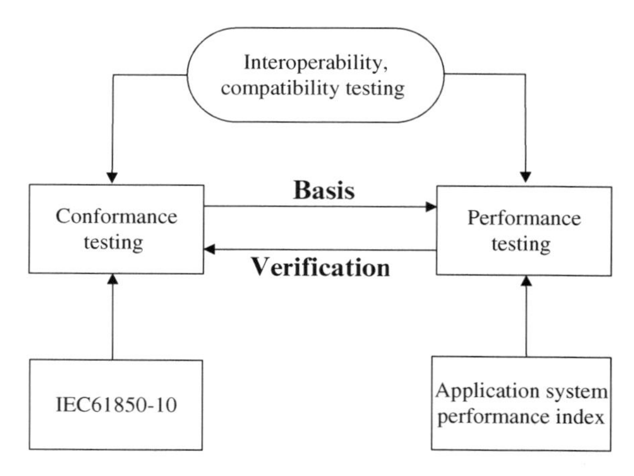

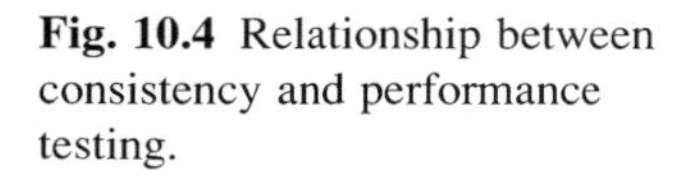
Fig. 10.4 Relationship between consistency and performance testing.

10.3.2 Active distribution network and SCADA system

SCADA systems are usually used by power companies for day-to-day management of the grids. The implementation of active distribution networks largely depends on the ability to efficiently and cost-effectively apply SCADA systems to network-managed communications and operational control. With the advancement of DG, there are some constraints of gird itself, such as voltage limitations, thermal overload, and the complexity of the hardware connections, these limitations and complexity have become major obstacles to DG development, therefore, in an active distribution network, applying SCADA system with an active control model will be a favorable solution.

10.3.2.1 Existing SCADA system

SCADA systems are responsible for coordinating communication and control actions between substations and control rooms. In the 1960s, first-generation SCADA systems were used in state-change alarming and manned substations. In the 1970s, the second generation of SCADA system adopted RTU technology with polling mode to provide information for the control room. The third generation of SCADA systems began in the 1980s with more advanced technologies that provide better bandwidth and efficiency than the second-generation systems.

The SCADA system is mainly used to control the power grid based on a local area network (LAN) with voltage levels higher than 6.6 kV, which is usually located in remote locations. System functions of SCADA system include data acquisition and processing, remote control, alarm processing, historical data, graphical human-machine interface (HMI), emergency control switch, demand-side management, etc. There are usually two control schemes for SCADA system, that is, centralized control and distributed control.

Some functions of the SCADA system require centralized control, such as scheduling of low-frequency load shedding and load side management, and the problem of the centralized control is the lack of reliable basic communication facilities. As some substations lack basic communication facilities such as RTUs, this results in a slower detection rate of switch changes in SCADA systems. Centralized SCADA systems have functions such as network topology management, asset databases, maintenance of hardware and software, and centralized configuration management. However, such systems tend to have some disadvantages such as the lack of economical communication infrastructure, slow operation, physical distance test bottlenecks, and single point of failure. Distributed SCADA consists of small SCADA systems located in different substations. Compared with centralized systems, distributed SCADA has some advantages such as modular repeatable logic function, low-cost wireless communication, and better switching response time. However, there are also some difficulties and challenges with this type of system: Distributed SCADA systems require additional maintenance facilities and management tools that are compatible with multiple distributed operating systems.

10.3.2.2 Architecture of SCADA system

The SCADA system consists of intelligent RTU and PLC that can perform simple, autonomous logic processes without the need for host intervention. SCADA system is mainly divided into four parts: RTU; PLC; master and HMI; and SCADA communication infrastructure. Fig. 10.5 shows the basic configuration of a master station. It is based on a LAN and has several servers and workstations connected to it.

The main function of the front-end processor (FEP) is to communicate with each RTU, sending the data collected by the RTU to the network, and receiving the commands sent by the dispatch workstation and sending them to the related RTUs. The FEP usually has the ability to communicate with multiple plant sites, each with its

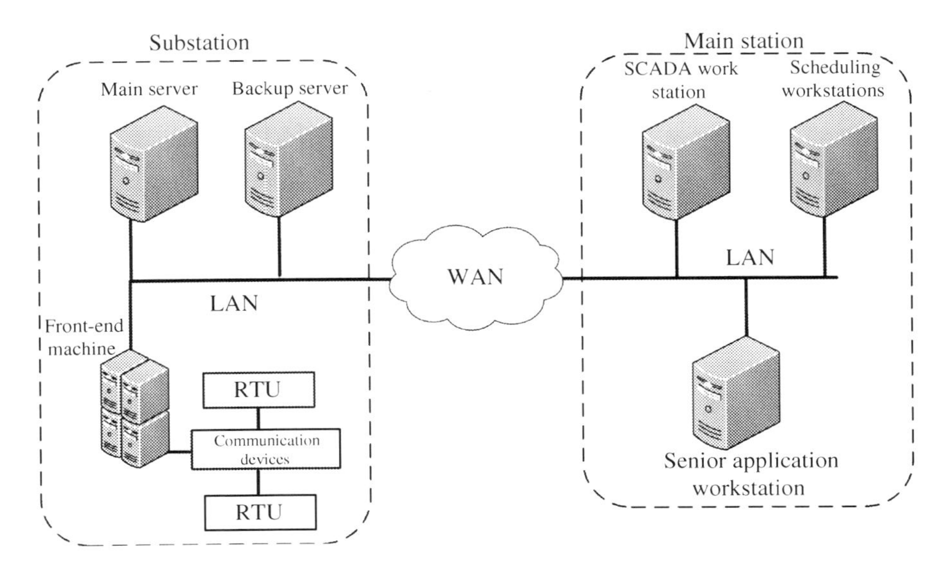

Fig. 10.5 System configuration of supervisory control and data acquisition (SCADA).

own independent port, and accommodates synchronous, asynchronous, and analog and digital communications.

A SCADA system usually consists of a master station and a plurality of substations. The master station is usually located in the dispatch control center, while the substation is installed in a power substation or a power plant. The master station communicates with the substations through a telecontrol channel or a wide area network to monitor and acquire the data. The main station of SCADA system can also be divided into front subsystem and back-end subsystem. The front subsystem is responsible for communication with the plant end and other scheduling control centers. The back-end subsystem is responsible for data processing tasks. The SCADA system hardware components and communications infrastructure are introduced briefly as follows:

1. Remote terminal. RTU is used to connect specific controlled objects in SCADA system. A typical RTU has a serial port, or Ethernet, or other types of communication interfaces, a microprocessor, sensors, control switches, etc., and it has certain impulse voltage protection ability. Sometimes a device bus or fieldbus can also be used to interconnect the RTU with the host system. Generally, RTUs could perform simple procedures autonomously without the intervention of a SCADA system, providing the necessary redundancy for safe operation.
2. PLC. PLC is a software-operated set of digital relays that, unlike general-purpose computers, are designed for use in environments with high electrical noise, vibration, and other impacts. There is dedicated I/O interface in PLC to connect sensors and actuators for reading limit switches, analog and digital variables. The PLC's inherent programming capabilities make it possible to replace thousands of relays with a single PLC, which greatly expand the operational flexibility by changing the code in the firmware. Generally, PLC provides RS232, RS485 port, and built-in TCP/IP protocol, and also provides commonly used industrial field bus protocol such as Modbus and other protocols.
3. Master station and HMI. At present, the RTU needs to operate autonomously without a master station or a human operator. The master station usually performs data analysis before issuing any command. Currently, SCADA system hardware needs to be robust enough to withstand extreme temperatures, vibrations, and voltages for demanding stringent safety standards for master and HMI software. The SCADA system master consists of the server and HMI workstations responsible for communicating with the field devices (RTU, PLC, etc.). For smaller systems, one computer may be sufficient, however, in larger SCADA systems, the master station may consists of multiple servers, distributed applications, and incident recovery sites. The function of HMI is to display graphically the data collected by the system, so that the operator can obtain the necessary control information.
4. SCADA communication infrastructure. SCADA systems generally use wireless, RS232, RS485, or modem to meet the communication needs. One of the major problems in the construction of communication systems in SCADA systems is the lack of standards at the beginning of design and construction. In addition, the design of system communication protocols has become increasingly compact, and many protocols send information to the master station through RTU polling only. In order to expand the customer, larger vendors of SCADA systems are always trying to promote self-developed communication protocols. Actually, typical present SCADA system communication protocols are proposed by device manufacturers such as Modbus and PR-570. However, these communication protocols have basically become the de facto standards and recognized by the major SCADA system vendors. Although the Internet is not used in SCADA system communication, TPC/IP protocol

can be integrated into the majority of these communication protocols. Since currently used RTUs, PLCs, and other automation equipment were developed prior to the promulgation of industry interoperability standards, therefore, a large number of communication protocols need to be developed for the proper communication between the SCADA system and numerous controller devices.

In order to develop international standards for communication between systems and equipment during power production, the International Electrotechnical Commission's 57th Technical Committee (IEC TC 57) was established in the early 1960s. In the process of setting standards, it is important to consider not only the equipment level but also the system parameters such as SCADA systems, EMSs, and distribution automation systems. Since US-based grid communications architectures already provide reference standards for real-time interfaces, protocols, and data models, IEC TC 75 most likely adopt these standards as IEC 61850 subsystems. The purpose of communication standardization is to ensure interoperability and interchangeability of functions, hardware and software interfaces, protocols, and data models among system manufacturers.

10.3.3 Information security issues in active distribution network

With the development of distributed energy and microgrid technology, all kinds of new loads are connected to the distribution network in the form of microgrids, thus more effective and flexible management is required. In this process, two-way electricity demand interaction including power flow and information flow is formed between the power operator and the user. In the active distribution network, the information interaction mainly occurs in the distribution network close to the user side. The terminal devices and some communication networks of the power users exist in an open form, thus significantly increases the possibility of information security incidents in the network. The information security incidents that are attacked maliciously from the distribution network happened frequently. In the power distribution terminal, once the load is maliciously occupied by unauthorized users or authorized to use illegally, it may cause the waste of power resources and instability of the grid (Tan et al., 2017).

The problem of information security of power system has been gradually paid attention in the industry, therefore, research institutions have carried out an extensive research on this subject, of which IEC 62351 is a typical safety standard. However, this standard is not a complete and specific solution since it only proposed that all levels in the communication system should take appropriate measures to ensure information security. The IEC also proposed specific safety requirements for two-way demand interactive power and also stressed the need to study solutions to safety requirements and other functions of the advanced metrology system of smart grids.

10.3.3.1 Information security requirements in active distribution network

In the active distribution network, information security usually needs to consider the confidentiality, integrity, availability, and certification in order to provide comprehensive security protection.

(1) Confidentiality. Confidentiality requirements mean that any sensitive data are not disclosed to third parties outside during the communication process. On the distribution side, this means that no third party shall have access to the user's metering equipment, except for the power operator and consumer, since such equipment often contains user consumption data that reflects the user's daily behavioral characteristics and may lead to user privacy leaks.

(2) Integrity needs. The requirement of integrity refers to preventing information from being modified during creation, transmission, and reception. In the process of power data communication, attackers can modify the original communication data through fake message injection, message replay, message delay, and so on, so as to destroy the key data exchange, thus causing the equipment damage and even the harm to the user. These data may be customer information such as price information, user account balance, etc., or network operation information such as voltage readings, equipment operation status, etc.

(3) Usability requirement. Usability requirement means that authorized users can access information even under the network attack. Denial-of-service (DoS) attack tries to delay the network traffic by delaying, blocking, and destroying the information transmission, thus disabling the communication nodes that need information exchange in the active distribution network. To launch such attacks, intruders only need to access to the communication channel without having to be authorized. As network users often use wireless communication to access to various devices, which is highly vulnerable to such attacks. Therefore, it is very important for power companies to put forward effective countermeasures to prevent such attacks.

10.3.3.2 Information security precautions

Under the environment of active distribution network, a large number of the above-mentioned information security problems arise due to the interactive power consumption caused by the access of distributed energy and information load. The problem of information security has a long history in the computer industry. Therefore, a few measures can be proposed by introducing various security measures from the information industry.

(1) Hierarchical control of user access. In the hierarchical control, different access rights can be set for users and control entities of different layers. The high-authority user can access the encrypted data information of the low-authority user by deducing the session key. When the low-authority users try to access system resources, the judgment on user rights should be made to control user access and to prevent users from occupying resources. Therefore, the power companies need to develop a safe and an effective hierarchical access control scheme to protect network information security and interests of users.

(2) Strengthen the detection of abnormal behavior. The communication network of the active distribution network environment is an open network. Even if the access control and secure communication mechanism are adopted, it is difficult to ensure that there is no security flaw in the communication system and it is difficult to resist completely malicious attacks. Unlike computer network virus detection, smart meters, if invaded by viruses and lurking, can hardly be detected by operators and users based on existing technologies. Under such security threats, active defense measures need to be studied, which is capable of real-time detection of abnormal behavior inside and outside the system, as such measures will effectively improve the system security and reduce network threats.

10.3.4 Active distribution network and big data

Active distribution network contains a large number of distributed power supply, usually with large-scale network and complex structure. The system usually possesses different characteristic feathers during the actual operation, such as strong interaction, multicoupling, and high random. The data generated during the operation may be featured in a variety of structures, complexity of sources, nonuniform timescales, and the differences in spatial scales. The main data types include the topology information of distribution network, distributed power status monitoring information, relevant regional meteorological information, equipment status monitoring information, distribution automation information, and user marketing information. It is estimated that a medium-sized distribution network will generate hundreds of terabytes of data each year. However, the current distribution network lacks the technologies related to analysis and processing of big data, and fails to make full use of the vast amounts of obtained data that can improve system operation and efficiency. Therefore, it is necessary to provide an effective management mechanism for massive data generated during the operation of the active distribution network. By establishing an efficient data processing platform, the operation status of the distribution network can be accurately predicted and evaluated, meanwhile, an efficient energy dispatch system can be constructed to put forward a reliable active control and protection strategy.

In recent years, big data technology has been widely used in the Internet industry, and its tremendous value has been reflected. Compared with the commercial fields and Internet industry, the application of big data in the smart grid is still in its infancy. For example, big data technology can be applied to the active distribution network for real-time detection and prediction, system operation status assessment, and system control optimization.

10.3.4.1 Features of the big data in the active distribution network

The data in the active distribution network can be roughly divided into three categories: active distribution network operating status and related monitoring information, such as network operating topology, distributed power, and equipment status information (He et al., 2017); regional weather information that will affect distributed power output, such as light, temperature, wind speed, and other information; state and marketing information of network users, such as the operation of electric vehicles, the user's electricity consumption information. The structure and time characteristics of the active distribution network data are such complex that it also reflects the significant features of the big data, which are mainly reflected in the following aspects:

(1) Huge data size. The national grid company of China deployed 353 smart meter information collection points in five residential areas in Beijing and collected 12,000 parameters, including frequency, voltage, and current, up to 34 GB of information can be generated in 1 day.
(2) Wide range of data types. From the data structure point of view,big data in the active distribution network can be divided into structured data (status operation information, etc.) and unstructured data(video surveillance, photos, overhauling document information, etc.). From the perspective of data applications, it covers electrical quantity information such

as current, voltage, and power, weather information for predicting distributed power output, and user behavior record information for analyzing the behavior of electric vehicle users, and so on.

(3) Rapid change. The operating status of the active distribution network is not constant. Due to the unstable output of the distributed power supply and the strong load fluctuation, the information about the operating electric quantities of the active distribution network and the topology data are constantly changing. Research relevant strategies should be studied about how to effectively store and utilize data that change so quickly.

10.3.4.2 The application of big data technology in the active distribution network

In the future, active distribution network will be panorama real-time intelligent distribution network based on big data processing and analysis technology. Distributed storage computing provides a platform for storage and analysis of heterogeneous, multisourced data in an active distribution network. With the elapse of time, platform operation will inevitably generate big data. Combined with big data storage, analysis and computing technology, advance active distribution network application will provide active control, active management, and active operation of power equipment. In the active distribution network, there are mainly following aspects of the application:

(1) Active distribution network energy optimization scheduling. Compared with the traditional power grid, the optimal scheduling model of active distribution network has undergone profound changes from the aspects of control variables, constraints, and objective functions. Active distribution networks include not only controlled DG units such as fuel cells and diesel generators, but also energy storage systems with charge-discharge characteristics and contact switches in the distribution network, all of which add difficulty to making energy scheduling strategy in the active distribution network. Due to the need to integrate multisource data information (such as distributed power start-stop information, electricity price information, contact switch status information, etc.) into the energy optimization scheduling model, the big data processing platform can store and extract multisource heterogeneous data, which greatly improve the data storage capacity and efficiency of the active distribution network energy scheduling system.

(2) Active distribution network operational status analysis and evaluation. One of the most significant features that distinguish active distribution network from the traditional distribution network is that the DG units, energy storage units, and microgrid unit for distribution network operators are controllable. The operation and control of active distribution network are based on the reasonable analysis of its status information. For example, in the case of output prediction for DG, data association rule mining can be used to establish the correlation between distributed power output and different types of weather data and weather stations in the surrounding area. Due to the uncertainty of distributed power output, the massive operation information in the active distribution network will also become the basis for mining the description of the probability of distributed power output.

On the power user side, due to the bidirectional flow of user energy in the active distribution network, based on the existing mass data, the big data cluster analysis method is used to classify the user load according to the stable mode, the fluctuation mode, and the random change mode. Second, by applying correlation analysis technology to setup the correlation between the weather factors and the load, finally, regression decision tree analysis method can be used to establish the intelligent forecasting system based on rule mining to improve the short-term load forecasting accuracy.

(3) Active distribution network protection and control. With the access of large-scale distributed renewable energy, the problem of random bilateral power flow arises in the distribution network, while the active management requirements also make the distribution network topology and operating mode more flexible and changeable, both of which poses a serious challenge in the realization of safe and reliable network protection. Large-scale access to distributed power supply and more flexible power load generated a lot of measurement information. Therefore, mass distribution network information can be effectively used to meet the new challenge posed by the protection of the active distribution network operation control.

Protection measurement information in the active distribution network is featured by multisource and multi-timescale, therefore, it will be an important research direction that automatically correlating the network protection measurement data and setting up the unified description of the method. At the same time, the efficient use of multisource and multi-timescale data needs to combine different tasks and taking into account the accuracy and timeliness of multiple data analysis tasks. In the aspect of fault feature analysis, it is necessary to dig deeply into massive and multisources historical fault recording data, by analyzing key failure characteristics of active distribution network in different scenarios, the protection and control scheme of distribution network based on fault information feedback could be proposed.

10.3.5 Applications of ICT technology in the active distribution network

In the active distribution network, the frequency of power flow changes is increased because of the access of large amount of distributed power supply. The impact on the power grid is increased and unpredictable, which brings even more difficulty in dispatching control of the power grid. In order to track the dynamic state, analysis process and make the right judgments, dispatchers must obtain large amount of practical information by using advanced analytical tools and decision-making methods and thus ensure the safety of the power system. The intellectualization of active distribution network mainly relies on a variety of automatic control, communication, and information management technology, the following is a brief introduction of the application of ICT in the field of active distribution network:

(1) Phasor measurement unit (PMU): PMU uses GPS system timing pulse as a synchronous clock and can be used in the fields of dynamic monitoring, protection, analysis, and prediction of power system. The deployment of PMU is the basic method to observe the dynamic process of power system. It not only plays an important role in the analysis of various dynamic behaviors of power system, but also works effectively in model checking and state estimation. The PMU uses high-precision GPS technology for measuring the current, voltage, and switching data of each node in the substation. With the GPS timing system, the PMU could achieve a precision of 0.1 μs and the phase error less than 0.018 degrees in 50-Hz power frequency. After collecting the data, the PMU can transmit the sampled data to the control center synchronously. The wide-area situational awareness (WASA) proposed in recent years is also based on the synchronous PMU technology. In the same reference time frame, the PMU can collect the real-time information of each node in the power grid, these information is encoded and sent to the data analysis center in a specific format via the high-speed data transmission system so that the power system steady-state and dynamic measurement and control center can obtain the accurate information, thus the corresponding regulation can be made timely.

(2) AMI system. AMI is referred to as the cornerstone of the smart grid. It is mainly composed of smart meters, bidirectional communication networks, metering data management system (MDMS), and HAN. AMI is an integrated system of smart meters and various bidirectional interactive terminals, communication networks, and computer software systems. It is not a single technology implementation, AMI is an integrated technology system used to measure, store, analyze, and operate data, and therefore, it can be regarded as a composite of ICT technology.

In addition to smart meters, communication system is crucial to the AMI implementation. AMI uses a two-way communication network that periodically reads smart meters and sends meter information and fault alarms over the LAN to the concentrator, which connects to the data center through a wide area network. At the power company's control end, commands and information is sent downstream to meters and power users via concentrators. Since there is less demanding on the communication rate in LAN, PLC and power line broadband (BPL) technology is often used. HAN is an important infrastructure in AMI. One of the most important devices in HAN is an intelligent interactive terminal in a user's room. The interactive terminal connects the smart meter and the user's electrical equipment through a gateway to form a LAN and can implement various energy management functions. At present, HAN networks mainly adopt zigbee or PLC as the basic communication technology. The basic function of MDMS is to confirm, edit, and estimate AMI data so as to ensure the integrity and accuracy of the data flow to the billing system when the communication network is interrupted or breakdown on the user side. The data provided by the MDMS can be used to implement time-of-day billing, peak tariffs, and other flexible billing methods.

(3) User-side energy management. In the active distribution network system, the client EMS realizes the intelligentialize and informationization of the distribution and electricity systems from the user side to meet the requirements of internal management in the client side, data exchange of internal subsystems, and connection of external demand-side management platforms. Its main objectives include: achieving the goal of user's energy monitoring and energy optimization, to improve the service efficiency of energy-consuming equipment through manual intervention or automatic control; providing user interaction with the grid company; through interaction with the power company, the client-side equipment enables the user equipment to achieve the goal of balancing the load in the client side; management of access to distributed energy equipment. Since access to energy storage equipment and renewable energy equipment in the grid may cause impact on the grid, at the same time, such equipment can also send electrical energy back to the grid and therefore require the management. To achieve the goal of energy management at the user side, several key ICT technologies below are required:

(a) Bidirectional real-time electricity price charging technology: including active energy and reactive energy metering; time-sharing measurement, the maximum demand measurement, electricity price program display; power quality and environmental parameters; information and data storage, transfer, output, and display functions.

(b) Network programmable control technology: including the fieldbus and industrial Ethernet communication protocol specification research and implementation, high-speed bus technology, control, and communication integration technology.
(c) Massive client data fusion and parallel computing technology: data collection, exchange, communication, storage, and sharing mechanism should be deeply researched.
(d) GIS (geographic information system) technology in outage management system (OMS). As the highest-level application in modern distribution management automation, a power OMS fully handles planned/incident power outages based on equipment management, maintenance staff scheduling and operational scheduling. The main task of the OMS is to analyze the data related to power outage, providing information about the scope of the power outage, analyzing the possible causes of the power outage, and making a power cut forecast. Specifically, it can be divided into three tasks.

There is a detailed network model of the power distribution system at the heart of a power OMS, GIS system will provides detailed information for the model. Through the power outage information reported by users, with the help of rules engine, OMS can predict the location of line breaks. In addition, an automatic meter reading (AMR) system can also provide circuit breaking detection and recovery functions. However, there is certain false alarm rate for the AMR system. Therefore, additional judgment logic needs to be provided to integrate the AMR system into the GIS system. In general, OMS is also integrated with SCADA systems to enable automatic monitoring of circuit breakers and other functions. Therefore, the distribution network data can be centrally and uniformly managed through GIS technology. The management objects of the GIS system include not only the location of power equipment (called space positioning data) such as substations, users, and towers, but also the information that cannot be directly linked with the geographical coordinates (known as nonspatial positioning data), such as properties of the power equipment, operation situation and power flow distribution and other. In the distribution network, the GIS system usually include distribution network connectivity rules and attributes information of the distribution facilities, with the network topology analysis function, GIS could conduct distribution network load transfer and power analysis to determine the scope of power outages, so as to determine the optimal fault isolation point.

10.4 Conclusion

Active distribution network can take advantage of information and communication technologies to manage proactively the access to the large-scale distributed energy distribution network, it can coordinate intermittent renewable energy and energy storage devices and other distributed energy units to achieve safety and economical operation for renewable energy. ICT technology can be said to be an important foundation for active distribution networks. This chapter summarizes and analyzes ICT technologies used in active distribution networks, and introduces several the ICT technology communication standards in smart grids and active distribution networks and the arisen problems in its application of ICT technologies.

References

D'Adamo, C., Jupe, S., Abbey, C., 2009. Global survey on planning and operation of active distribution networks-Update of CIGRE C6, 11 working group activities[C]//Electricity Distribution-Part 1. 20th International Conference and Exhibition on. IET, pp. 1–4.

Fang, X., Misra, S., Xue, G., et al., 2012. Smart grid-the new and improved power grid: a survey. IEEE Commun. Surv. Tutorials 14 (4), 944–980.

Gungor, V.C., Sahin, D., Kocak, T., et al., 2011. Smart grid technologies: communication technologies and standards. IEEE Trans. Ind. Inf. 7 (4), 529–539.

He, X., Ai, Q., Qiu, R.C., et al., 2017. A big data architecture design for smart grids based on random matrix theory. IEEE Trans. Smart Grid 8 (2), 674–686.

Panajotovic, B., Jankovic, M., Odadzic, B., 2011. ICT and smart grid[C]//. In: International Conference on Telecommunication in Modern Satellite Cable and Broadcasting Services, IEEE, pp. 118–121.

Rohjans, S., Uslar, M., Bleiker, R., et al., 2010. Survey of smart grid standardization studies and recommendations. In: Smart grid communications, 2010 first IEEE international conference on, pp. 583–588.

Tan, S., De, D., Song, W.Z., et al., 2017. Survey of security advances in smart grid: a data driven approach. IEEE Commun. Surv. Tutorials 19 (1), 397–422.

Yan, Y., Qian, Y., Sharif, H., et al., 2013. A survey on smart grid communication infrastructures: motivations, requirements and challenges. IEEE Commun. Surv. Tutorials 15 (1), 5–20.

Yu, Y.X., Luan, W.P., 2009. Review of smart grid. Proc. CSEE 29 (34), 1–8.

Further reading

Chai, W.K., Wang, N., Katsaros, K.V., et al., 2017. An information-centric communication infrastructure for real-time state estimation of active distribution networks. IEEE Trans. Smart Grid 6 (4), 2134–2146.

Yang, T., Zhao, R., Zhang, W., et al., 2017. On the modeling and analysis of communication traffic in intelligent electric power substations. IEEE Trans. Power Delivery 32 (3), 1329–1338.

Zhang, Y., Wang, L., Xiang, Y., et al., 2016. Inclusion of SCADA cyber vulnerability in power system reliability assessment considering optimal resources allocation. IEEE Trans. Power Syst. 31 (6), 4379–4394.

Virtual power plant communication system architecture

Matej Zajc, Mitja Kolenc†, Nermin Suljanović‡*
*Faculty of Electrical Engineering, University of Ljubljana, Ljubljana, Slovenia, †ELES, d.o.o., Ljubljana, Slovenia, ‡Milan Vidmar Electric Power Research Institute, Ljubljana, Slovenia

11.1 Introduction

The wide integration of renewable energy resources (RERs) into power systems is driven by environmental, commercial, and regulatory aims (Akorede et al., 2010). Besides the environmental impacts and commercial benefits (Dabbagh and Sheikh-El-Eslami, 2014; Kok, 2009), the massive deployment of distributed generation (DG) provides additional capacity for energy balancing and flexibility to power system operation (Fraunhofer, 2015). Conversely, generation in the distribution grid brings new challenges for power system operation such as voltage rises, power quality issues, complex protection, and new investments in grid infrastructure. The commercial and technical impacts of DG are significantly increased with their coordinated operation and aggregation using virtual power plants (VPPs) (Thavlov and Bindner, 2015).

There are a variety of geographically distributed energy resources (DERs) that can be aggregated to participate in the provision of ancillary services: commercial and industrial consumers such as steel mills, paper mills, and other factories and larger manufacturers; and producers such as photovoltaics (PV), battery energy storage systems (BESS), small-scale hydropower plants, gas turbines, electric vehicles (EVs), and combined heat and power (CHP) systems (Braun, 2007).

Ancillary services (ENTSO-E, 2009) are an important approach to balancing the power system and include different services, which transmission system operators (TSOs) activate to sustain power system stability and security. Ancillary services include scheduling and re-dispatching, reactive power and voltage control, congestion management, load-frequency control, balancing consumption and generation, and imbalance management (Giuntoli and Poli, 2013; Nikonowicz and Milewski, 2012; Palizban et al., 2014; Pudjianto et al., 2007). VPP is a tool that contributes to demand response programs and enables the active inclusion of DERs into ancillary service provision to the TSO or distribution system operator (DSO), most frequently load-frequency control (Ma et al., 2013). Therefore, VPPs can provide ancillary services, more specifically load-frequency control such as manual frequency restoration

Smart Power Distribution Systems. https://doi.org/10.1016/B978-0-12-812154-2.00011-0

reserve (mFRR) and automatic frequency restoration reserve (aFRR), by aggregating resources for generation or consumption.

The VPP, as an information and communication technology (ICT) infrastructure, can be deployed as a standalone or a cloud-based installation. Utilities, aggregators, retailers, and TSOs who own ICT infrastructure (servers and network equipment) for daily operations usually allocate their own resources for VPP installations to reduce investment costs and ensure security. However, cloud-based VPP deployment is more appropriate for smaller aggregators, retailers, and utilities with limited ICT assets (Samad et al., 2016).

The communication system inside a VPP has a hierarchical architecture and utilizes reliable and secure communication protocols providing reliability, performance, and security (Palizban et al., 2014). The use of TCP/IP-based infrastructure is one of the prominent trends in the smart grid domain (Ancillotti et al., 2013; Yang et al., 2011). The exchange of numerous operational and nonoperational data across different layers of the power system in specific time frames and the development of new ICT-based concepts requires the extension of existing and the development of new communication protocols based on TCP/IP transport.

In the downstream direction, the VPP communicates with numerous geographically scattered units (RER, DG, BESS, EVs, and flexible loads) enabling the timely response of DERs to provide the required capacity for ancillary services (Fig. 11.1). The VPP as an independent entity also communicates in the upstream direction, toward electricity retailers, aggregators, TSOs, DSOs, and the electricity market (Ancillotti et al., 2013; Etherden et al., 2015), integrating the VPP into the power system network's operation.

VPPs can be commercial (CVPP) or technical (TVPP). CVPPs facilitate the trading of DERs as a flexible resource on various energy markets (Braun, 2007; Pudjianto et al., 2007)—offering bids on ancillary service markets, for example—whereas

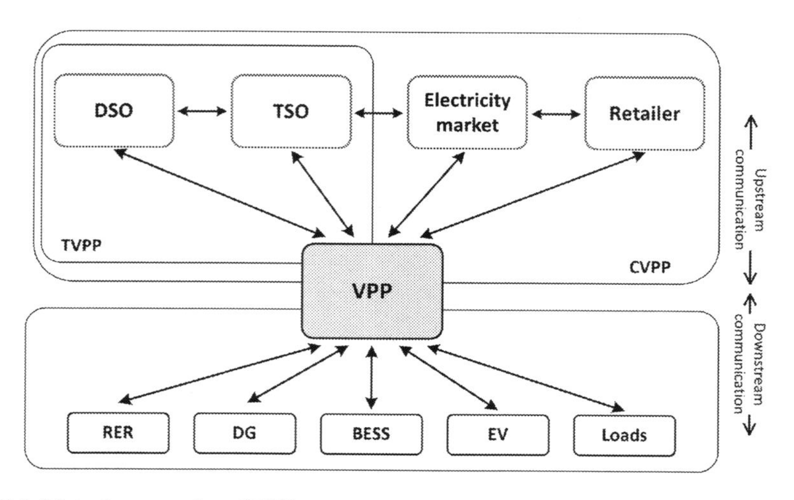

Fig. 11.1 Virtual power plant (VPP) concept.

TVPPs aggregate DERs from the same geographical area for technical purposes, focusing on solving local grid constrains (Li et al., 2016).

In the case of CVPPs, the aggregator or retailer offers flexible capacity on the electricity market. For example, the BESIC (Batterie-Elekttische Schwerlastfahrzeuge im Intelligenten Containerterminalbetrieb) pilot study demonstrated how the dynamic charging of battery stacks of automated guided vehicles at the container terminal in a port can be utilized to provide flexibility, with a total capacity of 640 kW, without having an impact on the port's operation (Wulff et al., 2016). The idea of dynamic charging is based on the possibility of shifting the automated guided heavy transport vehicles charging process to peaks in power production from renewable resources. The CVPP of a local utility monetized the available flexibility by providing load-frequency control as ancillary services to the TSO (Kolenc et al., 2017).

However, TVPPs are typically serving DSO for local system management (Giuntoli and Poli, 2013; Pudjianto et al., 2007). By exploiting TVPP capabilities, DSOs can include the active management of the distribution network among its roles and responsibilities. In this case, the ancillary services can be provided through DG and flexible load units at lower voltage levels at a large scale to balance the intermittent RERs and stabilize the grid. For instance, the Integrid project develops a TVPP to provide DSO with specific ancillary services and active grid management (voltage optimization, control, and forecasting algorithms) enabling the management of small-scale grid batteries and the optimization, control, and load forecasting of the distribution grid voltage (Integrid consortium, 2017). Furthermore, TVPP can also serve the TSO by providing system balancing, ancillary services, congestion management, and engaging its own emergency assets (Nikonowicz and Milewski, 2012). Therefore, TVPP can be used as part of the DSO or TSO system management infrastructure and is not directly related to the electricity market. In addition, several CVPPs with their available capacity can be coordinated by a TVPP to provide specific power system support services. CVPP and TVPP can act jointly (Nosratabadi et al., 2016) to achieve maximal technical and economic impact on the network.

11.2 VPPs in the smart grid concept

The complexity of the smart grid domain is captured in the smart grid architecture model (SGAM), manifesting the importance of ICT for communication between different components (Etherden et al., 2015; Trefke et al., 2013; Uslar and Trefke, 2014).

In this chapter, we propose a modified SGAM to illustrate how the VPP concept fits into smart grid architecture (Fig. 11.2). On the component layer, the VPP server is placed between the DSO and DER components to follow the general architecture shown in Fig. 11.1. In reference to zones of automation in the modified SGAM, VPP participates in operation, enterprise, and the market. The vertical axis corresponds to the interoperability layers illustrating the major protocols, data models, and services presented in this chapter.

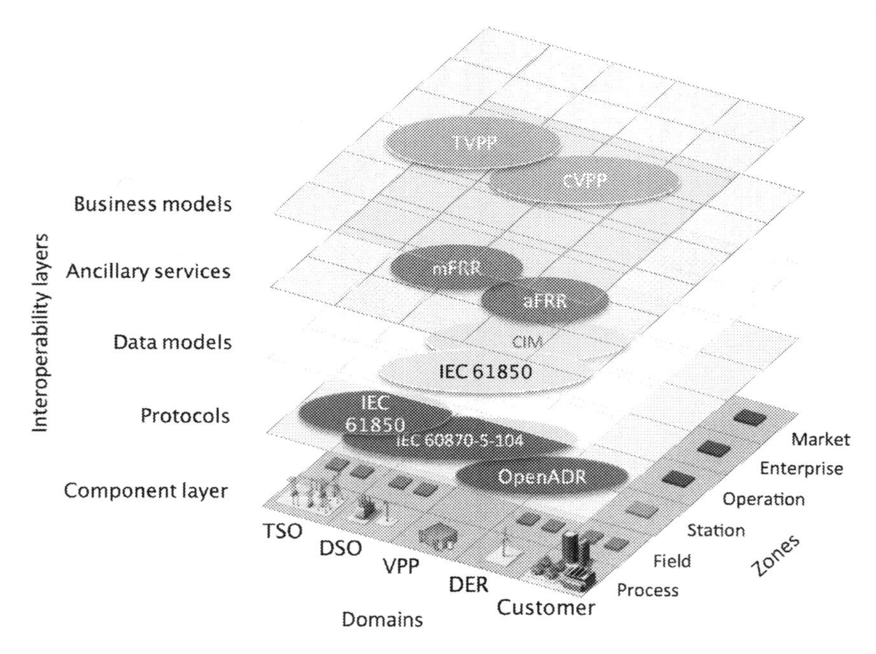

Fig. 11.2 Modified smart grid architecture model extended for VPP.

11.2.1 VPP system architecture

VPP is an entity that acts autonomously based on defined procedures and constrains to provide flexibility and balancing services to the TSO and DSO. VPP communicates in the upstream direction toward the DSO or TSO control center from where the VPP receives commands.

The role of TVPPs is to enable DERs to contribute to system management activities (Pudjianto et al., 2007). Since a TVPP influences the grid state, it communicates with the supervisory control and data acquisition (SCADA) system and energy management system (EMS) of the DSO to obtain operational parameters and data (power flows, voltage levels, network status, measurements, etc.). Information about market as well as DER data (operating parameters, marginal costs, and metering) are provided through one or more CVPPs. A CVPP solely utilizes market data without consideration of DER activation impact on the grid state.

In the upstream direction, VPP regularly reports aggregated VPP profiles including DER data to the TSO or DSO. In the control direction, the TSO or DSO's control center dispatches an activation signal to the VPP, which triggers the delivery of the requested capacity.

When a VPP is used for commercial and trading purposes as a CVPP, a retailer or aggregator offers the available aggregated capacity from the VPP to the electricity market. To exchange and forward relevant market-related data, that is, bids to the retailer, who offers services on the electricity market (e.g., intraday, day ahead, balancing, or ancillary service markets), the VPP needs to communicate with a retailer, who operates market platform applications. In relation to the electricity market, the VPP can receive pricing information utilized for internal resource optimization procedures.

Efficient VPP operation is closely connected with relevant market-related data being exchanged between the VPP and different market actors (e.g., retailors and aggregators). Retailer bids the capacity offered on the electricity market platform operated by the market organizer.

In order to utilize the flexibility potential of different DERs by the VPP technology or other flexibility platforms, comprehensive information (schedules, baseline data, market prices, etc.) needs to be exchanged. These communication protocols do not, however, provide sufficient mechanisms for additional information exchange and more advanced communication protocols are needed for VPP communication purposes.

Communication systems inside smart grids rely on TCP/IP infrastructure and facilitate web services as a mechanism to transfer messages. The VPP communication system is compliant with such architecture. Furthermore, new protocols must meet the VPP's operational requirements. These requirements led to the development of new application-specific protocols, such as Open ADR 2.0, that are primarily used for initiating automatic demand response programs, DER integration and VPP deployment. In parallel to web-based protocols, the interoperability of VPP with other power system components must be supported. The IEC 61850 protocol suite—the dominant communication protocol for data exchange inside power system automation—is also considered for VPP implementation.

The generalized architecture of a VPP is presented in Fig. 11.3. It consists of the main VPP core, a database, a reporting module, an additional functions module, and communication modules for downstream and upstream communications. Processing and decision making are done inside the VPP core via baseline calculation, optimization, and control modules.

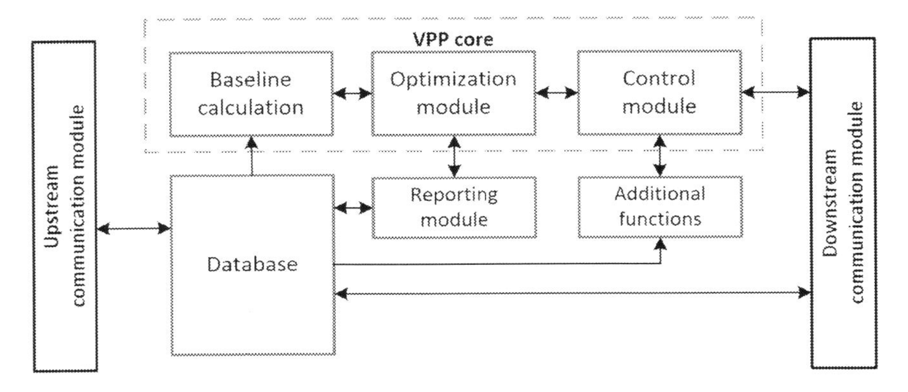

Fig. 11.3 VPP system architecture.

The VPP core performs decision-making processes based on data stored in the database (operational and nonoperational data)—for example, baseline value calculations; selecting DERs from the available portfolio based on preset optimization parameters such as price, availability, responsiveness, and type; and activation signal dispatch control.

The reporting module collects relevant operational data for individual report recipients for each reporting time period. Reporting templates are customizable for use as an input for SCADA, EMS, or other high-level external systems. Additional functions—such as weather forecasts and RER product forecasts—represent supporting modules contributing to the overall VPP operation and decision making. The communication modules ensure reliable communication upstream toward TSOs or DSOs, retailers, or other third-party systems, and downstream toward DERs.

11.2.2 VPP for ancillary services

VPPs provide ancillary services to the TSO and DSO to maintain reliable operation of the grid by offering a capacity of intermittent generation and flexible loads for load-frequency control. An activation request occurs when a TSO or DSO sends a set-point signal to a VPP to start an activation process to deliver the required capacity (Palizban et al., 2014).

The activation can be positive or negative regardless of load-frequency control action type (ENTSO-E, 2009). A positive activation is triggered when there is an electricity shortage in the grid and additional generation units need to be engaged or loads need to reduce consumption (Fig. 11.4A). During the negative activation (Fig. 11.4B), some DERs must reduce generation (generators) or increase consumption (loads) to follow the requested set point (the dashed line).

The Measurements line indicates the power consumption (load) or production (generator) for a particular DER. The available capacity is defined as a difference between availability (positive availability (Fig. 11.4A) and negative availability (Fig. 11.4B)) and the Measurements line. During the activation, the VPP calculates the provided capacity as the difference between the Baseline (dotted line) and Measurements line and reports this information together with other data to the TSO (or DSO).

11.3 Communication system architecture

The telecommunication systems in the smart grid are represented by a hierarchical communication architecture that comprises wide area networks (WANs), field area networks (FANs), neighbor area networks (NANs), and home area networks (HANs) (Fig. 11.5) (Ancillotti et al., 2013; Budka et al., 2014). Communication networks inside smart grids are classified according to the coverage range and smart grid applications, with different quality of service (QoS) requirements (Khan and Khan, 2013). By targeting a cost-effective solution, several communication technologies are dominating—for example, fiber optics, digital subscriber line (DSL) technologies, and cellular technologies (2G, 3G, and 4G).

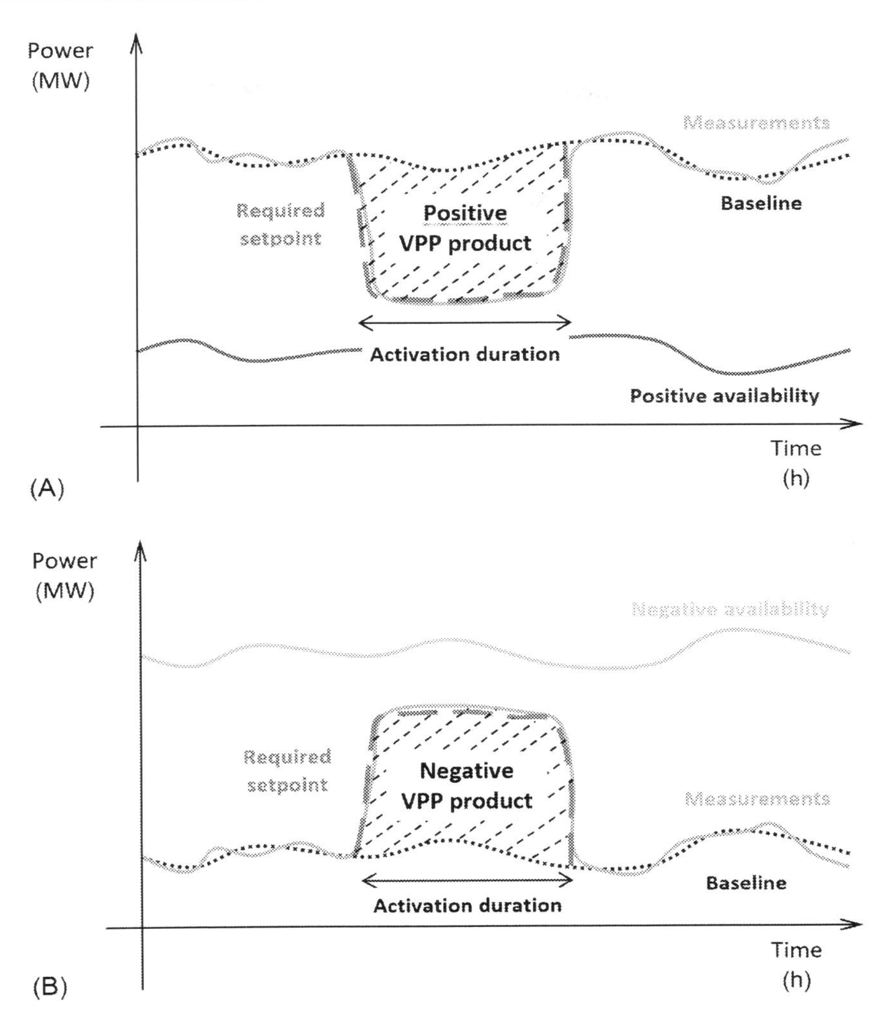

Fig. 11.4 Representation of positive and negative activations. (A) A positive activation, (B) a negative activation.

The VPP communication system is based on IP and Ethernet, and TCP is used as a reliable transport protocol. While both IP and Ethernet have a stochastic character, the QoS and security requirements are typically achieved through the implementation of a virtual private network (VPN) (Kolenc et al., 2017).

11.3.1 Communication requirements

All units providing ancillary services such as load-frequency control [frequency containment reserve (FCR), aFRR, and mFRR], regardless of whether they are conventional power plants or VPP systems used by aggregators, need to fulfil technical requirements [e.g., capacity, ramp-up times, full activation times (FATs), and

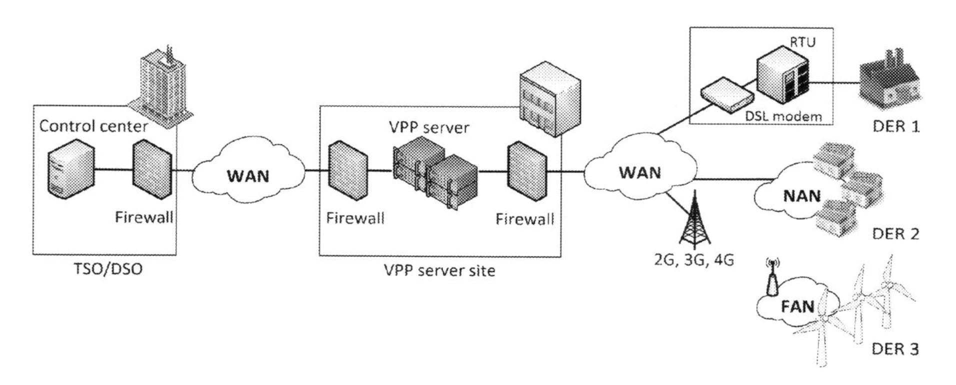

Fig. 11.5 VPP communication system architecture.

responsiveness] and communication requirements (e.g., data communication cycle time) as defined within the ENTSO-E operational handbook (ENTSO-E, 2009).

According to the guidelines for each load-frequency control action, the activation time (T_A) and cycle time (T_C) parameters are predefined (Table 11.1). The FAT, T_A, is defined as the time needed to transmit a set-point signal from the TSO or DSO to the VPP and downstream to DERs engaged in the procedure for a full activation of a particular load-frequency control action. The cycle time, T_C, indicates the time for collecting measurement data and transmitting it back to the TSO or DSO (ENTSO-E, 2009; Kolenc et al., 2018).

The load-frequency control is an important mechanism enabling TSOs to maintain stable power system frequency in the synchronous area. It follows a three-step procedure where FCR is followed by the activation of aFRR and mFRR actions. Each load-frequency control action has its own technical requirements in terms of activation time, cycle time, etc. Specified values in Table 11.1 are given as intervals since different national TSOs adjust values according to their local system characteristics (ENTSO-E, 2009).

Each producer/consumer has its own characteristic ramp-up. Fig. 11.6 illustrates ramp-up characteristics for selected producers (generators) and consumers (loads): (a) batteries, (b) industrial steam or gas turbines with a hot start, and (c) hydropower run-off river plant are presented in the producers group, followed by (d) CHP, (e) industrial loads such as cement mills, and (f) steel mills representing consumers group. The timescale indicates the approximate time needed for 100% activation of

Table 11.1 Load-frequency control technical requirements (ENTSO-E, 2009)

Load-frequency control action	Activation time T_A	Cycle time T_C
FCR	15 s–30 s	1–2 s
aFRR	5 min–15 min	1–5 s
mFRR	15 min	1 min

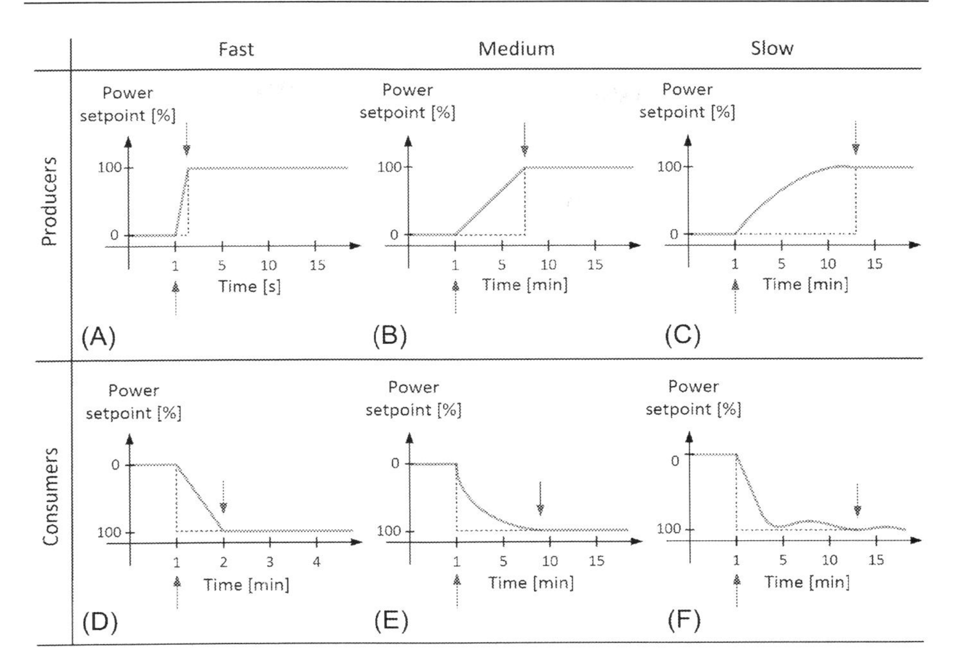

Fig. 11.6 Examples of different distributed energy resource (DER) ramp-up characteristics. (A) Batteries, (B) industrial steam or gas turbines with a hot start, (C) hydropower run-off river plant, (D) CHP, (E) industrial loads, and (F) steel mills.

an individual unit. Values on the abscise axis indicate the maximum level of flexible capacity at 100% and the minimum level that is on disposal for curtailment at 0%.

The operational costs of DERs and their technical characteristics (ramp-up time, offered capacity, reliability, etc.) impact upon the selection of DER for a particular ancillary service. A ramp-up time or ramping period is defined as a period when DER production (or consumption) will reach 100% of the set-point value requested by the VPP. Ramp-up times depend on the physical and technical characteristics of DER units (Table 11.2). As ramp-up time affects the overall latency of the VPP activation process, it can jeopardize reaching the target value T_A for a particular load-frequency control action. Therefore, DER units must be carefully considered within operation of the VPP, particularly for aFRR and FCR services where communication and control requirements are significantly more demanding (Table 11.1). Examples of DER types from Table 11.2 with different ramping characteristics and durations are illustrated in Fig. 11.6.

11.3.2 Communication technologies

One of the key components of the VPP concept is reliable, efficient, and safe communication infrastructure that enables two-way connectivity and reliable data exchange between all entities (Fig. 11.1) in the downstream direction, toward DERs, in the

Table 11.2 Response times of DER units used for aggregation (Klobasa et al., 2013; VDE-ETG, 2017)

Resource type	Ramp-up time	Classification
Consumers		
Steel mills, paper mills, cement mills, refineries	15 min	Slow
HVAC, chillers	5–15 min	Medium
Chillers	1 min	Fast
Producers		
Steam or gas turbine	10–15 min (cold start) 5 min (hot start)	Slow
Small-scale hydropower, wind mills, photo-voltaics (PVs), pump-storage, combined heat and power (CHP), diesel and gasoline generators	2–5 min	Medium
Batteries, battery energy storage systems (BESS)	1–10 s	Fast

upstream direction, toward TSOs, DSOs, electricity retailers, and the electricity market (Ancillotti et al., 2013; Etherden et al., 2015). VPPs need to communicate with numerous geographically scattered asset owners in the downstream direction where operating data, technical data, and cost-related data from each DER needs to be constantly exchanged with the VPP to meet service communication requirements. To ensure secure communication through public Internet access, VPNs are utilized to provide the necessary encryption and authentication mechanisms (Kolenc et al., 2017, 2018).

There are several wired and wireless technologies available to provide the required communications infrastructure on physical and link layers for the upstream and downstream directions (Della Giustina et al., 2013); fiber optics is the dominant technology used for the backbone network (WAN).

DERs are connected to VPPs via FANs and NANs with the wired and wireless technologies that are usually used for Internet access or are part of an advanced metering infrastructure (AMI). The most commonly used wired communication technologies are DSLs. Narrow-band power-line carrier (NB-PLC) technology offers an alternative to DER connectivity through smart metering infrastructure. For most cases, NB-PLC technology is not powerful, fast, and reliable enough to fulfil communication requirements of the VPP for providing demanding ancillary services, especially load-frequency control (e.g., aFRR) (Jansen et al., 2010).

Cellular communication systems are an attractive solution for FAN and NAN implementations, offering various advantages, but also facing challenges to be resolved for future successful deployments. Available technologies are 2G [General Packet Radio Service (GPRS)], 3G [Universal Mobile Telecommunication

System (UMTS)], and 4G [long-term evolution (LTE)] with typical data rates of 14.4 kbps for 2G (GSM), 56–171 kbps for 2G (GPRS), up to 3 Mbps for 3G (UMTS), and 50–100 Mbps for 4G (LTE) (Borovina et al., 2015; Mahmood et al., 2015). Even though 3G and 4G provide high data rates, end-to-end latency varies significantly and cannot always meet time-critical application requirements (Kalalas et al., 2016).

Latency and consequently packet loss in cellular communications can influence VPP performance, particularly when used for demanding load-frequency control services such as aFRR. To perform impact assessments of communication anomalies on VPP performance, the monitoring of selected QoS parameters (latency, packet loss, number of retransmissions, etc.) is crucial (Kolenc et al., 2016). The information is relevant for the VPP optimization module that can take into account an assessment of communication link stability and reliability, estimated on QoS parameter analysis in the downstream direction between the VPP and individual DERs. In the decision-making procedure, a VPP can therefore select resources with lower probability of communication link failures.

Currently, different wireline and wireless technologies are used in commercial and research deployments of VPPs. Here, we briefly overview selected projects relating to a VPP's service and the underlying technology.

Within the eBADGE project, a CVPP was used for providing capacity for cross-border balancing and demand-response purposes, aggregating commercial and industrial customers as well as residential customers (Kolenc et al., 2015). Heterogeneous telecommunication infrastructures (MPLS, xDSL, 3G, etc.) were used within the pilot deployment, using the proprietary HTTP-based communication protocol for the upstream and downstream direction.

The FutureFlow project proposed a common activation platform for providing aFRR for local and cross-border TSOs by engaging industrial DERs aggregated by CVPPs and other flexibility platforms (Beatens et al., 2016). Various wired communication media (xDSL and fiber optics) were used for successful pilot testing and demonstrations based on IEC 60870-5-104 and common information model (CIM).

As previously mentioned, the BESIC industrial research project considered a CVPP for providing mFRR services by engaging batteries from industrial transport vehicles to the local German TSO. Connectivity to public Internet access was achieved using xDSL and fiber optics. Data exchange was based on the OpenADR 2.0b communication protocol (Ihle et al., 2016).

For communications between the actors involved in the HybridVPP4DSO (Gutschi et al., 2017) and Integrid (Integrid consortium, 2017) projects, where VPP is used as a TVPP, wired (xDSL and fiber optics) and mobile (2G and 3G UMTS) technologies are used. In the scope of HybridVPP4DSO, a TVPP is utilized for solving local network constraints as well as CVPP for providing mFRR services to the TSO provided from locally located DERs. Similarly, the Integrid research project deals with applications of VPP technology for both technical and commercial services to the power distribution networks.

11.4 Communication protocols

This section is devoted to communication protocols used in VPP communication system. The activation process is outlined with the message exchange between TSO (or DSO), VPP, and DER. This is followed by an overview of commonly used communication protocols and selected QoS parameters (Section 11.5), crucial for reliable VPP operation.

11.4.1 The VPP message exchange

During VPP operation, continuous information exchange is needed between TSO (or DSO), VPP, and aggregated DERs. The VPP receives commands and set points from the top-level entities, but otherwise acts autonomously. The VPP is closely connected to the electricity market and market-related data is exchanged between the VPP and market actors (retailors and aggregators).

The VPP is receiving power measurements, curtailment capacity, and availability information from DERs. Information is periodically reported with the selected cycle time interval defined by a particular grid service type (Table 11.1). Simultaneously, VPP transmits aggregated pool measurements and calculated baseline values to the upstream entities.

Fig. 11.7 collects a protocol-agnostic message exchange between the TSO, VPP, and DER (Kolenc et al., 2018). Based on monthly, weekly, or daily tenders on the electricity market, the TSO sends a *BID activation* signal to the VPP. In the case of a valid bid, the VPP returns a *BID confirmation* back to the TSO. At the same time, the VPP dispatches a *START activation* signal with the required set-point value to the selected DER. The activation event starts when the DER sends an acknowledgment signal back to the VPP.

During the activation, the DER continuously sends measurements to the VPP to adjust production capacity to the selected set-point value. The VPP constantly sends reports to the TSO to monitor the provided capacity. If needed, the VPP can also adjust the set-point value during the activation event for the selected DER by sending the *CHANGE set-point value* signal. The activation event terminates with an *Activation END* signal sent to the DER. The VPP receives confirmation from the DER and terminates the activation event by sending the report to the TSO.

When the VPP acts as a TVPP to provide grid services on the distribution level, it needs to communicate with the DSO. In addition to the message exchange in Fig. 11.7, a VPP needs to receive operational data (power flows, voltage levels, network status, power quality measurements, etc.) from the DSO's SCADA or EMS systems. The exchange of operational data between the VPP and the DERs is similar to Fig. 11.7 as the VPP needs to continuously receive: power measurements, availability status, and available capacity in the monitoring direction; and dispatch set-point signals in the control direction. TVPP can also require services from CVPPs to exchange status, financial settlement, operational, and other relevant data and signals.

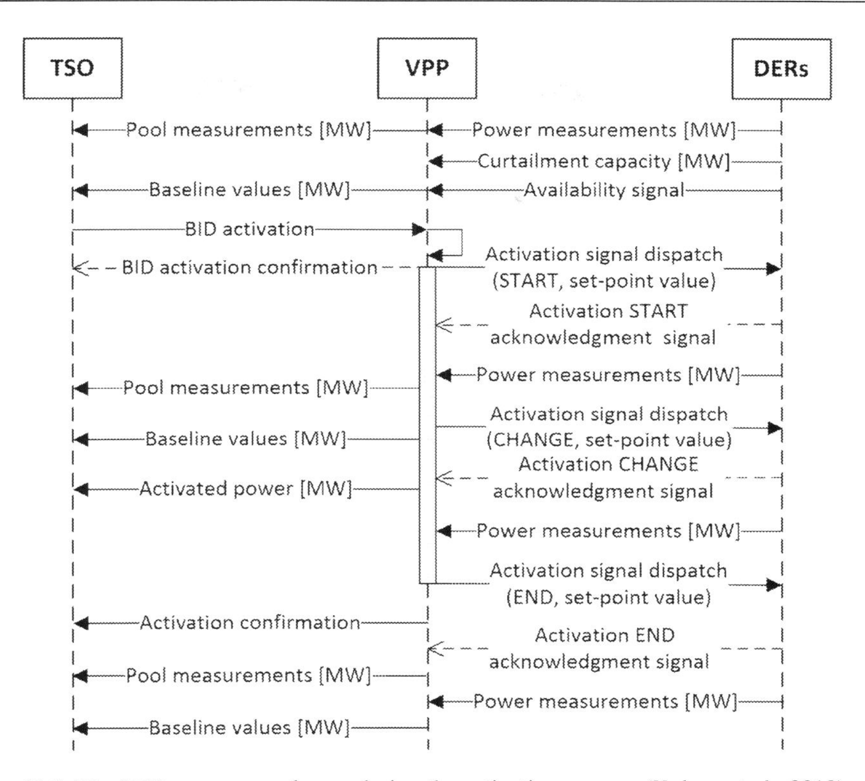

Fig. 11.7 The VPP message exchange during the activation process (Kolenc et al., 2018).

11.4.2 Communication system protocols for VPP

Communication system protocols for VPP system must adhere to several criteria: efficient and reliable communication, interoperability with other systems and stakeholders, and integration into the power system. For easier integration, it is usually desirable that the VPP system supports the communication protocols already in use by a particular TSO or DSO. In addition to standardized protocols, there are many proprietary protocols dominant in the power system automation domain that are currently in use by TSOs and DSOs.

Several communication protocols are used in current VPP systems; those frequently used are IEC 60870-5-104, OpenADR 2.0, IEC 61850, and Modbus (Ancillotti et al., 2013; Samad et al., 2016; Yang et al., 2011). The IEC 60870-5-104 protocol (IEC International Electrotechnical Commission, 2006), an extension of the IEC 60870-5-101 protocol, is widely used for telecontrol (SCADA). Due to its reliability, scalability, and efficiency, it is used in different communication systems in TSO and DSO domains (Etherden et al., 2015). The protocol has an open TCP/IP-based interface enabling connectivity to LAN and WAN.

Standards define modes of data transfer and device roles; downstream messages are exchanged in the control direction from the VPP (serving as a master station) to the DER (serving as a slave station) and in the monitoring direction from the DER to the VPP. In the upstream direction, the VPP serves as a slave station.

Legacy communication protocols such as IEC 60870-5-104 (Kolenc et al., 2018) and Modbus (Ancillotti et al., 2013) lack the support for exchange of flexibility-related data (schedules, baseline, market price data, etc.) required by the electricity market. For these services, new more advanced protocols are needed. While IEC 60870-5-104 provides basic VPP operation functionalities and connectivity to a TSO's SCADA and EMS, it lacks advanced services for DER integration and communication with the market. This is a light telecontrol protocol, which is simple to implement and compatible with SCADA and EMS systems.

There are several initiatives and standardization efforts to develop protocols suitable for exchanging specific data and parameters relevant for monetizing DER flexibility on electricity markets. The industry alliance VHPready e.V. (Industry Alliance VHPready e.V., 2017) is developing an industry standard for networking decentralized energy systems. The proposed VHPready 4.0 communication standard ensures interoperability and controllability of the system components of VPPs. The protocol stack is based on the TCP/IP protocol, whereas on the application level, protocols IEC 60870-5-104 or IEC 61850-4-420 (part of the IEC 61850 standard) are supported. The VHPready 4.0 protocol supports the integration of CHPs, wind and solar plants, batteries, heat pumps, etc.

To fulfil other technical and market requirements for the active engagement of DERs into flexibility provision on a pool level, more advanced protocols are needed. OpenADR 2.0 is an open standard with publicly available architecture, which is already in use in several commercial and research deployments providing a set of data models to facilitate ancillary services, demand response, market transactions dynamic pricing, and DER management (Samad et al., 2016). While OpenADR 1.0 was developed for US markets, the current version, OpenADR 2.0, enables wider use for different applications. It is also the main standard for demand response in Japan, while in Europe is still gaining momentum. It enables the exchange of measurement reports, forecasts, schedules, baseline, market price data, etc. As OpenADR 2.0 is XML based, the protocol has higher communication overheads resulting in higher bandwidth requirements (Kolenc et al., 2017).

The standard supports hierarchical architecture with two node types: virtual top node (VTN) and virtual end node (VEN). Similarly, as in the case of IEC 60870-5-104, VPP takes a dual role, serving as the VEN for the upstream direction where TSO (or DSO) serves as the VTN, and as the VTN in the downstream direction where the DER serves as the VEN (Kolenc et al., 2017).

As an alternative to OpenADR 2.0, the widely accepted IEC 61850 standard also shows promise for VPP communication and integration purposes. The standard comprises of communication architecture and information models for power system automation, bringing several advantages. The IEC 61850 standard, originally developed for applications in the transmission grid substation at the process level, is becoming a dominant standard for message exchanges inside the power utility. Another advantage of IEC 61850 is its object-oriented architecture and communication architecture, which carries out all traffic using IP and Ethernet protocols, taking into account specific QoS requirements set by time-critical applications. Implementation flexibility is achieved by virtual communication interfaces for measurement, control, and protection functions, named logical nodes, and various attributes assigned to data objects in the logical nodes. Such a flexible approach allows the modeling of all VPP components and the hierarchical architecture of the VPP communication system. Since all power grid components should be monitored and controlled remotely utilizing IEC 61850, it is expected that in the future the VPP ecosystem will also be capable to communicate using this protocol.

The current version of the IEC 61850 standard, however, does not provide all necessary capabilities for full VPP operation (Etherden et al., 2015). For example, it does not include flexible information exchange necessary for VPP functionalities such as aggregation, scheduling, pricing, and product prediction. However, substandard IEC 61850-7-420 (IEC International Electrotechnical Commission, 2016) provides standard communication interfaces and the ability for data exchange with SCADA systems and EMS. In particular, it sets standardized interfaces between VPPs and DERs, but without aggregation. Substandard IEC 61850–90-15 describes concepts regarding the integration of DERs into the grid (Etherden et al., 2015). It introduces DER centric energy services and a registry (database) that eases the access of flexibility aggregators to resource providers and DER technical capabilities. Such architecture is suitable for IEC 61850 message exchange utilizing web services.

It is obvious that IEC 61850 will appear in SGAM on both the protocol layer and the data model layer (Fig. 11.2). The IEC 61970 (CIM) standard is a standard for object-oriented information modeling for data exchange inside various power system components, which also fits on the data model layer of SGAM (Trefke et al., 2013; Yang et al., 2011). Data models defined by the IEC 61850 and CIM standards are overlapping, since they were originally designed for different purposes. Mapping from one model to another will require data format converters. For example, the IEC CIM 62325 series of standards define communications for deregulated energy markets, which is of significant importance for CVPP operation.

11.5 Communication system performance analysis

Communication system performance is a significant factor in various fields of power networks and evolving smart grids (Ancillotti et al., 2013; Khan and Khan, 2013). Monitoring QoS parameters enable insights into the communication network and equipment conditions (Hardy, 2001). Different services require specific levels of communication system performance for their undisturbed and reliable operation (Khan and Khan, 2013).

The main parameters that describe the state of the communication network and its performance are latency, packet loss, retransmissions, bandwidth, amount of traffic, and average data transmission speed (Hardy, 2001). Methodologies for monitoring

QoS parameters are well established in distributed smart grid communication networks (Kong, 2015; Vallejo et al., 2012).

QoS parameters in VPPs need to provide the necessary information for adequate VPP performance assessment (Kolenc et al., 2017, 2018). The most relevant QoS parameters in terms of their influence on the communication of system performance and consequently VPP operational reliability are latency, bandwidth, and packet loss.

For normal operation of the VPP communication system, latency is the most relevant parameter related to VPP service requirements, as defined for the load-frequency control in Table 11.1. Bandwidth is protocol dependent and defined by the amount of data needed for transmission in the selected time interval. A large bandwidth does not solve all latency issues, as latency can increase due to network congestion, network topology, etc. A higher rate of packet loss will result in numerous retransmissions on the transport layer and consequently the delayed delivery of messages, which further increases latency.

The selection of communication protocols influences the amount of data (exchanged between VPP, TSO, DSO, and DER) and packet size, both influencing the required bandwidth. To estimate required bandwidth for the IEC 60870-5-104 protocol, we can use its largest message of 286 bytes (I-format associated with the DER measurement reporting). If we assume that these messages are periodically reported every 200 ms, the required bandwidth can be approximated to 13.4 kbps with added Ethernet and TCP/IP overheads (Kolenc et al., 2018).

Similarly, in the BESIC project, we observed a case with a VEN, connecting one DER location, and reporting elementary parameters for VPP operation (Kolenc et al., 2017). In this particular case, OpenADR 2.0 uses messages with up to 2706 bytes for periodical measurement reports, activation events, and capacity forecasts. To estimate the bandwidth, we need to consider the message size together with the Ethernet and TCP/IP overheads, resulting in an estimated bandwidth of 111.4 kbps for messages delivered periodically within 200 ms. However, the size of an OpenADR 2.0 XML-based message can be larger if there are additional data tags integrated into the message. In the case of a VEN, which aggregates data from several DERs, the message size increases accordingly.

Latency influences T_A and T_C and impacts on the reliable delivery of load-frequency control actions. Time is needed to send a message from the TSO or DSO during the activation process; the activation of a DER adds several additional contributions besides the communication delays on communication networks (upstream, $T_{UPSTREAM}$ and downstream, $T_{DOWNSTREAM}$): the VPP server processing time (T_{VPP}), the remote terminal unit (RTU) controller processing time (T_{RTU}), and the DER unit ramp-up time ($T_{RAMP\text{-}UP}$).

$$T_A < T_{UPSTREAM} + T_{VPP} + T_{DOWNSTREAM} + T_{RTU} + T_{RAMP-UP} \tag{11.1}$$

In general, the ramp-up time of the DER units is one of major contributors to the activation time T_A (Table 11.2). In time critical services, for example, aFRR, all contributing elements need to be carefully optimized. In case of network problems, increased latency and packet loss leads to a loss of data or even a loss of the communication connection. Inappropriate communication technology that is not able to fulfil the latency requirements may hinder the VPP's operation.

As reported by Kalalas et al. (2016), the use of cellular communication technologies in smart grid domains brings several advantages as well as some disadvantages. Cellular systems enable wider coverage and easier connection of geographically dislocated resources where a direct Internet connection is not available. However, the network's stochastic nature, signal coverage, switching, and queuing problems may impact QoS levels from time to time, resulting in increased latency and data loss. For example, currently 4G LTE end-to-end latency is around 100 ms, which is still insufficient for delay critical applications (Kalalas et al., 2016). Therefore, time critical services such aFRR with demanding communication requirements (T_A and T_C) might be occasionally compromised if such mobile technologies are used.

11.6 Conclusions

VPPs are playing an important role in the evolving future power networks to cope with the rising share of distributed energy generation. Efficient massive integration of DERs offers solutions to many energy domain problems such as increasing peak demand. Currently, two types of VPPs are used for DER integration: CVPPs at the market level and TVPPs at the TSO and DSO level. Each type requires specific information exchange and efficient communication with different entities.

This chapter highlights VPP architecture with reference to the smart grid communication model. Communication requirements are considered for such VPP architectures to meet required QoS parameters to provide selected ancillary services. In this sense, we provide a protocol-agnostic sequence diagram for message exchange for VPP operation.

In the chapter, we provide an overview of available protocols and present their main concepts, advantages, and drawbacks. All protocols must rely on IP network and utilize public Internet infrastructure. The deployment of legacy SCADA protocols such as IEC 60870-5-104 proved that such a light protocol can be used for CVPP and to communicate with SCADA systems, but their shortages are addressed by the OpenADR 2.0 protocol, which is more complex to implement and more demanding for communication resources.

If VPP is considered as part of DSO, it must interoperate with other power system components. In this case, the IEC 61850 standard provides specifications for communication protocols and data structures that will meet these requirements. Furthermore, it is foreseen to include additional support for various DER units described with specific data models including flexibility parameters to be used for VPP communication purposes.

Current standardization efforts in the domain of communication protocols were outlined with insights into selected current research and industrial deployments. We described the importance of monitoring relevant QoS parameters to detect potential time-critical bottlenecks inside the VPP communication system.

References

Akorede, M.F., Hizam, H., Pouresmaeil, E., 2010. Distributed energy resources and benefits to the environment. Renew. Sust. Energ. Rev. 14, 724–734. https://doi.org/10.1016/j.rser.2009.10.025.

Ancillotti, E., Bruno, R., Conti, M., 2013. The role of communication systems in smart grids: architectures, technical solutions and research challenges. Comput. Commun. 36, 1665–1697. https://doi.org/10.1016/j.comcom.2013.09.004.

Beatens, R., Dierckxsens, C., Mijlemans, L., Joseph, P., Krisper, U., Lacko, R., Oštir, T., Paravan, D., Artač, G., Anže, P., Markočič, R., Sernec, R., Bregar, Z., Souvent, A., Nemček, P., Gutschi, C., Andolšek, A., Krmar, D., Gerbec, D., Panjan, G., Mohorko, M., Vukasovič, M., Stimmer, A., 2016. Future Flow Project, D1.1 Requirements for DR & DG Participation in aFRR Markets.

Borovina, D., Fahrioglu, M., Artnzen, A.A., Sernec, R., Suljanović, N., 2015. Communication aspects of smart metering systems.Proceedings of 20th International Conference on Transformative Science & Engineering, Business and Social Innovation (SDPS 2015). Fort Worth, Texas, USA.

Braun, M., 2007. Technological control capabilities of DER to provide future ancillary services. Int. J. Distrib. Energy Res. 3, 191–206. https://doi.org/10.1177/1350508407082265.

Budka, K.C., Deshpande, J.G., Thottan, M., 2014. Communication Networks for Smart Grids. Springer, London.

Dabbagh, S.R., Sheikh-El-Eslami, M.K., 2014. Risk-based profit allocation to DERs integrated with a virtual power plant using cooperative game theory. Electr. Power Syst. Res. 121, 368–378. https://doi.org/10.1016/j.epsr.2014.11.025.

Della Giustina, D., Ferrari, P., Flammini, A., Rinaldi, S., Sisinni, E., 2013. Automation of distribution grids with IEC 61850: a first approach using broadband power line communication. IEEE Trans. Instrum. Meas. 62, 2372–2383. https://doi.org/10.1109/TIM.2013.2270922.

ENTSO-E, 2009. P1: Load-frequency control and performance [WWW Document]. Cont. Eur. Oper. Handb. https://www.entsoe.eu/publications/system-operations-reports/operation-handbook/Pages/default.aspx.

Etherden, N., Vyatkin, V., Bollen, M., 2015. Virtual power plant for grid services using IEC 61850. IEEE Trans. Ind. Inf. 12, 437–447. https://doi.org/10.1109/TII.2015.2414354.

Fraunhofer, I.W.E.S., 2015. The European Power System in 2030: Flexibility Challenges and Integration Benefits. An Analysis with a Focus on the Pentalateral Energy Forum Region.

Giuntoli, M., Poli, D., 2013. Optimized thermal and electrical scheduling of a large scale virtual power plant in the presence of energy storages. IEEE Trans. Smart Grid 4, 942–955. https://doi.org/10.1109/TSG.2012.2227513.

Gutschi, C., Bleyl, J.W., Spreitzhofer, J., Taljan, G., Meissner, E., Esterl, T., Svetina, M., 2017. Economic Appraisal of Hybrid-VPP Use Cases: Substantial Savings Possible for New Generators Economic Appraisal.

Hardy, W.C., 2001. QoS Measurement and Evaluation of Telecommunications Quality of Service, first ed. Wiley Publishing, New York.

IEC International Electrotechnical Commission, (Ed.), 2006. IEC 60870-5-104: Telecontrol Equipment and Systems–Part 5-104: Transmission Protocols – Network Access for IEC 60870-5-101 Using Standard Transport Profiles. second ed. IEC International Electrotechnical Commission, Switzerland.

IEC International Electrotechnical Commission, (Ed.), 2016. IEC 61850-7-420: Communication Networks and Systems for Power Utility Automation–Part 7-420: Basic Communication Structure – Distributed Energy Resources Logical Nodes. second ed. IEC International Electrotechnical Commission, Switzerland.

Ihle, N., Runge, S., Gutschi, C., Gödderz, K., 2016. Data exchange format for demand response – evaluation and case study with OpenADR.14. Symposium Energieinnovation, Technische Universität Graz. Graz, Austria, pp. 1–12.

Industry Alliance VHPready e.V, 2017. VHPready (Virtual heat & power) [WWW Document]. Available from: https://www.vhpready.com.

Integrid consortium, 2017. Integrid project. [WWW Document]. Available from: https://integrid-h2020.eu/slovenia.

Jansen, B., Binding, C., Sundström, O., Gantenbein, D., 2010. Architecture and communication of an electric vehicle virtual power plant.2010 First IEEE International Conference Smart Grid Commun. (SmartGridComm)pp. 149–154. https://doi.org/10.1109/SMARTGRID.2010.5622033.

Kalalas, C., Thrybom, L., Alonso-Zarate, J., 2016. Cellular communications for smart grid neighborhood area networks: a survey. IEEE Access 4, 1469–1493. https://doi.org/10.1109/ACCESS.2016.2551978.

Khan, R.H., Khan, J.Y., 2013. A comprehensive review of the application characteristics and traffic requirements of a smart grid communications network. Comput. Netw. 57, 825–845. https://doi.org/10.1016/j.comnet.2012.11.002.

Klobasa, M., Angerer, G., Schleich, J., Buber, T., Gruber, A., Hünecke, M., von Roon, S., 2013. Lastmanagement als Beitrag zur Deckung des Spitzenlast- bedarfs in Süddeutschland.

Kok, K., 2009. Short-term economics of virtual power plants.20th International Conference on Electricity Distribution, pp. 1–4. https://doi.org/10.1049/cp.2009.1139.

Kolenc, M., Nemček, P., Gutschi, C., Suljanović, N., Zajc, M., 2017. Performance evaluation of a virtual power plant communication system providing ancillary services. Electr. Power Syst. Res. 149, 46–54.

Kolenc, M., Ihle, N., Gutschi, C., Nemček, P., Breitkreuz, T., Gödderz, K., Suljanović, N., Zajc, M., 2018. Virtual power plant using OpenADR 2. 0b for dynamic charging of automated guided vehicles. Int. J. Electr. Power Energy Syst. 104, 370–382.

Kolenc, M., Nemček, P., Suljanović, N., Zajc, M., 2016. Monitoring communication QoS parameters of distributed energy resources.IEEE International Energy Conference (ENERGYCON 2016). Leuven, Belgium, https://doi.org/10.1109/ENERGYCON.2016.7513900.

Kolenc, M., Zalaznik, N., Nemček, P., Turha, B., Mario, G., Sauer, I., Šterk, M., 2015. VPP and its role in the eBADGE project. Metering Int. 58–60.

Kong, P.-Y., 2015. Wireless neighborhood area networks with QoS support for demand response in smart grid. IEEE Trans. Smart Grid 7, 1913–1923.

Li, B., Shen, J., Wang, X., Jiang, C., 2016. From controllable loads to generalized demand-side resources: a review on developments of demand-side resources. Renew. Sust. Energ. Rev. 53, 936–944. https://doi.org/10.1016/j.rser.2015.09.064.

Ma, O., Alkadi, N., Cappers, P., Denholm, P., Dudley, J., Goli, S., Hummon, M., Kiliccote, S., MacDonald, J., Matson, N., Olsen, D., Rose, C., Sohn, M.D., Starke, M., Kirby, B., O'Malley, M., 2013. Demand response for ancillary services. IEEE Trans. Smart Grid 4, 1988–1995. https://doi.org/10.1109/TSG.2013.2258049.

Mahmood, A., Javaid, N., Razzaq, S., 2015. A review of wireless communications for smart grid. Renew. Sust. Energ. Rev. 41, 248–260. https://doi.org/10.1016/j.rser.2014.08.036.

Nikonowicz, L., Milewski, J., 2012. Virtual power plants – general review: structure, application and optimization. J. Power Technol. 92, 135–149.

Nosratabadi, S.M., Hooshmand, R.-A., Gholipour, E., Parastegari, M., 2016. A new simultaneous placment of distributed generation and demand response resources to determine virtual power plant. Int. Trans. Electr. Energy Syst. 26, 1103–1120. https://doi.org/10.1002/etep.2128.

Palizban, O., Kauhaniemi, K., Guerrero, J.M., 2014. Microgrids in active network management—part I: hierarchical control, energy storage, virtual power plants, and market participation. Renew. Sust. Energ. Rev. 1, 1–13. https://doi.org/10.1016/j.rser.2014.01.016.

Pudjianto, D., Ramsay, C., Strbac, G., 2007. Virtual power plant and system integration of distributed energy resources. IET Renew. Power Gener. 1, 10–16. https://doi.org/10.1049/iet-rpg.

Samad, T., Koch, E., Stluka, P., 2016. Automated demand response for smart buildings and microgrids: the state of the practice and research challenges. Proc. IEEE 104, 726–744. https://doi.org/10.1109/JPROC.2016.2520639.

Thavlov, A., Bindner, H.W., 2015. Utilization of flexible demand in a virtual power plant set-up. IEEE Trans. Smart Grid 6, 640–647. https://doi.org/10.1109/TSG.2014.2363498.

Trefke, J., Rohjans, S., Uslar, M., Lehnhoff, S., Nordström, L., Saleem, A., 2013. Smart grid architecture model use case management in a large European smart grid project.2013 4th IEEE/PES Innovative Smart Grid Technologies Europe (ISGT EUROPE). IEEE, pp. 1–5. https://doi.org/10.1109/ISGTEurope.2013.6695266.

Uslar, M., Trefke, J., 2014. Applying the smart grid architecture model SGAM to the EV domain.Proceedings of the 28th EnviroInfo 2014 Conference. BIS-Verlang, Oldenburg, Germany.

Vallejo, A., Zaballos, A., Selga, J.M., Dalmau, J., 2012. Next-generation QoS control architectures for distribution smart grid communication networks. IEEE Commun. Mag. 50, 128–134. https://doi.org/10.1109/MCOM.2012.6194393.

VDE-ETG, 2017. VDE-studie: Demand side integration-lastverschiebungspotenziale in Deutschland [WWW Document]. Available from: https://www.vde.com/de/etg/publikationen/studien/etg-vde-studie-lastverschiebungspotenziale.

Wulff, B., Thiele, M., Schmidt, J., 2016. Abschlussbericht: Verbundprojekt BESIC (Batterie-Elekttische Schwerlastfahrzeuge im Intelligenten Containerterminalbetrieb). Hamburg, Germany.

Yang, Q., Barria, J.A., Green, T.C., 2011. Communication infrastructures for distributed control of power distribution networks. IEEE Trans. Ind. Inf. 7, 316–327. https://doi.org/10.1109/TII.2011.2123903.

Inertia emulation from HVDC links for LFC in the presence of smart V2G networks

Sanjoy Debbarma, Rituraj Shrivastwa*[†]
*Electrical Engineering Department, NIT Meghalaya, Shillong, India, [†]Department of Electrical Power Engineering (IEE), Grenoble Institute of Technology, Grenoble, France

12.1 Introduction

The world is experiencing many challenges associated with the phasing out of natural resources such as coal and oil. One of the major solutions can be the integration of renewable power sources (RPS) such as solar photovoltaic (SPV) and wind energy to the grid. However, such intermittent sources will increase the unpredictability in the energy systems which in turn affect the grid frequency. It is well-known fact that any imbalance between total power generated and demanded in an interconnected power system affects the grid frequency and could results into blackouts if persist for a longer period of time. Power generation involving wind generator and SPV due to their variable output have more possibility of imposing stress on the power system. To maintain the equilibrium of power system, the balance between generation and load must be identical. Such balance can be maintained via automatic generation control (AGC) or load-frequency control (LFC) (Elgerd and Fosha, 1970). During abnormal situation, three tasks must be performed by the AGC mechanism: (1) grid frequency is to be maintained satisfactorily within permissible bounds, (2) any deviations in power flow through the tie-line between control areas must be zero, and (3) optimal generation must be done.

To deal with the frequency stability issue, speed governor is used through primary frequency control (PFC) loop. However, control action of such primary controller is inadequate to maintain the deviation and subsequently the secondary control technique is incorporated along with the governor. So far, many critical investigations pertaining to AGC or LFC have been studied. However, traditional way of controlling may not be suitable for modern power system with high penetration of RPS-based generation, DGs, and electric vehicle (EV) networks. It is interesting to mention that all the aforesaid sources and EV infrastructure involves the power electronic devices when connected to grid (Zhu et al., 2014). A typical modern power system is shown in Fig. 12.1 consisting of RPS, V2G networks and conventional units. It is well-known fact that massive integration of RPS, inspired by sustainability concerns, would present several challenges to power system control and operation.

Smart Power Distribution Systems. https://doi.org/10.1016/B978-0-12-812154-2.00012-2

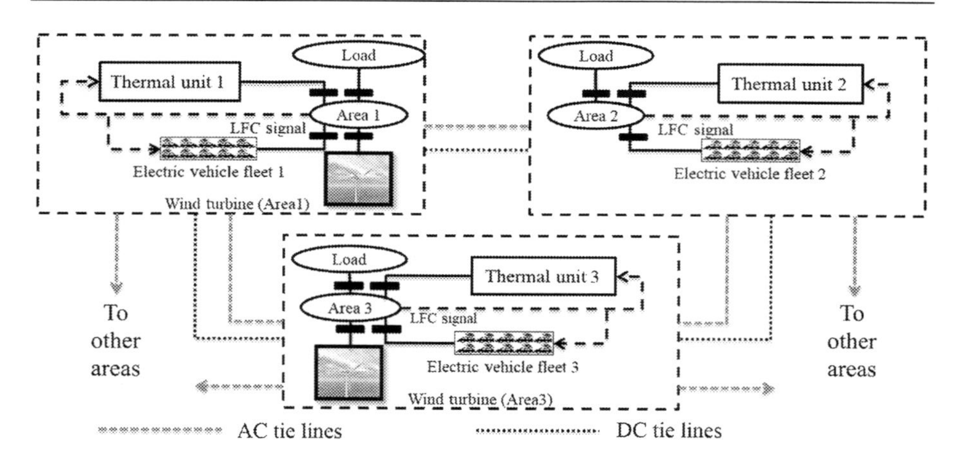

Fig. 12.1 Typical interconnected power system with RPS penetration and V2G network.

Also, the increasing penetration of RPS like wind power and PV generation may act to worsen the grid frequency stability due to lack of sufficient inertia. Traditionally rotational inertia to the power systems are provided by synchronous generators (SGs) (Castro and Acha, 2016). Undoubtedly, such low-inertia grid with power electronics converters (due to integration of RPS) presents a greater risk of instability requiring alleviation like changes to dispatch control systems, necessity for considerable amounts of spinning reserve and recurrent starting of conventional generators, which is often inefficient, challenging and costly from the business point of view. In many countries, RPS-based generations is interfaced with high-voltage direct current (HVDC) transmission lines due to attractive advantages such as fast frequency support, bidirectional controllability, and power oscillation damping. However, such modern power system with increasing electronically interfaced components due to HVDC systems and integration of RPS would suffer from lack of physical inertia and damping response which in turn put negative impacts on frequency regulation (FR) (Rakhshani et al., 2017). This is a trouble which demands crucial research investigation, as power networks with poor inertia are probably to become more common. To address these, many recent researchers designed methodology of emulating virtual inertia using the short-term energy storage system (ESS). In Zhu et al. (2014), inertia emulation controller (INEC) framework is implemented for contributing inertia by a DC link in voltage source converter (VSC)-based multiterminal DC (MTDC) system. Rakhshani et al. (2016a,b) focused on emulating the inertia using derivative of frequency signal for HVDC systems. But both the works used derivative of frequency signal for inertia emulation are very sensitive to noise and affect the system stability. Subsequently, concept of synchronous power controller (SPC) is implemented in converter station of HVDC links for inertia emulation (Rakhshani et al., 2017). The advantage of the SPC concept over other aforesaid methods is that it does not use phase-locked loop (PLL) for estimation of the signals. In other words, this method is a PLL-less concept for grid synchronization.

In most of the inertia emulation technique, DC link is treated as ideal battery source with immense storage which practically may not be (Ashabani and Mohamed, 2014). Therefore, an additional huge energy reserve is required to reduce the risk of instability on the power system. In smart grid environment, contribution from mobile vehicles can be one of the possible cost effective solutions for LFC problem (Pillai and Bak-Jensen, 2011). Recently, EVs are being adopted increasingly as it can alleviate effect to urban heat island to help local as well as global climates. It is found that, in practice, most of the EVs with onboard battery pack are parked in idle condition for several hours per day (Yilmaz and Krein, 2012, 2013). Besides acting as a load, EVs can also take part in PFC and secondary frequency control (SFC) through vehicle-to-grid (V2G) infrastructure (Debbarma and Dutta, 2017). More details on EV networks participating in LFC- and SPC-based HVDC transmission links are presented in the succeeding sections.

12.2 Inertia emulation from SPC-based HVDC systems for LFC

As discussed earlier, converter interfaced generators would invite negative impacts on future power systems by reducing the physical inertia, thereby imposing threat to power system stability. It is well-known fact that the kinetic energy stored in the rotating machines is responsible for counteracting any mismatch through inertial response until PFC activates. Now the question is how to compensate for reduced inertial response in future low-inertia power system. A probable solution is to improve the dynamics of the converters by specifying the properties of the grid-connected converters in such a way that it acts like a SG (Rakhshani et al., 2017). Although there exist different methods for inertia emulation, but all the techniques depends on PLL concept for measurement and estimation. Such amplified noise caused by frequency deviation measurements could bring instability problems (Rakhshani et al., 2016a,b). This limitation can be even worse, especially during abnormal faults and unsymmetrical system. In this section, the mathematical modeling of HVDC transmission system using SPC idea is presented for virtual inertia emulation. The principle idea is based on the control of VSC of HVDC link via active power loop and virtual admittance. Now, before modeling SPC-based HVDC transmission model, let us derive the conventional small signal model of HVDC link. With conventional small signal model of HVDC links incorporated in parallel with the HVAC transmission links, $\Delta P_{dci\text{-}j}(s)$ can be written as

$$\Delta P_{dci-j}(s) = \frac{K_{dci}}{1 + sT_{dci}} \Delta E_{dci}(s) \tag{12.1}$$

where K_{dci} represents feedback gain of HVDC tie-line, T_{dci} represents the time constant of HVDC tie-line, and ΔP_{dci-j} is direct current power deviation between area i and area j. Here, $\Delta E_{dci}(s)$ is the control error and can be represented as

$$\Delta E_{dci}(s) = \left[K_{fi}\Delta f_i(s) + K_{fj}\Delta f_j(s) + K_{ac}\Delta P_{tiei-j}\right] \quad (12.2)$$

where Δf_i and Δf_j are the frequency deviation in area i and area j where $i \neq j$. K_{fi}, K_{fj}, and K_{ac} are control gains of power modulation controller (PMC). With this new state in HVDC systems, area control error (ACE) will be reconstructed as follows:

$$ACE_{newi} = \left[\Delta f_i \beta_i + \Delta P_{totali-j}\right] \quad (12.3)$$

$$\Delta P_{totali-j} = \Delta P_{dci-j} + \Delta P_{tiei-j} \quad (12.4)$$

where $\Delta P_{totali-j}$is total tie-power line deviation. It is significant to mention that SPC loop in HVDC imitate the response of SG here but not fully; however, it demolishes the limitation of oscillatory response of SG (Rakhshani et al., 2017). The storage element such as capacitor between two converter stations in DC link can be modeled in the converter topology. The most advantageous feature of this SPC strategy is that any storage devices can be used alongside the converters to supply the deficit in the grid. Fig. 12.2 shows the active SPC loop used in modeling of HVDC systems. This loop consists of electromechanical block and virtual admittance-based converter block and can be obtained by reconstructing the second-order swing equation of SG.

In active SPC loop, P_i and P_o is the input power and the output power of converter, respectively, variation of which (ΔP) is the input to power loop controller (PLC). The mathematical model of PLC shown within the electromechanical block depicts the rotational behavior of SG and can be represented as

$$PLC(s) = \frac{\omega_n^2 / P_{Active}}{s + 2\zeta\omega_n} \quad (12.5)$$

where ω_n and ζ are the natural frequency and damping factor, respectively. P_{Active} is the delivered maximum active power. Here, PLC fixed a relative rotor frequency which append with synchronous speed of grid (ω_s) to give virtual rotor frequency of electromotive force (ω_r). Integration of ω_r enables the measurement of phase angle (θ_r) of virtual rotor which after subtracting from grid voltage angle (θ_{grid}) provide load

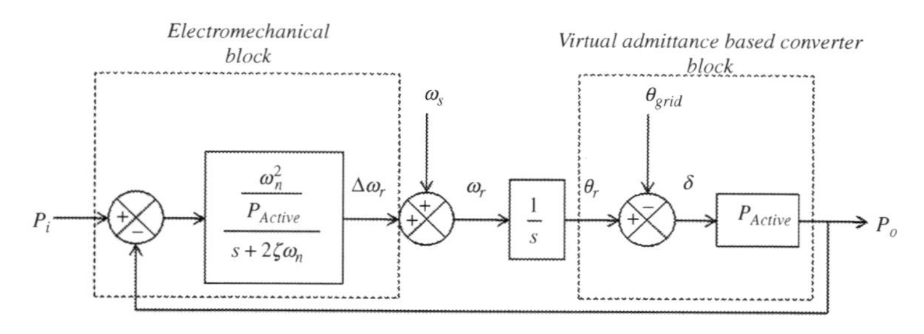

Fig. 12.2 SPC active loop for HVDC system.

angle (δ), that is, angle between grid voltage and virtual emf. Therefore, the final active power equation of the SPC is given as

$$P_o = P_{Active}{}^{*}\delta \tag{12.6}$$

where $P_{Active} = EV/X$, wherein E and V is virtual electromotive force and grid voltage, respectively. Here X represents reactance due to virtual admittance. Output power of SPC can be easily adjusted by changing appropriately the angle of output voltage of the converter. Thus the complete transfer function of the loop can be obtained as

$$\frac{P_o}{P_i} = \frac{\omega_n^2}{s^2 + 2\zeta\omega_n s + \omega_n^2} \tag{12.7}$$

where

$$\omega_n = \sqrt{\frac{P_{Active}}{J\omega_s}} \text{ and } \zeta = \frac{D}{2\sqrt{P_{Active}J.\omega_s}} \tag{12.8}$$

Here, D and J stand for damping constant parameter and moment of inertia, respectively. In Eq. (12.8), the proper selection of value of D and J could help to imitate the desired damping effect. Finally, for ith area power system scenario with SPC-based HVDC system, the control error in Eq. (12.2) can be rewritten as

$$\Delta E_{SPC,dci}(s) = \left[K_{SPC,fi}\Delta f_i(s) + K_{SPC,fj}\Delta f_j(s) + K_{SPC,ac}\Delta P_{tiei-j}\right] \tag{12.9}$$

where $\Delta E_{SPC,\ dci}(s)$ is the DC reference error signal and act as an input to SPC-based HVDC control whose output signal is $\Delta P_{SPC,\ dci-j}(s)$. For multiple HVDC links, the output power of DC link can be written as

$$\Delta P_{dci-j}(s) = \Delta P_{SPC,dci-j}(s) = \sum_{\substack{i=1 \\ i\neq j}}^{k} \Delta P_{SPC,dci-j} \tag{12.10}$$

where k is multiple SPC-based HVDC stations.

A SPC or virtual synchronous power (VSP) based concept is a PLL less concept for grid synchronization and considered as a good solution for the renewable generation systems with energy storage. Compared with the well-known virtual impedance structure, the virtual admittance structure emulates the output impedance without leading to the difficulties in implementation. The main advantages lay on the effectiveness for the complete range of harmonic frequencies and the simplicity in the inner loop implementation (Tamrakar et al., 2017). In this concept, the proposed control law reproduces basically the swing equation of a SG. The transmitted active power is increased or decreased by shifting the output voltage phase of the converter forward or backward. By regulating this power, the system automatically maintains the synchronization.

12.3 Introduction to V2G network

12.3.1 EV product and services for FR

Introduction of V2G technology offer EVs to participate in different ancillary services under competitive electric market. EVs provide an opportunity to grow new products and services for grid management. As a one type of distributed energy storage, EVs is highly expected to play a vital role for emergency reliability services such as FR (Liu et al., 2015). On the other hand, EVs are gaining popularity because of reduced dependence on petrol and greenhouse emissions thus contribute to clean environment. As per the report, China has planned to put 200 million EVs by 2050 (Vachirasricirikul and Ngamroo, 2014). It is found that average personal vehicles travel on the road is nearly 4%–5% of the time, sitting in home garages or parking lots the rest of the day (Yilmaz and Krein, 2012). In an area, thousands of EVs can be aggregated to discharge active power to grid if there is a change/mismatch in power demanded and generated and help conventional units to meet the requirements instantaneously. Thus, EVs can participate in PFC as well as in LFC through V2G infrastructure. The charging and discharging of battery of the EV owners during peak and normal hours can be smartly controlled by new for-profit entity called EV aggregator with proper strategy.

As analysis of EV capabilities, several products or services are recommended for initial deployment based on a combination of their potential usefulness to the ISO or regional transmission organization (RTO) and the likely response from aggregators and end consumers (IRC, 2010). In the phased implementation approach, the initial products and services are indicated by minimum infrastructure required and support of grid reliability. They include (as per IRC, 2010):

1. Emergency load curtailment (ELC)—EVs are able to provide a quick response load-curtailment resource for emergency events and may be aggregated for the maximum effect. Due to relatively simple mechanisms for engaging this resource, and the large benefit of doing so, ELC of EV charging is a likely near-term product. It could serve in a reliability-based or economic demand response (DR) capacity.
2. Dynamic pricing (DP)—DP might be a solution to accomplish charging of EV batteries in off-peak hours. Nonetheless, research on consumer behavior is important to understand how an EV owner will respond to retail price differentials. In addition, EV-specific DP may be one way to introduce DP to consumers while avoiding political sensitivities regarding DP for existing retail loads.
3. Enhanced aggregation (EA)—The potential for high concentrations of EV loads in the evening makes managing charging over the day a priority for the ISO/RTOs. Some aggregators, automakers, and information management groups appear to be proactive in developing scheduling capabilities, possibly using additional information provided by the ISO/RTO. This product would be complementary to planned time-of-use (TOU) programs typically offered by the retail utilities. It also could be potentially linked to a DP product.

When V2G infrastructure is installed in the existing distribution networks, the following market products can provide value to the ISO/RTOs and aggregators:

1. **Regulation**—Expected EV loads in the next few years will not likely have a large impact on the amount of total regulation in the ISO/RTO markets or on regulation market prices. However, the regulation market is attractive to EV stakeholders since it can generate fairly predictable revenues. In addition, the relatively simple but new communication requirements for this product make it a good trial for subsequent EV products and services.
2. **Reserves**—EVs are able to provide reserve resources with relatively simple control of EV charging. Furthermore, this product appears to complement upcoming developments in DR resources as a result of smart grid developments.

For EV aggregators to participate in wholesale electric markets, firstly, ISO must ensure that aggregators have the capability to identify EVs location. Then aggregator purchases the power by engaging themselves in the day ahead market, which later will resell to the EV owners at predetermined price. The function of aggregator is to collect the information of the status of EVs and send it to the control operator. EVs receive the control signal from the operator and update their data/information in real time. Such information may be state of charge (SOC), EVs' capacity, numbers of EVs plugged in to charging station and future driving demand. For bidirectional power flow, that is, for charging and discharging of EVs, an active grid connected bidirectional AC-DC converter that enforces active power factor correction, and a bidirectional DC-DC converter to regulate the battery charge or discharge current is required. When operating in charge mode, the charger should draw a sinusoidal current with a defined phase angle to control power and reactive power. In discharge mode, the charger should return current in a similar sinusoidal form. However, incorporating bidirectional power flow has many challenges as well; viz. battery degradation due to frequent charging and discharging regulation, extra costs for bidirectional converters, metering issues, proper forecasting of EV reserve and driving behavior, interface concerns, etc.

Fig. 12.3 shows a schematic diagram of three area multisource interconnected power systems where in each area EVs are connected to the grid using V2G network. In such power grid, communications must be bidirectional to report battery status and receive commands such that the grid as well as the EVs can send each other economic and control signals. This will help in tracking intermittent resources and alter charge rates to track power prices, frequency or power regulation, and spinning reserves. A variety of communication protocols have to be studied for this purpose, including ZigBee, Bluetooth, Z-Wave, etc. and choose the best one for the purpose. Furthermore, in order to maintain proper safety isolation is important. It is beneficial for EV functions, including the high-voltage battery, DC-DC converter, traction inverter, and charger. Galvanic isolation in EV supply equipment can be provided either with a line transformer or in the DC-DC converter stage with a high-frequency transformer. High-frequency transformer isolation supports voltage adjustment for better control, safety for load equipment, compactness, and suitability for varying applications.

12.3.2 Modeling of LFC system with V2G energy network and SPC HVDC links

For power pool operation, grid frequency must be managed in acceptable bound to support generation-load balance. Such mechanism is attained by centralized LFC/AGC. Any imbalance between supply demand is represented by ACE signal given by Eq. (12.11)

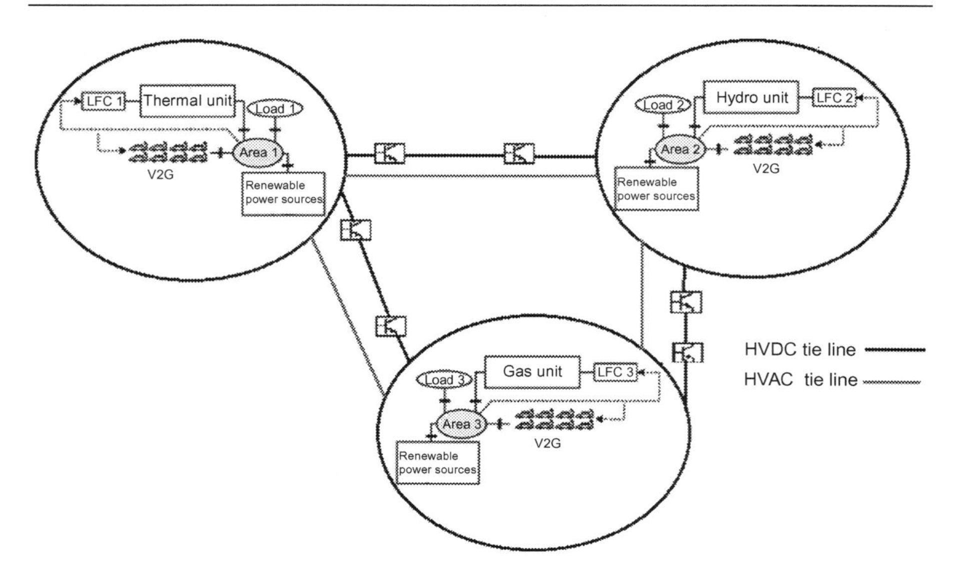

Fig. 12.3 Schematic diagram for three area system with V2G networks (Debbarma and Dutta, 2017).

$$ACE_i = \left[\Delta f_i \beta_i + \Delta P_{tiei-j}\right] \tag{12.11}$$

where β is bias factor, Δf_i is frequency deviation, and $\Delta P_{tiei\text{-}j}$ is net tie-line power between two control areas. Minimization of ACE signal would provide good dynamics and ensure the stability power system. Fig. 12.4 shows the basic frame of ith control area with EV penetration, thermal unit, and SPC HVDC links. The output power of the different sources is controlled via PFC loop and secondary loop. T_{gi} is steam governor time constant, T_{ti} is steam turbine time constant, R_i is the governor speed regulation parameter, T_{pi} and K_{pi} is the power system time constant and gain. *APF* is called ACE participation factor of EV fleet and conventional unit.

V2G energy network owing to its huge energy reserve can contribute to LFC of power grid. EVs act as a prosumer because they can charge their battery by taking energy from grid as well as inject energy to the grid. However, power availability of V2G network is uncertain due to random arrivals and departures of EVs. If arrivals and departures of EVs between any two districts are same, they are considered as symmetrical. However, practical scenario is generally asymmetrical between the districts especially during rush hours which in turn affect the battery SOCs. Thus, to realize the asymmetrical EVs movement in simulation study one must define the increasing and decreasing figure of EVs for different time interval. Based on the driving behaviors of EV owner and their expected SOC level, three different types of EV fleets can be categorized.

1. *Type I*: This type of fleet is very common where EVs randomly arrives and depart round the clock. Here injection of V2G power from aggregator following ACE is not fixed as the number of EVs kept changing with time.

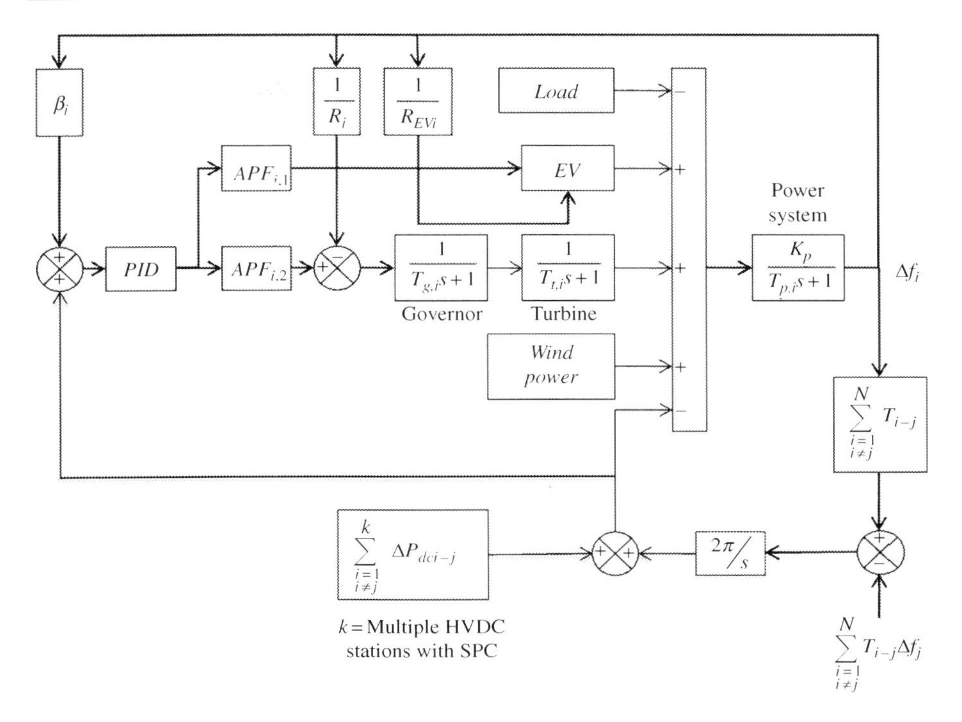

Fig. 12.4 Basic frame of *i*th control area with EV penetration and HVDC links.

2. *Type II*: In a lot of cases we may find that the EVs are parked in the garages for a longer period of time throughout the day. This type includes those EVs and the owners can sell the redundant power to the grid to earn revenues.
3. *Type III*: This type of the EVs includes those which come with the emergency requirement like charging and possess very low SOC.

EVs participating in FR can be defined based on smart balance SOC control technique as

$$P_{EVi} = \begin{cases} K_c{}^*ACE, & ACE \geq 0 \\ K_d{}^*ACE, & ACE < 0 \\ P_{AG}{}^{\max}, & K_c{}^*ACE \geq P_{AG}{}^{\max} \\ -P_{AG}{}^{\max}, & K_d{}^*ACE \leq -P_{AG}{}^{\max} \end{cases} \tag{12.12}$$

where P_{EVi} is output power of V2G fleet, $P_{AG}^{\max}$ and $P_{AG}^{\min}$ represents maximum and minimum value of V2G fleet power output, respectively. Here the charging (K_c) and discharging (K_d) coefficient are calculated based on Eqs. (12.13), (12.14), respectively.

$$K_c = K_{\max}\left\{1 - \left(\frac{SOC_i - SOC_{\text{low}}}{SOC_{\max} - SOC_{\text{low}}}\right)^n\right\} \tag{12.13}$$

$$K_d = K_{\max}\left\{1 - \left(\frac{(SOC_i - SOC_r) - SOC_{high}}{SOC_{\min} - SOC_{high}}\right)^n\right\} \quad (12.14)$$

where $K_{\max}$ is the maximum EV gain, SOC_{high} and SOC_{low} is the high and low SOC, respectively, $SOC_{\min}$ and $SOC_{\max}$ are the minimum and maximum SOC of the battery, respectively, and n is the specification of the designed battery SOC. In Eq. (12.14), SOC_r is defined as reserved SOC required by EV owners for their upcoming driving. Any EV owners can set SOC_r value keeping in the view their next travel distance.

12.4 Simulation studies

12.4.1 LFC of power system with V2G control

Here, a three area interconnected system is considered as a study system which comprises conventional thermal units and V2G networks having capacity of 2000, 4000, and 8000 MW in area1, area2 and area3, respectively. Modeling of the test model is based on the Fig. 12.4 where both HVAC- and SPC-based HVDC tie-lines are connected in parallel. Parameters of the thermal systems, EV fleets and other are taken from Debbarma et al. (2014) and provided in Table 12.1. The model of the wind turbine system is designed (Nandar, 2013) and the corresponding output is shown in Fig. 12.5. In our study, it is assumed that only first and third control areas are penetrated with high wind power. EV fleets in all the areas is modeled considering only Type I and Type II fleets as discussed above with their initial SOC as 65% and 85%, respectively. Load perturbation is assumed to be occurred in all the areas simultaneously.

Fig. 12.6A and B shows the frequency deviation taken place in area1 and tie-line power deviation between area1 and area2. It is clearly seen that when only Type II EV fleets is considered, fluctuations is more as the reserve capacity of fleets changes. However, when Type I and Type II fleets are considered in the networks, dynamic response of frequency and tie-power significantly improved. Following error and wind power fluctuations, V2G algorithm updates the value of K_c and K_d. For analysis, value of parameters in Eqs. (12.13), (12.14) are taken as $SOC_{\min} = 10\%$, $SOC_{\max} = 90\%$, $SOC_{\text{low}} = 20\%$, $SOC_{\text{high}} = 80\%$, $SOC_r = 20\%$, and $n = 2$ (Vachirasricirikul and Ngamroo, 2014). The charging and discharging rate of the EV battery is taken as ±5 kW.

Table 12.1 Parameters of the three area system (Debbarma et al., 2014; Debbarma and Dutta, 2017)

Parameter	Value
f, T_{gi}, T_{ri}, T_{ti}, K_r	60 Hz, 0.08 s, 10 s, 0.3 s, 0.5
H_i, T_w, D_i,	5 s, 1 s, 0.00833 p.u. MW/Hz
T_{pi}, K_{pi}	20 s, 120 Hz/p.u. MW
R_i, R_{EVi}	2.4 Hz/p.u. MW, 2.4 Hz/p.u. MW
β_i, K_{EV}, T_{Evi}	0.425 p.u. MW/Hz, 1, 1 s
a_{12}, a_{23}, a_{13}	−1/2, −1/2, −1/4

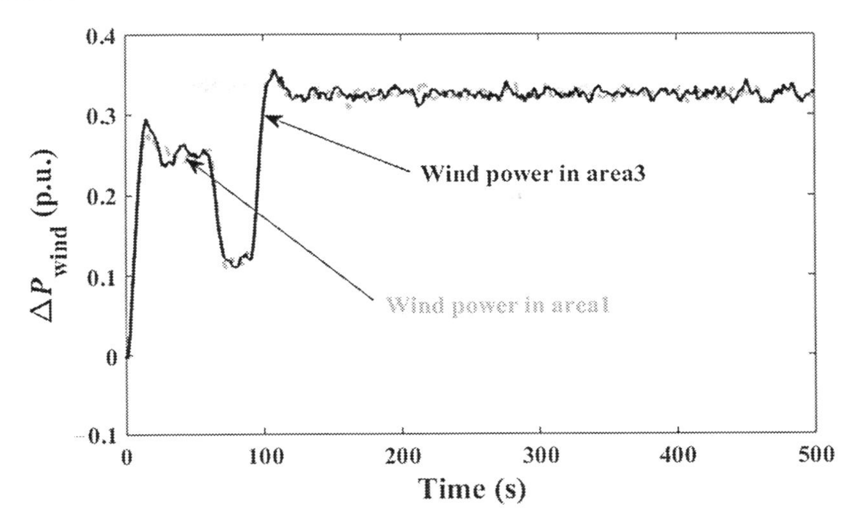

Fig. 12.5 Wind output power integrated to grid.

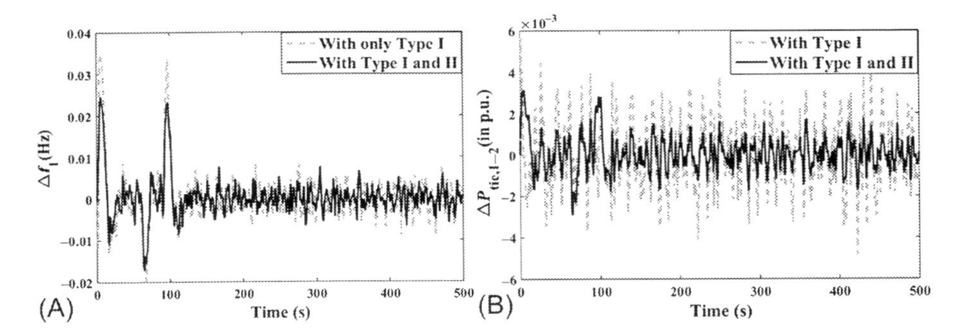

Fig. 12.6 (A) Frequency deviation in area1 and (B) tie-line power deviation between area1 and area2.

12.4.2 Inertia emulation from SPC HVDC transmission links of multiarea power system penetrated with V2G network and wind turbine systems

In this case, V2G network comprises all the types of EV fleets, that is, Types I, II, and III and control areas are interconnected via parallel HVAC- and SPC-based HVDC tie-lines. Same power system discussed above is considered for investigation. The main intension is to evaluate and show the positive impact of SPC-based HVDC links and EVs on grid frequency and tie-line power. The value of the parameters ζ and ω_n of SPC HVDC lines are taken as 1.31 and 6.87, respectively. For all the cases, parameters of the PMC and supplementary controller are tuned using cuckoo search algorithm.

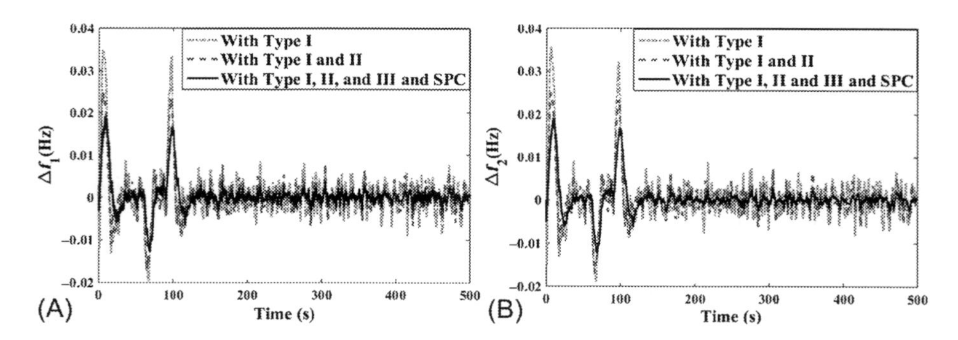

Fig. 12.7 (A) Frequency deviation in area1 and (B) frequency deviation in area2.

This is to mention further that any good optimization techniques can be used for selection of the optimum parameters of the controller. Eq. (12.8) can be used for calculating the related damping and the required inertia for SPC design. Fig. 12.7 represents the frequency deviation of a control area1 and control area2. It is clearly observed that the deviation corresponding to the case where all the EV fleets along with inertia emulation from SPC HVDC lines are not only suppressed and damped out to very low value but also remarkably curtailed the overshoot and undershoot. The quick recovery of the grid frequency is due to the contribution of the regulated V2G network and SPC-based HVDC transmission links. It is to be noted that proper selection of damping factor and inertia for HVDC system is very crucial for obtaining better result. More details on SPC HVDC can be studied from Rakhshani et al., 2017. In Fig. 12.8A, the dynamic response of emulated energy variation obtained from SPC-based DC system following load disturbance and wind variations is shown. The instantaneous response from HVDC links between the areas result into quick suppression of peak deviation and oscillations in dynamic response which undoubtedly shows the capability of SPC transmission system in emulating the virtual inertia. Fig. 12.8B shows the deviation in tie-line power exchanged between the area1 and area3.

In Fig. 12.9, SOC characteristics of V2G network in control area1 is shown. Type I fleet here is acting in a charging mode due to high surplus wind power. EVs in Type

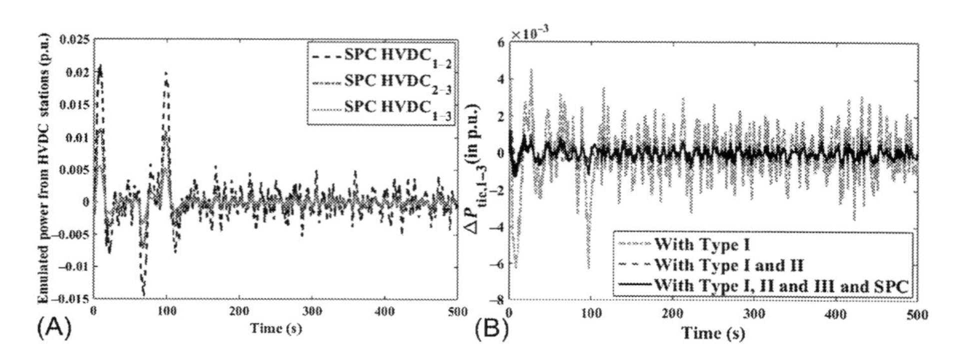

Fig. 12.8 (A) Variation in emulated power from SPC HVDC links and (B) tie-line power deviation.

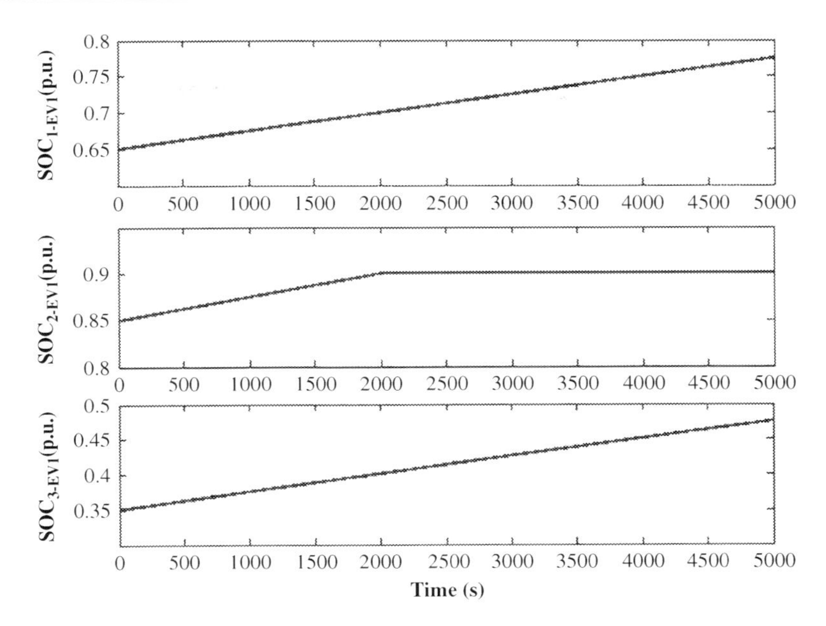

Fig. 12.9 SOC curve of Type I, II, and III fleets in area1.

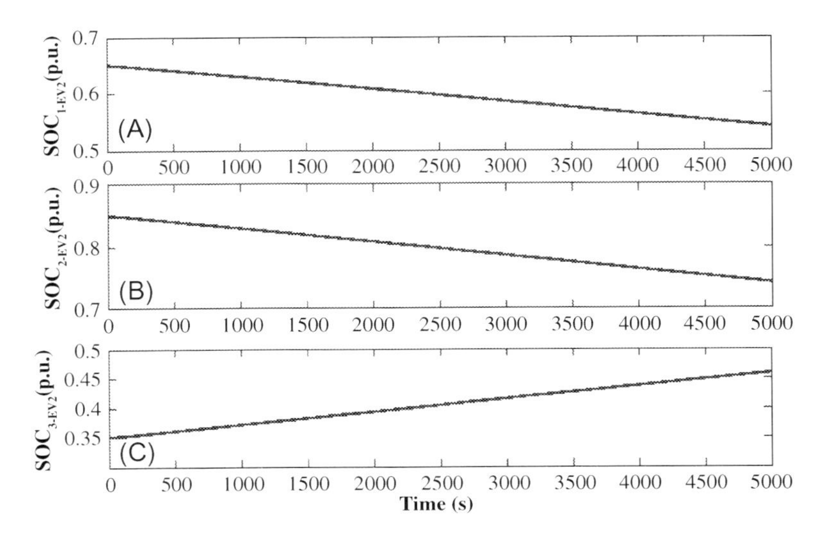

Fig. 12.10 SOC curve of (A) Type I, (B) Type II, and (C) Type III fleets in area2.

II fleet are charged up to their maximum limit, that is, 90% and ready for driving. Whereas Type III fleet owing to low SOC level continues to charge their batteries till the desired level is achieved. Fig. 12.10 is the SOC plots corresponding to the area2. It is seen that Type I and Type II EV fleets help to balance the power crisis (act in discharging mode) and assist thermal unit following perturbations. However, Type III is charging their batteries regardless of the grid condition for upcoming driving.

12.5 Conclusions

In this chapter, attempt has been made to analyze the cooperative control of V2G networks and SPC-based HVDC transmission systems for LFC in the presence of high wind penetration and conventional units. Based on the driving behavior of EV owners, V2G network for EVs participating in the LFC is designed which imitates real-time scenarios like random movement of EVs and upcoming charging demand. It has been found that the EV fleet responds fast unlike the conventional generation units. Modeling and contribution of SPC-based HVDC links is studied in details during load perturbations and variable wind power output. It is found that HVDC converter stations instantly provide inertia response following perturbation as well as wind output variation before thermal units react and diminish the peak deviations. Simultaneously based on ACE signal EVs react rapidly and assist in reducing the deviations and oscillations.

References

Ashabani, M., Mohamed, Y.A.I., 2014. Novel comprehensive control framework for incorporating VSCs to smart power grids using bidirectional synchronous-VSC. IEEE Trans. Power Syst. 29 (2), 943–957.

Castro, L.M., Acha, E., 2016. On the provision of frequency regulation in low inertia AC grids using HVDC systems. IEEE Trans. Smart Grid 7 (6), 2680–2690.

Debbarma, S., Dutta, A., 2017. Utilizing electric vehicles for LFC in restructured power systems using fractional order controller. IEEE Trans. Smart Grid 8 (6), 2554–2564.

Debbarma, S., Saikia, L.C., Sinha, N., 2014. Robust two-degree-of-freedom controller for automatic generation control of multi-area system. Int. J. Elect. Power Energy Syst. 63, 878–886.

Elgerd, O.I., Fosha, C.E., 1970. Optimum megawatt-frequency control of multi-area electric energy systems. IEEE Trans. Power Apparatus Syst. 89 (4), 556–563.

ISO/RTO Council (IRC), 2010. Assessment of plug-in electric vehicle integration with ISO/RTO systems.

Liu, H., Hu, Z., Song, Y., Wang, J., Xie, X., 2015. Vehicle-to-grid control for supplementary frequency regulation considering charging demands. IEEE Trans. Power Syst. 30 (6), 3110–3119.

Nandar, C.S.A., 2013. Robust PI control of smart controllable load for frequency stabilization of microgrid power system. Renew. Energy 56, 16–23.

Pillai, J.R., Bak-Jensen, B., 2011. Integration of vehicle-to-grid in the western Danish power system. IEEE Trans. Sustain. Energy 2 (1), 12–19.

Rakhshani, E., Remon, D., Cantarellas, A.M., Garica, J.M., Rodriguez, P., 2017. Virtual synchronous power strategy for multiple HVDC interconnections of multi-area AGC power systems. IEEE Trans. Power Syst. 32 (3), 1665–1677.

Rakhshani, E., Remon, D., Cantarellas, A.M., Rodriguez, P., 2016a. Analysis of derivative control based virtual inertia in multi-area high-voltage direct current interconnected power systems. IET Gener. Transm. Distrib. 10 (6), 1458–1469.

Rakhshani, E., Remon, D., Cantarellas, A.M., Garica, J.M., Rodriguez, P., 2016b. Modeling and sensitivity analyses of VSP based virtual inertia controller in HVDC links of interconnected power systems. Electr. Power Syst. Res. 141, 246–263.

Tamrakar, U., Shrestha, D., Maharjan, M., Bhattarai, B.P., Hansen, T.M., Tonkoski, R., 2017. Virtual inertia: current trends and future directions. Appl. Sci. 7 (7), 1–29.

Vachirasricirikul, S., Ngamroo, I., 2014. Robust LFC in a smart grid with wind power penetration by coordinated V2G control and frequency controller. IEEE Trans. Smart Grid 5 (1), 371–380.

Yilmaz, M., Krein, P.T., 2012. Review of benefits and challenges of vehicle-to-grid technology. IEEE Energy Conversion Congress and Exposition, pp. 3082–3089.

Yilmaz, M., Krein, P.T., 2013. Review of the impact of vehicle-to-grid technologies on distribution systems and utility interfaces. IEEE Trans. Power Electron. 28 (12), 5673–5689.

Zhu, J., Guerrero, J.M., Hung, W., Booth, C.D., Adam, G.P., 2014. Generic inertia emulation controller for multi-terminal voltage-source-converter high voltage direct current systems. IET Renew. Power Gener. 8 (7), 740–748.

Further reading

Debbarma, S., Saikia, L.C., Sinha, N., 2013. AGC of a multi-area thermal system under deregulated environment using a non-integer controller. Electr. Power Syst. Res. 95, 175–183.

Sasaki, T., Kadoya, T., Enomoto, K., 2004. Study on load frequency control using redox flow batteries. IEEE Trans. Power Syst. 19 (1), 660–667.

Yogarathinam, A., Kaur, J., Chaudhuri, N.R., 2017. Impact of inertia and effective short circuit ratio on control of frequency in weak grids interfacing LCC-HVDC and DFIG-based wind farms. IEEE Trans. Power Deliv. 32 (4), 2040–2051.

Zhong, Q.C., Weiss, G., 2011. Synchronverters: inverters that mimic synchronous generators. IEEE Trans. Ind. Electron. 58 (4), 1259–1267.

Internet of things application in smart grid: A brief overview of challenges, opportunities, and future trends

Qiang Yang
College of Electrical Engineering, Zhejiang University, Hangzhou, People's Republic of China

13.1 Introduction

Today, the electric power energy is mostly generated in centralized power plants and transported over a long-distance transmission network to a distribution network before reaching the end customers. Current power system operations rely mostly on supervisory control and data acquisition (SCADA) systems that were mostly designed with a centralized architecture in mind interconnecting a master terminal (at control center) and a large number of remote telemetry units (RTUs) located at geographically dispersed sites. The electric power networks have ubiquitously deployed physical devices in substations and fields that have sensing, actuating, and communication capabilities. It can be envisioned that more advanced computational and communication technologies will be integrated into the power system to carry the control system from local to wide-area scope to enable an intelligent power grid. As a result, in such electric cyber-physical system (ECPS), the smart grid will be operated and controlled firmly based on an underlying communication infrastructure as large amount of information will be exchanged among the power entities (Yang et al., 2012).

The Internet of things (IoT) is expected to grow to 50 billion connected devices by 2020 (Evans, 2011) providing valuable information to consumers, manufacturers. and utility providers. The IoT enables the devices to be interconnected through the Internet and peer-to-peer connections as well as closed networks, for example, smart grid infrastructure. Currently, much research effort has been made to adopt IoT technologies in smart grid throughout the power generation, transmission, distribution, and domestic (in-home and in-building) consumption. In particular, the demand response techniques can eventually regulate automatically by operating during off-peak energy hours and connect to sensors to monitor occupancy and energy conditions. On the other hand, the increasing connections among devices in the IoT infrastructure impose obvious cyber security challenges, for example, denial-of-service (DoS) attacks, Eavesdropping attacks, false information injection, to the critical infrastructures, like smart grid. The cyber attackers can cause substantial damage by breaching the IoT

Smart Power Distribution Systems. https://doi.org/10.1016/B978-0-12-812154-2.00013-4

infrastructure of a smart grid and manipulating the data transferred, thereby making the sensors take incorrect decisions. This can cause appliances to function in an undesired manner, resulting in widespread equipment damage as well as considerable financial loss. In addition, the IoT applications in smart grid are generally implemented and deployed in a large scale, which indicates that the modeling and simulation of large-scale IoT application in smart grid is a nontrivial task.

The rest of this chapter is organized as follows: Section 13.2 presents the domestic energy system model and demand response solutions based on IoT architecture; in Section 13.3, the issues of IoT cyber physical security are overviewed and discussed; followed by Section 13.4 providing a set of challenges and solutions of modeling and simulation of large-scale IoT in smart grid context. Finally, some conclusive remarks and future research work are given in Section 13.5.

13.2 Demand response opportunities in smart distribution systems

It is reported that the electric meter deployments in the US received early stimulus funding allowing regional utilities in 10 states to start pilot deployments in markets where educating customers was seen as a means to encourage adoption, and more importantly, real-energy savings. The massive adoption of electric smart plugs, in-home displays, and smart thermostats, as shown in Fig. 13.1, has given consumers a choice on which household devices they want to monitor. Simply plug the appliance into the smart plug and add it to the home network. Through ZigBee or Wi-Fi the user can then connect to the Internet to get information through a home gateway or allow direct connection via cloud connectivity with a smartphone or tablet. Consumers are adopting smart plugs more quickly than high-end appliances with smart technology since they are lower cost and allow retrofitting of existing appliances. It is making submeters less complex by packaging its complementary metrology, connectivity, and processors in an easy-to-use, low-power solution. In fact, the connection of devices together in building and homes is one of the next steps to reach the full benefits of the smart grid and many innovative solutions and convenient applications are already offered to the consumers. The introduction of dedicated home energy gateway, smart hub, or energy management system in mandated deployments like in the United Kingdom or Japan can significantly promote the IoT benefits for electric power consumers.

In recent decades, driven by the advances of distributed energy resources (DER) technologies, the renewable distributed power generators, for example, PVs and wind turbines, with the small-scale capacities from a number of kilowatts (kW) to megawatts (MW) and home-based energy storage (e.g., battery) become increasingly prevalent and have been deployed in the domestic scope. These renewable distributed generators can compensate the energy supply from the power utility to meet the domestic energy demand. Also, the real-time pricing (RTP) mechanism has been adopted in many nations as a leveraging tool aiming to enclose the customers into the loop of energy provision and adaptively manage their domestic loads to improve

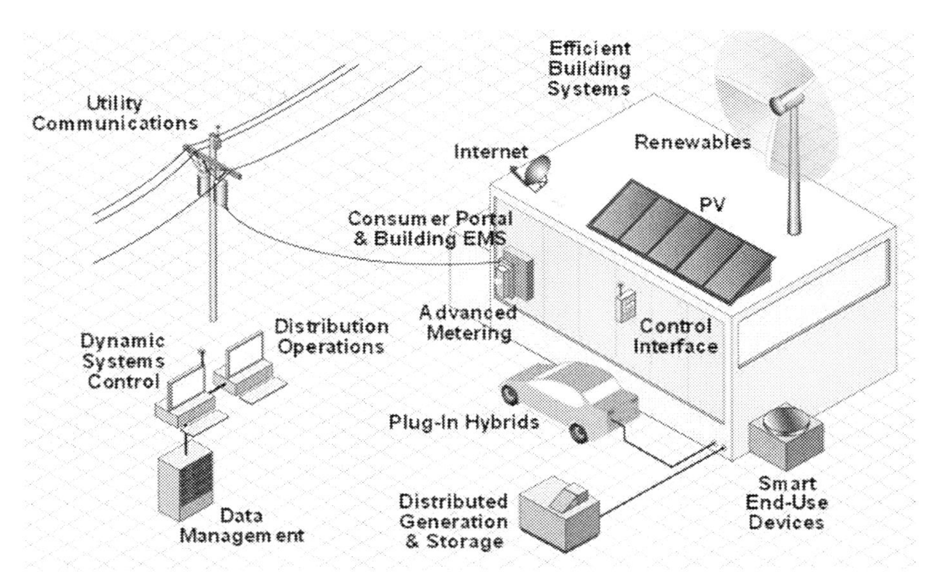

Fig. 13.1 Domestic smart grid technology using IoT-based technologies.

the global efficiency and reliability of the power supply. This brings obvious and direct benefit through appropriate demand response actions, however, the intermittency of renewable DGs and dynamical pricing make it a nontrivial task (Siano, 2014). To fully use the installed DGs, their intermittent energy must be replaced elsewhere in the supply/demand loop (Lujano-Rojasa et al., 2012; Finn and Fitzpatrick, 2014). This calls for a cost-effective demand response solution to efficiently manage the operations of appliances and allocate the domestic loads to the time slots with low costs in accordance to the DG outputs and the up-to-date RTP information. The installed energy storage can coordinate the intermittency of DGs and load pattern, which improves the overall resource utilization at the domestic scope.

Through the demand response techniques, residents have more opportunities to obtain a cost-effective energy management through appropriately controlling the schedulable demands and storage units based on the 1-day ahead predicted operation. A collection of novel energy scheduling algorithmic solution is utilized to exploit the potential economic benefits of matching power supply to the load demand while optimizing the DG and storage utilization efficiency at the domestic level in the context of RTP.

Recent investigations (e.g., Mohsenian-Rad et al., 2010; Zhao et al., 2013; Adika and Wang, 2014; Salinas et al., 2013; Wang et al., 2014; Beaudin et al., 2014; Delfino et al., 2014; Ziadi et al., 2014; Karami and Sanjari, 2014; Igualada et al., 2014; Marra et al., 2014) have been carried out to address the demand response challenges considering the availability of domestic DGs and storage in the IoT-enabled smart home environment. In Mohsenian-Rad et al. (2010), the authors proposed an optimal approach aiming to reduce the resident's payment, minimize the waiting time of appliances, and alleviate the peak-to-average ratio of power demand, by considering the RTP tariff together with the inclining block rates (IBR). In Zhao et al. (2013), two electricity price (EP) levels were set in IBR, and the EP can be switched to the higher price level once the power consumption exceeding the predefined threshold, and hence automatically scheduled the loads in a cost-effective fashion. In Adika and Wang (2014), the appliances are grouped into a set of clusters based on their operational preferences and a cluster-based scheduling approach was proposed to reduce the electricity cost. In Salinas et al. (2013), Lyapunov optimization theory was adopted to address the time-coupling energy management problem with the consideration of demand and DG uncertainties.

It should be noted that, however, the aforementioned solutions have not fully considered the impact of RTP as well as the energy storage on the domestic energy management, and hence the potential benefit needs to be further investigated. To this end (Wang et al., 2014; Beaudin et al., 2014; Delfino et al., 2014), a two-level control architecture considering domestic DGs and storage devices was presented to combine the long-term planning at the time scale of billing period (i.e., a month) and short-term corrective control to compensate the errors of DG and demand prediction. Those two-horizon algorithms could not only lessen computation time, but also improve computational accuracy. In Ziadi et al. (2014) and Karami and Sanjari (2014), the power reference schedule (look-up table) for the network elements, for example, DGs, combined heating and power (CHP) and energy storage system (ESS), was derived and generated at each time interval to optimize the energy management. The control of domestic battery (charging and discharging) was formulated as an mixed-integer linear programming (MILP) problem in the context of the vehicle-to-grid (V2G) system was studied in Igualada et al. (2014). In addition, a decentralized storage management solution was presented in Marra et al. (2014) to trigger the ESS activation only when the DG supply exceeded an optimized power threshold, to avoid overvoltage and adverse impact on the battery.

The aforementioned solutions have either not been able to thoroughly investigate the impact and potential benefits of correlative effect among DGs, storage, and demand, or not fully included the RTP variability into consideration. Also, the available solutions are firmly based on the underlying assumption of accurate prediction of RTP and DG dynamics, and hence the performance in the presence of prediction errors still remains open for further validation. To this end, the work presented in Yang and Fang (2017) addressed such outstanding technical challenges and presented a cost-effective demand response algorithmic solution in the context of domestic energy system with small-scale distributed renewable generators (e.g., PVs and wind turbines) and storage units while considering the impact of RTP. In details, this work presented a demand response approach for the domestic energy system, embedded with small-scale renewable DGs (solar and wind), and Li-ion battery acting as the storage unit, in the context of RTP environment. A collection of domestic appliances with diverse operational patterns and constraints are considered into the optimal load scheduling process. The proposed solution carries out the domestic energy management which combines the load control and storage-based scheduling can optimally match the domestic loads to the power supply as well as improving the utilization of DG resources and simultaneously reduce the electricity purchase cost from the power utility. The proposed algorithmic solution has been evaluated through a set of simulation experiments through a comparative study, and the robustness of the solution is also evaluated in the presence of prediction errors. It clearly demonstrates the effectiveness of the proposed solution for demand response under RTP. The key observations and results obtained from a comparative study are illustrated in Fig. 13.2.

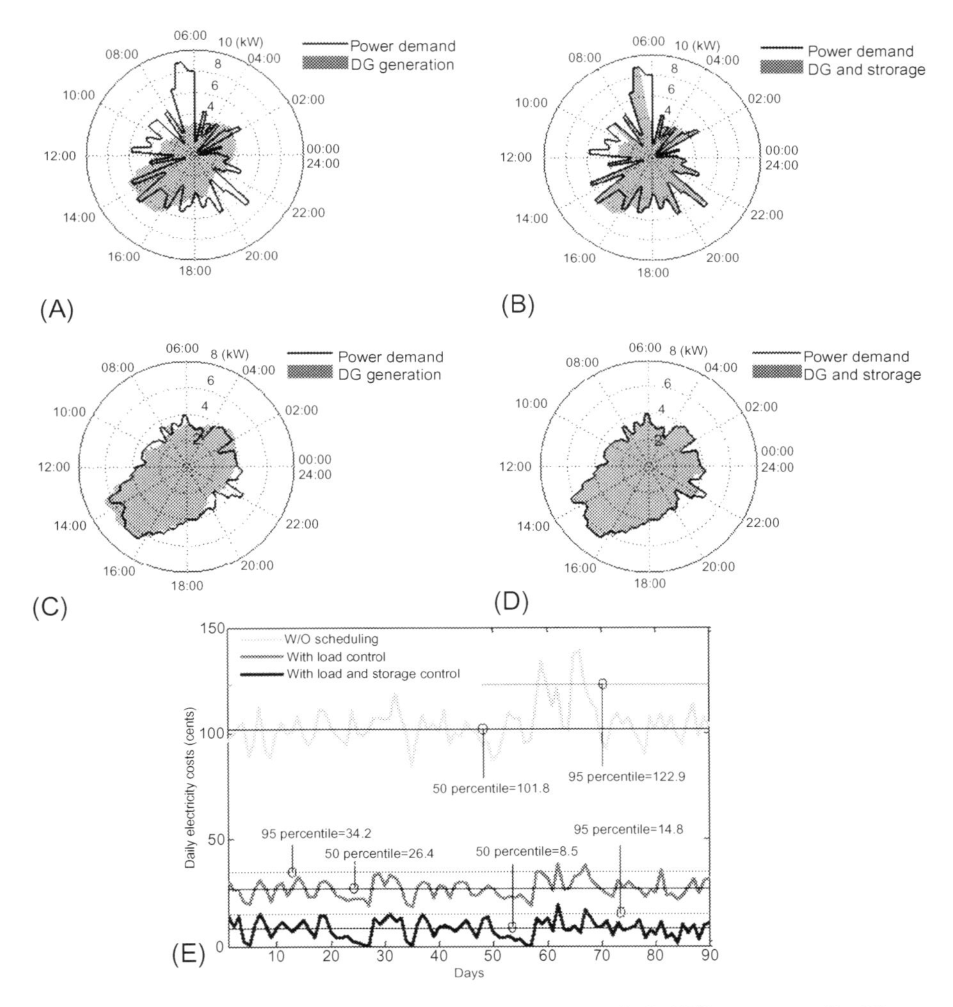

Fig. 13.2 Supply-demand matching performance (A): scenario I (W/O any control); (B): scenario II (storage-based scheduling only); (C): scenario III (load control only); (D): scenario IV (with load control and storage-based scheduling); and (E) daily electricity purchase under RTP and DG uncertainties (3 months).

13.3 IOT cyber physical security in smart grid

With the increasing adoption of IoT technologies in power utilities, the coupling of cyber and physical network as well as the operational complexity and diversity make the smarter grid vulnerable to different forms of attacks and much research effort has been carried out aiming to obtain better insights in cyber-physical system (CPS) cyberattacks, and hence to develop theories and tools to promote the cyber-security capability of CPS. In reality, unlike the attacks directly imposed on physical components, the attacks on the cyber system can often cost much less and more concealment, that is, difficult to be detected timely and accurately. The adverse impact induced by the cyberattack can significantly affect the operation of both cyber and physical system as a whole. The opponents may have the opportunities to inject false information or modify the available information at different parts of the smart grid, for example, the real-time electricity pricing (RTP) information in advanced metering infrastructure (AMI) that directly misleads the energy dispatch and energy consumption behaviors. This may also affect the decision-making process in the control center and result in sending the incorrect supervisory control signals to the field control and protection devices, which may lead to large-scale network cascading failures.

In addition, the opponents may hack the underlying communication infrastructure, including the communication devices or equipment (e.g., routers and switches) and the communication protocols. Currently, power utilities tend to adopt the standard packet-switching-based protocols (e.g., TCP/IP-based protocols) and information models (e.g., IEC61850 in substations), which makes the communication system more vulnerable to the intentional attacks compared with the private-vendor protocols. In addition, other types of cyberattacks, for example eavesdropping attacks, and malwares attacks, can also be applied to the communication system of smart grid, including smart meters, communication channels, and control center. In summary, it can be envisaged that the aforementioned cyberattacks can bring direct adverse or disastrous impact on the operation of smart energy networks. In the following sections, we look into the cyberattacks in a case study of PMU-based wide-area measurement system (WAMS) for electric transmission network.

The operation of power systems is firmly based on the underlying communication infrastructure, as illustrated in Fig. 13.3. There are many technical issues related to the architecture of the underlying communication infrastructure to facilitate the stable, reliable, and economical operation of transmission power grid. For example, the standards and protocols, the physical media and technologies, and the control algorithms in decision-making need to be further studied and validated. In addition, more attentions have been paid to both the power system itself and the characteristics of the communication infrastructure and their impact of degrading the underpinned functionalities. In Evans (2011), the recent study has looked into the security problem of state estimation in WAMS in the context of CPS which often exhibits complex structural characteristics and dynamic operational phenomenon. Typical attacks, for example, data tampering attack, on PMUs and adverse impact on network state estimation were explored and indicated that any malicious cyberattack will not just cause harm to the cyber system, but will also severely affect the normal operation

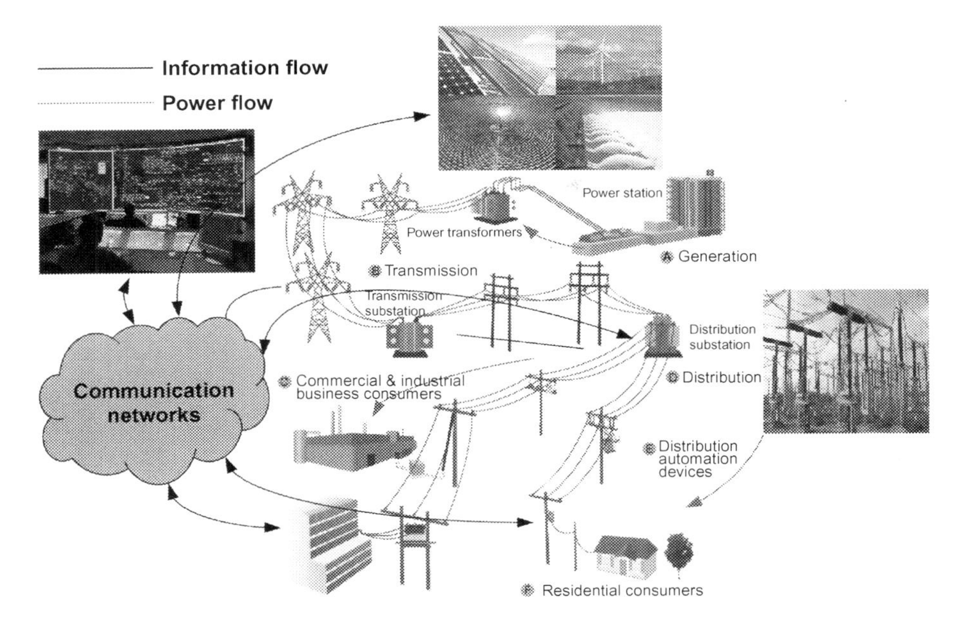

Fig. 13.3 IoT technologies adopted throughout the electric power systems.

of physical systems. This demand for efficient co-simulation tools to carry out the performance assessment, rather than only existing stand-alone power grid simulator or a communication system simulator. The performance of power network measurement and control functionalities can be evaluated, and the cost and benefit analysis can be carried out based on an effective, scalable, and efficient co-simulation platform for electrical cyber-physical systems (ECPS), for example, the configuration of PMU-based WAMS in transmission networks can be optimized and evaluated, and robustness performance under various operational faults, malfunctions, or attacks.

Existing studies have highlighted the importance of cyber security of ECPS against various forms of attacks from different aspects. This chapter exploits the cyberattack in ECPS with a focus on the false-data injection (FDI) attacks on PMU-based state estimation in power distribution networks. It is confirmed that FDI attacks can circumvent bad data detection programs and inject bias into the values of the estimated state in power systems (Deng et al., 2017). A number of recent studies have exploited the FDI attacks on power system state estimation and defense mechanisms, including detection-based methods (e.g., Kosut et al., 2010; Liu et al., 2014; Chaojun et al., 2015) and protection-based methods (e.g., Kim and Poor, 2011; Bi and Zhang, 2014; Mousavian et al., 2015; Lee and Kundur, 2014; Bi and Zhang, 2011; Liang et al., 2016; Yang et al., 2014; Liu et al., 2015; Chakhchoukh and Ishii, 2016).

The FDI attack detection approaches generally analyze the obtained measurements and detect the abnormal ones that cannot fit the expected distributions of historical measurements. A Kullback-Leibler distance (KLD)-based method was proposed for FDI attack detection by tracking and calculating the distance between two probability distributions derived from measurement variations (Chaojun et al., 2015). In Kim and Poor (2011), greedy algorithms were developed to strategically place secure PMUs at key buses in the power system to defend against data injection attacks with improved manageability and reduced cost. In Bi and Zhang (2014), the authors exploited the graphical defending algorithms against FDI attacks through protecting the state variables with the minimum number of measurements. In Mousavian et al. (2015), a risk mitigation model was presented for cyberattacks to PMU networks through solving a MILP problem to prevent cyberattack propagation and maintain the observability of the power system. A detection and identification of cyberattacks in PMU data based on the expectation-maximization algorithm was presented in Lee and Kundur (2014). The work in Bi and Zhang (2011) designed the defense mechanism of FDI attacks on state estimation based on a least-budget strategy. In Liang et al. (2016), an unobservable FDI attack on AC state estimation and its impacts on the physical system were studied. In Yang et al. (2014), the authors exploited the optimal FDI attack strategy to cause the maximum damage and developed efficient algorithms to identify the optimal meter set for protection. In Liu et al. (2015), the work proposed an efficient strategy for determining the optimal attacking region that requires reduced network information. Multiple robust estimators with different robustness properties to improve the overall cybersecurity of power state estimation were investigated with the consideration of investment reduction (Chakhchoukh and Ishii, 2016).

In addition, solutions (e.g., Beg Mohammadi et al., 2016; Philip and Jain, 2015; Rather et al., 2015) are available to address the issue of optimal PMU placement with different objectives and deployment constraints. The optimal PMU placement can be obtained through minimizing the cost of overall WAMS [PMUs, phasor data concentrators and communication infrastructures (CIs)] (Beg Mohammadi et al., 2016). An optimization formulation was proposed in Philip and Jain (2015) to improve the measurement redundancy with minimum number of PMUs while maintaining full-system observability. The work in Rather et al. (2015) focused on the minimization of total realistic cost with considering hidden but significant and integral part of PMU installation cost.

13.4 Modeling and simulation challenges of IoT in smart grid

Current studies of co-simulation platforms are mainly restricted to small-scale scenarios. This section continues to discuss the potential for simulation capability and highlights the key issue of realistic large-scale IoT infrastructure and operational scenarios for "what-if" evaluations.

13.4.1 Simulation, platforms, and complexity

Simulation is generally restricted to exploring a constructed, abstracted model of the real world and often interplays with measurement, experimentation, and analysis to capture various network behaviors. Most simulators have focused on the packet-level network simulations, that is, model the packet transmission and queuing as they are delivered through the network, and so forth. Alternatively, the fluid simulation (i.e., rate-based simulation) can be used which neglects the small-scale fluctuations in packet queue behavior and focus on source traffic rates and burst scale queuing. Many of them are developed restricted to address a specific facet of network behavior or target protocol. The key challenge is to simulation of realistic networks that are complex, heterogeneous, distributed, and continually changing, and to do this effectively, a general purpose tool requires an active community supporting and further developing the tool. There are two multi-protocol simulation tools, ns-2, and OPNET. However, both these tools are incapable of modeling large-scale networks, where typically OPNET is limited to hundreds of nodes (i.e., hosts, queues, etc.), and ns-2 to thousands, for practical simulation investigations.

13.4.2 Topology characterization and generation

The grid topology needs to adapt and shift from a centralized source to a distributed topology that can absorb different energy sources in a dynamic way. There is a need to track real-time energy consumption and demand to the energy supply: this goes with the deployment of more remote sensing equipment capable of measuring, monitoring, and communicating energy data that can be used to implement a self-healing grid, increase the overall efficiency, and increase the level of self-monitoring and decision making. The connected smart grid provides a communication network that will connect all the different energy-related equipment of the future.

As a result, the first basic question needs to be answered when simulating realistic large-scale ECPS is that how the network are physically interconnected, for example, topology, link properties (e.g., capacity, delay), etc. However, this basic question becomes hard to answer in large-scale scenarios, for example communication topology and properties for smart grids, which share several key properties make them extremely difficult to characterize, model and thus to simulate. In addition, there are no realistic communication topologies available for validating current emerging smart grid infrastructures. Research effort has been made to generate statistically correct or generally reasonable topological models for smart grids based on either statistical network characteristics (Wang et al., 2010) or theoretical graph models (Holmgren, 2006).

13.4.3 Traffic modeling and congestion

The development of current electric CPS is not only in terms of its size, but also in many other dimensions, for example, traffic types and volume. Currently many services and applications other than carried by TCP, for example, dispatch demands and management control signals, are delivered throughout current ECPSs to support the smart grid operation and management. In the meanwhile, the network traffic volume becomes fathomless in today's networks and bandwidth has been over-provisioned with hundreds of megabytes per second to accommodate thousands to millions of simultaneous data transmission in large networks. In ECPS co-simulation, the simulation of communication systems in fine-granularity, for example packet level, by simulating each individual source will lead to computationally prohibitive. Significant advances have been achieved for realistic traffic modeling and generation, from both simulation and analysis perspective. Here, we are more interested in traffic models for data communication in smart grid, for example, TCP elastic applications, like HTTP web browsing, FTP downloading, and UDP/RTP inelastic services, for example, VoIP and video conference.

However, it should be highlighted that there is a mismatch between the simulation paradigms of power system and communication system with respect to the fundamental simulation models (Perros, 2009): power system dynamic simulation uses time-driven methods while communication network simulation uses event-driven methods. These two simulation models must be seamlessly integrated to ensure a reliable synchronization between two coupling systems. In this chapter, the main technical contributions are summarized as follows: the ECPS consisting of electric power system and underlying communication system is discussed and the models of both systems are presented; the conceptual ECPS co-simulation platform is provided and discussed in details. A case study of assessing the effectiveness of ECPS simulator is presented with some preliminary results.

To address the challenges of co-simulation for ECPS, much research effort (e.g., Lin, 2012; Hopkinson et al., 2006; Nutaro et al., 2007; Zhu et al., 2011; Li et al., 2011; Liberatore and Al-Hammouri, 2011; Tong, 2010; Davis et al., 2006; Mallouhi et al., 2011) has been made to attempt to accurately model and simulate the interactive behavior between a power system and a communication network. The most notable existing co-simulation approaches and platforms are available in Ref. (Lin, 2012) and summarized in Table 13.1.

Table 13.1 ECPS co-simulation platforms/tools

	Component	Scalability	Real-time
EPOCHS Hopkinson et al. (2006)	PSCAD, PSLF, ns-2	Medium	No
ADEVS Nutaro et al. (2007)	Adevs, ns-2	Low	No
Zhu et al. (2011)	Simulink, OPNET	Medium	No
VPNET Li et al. (2011)	VTB, OPNET	Low	No
PowerNet Liberatore and Al-Hammouri (2011)	Modelica, ns-2	Low	No
Tong (2010)	OPNET	Low	No
Davis et al. (2006)	PowerWorld, RINSE	Medium	Yes
TASSCS Mallouhi et al. (2011)	PowerWorld, OPNET	Medium	Yes
GECO Lin (2012)	PSLF, ns-2	Medium	No

In literature, both EPOCHS (Hopkinson et al., 2006) and ADEVS (Nutaro et al., 2007) are the well-known approaches for co-simulation of ECPS. EPOCHS (Hopkinson et al., 2006) proposed a power system modeling tool with consideration of the underlying communication network based on federated a dynamic simulation which integrates subcomponents under the high-level architecture (HLA) framework. Similar to EPOCHS, ADEVS (Nutaro et al., 2007) was proposed to promote the synchronization method through modeling the power system using DEVS formalism and the communication network is modeled in ns-2. In Zhu et al. (2011), an integration of MATLAB Simulink and OPNET was reported to study the impact of Information & Communication Technology (ICT) architecture on the reliability of WAMS applications. In Li et al. (2011), an integration of virtual test bed (VTB) software and OPNET called VPNET was introduced for simulating remotely controlled power electronic devices in the power system. The synchronization method used in this chapter is similar to the EPOCHS's method. The tool of PowerNet (Liberatore and Al-Hammouri, 2011) was presented which was designed to integrate Modelica and NS2 with the same synchronization approach. In Tong (2010), the OPNET was extended to simulate wide-area communication networks in power system where the power system dynamic simulation was simplified as a virtual demander.

In Davis et al. (2006), a SCADA cyber security test bed was designed with the focus to assess the robustness of the underlying communication infrastructure of the power system under cyberattacks. Another SCADA cyber security test bed was proposed in Mallouhi et al. (2011) which integrated PowerWorld and OPNET to verify power system performance under cyberattacks on SCADA. In Lin (2012), the author presented a power system and communication network co-simulation framework GECO (global event-driven co-simulation) which can be considered as a universal design pattern for hybrid system co-simulation based on PSLF and ns-2. Recently, functional mock-up interface becomes popular as a standard for implementation of co-simulation platform for hybrid energy systems (Du et al., 2014; Chen et al., 2016). Also, the real-time digital simulator (RTDS) is an efficient tool for power grid simulation (Luo et al., 2015).

13.4.4 IoT-based smart grid co-simulation model and design

The key issue of the design of the power system and communication network co-simulation framework is that it has to accurately synchronize the simulation time in two distinct simulation models. The power system simulation usually uses time-driven method while the communication network simulation usually uses event-driven method. Between the simulators, the OLE for process control (OPC) is used to as an interface for data exchange. It specifies an interface between process data client applications and servers. The standard was purposely limited to the reading and writing of process values. Alarm handling, process events, security, batch structures, and historical data access were all deferred to subsequent releases. OPC provides for a high degree of interoperability between client and server applications supplied by different vendors. OPC allows software modules from a variety of vendors to interoperate through a standard interface.

The work presented in Lu et al. (2015) discussed the issue of co-simulation for IoT-based ECPS consisting of electric power system and underlying communication system, including the traffic models as well as a generic framework of co-simulator. The coupled two systems, that is, physical power system and cyber communication system, are characterized as continuous system and discrete system, respectively. As a result, the coordinated simulation of these two systems need to be implemented through integrates the time-driven and event-driven simulators. Fig. 13.4 illustrates a generic framework of co-simulation platform which can carry out the ECPS simulations through a networked structure.

It can be seen that this networked co-simulation platform basically can be broken down into three simulation blocks: (1) power network simulator which could adopt different simulators of power systems (running at computer 1); communication network simulator/emulator (running at computer 2); and control center simulator (running at computer 3). In the computers which carry out the simulation of power systems and control center functionalities, different simulators can be used in respect to different simulation purposes. In the middle, the communication simulators (e.g., ns-2, MATLAB/TRUETIME toolbox) can be adopted to simulate the communication systems interconnecting the power system and control center or local controllers. The network-introduced delay which occurs during the data exchange through the shared medium can be either constant or time varying. Some information not only suffers from delay (i.e., latency) but also can be lost during transmission. The delay and loss can degrade the performance of the control system.

13.5 Conclusions

Regulations and standards continue to drive the adoption of connected devices all across the smart grid industry from grid infrastructure to smart meters down to homes and buildings. While migrating to smart meters adds a new layer of complexity, the return on investment, such as improved customer experience and energy efficiency, is becoming more apparent. The grid itself is also changing and moving toward a fully automated substation network with connected data concentrators already being deployed. The connectivity and accessibility that the IoT brings further enhance the customer experience and efficiencies allowing greater interaction and control for consumers. Additionally the IoT delivers more data for manufacturers and utility providers to reduce costs through diagnostics and neighborhood-wide meter reading capabilities. Ultimately, the IoT will be instrumental in building a more connected, cost effective, and smarter smart grid.

With the industry's broadest IoT-ready portfolio of wired and wireless connectivity technologies, microcontrollers, processors, sensors and analog signal chain, and power solutions, TI offers cloud-ready system solutions designed for IoT accessibility. From high-performance home, industrial, and automotive applications to battery-powered wearable and portable electronics or energy-harvested wireless sensor nodes, TI makes developing applications easier with hardware, software, tools and support to get anything connected within the IoT.

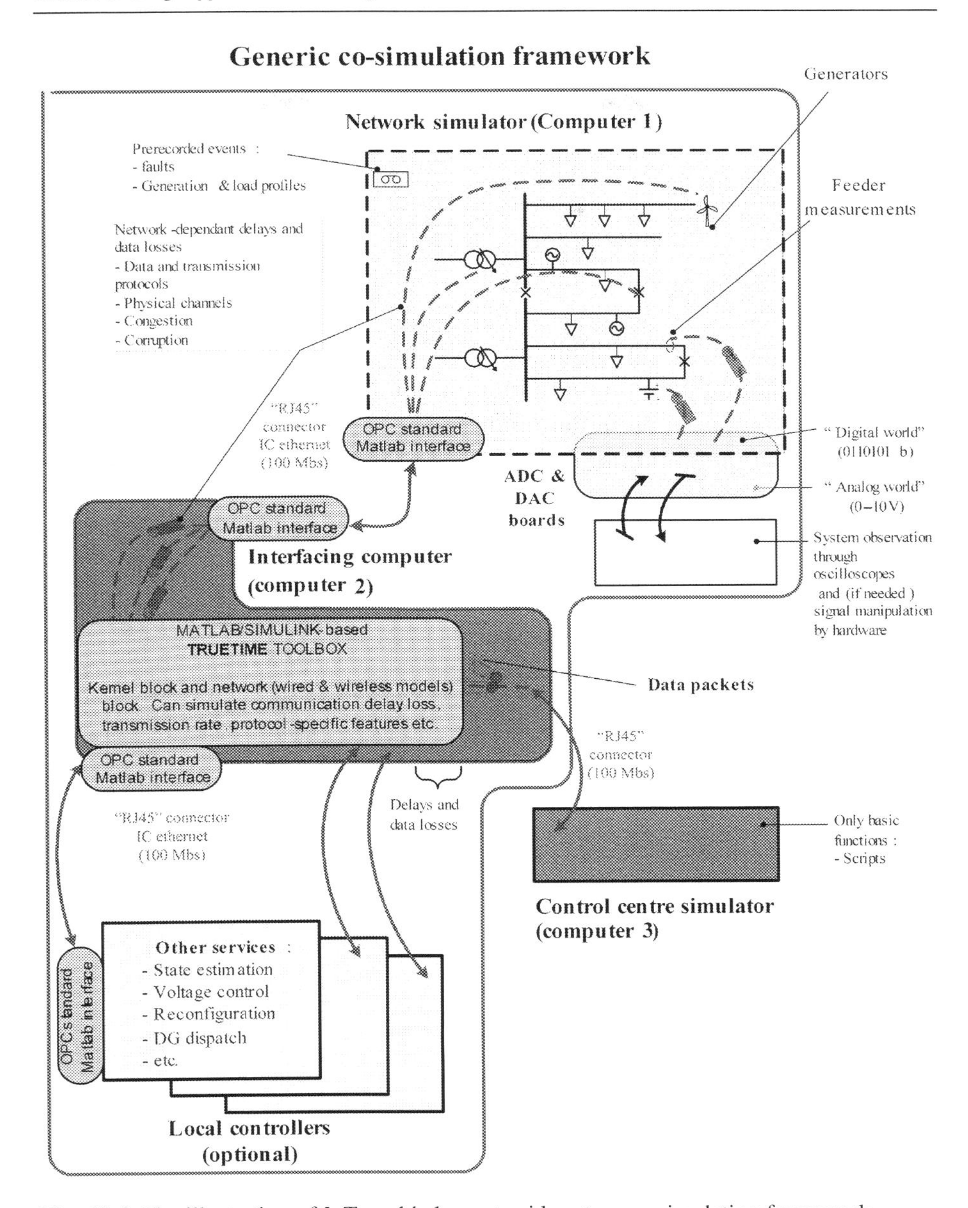

Fig. 13.4 The illustration of IoT-enabled smart grid system co-simulation framework.

In respect to the future work, two directions are considered worth further research effort. The advanced prediction tools and the error correction methods need to be explored to improve the prediction accuracy of DG generation and RTP, to guarantee the performance of the proposed solution; also the potential benefit of cooperative operation of multiple domestic energy management systems covering a large number of residents needs to be studied. In addition, it should be noted that this co-simulation platform has the following limitations which required further exploitation and extensions: the tool simulates communication systems at the packet level. There are no detailed models for simulation of lower levels of the protocol stack, for example, MAC layer, physical layer; the simulation tool is not able to support comprehensive reliability analysis and communication system redundancy assessments; current communication performance analysis focuses mainly on analyzing communication latency and loss probability, and simplified attacks during data transmission.

References

Adika, C.O., Wang, L.F., 2014. Autonomous appliance scheduling for household energy management. IEEE Trans. Smart Grid 5 (2), 673–682.

Beaudin, M., Zareipour, H., Bejestani, A.K., Schellenberg, A., 2014. Residential energy management using a two-horizon algorithm. IEEE Trans. Smart Grid 5 (4), 1712–1723.

Beg Mohammadi, M., Hooshmand, R.A., Haghighatdar Fesharaki, F., 2016. A new approach for optimal placement of PMUs and their required communication infrastructure in order to minimize the cost of the WAMS. IEEE Trans. Smart Grid 7 (1), 84–93.

Bi, S., Zhang, Y.J., 2011. Defending mechanisms against false-data injection attacks in the power system state estimation. 2011 IEEE GLOBECOM Workshops (GC Wkshps), Houston, TX, 1162–1167.

Bi, S., Zhang, Y.J., 2014. Graphical methods for defense against false-data injection attacks on power system state estimation. IEEE Trans. Smart Grid 5 (3), 1216–1227.

Chakhchoukh, Y., Ishii, H., 2016. Enhancing robustness to cyber-attacks in power systems through multiple least trimmed squares state estimations. IEEE Trans. Power Syst. 31 (6), 4395–4405.

Chaojun, G., Jirutitijaroen, P., Motani, M., 2015. Detecting false data injection attacks in AC state estimation. IEEE Trans. Smart Grid 6 (5), 2476–2483.

Chen, J., Garcia, H.E., Kim, J.S., Bragg-Sitton, S.M., 2016. Operations optimization of nuclear hybrid energy systems. Nucl. Technol. 195 (2), 143–156.

Davis, C.M., Tate, J.E., Okhravi, H., Grier, C., Overbye, T.J., Nicol, D., 2006. SCADA cyber security testbed development. In: Proceedings of 38th North American Power Symposium (NAPS 2006), pp. 483–488.

Delfino, F., Minciardi, R., Pampararo, F., Robba, M., 2014. A multilevel approach for the optimal control of distributed energy resources and storage. IEEE Trans. Smart Grid 5 (4), 2155–2162.

Deng, R., Xiao, G., Lu, R., Liang, H., Vasilakos, A.V., 2017. False data injection on state estimation in power systems—attacks, impacts, and defense: a survey. IEEE Trans. Ind. Inform. 13 (2), 411–423.

Du, W., Garcia, H.E., Paredis, C.J.J., 2014. An optimization framework for dynamic hybrid energy systems. In: Proceedings of 10th International Modelica Conference, pp. 767–776.

Evans, D., April 2011. The Internet of Things: How the Next Evolution of the Internet is Changing Everything. CISCO. Retrieved from http://www.cisco.com/web/about/ac79/docs/innov/IoT_IBSG_0411FINAL.pdf.

Finn, P., Fitzpatrick, C., 2014. Demand side management of industrial electricity consumption: promoting the use of renewable energy through real-time pricing. Appl. Energy 113, 11–21.

Holmgren, A.J., 2006. Using graph models to analyze the vulnerability of electric power networks. Risk Anal. 26 (4), 955–969.

Hopkinson, K., Wang, X., Giovanini, R., Thorp, J., Birman, K., Coury, D., 2006. EPOCHS: a platform for agent-based electric power and communication simulation built from commercial off-the-shelf components. IEEE Trans. Power Syst. 21 (2), 548–558.

Igualada, L., Corchero, C., Zambrano, M.C., Heredia, F.-J., 2014. Optimal energy management for a residential microgrid including a vehicle-to-grid system. IEEE Trans. Smart Grid 5 (4), 2163–2172.

Karami, H., Sanjari, M.J., 2014. et al., An optimal dispatch algorithm for managing residential distributed energy resources. IEEE Trans. Smart Grid 5 (5), 2360–2367.

Kim, T.T., Poor, H.V., 2011. Strategic protection against data injection attacks on power grids. IEEE Trans. Smart Grid 2 (2), 326–333.

Kosut, O., Jia, L., Thomas, R., Tong, L., 2010. On malicious data attacks on power system state estimation. In: Proceedings of 45th International Universities Power Engineering Conference (UPEC), Cardiff, UK, pp. 1–6.

Lee, D., Kundur, D., 2014. Cyber-attack detection in PMU measurements via the expectation-maximization algorithm. 2014 IEEE Global Conference on Signal and Information Processing (GlobalSIP), Atlanta, GA, 223–227.

Li, W., Monti, A., Luo, M., Dougal, R.A., 2011. VPNET: A co-simulation framework for analyzing communication channel effects on power systems. Proceedings of IEEE Electric Ship Technologies Symposium (ESTS), 143–149.

Liang, J., Sankar, L., Kosut, O., 2016. Vulnerability analysis and consequences of false data injection attack on power system state estimation. IEEE Trans. Power Syst. 31 (5), 3864–3872.

Liberatore, V., Al-Hammouri, A., 2011. Smart grid communication and co-simulation. In: Proceedings of IEEE Energy Tech, pp. 1–5.

Lin, H., 2012. Communication infrastructure for the smart grid: a co-simulation based study on techniques to improve the power transmission system functions with efficient data networks. PhD dissertation, Virginia, September.

Liu, L., Esmalifalak, M., Ding, Q., Emesih, V., Han, Z., 2014. Detecting false data injection attacks on power grid by sparse optimization. IEEE Trans. Smart Grid 5 (2), 612–621.

Liu, X., Bao, Z., Lu, D., Li, Z., 2015. Modeling of local false data injection attacks with reduced network information. IEEE Trans. Smart Grid 6 (4), 1686–1696.

Lu, Y., Yang, Q., Xu, W., Lin, Z., Yan, W., 2015. Cyber security assessment in PMU-based state estimation of smart electric transmission networks. Proceedings of 27th Chinese Control and Decision Conference (CCDC), Qingdao. .

Lujano-Rojasa, J.M., Monteirob, C., Dufo-Lópeza, R., Bernal- Agustín, J.L., 2012. Optimum residential load management strategy for real time pricing (RTP) demand response programs. Energy Policy 45, 671–679.

Luo, Y., Wang, C., Tan, L., Liao, G., Zhou, M., et al., 2015. Application of generalized predictive control for charging super capacitors in microgrid power systems under input constraints. IEEE International Conference on Cyber Technology in Automation, Control, and Intelligent Systems (CYBER), 1708–1713, 2015.

Mallouhi, M., Al-Nashif, Y., Cox, D., Chadaga, T., Hariri, S., 2011. A testbed for analyzing security of SCADA control systems (TASSCS). Second IEEE PES Innovative Smart Grid Technologies Conference.

Marra, F., Yang, G., Traeholt, C., Ostergaard, J., Larsen, E., 2014. A decentralized storage strategy for residential feeders with photovoltaics. IEEE Trans. Smart Grid 5 (2), 974–981.

Mohsenian-Rad, A.-H., et al., 2010. Optimal residential load control with price prediction in real-time electricity pricing environments. IEEE Trans. Smart Grid 1 (2), 120–133.
Mousavian, S., Valenzuela, J., Wang, J., 2015. A probabilistic risk mitigation model for cyber-attacks to PMU networks. IEEE Trans. Power Syst. 30 (1), 156–165.
Nutaro, J., Kuruganti, P.T., Miller, L., Mullen, S., Shankar, M., 2007. Integrated hybrid simulation of electric power and communications systems. Proceedings of IEEE Power Engineering Society General Meeting, 1–8.
Perros, H., 2009. Computer Simulation Techniques: The Definitive Introduction. North Carolina State University, Raleigh, NC.
Philip, J.G., Jain, T., 2015. Optimal placement of PMUs for power system observability with increased redundancy. In: 2015 Conference on Power, Control, Communication and Computational Technologies for Sustainable Growth (PCCCTSG), Kurnool, pp. 1–5.
Rather, Z.H., Chen, Z., Thøgersen, P., Lund, P., Kirby, B., 2015. Realistic approach for phasor measurement unit placement: consideration of practical hidden costs. IEEE Trans. Power Deliv. 30 (1), 3–15.
Salinas, S., Li, M., Li, P., Fu, Y., 2013. Dynamic energy management for the smart grid with distributed energy resources. IEEE Trans. Smart Grid 4 (4), 2139–2151.
Siano, P., 2014. Demand response and smart grids – a survey. Renew. Sustain. Energy Rev. 30, 461–478.
Tong, X., 2010. The co-simulation extending for wide-area communication networks in power system. Proceedings of Asia-Pacific Power and Energy Engineering Conference (APPEEC), 1–4.
Wang, Y., Lin, X., Pedram, M., 2014. Adaptive control for energy storage systems in households with photovoltaic modules. IEEE Trans. Smart Grid 5 (2), 992–1001.
Wang, Z., Scaglione, A., Thomas, R., 2010. Generating statistically correct random topologies for testing smart grid communication and control networks. IEEE Trans. Smart Grid, 28–39.
Yang, Q., Barria, J., Laurenson, D., 2012. On the use of LEO satellite constellation for active network management in power distribution networks. IEEE Trans. Smart Grid 3 (3), 1371–1381.
Yang, Q., Fang, X., 2017. Demand response under real-time pricing for domestic households with DGs and energy storage. IET Gener. Transm. Distrib.
Yang, Q., Yang, J., Yu, W., An, D., Zhang, N., Zhao, W., 2014. On false data-injection attacks against power system state estimation: modeling and countermeasures. IEEE Trans. Parallel Distrib. Syst. 25 (3), 717–729.
Zhao, Z., Lee, W.C., Shin, Y., Song, K.-B., 2013. An optimal power scheduling method for demand response in home energy management system. IEEE Trans. Smart Grid 4 (3), 1391–1400.
Zhu, K., Chenine, M., Nordstrom, L., 2011. ICT architecture impact on wide area monitoring and control systems' reliability. IEEE Trans. Power Deliv. 26 (4), 2801–2808.
Ziadi, Z., Taira, S., Oshiro, M., Funabashi, T., 2014. Optimal power scheduling for smart grids considering controllable loads and high penetration of photovoltaic generation. IEEE Trans. Smart Grid 5 (5), 2350–2359.

Glossary

IoT Internet of things is considered as a network comprised of physical objects capable of gathering and sharing electronic information.

Active distribution network A distribution network with systems in place to control a combination of distributed energy resources comprising of generators and storage.

Distributed control The control architecture adopts decentralized elements or functionalities to control the distributed components in systems.

Distributed generation (DG) It is an approach that employs small-scale technologies to produce electricity close to the end users of power. DG technologies often consist of modular (and sometimes renewable-energy) generators, and they offer a number of potential benefits. In many cases, distributed generators can provide lower-cost electricity and higher power reliability and security with fewer environmental consequences than can traditional power generators.

Supervisory control and data acquisition (SCADA) It is a system for remote monitoring and control that operates with coded signals over communication channels (using typically one communication channel per remote station).

Demand response Changes in electric usage by end-use customers from their normal consumption patterns in response to changes in the price of electricity over time, or to incentive payments designed to induce lower electricity use at times of high wholesale market prices or when system reliability is jeopardized. DR includes all intentional modifications to consumption patterns of electricity to induce customers that are intended to alter the timing, level of instantaneous demand, or the total electricity consumption.

Cyber-physical system It is a mechanism controlled or monitored by computer-based algorithms, tightly integrated with the internet and its users. In cyber-physical systems, physical and software components are deeply intertwined, each operating on different spatial and temporal scales, exhibiting multiple and distinct behavioral modalities, and interacting with each other in a myriad of ways that change with context.

H-infinity-based microgrid state estimations using the IoT sensors

M.M. Rana, Wei Xiang, Eric Wang
College of Science and Engineering, James Cook University, Townsville, QLD, Australia

14.1 Introduction

The Internet of things (IoT) is a promising area of research topic both in academia and industrial domains (Haga et al., 2017). Basically, it can be applied to many areas such as smart grid, air conditioning system, smart agriculture, and transportation systems (Pacheco and Hariri, 2016; Shinohara and Namerikawa, 2017). The main component of IoT framework is the wireless sensor networks. In fact, the sensors are deployed in the physical space, and the sensing information is communicated to the cyber space to make the world better for living, monitoring, servicing, and enjoying (Aref and Tran, 2017; Kaur and Kalra, 2018). Driven by this motivation, this chapter proposes an H-infinity-based microgrid state estimation algorithm using the IoT-embedded sensors. Technically, the energy management system uses the sensing information to estimate the microgrid states so that the power system situation awareness and security will be enhanced (Zhou and Miao, 2016). It is believed that the IoT can mitigate the global warming at an acceptable level after encouraging to use the environment friendly renewable microgrid.

In order to estimate the power system states such as bus voltage, many methods have been proposed in the literature. To begin with, the weighted least-squared-based AC state estimation is presented in Hug and Giampapa (2012). In Salinas and Li (2016), proposes a Kalman filter (KF)-based centralized energy theft detection algorithm. Basically, it can effectively identify the energy thieves by decomposing the KF into two parallel and loosely coupled filters. Furthermore, the KF-based decentralized fault detection of multiple cyber attacks in power systems is explored in Irita and Namerikawa (2015). In this scheme, each sensor node can exchange information in the neighboring nodes; finally, it uses the fault detection and fault distinction matrices. In Keller et al. (2012), proposes a modified unknown input KF algorithm for the state filtering of network controlled systems subject to random cyber attacks. In Zhao et al. (2017), develops a robust iterative extended KF-based maximum likelihood approach for estimating power system states. Moreover, a class of malicious attacks against remote state estimation in cyber-physical systems is demonstrated in Li et al. (2017). Finally, the linear minimum variance sense-based distributed KF state estimation algorithm is proposed in Chen et al. (2017). All the KF-based approaches are required to know the exact noise statistics which are generally unknown at the energy management system. In fact, the H-infinity-based microgrid state estimation is proposed in Rana (2017a), but it does not analysis the estimation performance

Smart Power Distribution Systems. https://doi.org/10.1016/B978-0-12-812154-2.00014-6

considering sensor faults and large disturbances. Additionally, the dynamic state estimation using the H-infinity algorithm is presented in Zhao (2018). Furthermore, the H-infinity for power system stability is analyzed in Zadehbagheri et al. (2017), Rusu and Livint (2017), and Sakthivel and Arun (2016). Motivated by the aforementioned analysis, this article proposes an H-infinity-based microgrid state estimation using the IoT-embedded sensors. In fact, the main contributions of this article are as follows:

- The microgrid is expressed as an user friendly discrete-time state-space framework. For doing this, the renewable distributed energy resource (DER) is connected to the distribution test feeder. After applying the Kirchhoff's laws, the system is represented to the continues-time state-space model. Using discretization technique, the continues-time system is represented to the discrete-time framework.
- The IoT-embedded smart sensors are used to sense the microgrid states. For doing this, the IoT-enabled smart sensors are deployed around the test feeder circuit to get the noisy measurement for state estimations.
- For estimating the system states in the energy management system, an H-infinity-based microgrid state estimation algorithm is proposed. This algorithm does not require the process and measurement noise statistics.
- Extensive numerical simulations are conducted considering the sensor faults and large disturbance conditions. It can be seen that the proposed algorithm can effectively estimate the system states in approximately 200 iterations.

Organization: The manuscript is organized as follows. Section 14.2 describes the IoT-based observation model. Furthermore, the H-infinity-based state estimation algorithm is presented in Section 14.3. Moreover, the effectiveness of the proposed algorithm is demonstrated in Section 14.4, where we are firstly described the microgrid model. The paper is wrapped up in Section 14.5.

Notation: Upper case letters and bold face lower are used to represent matrices and vectors, respectively; superscripts x' denotes the transpose of x and I is the identity matrix.

14.2 Observation model

The discrete-time state-space framework is written as follows:

$$\mathbf{s}(t+1) = \mathbf{F}\mathbf{s}(t) + \mathbf{B}\mathbf{u}(t) + \mathbf{v}(t), \tag{14.1}$$

where $\mathbf{F}$ is system state matrix, $\mathbf{s}(t)$ is the system state, $\mathbf{B}$ is the input matrix, $\mathbf{u}(t)$ is the system control effort, and $\mathbf{v}(t)$ is the system noise whose covariance matrix is $\mathbf{Q}(t)$. The measurement from the IoT embedded sensor is given by

$$\mathbf{z}(t) = \mathbf{H}\mathbf{s}(t) + \mathbf{w}(t), \tag{14.2}$$

where $\mathbf{z}(t)$ is the measurement information, $\mathbf{H}$ is the sensing matrix and $\mathbf{w}(t)$ is the noisy measurement whose covariance matrix is $\mathbf{R}(t)$. Basically, the sensing measurement is used for state estimation at the energy management system.

14.3 H-Infinity for microgrid state estimation

Basically, H-infinity is an iterative process to estimate the system state without knows the exact noise statistics. The followings are the steps for H-infinity filter (Rana, 2017a; Simon, 2006; Allen et al., 2010):

$$\hat{s}(t+1) = F\hat{s}(t) + Bu(t) + K(t)[z(t) - H\hat{s}(t)].$$

Here, $\hat{s}(t)$ and $\hat{s}(t+1)$ are the prior and posterior state estimation, and the H-infinity filter gain $\mathbf{K}(t)$ is given by

$$K(t) = P(t)\left[I - \theta\overline{M}(t)P(t) + H'R^{-1}(t)HP(t)\right]^{-1}H'R^{-1}(t).$$

Here, $\mathbf{P}(t)$ is the prior error covariance, θ is the user defined bound, and $\overline{M}(t) = L(t)M(t)L(t)$, where $\mathbf{L}(t)$ as well as $\mathbf{M}(t)$ are the user-defined performance variables. The estimated error covariance matrix $\mathbf{P}(t+1)$ is computed by

$$P(t+1) = FP(t)\left[I - \theta\overline{M}(t)P(t) + H'R^{-1}(t)HP(t)\right]^{-1}F' + Q(t).$$

In each step, the following condition is hold:

$$P^{-1}(t) - \theta\overline{M}(t) + H'R^{-1}(t)H > 0.$$

Using the microgrid, the performance of the H-infinity is demonstrated in the following section.

14.4 Microgrid modeling and simulation results

This section describes the microgrid state-space framework and simulation results. It shows that the proposed algorithm can effectively estimate the system states under the condition of sensor faults and large disturbances.

14.4.1 Microgrid modeling

Generally speaking, the microgrid provides clean, green, and sustainable energy to the user (Myeong-Jun and Kyeong-Hwa, 2017; Salas-Puente et al., 2017). Sometimes, the microgrid incorporating DER such as solar cell is connected the electric grid to supply energy. In the distribution power system, the microgrid is connected to the feeder circuit. To illustrate, Fig. 14.1 shows a distribution test feeder where the DER is connected to the point of common coupling (PCC) through coupling inductor.

Basically, the DER is represented by voltage source V_s which is connected to the PCC whose bus voltage is denoted by V_b. The coupling inductor exists between them.

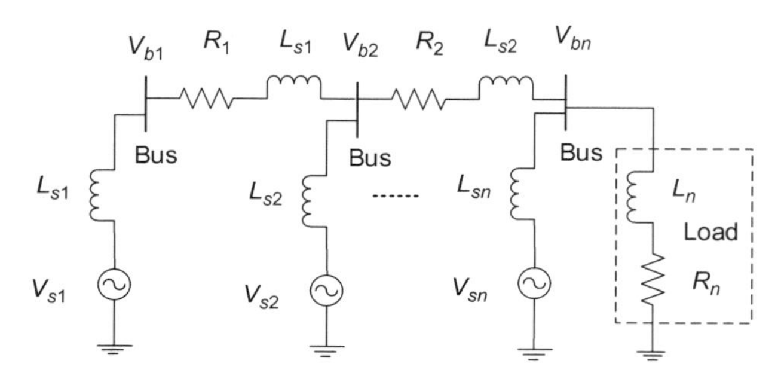

Fig. 14.1 The n-bus system connected to DERs (Korukonda et al., 2017; Li et al., 2012).

For simplicity, this chapter considers that there are 4-bus in the power network. The continuous-time state-space framework is expressed as follows:

$$\Delta \dot{V}_b = F_c \Delta V_b + B_c \Delta V_s, \tag{14.3}$$

where $\Delta V_b = V_b - V_{bref}$ is the PCC bus voltage deviation, $V_b = [V_{b1}V_{b2}V_{b3}V_{b4}]'$, V_{bref} is the bus reference voltage, $\Delta V_s = V_s - V_{sref}$ is the DER voltage deviation, $V_s = [V_{s1}V_{s2}V_{s3}V_{s4}]'$, and V_{sref} is the reference voltage source. Using the discretization step size parameter μ, Eq. (14.3) is transformed to discrete-time state-space framework in Eq. (14.1), where $F = I + \mu F_c$ is the system state matrix, $s = V_b - V_{bref}$ is the state/PCC voltage to be estimated, $B = \mu B_c$ is the system input matrix, $u = V_s - V_{sref}$ is the control effort, and **v** is the process uncertainty. The continuous-time system state and input matrices are given in Korukonda et al. (2017).

14.4.2 Simulation results

For simulation, the discretization step size parameter is 0.01 s. The process and measurement noises are considered the Gaussian distribution with zero means and 0.00001**I** and 0.0001**I** covariance matrices. The simulations are conducted considering three scenarios. Firstly, it is considered that sensors can sense the system states directly, that is, sensing matrix **C** is the identity matrix (Korukonda et al., 2017; Li et al., 2012; Rana, 2017b). Considering this scenario, the simulation results are depicted in Figs. 14.2–14.5. It can be seen that the H-infinity filter can able to estimate the system state within the maximum of 30 iterations. Specifically, Fig. 14.2 shows the PCC voltage deviation at bus 1 and its estimation result. It is observed that the H-infinity can able to estimate the system state within 20 iterations. Other system states have almost similar estimation performance.

Sometimes, the sensor cannot sense the system states directly, even sensors are completely dead. In this scenario, it is considered that strength of the first three sensors are reduced to 90%, while the last sensor is completely dead. Considering this scenario, the simulation results are demonstrated in Figs. 14.6–14.9. It can be seen that the proposed algorithm can effectively estimate the system states within 70 iterations.

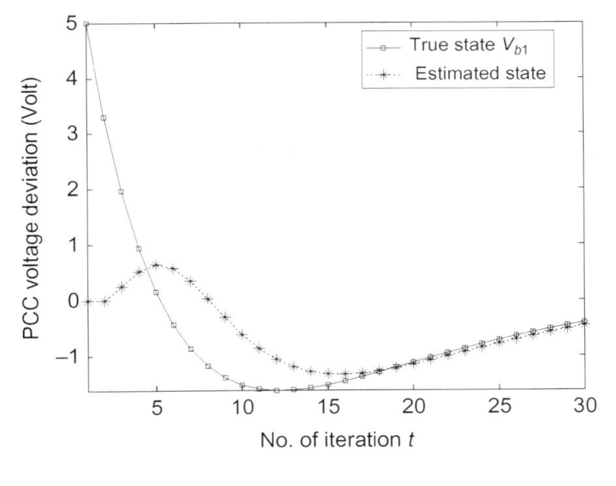

Fig. 14.2 PCC voltage deviation at bus 1 when sensors sense system state directly.

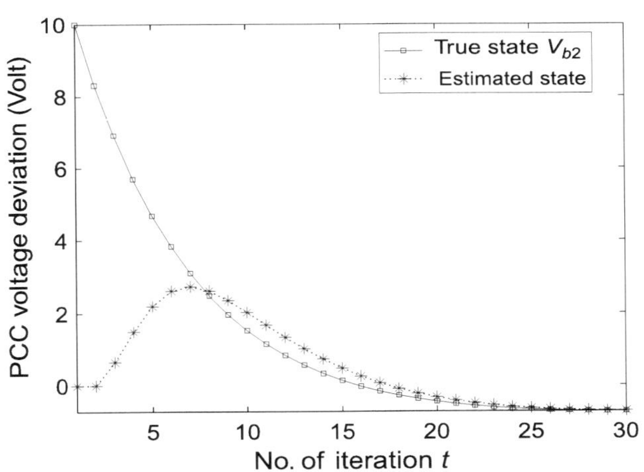

Fig. 14.3 PCC voltage deviation at bus 2 when sensors sense system state directly.

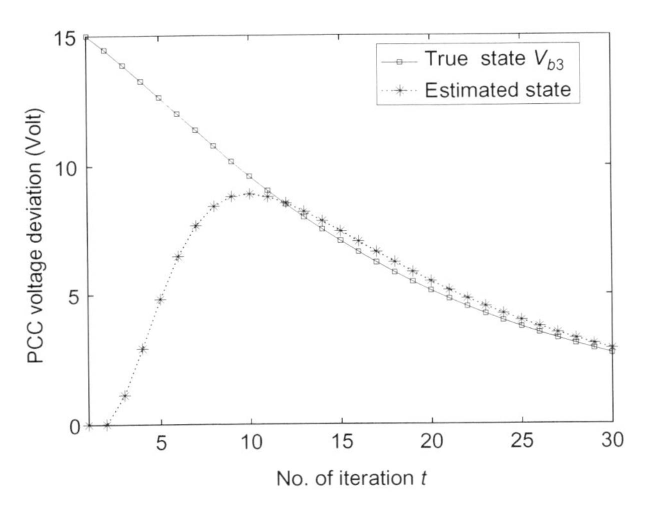

Fig. 14.4 PCC voltage deviation at bus 3 when sensors sense system state directly.

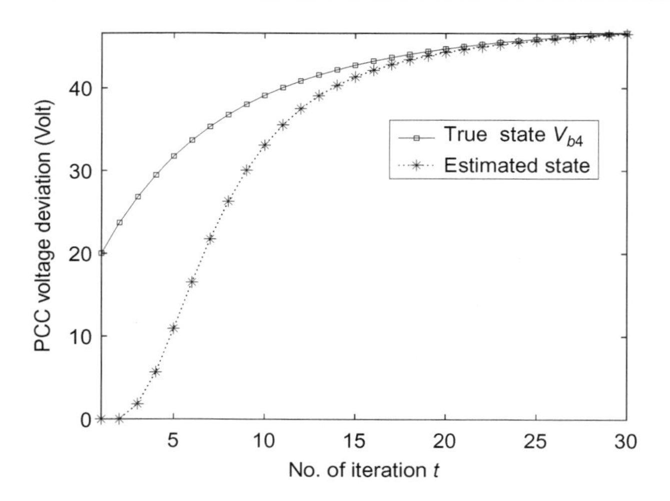

Fig. 14.5 PCC voltage deviation at bus 4 when sensors sense system state directly.

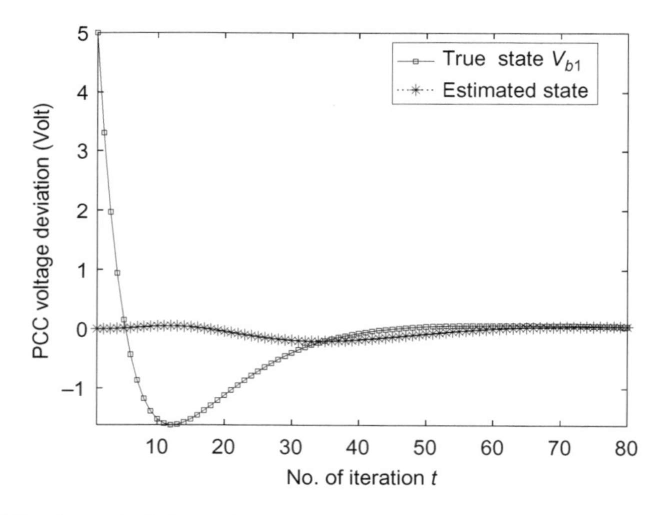

Fig. 14.6 PCC voltage deviation at bus 1 when sensors are not working properly.

Particularly, Fig. 14.9 shows the PCC voltage deviation at bus 4 and its estimation result. It can be observed that the proposed algorithm requires more iteration to estimate the system state as expected (Figs. 14.10–14.13).

14.5 Conclusion and future work

This article proposes an H-infinity-based microgrid state estimation algorithm using the IoT-based sensors. After representing the microgrid in a state-space framework, the IoT-enabled smart sensors are deployed to obtain the sensing state information.

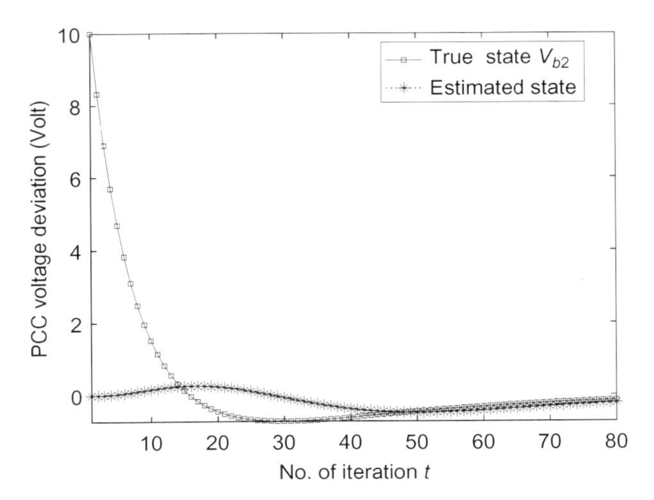

Fig. 14.7 PCC voltage deviation at bus 2 when sensors are not working properly.

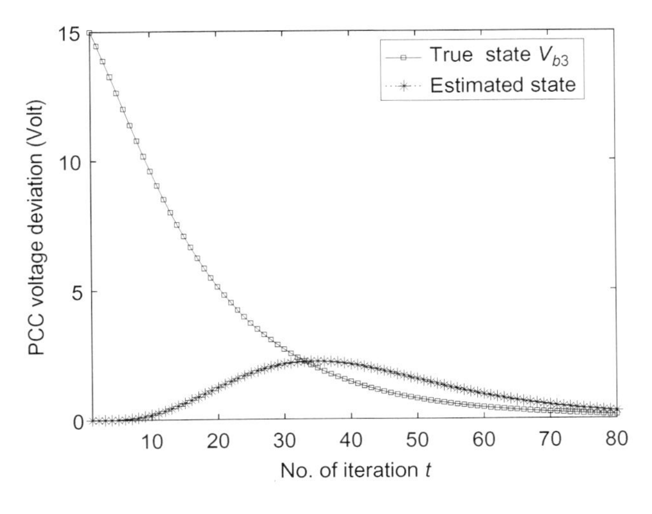

Fig. 14.8 PCC voltage deviation at bus 3 when sensors are not working properly.

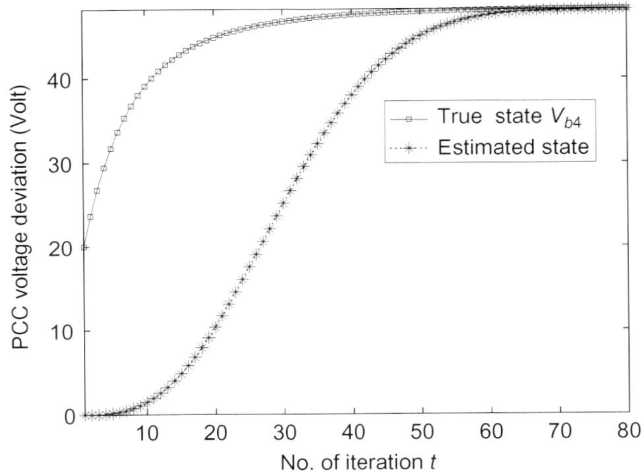

Fig. 14.9 PCC voltage deviation at bus 4 when sensors are not working properly.

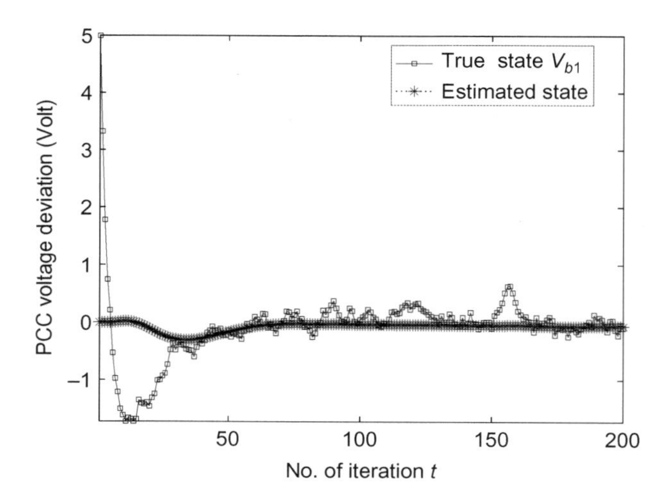

Fig. 14.10 PCC voltage deviation at bus 1 with sensors damage and large disturbances conditions.

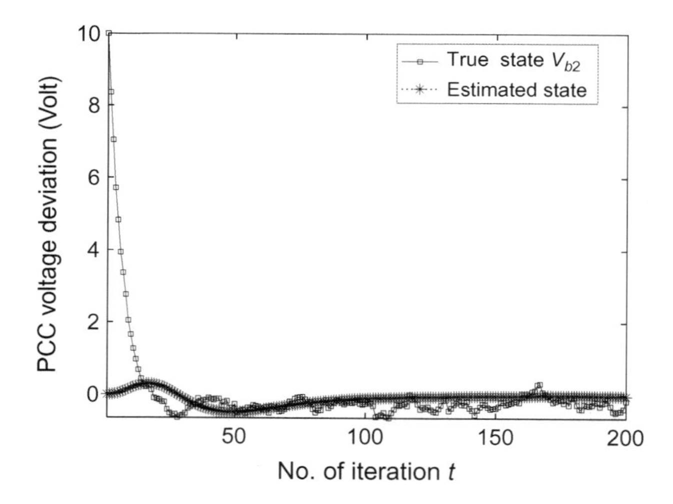

Fig. 14.11 PCC voltage deviation at bus 2 with sensors damage and large disturbances conditions.

In the energy management system, an H-infinity-based optimal filtering approach is adopted and verified through numerical simulations. It shows that H-infinity algorithm can able to estimate the microgrid states even there are large disturbance and sensor faults. Consequently, this work is valuable to design the IoT-based smart energy management system. Based on the analysis, we will implement Chi-squared detector to detect cyber attacks. In future, we will also apply the control algorithm to stabilize the microgrid states.

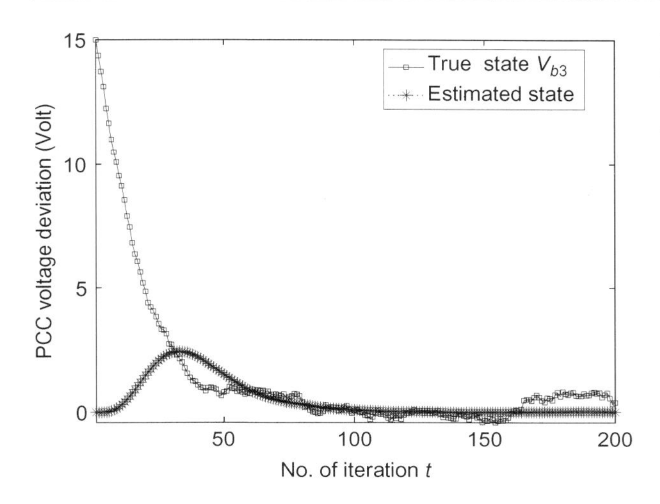

Fig. 14.12 PCC voltage deviation at bus 3 with sensors damage and large disturbances conditions.

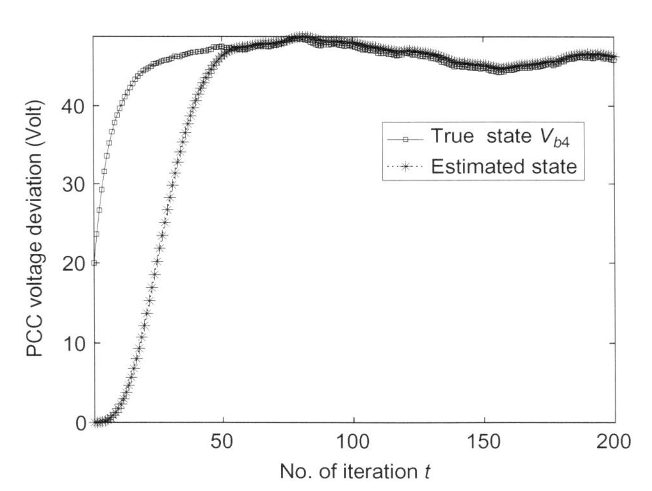

Fig. 14.13 PCC voltage deviation at bus 4 with sensors damage and large disturbances conditions.

References

Allen, R., Lin, K.-C., Xu, C., 2010. Robust estimation of a maneuvering target from multiple unmanned air vehicles' measurements. In: Proceedings of the International Symposium on Collaborative Technologies and Systems, pp. 537–545.

Aref, A., Tran, T., 2017. Multi-criteria trust establishment for internet of agents in smart grids. Multiagent Grid Syst. 13 (3), 287–309.

Chen, B., Ho, D.W., Hu, G., Yu, L., 2017. Secure fusion estimation for bandwidth constrained cyber-physical systems under replay attacks. IEEE Trans. Cybern. 48 (6), 1862–1876.

Haga, J.H., Fernandez, C., Takefusa, A., Ikeda, T., Tanaka, J., Belter, B., Kudoh, T., 2017. Building intelligent future internet infrastructures: FELIX for federating software-defined networking experimental networks in Europe and Japan. IEEE Syst. Man Cybern. Mag. 3 (4), 35–42.

Hug, G., Giampapa, J.A., 2012. Vulnerability assessment of AC state estimation with respect to false data injection cyber-attacks. IEEE Trans. Smart Grid 3 (3), 1362–1370.

Irita, T., Namerikawa, T., 2015. Decentralized fault detection of multiple cyber attacks in power network via Kalman filter. European Control Conference, pp. 3180–3185.

Kaur, M., Kalra, S., 2018. Security in iot-based smart grid through quantum key distribution. In: Advances in Computer and Computational Sciences. Springer, Singapore, pp. 523–530.

Keller, J.-Y., Sauter, D., Chabir, K., 2012. State filtering for discrete-time stochastic linear systems subject to random cyber attacks and losses of measurements. In: Mediterranean Conference on Control and Automation. pp. 935–940.

Korukonda, M.P., Mishra, S.R., Shukla, A., Behera, L., 2017. Handling multi-parametric variations in distributed control of cyber-physical energy systems through optimal communication design. IET Cyber-Phys. Syst. Theory Appl. 2 (2), 90–100.

Li, H., Lai, L., Poor, H.V., 2012. Multicast routing for decentralized control of cyber physical systems with an application in smart grid. IEEE J. Sel. Areas Commun. 30 (6), 1097–1107.

Li, Y., Quevedo, D.E., Dey, S., Shi, L., 2017. A game-theoretic approach to fake-acknowledgment attack on cyber-physical systems. IEEE Trans. Signal Inf. Process. Netw. 3 (1), 1–11.

Myeong-Jun, J., Kyeong-Hwa, K., 2017. Control of power electronic converters for efficient operation of microgrid consisting of multiple DG units and ESS. DEStech Trans. Eng. Technol. Res. https://doi.org/10.12783/dtetr/amsm2017/14875.

Pacheco, J., Hariri, S., 2016. IoT security framework for smart cyber infrastructures. In: International Workshops on Foundations and Applications of Self Systems, pp. 242–247.

Rana, M., 2017a. Architecture of the internet of energy network: an application to smart grid communications. IEEE Access 5, 4704–4710.

Rana, M.M., 2017b. Internet of things for smart grid automation. Int. Robot. Autom. J. 3 (5), 1–2.

Rusu, F.A., Livint, G., 2017. H-infinity controller for power sources applied to electrical vehicles. In: International Carpathian Control Conference, pp. 488–492.

Sakthivel, R., Arun, M., 2016. Design of H-infinity robust power system stabilizer using fuzzy and ANN controllers based on EABC optimal power system. Adv. Nat. Appl. Sci. 10 (16), 55–66.

Salas-Puente, R., Marzal, S., Gonzalez-Medina, R., Figueres, E., Garcera, G., 2017. Experimental study of a centralized control strategy of a DC microgrid working in grid connected mode. Energies 10 (10), 1627.

Salinas, S.A., Li, P., 2016. Privacy-preserving energy theft detection in microgrids: a state estimation approach. IEEE Trans. Power Syst. 31 (2), 883–894.

Shinohara, T., Namerikawa, T., 2017. Manipulative zero-stealthy attacks in cyber-physical systems: existence space of feasible attack objectives. In: Conference on Control Technology and Applications, pp. 1123–1128.

Simon, D., 2006. Optimal State Estimation: Kalman, H Infinity, and Nonlinear Approaches. John Wiley and Sons, New Jersey.

Zadehbagheri, M., Pishavaie, M., Ildarabadi, R., Sutikno, T., 2017. The coordinated control of FACTS and HVDC using H-infinity robust method to stabilize the inter-regional oscillations in power systems. Int. J. Power Electron. Drive Syst. 8 (3), 1274.

Zhao, J., 2018. Dynamic state estimation with model uncertainties using h-infinity extended Kalman filter. IEEE Trans. Power Syst. 33 (1), 1099–1100.
Zhao, J., Netto, M., Mili, L., 2017. A robust iterated extended Kalman filter for power system dynamic state estimation. IEEE Trans. Power Syst. 32 (4), 3205–3216.
Zhou, Y., Miao, Z., 2016. Cyber attacks, detection and protection in smart grid state estimation. In: North American Power Symposium.pp. 1–6.

Part Three

Optimization models/methods in smart distribution networks (optimization aspects)

Management of renewable energy source and battery bank for power losses optimization

Rajeev Kumar Chauhan, Kalpana Chauhan*†

*School of Computing and Electrical Engineering, Indian Institute of Technology, Mandi, India, †Department of Electrical and Electronics Engineering, Galgotias College of Engineering and Technology, Greater Noida, India

15.1 Introduction

This chapter is devoted to design an intelligent energy management system (IEMS) for a grid connected DC microgrid. The photovoltaic (PV) plant and multibattery bank (BB) are the internal power source of the DC microgrid. The control objectives of the proposed IEMS are (i) to ensure the load sharing (according to the source capacity) among sources, (ii) the selection of the closest source to the load, (iii) to reduce the power losses in the system, and (iv) to enhance the system reliability and (v) power quality. The IEMS is novel because it is following the ideal characteristics of the battery for the power sharing and the selection of the closest source to minimize the power losses in the lines. The IEMS allows continuous and accurate monitoring with intelligent control of distribution system operations such as battery bank energy storage (BBES) system, PV system, and customer utilization of electric power. The proposed IEMS is showing the better operational performance for operating conditions in terms of load sharing, loss minimization, and reliability enhancement of the DC microgrid.

The microgrid consists of interconnected distributed energy resources (DERs), capable of providing sufficient and continuous energy for a significant portion of internal load demand (Engelen et al., 2006; Chauhan et al., 2013). This concept easily works in both the AC and DC systems, but the implementation of the DC microgrid provides several advantages in terms of redundancy, modularity, fault tolerance, high efficiency, high reliability, easy maintenance, smaller size, and lower design cost. Besides from reducing resource and financial costs, DC microgrid simplifies and provides the opportunity to integrate renewable energy source (RES) that are intrinsically DC source (i.e., PV, small wind turbines, or fuel cells) at higher efficiency (Starke et al., 2008; Chauhan et al., 2016a). Most of the loads in the residential/commercial buildings are DC loads or easily DC convertible AC loads. Hence, the DC output of the RES can be directly used within the buildings without converting to the AC to feed to the electric grid (Salomonsson and Sannino, 2007; Chauhan et al., 2015). The PV arrays inherently generate DC power and a DC microgrid is a suitable solution to interconnect PV generation systems with public utility (PU) and loads via an inverter

Smart Power Distribution Systems. https://doi.org/10.1016/B978-0-12-812154-2.00015-8

(Salomonsson and Sannino, 2007). The stochastic nature of PV power and dynamic nature of demand, force the energy storage devices connected in the DC microgrid to smooth the power flow with high quality. At present, low-voltage direct current (LVDC) distribution systems are used only in some applications, such as telecommunication equipment with 48 V DC (Mizuguchi et al., 1990), traction systems (Hill, 1994), and shipboard systems (Ciezki and Ashton, 2000).

The use of BBES system is a popular alternative to maximize the penetration level of variable power coming from RES into DC microgrid. The optimal operation of DC microgrid requires proper control strategies such as battery control and monitoring system (BCMS). It monitors critical conditions of BB and allows an early detection of outage conditions. It implies that the monitoring system of each source can play an important role in reliability and security of the DC microgrid (Chauhan and Rajpurohit, 2015). The control of load sharing between the PV plant and the BBES is a critical issue in DC microgrid, which should be based on the sources and storage power rating and characteristics. The last one involves complexity related to the charging/discharging processes of batteries. Many load-sharing techniques have been reported in the literature during the last decade. Energy management system is playing an important role in DC microgrid. There are many literatures with considering the different problems associated with this system. In some literature (Abbey and Joos, 2005; Alawi et al., 2007; Cellura et al., 2008; Goa et al., 2008; Hajizadeh and Golkar, 2007; Jeong et al., 2005), genetic algorithm or neural network-based optimization have been used for microgrid. However, the drawback of these types of techniques is that they do not have the ability to welcome the new source or load. They were designed for a fixed number of sources and loads. Moreover, these systems are showing the centralized control algorithm. The proposed IEMS is a solution for some of these problems. The most obvious features such as voltage stability, selection of the closest source to load and approaching the ideal characteristics of the source (battery) makes proposed IEMS more advance. The use of battery keeps the power balance in DC microgrid. The current and voltage-based power sharing approach of IEMS keeps the source characteristics very close to its ideal characteristics. It minimizes the voltage fluctuation and keeps the voltage closer to a reference voltage in the DC microgrid. The selection of the closest source to the load feature of the IEMS minimizes the power losses in the lines. On the other hand, the traditional approach not having all these features such as the selection of the nearest source according to the requirement and availability of extra power simultaneously. The proposed IEMS can work with multisources and loads for lumped and distributed load. The proposed IEMS can work with multisources and loads for lumped and distributed load. To validate the proposed control strategy a real-time GUI has been configured and results are shown for different operating time intervals of a typical day.

15.2 Energy management system for DC microgrid

The two types of DC microgrid configurations (i) distributed energy resources with lumped load (DERLL) and (ii) distributed energy resources with distributed load (DERDL) are considered for development of the IEMS. The DERLL configuration is designed so that BBs connected to the DC microgrid will follow its ideal charging/discharging characteristic during power sharing. On the other hand the DERDL configuration is designed for the selection of the closest source to the load to minimize the power losses in the DC microgrid. Three energy sources PU, PV, and BB are considered in this chapter for both configurations.

15.2.1 Electrical circuit of battery

In renewable energy systems, the gel type and maintenance free lead acid battery is used widely. However, other types of batteries such as nickel-cadmium and nickel-metal-hydride also available for use, but they are very expensive and not easily available (Duryea et al., 2001). It can be seen that the daily energy demand is approximately constant. When the weather is overcast, less energy is supplied by the PV array. This causes the battery state of charge (SOC) to reduce accordingly. Fig. 15.1 showing a simplified equivalent circuit for battery with limits of no diffusion rate and accessible active site with large fraction. The resistances between the active electrode and external battery terminal are represented by resistors R_{CC_c} and R_{CC_a}, respectively. However, R_{El} is the electrolyte resistance. At anode and cathode interface, there are double-layer capacitances C_{DL_a} and C_{DL_c}, respectively. The series combination of resistors R_F and capacitors C_ϕ creates Faradic reaction. The values of capacitors and resistors are the function of operating parameters such as DC current, temperature, and SOC (Sritharan, 2012; Conway, 1991).

There are some phenomena which may affect the battery operation during charging mode such as fully charging of batteries with constant charging current, application of high charging currents due to which the voltage reaches to the gassing threshold so quickly, internal resistors increase the power loss and the thermal effect increases at the high level SOC (Dizqah et al., 2015). At the negative plate there are exothermic oxygen recombination reactions as well as I^2R losses during the charging of battery which creates significant heating. The heating is directly proportional to the current therefore the generated heat increases the current and the excessive current creates a heating effect on the battery cell.

The depth of previous stage discharge as well as available current for charging decides the magnitude and charging rate of the battery. The available and allowed charging current should be a function of discharge depth and limited maximum values keep the temperature below 10°C during battery recharge (C and D Technologies, 2012). The relation is shown in the example as in Table 15.1.

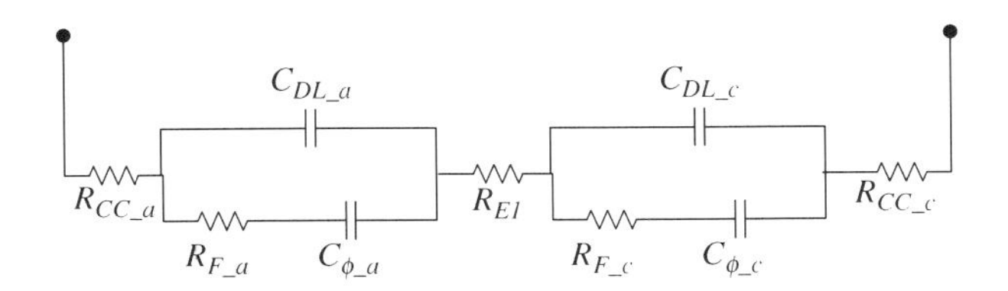

Fig. 15.1 Electrical equivalent circuit diagram of battery.

Table 15.1 Maximum allowable recharge current

Approximate discharge rate and duration	Depth of previous discharge (20 h rate) (%)	Maximum allowable recharge current per 100 Ah rated capacity (A)
15 min	45	100
30 min	55	50
60 min	62	40
3 h	75	33
5 h	81	25
8 h	88	20
20 h	100	10

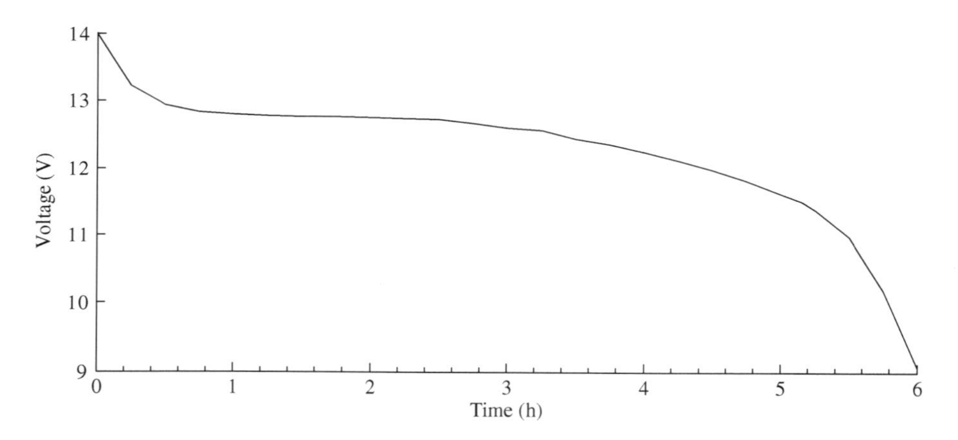

Fig. 15.2 Typical charging-discharging characteristics of battery bank.

Whenever a battery is connected to a PV system, it is showing very different characteristics with the one which a normal battery has its own charging/discharging characteristics, as shown in Fig. 15.2. The reason is that the PV system is neither showing constant current nor constant voltage as the weather change occurs even in a single day (Lalwani et al., 2011). The charging rate of the battery from the PV is depending upon the power generated due the sun radiations on the PV. The unbalanced charging/discharging of battery, changes its SOC, as SOC is the cumulative sum of the daily charge/discharge energy transfers. If the battery charging/discharging rate does not follow its own ideal characteristics, then the battery backup time as well as the battery life may decrease. The load sharing of BB is based on the battery rating and its own designed charging/discharging so that BBs will not empty at different instant. At that point, empty battery will behave as a load and start charging from the DC microgrid.

On the other hand, the series connection of BBs will charge the discharged batteries with reverse voltage, which may damage the battery because batteries are not designed for reverse charging (Gu et al., 1999). The no load terminal voltage of the battery can be expressed as

$$V_t = V_0 - \lambda_p \frac{\mu_n}{\mu_n - \mu_e} + \alpha e^{-\beta \mu_e} \tag{15.1}$$

where V_t is no load voltage (V), V_0 is constant voltage (V), λ_p is polarization voltage (V), μ_n is the battery nominal capacity (Ah), α is exponential voltage (V), μ_e is the battery extracted capacity (Ah), and β is the battery exponential capacity $(\mathrm{Ah})^{-1}$.

15.2.2 Battery SOC

In general, the SOC of the battery is the ratio of its present capacity $[\mu(t)]$ to the nominal capacity (μ_n) of the battery (Blaine and Newman, 1997; Phurailatpam et al., 2016). The SOC can be defined as

$$SOC(t) = \frac{\mu(t)}{\mu_n} \tag{15.2}$$

The coulomb counting method measures the charging/discharging current of a battery and integrates the charging/discharging current over time in order to estimate SOC (Phurailatpam et al., 2016; Chauhan et al., 2017). The SOC can be calculated as

$$SOC(t) = SOC(t-1) + \frac{i(t)}{\mu_n} \Delta t \tag{15.3}$$

15.2.3 Battery reference currents

The charging/discharging reference current of the battery depends on their nominal capacity and SOC. The reference charging current $[I_{rcb}(t)]$ of battery can be expressed as (Chauhan et al., 2017, 2016c):

$$I_{rcb}(t) = \left(\frac{1 - \psi_b(t)^b}{100} \right) \mu_b \tag{15.4}$$

where ψ_b is the SOC of the battery at time instant t, while μ_b is the nominal power capacity of the battery.

The reference charging power (P_{rcb}) of the battery is the function of the charging current and terminal voltage (V_b) of the battery and can be expressed as (Chauhan et al., 2017, 2016c; Chauhan and Chauhan, 2017):

$$P_{rcb}(t) = \left(\frac{1 - \psi_b(t)^b}{100} \right) \mu_b V_b \tag{15.5}$$

Similarly, the discharging reference current [$I_{rdb}(t)$] of battery can be expressed as

$$I_{rdb}(t) = \left(\frac{\psi_b(t)^b}{100}\right)\mu_b \tag{15.6}$$

While the discharging reference power (P_{rdb}) can be expressed as

$$P_{rdb}(t) = \left(\frac{\psi_b(t)^b}{100}\right)\mu_b V_b \tag{15.7}$$

The voltage control of DC bus and the current control of battery (i.e., charging/discharging) are the great challenge. The minimum selector is a solution for this problem.

15.2.4 *Minimum selector*

The idea behind the minimum selector is to select the minimum value of the current/voltage from all the input at particular time instant those obtained from simulation and send the digital input to the pulse width modulation (PWM) to actuate the power electronics converter (PEC) to achieve the (standard) desired value. An example of minimum selector is given here (Fig. 15.3). The four inputs have been taken for example the ICs 1A, 1B, and 1C are the LM 324 operational amplifiers whereas the ICs 2A, 2B, and 2C are the 74C85 exclusive OR gate and ICs 3A, 3B, and 3C are the CA4066B analog switch.

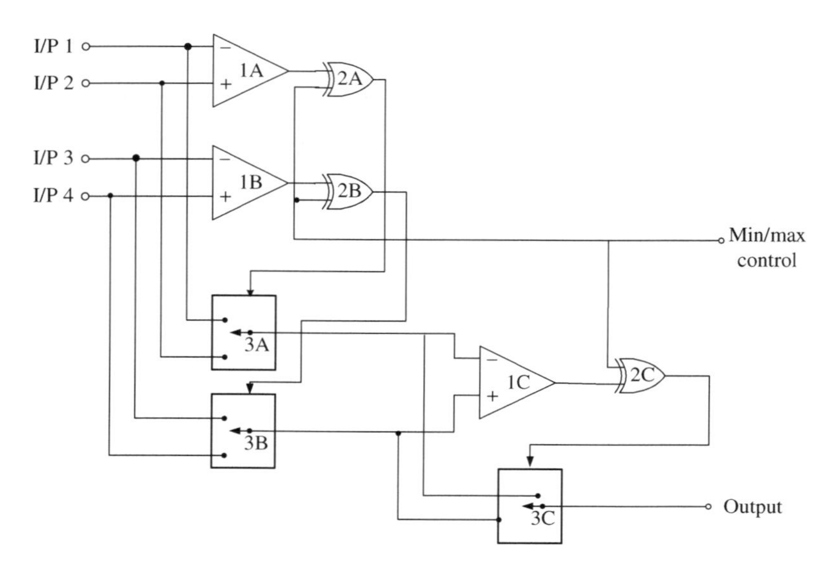

Fig. 15.3 Minimum selector circuit.

15.2.5 Distributed energy resource with lumped load (DERLL)

The first configuration considered for designing IEMS is the DERLL configuration. In this scheme, the PV and PU are connected to the DC microgrid via DC-DC and AC-DC converters, respectively. The power sharing of BB-1 and BB-2 has been controlled using a power electronic converter (PEC-1) and PEC-2, respectively. The voltage stability of DC microgrid could be achieved by power balancing in the DC microgrid. This power balance can be attained by a correct adjustment of power contribution by AC-DC converter of PU, DC-DC converters of the PV plant, BB-1, and BB-2. It adjusts the voltage of DC microgrid at an acceptable limit during transient as well as in steady-state conditions. Fig. 15.4 shows the DERLL system having a controlling strategy of DC microgrid voltage and battery characteristics. The Minimum Selector-1 output is the function of output of the Controller-1 (voltage control loop) and Controller-2 (current control loop).

Controller-1 output signal (Γ_1) is the function of the error generated due to reference DC microgrid voltage (U_{rg}) and actual (measured) DC microgrid voltage (U_{mg}):

$$\Gamma_1 = f_1(V_{eg}) \tag{15.8}$$

where V_{eg} is the error in microgrid voltage which is expressed as

$$V_{eg} = V_{rg} - V_{mg} \tag{15.9}$$

Controller-2 output signal (Γ_2) is the function of the error generated due to the standard charging/discharging current of BB-1 and the actual (measured) current (flowing between BB-1 and DC microgrid):

$$\Gamma_2 = f_2(I_{eb1}) \tag{15.10}$$

where I_{eb1} is the error in BB-1 charging/discharging current and can be expressed as

$$I_{eb1} = I_{rb1} - I_{mb1} \tag{15.11}$$

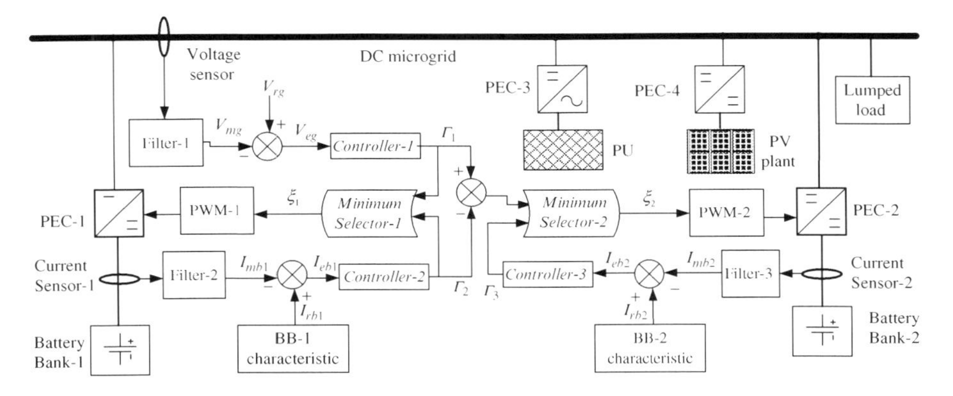

Fig. 15.4 Schematic diagram of IEMS for DERs with lumped load.

where I_{rb1} is the standard charging/discharging current of the BB-1 and I_{mb1} is the actual measured current sharing between the DC microgrid and BB-1.

Minimum Selector-1 output signal (ξ_1) is the minimum value of both of these controller outputs (Γ_1, Γ_2):

$$\xi_1 = \lim_{x \to 0} \left[\frac{d}{dx} (\Gamma_1(x), \Gamma_2(x)) \right] \tag{15.12}$$

The ξ_1 signal actuates the PWM-1 to operate the PEC-1 for adjusting the minimum value to its standard value, that is, the value of I_{eb1} and V_{eg} near to zero. The Minimum Selector-2 output signal (ξ_2) is the minimum value of the difference of output of the Controller-1 and Controller-2 and the output of the *Controller-3*.

Controller-3 output signal (Γ_3) is the function of the error generated due to the standard charging/discharging current of BB-2 and the actual (measured) current (flows between BB-2 and DC microgrid):

$$\Gamma_3 = f_3(I_{eb2}) \tag{15.13}$$

The I_{eb2} is the error in BB-2 charging/discharging current and can be expressed as

$$I_{eb2} = I_{rb2} - I_{mb2} \tag{15.14}$$

where I_{rb2} the standard charging/discharging current of the BB-2 and I_{mb2} the actual (measured) current sharing between the DC microgrid and BB-2.

The output of Minimum Selector-2 can be expressed as

$$\xi_2 = \lim_{x \to 0} \left[\frac{d}{dx} [\{\Gamma_1(x) - \Gamma_2(x)\}, \Gamma_3(x)] \right] \tag{15.15}$$

The ξ_2 signal actuates the PWM-2 to operate the PEC-2 for adjusting the minimum value to its standard value, that is, to set the I_{eb2} near to zero.

15.2.6 Distributed energy resource with distributed load (DERDL)

The second configuration considered for designing IEMS is DERDL shown in Fig. 15.5. There are four homes (*Home-1*, *Home-2*, *Home-3*, and *Home-4*) supplied by PU, DER, and the BBs via DC microgrid. This is assumed that each home has an own PV plant. The output of current sensors mounted between power source and homes as shown in Fig. 15.5. It also acts as a feedback to the controllers (PID controllers) of power source. The feedback signal (ζ_{dy}) from yth current sensor is as follows:

$$\zeta_{dy} = i_{hy} - i_{csy} \tag{15.16}$$

where $y = 1, 2, 3, 4$. i_{hy} is the current consumption in yth home and i_{csy} is the current measured by the yth current sensor.

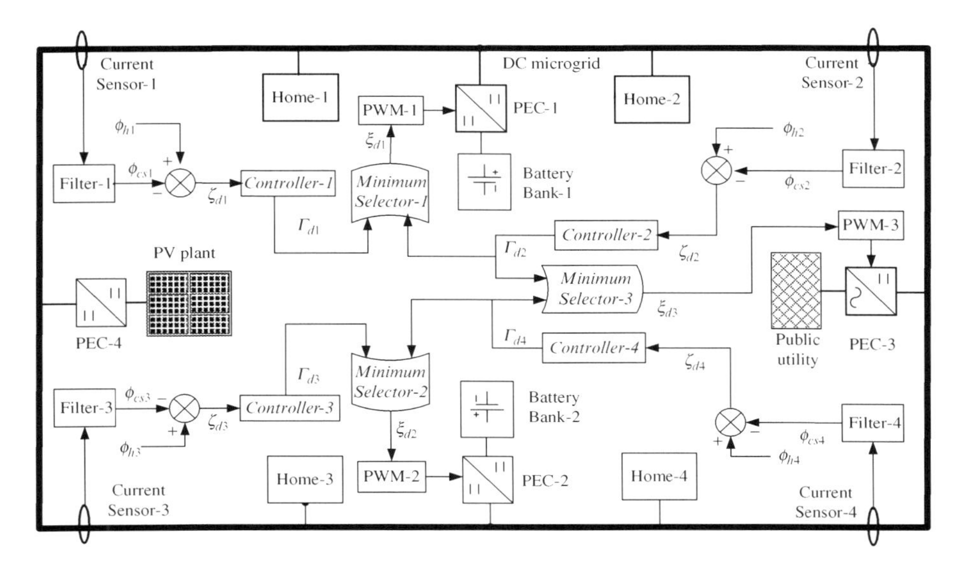

Fig. 15.5 Schematic diagram of IEMS for DERs with distributed load.

The ζ_{d1}, ζ_{d2}, ζ_{d3}, and ζ_{d4} are the error signals to active the *Controller-1*, *Controller-2*, *Controller-3*, and *Controller-4*, respectively, to select the power source for the fulfillment of the required power. In this case, the controller output depends on three parameters: (i) the error signal which is basically the feedback to the controller, (ii) minimum distance between slave home (which require borrowing power from the nearest source) and master plant (which require to donate the power for other nearest homes), and (iii) power availability of the selected power source.

Mathematically, the yth controller output Γ_{dy} can be expressed as

$$\Gamma_{dy} = f\left(\zeta_{dy}, \Delta, \hbar\right) \tag{15.17}$$

where Δ is the minimum distance between *slave* and *master plant* and expressed as

$$\Delta = \lim_{x \to 0} \frac{d}{dx} \{D(x)\} \tag{15.18}$$

And $\hbar$is the available load-sharing capacity of master plant after fulfillment of the self-requirement, and satisfies the condition

$$\hbar \geq \zeta_d \geq \Psi_h - \Psi_s \tag{15.19}$$

where Ψ_h is the master home/plant feeding power to the DC microgrid and Ψ_s is Slave home power consumption from the DC microgrid.

The controller helps to select the nearest source that has the sufficient power, which is more, or equal to the required power by the slave home. The controller checks all three parameters, discussed earlier in this section and continues checking until the three conditions will be satisfied.

There is Minimum Selector-1 for detecting the home, which requires the extra power from other nearest home. The Minimum Selector-1 output (ξ_{d1}) is the minimum value between these controller outputs (Γ_{d1}, Γ_{d2}):

$$\xi_{d1} = \lim_{x \to 0} \left[\frac{d}{dx} \{ \Gamma_{d1}(x), \Gamma_{d2}(x) \} \right] \tag{15.20}$$

In a similar way, the *Minimum Selector-2* output (ξ_{d2}) is the minimum value between these controller outputs (Γ_{d3}, Γ_{d4}):

$$\xi_{d2} = \lim_{x \to 0} \left[\frac{d}{dx} \{ \Gamma_{d3}(x), \Gamma_{d4}(x) \} \right] \tag{15.21}$$

The ξ_{d1} and ξ_{d2} actuates the PWM-1 and PWM-2 to operate the PEC-1 and PEC-2, respectively, to find out the minimum value shown from both controller and helps to find out the nearest source to the power desiring home. On the other hand, the Minimum Selector-3 sends the control signal to the PU converter (PEC-3) to supply desired minimum demand. The Minimum Selector-3 output (ξ_{d3}) is the minimum value of both of these controller outputs (Γ_{d2}, Γ_{d4}):

$$\xi_{d3} = \lim_{x \to 0} \left[\frac{d}{dx} \{ \Gamma_{d2}(x), \Gamma_{d4}(x) \} \right] \tag{15.22}$$

In case of distributed load in DC microgrid, there may arise a voltage difference, which varies when the power flow across the interconnecting cable changes according to distributed load (Chauhan and Chauhan, 2017; Anand et al., 2013). Due to dependency of the source currents on the node voltages for droop control they also depend on load distribution in the presence of connecting cable resistance. If the droop gain increases the deviation in source currents can be reduced, but the voltage variation also increased with this. For example, a two-source (source u_1 and u_2) and two-load (load r_{L1} and r_{L2}) system is shown in Fig. 15.6 There are two-voltage sources connected by a resistive cable and local loads connected to each source. The analysis of this circuit for distributed load is presented in the results section.

15.3 Results and discussions

This section presents simulation results of the proposed IEMS for the DC microgrid for a typical day in the first subsection. In the next subsection the performance of the proposed IEMS (in case of lumped load) has been explained in terms of comparison with other IEMS. The charging/discharging characteristic of battery in the proposed IEMS is compared with ideal characteristics of the battery. The system is compared in terms of power consumed from the PU, PV, and BBs. The proposed IEMS is showing much better and managed power sharing between the power sources. After this,

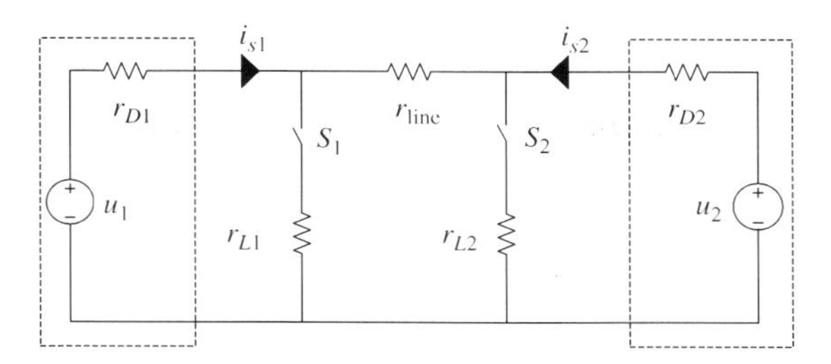

Fig. 15.6 Conceptual diagram of two-source and two-load DC microgrid.

the next section is describing the performance of the proposed IEMS (in case of distributed load). The load-sharing characteristics of the proposed IEMS have been described for voltage drop and power loss due to distributed load. The parameters of the DC bus of the microgrid can be found in the Table 15.2.

15.3.1 Real-time GUI of IEMS for DC microgrid

A real-time application using the proposed IEMS for DC microgrid has been simulated using MATLAB™. It includes a friendly graphical user interface (GUI) to allow a very flexible management of input data and very detailed representation of output results. Manual operated PV input tab has been used to connect or disconnect the PV to the DC microgrid. The user can change the ON/OFF state of the appliances any time by the corresponding buttons. The loads, including BB and PV panel (of maximum 800 W each) are plugged into the DC microgrid. The PV plant power production is based on real experimental data, recorded from the real-residential plant. This data are collected on the PV plant by recording in a typical day and representing the maximum power generation during the afternoon. The hourly power consumptions for all loads (appliances) in each household are depicted in the GUI. Typical conditions of DC microgrid located in India have been used for simulation purpose in this chapter. Environmental conditions and load values have been randomly changed day by day to test the behavior of the proposed system. The simulation results presented in this section are based on a discrete time step during a 24 h of a typical day. There is a "start system" switch in the GUI to start and stop the simulation. The average computational time for all the off-line simulations presented in this section is 180 s (as shown on Fig. 15.7). The GUI has an output window shows the simulation results after pressing the "start system" switch. To reset the graph there is a push-up switch in the GUI

Table 15.2 DC bus parameters

System parameters	
Cable cross-sectional area (A)	241.9 mm^2
Unit resistance of bus (r_{ul})	121 mΩ/km
Segment length of bus (l)	100 m
System voltage	12 V

named as "clear graph". Fig. 15.7 shows six trend lines in the GUI. The illustrative results presented in Fig. 15.7 are based on the assumption of two households at constant power load during the whole simulation period (24 h). During the operation of the IEMS for DC microgrid simulator four-operational modes are found.

Mode I—Stand-alone period: During dark periods in the night, the PV plant is not capable to produce electricity and also, the DC microgrid is assumed to be disconnected from the PU. As consequence between 00:00 and 06:00 h the total power demanded by the loads is supplied by the BBs. The BB-1 has started load sharing since 00:00 h and BB-2 starts load sharing when the BB-1 reached at its 90% capacity. The energy balancing equation for this operating mode can be expressed as

$$\sum_{j=1}^{n_b} P_{bb\mathrm{j}}(kt, \mu_s) = \sum_{j=1}^{n_l} P_{load\mathrm{j}}(kt) \tag{15.23}$$

where $P_{loadj}(kt)$ represents load demand and $P_{bbj}(kt,\mu)$ represent the power sharing between jth battery and DC microgrid, n_b and n_l are the number of BBs and loads. μ_s is the stored energy in BBs.

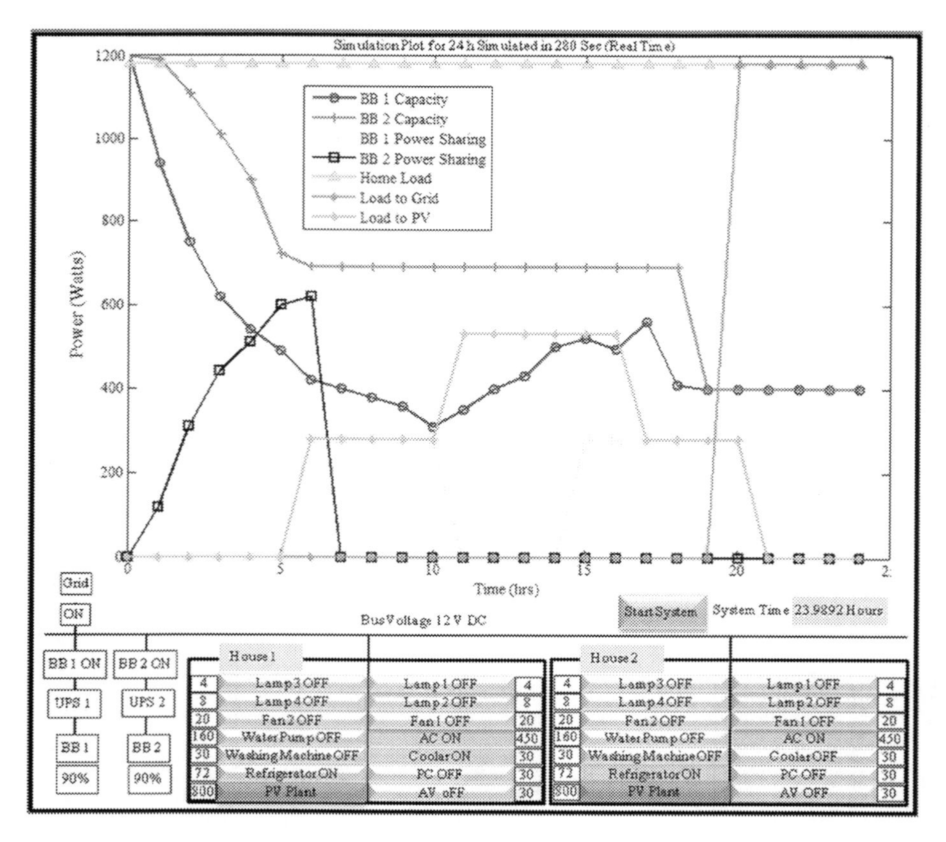

Fig. 15.7 Graphic user interface for DC microgrid.

The results of the IEMS for DC microgrid simulator have been shown in the output window depicts that during the time period of this operating mode the BB-1 and BB-2 are providing energy (discharging process). Trend line representing the power imported from the PU is located at zero level because during the stand-alone operational mode, the power supply from the grid is not providing energy.

Mode II—PV and BB discharge: When the morning time reached, around 06:00 h, PV plant starts power generation, but not giving the rated power due to dull sunshine. During 06:00–10:10 h and 14:40–18:30 h time interval, the PV power is less than the load. Moreover, the demand is higher than the PV generation in this scenario. Besides that the surplus demand is less than the power supply capacity of BB-1. Therefore, the BB-1 starts discharging to balance the surplus demand (i.e., difference of PV power and demand) while the BB-2 remains in isolated mode. The DC microgrid power is balanced by the PV and BBs and can be expressed as

$$\sum_{j=1}^{n_b} P_{bbj}(kt, \mu_s) + \sum_{j=1}^{n_{pp}} P_{pvj}(kt) = \sum_{j=1}^{n_l} P_{loadj}(kt) \tag{15.24}$$

where $P_{pvj}(kt)$ is the PV output power and n_{pp} is the number of PV plants.

During 06:00–10:10 h, there is availability of PV plant power. Hence, one battery bank (BB-2) stops discharging while another (BB-1) is still going to discharge. BB-1 again starts discharging due to low power from the PV-plant during 14:40–18:30 h as shown in the GUI. So load has been taken by the BB-1.

Mode III—PV and BB charging: During 10:10–14:40 h, the sun is brightest. Therefore the PV starts the power generation closer to its rated capacity. Moreover, the PV power is greater than the load. As the PV surplus power is less than the charging rate of the BB-1, therefore BB-1 start charging with the surplus PV power and BB-2 remains in isolated mode as shown in Fig. 15.7. The power balancing equation can be expressed as

$$\sum_{j=1}^{n_{pp}} P_{pvj}(kt) = \sum_{j=1}^{n_l} P_{loadj}(kt) + \sum_{j=1}^{n_b} P_{bbj}(kt, \mu_s) \tag{15.25}$$

During the GUI, the power from a PV plant is higher than the load so it starts charging the BB-1 after fulfillment the load requirement and BB-1 sharing the power. Thick representation of grid trend line and BB-1 sharing line are showing the peak hours of PV-plant power.

Mode IV—Importing from PU: During 18:30–24:00 h, there is sunset (night hours). Besides that the PU is available and BBs behaves as an outage source. Therefore the demand is supplied by PU. The power balancing equation can be expressed as

$$P_{pu}(\mu_s, i_l) = \sum_{j=1}^{n_l} P_{loadj}(kt) \tag{15.26}$$

where $P_{pu}(\mu_s, i_l)$ represents the PU power supplied to the DC microgrid and i_l represents the current of the connected load to the DC microgrid and expressed as

$$i_l = \sum_{j=1}^{n_l} i_j \tag{15.27}$$

In this duration, the PU supply is available and takes the complete load as shown in Fig. 15.7. The trend line associated with load to grid is approaching the same level as the load. Trend lines of other sources become constant at this time. The four-operational modes allow the IEMS for DC microgrid to fulfillment of the load demand. When the BBs are fully charged, then the curve goes on decreasing from its maximum value to a null value. The BBs recharging management, operational model III, is one of the most important tasks in order to get uninterrupted power supply in the critical hours of the day. In the simulation results of the proposed IEMS for DC microgrid, the BBs recharging has been performed during the peak hours of the PV plant, that is, during afternoon time. It can be easily seen in Fig. 15.7 that grid power is not required during 00:00–19:30 h, total load demand is covered by PV plants and BBs. When the BB of one home is completely charged after fulfilling its own demand, the excess of power may be sold to the other home or grid.

15.3.2 Performance results of IEMS in case of lumped load

Similar to Chauhan et al. (2016b), the available set values were randomly mixed to defer the day-to-day load for testing the proposed system. A discrete time simulation is performed and all parameters are set at their initial values. The set of data for simulation is as there were two PV arrays (40 kW each), two wind turbines (40 kW each), two batteries (30 kW nominal each), and the load data are collected in real time from industrial and residential area. The DC microgrid peak power was 30 kW. In the proposed IEMS, there is a community PV plant of 10 kW and, two batteries (1200 Ah, 12 V each) and four homes including captive PV plant. The load data are collected in real time from the Pecan Street Project, University of Texas at Austin, USA (home 1 demand: 5 kW, PV plant 5 kW; home 2: 7.5 kW, PV plant 5 kW; home 3: 7 kW, PV plant 5 kW; and home 4: 6 kW PV plant 5 kW). The microgrid peak power is 20 kW. The demand curve obtained from both systems is shown in Fig. 15.8.

Fig. 15.9 shows a comparative analysis of proposed IEMS for DC microgrid in the form of BB performance. The data collected as the combined performance of the both BB used in the proposed IEMS for DC microgrid. As targeted, the response of charging/discharging characteristics of BBs in the proposed IEMS for DC microgrid following the standard characteristics of the battery, shown in Fig. 15.2, with near to zero error, as the error plot in Fig. 15.9 is very near to zero value. Comparative to previous latest IEMS for DC microgrid, for example (Cirrincione et al., 2011; Jiang and Zhang, 2011), the proposed IEMS shows better performance in terms of stability, efficiency, power losses, energy saving. In Cirrincione et al. (2011) and

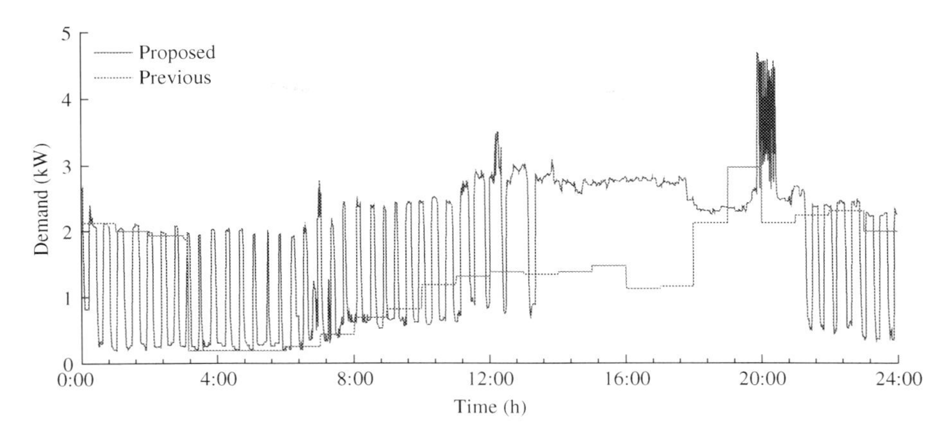

Fig. 15.8 Demand curve of DC microgrid.

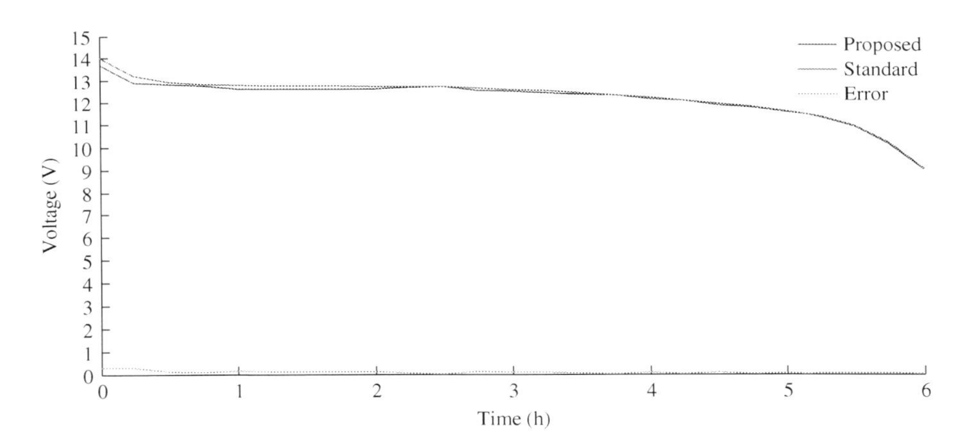

Fig. 15.9 Comparative analysis of battery bank characteristics.

Jiang and Zhang (2011), a self-regulate system with a coherent arrangement of power sources and loads are designed to enhance the efficiency. Each microgrid is having the ability to take its own decisions. There is an optimization of utilization of source according to load.

Fig. 15.10 shows plots of power imported from the PU during a typical 24-h day. The plot concerned with previous IEMS (Chauhan et al., 2016c) for DC microgrid shows that there is a need to consume power during the whole day time to time. There are more fluctuations, however the plot related to the proposed IEMS for DC microgrid having a good power balancing (stability) in the DC microgrid. It shows that the proposed IEMS for DC microgrid consumes power from the grid only at the night time.

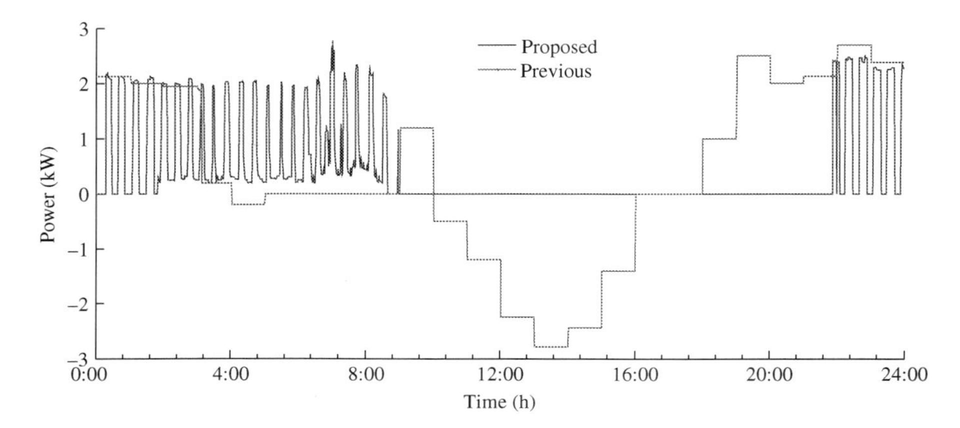

Fig. 15.10 Analysis of IEMS for DC microgrid in terms of public utility power.

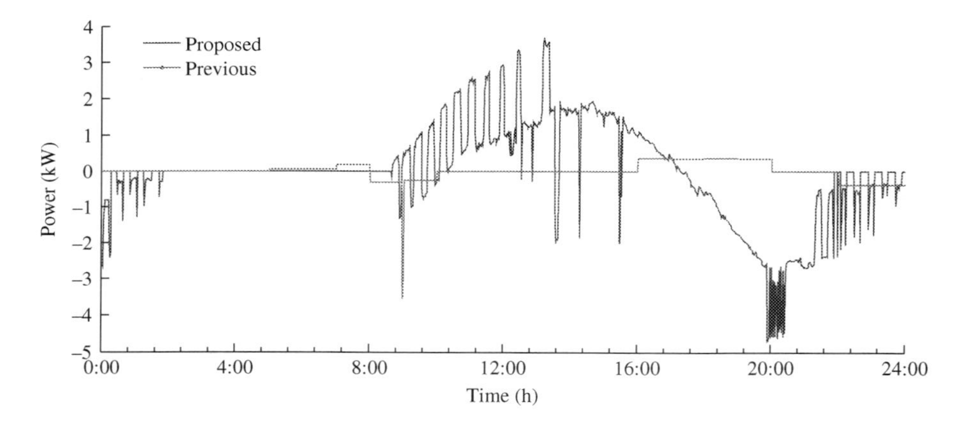

Fig. 15.11 Analysis of IEMS for DC microgrid in terms of battery bank power.

Fig. 15.11 shows the plots of power consumed from BBs in a typical day. The plot concerned with the previous IEMS showing that there is less utilization of BBs in the typical day. The consumer has to purchase electricity from the grid according to the demand. The plot of the proposed IEMS shows an efficient utilization of BBs in the whole day. However, in the day time BBs gets charged from the surplus PV power during 08:00–17:00 h time interval. It shows that the proposed IEMS is managed by BB and it does not need continuous (24*7 h) PU connectivity. While in the previous scheme, the surplus PV production is balanced by the PU and BB get charged by PU during the 16:00–20:00 h time interval. It verifies that the previous scheme is more dependent on the PU and needs the continuous (24*7 h) PU connectivity.

The plot of power consumed from PV plant in a typical day is depicted in Fig. 15.12. The plot concerned with the previous IEMS shows that there is an uncontrolled charging and consumption of PV power. During the night time and the starting

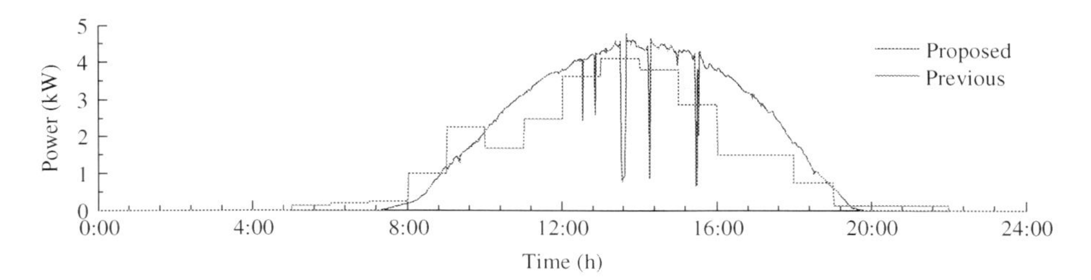

Fig. 15.12 Analysis of IEMS for DC microgrid in terms of PV power.

of the day, the PV is unavailable in the proposed IEMS. As the day proceeds, the PV power also increases in a right manner and ends also in the same manner. The plot of the proposed IEMS shows an efficient utilization of PV power during the whole day.

The above analysis shows the high efficiency and stability of the proposed IEMS for DC microgrid in comparison to the previous available IEMS for DC microgrid (Chauhan et al., 2016c).

15.3.3 Performance results of IEMS in case of distributed load

The DC microgrid consists of four homes with captive PV power plant, including with one PV and two BBs of the community and PU as shown in Fig. 15.5. The demand and power consumption of BB power, PU power, and PV power of four homes of the DC microgrid for a typical day is shown in Fig. 15.13.

In Fig. 15.14, there is a voltage-current characteristic of the circuit shown in Fig. 15.6 by considering the analysis in Refs (Chauhan and Chauhan, 2017; Anand et al., 2013). The voltage axis on the graph is the voltage across load-2 (r_{L2}) and the current axis on the currents i_{s1} and i_{s2} drawn from source-1 and source-2, respectively. The characteristics obtained are having a slope which depends on the load to source resistance and droop gain. In this case, the low value of droop gain r_{D1} and r_{D2} leads to good voltage regulation, that is, small voltage drop; however, the current variation increases. If the value of droop gains r_{D1} and r_{D2} is high, then the current deviation from their ideal values is significantly lower than the previous case. However, there has an increment in voltage regulation; it is worth to say the selection of the DC voltage droop can be optimized. Moreover, the voltage droop increases as the load is supplied by the far source, for example, when the *load-1*(r_{L1}) is switched off (S_1 off i.e., $i_{L1} = 0$) and *load-2* (r_{L2}) supplied by only *source-2* ($i_{L2} = i_{s2}$ and $i_{s1} = 0$) the voltage drop is minimum. On the other hand, when the *load-2* is supplied by *source-1* ($i_{L2} = i_{s1}$ and $i_{s2} = 0$) then the voltage drop is maximum.

Fig. 15.15 shows the power-current characteristic of the circuit shown in Fig. 15.6. The power axis on the graph having the addition of power loss in the source gains r_{D1} and r_{D2} and cable resistance r_{line}. The current axis having the current i_1 and i_2 drowns from the *source-1* and *source-2*, respectively. An exponential sloppy characteristic has been obtained. In this case, the power losses increase exponentially as the load is supplied by the far source. For example, when the *load-1* is switched

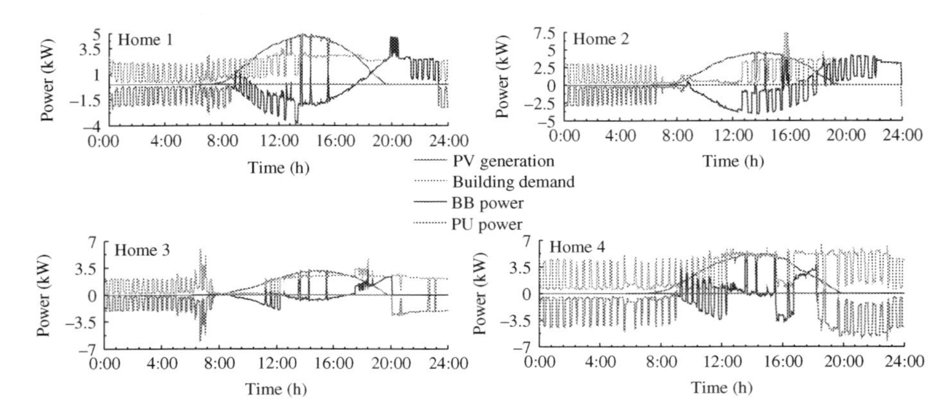

Fig. 15.13 Homes demand and power consumption from photovoltaic, battery bank, and public utility.

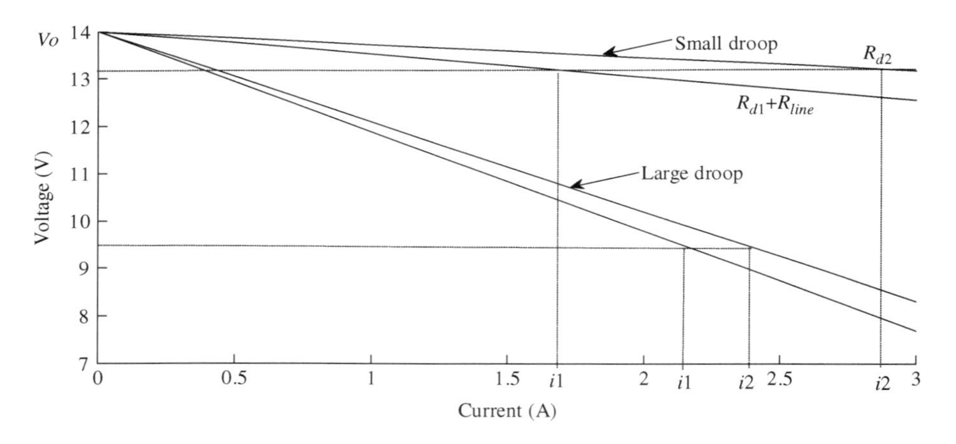

Fig. 15.14 Load-sharing characteristics for voltage drop due to distributed loads in DC microgrid.

"OFF" (S_1 "OFF", i.e., $i_{L1} = 0$) and load-2 supplied by only source-2 ($i_{L2} = i_{s2}$ and $i_{s1} = 0$) the power loss is minimized. On the other hand, when the Load-2 is supplied by the source-1 ($i_{L2} = i_{s1}$ and $i_{s2} = 0$), then the power loss is maximized in the system. Furthermore, the power losses are also varying with the source droop gain value r_{D1} and r_{D2}. As the droop gains increase the system power loss exponentially increases. In this case the power loss increases exponentially with the increase in droop gain values r_{D1} and r_{D2} and current deviation decreases in comparison to small droop values.

The DC bus is divided into four links and four sub-links. The link1 and link3 connects the home-1 and home-3 to the PEC-4. Similarly, the link2 and link4 connects the home-2 and home-4 to the PEC-3. Moreover the sub-link1 and sub-link3 connects the home-1 and home-2 to the PEC-1. Similarly, the sub-link2 and sub-link4 connects the home-3 and home-4 to the PEC-2. The length of the links and sub-links

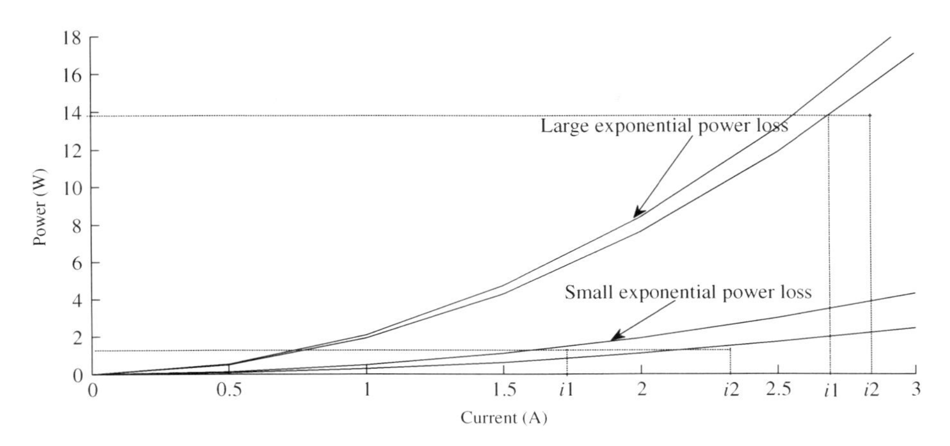

Fig. 15.15 Load-sharing characteristics for power loss due to distributed loads in DC microgrid.

is 20 and 5 m, respectively. The microgrid power loss (P_{loss}) in the DC bus for the proposed and previous approach is calculated using Eq. (15.22) and shown in Fig. 15.16.

$$P_{loss} = \sum_{j=1}^{n_{lk}} \left(i_{csj}\right)^2 r_{1j} + \sum_{j=1}^{n_{slk}} \left(i_{csj} - i_{hj}\right)^2 r_{2j} \tag{15.28}$$

where r_{1j} and r_{2j} are the resistance of the jth link and sub-link respectively, n_{lk} and n_{slk} are the number of link and sublink respectively, ϕ_{csj} is the current measured by the current sensor mounted on the jth link, ϕ_{hj} is the current of the jth home.

When the microgrid is in the PU connected mode, all the homes are feed by the PU via same path for both the approaches. During 0:00–08:40 h, the load is feed by the PU and the power loss in the DC bus remains same for both the approaches. During

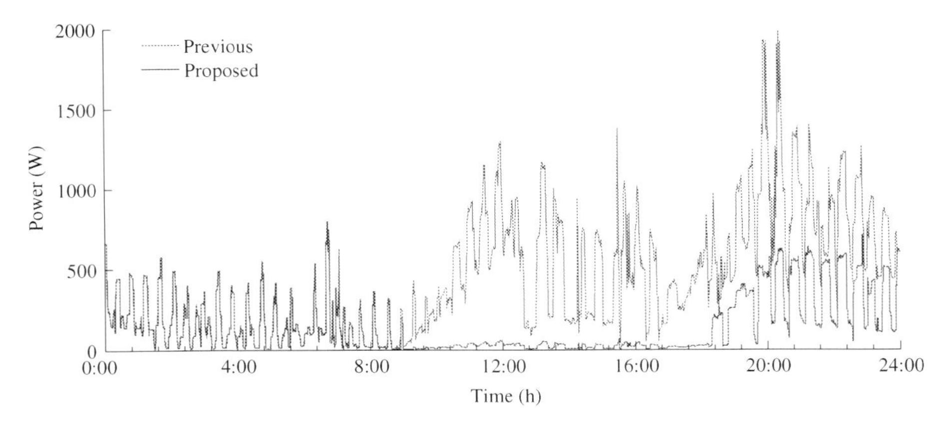

Fig. 15.16 Power losses in the lines for distributed load with distributed generation.

08:41–18:20 h, the captive power plants (home own PV plant) generation is higher than the demand and BB acts as a load and charged by the closest PV plant. The closest BBs changing current circulates in the DC bus of the microgrid and power loss in the bus remains at the least level. During 18:21–23:59 h, the PV generation becomes zero and loads are supplied by the BBs. The load currents are flowing in the DC bus and following the different path that is why the power losses remain at the highest level during this time interval. The proposed system reduces the power losses in the bus (i.e., improve the system efficiency) that is why the running cost of the proposed system is reduced.

15.4 Conclusion

In this chapter, an IEMS is proposed for DC microgrids that have the ability to provide a source selection based on the priority of the closest source to the load. The distant sources have been approached only if the nearer source is not able to supply the load. This criterion will help to decrease the power loss and voltage droop in DC microgrid, as consequence, there is an intrinsic increase in the systems efficiency. The proposed IEMS for DC microgrid has been designed to keep the BB measured sharing current and ideal current ratio less than or equal to one and follow the charging/discharging characteristics of the BB. It helps to ensure a longer life time and battery's performance. The used approach is very close to the BB ideal characteristic during 00:00–06:00 h. Real-time monitoring of the stored energy in BB is an additional advantage of the proposed IEMS. This may be helpful to predict the future load-sharing capacity of BB. At the same time, it increases the reliability and security of DC microgrid as having the knowledge about the available energy to use.

The proposed IEMS for DC microgrid has several advantages over other previous IEMS's presented in the literature. First, there is no restriction for the number of sources and loads, that is, it is designed as a universal system. The proposed IEMS is distributed in nature. Third, there is a smart management policy exists for the BBs recharging and discharging which makes the system much efficient to fulfill uninterrupted load demand. The microgrid is very useful for the poorly grid connected areas. It is well seen that the distributed concept shows a much better solution of energy demand than the lumped system. Utilizing intelligent sensors may improve the system accuracy. The future research includes the implementation and testing of the IEMS for a multiterminal microgrid with batteries, PV panels in the field. The more intelligent instrumentation may be utilized to get an optimized system in terms of efficiency and cost.

References

Abbey, C., Joos, G., 2005. Energy management strategies for optimization of energy storage in wind power hybrid system. In: Proceedings of IEEE 36th Power Electronics Specialists Conference, June, pp. 2066–2072.

Alawi, A.A., Alawi, S.M.A., Islam, S.M., 2007. Predictive control of an integrated pv-diesel water and power supply system using an artificial neural network. Renew. Energy 32 (8), 1426–1439.

Anand, S., Fernandes, B.G., Guerrero, J.M., 2013. Distributed control to ensure proportional load sharing and improve voltage regulation in low voltage dc micro-grids. IEEE Trans. Power Electron. 28 (4), 1900–1913.

Blaine, P., Newman, J., 1997. Modeling of nickel-metal hydride batteries. J. Electrochem. Soc. 144 (11), 3818–3831.

C and D Technologies, 2012. Thermal Runaway in VRLA Batteries-Its Cause and Prevention. Technical Bulletin 41-7944, C and D Technologies, Blue Bell, PA, pp. 1–14.

Cellura, M., Cirrincione, G., Marvuglia, A., Miraoui, A., 2008. Wind speed spatial estimation for energy planning in Sicily: a neural kriging application. Renew. Energy 33 (6), 1251–1266.

Chauhan, K., Chauhan, R.K., 2017. Optimization of grid energy using demand and source side management for DC microgrid. J. Renew. Sustain. Energy. 9(3). 035101. https://doi.org/10.1063/1.4984619.

Chauhan, R.K., Phurailatpam, C., Rajpurohit, B.S., Gonzalez-Longatt, F.M., Singh, S.N., 2017. Demand side management system for autonomous DC microgrid. Technol. Econ. Smart Grids Sustain. Energy 2 (4), 1–11.

Chauhan, R.K., Rajpurohit, B.S., 2015. DC distribution system for energy efficient buildings. In: Proceedings of IEEE, 18th National Power System Conference, IIT Guwhati, India, 18–20 December. pp. 1–6.

Chauhan, R.K., Rajpurohit, B.S., Hebner, R.E., Singh, S.N., Gonzalez-Longatt, F.M., 2016a. Voltage standardization of DC distribution system for residential buildings. J. Clean Energy Technol. 4 (3), 167–172.

Chauhan, R.K., Rajpurohit, B.S., Hebner, R.E., Singh, S.N., Longatt, F.M.G., 2015. Design and analysis of PID and fuzzy-PID controller for voltage control of DC microgrid. In: Proceedings of IEEE PES Innovative Smart Grid Technologies (ISGT) Asian Conference, Bangkok, 4–6 November. pp. 1–6.

Chauhan, R.K., Rajpurohit, B.S., Singh, S.N., Gonzalez-Longatt, F.M., 2013. DC grid interconnection for conversion losses and cost optimization. In: Renewable Energy Integration: Challenges and Solutions. Springer, Singapore, pp. 327–345.

Chauhan, R.K., Rajpurohit, B.S., Singh, S.N., Gonzalez-Longatt, F.M., 2016b. Real time energy management system for smart buildings to minimize the electricity bill. Int. J. Emerg. Electr. Power Syst. 18(3). https://doi.org/10.1515/ijeeps-2016-0238.

Chauhan, R.K., Rajpurohit, B.S., Singh, S.N., Gonzalez-Longatt, F.M., 2016c. Intelligent energy management system for PV-battery-based microgrids in future DC homes. Int. J. Emerg. Electr. Power Syst., De Gruyter 17 (2), 173–189.

Ciezki, J.G., Ashton, R.W., 2000. Selection and stability issues associated with a navy shipboard dc zonal electric distribution system. IEEE Trans. Power Deliv. 15 (2), 665–669.

Cirrincione, M., Cossentino, M., Gaglio, S., Hilaire, V., Koukam, A., Pucci, M., 2011. Intelligent energy management system. In: Proceedings of IEEE 9th International Conference on Industrial Informatics, July. pp. 232–237.

Conway, B.E., 1991. Transition from super capacitor to battery behavior in electrochemical energy storage. J. Electrochem. Soc. 138 (6), 1539–1548.

Dizqah, A.M., Maheri, A., Busawon, K., Kamjoo, A., 2015. A multivariable optimal energy management strategy for standalone dc micro-grids. IEEE Trans. Power Syst. 30 (5), 2278–2287.

Duryea, S., Islam, S., Lawrence, W., 2001. A battery management system for standalone photovoltaic energy system. IEEE Ind. Appl. Magaz. 7 (3), 67–72.

Engelen, K., Shun, E.L., Vermeyen, P., Pardon, I., Hulst, R.D., Driesen, J., Belmans, R., 2006. The feasibility of small scale residential DC distribution systems. In: Proceedings of IEEE 32nd Annual Conference on Industrial Electronics, November. pp. 2618–2623.

Goa, D., Jin, Z., Lu, Q., 2008. Energy management strategy based on fuzzy logic for a fuel cell hybrid bus. J. Power Sources 185 (1), 311–317.

Gu, W.B., Wang, C.Y., Li, S.M., Geng, M.M., Liaw, B.Y., 1999. Modeling discharge and charge characteristics of nickel-metal hydride batteries. Electrochim. Acta 44 (25), 4525–4541.

Hajizadeh, A., Golkar, M.A., 2007. Intelligent power management strategy of hybrid distributed generation system. Int. J. Electr. Power Energy Syst. 29 (10), 783–795.

Hill, R., 1994. Electric railway traction part 3 traction power supplies. Power Eng. J. 8, 275–286.

Jeong, K.S., Lee, W.Y., Kim, C.S., 2005. Energy management strategies of a fuel cell/battery hybrid system using fuzzy logics. J. Power Sources 145 (2), 319–326.

Jiang, W., Zhang, Y., 2011. Load sharing techniques in hybrid power systems for dc micro-grids. In: Proceedings of IEEE Asia Pacific Power and Energy Engineering Conference, March. pp. 1–4.

Lalwani, M., Kothari, D.P., Singh, M., 2011. Size optimization of stand-alone photovoltaic system under local weather conditions in India. Int. J. Appl. Eng. Res. 1 (4), 952–961.

Mizuguchi, K., Muroyama, S., Kuwata, Y., Ohashi, Y., 1990. A new decentralized dc power system for telecommunications systems. In: Proceedings of IEEE 12th International Telecommunications Energy Conference, October. 22, pp. 55–62.

Phurailatpam, C., Chauhan, R.K., Rajpurohit, B.S., Gonzalez-Longatt, F.M., Singh, S.N., 2016. Demand side management system for future buildings. In: Proceedings of IEEE International Conference on Sustainable Green Buildings and Communities, IIT Madras, India, 18–20 December. pp. 1–6.

Salomonsson, D., Sannino, A., 2007. Low voltage DC distribution system for commercial power systems with sensitive electronic loads. IEEE Trans. Power Deliv. 22 (3), 1620–1627.

Sritharan, T., 2012. Impact of Current Waveforms on Battery Behavior. Department of Electrical and Computer Engineering, University of Toronto, pp. 1–95.

Starke, M., Tolbert, L.M., Ozpineci, B., 2008. AC vs. DC distribution: a loss comparison. In: Proceedings of IEEE Conference and Exhibition on Transmission and Distribution, April. pp. 1–7.

Scenario-based methods for robust electricity network planning considering uncertainties

16

Jinghua Li, Yujin Huang, Bo Lu, Shanyang Wei, Zhibang Wang
Department of Electrical Engineering, Guangxi University, Nanning, People's Republic of China

16.1 Introduction

As the rapid development of renewable power, the network expansion planning problem becomes more and more complex (Zhao et al., 2011). It is because that the renewable power is stochastic variable, and what is worse, its stochastic features could not be accurately formulated by a tractable mathematical equation. Hence, optimal planning for a network scheme that can satisfy the power transmission demand with stochastic renewable power integration is an intractable issue.

Simulation of scenario is a common method to consider uncertainties of renewable power (Javadi et al., 2013; Sánchez-Martín et al., 2005; Orfanos et al., 2013; Oliveira et al., 2007; Roh et al., 2009; Qiu et al., 2015; Carrión et al., 2007). The Monte Carlo method is a typical scenario generation method, which is popularly used in TNEP problem (Javadi et al., 2013; Sánchez-Martín et al., 2005; Orfanos et al., 2013; Oliveira et al., 2007). However, to accurately approximate the uncertain features, a large number of scenarios are required by Monte Carlo method, which will burden the computation heavily. Some methods have been proposed to reduce the number of scenarios to alleviate the computational burden (Roh et al., 2009; Qiu et al., 2015). However, the efficiency and accuracy of the reduction method are required to be improved further.

At present, there are several available methods for scenario reduction. The clustering (C) method (Baringo and Conejo, 2013) is the simplest method to reduce the number of original scenarios (OS) with the objective of minimizing the Euclidean distance (ED) between the reduced scenarios (RS) and OS. However, the C method intrinsically assumes that the probabilities of all OS are the same, which is not suitable for the conditions of the OS with different probabilities. As an improvement method, Backward or forward method (BF) (Chow and Liu, 1968; Kuska et al., 2003; Razali and Hashim, 2010; Sumaili et al., 2011) is a traditional reduction method considering the influences of the different probabilities. However, the computational burden of the BF method is too heavy to apply in the reduction of large- or even middle-scale scenarios. For improving the computational efficiency, a new algorithm based on particle swarm optimization (PSO) has been proposed in Pappala et al. (2009). However, about

Smart Power Distribution Systems. https://doi.org/10.1016/B978-0-12-812154-2.00016-X

all of the above reduction-based methods, only the space distance (SD) between reduced discrete probability distribution (RDPD) and original discrete probability distribution (ODPD) is considered while ignoring the stochastic features of ODPD, which will decrease the approximation quality of the RDPD.

Moment-matching (MM) method (Ross, 2007) is a preferable approach to generating RDPD to match the stochastic features of ODPD. However, the MM method involves a serious NP-Hard problem. So, an improved MM (IMM) method is proposed by (Høyland et al., 2003) to solve the referred problem. According to Høyland et al. (2003), matrix transformation and cubic transformation are adopted to make the generated scenarios matching the objective mean, variance, skewness, kurtosis, and correlation, which avoids the complex computation of traditional MM method.

Alternative, Taguchi's orthogonal array (OA) testing (Yu et al., 2011) are popularly used to generate extreme scenarios for testing the robustness of planning schemes. However, the scenarios generated by Taguchi's OA testing are too extreme, and occurrence possibilities of the scenarios are very small. So the planning scheme based on OA is too conservative to achieve a trade-off between economy and robustness.

With the generated scenarios, a robust planning can be presented and formulated, which is robust against all the scenarios that represent the stochastic features of renewable generation power. In this chapter, we aim to introduce a kind of scenario-based robust optimization methods for solving the electricity distribution network planning with uncertainties.

The remainder of this chapter is organized as follows: Section 16.2 presents the robust TNEP problem based on the scenarios of wind and load. Section 16.3 proposes the IMM and OA methods for generation of representative scenarios. Section 16.4 details the mathematic formulation of the robust TNEP problem based on the generated scenarios. Section 16.5 presents the case studies and verifies the method by comparing with the IMM and OA method. The conclusions are drawn in Section 16.6.

16.2 The mathematical model of network planning problem

16.2.1 The model of network planning problem with renewable power

The purpose of transmission network planning is to find an optimal economic construction scheme to ensure that the power system can operate safely and reliably in different operating environments. Therefore, a robust transmission network planning model based on wind power scenarios is established, as shown in Eqs. (16.1)–(16.9) (Yu et al., 2011). The objective function is the sum of construction cost of the lines, penalty cost of load curtailment, and penalty cost of wind power curtailment and the objective function is aimed at obtaining the robust grid construction scheme which satisfies the given scenarios without load curtailment and wind power curtailment as much as possible.

16.2.1.1 Objective function

In general, the objective function of the transmission network planning is to minimize the construction cost when the system is without load curtailment and wind power curtailment. Thus, the objective function contains two parts, as shown in Eq. (16.1). The left side of the plus sign is the first part, which means the construction cost; the right side of the plus sign is the second part, which is the penalty cost of load curtailment and the penalty cost of wind power curtailment.

$$\min \sum_{(i-j)\in\Omega_L} c_{ij}n_{ij} + \mu \sum_{i\in\Omega_B} (r_i + w_i) \tag{16.1}$$

where c_{ij} is the construction cost of one line added to the $i-j$ right of way, whose unit is \$; n_{ij} is the number of lines added between node i and node j, which is an integer variable; μ is the penalty factor for load curtailment and wind power curtailment, whose unit is \$/MW; r_i is the load curtailment of node i, MW; w_i is the wind power curtailment of node i, whose unit is MW; and Ω_L, Ω_B, and Ω_H are the set of lines, the set of system nodes, and the set of scenarios, respectively.

16.2.1.2 Constraints

The robust transmission network planning problem (NPP) is subject to the following constraints.

Nodal power balance

The power balance constraint is that the sum of output of thermal power units and wind power must be equal to the sum of the load and the loss of the network, and considers load curtailment and wind power curtailment as

$$\sum_{l=1}^{N_L} S_{il} \times p_{ij} + g_i + u_i + r_i = d_i + w_i \quad i \in \Omega_B \tag{16.2}$$

where S_{il} is the correlation coefficient of node i and branch l; p_{ij} is the power of branch $i-j$, whose unit is MW; g_i is the output of thermal power units at node i, whose unit is MW; u_i is the wind power at node i, whose unit is MW; and d_i is the forecasted load at node i, whose unit is MW.

Branch power

In the study of transmission network planning, the power flow constraint of the branch is usually in the form of DC power flow as

$$p_{ij} - \chi_{ij}\left(n_{ij}^0 + n_{ij}\right)\left(\theta_i - \theta_j\right) = 0 \quad (i-j) \in \Omega_L \tag{16.3}$$

where χ_{ij} is the reactance of line $i-j$; n_{ij}^0 is the initial number of lines between $i-j$; θ_i is the phase angle of voltage at node i, whose unit is rad; and θ_j is the phase angle of voltage at node j, whose unit is rad.

The branch power limits

The transmission capacity constraint of the branch is that the transmission capacity of each branch must be within a certain limit as

$$|p_{ij}| \leq \left(n_{ij}^0 + n_{ij}\right) \overline{p}_{ij} \quad (i-j) \in \Omega_L \tag{16.4}$$

where $\overline{p}_{ij}$ is the capacity of line $i-j$, whose unit is MW.

Power output limits

Each of the thermal power units is subject to its operating constraints. The output must be among the minimum and maximum production levels as

$$\underline{g}_i \leq g_i \leq \overline{g}_i \quad i \in \Omega_B \tag{16.5}$$

where $\overline{g}_i$ is the maximum output of units at node i, whose unit is MW and $\underline{g}_i$ is the minimum output of units at node i, whose unit is MW.

Load curtailment limits

The load curtailment constraint is that the system's load curtailment must be less than or equal to the load of the node and greater than or equal to 0 as

$$0 \leq r_i \leq d_i \quad i \in \Omega_B \tag{16.6}$$

Wind curtailment limits

The wind curtailment constraint is that the system's wind curtailment must be less than or equal to the wind power of the node and greater than or equal to 0 as

$$0 \leq w_i \leq u_i \quad i \in \Omega_B \tag{16.7}$$

Number of line limits

The number of lines must be less than or equal to the maximum number of lines allowed to build as

$$0 \leq n_{ij} \leq \overline{n}_{ij} \quad (i,j) \in \Omega_L \tag{16.8}$$

$$n_{ij} \text{ is an integer} \tag{16.9}$$

where $\overline{n}_{ij}$ is the maximum number of lines allowed to build between node i and node j.

Eqs. (16.1)–(16.9) form the mathematical model of the transmission NPP. The model is a mixed integer nonlinear programming problem with stochastic variables, a large number of constraints, and discrete variables. Compared with other methods of dealing with stochastic variables, the scenario-based method is simpler and more operable. Therefore, the scenario-based method is used to deal with the wind power randomness in the proposed model. In addition, the proposed model is transformed from the uncertain planning problem to the determining planning problem by using the generated scenarios to replace the wind power.

16.2.2 The robust model of NPP based on scenarios

The aim of robust TNEP is to plan an optimal transmission network to satisfy the stochastic variation of wind power, at the lowest cost of construction, load curtailment and wind power curtailment. The robust TNEP problem based on the representative scenarios of wind power and load is formulated as the following.

16.2.2.1 Objective function

Minimize

$$\sum_{(i-j)\in\Omega_L} c_{ij}n_{ij} + \sum_{h\in\Omega_H}\sum_{i\in\Omega_B}\left(\mu_{r,h}r_{i,h} + \mu_{w,h}w_{i,h}\right) \quad (16.10)$$

The objective function includes three parts: construction cost of the lines, penalty cost of load curtailment, and penalty cost of wind power curtailment. In this chapter, the penalty factors are decided by the electricity price and the duration time of scenario h in the service time of the line, given by

$$\mu_{r,h} = c_r \times T_h \quad (16.11a)$$

$$\mu_{w,h} = c_w \times T_h \quad (16.11b)$$

where c_r and c_w are the prices of load and wind power, whose unit is \$/MWh and T_h is the duration time of scenario h.

16.2.2.2 Constraints

The robust TNEP problem is subject to the following constraints, including nodal power balance, branch power, the branch power limits, power output limits, load curtailment limits, wind curtailment limits, number of line limits, as well as the node angle limits, which are shown by Eqs. (16.12)–(16.19), respectively.

Nodal power balance

In each node, the supplies of the power are required to the demands. The supplies include branch power, generation power of the traditional generator, and wind power. The wind power equals to the available power minus the wind curtailment. The demands equal to the load minus the load curtailment. The nodal power balance constraint is formulated as

$$\sum_{l=1}^{N_L} S_{il} \times p_{ij,h} + g_{i,h} + (u_{i,h} - w_{i,h}) = (d_i - r_{i,h}) \tag{16.12}$$

Branch power

For simplified calculation, the power flowing in a branch equals to the product of reactance and phase angle of voltage, formulated as

$$p_{ij,h} - \chi_{ij}\left(n_{ij}^0 + n_{ij}\right)\left(\theta_{i,h} - \theta_{j,h}\right) = 0 \tag{16.13}$$

The branch power limits

Due to the physical property of electrical line, the branch power is restricted to an upper limitation, shown as

$$\left|p_{ij,h}\right| \leq \left(n_{ij}^0 + n_{ij}\right)\overline{p}_{ij} \tag{16.14}$$

Power output limits

Each of the generating units is subject to its operating constraints. Among the minimum and maximum production levels ($\underline{g}_i$ and $\overline{g}_i$, respectively) for unit i are such that

$$\underline{g}_i \leq g_{i,h} \leq \overline{g}_i \tag{16.15}$$

Load curtailment limits

The load curtailment of a node cannot exceed the load connecting with the node, shown as

$$0 \leq r_{i,h} \leq d_i \tag{16.16}$$

Wind curtailment limits

The wind curtailment should less than the available wind power, shown as

$$0 \leq w_{i,h} \leq u_{i,h} \tag{16.17}$$

Number of Line limits

The number of lines constructed between two nodes should be integer and it is limited by

$$0 \leq n_{ij} \leq \overline{n}_{ij}, \quad n_{ij} \text{ is an integer} \tag{16.18}$$

Nodal angle limits

In each scenario, the nodal angel at each bus is limited in the range of $[-2\pi, 2\pi]$, which is formulated by

$$-2\pi \leq \theta_{i,h} \leq 2\pi \tag{16.19}$$

Eqs. (16.10)–(16.19) form the mathematical model of the TNEP problems. Seen from the model, $n_{ij}((i-j) \in \Omega_L)$ is the construction planning of network, which satisfies all the given scenarios of wind power $u_{i,h}(i \in \Omega_B, h \in \Omega_H)$. Substitute the generated scenarios into the Eqs. (16.10)–(16.19) and solve by modern interior point method (Wei et al., 2000). Then the robust construction planning is obtained.

16.3 Scenario generation methods

16.3.1 MM scenario generation method

16.3.1.1 Basic idea of MM method

Compared with other scenario generation methods, the MM method has no requirement for the distribution of the original data, and it is not necessary to obtain the marginal distribution of the variables in advance to facilitate the generation of power scenarios of multiple wind farms. Its main idea is to generate a small number of scenarios that coincide with the specified moments of the original data (usually the first four moment: mean, variance, skewness, and kurtosis) and the correlation matrix. The MM method is simple and easy to implement, and it can get a few scenarios that reflect the statistical properties of multiple stochastic variables.

MM is a common method of scenario generation, its main idea is to make the generated scenario with a given moment to match. In general, the higher the order of the selected moments, the more accurate the calculation, but the greater the amount of computation. The results show that the calculation accuracy of the first four moments has been satisfied, and the computational complexity is relatively small. Therefore, this chapter uses the first four moments, mean, variance, skewness, kurtosis, and correlation matrix as the matching term. Take the three variables as an example, the basic idea of the MM method is shown in Fig. 16.1.

In Fig. 16.1, $M_{1,k}$, $M_{2,k}$, and $M_{3,k}$ represent the k-order moments of the original wind-power scenario of three wind farms, $k = 1, 2, 3, 4$; R_{12}, R_{23}, and R_{13} represent correlation coefficients between the three wind farms. The purpose of the MM method is to generate a small number of scenarios representing the three wind-farm power sequences, which satisfy both $M_{1,k}$, $M_{2,k}$, and $M_{3,k}$ and can satisfy the correlation coefficients R_{12}, R_{23}, and R_{13}.

The correlation matrix R in the MM method needs to satisfy two conditions. The first condition is that R should be a symmetric positive semi-definite matrix with diagonal elements of 1. To determine whether R satisfies the condition, it is only necessary to perform Cholesky decomposition on the matrix R for pretreatment. If the Cholesky decomposition cannot be performed, it indicates that R is not a semi-definite matrix.

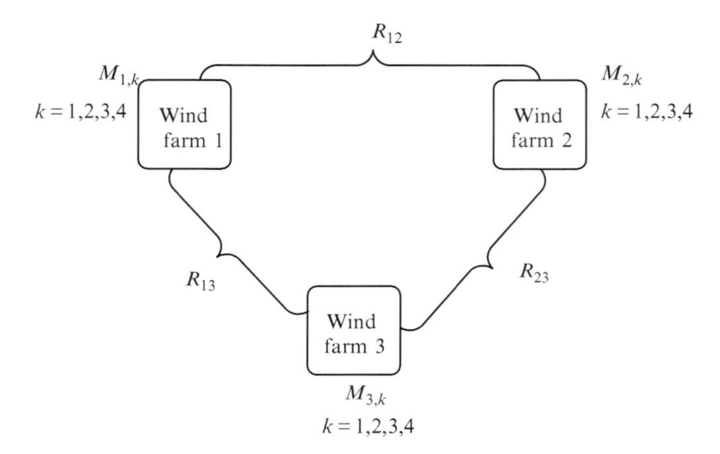

Fig. 16.1 The schematic diagram of MM.

If R is not a semi-definite matrix, then the data need to be reanalyzed. Alternatively, the method proposed by Higham (2000) and Lurie and Goldberg (1998) can be used to find a correlation matrix R that satisfies the requirements to some extent. Among them, the method in Lurie and Goldberg (1998) is more flexible and adaptable; it can be based on the requirements of the researchers to weight the correlation coefficient, the possible value of the constraints. Another condition is that R should be a nonsingular matrix, that is, each random variable cannot be a parallel vector. The Cholesky decomposition method can also be used to determine the condition, because for the collinear vector correlation matrix Cholesky decomposition, the resulting lower triangular matrix will appear 0, then the matrix Cholesky decomposition failure.

The condition for the correlation matrix R is not very strict because if the vectors of the stochastic variables are collinear, it means that at least one of the variables can be calculated from other variables. From this point of view, the dimension of the problem is reduced number.

16.3.1.2 Matching statistical moments

The most common statistical properties to be considered are the moments of the stochastic processes. Usually, the first four moments (mean, variance, skewness, and kurtosis) and correlation are taken into consideration as the stochastic features. There are several approaches to calculate the moments either based on sample or based on a mathematical definition. Next, we give the definitions of the expectation, variance, skewness, kurtosis, and correlation matrix.

(1) Mean

The mean is the first central moment. It simply represents the weighted sum of the stochastic variable values, that is, the arithmetical mean. Its mathematical definitions are:

For continuous variable:

$$E(\mathbf{X}) = \int_{-\infty}^{+\infty} x \cdot f(x)dx \tag{16.20}$$

X is considered as a stochastic variable; $f(x)$ is considered as the probability density function of the stochastic variable X.

For discrete variables:

$$E(\mathbf{X}) = \sum_{i=1}^{\infty} x_i p_i \tag{16.21}$$

The notation x_i is considered as the ith possible value of the stochastic variable **X**; and p_i is considered as the probability for each stochastic variable value x_i.

For historical samples:

$$\overline{x} = \frac{\sum_{i=1}^{N} \mathbf{X}_i}{N} \tag{16.22}$$

(2) Variance

The variance is the root-mean-square (RMS) deviation of the values from their means, it is the most common measure of statistical dispersion, measuring how widely spread the values in a data set are. Its mathematical definitions are:

For continuous variable:

$$\sigma(\mathbf{X}) = \sqrt{E(\mathbf{X}^2) - [E(\mathbf{X})]^2} = \sqrt{\int_{-\infty}^{+\infty} [x - E(\mathbf{X})]^2 f(x)dx} \tag{16.23}$$

For discrete variable:

$$\sigma(\mathbf{X}) = \sqrt{E(\mathbf{X}^2) - [E(\mathbf{X})]^2} = \sqrt{\sum_{i=1}^{\infty} [x_i - E(\mathbf{X})]^2 p_i} \tag{16.24}$$

For historical samples:

$$s = \sqrt{\frac{\sum_{i=1}^{N} (\mathbf{X}_i - \overline{x})^2}{N - 1}} \tag{16.25}$$

(3) Skewness

Skewness is the measure of the asymmetry of the probability distribution. Theoretically speaking, a positive skewness represents a longer right tail in comparison with the left tail, while a negative skewness represents the opposite situation. Its mathematical definitions are:

For continuous variable:

$$\text{skew}(\mathbf{X}) = \frac{E[\mathbf{X} - E(\mathbf{X})]^3}{\sigma(\mathbf{X})^3} = \frac{\int_{-\infty}^{+\infty} [x - E(\mathbf{X})]^3 f(x)dx}{\left(\int_{-\infty}^{+\infty} [x - E(\mathbf{X})]^2 f(x)dx\right)^{\frac{3}{2}}} \tag{16.26}$$

For discrete variable:

$$\text{skew}(\mathbf{X}) = \frac{E[\mathbf{X} - E(\mathbf{X})]^3}{\sigma(\mathbf{X})^3} = \frac{\sum_{i=1}^{\infty} [x_i - E(\mathbf{X})]^3 p_i}{\left(\sum_{i=1}^{\infty} [x_i - E(\mathbf{X})]^2 p_i\right)^{\frac{3}{2}}} \tag{16.27}$$

For historical samples:

$$\text{skew} = \frac{\frac{1}{N}\sum_{i=1}^{N} (x_i - \overline{x})^3}{\left[\frac{1}{N}\sum_{i=1}^{N} (x_i - \overline{x})^2\right]^{\frac{3}{2}}} \tag{16.28}$$

(4) Kurtosis

The kurtosis is the forth standardized central moment. The kurtosis is a measure of the peakedness of the probability distribution. The kurtosis of the normal distribution is 3. Therefore, in many cases the kurtosis is defined as Kurtosis—3, in order to easily compare the peakedness to the one of normal distribution. A high kurtosis occurs when a high percentage of the variance is due to infrequent extreme deviations from the mean. On the other hand, a low kurtosis occurs if the variance is mostly due to frequent modestly sized deviations for the mean. Its mathematical definitions are:

For continuous variable:

$$\text{kure}(\mathbf{X}) = \frac{E[\mathbf{X} - E(\mathbf{X})]^4}{\sigma(\mathbf{X})^4} = \frac{\int_{-\infty}^{+\infty} [x - E(\mathbf{X})]^4 f(x)dx}{\left(\int_{-\infty}^{+\infty} [x - E(\mathbf{X})]^2 f(x)dx\right)^{2}} \tag{16.29}$$

For discrete variable:

$$\text{kure}(\mathbf{X}) = \frac{E[\mathbf{X} - E(\mathbf{X})]^4}{\sigma(\mathbf{X})^4} = \frac{\sum_{i=1}^{\infty} [x_i - E(\mathbf{X})]^4 p_i}{\left(\sum_{i=1}^{\infty} [x_i - E(\mathbf{X})]^2 p_i\right)^{2}} \tag{16.30}$$

For historical samples:

$$\text{kurt} = \frac{\frac{1}{N}\sum_{i=1}^{N}(x_i - \overline{x})^4}{\left[\frac{1}{N}\sum_{i=1}^{N}(x_i - \overline{x})^2\right]^2} \tag{16.31}$$

(5) Correlation matrix

In probability theory and statistics, correlation—also called correlation coefficients—indicates the strength and direction of a linear relationship between two stochastic variables. In general statistical usage, correlation refers to the departure of two variables from independence although correlation does not imply causation. In this broad sense there are several coefficients, measuring the degree of correlation, adapted to the nature of data.

A number of different coefficients are used for different situations. The best known is the Pearson product-moment correlation coefficient, which is obtained by dividing the covariance of the two variables by the product of their variances. Its mathematical definitions are:

For continuous variable:

$$\begin{aligned}\rho_{XY} &= \frac{\text{cov}(\mathbf{X},\mathbf{Y})}{\sigma_{\mathbf{X}}\sigma_{\mathbf{Y}}} = \frac{E\{[\mathbf{X}-E(\mathbf{X})][\mathbf{Y}-E(\mathbf{Y})]\}}{\sigma_{\mathbf{X}}\sigma_{\mathbf{Y}}} \\ &= \frac{\int_{-\infty}^{+\infty}\int_{-\infty}^{+\infty}[x-E(\mathbf{X})][y-E(\mathbf{Y})]f(x,y)dxdy}{\sqrt{\int_{-\infty}^{+\infty}[x-E(\mathbf{X})]^2 f(x)dx \cdot \int_{-\infty}^{+\infty}[y-E(\mathbf{Y})]^2 f(y)dy}}\end{aligned} \tag{16.32}$$

$\mathbf{Y}$ is considered as a stochastic variable; $f(y)$ is considered as the probability density function of the stochastic variable $\mathbf{Y}$. The notation ρ_{XY} is considered as the correlation coeffcient between two stochastic variables $\mathbf{X}$ and $\mathbf{Y}$, $f(x,y)$ is considered as the joint probability distribution function between two stochastic variables $\mathbf{X}$ and $\mathbf{Y}$.

For discrete variable:

$$\begin{aligned}\rho_{XY} &= \frac{\text{cov}(\mathbf{X},\mathbf{Y})}{\sigma_{\mathbf{X}}\sigma_{\mathbf{Y}}} = \frac{E\{[\mathbf{X}-E(\mathbf{X})][\mathbf{Y}-E(\mathbf{Y})]\}}{\sigma_{\mathbf{X}}\sigma_{\mathbf{Y}}} \\ &= \frac{\sum_{i}^{\infty}\sum_{j}^{\infty}[x_i-E(\mathbf{X})]\left[y_j-E(\mathbf{Y})\right]p_{ij}}{\sqrt{\sum_{i=1}^{\infty}[x_i-E(\mathbf{X})]^2 p_i \cdot \sum_{j=1}^{\infty}\left[y_j-E(\mathbf{Y})\right]^2 p_j}}\end{aligned} \tag{16.33}$$

The notation p_{ij} is considered as joint probabilities between two stochastic variables values x_i and y_i.

For historical samples:

$$r_{XY} = \frac{\sum_{i}^{N}(x_i - \overline{x})(y_i - \overline{y})}{\sqrt{\sum_{i=1}^{N}(x_i - \overline{x})^2 \cdot \sum_{j=1}^{N}(y_i - \overline{y})^2}} \tag{16.34}$$

16.3.1.3 Mathematical model of MM method

The objective of the MM method is to minimize the bias of the stochastic features between the RDPD and ODPD, subject to the sum of the probabilities of all RS are 1. Usually, the first four moments (mean, variance, skewness, and kurtosis) and correlation are taken into consideration as the stochastic features (Ross, 2007; Høyland et al., 2003; Xu et al., 2012). Accordingly, the problem of MM method is formulated as Eqs. (16.35)–(16.36), named as Problem 1.

$$\min \sum_{n=1}^{N}\sum_{k=1}^{4}\omega_k(m_{nk} - M_{nk})^2 + \sum_{n,l\in\{1,\cdots N\},n\leq l}\omega_r(c_{nl} - C_{nl})^2 \tag{16.35}$$

$$m_{n1} = \sum_{s=1}^{S} p_s w_{ns} \tag{16.35a}$$

$$m_{n2} = \sqrt{\sum_{s=1}^{S}(w_{ns})^2 p_s - m_{n1}^2} \tag{16.35b}$$

$$m_{n3} = \frac{\sum_{s=1}^{S}(w_{ns} - m_{n1})^3 \cdot p_s}{m_{n2}^3} \tag{16.35c}$$

$$m_{n4} = \frac{\sum_{s=1}^{S}(w_{ns} - m_{n1})^4 \cdot p_s}{m_{n2}^4} \tag{16.35d}$$

$$c_{nl} = \frac{\sum_{s=1}^{S}(w_{ns} - m_{n1})(w_{ls} - m_{l1})p_s}{\sqrt{\left(\sum_{s=1}^{S}(w_{ns} - m_{n1})(w_{ns} - m_{n1})p_s\right)\left(\sum_{s=1}^{S}(w_{ls} - m_{l1})(w_{ls} - m_{l1})p_s\right)}}$$

$n,l \in N_W$ and $n < l$

(16.35e)

s.t

$$\sum_{s=1}^{S} p_s = 1, \quad p_s \geq 0 \tag{16.36}$$

The notations in Problem 1 are explained as follows, including the indices, parameters, known variables as well as unknown variables.

- Indices
  k is the order of the moment, $k = 1, 2, 3, 4$. n and l are the number of wind farm, $n, l \in \{1, \cdots, N\}$.
- Parameters
  ω_k and ω_r are the penalty factors of the unsatisfied degrees of the first four moments and correlation matrix, respectively. N is the number of wind farms and S is the scenario number of the RDPD.
- Known variables
  M_{nk} is kth-order moment of ODPD of the wind farm n. C_{nl} is the correlation of wind farm n and l of ODPD. M_{nk} and C_{nl} are calculated by Eqs. (A.1)–(A.5) in the Appendix based on the known ODPD.
- Unknown variables
  $m_{nk}(k = 1, 2, 3, 4 \;\; n = 1, \cdots, N)$ is the kth-order moment of RDPD of the wind farm n, which is calculated by Eqs. (16.35a)–(16.35d). c_{nl} is the correlation of wind farm n and l, which is calculated by Eq. (16.35e). w_{ns} is the value of wind farm n in the scenario s of RDPD. p_s is the probability of scenario s.

Based on the descriptions of symbols above, the computational complex of the Problem 1 is detailed as follows:

As seen from Eq. (16.35), the objective function includes the second-order items $(m_{nk})^2$ $(k = 1, 2, 3, 4)$. In addition, the forth-order moment m_{n4} has the item of $p_s(p_s w_{ns})^4$ as shown by Eq. (16.35d). So, Eq. (16.35) includes the item $(p_s(p_s w_{ns})^4)^2$, which causes the Problem 1 is hard to solve. Besides, the SD between RDPD and ODPD has not been considered in Problem 1, which may decrease the approximation accuracy of RDPD (Hochreiter and Pflug, 2007).

Hence, it is necessary to reduce the computational complexity of MM method and add the consideration of SD to enhance the approximation accuracy of RDPD.

16.3.2 IMM method

16.3.2.1 The basic idea of improvement

The aim of the MM method is to generate the scenarios that satisfy the target moments and correlation. However, the traditional MM method involves in the complex calculation with intractable high-dimensional discrete variable (Ross, 2007). Hence, in this chapter, we apply a heuristic moment-matching (HMM) method (Høyland et al., 2003) to generate wind power scenarios, which avoids the complex calculation.

The basic process of the HMM method is shown in Fig. 16.2. Seen from the Fig. 16.2, the proposed method includes three steps.

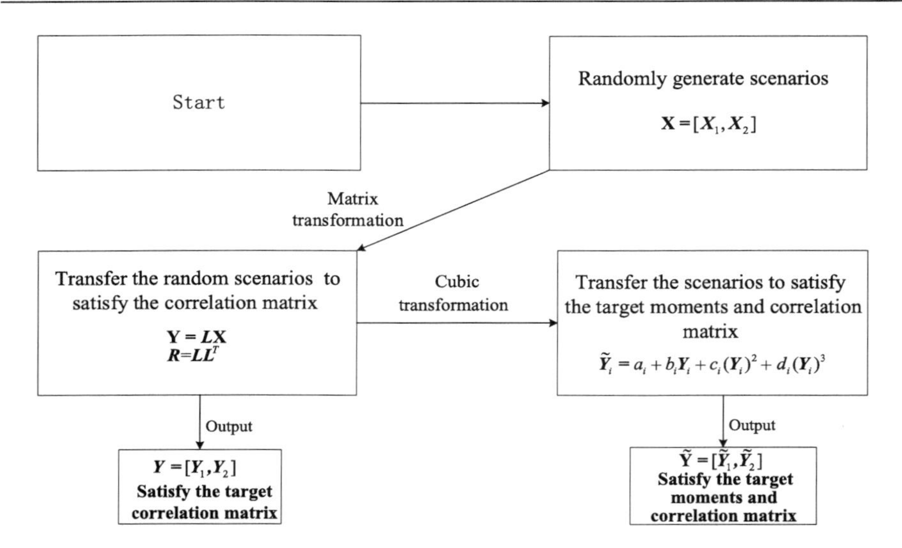

Fig. 16.2 The main steps of generating wind power scenarios based on HMM.

Step 1: Randomly generate scenarios X. In Fig. 16.2, there are two wind farms and the solid points represent the random scenarios. $X=[X_1,X_2]$, $\boldsymbol{X}_1$ and $\boldsymbol{X}_2$ are the scenarios of wind farm 1 and wind farm 2, respectively.

Step 2: Transfer the random scenarios X to satisfy the correlation matrix R, using the matrix transformation (Høyland et al., 2003). Denote the matrix transformed scenarios as Y.

Step 3: Transfer the scenarios Y to satisfy the target moments, using the cubic transformation (Wei et al., 2000). The cubic transformed scenarios are denoted as Y, which are the target scenarios.

Seen from the steps introduced above, matrix transformation and cubic transformation are important in the HMM method. Both are introduced below.

16.3.2.2 Two important transformations

Matrix transformation

The matrix transformation (Høyland et al., 2003) is formulated as

$$Y=LX \tag{16.37}$$

where $\boldsymbol{x}_i \in \boldsymbol{X}$ and $\boldsymbol{x}_1, \cdots, \boldsymbol{x}_N$ are independence; and $\boldsymbol{L}$ is the lower triangular matrix of R, shown in Fig. 16.2. The lower triangular matrix L can be obtained by Cholesky decomposition (Høyland et al., 2003).

The matrix L always comes from a Cholesky decomposition of the correlation matrix L, so we have $LL^T=R$.

From theory we know that if X is an n-dimensional N (0, 1) stochastic variable with correlation matrix $\boldsymbol{I}$ (and therefore with $\boldsymbol{x_i}$ mutually independent), then the $\boldsymbol{Y} = \boldsymbol{LX}$ is an n-dimensional N (0, 1) stochastic variable with correlation matrix.

The input phase

In this phase we work only with the target moments and correlations, we do not yet have any outcomes. This means that all operations are fast and independent of the number of scenarios s.

Our goal is to generate a discrete approximation z of an n-dimensional stochastic variable Z with matrix of target moments (***TARMOM***) and correlation matrix R. Since the matrix transformation needs zero means and variances equal to 1, we have to change the targets to match this requirement. Instead of z we will thus generate stochastic variables y with matrix of normalized moments (***MOM***) (and correlation matrix R), such that $MOM_1 = 0$, and $MOM_2 = 1$. z is then computed at the very end of the algorithm as

$$z = \alpha y + \beta \tag{16.38}$$

It is easily shown that the values leading to the correct z are:

$$\alpha = TARMOM_2^{\frac{1}{2}} \quad MOM_3 = \frac{TARMOM_3}{\alpha^3} \tag{16.39}$$

$$\beta = TARMOM_1 \quad MOM_4 = \frac{TARMOM_4}{\alpha^4} \tag{16.40}$$

The final step in the input phase is to derive moments of independent univariate stochastic variables $\boldsymbol{x_i}$, by $\boldsymbol{Y} = \boldsymbol{LX}$, the target moments and correlations can be obtained then. To do this need to find the Cholesky-decomposition matrix L, that is, a lower-triangular matrix L satisfying $R = LL^T$.

Cubic transformation

The cubic transformation (Høyland et al., 2003) is formulated as

$$\widetilde{\boldsymbol{Y}}_i = a_i + b_i\boldsymbol{Y}_i + c_i\boldsymbol{Y}_i^2 + d_i\boldsymbol{Y}_i^3 \tag{16.41}$$

where $\boldsymbol{Y}_i \in \boldsymbol{Y}, \widetilde{\boldsymbol{Y}}_i \in \widetilde{\boldsymbol{Y}}$. a_i, b_i, c_i, and d_i are the transforming coefficients, which are calculated as follows.

Suppose the moments of $a + b\boldsymbol{Y}_i + c\boldsymbol{Y}_i^2 + d\boldsymbol{Y}_i^3$ equal to the target moments, formulated as

$$M_{i,k}\left(a_i + b_i\boldsymbol{Y}_i + c_i\boldsymbol{Y}_i^2 + d_i\boldsymbol{Y}_i^3\right) = \overline{M}_{i,k} \quad k = 1,2,3,4 \tag{16.42}$$

where $M_{i,k}$ is the kth moment and $\overline{M}_{i,k}$ is the target moment of the wind farm i. Usually, the first four moments, namely mean, variance, skewness, and kurtosis, are regarded to represent the stochastic features of the wind power (Ross, 2007). Hence, we set $k = 1, 2, 3, 4$. $\overline{M}_{i,k}$ is the known parameter calculated from historical scenarios. The definitions of $M_{i,k}$ and $\overline{M}_{i,k}$ are also given by Ross (2007).

After solving (16.42), the transforming coefficients a_i, b_i, c_i, and d_i are obtained. Then the target scenarios $\widetilde{\boldsymbol{Y}}$ are generated by using (16.41).

This transformation comes from Fleishman (1978), where a method to generate a univariate non-normal stochastic variable $\widetilde{\boldsymbol{Y}}_i$ with given first four moments is introduced. It takes a N (0, 1) variable $\widetilde{X}_{\mathrm{i}}$ and uses a cubic transformation

$$\widetilde{\boldsymbol{Y}}_{\mathrm{i}} = \mathrm{a} + \mathrm{b}\widetilde{X}_{\mathrm{i}} + \mathrm{c}\widetilde{X}_{\mathrm{i}}^{2} + \mathrm{d}\widetilde{X}_{\mathrm{i}}^{3} \tag{16.43}$$

to obtain $\widetilde{\boldsymbol{Y}}_i$ with the target moments. Parameters a, b, c, and d are found by solving a system of nonlinear equations. Those equations utilize normality of the variable $\widetilde{X}_{\mathrm{i}}$.

The problem of this approach is that $\widetilde{X}_{\mathrm{i}}$ must have the first 12 moments equal to those of N (0, 1) in order to get exactly the target moments of $\widetilde{\boldsymbol{Y}}_i$. Since this is hard to achieve, either by sampling or discretization, the results with those of formulas are approximate. We have thus dropped the assumption of normality and derived formulas that work with an stochastic variable $\widetilde{X}_{\mathrm{i}}$—see Appendix A. Parameters a, b, c, and d are now given by a system of four implicit equations.

We have used a nonlinear mathematical-programming (least-squares) model to solve the system. In this model we have a, b, c, and d as decision variables, and we express the moments of $\widetilde{y}_{\mathrm{i}}$ as functions of these variables and the first 12 moments of $\widetilde{x}$. We then minimize the distance between those moments and their target values. We do not need to assume that the system has solution—if the solution does not exist, the model gives us the $\widetilde{y}_{\mathrm{i}}$ with the closest possible moments.

16.3.2.3 The steps of generating scenarios based on IMM methods

The steps of HMM method for generating wind power and load scenarios are summarized as the following. N_W is the number of wind farms and N_H is the number of scenarios. In addition, the dimension of the stochastic variables is $N_W + 1$, which is the sum of the number of wind farms and load.

Step 1: *Initialization.* Calculate the target moments ($\overline{M}_{i,k}$, $i = 1, \cdots, N_W + 1, k = 1,2,3,4$) and the correlation matrix ($\boldsymbol{R}$) of the N_W wind farms and load. Give the threshold value of moment error (ε_m) and correlation error (ε_r).

Step 2: *Normalization.* For convenient calculation, normalize the target moments to let the mean and variance to be 0 and 1, respectively. The normalization is formulated as (Høyland et al., 2003)

$$\overline{M}_{i,1}^{N} = 0,\ \overline{M}_{i,2}^{N} = 1,\ \overline{M}_{i,3}^{N} = \frac{\overline{M}_{i,3}}{\left(\sqrt{\overline{M}_{i,2}}\right)^{3}},\ \overline{M}_{i,4}^{N} = \frac{\overline{M}_{i,4}}{\left(\overline{M}_{i,4}\right)^{2}} \tag{16.44}$$

where $\overline{M}_{i,k}^{N}$ is the moment after normalizing.

Step 3: *Randomly generate scenarios.* Randomly generate NH scenarios of N_W wind farms and load, which satisfy N (0, 1), respectively. Denote the generated scenarios as $\boldsymbol{X}_{N_H \times (N_W+1)}$.

Step 4: *Matrix transformation.* Do Cholesky decomposition $\boldsymbol{R} = \boldsymbol{L}\boldsymbol{L}^T$ and transfer $\boldsymbol{X}_{N_H \times (N_W+1)}$ to satisfy the target correlation matrix using (16.37). The output of this step is denoted as $\boldsymbol{Y}_{N_H \times (N_W+1)}$.

Step 5: *Cubic transformation.* Calculate the cubic transforming coefficients a_i, b_i, c_i, and d_i by Eq. (16.42). Then transfer $\boldsymbol{Y}_{N_H \times (N_W+1)}$ to satisfy the normalized moments using Eq. (16.41). The output of this step is denoted as $\widetilde{\boldsymbol{Y}}^N_{N_S \times N_H}$.

Step 6: *Judge.* See whether $\varepsilon_r \leq \underline{\varepsilon}_r$ and $\varepsilon_m \leq \underline{\varepsilon}_m$ or not? If Yes, go next; Else, go to step 3.

$$\varepsilon_r = \left\| \left(R^G_{il} - R_{il} \right) \right\|_2 = \sqrt{\sum_{i=1}^{N_H} \sum_{l=1}^{N_H} \left(R^G_{il} - R_{il} \right)^2} \tag{16.45}$$

$$\varepsilon_m = \sum_{i=1}^{N_H} |M^G_{ik} - \overline{M}^N_{ik}| / \overline{M}^N_{ik} \tag{16.46}$$

where R^G_{il} and R_{il} are the correlations of generated and target scenarios. M^G_{ik} and $\overline{M}^N_{ik}$ are the moment of generated scenarios and the normalized moment, respectively.

Step 7: *Inversion of normalized scenarios.* Invert the normalized scenarios $\widetilde{\boldsymbol{Y}}^N_{N_H \times (N_W+1)}$ to satisfy the target moments by (Høyland et al., 2003)

$$\widetilde{\boldsymbol{Y}}_i = \sqrt{\overline{M}_{i,2}}\widetilde{\boldsymbol{Y}}^N_i + \overline{M}^N_{i,3} \tag{16.47}$$

where $\widetilde{\boldsymbol{Y}}_i \in \widetilde{\boldsymbol{Y}}$, $\widetilde{\boldsymbol{Y}}^N_i \in \widetilde{\boldsymbol{Y}}^N$.

$\widetilde{\boldsymbol{Y}}$ is the target scenarios of N_W wind farms and load. Based on the generated scenarios, the robust TNEP problem is formulated in the next section.

16.3.3 Taguchi's OA testing method

For the combination of wind power and load, the number of stochastic variables is very large. As a result, it is difficult to solve the problem, and reduce the scenarios by using Taguchi's OA, while maintaining the maximum degree of stochastic variables statistical information.

(1) Taguchi's OA

The Taguchi's OA was presented by Dr. Genichi Taguchi's in Japan in the late 1970s, it was originally used to predict the quality of the product from an engineering point of view. Based on the social loss of the product, in the early product development and design, it can effectively guarantee the quality of the product (Lin, 2004). Taguchi's

Table 16.1 Scenarios determined by OA $L_4(2^3)$

Scenario	Variable level		
	X_1	X_2	X_3
1	1	1	1
2	1	2	2
3	2	1	2
4	2	2	1

OA was popular in Japan at the beginning, and then swept the United States, finally developed to global scale. It has experienced a rapid development and now is widely used in multiple fields such as motor, automobile, optical, chemical, and computer products. Compared with the Monte Carlo sampling method, the number of scenarios in the Taguchi's OA is much smaller, which ensures that the scenarios provide good statistical information with the smallest number of scenarios in the uncertain operating space and dramatically decreases the difficulty of testing.

OA is formed as follows (Yu et al., 2011):

Suppose there is a system $\boldsymbol{X}$ consisting of F variables, denote $\boldsymbol{X} = X(x_1,x_2,\ldots,x_F)$, $x_1,x_2,\ldots,x_F$ as system variables. When the level number of all variables is the same (the level number represents the number of variable values, and each level corresponds to the individual values of the variables), assuming that they are O, then OA is $L_H(O^F)$, and H is the number of scenarios. When the level number of all variables is not the same, assuming that n variables have O_1 kinds of levels, and n variables are combined to get O_1^n combinations, m variables have O_2 level, and n variables are combined to get O_2^n combinations, all combinations of $O_1^n \times O_2^n$, then the OA form of $L_H(O_1^n \times O_2^n)$.

Take OA $L_4(2^3)$ as an example, Table 16.1 shows the scenarios that are determined by OA $L_4(2^3)$. Among them, "1" and "2" are the levels, representing the corresponding values of this variable. For example, when a variable is valued at 10 and 20 MW, the levels "1" and "2" can correspond to 10 and 20 MW, respectively, or 20 and 10 MW, respectively. Table 16.1 shows that there are three variables in this system, each of which has two levels, and only four of the system's composite scenarios are formed by OA. Tables B.1 and B.2 in the Appendix give OA $L_8(2^7)$ and OA $L_{36}(2^{11} \times 3^{12})$, respectively.

There are two main characteristics of Taguchi's OA. One is the reduction of the scenario, the other is the limit scenario.

(1) The reduction of scenarios

In general, H is less than O^F or $O_1^n \times O_2^n$. For example, OA $L_4(2^3)$, $H = 4$, $O^F = 2^3 = 8$, 4 less than 8. Also, in OA $L_{36}(2^{11} \times 3^{12})$, H = 36, 36 is much smaller than $O^F = 2^{11} \times 3^{12}$. It can be seen that after the use of OA, the number of scenarios was significantly reduced.

(2) The limit scenarios

OA has the following properties:

(i) Each column is self-balanced, which means each level appears H/O times. As shown in Table 16.1, the levels of "1" and "2" in columns 1, 2, and 3 appear $4/2 = 2$ times;
(ii) Any two columns are mutual-balanced, which means for any two columns, each two variable levels of the combination is with the same frequency. As shown in Table 16.1, for variable X_1, X_2, the combination of "11," "12," "21," and "22" appear once each;
(iii) The combination of OA is evenly distributed in all possible spaces. As shown in Fig. 16.3, OA $L_4(2^3)$, the solid dots are OA combinations, which are evenly distributed across the vertices in the three-dimensional cube.

Thus, the resulting OA guarantees the limit of scenarios.

In addition, OA also has an important nature: when ignoring a column or some columns, the remaining table still meet the characteristics of OA.

As a result, OA can obtain a few limit scenarios, and when the system satisfies these scenarios, the rest of the scenario is often satisfied.

(2) Get the limit scenario

Assume that the wind power has three scenarios, which is 0, the mean and the rated value, respectively. Also assume that the load has two scenarios, namely $\mu + \sigma$ and $\mu - \sigma$. Therefore, it is necessary to select a mixed OA with two levels and three levels. Assume that the power system has n_u fan access nodes and n_d load access nodes, then the OA can be expressed as $L_H(2^{n_d} \times 3^{n_u})$. By checking the table (The University of York, 2004) or the calculation (Yang, 1978), the wind power and load composed of OA can be obtained, and the limit scenarios of the system can be obtained consequently.

Firstly, determine the number of system variables and the number of horizontal variables per variable. Take a system as an example, in this system, the variables include load and fan power, which possess five load nodes. Each load node has two kinds of levels ($\mu + \sigma$, $\mu - \sigma$), one fan access node and three kinds of level (0, mean, rated). Since the numbers of levels of wind power and load are not the same, so the system's Taguchi's OA goes to mixed OA. With $O_1 = 2$ and $O_2 = 3$, the system

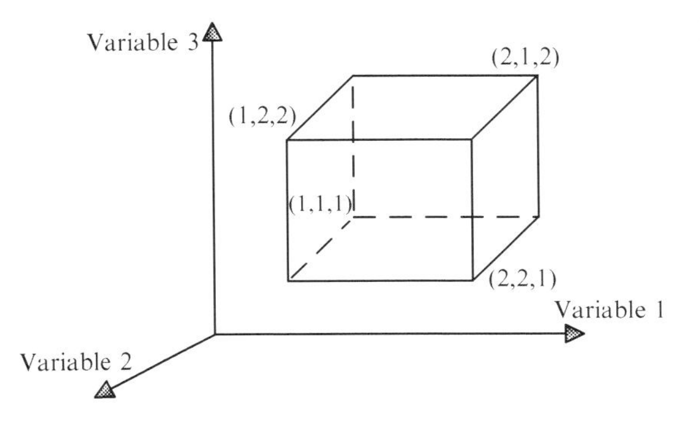

Fig. 16.3 Schematic diagram of OA L_4 (2^3).

should be $L_H(2^5 \times 3^1)$. For standard OA, there is no $L_H(2^5 \times 3^1)$ but $L_{36}(2^{11} \times 3^{12})$. According to the properties of the array, ignoring any column of OA does not change the nature of the array. Therefore, we can choose $L_{36}(2^{11} \times 3^{12})$, and then select "$2^{11}$" in the first five columns, "3^{12}" in the first one column. As a result, a system with the Taguchi's OA can be formed, which is referred to as $L_{36}(2^5 \times 3^1)$.

16.4 The solving process of the robust network planning

The robust optimization process of transmission network planning is divided into two parts:

(1) The transformation of deterministic robust optimization model. Consider the wind power transmission network planning model as the wind power, which is a stochastic programming model, and the model is transformed into a deterministic robust optimization model.

(2) Solving robust optimization model. The established robust optimization model is a mixed integer nonlinear optimization problem, in this chapter, the model is solved by using GAMS software, which can overcome the shortcomings such as the huge number of iterations, large calculation and unstable results.

The basic idea of solving is shown in Fig. 16.4. Firstly, a robust power grid optimization planning model is established to satisfy all the representative scenarios. Secondly, use the MM method to generate a small number of representative scenarios that reflect the stochastic characteristics of wind power in a number of wind farms, then replace the wind power stochastic variables in the model with the generated

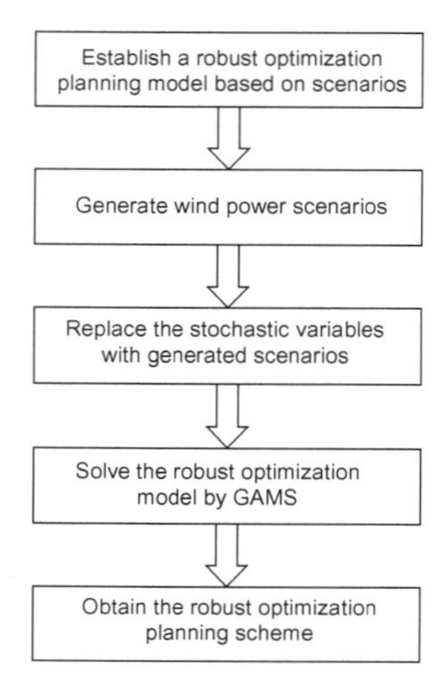

Fig. 16.4 Basic idea of robust optimal TNEP based on scenario.

scenarios; next, transform the uncertainty planning into deterministic planning. Finally, the DICOPT solver in GAMS software is used to solve the model and then obtain a robust optimal programming scheme. The obtained planning scheme can cope with the wind power with all possible statistical scenarios and possesses a good robustness, which provides a novel solution for the power system with access nodes of plurality in transmission network planning.

16.5 Case studies

In this section, the proposed method is tested on three cases: Garver 6-bus system, IEEE24-bus system, and IEEE RTS-96 system. The wind data here are from 50 Hz, Amprion, and APG, which are downloaded from the (Amprion in Germany, n.d.). They are numbered as Wind Farm1 (WF1), Wind Farm2 (WF2), and Wind Farm3 (WF3), respectively. The capacities of the three wind farms are 12,200, 5200, and 1080 MW, respectively. The wind data used in the test is recorded every 15 min and from 2012-1-1 and 2012-12-31. Hence, there are altogether 35,136 historical wind power points, which are used as OG. Firstly, we use the HMM method to generate the scenarios of wind farms. Then, the generated scenarios are applied into the TNEP of the three cases. The proposed TNEP is verified by comparison with the Taguchi's OA method proposed by Yu et al. (2011).

16.5.1 Test of the generated scenarios

Firstly, normalize the wind data, next set the $\underline{\varepsilon}_{r} = 0.01$ and $\underline{\varepsilon}_{r} = 0.15$. Then, respectively generate 20, 30, 40, 50, 60, 70, 80, 90, and 100 scenarios using the HMM method. At last, compare the stochastic features with those of the original scenarios. The comparison results are shown in Fig. 16.5. The errors of correlation in Fig. 16.5 are calculated by Eq. (16.46).

Seen from the Fig. 16.5A, the errors of the correlation do not decrease with the increasing of the scenarios number. It is because the scenarios are randomly generated in the Step3 (Section 16.3.2.3) of the scenario generation method, which brings irregular results. However, all the errors of correlation are less than the preset threshold value 0.01.

Seen from the Fig. 16.5B, as a whole, the errors of the first four moments decrease with the increasing of the scenarios number. The errors are less than 2% when the number is larger than 40. The errors of mean and variance are larger than that of skewness and kurtosis. It is because the normalized wind power is used here and their means and variance are less than 1. The errors of skewness and kurtosis are nearly 0 in the Fig. 16.5B. By the comparison results, the stochastic features of generated scenarios are close to those of original scenarios.

Seen from the Fig. 16.5, the HMM method has good performance in approximated the stochastic features of original scenarios, which uses much fewer scenarios number.

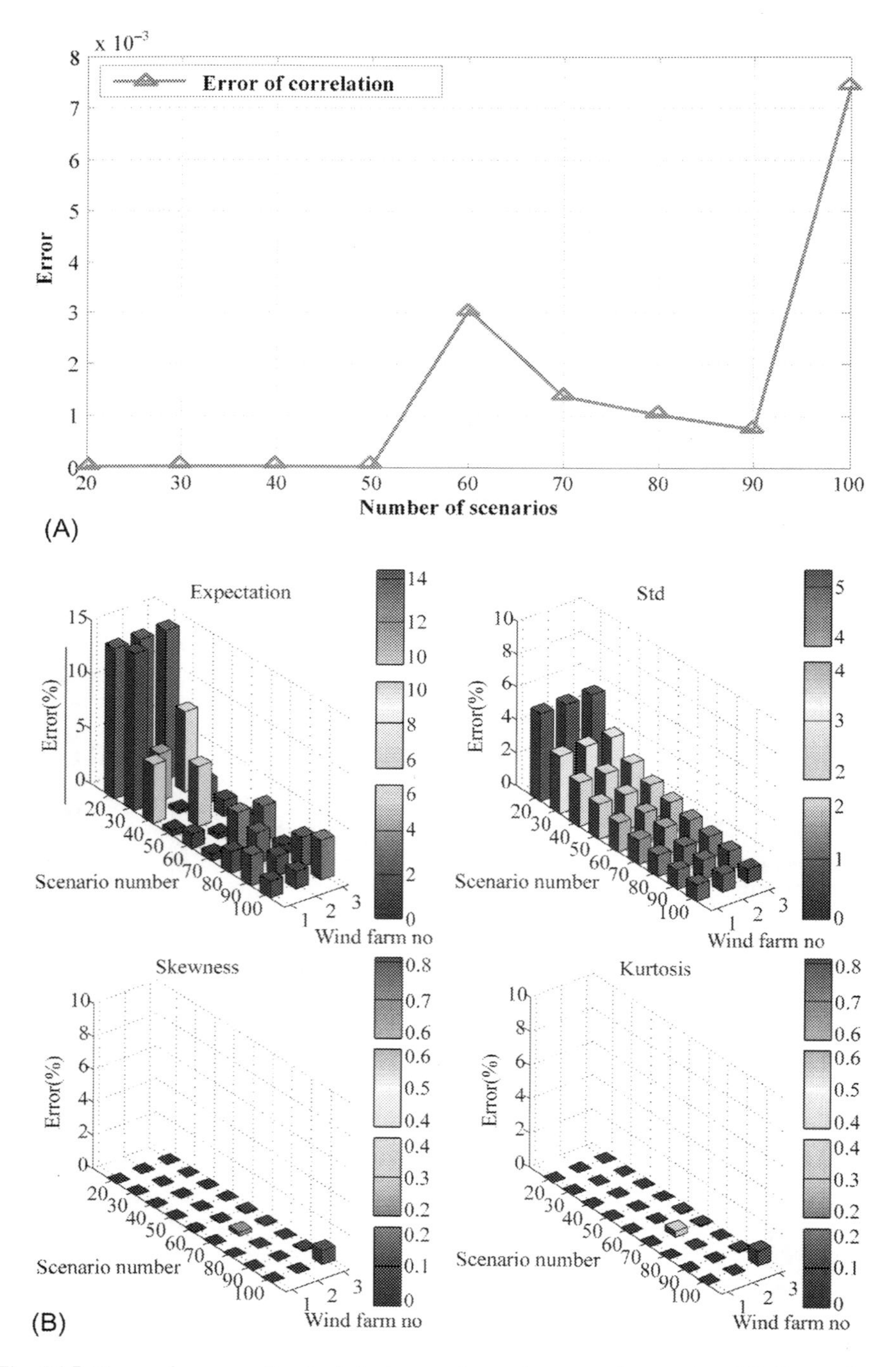

Fig. 16.5 Comparisons results of original scenarios and generated scenarios. (A) Comparisons of correlation matrix. (B) Comparisons of first four moments.

We can choose the scenarios from 20 to 100 to represent the original scenarios, according to the requirements of calculation and precision.

16.5.2 Robust planning schemes of three cases

16.5.2.1 Garver 6-bus system

The initial topology of Garver 6-bus is shown in Fig. 16.6. G1–G3 are the thermal units and W1–W3 are the wind farms. The parameters of the branches, generators, and nodal loads are from Yu et al. (2011). The capacities of the three wind farms are 30, 30, and 300 MW, respectively. The three wind farms' scenarios are obtained from Section 16.5.1. The actual value of the wind power scenarios equals to the normalized value multiply with the capacities of wind farms, namely, 30, 30, and 300 MW.

In this test, we choose the number of scenarios as 40. Set the operating hours of the three wind farms is 2000 h a year and the operating hour of each scenario is $T_h = 2000/40 = 50\,\text{h}$. If the electrical price is $c_r = c_w = 500\$/\text{MWh}$, then the penalty factors of the load curtailment and wind curtailment are $\mu_{r,h} = \mu_{w,h} = 500 \times 50 = 2500\$$. For comparisons, we calculate the planning schemes by HMM method and Taguchi's OA method proposed by Yu et al. (2011). The extreme scenarios based on OA method are shown in Table 16.2.

The calculation results are shown in Table 16.3. Seen from the Table 16.3, the scheme of OA method is needed to add two lines between node 2 and node 6. While the proposed method only need to add one line from node 4 to node 6, which saves half of construction cost than OA method. The penalty costs of both methods are 0$.

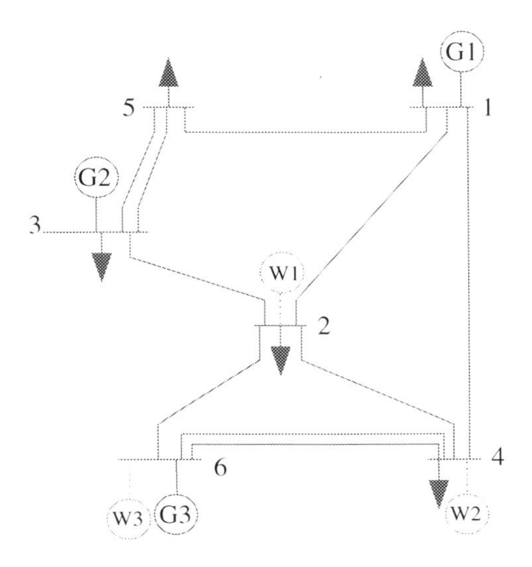

Fig. 16.6 Topology of Garver6-bus system.

Table 16.2 Scenarios chose from Taguchi's orthogonal array

	W1	W2	W3
1	0	0	0
2	0	1	1
3	1	0	1
4	1	1	0

Table 16.3 Construction planning of Garver 6-bus system

	OA	Proposed method
Planning	Lines 2–6 = 2	Line 4–6 = 1
Construction ost ($)	6000	3000
Penalty cost ($)	0	0

Table 16.4 Verified results of Garver 6-bus system

	Initial	OA	HMM
Load_C (MWh)	0	0	0
Load_T	0	0	0
Wind_C (MWh)	6200	0	247
Wind_T	137	0	13

Furthermore, we carry on the experiments to test the robustness of the planning schemes from both methods. The experiments are as follows. First, we generate 1000 scenarios of wind power by Monte Carlo method, and then calculate the load curtailment and wind curtailment of the network schemes, when facing the above 1000 scenarios. The verified results are shown in Table 16.4.

In Table 16.4, "Load_C" and "Wind_C" represent the load curtailment and wind curtailment. "Load_T" and "Wind_T" represent the times of load curtailment and wind curtailment. "Initial" means the initial topology without adding lines. Seen from the Table 16.4, the probabilities of load curtailment of the three methods are 0. As for wind curtailment, although there is wind curtailment of the HMM method, its value and probability is small. In the HMM method, the probability of wind curtailment is 0.013 (13/1000) and the wind curtailment at every time is 19MW (247/13), which are much less than initial topology.

The wind curtailment of OA method is 0, but its construction cost is double the cost of HMM method. Hence, the construction planning of the HMM method is better in trade-off robustness and economy.

Table 16.5 Construction planning of IEEE 24-bus system

	OA	HMM
Planning	Line 21–22 = 1, Line 6–7 = 1	Lines 7–8 = 2, Line 21–22 = 1
Construction cost ($)	144,000	126,000
Penalty cost ($)	0	0

Table 16.6 Verified results of IEEE 24-bus system

	Initial	OA	HMM
Load_C (MW)	0	0	0
Load_T	0	0	0
Wind_C (MW)	16,733	0	0
Wind_T	116	0	0

16.5.2.2 IEEE 24-bus system

The parameters of the branches, generators, buses, and nodal loads of IEEE 24-bus system are from Romero et al. (2005). Also, three wind farms are considered in the test. The capacities of the three farms are 1000, 200, and 1000 MW, and the wind farms connect with node 7, node 16, and node 22, respectively. The scenarios of wind power are obtained from Section 16.5.1 and we choose the scenarios number as 40. The parameters of penalty factors are also as same as the ones in the Garver 6-bus system.

The results of the planning schemes of IEEE 24-bus system are shown in Table 16.5. Table 16.6 shows the verified results of the planning schemes from initial topology, OA method, and HMM method.

As seen from Table 16.5, the construction cost of HMM method is less than OA method. In Table 16.6, both the OA and proposed methods have not load curtailment and wind curtailments during the 1000 tests. While as for the initial scheme, the probability of wind curtailment is 0.116. Hence, compared with the OA method, the HMM method can have the same robust with less cost in the IEEE 24-bus system. Compare with the initial schemes, the HMM method is much more robust.

16.5.2.3 IEEE RTS-96 system

The parameters of the branches, generators, bus, and node loads of IEEE RTS-96 system are from Wong et al. (1999). Three wind farms are also considered and their capacities are 1200, 1200, and 800 MW. The wind farms connect with node 7, node

Table 16.7 Construction planning of IEEE RTS-96 system

	OA	Proposed method
Planning	Line 7–8 = 1, Lines 7–27 = 3, Line 13–39 = 1, Line 15–16 = 1, Line 15–21 = 1, Line 16–17 = 1, Lines 16–19 = 2, Line 16–22 = 1, Lines 21–22 = 2, Line 23–41 = 1, Line 39–40 = 1, Line 39–45 = 1, Line 40–41 = 1, Line 40–43 = 1, Line 41–42 = 1, Lines 45–46 = 2	Line 7–8 = 1, Line 7–27 = 1, Line 15–21 = 1, Line 16–17 = 1, Line 16–19 = 1, Line 17–18 = 1, Lines 21–22 = 2, Line 39–40 = 1, Line 39–45 = 1, Line 40–43 = 1, Lines 45–46 = 2
Construction cost ($)	692,000	386,000
Penalty cost ($)	0	0

Table 16.8 Verified results of IEEE RTS-96 system

	Initial	OA	Proposed method
Load_C (MW)	0	0	0
Load_T	0	0	0
Wind_C (MW)	291,993	0	1137
Wind_T	984	0	19

22, and node 47, respectively. All the other parameters are set as the same as the ones in Garver 6-bus system and IEEE 24-bus system.

The planning results of IEEE RTS-96 system are shown in Tables 16.7 and 16.8. Seen from the Table 16.7, the adding lines of HMM method are much less than that of the OA method. In addition, the construction cost of the OA is almost 1.8 times of proposed method. In Table 16.8, during the 1000 tests, there is no load curtailment in the three methods. The wind curtailment occurs 984 in the initial scheme, which means the initial could not accommodate such large scale of wind power. The wind curtailment probability of the HMM method is 0.019, and the probability of OA method is 0. From Table 16.8, we can see that the planning scheme of OA is the most robust one among all schemes. However, its construction cost is too high. The results show the HMM method can have good performance on the coordination of robustness and economy.

16.6 Conclusion

This chapter proposed a robust transmission network expansion planning method, which considers the wind power uncertainties by using scenarios. A HMM method is proposed to approximate the stochastic features of wind power and load. It proves that the HMM method has good performance in capturing the features of mean, variance, skewness, kurtosis, and correlation matrix by using a few number of scenarios. Correspondingly, with the HMM method, the calculation burden of the TNEP problem based on scenarios will greatly be reduced. Shown by the comparisons results of Taguchi's OA method, the proposed method has better performance in trade-off robustness and economy. At last, the analysis of the influencing factors shows that the number of scenarios and the penalty factors are not very sensitivity to the planning scheme.

Appendix

Appendix A

The first four moments and correlation matrix are calculated by (A.1)–(A.5), respectively.

$$M_{n1} = \frac{\sum_{s_o=1}^{S_o} w_{ns_o}}{S_o} \tag{A.1}$$

$$M_{n2} = \sqrt{\frac{\sum_{s_o=1}^{S_o} (w_{ns_o} - M_{n1})^2}{S_o}} \tag{A.2}$$

$$M_{n3} = \frac{S_o}{(S_o - 1)(S_o - 2)} \sum_{s_o=1}^{S_o} \left(\frac{w_{ns} - M_{n1}}{M_{n2}} \right)^3 \tag{A.3}$$

$$M_{n4} = \left(\frac{S_o(S_o + 1)}{(S_o - 1)(S_o - 2)(S_o - 3)} \sum_{s_o=1}^{S_o} \left(\frac{w_{ns} - M_{n1}}{M_{n2}} \right)^4 \right) - 3\frac{(S_o - 1)}{(S_o - 2)(S_o - 3)} + 3 \tag{A.4}$$

$$C_{il}=\frac{\frac{1}{S_o}\sum_{s_o=1}^{S_o}(w_{ns_o}-M_{n1})(w_{ls_o}-M_{l1})}{\sqrt{\left\{\frac{1}{S_o}\sum_{s_o=1}^{S_o}(w_{ns_o}-M_{n1})(w_{ns_o}-M_{n1})\right\}\left\{\frac{1}{S_o}\sum_{s_o=1}^{S_o}(w_{ls_o}-M_{l1})(w_{ls_o}-M_{l1})\right\}}},\quad \forall n\neq l \tag{A.5}$$

where S_o is the scenario number of the ODPD and w_{ns_o} is the $(s_o)^{\text{th}}$ OS of the wind farm n.

Appendix B

See Tables B.1 and B.2.

Table B.1 Scenarios determined by OA $L_8(2^7)$

Scenario	Variable level						
	X_1	X_2	X_3	X_4	X_5	X_6	X_7
1	1	1	1	1	1	1	1
2	1	1	1	2	2	2	2
3	1	2	2	1	1	2	2
4	1	2	2	2	2	1	1
5	2	1	2	1	2	1	2
6	2	1	2	2	1	2	1
7	2	2	1	1	2	2	1
8	2	2	1	2	1	1	2

Appendix C

Garver 6 bus

See Fig. C.1 and Tables C.1–C.3.

IEEE 24-bus system

See Fig. C.2 and Tables C.4 and C.5.

IEEE RTS-96 system

See Fig. C.3 and Tables C.6–C.8.

Table B.2 Scenarios determined by OA $L_{36}(2^{11} \times 3^{12})$

Scenario	Variable level											
	X_1	X_2	X_3	X_4	X_5	X_6	X_7	X_8	X_9	X_{10}	X_{11}	X_{12}
1	1	1	1	1	1	1	1	1	1	1	1	1
2	1	1	1	1	1	1	1	1	1	1	1	2
3	1	1	1	1	1	1	1	1	1	1	1	3
4	1	1	1	1	1	2	2	2	2	2	2	1
5	1	1	1	1	1	2	2	2	2	2	2	2
6	1	1	1	1	1	2	2	2	2	2	2	3
7	1	1	2	2	2	1	1	1	2	2	2	1
8	1	1	2	2	2	1	1	1	2	2	2	2
9	1	1	2	2	2	1	1	1	2	2	2	3
10	1	2	1	2	2	1	2	2	1	1	2	1
11	1	2	1	2	2	1	2	2	1	1	2	2
12	1	2	1	2	2	1	2	2	1	1	2	3
13	1	2	2	1	2	2	1	2	1	2	1	1
14	1	2	2	1	2	2	1	2	1	2	1	2
15	1	2	2	1	2	2	1	2	1	2	1	3
16	1	2	2	2	1	2	2	1	2	1	1	1
17	1	2	2	2	1	2	2	1	2	1	1	2
18	1	2	2	2	1	2	2	1	2	1	1	3
19	2	1	2	2	1	1	2	2	1	2	1	1
20	2	1	2	2	1	1	2	2	1	2	1	2
21	2	1	2	2	1	1	2	2	1	2	1	3
22	2	1	2	1	2	2	2	1	1	1	2	1
23	2	1	2	1	2	2	2	1	1	1	2	2
24	2	1	2	1	2	2	2	1	1	1	2	3

Continued

Table B.2 Continued

Scenario	Variable level											
	X_1	X_2	X_3	X_4	X_5	X_6	X_7	X_8	X_9	X_{10}	X_{11}	X_{12}
25	2	1	1	2	2	2	1	2	2	1	1	1
26	2	1	1	2	2	2	1	2	2	1	1	2
27	2	1	1	2	2	2	1	2	2	1	1	3
28	2	2	2	1	1	1	1	2	2	1	2	1
29	2	2	2	1	1	1	1	2	2	1	2	2
30	2	2	2	1	1	1	1	2	2	1	2	3
31	2	2	1	2	1	2	1	1	1	2	2	1
32	2	2	1	2	1	2	1	1	1	2	2	2
33	2	2	1	2	1	2	1	1	1	2	2	3
34	2	2	1	1	2	1	2	1	2	2	1	1
35	2	2	1	1	2	1	2	1	2	2	1	2
36	2	2	1	1	2	1	2	1	2	2	1	3

Scenario	Variable level										
	X_{13}	X_{14}	X_{15}	X_{16}	X_{17}	X_{18}	X_{19}	X_{20}	X_{21}	X_{22}	X_{23}
1	1	1	1	1	1	1	1	1	1	1	1
2	2	2	2	2	2	2	2	2	2	2	2
3	3	3	3	3	3	3	3	3	3	3	3
4	1	1	1	2	2	2	2	3	3	3	3
5	2	2	2	3	3	3	3	1	1	1	1
6	3	3	3	1	1	1	1	2	2	2	2
7	1	2	3	1	2	3	3	1	2	2	3
8	2	3	1	2	3	1	1	2	3	3	1
9	3	1	2	3	1	2	2	3	1	1	2
10	1	3	2	1	3	2	3	2	1	3	2
11	2	1	3	2	1	3	1	3	2	1	3

Scenario	Variable level										
	X_{13}	X_{14}	X_{15}	X_{16}	X_{17}	X_{18}	X_{19}	X_{20}	X_{21}	X_{22}	X_{23}
12	3	2	1	3	2	1	2	1	3	2	1
13	2	3	1	3	2	1	3	3	2	1	2
14	3	1	2	1	3	2	1	1	3	2	3
15	1	2	3	2	1	3	2	2	1	3	1
16	2	3	2	1	1	3	2	3	3	2	1
17	3	1	3	2	2	1	3	1	1	3	2
18	1	2	1	3	3	2	1	2	2	1	3
19	2	1	3	3	3	1	2	2	1	2	3
20	3	2	1	1	1	2	3	3	2	3	1
21	1	3	2	2	2	3	1	1	3	1	2
22	2	2	3	3	1	2	1	1	3	3	2
23	3	3	1	1	2	3	2	2	1	1	3
24	1	1	2	2	3	1	3	3	2	2	1
25	3	2	1	2	3	3	1	3	1	2	2
26	1	3	2	3	1	1	2	1	2	3	3
27	2	1	3	1	2	2	3	2	3	1	1
28	3	2	2	2	1	1	3	2	3	1	3
29	1	3	3	3	2	2	1	3	1	2	1
30	2	1	1	1	3	3	2	1	2	3	2
31	3	3	3	2	3	2	2	1	2	1	1
32	1	1	1	3	1	3	3	2	3	2	2
33	2	2	1	2	1	1	3	1	1	3	3
34	3	1	2	3	2	3	1	2	2	3	1
35	1	2	3	1	3	1	2	3	3	1	2
36	2	3	1	2	1	2	3	1	1	2	3

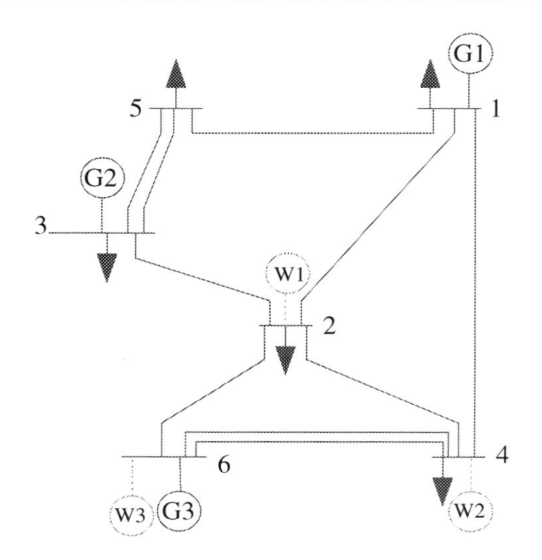

Fig. C.1 Topology of Garver 6-bus system.

Table C.1 Parameters of lines

No	From	To	n_{ij}^0	n_{ij}^0	($)	$\overline{n}_{ij}$
1	1	2	1	100	40,000	3
2	1	3	0	100	38,000	3
3	1	4	1	80	60,000	3
4	1	5	1	100	20,000	3
5	1	6	0	70	68,000	3
6	2	3	1	100	20,000	3
7	2	4	1	100	40,000	3
8	2	5	0	100	31,000	3
9	2	6	1	100	30,000	3
10	3	4	0	82	59,000	3
11	3	5	2	100	20,000	3
12	3	6	0	100	48,000	3
13	4	5	0	75	63,000	3
14	4	6	2	100	30,000	3
15	5	6	0	78	61,000	3

Table C.2 Parameters of generators and load

Bus	$\overline{g}_i$	$\underline{g}_i$	d_i
1	150	30	80
2	0	0	240
3	360	165	40
4	0	0	160
5	0	0	240
6	300	150	0

Table C.3 Scenarios chosen from Taguchi's orthogonal array

	W1	W2	W3
1	0	0	0
2	0	1	1
3	1	0	1
4	1	1	0

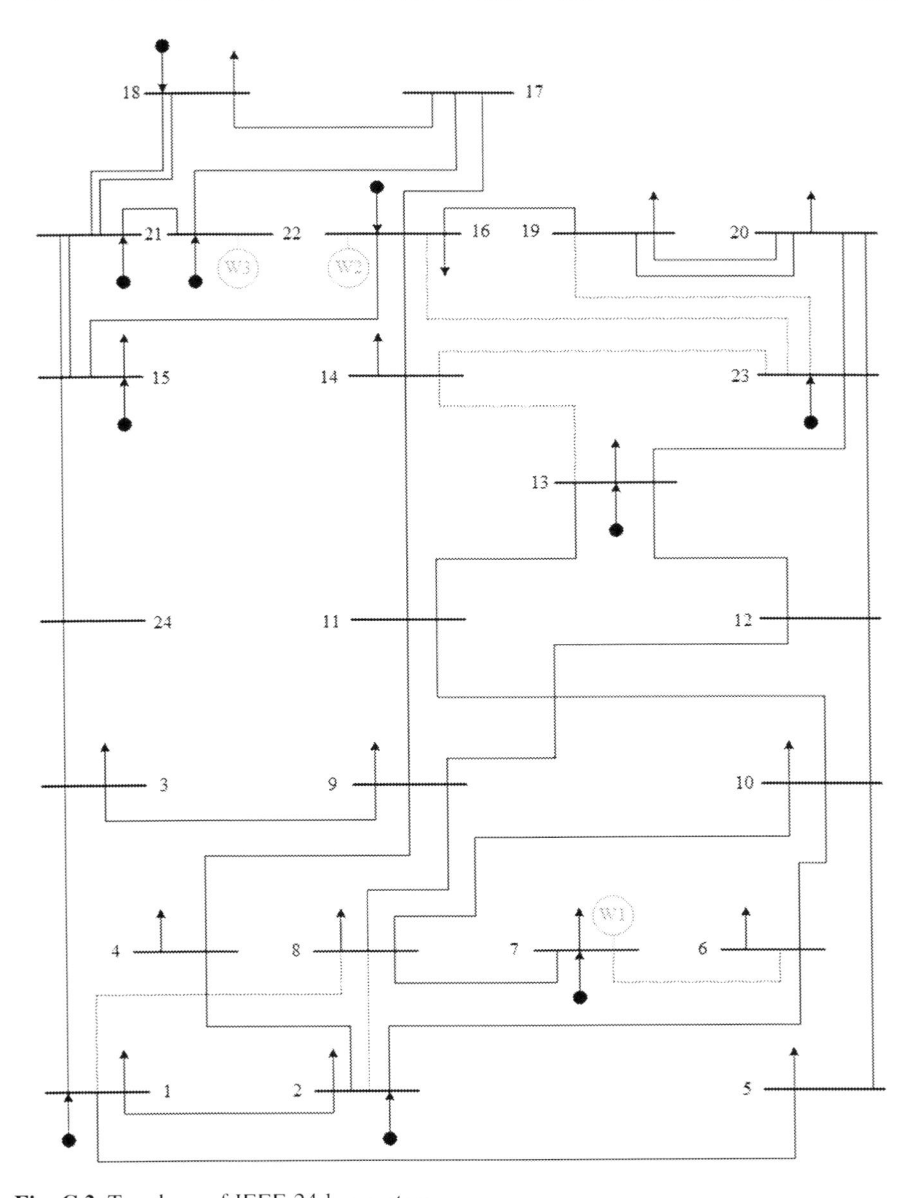

Fig. C.2 Topology of IEEE 24-bus system.

Table C.4 IEEE 24 system generation and load in MW

Bus	G_0	G_1	G_2	G_3	G_4	D
1	576	576	465	576	520	324
2	576	576	576	576	520	291
3	0	0	0	0	0	540
4	0	0	0	0	0	222
5	0	0	0	0	0	213
6	0	0	0	0	0	408
7	900	900	722	900	812	375
8	0	0	0	0	0	513
9	0	0	0	0	0	525
10	0	0	0	0	0	585
11	0	0	0	0	0	0
12	0	0	0	0	0	0
13	1773	1773	1424	1457	1599	0
14	0	0	0	0	0	795
15	645	645	645	325	581	582
16	465	465	465	282	419	951
17	0	0	0	0	0	300
18	1200	1200	1200	603	718	999
19	0	0	0	0	0	543
20	0	0	0	0	0	384
21	1200	1200	1200	951	1077	0
22	900	900	900	900	900	0
23	1980	315	953	1980	1404	0
24	0	0	0	0	0	0

Table C.5 IEEE 24 system circuit data

Bus	Bus	n_{ij}^0	res. (p.u.)	reat. (p.u.)	f_{ij} MW	C_{ij} 10^3 US $
1	2	1	0.0026	0.0139	175	3000
1	3	1	0.0546	0.2112	175	55000
1	5	1	0.0218	0.0845	175	22000
2	4	1	0.0328	0.1267	175	33000
2	6	1	0.0497	0.1920	175	50000
3	9	1	0.0308	0.1190	175	31000
3	24	1	0.0023	0.0839	400	50000
4	9	1	0.0268	0.1037	175	27000
5	10	1	0.0228	0.0883	175	23000
6	10	1	0.0139	0.0605	175	16000
7	8	1	0.0159	0.0614	175	16000
8	9	1	0.0427	0.1651	175	43000
8	10	1	0.0427	0.1651	175	43000
9	11	1	0.0023	0.0839	400	50000
9	12	1	0.0023	0.0839	400	50000

Table C.5 Continued

Bus	Bus	n_{ij}^0	res. (p.u.)	reat. (p.u.)	f_{ij} MW	C_{ij} 10^3 US $
10	11	1	0.0023	0.0839	400	50000
10	12	1	0.0023	0.0839	400	50000
11	13	1	0.0061	0.0476	500	66000
11	14	1	0.0054	0.0418	500	58000
12	13	1	0.0061	0.0476	500	66000
12	23	1	0.0124	0.0966	500	134000
13	23	1	0.0111	0.0865	500	120000
14	16	1	0.0050	0.0389	500	54000
15	16	1	0.0022	0.0173	500	24000
15	21	2	0.0063	0.0490	500	68000
15	24	1	0.0067	0.0519	500	72000
16	17	1	0.0033	0.0259	500	36000
16	19	1	0.0030	0.0231	500	32000
17	18	1	0.0018	0.0144	500	20000
17	22	1	0.0135	0.1053	500	146000
18	21	2	0.0033	0.0259	500	36000
19	20	2	0.0051	0.0396	500	55000
20	23	2	0.0028	0.0216	500	30000
21	22	1	0.0087	0.0678	500	94000
1	8	0	0.0348	0.1344	500	35000
2	8	0	0.0328	0.1267	500	33000
6	7	0	0.0497	0.1920	500	50000
13	14	0	0.0057	0.0447	500	62000
14	23	0	0.0080	0.0620	500	86000
16	23	0	0.0105	0.0822	500	114000
19	23	0	0.0078	0.0606	500	84000

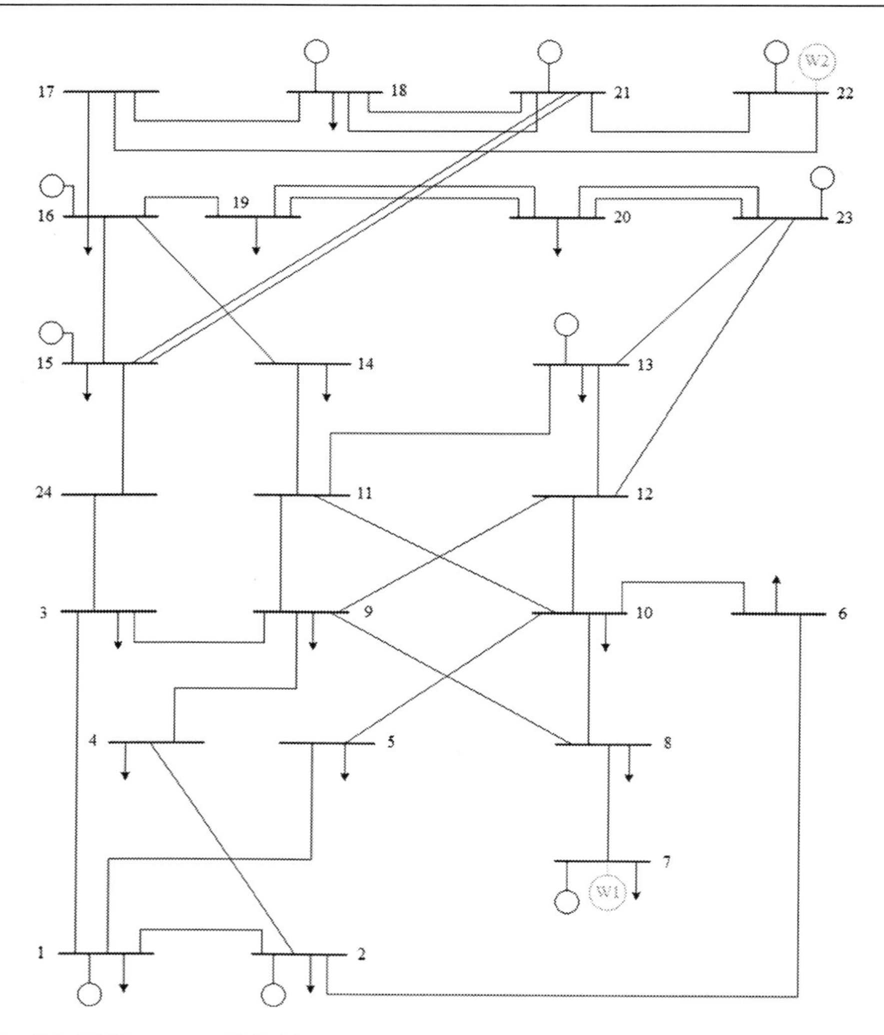

Fig. C.3 IEEE one area RTS-96.

Table C.6 RTS-96 system bus load data

Bus number	Bus load	Load		If Peak load 10% higher	
	% of system load	MW	MVar	MW	MVar
101,201,301	3.8	108	22	118.8	24.2
102,202,302	3.4	97	20	106.7	22.0
103,203,303	6.3	180	37	198.0	40.7
104,204,304	2.6	74	15	81.4	16.5
105,205,305	2.5	71	14	78.1	16.4
106,206,306	4.8	136	28	149.6	30.8
107,207,307	4.4	125	25	137.5	27.5
108,208,308	6.0	171	35	188.1	38.5
109,209,309	6.1	175	36	192.5	39.6
110,210,310	6.8	195	40	214.5	44.0
113,213,313	9.3	265	54	291.5	59.4
114,214,314	6.8	194	39	213.4	42.9
115,215,315	11.1	317	64	348.7	70.4
116,216,316	3.5	100	20	110.0	22.0
118,218,318	11.7	333	68	366.3	74.8
119,219,319	6.4	181	37	199.1	40.7
120,220,320	4.5	128	26	140.8	28.6
	Total 100.0	2850	580	3135	638

Table C.7 RTS-96 system generator data

Unit group	Unit size (MW)	Unit type	Force outage rate	MTTF (hour)	MTTR (hour)	Scheduled maint. wks/year
U12	12	Oil/steam	0.02	2940	60	2
U20	20	Oil/CT	0.10	450	50	2
U50	50	Hydro	0.01	1980	20	2
U76	76	Coal/steam	0.02	1960	40	3
U100	100	Oil/steam	0.04	1200	50	3
U155	155	Coal/steam	0.04	960	40	4
U197	197	Oil/steam	0.05	950	50	4
U350	350	Coal/steam	0.08	1150	100	5
U400	400	Nuclear	0.12	1100	150	6

Table C.8 RTS-96 system branch parameter

Branch ID	i	j	X (p.u)	n_0	p_{max} (MW)	Cost (1000US$)	n_{max}
1	1	2	0.014	1	175	3	3
2	1	3	0.211	1	175	55	3
3	1	5	0.085	1	175	22	3
4	2	4	0.127	1	175	33	3
5	2	6	0.192	1	175	50	3
6	3	9	0.119	1	175	31	3
7	3	24	0.084	1	400	50	3
8	4	9	0.104	1	175	27	3
9	5	10	0.088	1	175	23	3
10	6	10	0.061	1	175	16	3
11	7	8	0.061	1	175	16	3
12	7	27	0.161	1	175	42	3
13	8	9	0.165	1	175	43	3
14	8	10	0.165	1	175	43	3
15	9	11	0.084	1	400	50	3
16	9	12	0.084	1	400	50	3
17	10	11	0.084	1	400	50	3
18	10	12	0.084	1	400	50	3
19	11	13	0.048	1	500	33	3
20	11	14	0.042	1	500	29	3
21	12	13	0.048	1	500	33	3
22	12	23	0.097	1	500	67	3
23	13	23	0.087	1	500	60	3
24	13	39	0.075	1	500	52	3
25	14	16	0.059	1	500	27	3
26	15	16	0.017	1	500	12	3
27	15	21	0.049	2	500	34	3
28	15	24	0.052	1	500	36	3
29	16	17	0.026	1	500	18	3
30	16	19	0.023	1	500	16	3
31	17	18	0.014	1	500	10	3
32	17	22	0.105	1	500	73	3
33	18	21	0.026	2	500	18	3
34	19	20	0.04	2	500	27.5	3
35	20	23	0.022	2	500	15	3
36	21	22	0.068	1	500	47	3
37	23	41	0.074	1	500	51	3
38	25	26	0.014	1	175	3	3
39	25	27	0.211	1	175	55	3
40	25	29	0.085	1	175	22	3
41	26	28	0.127	1	175	33	3
42	26	30	0.192	1	175	50	3
43	27	33	0.119	1	175	31	3
44	27	48	0.084	1	400	50	3

Table C.8 Continued

Branch ID	i	j	X (p.u)	n_0	p_{max} (MW)	Cost (1000US$)	n_{max}
45	28	33	0.104	1	175	27	3
46	29	34	0.088	1	175	23	3
47	30	34	0.061	1	175	16	3
48	31	32	0.061	1	175	16	3
49	32	33	0.165	1	175	43	3
50	32	34	0.165	1	175	43	3
51	33	35	0.084	1	400	50	3
52	33	36	0.084	1	400	50	3
53	34	35	0.084	1	400	50	3
54	34	36	0.084	1	400	50	3
55	35	37	0.048	1	500	33	3
56	35	38	0.042	1	500	29	3
57	36	37	0.048	1	500	33	3
58	36	47	0.097	1	500	67	3
59	37	47	0.087	1	500	60	3
60	38	40	0.059	1	500	27	3
61	39	40	0.017	1	500	12	3
62	39	45	0.049	2	500	34	3
63	39	48	0.052	1	500	36	3
64	40	41	0.026	1	500	18	3
65	40	43	0.023	1	500	16	3
66	41	42	0.014	1	500	10	3
67	41	46	0.105	1	500	73	3
68	42	45	0.026	2	500	18	3
69	43	44	0.04	2	500	27.5	3
70	44	47	0.022	2	500	15	3
71	45	46	0.068	1	500	47	3

References

Amprion in Germany. (n.d.) Transparency in energy markets. Available from: http://www.transparency.eex.com/en/ (Accessed May 2010)

Baringo, L., Conejo, A.J., 2013. Correlated wind-power production and electric load scenarios for investment decisions. Appl. Energy 101, 475–482.

Chow, C.K., Liu, C.N., 1968. Approximating discrete probability distributions with dependence trees. IEEE Trans. Inf. Theory IT-14 (3), 462–467.

Carrión, M., Arroyo, J.M., Alguacil, N., 2007. Vulnerability-constrained transmission expansion planning: a stochastic programming approach. IEEE Trans. Power Syst. 22 (4), 1436–1445.

Fleishman, A.I., 1978. A method for simulating nonnormal distributions. Psychometrika. 43, 521–532.

Higham, N.J., 2000. Computing the nearest correlation matrix—A problem from finance. Manchester Centre for Computational Mathematics. Report number: 369.

Høyland, K., Kaut, M., Wallace, S.W., 2003. A heuristic for moment-matching scenario generation. Comput. Optim. Appl. 24 (2-3), 169–185.

Hochreiter, R., Pflug, G.C., 2007. Financial scenario generation for stochastic multi-stage decision processes as facility location problems. Ann. Oper. Res. 152 (1), 257–272.

Javadi, M.S., Saniei, M.H., Mashhadi, R., Gutiérrez-Alcaraz, G., 2013. Multi-objective expansion planning approach: distant wind farms and limited energy resources integration. IET Renew. Power Gener. 7 (6), 652–668.

Kuska, N.G., Heitsch, H., Römisch, W., 2003. In: Scenario reduction and scenario tree construction for power management problems.Proceedings of 2003 IEEE Bologna Power ECH Conference, Bologna, Italy, 23–26 June.

Lurie, P.M., Goldberg, M.S., 1998. An approximate method for sampling correlated random variables from partially-specified distributions. Manag. Sci. 44 (2), 203–218.

Lin, X.X., 2004. Actual Combat Technique of Taguchi Method. Haitian Publishing House, Shenzhen.

Oliveira, G.C., Binato, S., Pereira, M.V.F., 2007. Value-based transmission expansion planning of hydrothermal systems under uncertainty. IEEE Trans. Power Syst. 22 (4), 1429–1435.

Orfanos, G.A., Georgilakis, P.S., Hatziargyriou, N.D., 2013. Transmission expansion planning of systems with increasing wind power integration. IEEE Trans. Power Syst. 28 (2), 1355–1362.

Pappala, V.S., Erlich, I., Rohrig, K., Dobschinski, J., 2009. A stochastic model for the optimal operation of a wind-thermal power system. IEEE Trans. Power Syst. 24 (2), 940–950.

Qiu, J., Dong, Z.Y., Zhao, J.H., Xu, Y., Zheng, Y., Li, C.X., Wong, K.P., 2015. Multi-stage flexible expansion co-planning under uncertainties in a combined electricity and gas market. IEEE Trans. Power Syst. 30 (4), 2119–2129.

Romero, R., Rocha, C., Mantovani, J.R.S., Sanchez, I.G., 2005. Constructive heuristic algorithm for the DC model in network transmission expansion planning. IEEE Proc. Gener. Transm. Distrib. 52 (2), 277–282.

Ross, O., 2007. Interest rate scenario generation for stochastic programming. M.Sc. thesis, Department of Informatics and Mathematical Modelling, The Technical University of Denmark (DTU), Lyngby, Copenhagen, Denmark.

Roh, J.H., Shahidehpour, M., Wu, L., 2009. Market-based generation and transmission planning with uncertainties. IEEE Trans. Power Syst. 24 (3), 1587–1598.

Razali, N.M.M., Hashim, A.H., 2010. In: Backward reduction application for minimizing wind power scenarios in stochastic programming.4th International Power Engineering and Optimization Conference (PEOCO 2010), Shah Alam, Selangor, Malaysia, 23–24 June.

Sánchez-Martín, P., Ramos, A., Alonso, J.F., 2005. Probabilistic midterm transmission planning in a liberalized market. IEEE Trans. Power Syst. 20 (4), 2135–2142.

Sumaili, J., Keko, H., Miranda, V., Zhou, Z., Botterud, A., Wang, J.H., 2011. In: Finding representative wind power scenarios and their probabilities for stochastic models.2011 16th International Conference on Intelligent System Application to Power Systems (ISAP), Hersonissos, Greece, 25–28 September.

The University of York, 2004. Orthogonal arrays (Taguchi Designs). Available from: https://www.york.ac.uk/depts/maths/tables/orthogonal.htm/. Accessed May 2004.

Wong, P., Albrecht, P., Allan, R., Billinton, R., Chen, Q., Fong, C., Haddad, S., Li, W., Mukerji, R., Patton, D., Schneider, A., Shahidehpour, M., Singh, C., 1999. The IEEE reliability test system-1996. A report prepared by the reliability test system task force of the application of probability methods subcommittee. IEEE Trans. Power Syst. 14 (3), 1010–1020.

Wei, H., Sasaki, H., Kubokawa, J., Yokoyama, R., 2000. Large scale hydrothermal optimal power flow problems based on interior point nonlinear programming. IEEE Trans. Power Syst. 15 (1), 396–403.

Xu, D.B., Chen, Z.P., Yang, L., 2012. Scenario tree generation approaches using K-means and LP moment matching methods. J. Comput. Appl. Math. 236 (17), 4561–4579.

Yang, Z.X., 1978. Construction of Orthogonal Arrays. People's Publishing House, Jinan.

Yu, H., Chung, C.Y., Wong, K.P., 2011. Robust transmission network expansion planning method with Taguchi's orthogonal array testing. IEEE Trans. Power Syst. 26 (3), 1573–1580.

Zhao, J.H., Foster, J., Dong, Z.Y., Wong, K.P., 2011. Flexible transmission network planning considering distributed generation impacts. IEEE Trans. Power Syst. 26 (3), 1434–1443.

Scenarios/probabilistic optimization approaches for network operation considering uncertainties

Jinghua Li, Bo Chen, Yuhong Mo, Jiasheng Zhou
Department of Electrical Engineering, Guangxi University, Nanning, People's Republic of China

17.1 Introduction scenario

Large-scale stochastic renewable energy (such as wind power) is integrating into the grid, which leads to essential changes in power system planning and operation. Wind power possesses two characteristics, fluctuation and uncertainty, which are the core issues in large-scale wind power integration to keep the operation of power systems safe and stable.

In terms of mathematical essence, power system optimization with wind power integration belongs to complex multistage stochastic decision problem. From that point, it is sophisticated to be solved accurately and conventionally, so it is necessary to seek a suitable method for power system optimization planning and operation with renewable energy integration.

Since 2012, the capacity of wind power integration of China has outnumbered the capacity of the United States, which made China become the largest country in the field of wind power. With the development of large-scale wind power integration, the fluctuation and uncertainty of wind power itself have made the power system change profoundly. Because power system has evolved to complex load-power multistage stochastic decision coupling system, the safe and stable operation of power system faces unprecedented difficulties and challenges. Under this circumstance, conventional optimization approaches for network operation no longer suit. Therefore, new methods are required to handle the influence the uncertainties of wind power bring. To demonstrate the meaning of this research, the stochastic properties of wind power and their influences to power system operation are introduced below, respectively.

Smart Power Distribution Systems. https://doi.org/10.1016/B978-0-12-812154-2.00017-1

17.1.1 Stochastic properties of wind power

The stochastic properties of wind power include two aspects: the intermittency/fluctuation and the uncertainty/unpredictable. The detailed introductions of those two aspects are as followed.

17.1.1.1 Fluctuation

The wind varies vastly in the nature, which leads to the intermittency and fluctuation of wind power. Hence, the output of wind power cannot be effectively controlled by power system operators. As a result, extra energy is required to deal with the sudden change of wind power for keeping the power balanced. Such ancillary services include frequency regulation, voltage support, etc. Fig. 17.1 illustrates the fluctuation of wind power. From the figure, wind power fluctuates violently, which increases the difficulty of power system operation and dispatch.

To reveal the short-term fluctuation of wind farm, take the data of a wind farm, whose installed capacity is 49.3 MW, for statistical analysis. The results are shown in Table 17.1.

From Table 17.1, with the increase of time scale, the maximum fluctuation and the maximum fluctuation ratio of wind power output increase accordingly. The statistical analysis results show that the maximum fluctuation is 47.36 MW, the maximum fluctuation ratio is 96.07%, which appear when the time scale is 60 s.

From Fig. 17.1 and Table 17.1, wind power has a certain level of fluctuation, which means if not stabilized on time, the fluctuation of frequency and voltage will raise.

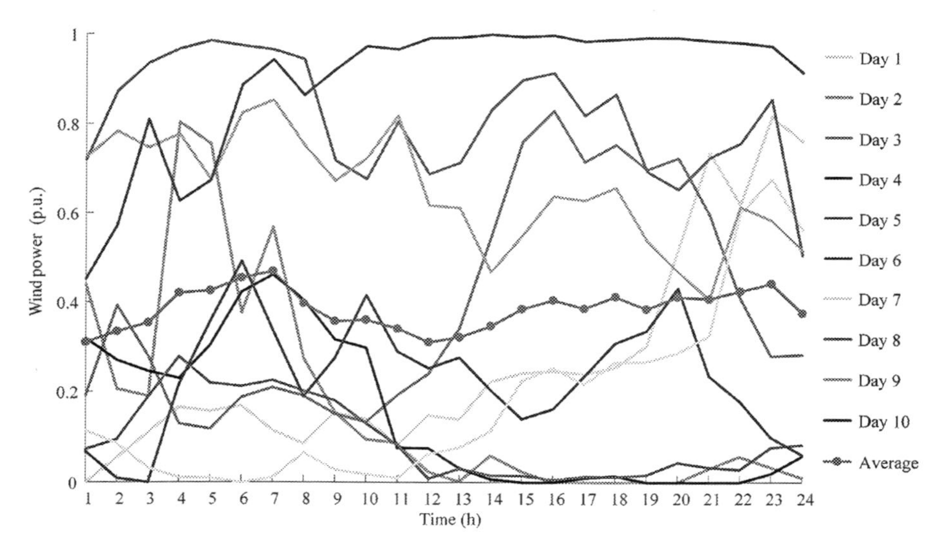

Fig. 17.1 Wind power output curves.

Table 17.1 Analysis of wind power short-term fluctuation

	6 s	12 s	24 s	30 s	60 s
Maximum fluctuation (MW)	33.36	45.09	46.04	46.57	47.36
Maximum fluctuation ratio (%)	67.67	91.46	93.38	94.46	96.07

Worse, it may lead to power system failure. Also, with the increase of wind power penetration, its output fluctuation becomes more and more prominent, and the operation and dispatch of power system will become even more complex and difficult accordingly.

17.1.1.2 Uncertainty

Uncertainty, also known as unpredictable, represents the ability of prediction for wind power. Different from fluctuation, the unpredictable of wind power can be eliminated, but the fluctuation exists eternally instead. In other words, the fluctuation of wind power still exists even though the accuracy of wind power prediction achieves the level of 100% because wind power output depends on changeable wind energy entirely.

Wind power cannot be predicted precisely at present. Fig. 17.2 illustrates the comparison of realized wind power curve and predicted wind power curve. It can be seen from the figure that forecasted values differ from realized values quite much.

The uncertainty of wind power has much influence on day-ahead unit commitment (UC) problem. Meanwhile, day-ahead unit dispatch plan and spinning reserve (SR)

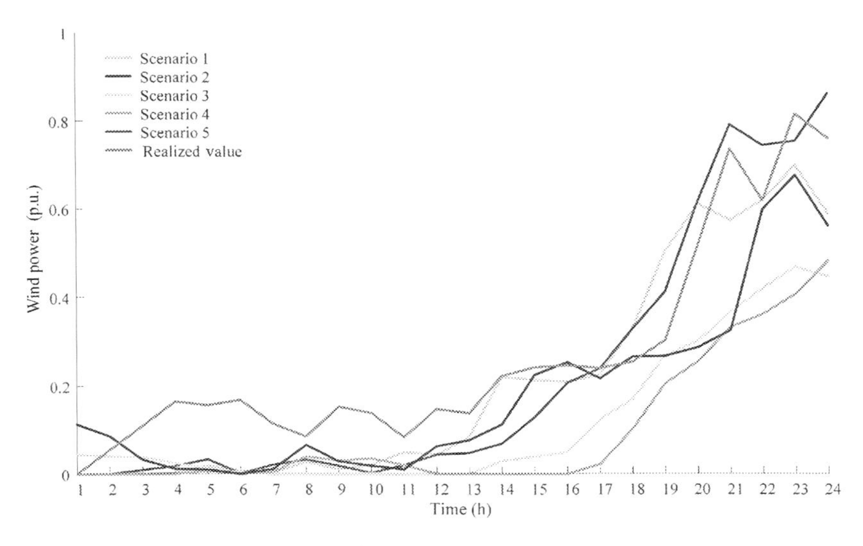

Fig. 17.2 Day-ahead prediction scenarios for wind power.

should take multiple randomness of load and power into consideration, in that case, real-time operation can be ensured to cope with emergencies, also will not cause excessive SR allocation. As a result, the goal of safe, reliable, economical, and environment-friendly power supply can be achieved.

17.1.2 Analysis of the influence of wind power properties on power system operation

17.1.2.1 Peak-valley difference increases, day-ahead UC is hard to set

Large-scale integration of wind power will increase the peak-valley difference of power system operation, which makes day-ahead UC is hard to set. That is so because with large-scale integration of wind power, the peak-valley difference of system net load (the difference between system load and wind power output) increases, the load regulation capacity of traditional thermal power unit is insufficient. Under this circumstance, it is difficult to set up an on-off dispatch plan to meet the demand of load regulation capacity of system.

As illustrated in Fig. 17.3, P1 and P2 represent two different UCs, $P1_{max}$, $P2_{max}$, $P1_{min}$, and $P2_{min}$ represent the sum of maximum unit output, the sum of minimum unit output of two plans, respectively. From Fig. 17.3, $P1_{min}$ can cover the demand of net load valley while $P1_{max}$ exceeds the demand of net load peak. However, $P2_{max}$ can meet the demand of net load peak while $P2_{min}$ is lower than the demand of net load valley. To conclude, when the peak-valley difference of net load reaches a certain level, it is hard to set up a UC to meet both the demands of load peak and valley.

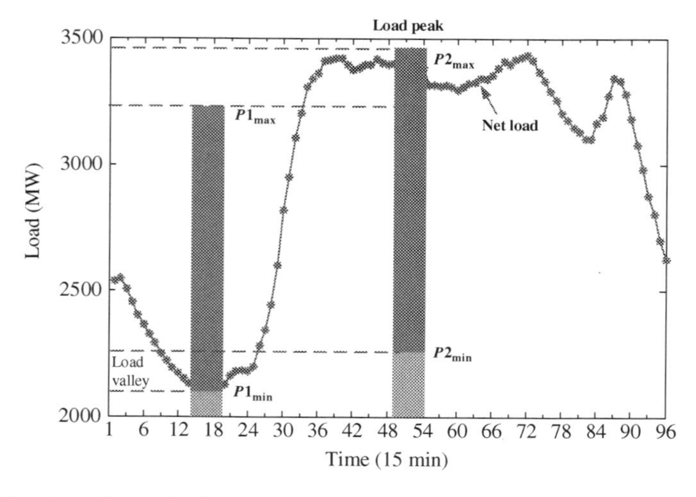

Fig. 17.3 Situation of oversized peak-valley difference.

After the large-scale wind power integration, the peak-valley difference increased. The load regulation capacity can hardly be met by regular thermal units. Therefore, the load regulation capacity of system with stochastic wind power integration should be estimated to set other flexible units in advance. Consequently, scientific and reasonable generation dispatch plan can be made.

17.1.2.2 High ramp events appear, the ramp of unit output is insufficient

The instability and fluctuation of wind power sometimes largely raise the fluctuation of net load at adjacent time periods. When the ramp ability of system units cannot deal with the fluctuation of net load at adjacent time periods, the system can only use the method of load curtailment or wind power curtailment to maintain the active power balance of the system.

Fig. 17.4 illustrates the situation of upward ramp of system units falling short. From the figure, an upward ramp event appears at time t. Limited by the ramp of unit output, the maximum output power of the unit is less than the increase of the net load from time t to time $t + 1$. The active power of the system falls short consequently, as a result, the system needs to maintain power balance by cutting off load or implementing flexible power supply (such as energy storage (ES)).

Fig. 17.5 illustrates the situation of downward ramp of system units falling short. From the figure, a large downward ramp event appears at time t. Limited by the ramp of unit output, the maximum output power of the unit is less than the decrease of the net load from time t to time $t + 1$. The active power of the system is surplus consequently, as a result, the system needs to maintain power balance by abandoning wind power or saving extra active power.

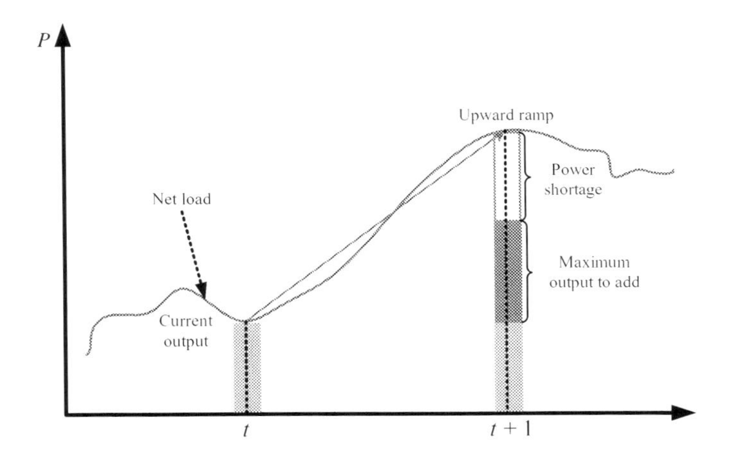

Fig. 17.4 Deficiency of upward ramp.

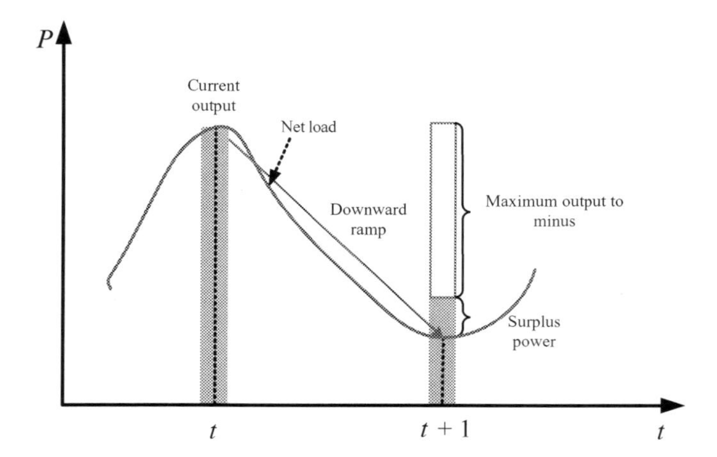

Fig. 17.5 Deficiency of downward ramp.

17.1.2.3 The prediction accuracy is not high, the prediction curve differs much from the realized curve, the pressure of real-time dispatch is huge

Due to the uncertainty of wind power, prediction values differ much from realized values, which leads to large deviation between the dispatch plan based on prediction values and actual operation. As a result, the regulation of system is of large quantity. Also, large-scale flow may transfer easily, the system security and stability are threatened, the difficulty of dispatch increase, worse, the regulation cannot be achieved. As illustrated in Fig. 17.6, realized net load curve deviates largely from predicted net load

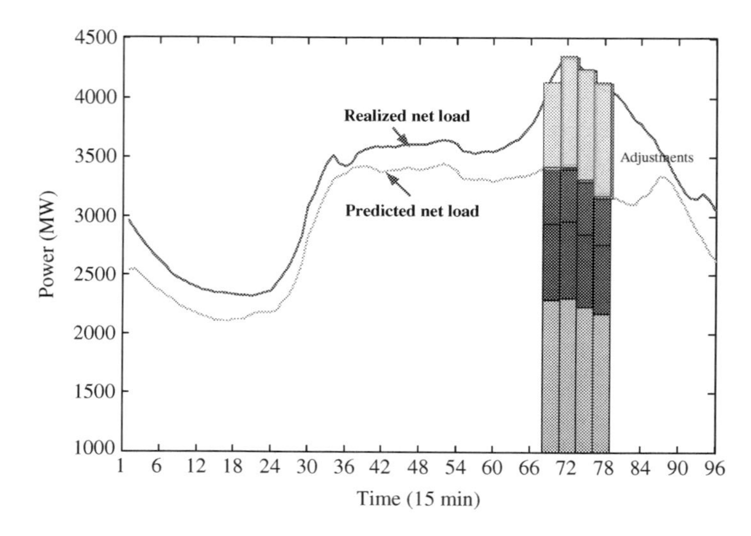

Fig. 17.6 Prediction with large deviation.

curve at time period 72, so the unit output should be increased. If it is not adjusted properly, line overload may be raised, which makes the power of some nodes cannot be balanced, causing power system failure.

To sum up, the fluctuation and uncertainty of wind power have much influence to power system operation. Deep analysis of the power system operation problem with wind power integration is required, as a result, the optimization operation model reflecting the characteristics of wind power is established. Researching on effective solving method, safe, stable, economical, and green operation of power system with strong stochastic power can be achieved.

17.2 Basic problems of power system optimization with large-scale wind power integration

The quantitative description of the stochastic characteristics of wind power, and the optimization operation theory and method of power system to deal with the randomness of wind power are two basic problems to realize the safe and stable operation of power system with large-scale wind power integration.

17.2.1 Traditional power system optimization operation problem

The optimization variables of traditional power system optimization operation include two parts: control variables and state variables. The control variables refer to the independent variables which can be controlled, including active power output of unit, reactive power output of each generator/synchronous compensator, the position of phase shifter, the position of adjustable transformer tap, the capacity of paralleling reactor/capacitor; in some emergency situations, rapid start-up of hydropower generating units, some load unload can also be used as a means of control. The state variables are the dependent variables of the control variables, usually including the voltage of each node, the power of each branch, and so on.

The constraints of traditional power system optimization problem include equality constraints and inequality constraints, which are as follows (Wang, 2003):

- Balance constraint of active and reactive power of each node.
- Upper and lower bound constraint of active output of each generator.
- Upper and lower bound constraint of reactive output of each generator/synchronous compensator.
- Capacity constraints of paralleling reactor/capacitor.
- Position constraints of phase shifter.
- Position constraints of adjustable transformer tap.
- Upper and lower bound constraint of the voltage of each node.
- Transmit power constraint of each branch.

The objective functions of traditional power system optimization probleminclude:

- Minimum system operation cost, which commonly refers to minimum cost of fuel for thermal power unit, including the cost of unit start-up and shut down.

- Minimum active power transmission loss. With minimum active power transmission loss as the objective function, improve the voltage quality while reducing the system active power loss.

For traditional power system, variables consist of thermal unit, nuclear unit, and hydroelectric unit such controllable deterministic variables. This belongs to deterministic optimization problem, whose mathematic model can be written as

$$\begin{aligned} &\min f(x_G) \\ &s.t. h(x_G)=0 \\ &\quad g(x_G)\leq 0 \\ &x_G: \text{common variable} \end{aligned} \tag{17.1}$$

For model (Eq. 17.1), it can be solved by nonlinear programming method, quadratic programming method, linear programming method, hybrid programming method, modern interior point method, and artificial intelligence method such mature optimization algorithms.

17.2.2 New power system optimization operation problem

For new power system, besides the traditional control variables and state variables, the optimization variables also include wind power, photovoltaic power such uncontrollable stochastic variables, which makes the optimization problem of power system convert from deterministic programming model into uncertain programming model.

The constraints of the new power system optimization model are based on the traditional model constraints, and the probabilistic function constraints and mean function constraints that characterize the stochastic characteristics of wind power are added. In addition, it may include constraints such as output characteristics, fluctuation characteristics, and probability distributions of related characteristics describing stochastic characteristics.

Besides the minimum cost of system operation, the minimum active power transmission loss and other traditional objective function and the objective function considering environmental factors, a new power system function will be more flexible and various. Due to the stochastic variables of wind power output, the objective function can also be minimum means of system operation cost, minimum active power transmission loss, etc.

To sum up, the mathematic model of new power system optimization problem can be written as

$$\begin{aligned} &\min E(f(x_G, x_W)) \\ &s.t. h(x_G, x_W)=0 \\ &\quad g_1(x_G, x_W)\leq 0 \\ &\quad \Pr\{g_2(x_G, x_W)\}\geq a \\ &\quad E\{g_2(x_G, x_W)\}\leq \beta \\ &\quad x_W \text{ or } T(x_W) \text{ obeys to a certain distribution} \\ &x_G: \text{ common variable} \\ &x_W: \text{ stochastic varibale} \\ &T(x_W): \text{ eigenfunction of stochasitc variable} \\ &\Pr: \text{ probability function} \end{aligned} \tag{17.2}$$

Table 17.2 Comparison of operation optimization problem between traditional and new power system

	Traditional power system optimization	**New power system optimization**
Basic problem	Deterministic programming	Uncertain programming
Stochastic variable	Weak randomness	Strong randomness
Constraint condition	Deterministic function	Deterministic function constraint/mean constraint/probabilistic constraint
Objective function	Deterministic function	Deterministic function/mean function/ probabilistic function
Solving method	Regular optimization calculating method	Stochastic programming calculation method

For model (Eq. 17.2), it can be solved by Monte Carlo method, scenarios-based method, probability/mean function with analytic transformation, and so on. However, the mathematical methods mentioned above require much about the performance of the problem to be solved. For instance, the methods require stochastic variables obey a common distribution, which leads to difficulties of applying to practical problems. In order to adapt practical problems, it requires the actual power system optimization problems or mathematical methods to be improved.

To sum up, comparing with traditional power system, the characteristics of new system operation optimization with large-scale wind power integration are listed in Table 17.2.

From Table 17.2, after integration of wind power, the mathematical nature of power system optimization problem has changed, and how to deal with the stochastic variables in the optimization model has become the key of solving the problem.

17.3 Research status of power system optimization with large-scale wind power integration

In recent years, the optimization operation of power systems with large-scale wind power integration has been widely studied. The following part is summarized from the model and method two aspects.

17.3.1 Research status of model

From the view of model, there are three typical models: traditional deterministic model, fuzzy model, and stochastic model. Besides, the research direction of a class of models is extended based on the traditional optimization model, adding more practical factors, such as environmental factors, market factors, and so on.

17.3.1.1 Traditional deterministic model

Instead of introducing wind power as a stochastic variables into the optimization model, traditional optimization modeling methods (Chen, 2008; Chen et al., 2006a; Lee, 2007) add SR capacity to deal with the uncertainty of wind power, and take a certain percentage of the total wind power as the additional spare demand reference value. This kind of modeling method deals with the stochastic characteristic of wind power to ensure safe and reliable operation of the system by increasing the reserve capacity to in a certain degree. However, lacking of analyzing and using uncertainty characteristics of the wind power, exorbitant reserve high additional methods is often attached to ensure safe and reliable operation of the system. Therefore, deterministic analysis model is too conservative, which will cause the waste of spare capacity, and indirectly increase the cost of power generation economic dispatch. Moreover, when the wind power integration capacity reaches a certain scale, this kind of methods which relies on the high proportion of the allocation or even the case of equal standby cannot be satisfied.

17.3.1.2 Fuzzy theory-based model

This kind of model is based on fuzzy theory to deal with unknown wind power. Generally speaking, this kind of method is defined as membership function to describe the system of wind power and the possibility of objective function belonging to a certain range, which select the minimum membership degree as the satisfaction index for decision makers to optimize by transforming the objective function into maximum satisfaction index.

Take each period of active power output as fuzzy variables, based on the active power output of the wind farm and the total consumption amount of the cost of membership function, the generation cost minimization problem was transformed into solving the maximum satisfaction index, and resulted in adapting the randomness of wind power better (Chen et al., 2006b). Stochastic fuzzy variables were used to describe the wind power, which considered the minimum cost of purchasing electricity and pollution emissions minimum as double objective optimization problem, and transform it into single-objective optimization problem of the decision maker with the greatest degree, the optimization results reflect the randomness of wind power to a certain extent (Ma et al., 2013). The usage of fuzzy variables was presented to describe the power of wind power, and applies fuzzy theory to the optimization operation problem with uncertain wind power (Miranda and Hang, 2005; Hong and Li, 2006; Wang et al., 2006).

17.3.1.3 Stochastic optimization model

In this kind of model, the wind speed/wind power is taken as the stochastic variable, and the optimization model is established, which contains the mean function or probability constraints of wind speed/wind power.

The uncertainty of wind farm output was considered, and the overestimation and underestimation of the expected cost caused by the wind farm output was added to the objective function (Hetzer et al., 2008; Zhao et al., 2010; Dong et al., 2012; Zhang et al., 2011; Villanueva et al., 2012). The influence of the uncertainty of wind power on the dispatch plan was also considered, and the chance constraint was also introduced into the model, which allowed the decision not satisfied the constraints to a certain degree (Sun et al., 2009; Liu, 2010; Tian et al., 2011). A wait-and-see (WS) model was established, which considered the randomness of wind power in the model, and evaluated the influence of stochastic wind speed on power generation (Liu, 2010). The UC model was built as a two-stage stochastic process with chance constraints (CCTS) (Wang et al., 2012c).

Some references considered the risk wind power brought to the system operation, and the risk constraints were added into the models. A risk reserve constrained dynamic economic dispatch model was presented, the additional wind power output and the actual output plan deviation were introduced as a penalty term into the objective function for optimization (Zhou et al., 2012). With the constraint of the limitation of operation risk rate, the risk rate can be limited to a certain degree, and the risk cost caused by the uncertain factors of wind power can be reduced while minimizing the operation cost (Ren et al., 2010b). The operation risk was taken as the reliability index, which was included in the constraints. And then the optimization dispatch model considering operation reliability with wind power integration was established to restrain the wind power impact on the power system reliability (Hu et al., 2013).

In some stochastic models, the stochastic factors such as load and unit failure rate were considered in addition to the stochastic factors of wind power.

With both the randomness of wind power and load prediction errors being considered, an energy shortage expectation was introduced into the objective function, and a UC model to optimize the model was established accordingly (Liu et al., 2012). Yuan et al. (2013) comprehensively considered the uncertain factors such as forced fault rate, load and wind power output prediction error to established a optimization model of power system generation and regulation with large-scale wind power integration. Consequently, the coordination between wind power and thermal power can be realized, and the optimization allocation scheme of thermal units meeting the requirements of reliability can be obtained.

17.3.1.4 Extended model considering multiple factors

With the development of wind power integration technology, more and more attention has been paid to the environmental problems. In recent years, the main environmental problems are caused by energy conversion process. Among all over the world, many countries have issued a series of environmental protection laws, the requirements for generation enterprises aim at the emission of SO_2 and NO_x. Therefore, many researchers add the environmental benefits of wind power in the optimization operation model, which guarantees the dispatch decisions meet the requirements of national and total emissions. For instance, the environmental costs of thermal power was considered in the objective function (Ren et al., 2010a). Pollution emissions and generation cost were taken at the same time as the objective function (Liu and Xu,

2010). The price penalty factor of emissions was proposed and embedded into the generator fuel cost function (Chen and Chen, 2013).

On the one hand, large-scale wind power integration replaces the traditional energy consuming unit, which makes the total cost of the system decreased; on the other hand, because of the fluctuation and intermittency of wind power, capacity credit decreases, which leads to the decline of the reliability of the system operation. Therefore, many scholars combine reliability research with the economic dispatch problem of wind power. In order to restrain the negative influence of wind power integration (Hu et al., 2013), the optimal operation model of wind power system considering the operation reliability was established by taking the operation risk index as the operation reliability index. A dynamic economic dispatch model with wind power considerting reliability constraints was proposed. This model contained reliability constraints besides regular operation constraints and took the cost of interruption load into the objective function, which made the model conform to the requirements of practical operation more (Jiang et al., 2013).

17.3.2 Research status of solving method

For traditional deterministic models, there are already mature optimization algorithms, which mainly include two categories: analytic algorithm and meta-heuristic algorithm. The solution of the uncertain programming model with wind power is still under study. At present, there are two kinds of solving methods, which are analytic method and simulation method.

17.3.2.1 Analytic algorithm

The analytic method assumes that the stochastic variable obeys some common distribution, and converts the mean function or probability constraint function into ordinary deterministic function by mathematical operation. This kind of conversion processing method process is relatively accurate, but need to assume that the wind speed or wind power distribution obeys a common probability. However, recent studies found that wind speed and wind power hardly obey certain distributions.

With the assumption of wind speed obeying Weibull distribution, the distribution function of wind power was obtained by using the relationship between wind speed and wind power (Qiu et al., 2011; Liu, 2011; Shi et al., 2012a). Based on this, the probability constraint or the mean function with stochastic variables was transformed into deterministic constraints to solve the power balance constraints and the SR constraints. With the assumption of wind speed obeying Weibull distribution, the analytic differential formula of optimization objective function with approximation error distribution was obtained to deal with the uncertain economic dispatch problem (Zhang et al., 2011). With the assumption of wind power obeying beta distribution, the probability constraint and mean function of the stochastic wind power in the model were equivalently transformed into deterministic constraints to solve the problem (Zhang et al., 2013). Based on the probabilistic constraints in the sequence computation theory, combined with genetic algorithm, the dynamic economic dispatch problem of the power system with wind farms was solved (Liu et al., 2013). With the assumption of load obeying normal distribution and wind speed prediction error obeying normal distribution, load-wind power prediction error distribution was used to solve the chance constraints of SR in the model (Xia et al., 2013). Gao et al. (2013) directly assumed that short-term wind power obeyed normal distribution to solve the problem.

17.3.2.2 Scenario simulation method

Scenario simulation method can be divided into two categories. One is based on Monte Carlo method/time series method to sample the stochastic variables according to the probability distribution, and simulate the data input and output of probability function or mean function, then take the simulation results as the sample to approximate mean function or probability constraint function. This method is with simple calculation, and can deal with stochastic variables obeying various distribution. However, this method requires multiple times of sampling and huge number of scenarios, which leads to slow calculation speed.

The other one is based on the historical samples or the samples generated by Monte Carlo method/time series method, then uses reduction technology to generate a small number of samples or extreme samples, and uses the scenario as the value of the stochastic variables in the model to realize the transformation from stochastic model to deterministic model. How to generate a small number of samples that can reflect the probability distribution of stochastic variables is the difficulty and key of this kind of solving method.

17.3.3 Existing problems

Based on the analysis of wind power system optimization mentioned above, in order to meet the needs of security, stability, economy, and green operation of the power system with large-scale wind power integration, continuous improvement of new power system optimization model and the progress of solving method are required, which play a positive role to alleviate the risk of wind power and the uncertainty brought to power system operation. However, there are still some fundamental problems worthy of study.

17.3.3.1 Lack of scientific consideration for the randomness of wind power

From the current research, most references are only based on the assumption that the wind speed/wind power or its prediction error obeys normal distribution, Weibull distribution or beta distribution. But the studies of the actual system show that, wind power hardly obeys to common distribution, so it is necessary to study more effective and scientific optimization method to deal with the randomness of wind power and get rid of the dependence of assumption in solving process.

17.3.3.2 Current researches concentrate on some sub problems, such as power system economic dispatch, which leads to the lack of system and framework research, and the progresses of various researches are uneven

At present, most studies focus on the economic dispatch problem, while the optimization of the system capacity, the capacity of ancillary services and the ES capacity are relatively falling behind. It is of great significance to establish and develop a scientific research system and framework, and then fully solve large-scale wind power integration problem.

17.4 *p*-Efficient point theory

To deal with the characteristic of anti-peak regulation, unpredictable and strong randomness of large-scale wind power, the ability of rapid positive/negative power regulation is required for the system. The rapid response of ES and the ability of positive/negative bidirectional regulation make it have unique advantages in peak load regulation, frequency regulation, reducing reserve capacity, and improving unit operation efficiency. However, limited by geographical location, technical and economic constraints, the use of ES capacity is constrained. Therefore, how to set up the power/capacity of ES has become the focused problem in the field of economy and reliability.

At present, relevant references have studied the optimization allocation of ES capacity in wind power system (Zhang et al., 2012; Xu et al., 2012; Chen et al., 2012). Most of the existing ES capacity optimization model took economy as the objective function, and took the operation requirement (including the charge/discharge process and the stage of charge of the ES system) of the system (microgrid (Xu et al., 2012; Chen et al., 2012; Wang et al., 2012a,b)/isolated system (Shi et al., 2012b)/large-scale power system) as the constraints to establish mathematical model, to solve the power/capacity allocation problem of a single ES device. This chapter aims to solve the problem of minimum ES power and SR which are required to deal with the error of wind power and load prediction for power system with large-scale wind power integration, establish minimum optimization allocation ES power with chance constraints. The results can provide theoretical basis for deciding the ES device capacity and the index energy should meet or reach, also can guide the storage capacity allocation to make it meet the requirements of system operation.

Transforming chance constraints into deterministic constraints is the difficulty and significance of solving uncertain programming problems with chance constraints. Usually, there are two methods for power system to deal with chance constraints. First, used simulation-approximation chance-constrained function (Qiao et al., 2008; Wang et al., 2006, 2011), which uses Monte Carlo to simulate chance-constrained function, to generate input and output data, then used intelligent algorithm (neural network, genetic algorithm) to fit the data generated to deterministic function. This method is easy to achieve, which leads to wide use. However, the existences of computational complexity and fitting inaccuracy result in the conversion error between the original chance-constrained function and the deterministic function; Second, analytic method (Wang, 1994) which transformed chance constraints into deterministic equivalent reservation for calculation. In early stage, using analytic method to solve probabilistic reserve constraints is based on the situation that the load prediction errors obey the normal distribution. However, the prediction errors may not obey the normal distribution when wind power is integrated (Liu et al., 2012). Also, for the chance constraints with more than one stochastic variables (wind power prediction error and load prediction error), analytic method is no longer suitable because of the complexity and difficulty.

In the view of that, this chapter solves chance constraints based on stochastic programming, p-efficient point theory. Precise conversion of chance constraints with double stochastic variables (wind power prediction error and load prediction error) which do not obey common probability distributions (such as normal distribution, t-distribution) is also deduced in this chapter. First, transform two stochastic variables chance constraints into single variable chance constraints; by wind power and load forecast errors of the historical sample, estimate the probability distribution function of stochastic variable with statistical characteristics, so to realize the precise conversion from chance constraints to deterministic constraints. It provides new ideas and approaches for the application of analytic method to power system optimization problems with chance constraints.

17.4.1 Minimum ES power allocation problem in power system

In this section, two optimization models for ES are described. Model-I is a planning model which is used to minimize ES for power system with large-scale wind power integrated. Model-II is used to verify the effectiveness of the results obtained with the Model-I by different scenarios.

For day-ahead dispatch of power system with high wind power penetration, the traditional deterministic UC of conventional generators is first processed by a priority list method (Senjyu et al., 2003) based on forecasted wind power and load demand of the next day. This can fully utilize the conventional power resources with relative low cost. In this stage, the forecast errors are not considered. As it is well known that precise prediction of the wind power is difficult to realize, the scheduled units including the SR may not be able to balance the fluctuation in actual wind power and load demand. ES is then scheduled to enhance the SR to cope with such unbalance. Model-I is used for this process.

17.4.1.1 Model-I: For minimum ES dispatch

The ES power in the dispatch period is represented as a variable to be minimized. The SRs are considered as the chance constraints. Both the wind power and the load demand forecast errors are considered as stochastic variables. It is assumed here that the total load of the system is always greater than the wind power generation (Pappala et al., 2009). To minimize the ES, the following model can be identified, where the ES capacity limitation is not considered, and power system network topology and the node power balance are not considered as usual (Senjyu et al., 2003; Pappala et al., 2009; Hadji and Vahidai, 2012; Chen, 2008).

(1) Objective function

$$\min\left[C_1^{ES},\ C_2^{ES},\ldots,C_t^{ES},\ldots,C_T^{ES}\right] \tag{17.1}$$

(2) Constraints

$$\Pr\left\{P_t^{L,A}-P_t^{W,A}\leq C_t^{ES}+\sum_{i=1}^{N_G}\overline{P}_i^G\right\}\geq a^{rs} \tag{17.2}$$

$$\Pr\left\{P_t^{L,A}-P_t^{W,A}\geq \sum_{i=1}^{N_G}\underline{P}_i^G-C_t^{ES}\right\}\geq a^{rs} \tag{17.3}$$

$$P_t^{L,A}-P_t^{W,A}\geq 0 \tag{17.4}$$

$$C_t^{ES}+\sum_{i=1}^{N_G}\overline{P}_i^G\geq P_t^{L,F}-P_t^{W,F} \tag{17.5}$$

$$\sum_{i=1}^{N_G}\underline{P}_i^G-C_t^{ES}\leq P_t^{L,F}-P_t^{W,F} \tag{17.6}$$

$$P_t^{L,A}=P_t^{L,F}+\Delta P_t^L \tag{17.7}$$

$$P_t^{W,A}=P_t^{W,F}+\Delta P_t^W \tag{17.8}$$

Eq. (17.2) describes the constraint of the upper SR constraint. It corresponds to the high load level operating condition. In this condition, ES operates at the discharging mode. This equation means that the sum of all the upper limits of scheduled units plus the discharged power from ES should be greater than the actual net load ($P_t^{L,A}-P_t^{W,A}$) at a confidence level of a^{rs}.

Eq. (17.3) describes the constraint of the lower SR constraint. It corresponds to the low load level operating condition. In this condition, ES operates at the charging mode. This equation means that the sum of all the lower limits of scheduled units minus the charged power into ES should be smaller than the actual net load ($P_t^{L,A}-P_t^{W,A}$) at a confidence level of a^{rs}.

Eq. (17.4) assures that system load always exceeds the wind power.

Eq. (17.5) states that the sum of all the upper limits of scheduled units plus the discharged power from ES must be greater than the forecast net load ($P_t^{L,F}-P_t^{W,F}$) to ensure power balance.

Eq. (17.6) states that the sum of all the lower limits of scheduled units minus the charged power into ES should be smaller than the forecast net load ($P_t^{L,F} - P_t^{W,F}$) to ensure power balance.

The above model is a chance-constrained programming model. In this model, $P_t^{L,A}$ (ΔP_t^L) and $P_t^{W,A}$ (ΔP_t^W) are introduced as the stochastic variables. In order to solve this model, the p-efficient point theory is used to change the chance-constraint model to a deterministic constraint model.

17.4.1.2 Model-II: For ES dispatch effectiveness verifying

This model is used to evaluate the effectiveness of the scheduled ES obtained by Model-I. The conventional UC of the power system is the same as that used in Model-I. Evaluation is carried out for different system operation scenarios, including some extreme scenarios. The effectiveness verifying is performed by comparing the actually required ES power C_t ($t = 1, 2, \ldots, T$) obtained from solving Model-II with the scheduled C_t^{ES} ($t = 1, 2, \ldots, T$) obtained from solving Model-I.

(1) Objective function

$$\min \sum_{t=1}^{T} C_t \tag{17.9}$$

(2) Constraints

$$u_t C_t + \sum_{i=1}^{N_G} P_{i,t}^G + \left(P_t^{W,A}\right)_s = \left(P_t^{L,A}\right)_s \tag{17.10}$$

$$\underline{P}_i^G \le P_{i,t}^G \le \overline{P}_i^G \tag{17.11}$$

$$P_{i,t-1}^G - P_{i,t}^G \le T_{\Delta t} \Delta \underline{P}_i^G \tag{17.12}$$

$$P_{i,t}^G - P_{i,t-1}^G \le T_{\Delta t} \Delta \overline{P}_i^G \tag{17.13}$$

Eq. (17.10) assures active power balance of the system, it is carried out in every 15 min.

Eq. (17.11) shows that the minimum and maximum power output of every generation unit should not be violated.

Eqs. (17.12), (17.13) show that the power output ramps up/down rates of every generation unit should not exceed their limits.

17.4.2 p-Efficient point theory of solving chance constraints

p-Efficient point theory is originally proposed to solve the separable chance-constraints problems. The chance-constrained optimization model can be written in the following generic form (Birge and Louveaux, 1997):

$$\begin{aligned}&\underset{x}{\text{Min}}\ c(\boldsymbol{x})\\&s.t:\ \Pr\{\boldsymbol{g}(x)\geq\boldsymbol{Z}\}\geq p\\&\boldsymbol{x}\in\boldsymbol{X}\end{aligned} \tag{17.14}$$

Converting the chance-constrained model Eq. (17.14) to a deterministic model includes the following steps.

Denote the cumulative distribution function (CDF) of stochastic variable vector $\mathbf{Z}$ as $\boldsymbol{F}_{\mathrm{Z}}(\mathbf{z})=\Pr\{\mathbf{Z}\leq\mathbf{z}\}$. Then, Eq. (17.14) is equal to

$$\begin{aligned}&\underset{x}{\text{Min}}\ c(\boldsymbol{x})\\&F_z(\mathbf{g}(\mathbf{x}))\geq p\\&\boldsymbol{x}\in\boldsymbol{X}\end{aligned} \tag{17.15}$$

Define the p-level set of variable $\boldsymbol{Z}$ by

$$\boldsymbol{Z}_p=\{\mathbf{z}\in R^m:\boldsymbol{F}_Z(\mathbf{z})\geq p\} \tag{17.16}$$

Eq. (17.14) is then compactly rewritten as follows:

$$\begin{aligned}&\underset{x}{\text{Min}}\, c(x)\\&s.t.\boldsymbol{g}(\boldsymbol{x})\in\boldsymbol{Z}_p\\&\boldsymbol{x}\in\boldsymbol{X}\end{aligned} \tag{17.17}$$

Here, we introduce the p-efficient point. For a given $p\in(0,1)$, a point $v\in R^m$ is called a p-efficient point of the probability distribution function F if $F(v)\geq p$ and there is no $z\leq v,\ z\neq v$ such that $F(z)\geq p$.

This definition suggests that a p-efficient point is the minimum point of the level set Z_p. Using the p-efficient point, Eq. (17.17) can be changed to the following form:

$$\begin{aligned}&\underset{x}{\text{Min}}\, c(x)\\&s.t.\boldsymbol{g}(\boldsymbol{x})\geq\mathbf{v}\\&\boldsymbol{x}\in\boldsymbol{X}\end{aligned} \tag{17.18}$$

Specially, if $\mathbf{Z}$ is a scalar stochastic variable, then, there is only one p-efficient point which is equal to $F_Z^{(-1)}(p)$ for every $p\in(0,1)$. Eq. (17.18) is further simplified to

$$\begin{aligned}&\underset{x}{\text{Min}}\, c(\boldsymbol{x})\\&s.t.g(\boldsymbol{x})\geq v\\&\boldsymbol{x}\in\boldsymbol{X}\end{aligned} \tag{17.19}$$

Eq. (17.19) is a deterministic constrained model equivalent to the chance-constrained Eq. (17.14). It can be easily solved by some conventional mathematic method, such as linear programming.

17.4.3 Conversion of the chance constraints to deterministic constraints

Substituting the explicitly representation of the two stochastic variables, ΔP_t^L and ΔP_t^W into the chance constraints (Eqs. 17.2 and 17.3), the following equation can be obtained:

$$\Pr\left\{\Delta P_t^L - \Delta P_t^W \le C_t^{ES} + \sum_{i=1}^{N_G} \overline{P}_i^G - P_t^{L,F} + P_t^{W,F}\right\} \ge a^{rs} \tag{17.20}$$

$$\Pr\left\{\Delta P_t^L - \Delta P_t^W \ge \sum_{i=1}^{N_G} \underline{P}_i^G - C_t^{ES} - P_t^{L,F} + P_t^{W,F}\right\} \ge a^{rs} \tag{17.21}$$

To simplify solving of Model-I, forecast error of the net load is introduced to reduce the number of stochastic variables. Define the forecast error of the net load as

$$\left(\Delta P_t^{L-W}\right)^* = \left|\frac{\Delta P_t^L - \Delta P_t^W}{P_t^{L,F} - P_t^{W,F}}\right| \tag{17.22}$$

It can be seen from this equation that $(\Delta P_t^{L-W})^* \ge 0$ is always true. Substituting Eq. (17.22) into Eqs. (17.20), (17.21), the constraints can be converted to the following form:

$$\Pr\left\{\left(\Delta P_t^{L-W}\right)^* \le \left[\begin{array}{l}\left(C_t^{ES} + \sum_{i=1}^{N_G} \overline{P}_i^G - P_t^{L,F} + P_t^{W,F}\right) \\ /\left(P_t^{L,F} - P_t^{W,F}\right)\end{array}\right]\right\} \ge a^{rs} \tag{17.23}$$

$$\Pr\left\{\left(\Delta P_t^{L-W}\right)^* \le \left[\begin{array}{l}\left(C_t^{ES} - \sum_{i=1}^{N_G} \underline{P}_i^G + P_t^{L,F} - P_t^{W,F}\right) \\ /\left(P_t^{L,F} - P_t^{W,F}\right)\end{array}\right]\right\} \ge a^{rs} \tag{17.24}$$

Denote $F_{(\Delta P_t^{L-W})^*}(z) = \Pr\{(\Delta P_t^{L-W})^* \le z\}$ as the CDF of $(\Delta P_t^{L-W})^*$ and $F^{-1}_{(\Delta P_t^{L-W})^*}(z)$ as the inverse function of $F_{(\Delta P_t^{L-W})^*}(z)$. Then, Eqs. (17.23), (17.24) can be converted to the following form:

$$F_{(\Delta P_t^{L-W})^*}\left[\left(C_t^{ES} + \sum_{i=1}^{N_G} \overline{P}_i^G - P_t^{L,F} + P_t^{W,F}\right) / \left(P_t^{L,F} - P_t^{W,F}\right)\right] \ge a^{rs} \tag{17.25}$$

$$F_{(\Delta P_t^{L-W})^*}\left[\left(C_t^{ES} - \sum_{i=1}^{N_G} \underline{P}_i^G + P_t^{L,F} - P_t^{W,F}\right) / \left(P_t^{L,F} - P_t^{W,F}\right)\right] \ge a^{rs} \tag{17.26}$$

Let $Z^{U,a}$ be the p-efficient point of $F_{(\Delta P_t^{L-W})^*}$, the deterministic form of the chance constraints are given by Eqs. (17.27), (17.28)

$$C_t^{ES} + \sum_{i=1}^{N_G} \overline{P}_i^G - P_t^{L,F} + P_t^{W,F} \geq \left(P_t^{L,F} - P_t^{W,F}\right) Z^{U,a} \tag{17.27}$$

$$C_t^{ES} - \sum_{i=1}^{N_G} \underline{P}_i^G + P_t^{L,F} - P_t^{W,F} \geq \left(P_t^{L,F} - P_t^{W,F}\right) Z^{U,a} \tag{17.28}$$

To obtain the SR coefficient explicitly, Eqs. (17.27), (17.28) are rewritten as

$$C_t^{ES} + \sum_{i=1}^{N_G} \overline{P}_i^G \geq \left(P_t^{L,F} - P_t^{W,F}\right)\left(1 + Z^{U,a}\right) \tag{17.29}$$

$$\sum_{i=1}^{N_G} \underline{P}_i^G - C_t^{ES} \leq \left(P_t^{L,F} - P_t^{W,F}\right)\left(1 - Z^{U,a}\right) \tag{17.30}$$

Thus, $Z^{U,\ a}$ and $-Z^{U,\ a}$ are the required SR coefficients. Correspondingly, $(P_t^{L,\ F} - P_t^{W,\ F})Z^{U,\ a}$ and $-(P_t^{L,\ F} - P_t^{W,\ F})Z^{U,\ a}$ are the required SR for the system. The required SR and the ES power are determined by the p-efficient point $Z^{U,a}$.

17.4.4 Calculation of the p-efficient point $Z^{U,a}$

In order to obtain the p-efficient point $Z^{U,a}$, it is necessary to obtain the probability distribution function of the net load forecast error $(\Delta P_t^{L-W})^*$.

Although some papers assume that the net load forecast errors are stochastic variables with the normal probability distribution function, such kind assumption is not always true in the practice. This chapter uses the Kernel estimation method to obtain the probability distribution function of the stochastic variable directly from the history data.

Let $x_1, x_2, \ldots, x_i, \ldots, x_N$ be the historical net load samples. The probability density function (PDF) $f(x)$ and the CDF $F(x)$ at x can be calculated as follows (Wei and Zhang, 2008):

$$f(x) = \frac{1}{Nh} \sum_{i=1}^{N} K\left(\frac{x - x_i}{h}\right) \tag{17.31}$$

$$F(x) = \frac{1}{N} \sum_{i=1}^{N} K\left(\frac{x - x_i}{h}\right) \tag{17.32}$$

$$K\left(\frac{x - x_i}{h}\right) = \frac{1}{\sqrt{2\pi}} \exp\left(-\frac{1}{2} \times \left(\frac{x - x_i}{h}\right)^2\right) \tag{17.33}$$

$$h = 1.06SN^{-0.2} \tag{17.34}$$

The total wind power of EirGrid from January to April in 2013 with the time period of 15 min are used to test the efficiency of the method. The data are from the website http://www.eirgrid.com/operations/systemperformancedata/windgeneration/. By using the Kernel estimation method and the normal distribution estimation method, respectively, the corresponding PDFs are obtained and given in Fig. 17.7 together with the histogram of the original dada, where the parameter of the normal distribution is estimated using the maximum likelihood estimation method (Wei and Zhang, 2008). It can be seen from Fig. 17.7 that more precise PDF can be obtained by the Kernel estimation. The effectiveness of the Kernel estimation method is also investigated by comparing the following parameter values of the Kernel PDF and the history data, including the mean, the variance, the skewness, the kurtosis (Hodge and Milligan, 2011), and the Rényi entropy (Hodge et al., 2012) as that given in Table 17.1. It can be seen that the original data distribution characteristics can be much better depicted by the Kernel estimation method (Table 17.3).

Based on the distribution function $F_{(\Delta P_t^L - W)^*}$ obtained by the Kernel estimation, it is easy to obtain the p-efficient point $Z^{U,a}$ by solving the following equation:

$$Z^{U,a} = F^{-1}_{(\Delta P_t^L - W)^*}(z) \tag{17.35}$$

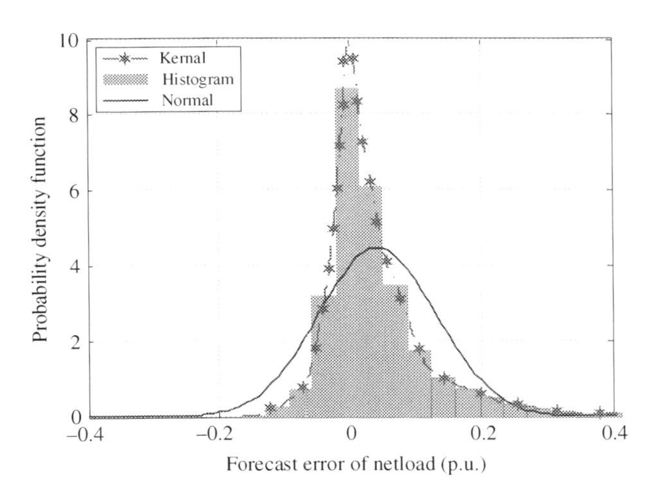

Fig. 17.7 Comparison of probability density functions.

Table 17.3 The deviation between Kernel PDF and history data

	Mean	Variance	Skewness	Kurtosis	*R*-entropy
Kernel PDF	0.0438	0.0534	17.8704	15.3821	14.4400
History data	0.0444	0.0548	3.0131	16.7462	14.4361

17.4.5 Case study of p-efficient point theory

17.4.5.1 Description of the case

Employ wind farm-thermal unit-storage system for simulation, the data of wind farm are taken from the historical data, which include both predicted values and realized values of 8640 points at intervals of 15 min, of EirGrid from January to March 2017. Similarly, the data of load are taken from historical data after system being reconstructed at intervals of 15 min, including both predicted values and realized values. In this chapter, the SR and the ES of the system at April 1, 2012 are taken to demonstrate, the data of load and wind power at the very day are shown in Fig. 17.8.

Fig. 17.8 also shows the maximum and minimum curve of thermal units output at April 1, 2017. It can be figured out that current thermal units can balance the net load of prediction.

17.4.5.2 p-Efficient point $Z^{U,a}$

At the confidence level of 0.95, it can be calculated by inverse operation of cumulative probability distribution function that $Z^{U,a} = 0.106$ (which means the relative error is $\pm 10.6\%$ at the confidence level of 95%).

17.4.5.3 Allocation plan for SR and ES

Calculate the positive/negative SR the system needs and the coefficient of positive/negative SR (1.106, 0.894). The SR and the ES power the system needed at April 1, 2012 are shown in Table 17.4.

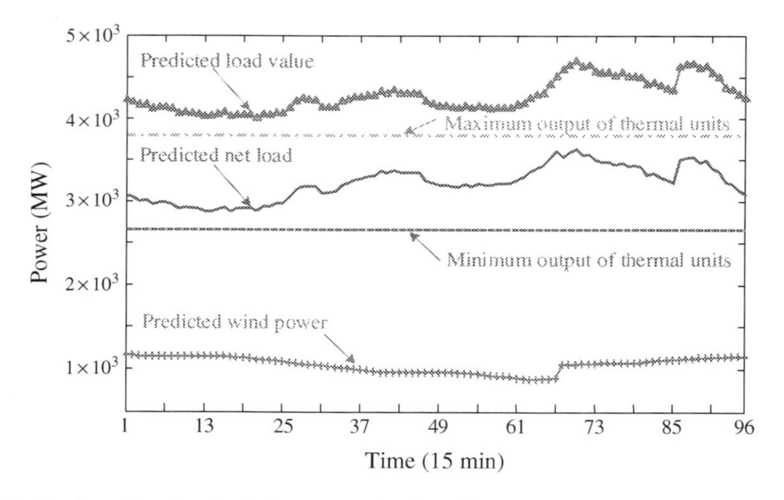

Fig. 17.8 Predicted load and wind power at April 1, 2012.

Table 17.4 Allocation plan for SR and ES

Time	SR	ES	Time	SR	ES	Time	SR	ES
1	326	0	33	331	0	65	364	0
2	323	0	34	339	0	66	374	102
3	319	0	35	342	0	67	383	198
4	320	0	36	346	0	68	375	115
5	314	9	37	347	0	69	380	165
6	317	0	38	350	0	70	385	217
7	317	0	39	351	0	71	377	138
8	315	5	40	357	0	72	379	151
9	309	53	41	354	0	73	371	71
10	311	36	42	359	0	74	369	47
11	309	51	43	355	0	75	370	56
12	308	61	44	355	0	76	365	5
13	305	86	45	355	0	77	369	46
14	306	80	46	355	0	78	363	0
15	308	66	47	345	0	79	364	0
16	311	38	48	342	0	80	362	0
17	306	80	49	339	0	81	352	0
18	309	57	50	339	0	82	354	0
19	309	50	51	338	0	83	349	0
20	310	45	52	337	0	84	346	0
21	307	70	53	340	0	85	342	0
22	312	26	54	338	0	86	371	75
23	312	29	55	341	0	87	374	101
24	316	0	56	339	0	88	375	112
25	316	0	57	339	0	89	368	42
26	324	0	58	341	0	90	370	65
27	334	0	59	341	0	91	361	0
28	338	0	60	342	0	92	356	0
29	338	0	61	346	0	93	343	0
30	338	0	62	350	0	94	340	0
31	329	0	63	356	0	95	333	0
32	330	0	64	362	0	96	328	0

Fig. 17.9 shows the comparison curve of the SR of the system and planned SR that the ES undertakes. It can be seen from the figure, in the nonpeak/valley period of load, thermal units can meet the requirements of system reserve without the allocation of storage; in peak period of load (such as period 66–72), most of the positive SR are undertaken by ES, the thermal units bear low adjustment ability due to undertaking most of the load; in low period of load (such as period 8–23), ES is also needed to undertake the negative SR.

17.4.5.4 System running test

Using Model-II, calculate two possible extreme scenarios (scenarios 1 and 2) and realized net load (scenario 3) to verify whether the configured ES can cope with the prediction error of net load.

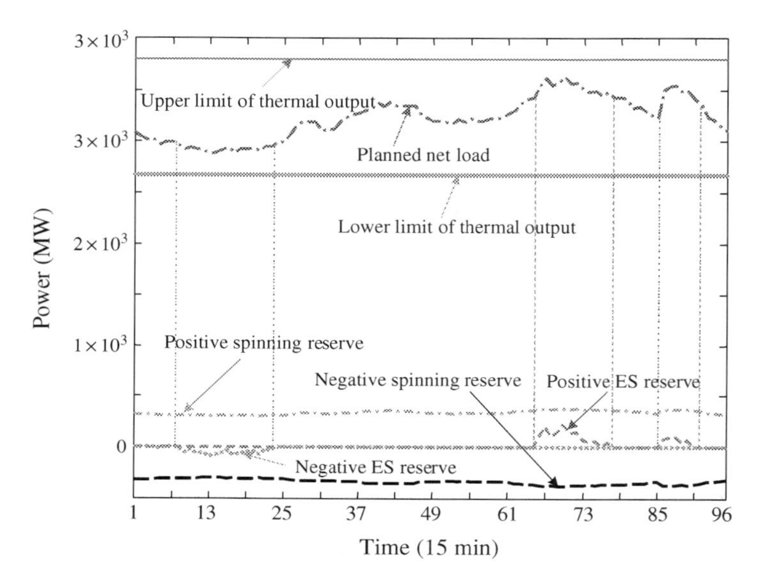

Fig. 17.9 Comparison of spinning reserve of system and ES allocation.

Scenario 1(S1): assuming that the load prediction is at the maximum positive deviation (103%), the wind power prediction is at the maximum negative deviation (80%).
Scenario 2(S2): assuming that the load prediction is at the maximum negative deviation (97%), the wind power prediction is in the maximum positive deviation (120%).
Scenario 3(S3): realized wind power and load.

The results of three scenarios are shown in Fig. 17.10, it can be seen that regular generator units (the top line represents the upper limit of generator units and the low line represents the lower limit of generator units) cannot balance the power of scenarios 1 and 2. When realized net load exceeds the upper or lower limit of the thermal power units, ES is needed to keep the balance of power system.

Introduce the net load of each period into $P_t^{W,A}$, $P_t^{L,A}$ in Model-II to verify, the calculated output curve of ES is shown in Fig. 17.11. It can be seen that in scenario 1, the system only needs positive SR which is close to and less than the allocation storage capacity; in scenario 2, the system requires only negative SR whose absolute value is close to and smaller than the configured storage capacity; from Fig. 17.10, realized net load basically operate in the range of thermal units adjustment, small amount of ES are only needed in the rush hour. It can be seen that, the configured ES can meet the need of both the actual operation of the system and the requirements of the preset extreme scenarios.

17.5 Moment matching theory

It is a common method to deal with the randomness of wind power generation by generating representative scenarios that can accurately reflect the stochastic characteristics of wind power. In practical operation of the system, analyzing the stochastic

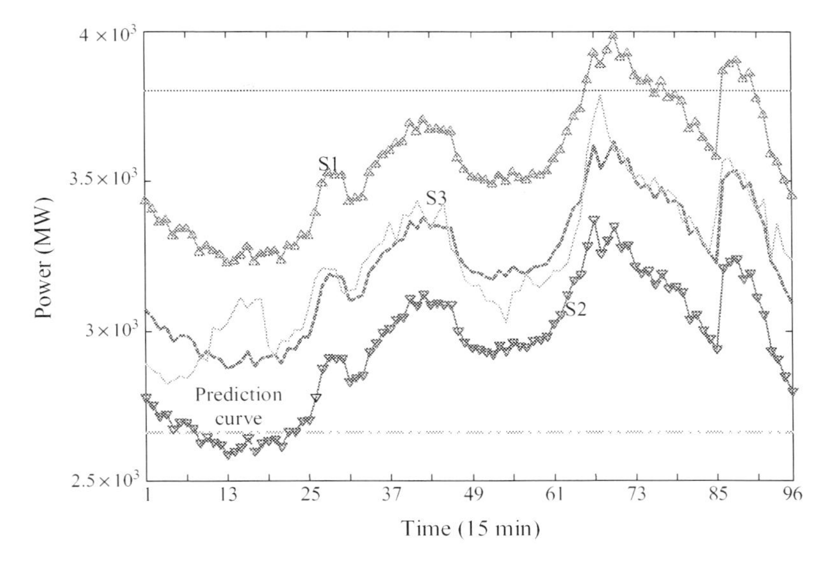

Fig. 17.10 Scenarios calculation.

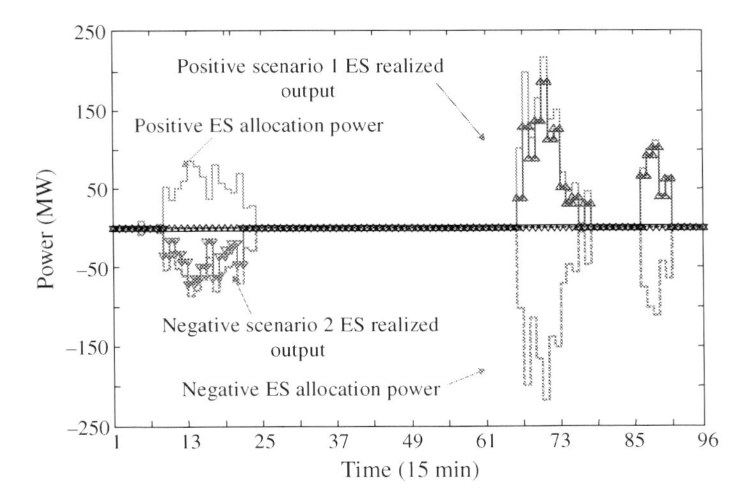

Fig. 17.11 Realized situation of ES.

characteristics of wind power, constructing typical wind power generation scenarios, making measures to deal with stochastic change of wind power for typical scenarios are other ways to deal with optimal operation of power system with wind power. The mathematical model established in this way is called the WS model. The key to solve the WS model is to obtain the discrete probability distribution which is similar to the probability distribution of the original stochastic variables, that is to say, to generate the scenarios.

This scenario analysis method can be used in power system operation. This section takes dispatch of power system with wind power as an example to introduce the application of this method in power system.

17.5.1 Traditional moment matching method

The basic idea of moment matching method to generate scenarios is to make the statistical properties of the scenario close to the statistical properties of stochastic variables, which is a method of matching statistical characteristics. If the statistical property is the moment, then it is called the moment matching. The whole framework is based on a mathematical optimization model, which was first proposed by Hoyland and Wallace (Høyland and Wallace, 2001). Moment matching method is easy to apply and can be applied to many uncertain decision problems, such as financial portfolio, power system dispatch. By adding statistical properties such as moments and covariance to the optimization objective of the scenario model, these statistical properties can be matched under the constraints of the problem. In general, the first four central moments are used as the statistical properties to be matched to meet the requirements of the problem. Next, we introduce an optimal model for generating single-period scenarios using moment matching method (Gülpınar et al., 2004; Wei, 2008). The formula for the central moment of each order can be referred in Ross (2007).

(1) Objective function

$$\min \sum_{i=1}^{w}\sum_{k=1}^{4}\omega_{ik}(m_{ik}-M_{ik})^2+\sum_{i,l\in I,i<l}\omega_{il}(c_{il}-C_{il})^2 \tag{17.36}$$

w is the number of stochastic variables; i is a set of stochastic variables; ω_{ik} is the weight of the stochastic variables i at order central moment k; ω_{ik} is the weight of central moment of order k of the stochastic variable i; m_{ik} is the k order central moment of stochastic scenario i; M_{ik} is the k order central moment of stochastic scenario i; ω_{il} is the covariance weight between stochastic variable i and stochastic variable l; and c_{il} is the covariance between stochastic variable i and stochastic scenario l.

The physical meaning of (Eq. 17.36): four order center moment of stochastic variables w and the distance of top four center order moment of original scenario generation should be minimized, the statistical characteristics of such scenario as close as possible to the original statistical characteristics.

(2) Constraint condition

$$\sum_{j=1}^{N}p_j=1 \tag{17.37}$$

$$m_{i1}=E(X_i)=\sum_{j=1}^{N}x_{ij}p_j,\quad i\in I \tag{17.38}$$

$$m_{i2} = \sigma_i = \sqrt{E\left(X_i^2\right) - E(X_i)^2} = \sqrt{\sum_{j=1}^{N} x_{ij}^2 p_j - m_{i1}^2} \tag{17.39}$$

$$m_{i3} = \frac{E\left[\left(X_i - E(X_i)\right)^3\right]}{\sigma_i^3} = \frac{\sum_{j=1}^{N} \left(x_{ij} - m_{i1}\right)^3 p_j}{m_{i2}^3} \tag{17.40}$$

$$m_{i4} = \frac{E\left[\left(X_i - E(X_i)\right)^4\right]}{\sigma_i^4} = \frac{\sum_{j=1}^{N} \left(x_{ij} - m_{i1}\right)^4 p_j}{m_{i2}^4} \tag{17.41}$$

$$c_{il} = E((X_i - m_{i1})(X_l - m_{l1})) = \sum_{j=1}^{N} \left(x_{ij} - m_{i1}\right)\left(x_{lj} - m_{l1}\right)p_j, \quad i,l \in I \text{ and } i < l \tag{17.42}$$

$$p_j \geq 0, \quad j = 1, \ldots, N \tag{17.43}$$

N is the number of scenarios; E is the mean in statistics; X_i is the scenario set of stochastic variables i; x_{ij} is scenario j of stochastic variables i; and p_j is the probability of scenario j.

The physical meaning of Eq. (17.37) is that the sum of the probability of each stochastic variable equals to 1.

The physical meanings of Eqs. (17.38–17.41) are mean, variance, skewness, and kurtosis, respectively.

The physical meaning of Eq. (17.42) is the covariance of the stochastic scenarios, and the correlation among the stochastic variables.

The physical implication of Eq. (17.43) is that the probability of each scenario of a stochastic variable must be nonnegative.

Among them, the estimation formula of the matching target moment is as follows:

$$M_{i1} = \frac{\sum_{h=1}^{n} y_h}{n} \tag{17.44}$$

$$M_{i2} = \sqrt{\frac{\sum_{h=1}^{n} \left(y_h - M_{i1}\right)^2}{n}}, \quad s = \sqrt{\frac{\sum_{h=1}^{n} \left(y_h - M_{i1}\right)^2}{n-1}} \tag{17.45}$$

$$M_{i3} = \frac{n}{(n-1)(n-2)} \sum_{h=1}^{n} \left(\frac{y_h - M_{i1}}{s}\right)^3 \tag{17.46}$$

$$M_{i4} = \left(\frac{n(n+1)}{(n-1)(n-2)(n-3)} \sum_{h=1}^{n} \left(\frac{y_h - M_{i1}}{s} \right)^4 \right) - 3\frac{(n-1)}{(n-2)(n-3)} + 3, \quad \forall i \in I \tag{17.47}$$

$$C_{il} = \frac{1}{n-1} \sum_{h=1}^{n} (y_{ih} - M_{i1})(y_{lh} - M_{l1}), \quad \forall i \neq l \in I \tag{17.48}$$

where n is the observational sample number of stochastic variables; y_h is the observation value of sample h of stochastic variables; and M_{i1}, M_{i2}, M_{i3}, and M_{i4} is the first-, second-, third-, and the fourth-order target moments of stochastic variables i.

The empirical results show that the moment matching method can make the first-order moments of the scenario and the stochastic variables close enough with a small number of scenarios.

17.5.2 The improved moment matching method

17.5.2.1 The shortcomings of the traditional moment matching method

The core of the traditional moment matching method is a nonlinear, nonconvex, and multivariable optimization model. In the process of solving, the scenario value and the corresponding probability are taken as variables simultaneously. However, it is difficult to solve the problem and find the global optimal solution. Next, the complexity of the model will be explained more clearly. For example, put Eqs. (17.38)–(17.41) into $\sum_{k=1}^{4} \omega_{ik}(m_{ik} - M_{ik})^2$, this equation can be converted to Eq. (17.49).

$$\begin{aligned}
&\sum_{k=1}^{4} \omega_{ik}(m_{ik} - M_{ik})^2 \\
&= \omega_{i1}(m_{i1} - M_{i1})^2 + \omega_{i2}(m_{i2} - M_{i2})^2 + \omega_{i3}(m_{i3} - M_{i3})^2 + \omega_{i4}(m_{i4} - M_{i4})^2 \\
&= \omega_{i1}(x_{i1}p_1 + x_{i2}p_2 + \cdots x_{iN}p_N - M_{i1})^2 + \omega_{i2}(m_{i2} - M_{i2})^2 + \omega_{i3}(m_{i3} - M_{i3})^2 \\
&+ \omega_{i4} \left(\frac{\sum_{j=1}^{N} \left[x_{ij} - (x_{i1}p_1 + x_{i2}p_2 + \cdots x_{iN}p_N) \right]^4 \cdot p_j}{\left[(x_{i1}^2 p_1 + x_{i2}^2 p_2 + \cdots x_{iN}^2 p_N) - (x_{i1}p_1 + x_{i2}p_2 + \cdots x_{iN}p_N)^2 \right]^2} - M_{i4} \right)^2
\end{aligned} \tag{17.49}$$

Eq. (17.49) is a polynomial of high degree and complexity where $x_{i1}, x_{i2}, \ldots, x_{iN}$ (1, 2, …, w) and $p_1, p_2, \ldots, p_N$ are its variables. If we try to bring Eqs. (17.38)–(17.42) into the whole objective function, and make Eqs. (17.37), (17.43) to be the constraint condition, then this model is quite difficult to solve.

Although the obtained scenarios satisfies the requirements of statistical characteristics, they do not satisfy the original probability distribution. This is another drawback of the model. For instance, the following four distributions are consistent in all their top four order moments:

(1) Uniform distribution on interval [−2.44949, 2.4949].
(2) The weight of mixed normal distribution N (1.244666, 0.450806) and N (−1.244666; 0.450806) is 0.5.
(3) The discrete distribution of equal probability, such as Table 17.5.
(4) Discrete distribution of unequal probabilities, such as Table 17.6.

These are shown in Fig. 17.11. On the left is the schematic diagram of distribution 1, 2, and 4; on the right is schematic diagram of distribution 1, 2, and 3. Visual inspection shows that these distributions do not have much in common (Fig. 17.12).

Although the results shown above raise doubts about the matching method, it is crucial that the scenario generation is used to provide input to the optimization problem. This result does not necessarily mean that the scenario is not valid. This shows that, in addition to scenario generation methods, the stability analysis should be done to check whether the accuracy criteria are met.

Table 17.5 The discrete probability distribution with equal probability

Value	−2.0395	−0.91557	0	0.91557	2.0395
Probability	0.2	0.2	0.2	0.2	0.2

Table 17.6 The discrete probability distribution with unequal probability

Value	−3.5	−1.4	0	1.4	3.5
Probability	0.013	0.429	0.1162	0.429	0.013

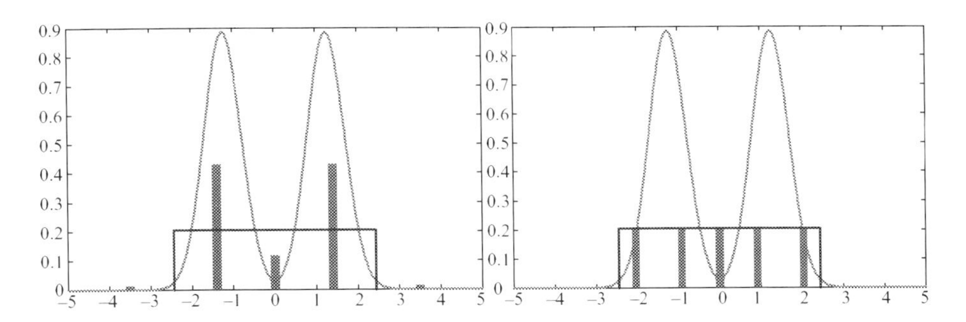

Fig. 17.12 The first four moments of the same four distributions.

For the above shortcomings to analyze the reasons, the traditional moment matching method generates the scenario completely according to the statistical properties of stochastic variables, and the result may be too far away from the original value. On the other hand, the corresponding probability is too far away from the possibility of actual occurrence, and the distribution of the scenario deviates from the original distribution of stochastic variables.

17.5.2.2 Improvement of moment matching method based on k-means clustering

1. Clustering method

Clustering is a set of samples that have no class labels. According to a criterion, the similarity between the samples is measured, and then the similarity is divided into several subsets, so that the similar samples can be classified into one class, and the dissimilarity samples are divided into different classes. From a mathematical point of view, there is the following expression:

Let $\boldsymbol{X} = \{\boldsymbol{x}_1, \boldsymbol{x}_2, \ldots, \boldsymbol{x}_n\}$ be the sample set of clustering analysis, and use the finite parameter value x_{kj} ($j = 1, 2, \ldots, s$) to describe the feature of $\boldsymbol{x}_k$ ($k = 1, 2, \ldots, n$), and then constitute a feature to $(x_{k1}, x_{k2}, \ldots, x_{ks})$ as the judgment of $\boldsymbol{x}_k$ similarity mark. According to the similarity between the eigenvectors corresponding to the n samples in the set X, the $\boldsymbol{x}_1, \boldsymbol{x}_2, \ldots, \boldsymbol{x}_n$ is divided into multiple disjoint subsets $X_1, X_2, \ldots, X_c$ and the following conditions are required:

$$X_1 \cup X_2 \cup \ldots \cup X_c = \mathbf{X}, X_i \cup X_j = \varnothing \\ 1 \leq i \neq j \leq c \tag{17.50}$$

In clustering algorithm, the similarity between samples is usually described by the distance between feature vectors. The smaller the distance is, the greater the similarity is, and the more likely the classification is. For samples with n variables, $d\,(i,j)$ is used to represent the distance between the i sample and the j sample, and the common distance formulas are as follows:

(a) The Ming's distance

$$d_{ij}(q) = \left(\sum_{k=1}^{p} |x_{ik} - x_{jk}|^q \right)^{\frac{1}{q}}, \quad q > 0 \tag{17.51}$$

When the values of q are different, the corresponding distances are different:

(i) Absolute distance

$$d_{ij}(1) = \sum_{k=1}^{p} |x_{ik} - x_{jk}| \tag{17.52}$$

(ii) Euclidean distance

$$d_{ij}(2) = \left(\sum_{k=1}^{p} \left| x_{ik} - x_{jk} \right|^2 \right)^{\frac{1}{2}} \tag{17.53}$$

(iii) Chebyshev distance

$$d_{ij}(\infty) = \max \left| x_{ik} - x_{jk} \right| (1 \le k \le p) \tag{17.54}$$

(b) Mahalanobis distance

The observed values of each sample variable are usually stochastic variables, and the stochastic vectors (composed of stochastic variables) are distributed according to certain rules. There are also related to possible between the components, the Mahalanobis distance between i vector and j vector is

$$d_{ij}^2(m) = \left(x_i - x_j\right)^T S^{-1} \left(x_i - x_j\right) \tag{17.55}$$

Among them, S is the covariance matrix of stochastic variables, S^{-1} is the inverse matrix of S, and there is a defect in Mahalanobis distance, that is, S is difficult to determine.

(c) Manhattan distance

Manhattan distance, also known as the distance between urban blocks, is the sum of the projection of the line segments generated by two points on the fixed rectangular coordinate system in Euclidean space. Distance expression is

$$D = \left| x_{i1} - x_{j1} \right| + \left| x_{i2} - x_{j2} \right| + \cdots + \left| x_{ip} - x_{jp} \right| \tag{17.56}$$

2. k-means clustering

k-clustering (k-means clustering) is an unsupervised real-time clustering algorithm proposed by Mac Queen. It is one of the most common clustering methods. Its basic idea is: the distance to the sum of all the data points in the sample space and the cluster center as the objective function to measure the error, based on data partition for minimizing the k reservation, the principle is simple and easy to handle large amounts of data.

The objective function directly reflects the effect of clustering. The smaller the value is, the more compact and independent the clustering is. Therefore, it is necessary to improve and optimize the clustering scheme by decreasing the value of the objective function, and when the minimum value is taken, the clustering is the optimal clustering scheme. k-mean clustering uses Euclidean distance as similarity index to measure similarity, whose expression is

$$J = \sum_{j=1}^{k} \left(\sum_{i=1}^{n_j} \left\| x_i^j - c_j \right\|^2 \right) \tag{17.57}$$

k is the number of classification; x_i^j denotes the i sample of group j data; c_j denotes the clustering center of group j sample data; and n_j denotes the number of sample data in group j.

It can be known from formula (17.57) that the clustering process is to find the best center c_j, so that the objective function is minimum.

$$\frac{\partial J}{\partial c_j} = \frac{\partial}{\partial c_j} \sum_{i=1}^{n_j} \left(x_i^j - c_j \right)^2 = -2 \sum_{i=1}^{n_j} \left(x_i^j - c_j \right) = 0 \tag{17.58}$$

So the clustering center is

$$c_j = \frac{1}{n_j} \sum_{i=1}^{n_j} x_i^j \tag{17.59}$$

The running process of k-mean clustering algorithm:

Step 1: specify the number of clusters k, randomly select k points as the initial clustering center, and set the number of iterations or iterative stop threshold.
Step 2: according to Eq. (17.53) calculating the distance between the sample and the cluster center, each sample is assigned to the nearest cluster center to form the class.
Step 3: according to Eq. (17.59) calculate the average vector of each class, update the cluster center of this class, repeat Step 2, iterate until the formula (17.57) is smaller than the stop threshold or reach the maximum number of iterations.

3. The whole process of scenario creation is improved by moment matching method, as shown in Fig. 17.13.
Step 1: according to the historical observation samples of stochastic variables, using Eqs. (17.44)–(17.48) to estimate the statistical characteristics of the original data, that is, the average value, standard deviation, skewness, kurtosis, covariance, used as target moment matching.
Step 2: through the operation of k-mean algorithm, the historical observation samples are clustered, and the clustering center is used as the scenario value of stochastic variables.
Step 3: after the first and the second step, the estimated mean, variance, skewness, kurtosis, covariance and the scenario values are used in Eqs. (17.36)–(17.43), calculate the probability of the corresponding scenarios.

Through the above three steps, the generation of the stochastic scenarios is completed.

17.5.3 Superiority analysis of improved moment matching method

The traditional k-mean clustering method generates the scenario by giving the number of pre-generated scenarios, that is to say, the number of classes. Then initialize the center of the class of the original sample, and then iterate to update the location of

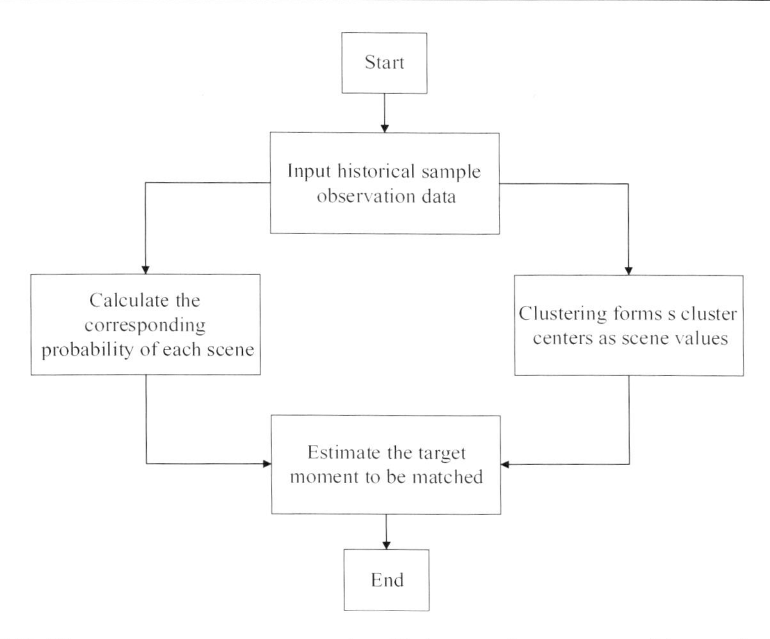

Fig. 17.13 The process to generate scenario with improved moment matching method.

the center. When the sum of the Euclidean distances between the center point and the sample in the class is smaller, the iteration is stopped. In order to determine the possibility of the occurrence of various scenarios, we can use the original sample to be divided into the number of samples in each kind to describe. The more the number is, the more likely the occurrence of the scenario is, the ratio of the number of samples in each class to the total sample will be the probability of occurrence of each scenario.

The principle of generating scenario by traditional moment matching method is to take the scenario value and corresponding probability as variables. Then, the statistical characteristics of the two variables are compared with the statistical characteristics of the original samples, and the optimization model is constructed by minimizing the error. Finally, we can get the scenario by solving this model. The comparison of two traditional methods are shown as Table 17.7.

The improved moment matching method is similar to the traditional moment matching method. The scenario value is classified by the k-means clustering method to classify the original samples, and the representative center points are generated. Finally, the scenario probability is taken as the variable to solve the optimization model. The scenario value is obtained by k-means clustering, which make the distances between the generated scenario and the sample close enough. This can be consistent with its original distribution as much as possible. Not only that, the scenario value and probability of the two variables can be separated, not at the same time involved in the optimization of the solution, which greatly reduces the difficulty of solving. The scenario generated by the proposed method is also matched with the original first four moments and covariance. Compared with the above two traditional

Table 17.7 Compared traditional *k*-means clustering method with traditional moment matching method

Method	Advantage	Disadvantage
Traditional *k*-means clustering method	The principle is simple and easy to implement, and the generated scenario has the class representation of the original sample	The probability of the occurrence of the scenario is simply handled, which does not guarantee the statistical rules of the original sample
Traditional moment matching method	The mean, variance and other statistical properties of the generated scenario are consistent with the original sample	The model is complex and difficult to solve, and the scenario cannot be consistent with the original probability distribution

methods, the improved moment matching method just takes the advantage of the two methods. The *k*-means clustering scenario ensures the typicality of the original sample, while the moment matching modifies the statistical characteristics, thus a small number of scenarios can be generated to meet the needs of the problem.

17.5.4 Scenario generation example based on improved moment matching method

17.5.4.1 Example description

In this part, the output power of wind farm is taken as a stochastic variable, so the scenario of wind power generation is needed. The wind power historical sample semploy the actual wind power output of a wind farm in EirGrid, Ireland, between January 1, 2013 and December 31, 2014, 12 times a day, a total of 730 historical samples. The empirical cumulative distribution map and frequency histogram are shown in Figs. 17.14 and 17.15.

The empirical cumulative distribution curves of wind power are similar to those of parabola in Figs. 17.14 and 17.15, and there is a certain rule to follow. The sample size ranged from 0 to 120 MW.

17.5.4.2 Statistical properties of the original scenario

According to the formulas (17.44)–(17.47), the first four statistical characteristics of the original wind power in this period are calculated, as shown in Table 17.8.

As shown in Table 17.8, the original mean wind power is 113.347 MW, standard deviation is 88.842 MW, the skewness kurtosis was 0.602, 2.227, thus it can be seen that the original statistical regularity of wind power. The mean 113.347 MW shows that the average output power of the wind farm is 113.347 MW during the period, and the standard deviation (for better illustration, standard deviation is employed here instead of variance) 88.842 MW shows that the output power is unstable. The output power is not kept at 113.347 MW, but seriously deviates from the 113.347 MW

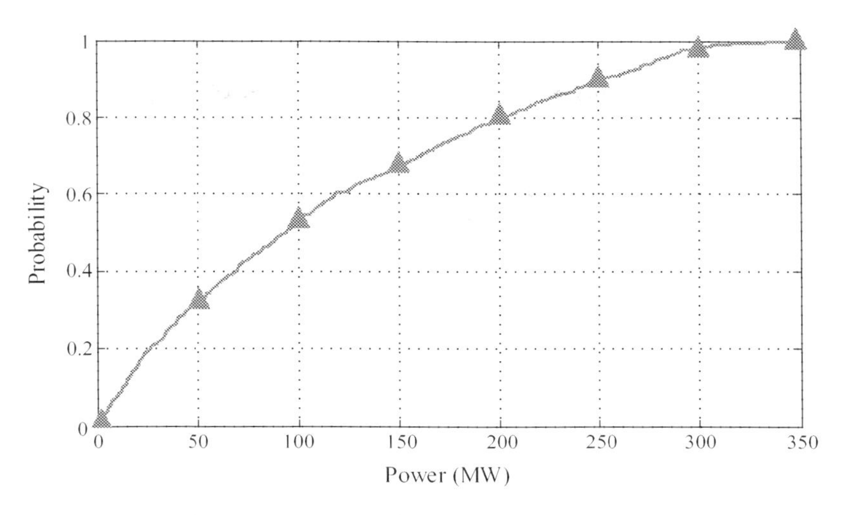

Fig. 17.14 The curve for wind power experience cumulative distribution.

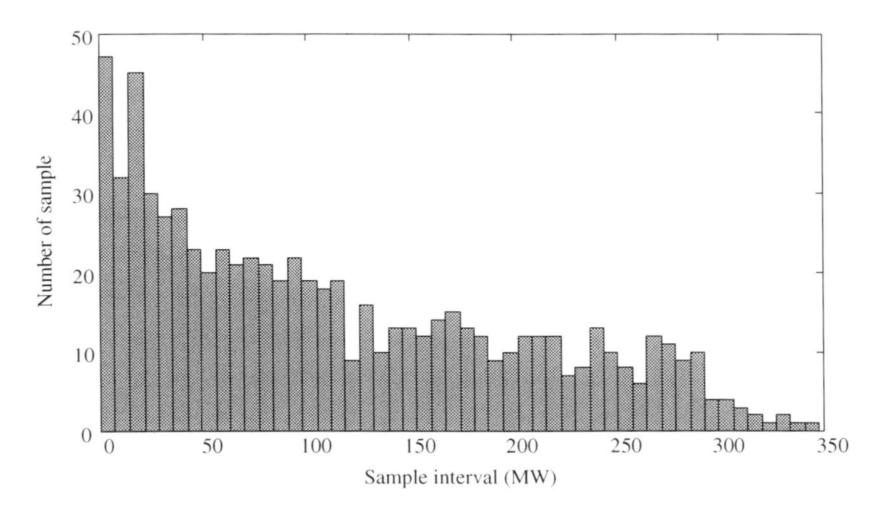

Fig. 17.15 The sample frequency histogram for wind power.

Table 17.8 The first four statistical properties for original wind power scenario

Mean value	Standard deviation	Skewness	Kurtosis
113.347	88.842	0.602	2.227

variation output, and the fluctuation is larger. Skewness 0.602 indicates that the probability distribution curve of the output power is asymmetric, and the right tail is longer than the left one. The 2.227 shows that the probability distribution curve of the peak output power.

17.5.4.3 Improved scenario generated by moment matching method

Using improved moment matching method to generate 10 and 50 scenarios.

First, according to the *k*-means clustering algorithm, the original wind power is clustered into 10 and 50 categories, respectively, and the clustering centers are shown in Tables 17.9 and 17.10.

Then, the clustering centers in Tables 17.9 and 17.10 are used as the scenario values. They are substituted into the original moment matching model to solve the corresponding probability, respectively. The results are shown in Tables 17.11 and 17.12.

17.5.4.4 Comparison and analysis of various scenarios generation methods

First, the comparison of statistical properties. The original wind power sample (original scenario) and the improved first moment matching method are used to generate the first four-order statistical characteristics of the scenario, as shown in Table 17.13.

The first four-order statistical properties of the original scenario and the *k*-mean clustering method are shown in Table 17.14.

We can see from Tables 17.13 and 17.14 that the improved moment matching method produces the first four-order statistical properties of the scenarios, which is consistent with the statistical properties of the original samples. The traditional *k*-means clustering method generates the first four statistical characteristics of the scenario, and there is a small deviation between the index value and the original sample.

Cumulative distribution comparisons are presented next. When 10 scenarios are generated, the cumulative distribution of the generated scenario and the original empirical cumulative distribution curve are shown in Figs. 17.16 and 17.17.

Fig. 17.16 is the result comparison of the improved moment matching method generating 10 scenarios. Fig. 17.17 is the result comparison of the traditional *k*-means

Table 17.9 The clustering centers of 10 scenarios generated

Category	Clustering center	Category	Clustering center
1	41.239	6	244.004
2	202.691	7	125.066
3	162.543	8	289.397
4	66.490	9	20.881
5	6.618	10	95.219

Table 17.10 The clustering centers of 50 scenarios generated

Category	Clustering center	Category	Clustering center
1	43.700	26	101.400
2	12.263	27	24.344
3	146.123	28	118.046
4	64.583	29	270.364
5	206.533	30	70.267
6	239.538	31	204.600
7	59.950	32	54.922
8	251.678	33	3.989
9	33.942	34	126.629
10	15.576	35	20.257
11	304.460	36	331.900
12	169.350	37	9.400
13	286.790	38	111.378
14	97.957	39	209.967
15	230.350	40	152.817
16	88.494	41	82.375
17	107.457	42	1.757
18	140.311	43	202.520
19	177.707	44	49.095
20	114.543	45	187.711
21	6.541	46	39.114
22	93.987	47	196.600
23	216.514	48	161.221
24	133.000	49	223.444
25	76.493	50	28.969

Table 17.11 The corresponding probability of 10 scenarios generated

Scenario number	Probability	Scenario number	Probability
1	0.10643841	6	0.06091200
2	0.07507444	7	0.15328766
3	0.09176213	8	0.09267591
4	0.08125177	9	0.15188342
5	0.09459865	10	0.09211561

clustering method generating 10 scenarios. It can be seen that the cumulative distribution curve generated by the improved moment matching method is the same as that of the original empirical cumulative distribution, and there are several points falling, and the curve has jitter. In contrast, k-means clustering method is a little smoother.

The following is a quantitative description of the difference in size. The cumulative probability corresponding to each scenario spot, such as Table 17.15, is measured by

Table 17.12 The corresponding probability of 50 scenarios generated

Scenario number	Probability	Scenario number	Probability
1	0.02783355	26	0.01796705
2	0.02421719	27	0.02621278
3	0.01505747	28	0.01642410
4	0.02423905	29	0.02498481
5	0.01591301	30	0.02300211
6	0.01925186	31	0.01580543
7	0.02525697	32	0.02628540
8	0.02168921	33	0.02310604
9	0.02755251	34	0.01586148
10	0.02473992	35	0.02552395
11	0.01758251	36	0.01120875
12	0.01481664	37	0.02379875
13	0.02349032	38	0.01696467
14	0.01837317	39	0.01612112
15	0.01794359	40	0.01491386
16	0.01967002	41	0.02066290
17	0.01732921	42	0.02285844
18	0.01523449	43	0.01569698
19	0.01490081	44	0.02726623
20	0.01669609	45	0.01511930
21	0.02341625	46	0.02790141
22	0.01888379	47	0.01542694
23	0.01658533	48	0.01481995
24	0.01553091	49	0.01719123
25	0.02173866	50	0.02693380

Table 17.13 Contrast first four moments between original scenarios and generate scenarios

	Mean value	Standard deviation	Skewness	Kurtosis
Original	113.347	88.842	0.602	2.227
10 Scenarios	113.347	88.842	0.602	2.227
50 Scenarios	113.347	88.842	0.602	2.227

Table 17.14 Contrast first four moments between original scenarios and generate scenarios

	Mean value	Standard deviation	Skewness	Kurtosis
Original	113.347	88.842	0.602	2.227
10 Scenarios	113.347	88.246	0.581	2.146
50 Scenarios	113.347	88.755	0.599	2.217

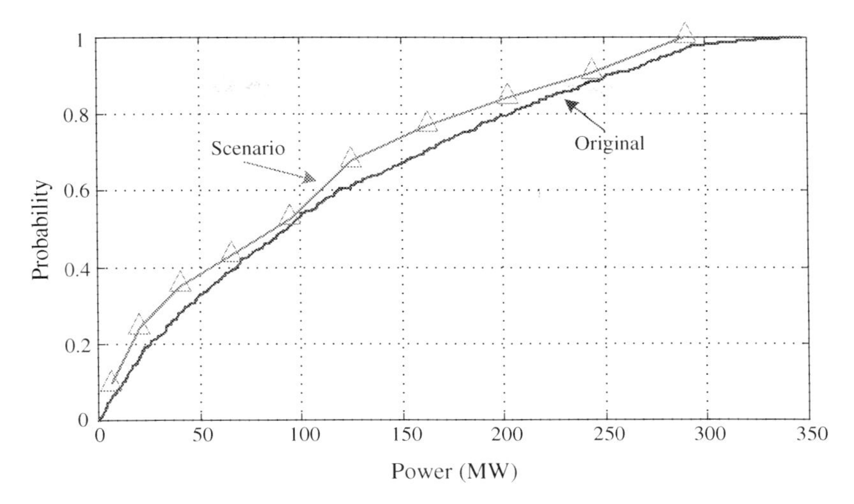

Fig. 17.16 Compare the scenarios generated by improved moment matching method with the original scenarios distribution (10 scenarios).

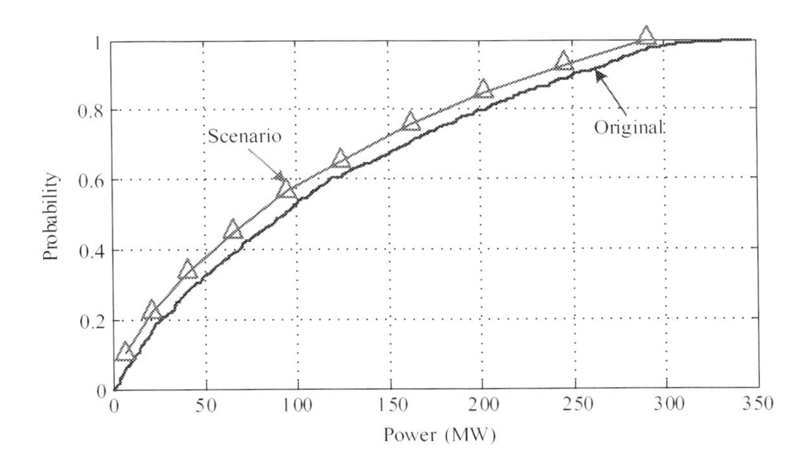

Fig. 17.17 Compare the scenarios generated by *k*-means clustering with the original scenarios distribution (10 scenarios).

the sum of the distance difference between two points at the scenario spots with different distribution curves. Then the improved moment matching method of scenario distribution curve and distribution curve of the original scenario the distance difference is 0.48239833, *k*-means clustering method of scenario distribution curve and distribution curve of the original scenario the distance difference is 0.48356164, the less the scenario situation, improved moment matching method of scenario distribution curve and distribution curve is closer to the original scenario.

Table 17.15 The cumulative probability in the point of scenarios generated with different methods (10 scenarios)

Scenario spot	Improved moment matching method	*k*-means cluster	Original
6.618	0.09459865	0.106849315	0.057534247
20.881	0.24648207	0.221917808	0.168493151
41.239	0.35292048	0.335616438	0.284931507
66.490	0.43417225	0.450684932	0.390410959
95.219	0.52628786	0.567123288	0.508219178
125.066	0.67957552	0.654794521	0.609589041
162.543	0.77133765	0.756164384	0.705479452
202.691	0.84641209	0.846575342	0.798630137
244.004	0.90732409	0.920547945	0.884931507
289.397	1.00000000	1.000000000	0.968493151

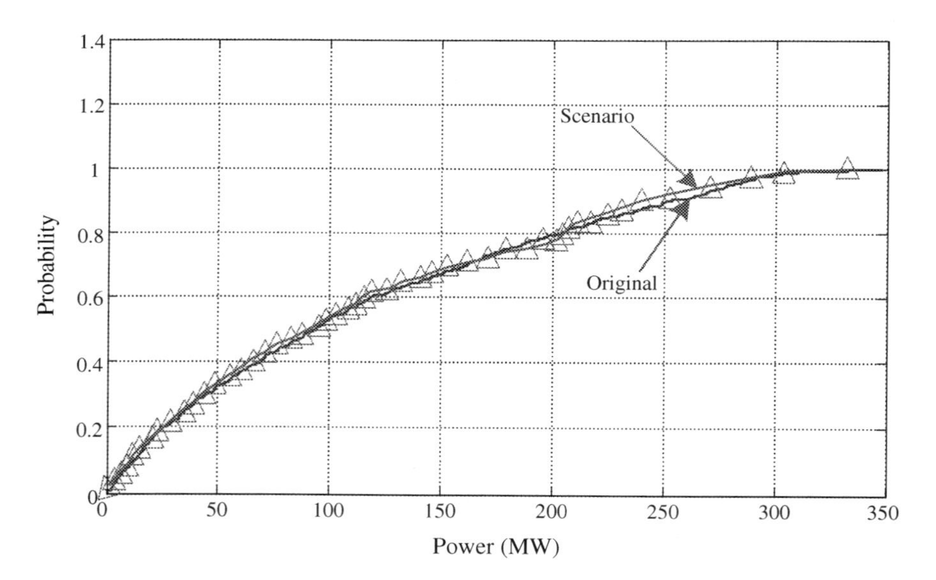

Fig. 17.18 Compare the scenarios generated by improved moment matching method with the original scenarios distribution (50 scenarios).

When 50 scenarios are generated, the cumulative distribution of the generated scenario and the original empirical cumulative distribution curve are shown in Figs. 17.17 and 17.18.

Fig. 17.18 is the result comparison of the improved moment matching method to generate the 50 scenarios. Fig. 17.19 is the result comparison of the traditional *k*-means clustering method to generate the 50 scenarios. It can be seen that the two curves are close to each other when the number of scenarios increases. Because the

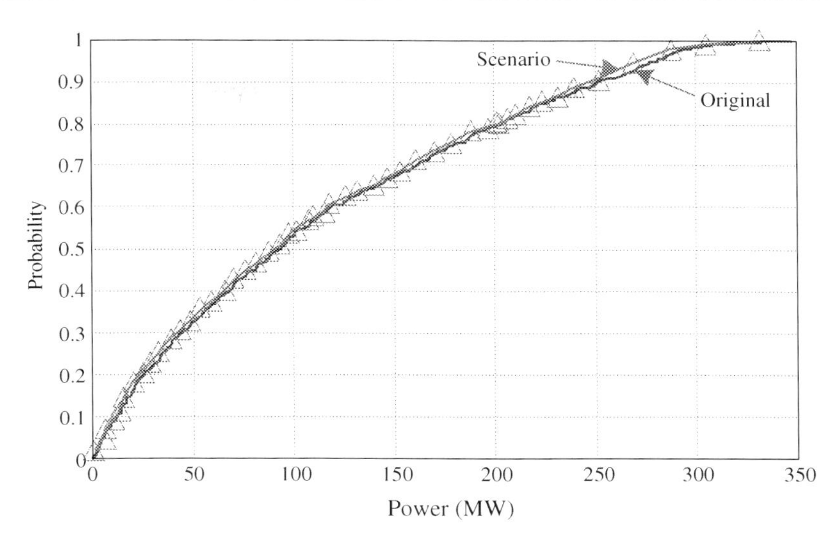

Fig. 17.19 Compare the scenarios generated by *k*-means clustering with the original scenarios distribution (50 scenarios).

number of scenarios increases, the loss of historical data is less. But the more the scenario is, the harder it is to calculate, which contradicts the application. You can choose the number of scenarios that meet the requirements by repeated calculations.

The cumulative probability corresponding to each scenario spot is shown in Table 17.16. The distance difference between the scenario distribution curve and the original scenario distribution curve is 0.64224024 by using the improved moment

Table 17.16 The cumulative probability in the point of scenarios generated with different methods (50 scenarios)

Scenario spot	Improved moment matching method	*k*-means cluster	Original
1.757	0.02285844	0.01917808	0.00684932
3.989	0.04596448	0.04520548	0.03287671
6.541	0.06938073	0.06849315	0.05616438
9.400	0.09317948	0.08630137	0.07808219
12.263	0.11739667	0.10821918	0.10000000
15.576	0.14213659	0.14246575	0.12602740
20.257	0.16766054	0.18082192	0.16301370
24.344	0.19387332	0.20547945	0.19315068
28.969	0.22080712	0.22328767	0.21369863
33.942	0.24835963	0.25616438	0.24109589
39.114	0.27626104	0.28493151	0.26849315

Continued

Table 17.16 Continued

Scenario spot	Improved moment matching method	*k*-means cluster	Original
43.700	0.30409459	0.30410959	0.29452055
49.095	0.33136082	0.33013699	0.31917808
54.922	0.35764622	0.35479452	0.34246575
59.950	0.38290319	0.37671233	0.36575342
64.583	0.40714224	0.39315068	0.38493151
70.267	0.43014435	0.42602740	0.41095890
76.493	0.45188301	0.44657534	0.43424658
82.375	0.47254591	0.46849315	0.45753425
88.494	0.49221593	0.49178082	0.47671233
93.987	0.51109972	0.51232877	0.50000000
97.957	0.52947289	0.53150685	0.52191781
101.400	0.54743994	0.54657534	0.54109589
107.457	0.56476915	0.56575342	0.55616438
111.378	0.58173382	0.57808219	0.57123288
114.543	0.59842991	0.58767123	0.58219178
118.046	0.61485401	0.60547945	0.59726027
126.629	0.63071549	0.62465753	0.61643836
133.000	0.64624640	0.64109589	0.63287671
140.311	0.66148089	0.65342466	0.64657534
146.123	0.67653836	0.67123288	0.66301370
152.817	0.69145222	0.68767123	0.67945205
161.221	0.70627217	0.71369863	0.70136986
169.350	0.72108881	0.73561644	0.72465753
177.707	0.73598962	0.75616438	0.74657534
187.711	0.75110892	0.78082192	0.77123288
196.600	0.76653586	0.79589041	0.78767123
202.520	0.78223284	0.80273973	0.79726027
204.600	0.79803827	0.80547945	0.80410959
206.533	0.81395128	0.81369863	0.80958904
209.967	0.83007240	0.82191781	0.81780822
216.514	0.84665773	0.84109589	0.83150685
223.444	0.86384896	0.85342466	0.84657534
230.350	0.88179255	0.86438356	0.85753425
239.538	0.90104441	0.88630137	0.87534247
251.678	0.92273362	0.91095890	0.90273973
270.364	0.94771843	0.94931507	0.92739726
286.790	0.97120875	0.97808219	0.96438356
304.460	0.98879126	0.99178082	0.98630137
331.900	1.00000001	1.00000000	0.99589041

matching method, and the distance difference between the scenario distribution curve of k-mean clustering method and the original scenario distribution curve is 0.49726027. It can be seen that the scenario distribution curve of k-mean clustering method is closer to the original scenario distribution curve.

On the whole, the improved moment matching method has absolute advantages in generating the first four-order moments statistics, and the cumulative distribution has a little fluctuation compared with the k-means clustering method, so it can be used to make the dispatch decision effectively.

17.6 Conclusion

This chapter proposes two approaches for network operation with consideration of stochastic renewable generation. Based on the content above, conclude two approaches as followed, respectively.

Probability-based approach: This chapter establishes the minimum power optimization model based on stochastic programming p-efficient point theory, the conversion from uncertain programming model to deterministic model is deduced, the positive/negative SR coefficient dealing with the prediction error of net load is obtained, required SR and minimum storage energy power are calculated, the maximum wind power consumption is achieved at the given confidence level without wind curtailment and load curtailment. In addition, extreme scenarios are set up to verify the proposed method, which provides ideas, methods, and approaches for the verification of ES capacity and ES power allocation. The calculation results provide theoretical basis for determining the required energy for the ES device, and thus guide the allocation of ES capacity.

Scenario-based approach: this chapter is based on analyzing the domestic and foreign research on economic dispatch in wind power integrated system on the uncertain information and an improved method to mining scenario of wind power, and based on the realization of solving the stochastic economic dispatch of power system with wind farms. Combining the advantages of the traditional moment matching method and clustering method, a wind power scenario simulation method based on improved moment matching is proposed. This method generates a set of representative scenarios close to the original scenarios of wind power generation in terms of statistical indices and probability distributions of the first four moments. Example verification shows that compared with the k-mean clustering method, the improved method of moment matching can effectively mine the uncertain information of wind power, and provide decision-making information for the economic dispatch of power system with wind farms. Also, the uncertainty of wind farm output power is represented by scenarios, and the stochastic economic dispatch problem of power system with wind farm is transformed into the deterministic problem in each scenario, which is easy to solve.

References

Birge, J., Louveaux, F., 1997. Introduction to Stochastic Programming. Springer, New York, pp. 103–107.

Chen, C.L., 2008. Optimal wind-thermal generating unit commitment. IEEE Trans. Energy Convers. 23 (1), 273–280.

Chen, G.G., Chen, J.F., 2013. Environmental/economic dynamic dispatch modeling and method for power systems integrating wind farms. Proc. CSEE 33 (10), 27–35.

Chen, C.L., Lee, T.Y., Jan, R.M., 2006a. Optimal wind-thermal coordination dispatch in isolated power systems with large integration of wind capacity. Energy Convers. Manag. 47, 3456–3472.

Chen, Y.H., Chen, J.F., Duan, Z.X., 2006b. Fuzzy modeling and optimization algorithm on dynamic dispatch in wind power integrated system. Autom. Electr. Power Syst. 30 (2), 22–26.

Chen, S.X., Gooi, H.B., Wang, M.Q., 2012. Sizing of energy storage for micro-grids. IEEE Trans. Smart Grid 3 (1), 142–151.

Dong, X.T., Yan, Z., Feng, D.H., et al., 2012. Power system economic dispatch considering penalty cost of wind farm output. Power Syst. Technol. 36 (8), 76–80.

Gao, H.J., Liu, J.Y., Wei, Z.B., et al., 2013. A security-constrained dispatching model for wind generation units based on extreme scenario set optimization. Power Syst. Technol. 37 (6), 1590–1595.

Gülpınar, N., Rustem, B., Settergren, R., 2004. Simulation and optimization approaches to scenario tree generation. J. Econ. Dyn. Control 28, 1291–1315.

Hadji, M.M., Vahidai, B., 2012. A solution to the unit commitment problem using imperialistic competition algorithm. IEEE Trans. Power Syst. 27 (1), 117–124.

Hetzer, J., Yu, D.C., Bhattarai, K., 2008. An economic dispatch model incorporating wind power. IEEE Trans. Energy Convers. 23 (2), 603–611.

Hodge, B.M., Milligan, M., 2011. Wind power forecasting error distributions over multiple timescales. Power & Energy Society General Meeting, Detroit Michigan, USA, July 24–29.

Hodge, B.M., Orwig, K., Milligan, M., 2012. Examining information entropy approaches as wind power forecasting performance metrics.12th International Conference on Probabilistic Methods Applied to Power System, Istanbul, Turkey, June 10–14.

Hong, Y.Y., Li, C.T., 2006. Short-term real-power scheduling considering fuzzy factors in an autonomous system using genetic algorithms. IEE Proc. Generat. Transm. Distrib. 153 (6), 684–692.

Høyland, K., Wallace, S.W., 2001. Generating scenario trees for multistage decision. Manag. Sci. 47 (2), 295–307.

Hu, G.W., Bie, C.H., Wang, X.F., 2013. Optimal dispatch in wind integrated system considering operation reliability. Trans.China Electrotech. Soc. 28 (5), 58–65.

Jiang, W., Cheng, Y.X., Yan, Z., et al., 2013. Reliability-constrained dynamic economic dispatch of power system with wind farms. Electr. Power Autom. Equip. 33 (7), 27–33.

Lee, T.Y., 2007. Optimal spinning reserve for a wind-thermal power system using EIPSO. IEEE Trans. Power Syst. 22 (4), 1612–1621.

Liu, X., 2010. Economic load dispatch constrained by wind power availability: a wait-and-see approach. IEEE Trans. Smart Grid 1 (3), 347–355.

Liu, X., 2011. Emission minimisation dispatch constrained by cost and wind power. IET Gener. Transm. Distrib. 5 (7), 735–742.

Liu, X., Xu, W., 2010. Economic load dispatch constrained by wind power availability: a here-and-now approach. IEEE Trans. Sustainable Energy 1 (1), 2–9.
Liu, B., Zhou, J.Y., Zhou, H.M., et al., 2012. An improved model for wind power forecast error distribution. East China Electr. Power 40 (2), 286–291.
Liu, D.W., Guo, J.B., Huang, Y.H., et al., 2013. Dynamic economic dispatch of wind integrated power system based on wind power probabilistic forecasting and operation risk constraints. Proc. CSEE 33 (16), 9–15.
Ma, R., Kang, R., Jiang, F., et al., 2013. Multi-objective dispatch planning of power system considering the stochastic and fuzzy wind power. Power Syst. Protect. Control 41 (1), 150–156.
Miranda, V., Hang, P.S., 2005. Economic dispatch model with fuzzy wind constraints and attitudes of dispatchers. IEEE Trans. Power Syst. 22 (4), 2143–2145.
Pappala, V.S., Erlich, I., Rohrig, K., et al., 2009. A stochastic model for the optimal operation of a wind-thermal power system. IEEE Trans. Power Syst. 24 (2), 940–950.
Qiao, J.G., Xu, F., Lu, Z.X., et al., 2008. Optimization analysis model of grid-connected wind capacity based on dependent chance programming. Autom. Electr. Power Syst. 32 (10), 84–103.
Qiu, W., Zhang, J.H., Liu, N., 2011. Model and solution for environmental/economic dispatch considering large-scale wind power penetration. Proc. CSEE 31 (19), 8–16.
Ren, B.Q., Peng, M.H., Jiang, C.W., et al., 2010a. Short-term economic dispatch of power system modeling considering the cost of wind power. Power Syst. Protect. Control 38 (14), 67–72.
Ren, B.Q., Peng, M.H., Jiang, C.W., et al., 2010b. Short-term economic scheduling model including wind park based on improved genetic algorithm and risk management. Modern Electr. Power 2 (1), 76–80.
Ross, O., 2007. Interest Rate Scenario Generation for Stochastic Programming. The Technical University of Denmark.
Senjyu, T., Shimabukuro, K., Uezato, K., et al., 2003. A fast technique for unit commitment problem by extended priority list. IEEE Trans. Power Syst. 18 (2), 882–888.
Shi, L.B., Wang, C., Yao, L.Z., et al., 2012a. Optimal power flow solution incorporating wind power. IEEE Syst. J. 6 (2), 233–241.
Shi, Q.J., Geng, G.C., Jiang, Q.Y., 2012b. Real-time optimal energy dispatch of standalone microgrid. Proc. CSEE 32 (16), 26–35.
Sun, Y.Z., Wu, J., Li, G.J., et al., 2009. Dynamic economic dispatch considering wind power penetration based on wind speed forecasting and stochastic programming. Proc. CSEE 29 (4), 41–47.
Tian, K., Zeng, M., Yan, F., et al., 2011. A dynamic economic scheduling model considering environmental protection cost and impact of connecting wind power to power grid. Power Syst.Technol. 35 (6), 55–59.
Villanueva, D., Feijoo, A., Pazos, J.L., 2012. Simulation of correlated wind speed data for economic dispatch evaluation. IEEE Trans. Sustainable Energy 3 (1), 142–149.
Wang, 1994. Basic of Power System Planning. China Electric Power Press, Beijing, pp. 41–57.
Wang, X.F., 2003. Modern Power System Analysis. Science Press, Beijing.
Wang, L., Yu, Z.W., Wen, F.S., 2006. A chance-constrained programming approach to determine requirement of optimal spinning reserve capacity. Power Syst. Technol. 30 (20), 14–19.
Wang, C.F., Liang, J., Zhang, L., et al., 2011. Impact of standby transmission components on the reliability of generation and transmission systems. Autom. Electr. Power Syst. 35 (17), 14–19.

Wang, C.S., Yu, B., Xiao, J., et al., 2012a. Sizing of energy storage systems for output smoothing of renewable energy systems. Proc. CSEE 32 (16), 1–8.

Wang, H.Y., Bai, X.M., Xu, J., 2012b. Reliability assessment considering the coordination of wind power, solar energy and energy storage. Proc. CSEE 32 (13), 13–20.

Wang, Q.F., Guan, Y.P., Wang, J.H., 2012c. A chance-constrained two-stage stochastic program for unit commitment with uncertain wind power output. IEEE Trans. Power Syst. 27 (1), 206–215.

Wei, F.M., 2008. Scenario Generation in Dynamic Investment Portfolio Based on Stochastic Programming. Tongji University.

Wei, Y.H., Zhang, S.Y., 2008. Copula Theory and Its Application in Financial Analysis. China Environmental. Science Press, Beijing, pp. 10–23.

Xia, S., Zhou, M., Li, G.Y., 2013. A coordinated active power and reserve dispatch approach for wind power integrated power systems considering line security verification. Proc. CSEE 33 (13), 18–26.

Xu, L., Ruan, X.B., Zhang, B.H., et al., 2012. An improved optimal sizing method for wind-solar-battery hybrid power system. Proc. CSEE 32 (25), 88–98.

Yuan, B., Zhou, M., Li, G.Y., et al., 2013. A coordinated dispatching model considering generation and operating reserve for wind power integrated power system based on ELNSR. Power Syst. Technol. 37 (3), 800–807.

Zhang, Z.S., Sun, Y.Z., Li, G.J., et al., 2011. A solution of economic problem considering wind power uncertainty. Autom. Electr. Power Syst. 35 (22), 125–130.

Zhang, K., Mao, C.X., Xie, J.W., et al., 2012. Optimal design of hybrid energy storage system capacity for wind farms. Proc. CSEE 32 (25), 79–87.

Zhang, H.F., Gao, F., Wu, J., et al., 2013. A dynamic economic dispatching model for power grid containing wind power generation system. Power Syst. Technol. 37 (5), 1298–1303.

Zhao, J.H., Wen, F.S., Xue, Y.S., et al., 2010. Power system stochastic economic dispatch considering uncertain outputs from plug-in electric vehicles and wind generators. Autom. Electr. Power Syst. 34 (20), 22–29.

Zhou, W., Sun, H., Gu, H., et al., 2012. Dynamic economic dispatch of wind integrated power systems based on risk reserve constraints. Proc. CSEE 32 (1), 47–55.

Further reading

Abbey, C., Joos, G., 2009. A stochastic optimization approach to rating of energy storage systems in wind-diesel isolated grids. IEEE Trans. Power Syst. 24 (1), 418–426.

Chakraborty, S., Senjyu, T., Toyama, H., et al., 2009. Determination methodology for optimising the energy storage size for power system. IET Gener. Transm. Distrib. 3 (11), 987–999.

Gulpinar, N., Rustem, B., Settergren, R., 2004. Simulation and optimization approaches to scenario tree generation. J. Econ. Dyn. Control, 1291–1315.

Han, T., Lu, J.P., Qiao, L., et al., 2011. Optimized scheme of energy-storage capacity for grid-connected large-scale wind farm. Power Syst. Technol. 34 (1), 169–173.

IEC, 2012. Grid Integration of Large-Capacity Renewable Energy Sources and Use of Large-Capacity Electrical Energy Storage. IEC, Oslo.

Korpaas, M., Holen, A.T., Hildrum, R., 2003. Operation and sizing of energy storage for wind power plants in a market system. Int. J. Electr. Power Energy Syst. 25 (8), 599–606.

Le, H.T., Nguyen, T.Q., 2008. Sizing energy storage systems for wind power firming: an analytical approach and a cost- benefit analysis.Power and Energy Society General Meeting-Conversion and Delivery of Electrical Energy in the 21st Century, IEEE, pp. 1–8.

Li, B.H., Shen, H., Yang, Y., et al., 2011. Impacts of energy storage capacity configuration of HPWS to active power characteristics and its relevant indices. Power Syst. Technol. 35 (4), 123–128.

Liu, J., Luo, X.J., 2010. Environmental economic dispatching adopting multiobjective random black-hole particle swarm optimization algorithm. Proc. CSEE 30 (34), 105–111.

Liu, G.D., Tomsovic, K., 2012. Quantifying spinning reserve in systems with significant wind power penetration. IEEE Trans. Power Syst. 27 (4), 2385–2393.

Lu, N., Chow, J., Desrochers, A., 2004. Pumped-storage hydro-turbine bidding strategies in a competitive electricity market. IEEE Trans. Power Syst. 19 (2), 834–841.

Ma, X.Y., Wu, Y.W., Fang, H.L., et al., 2011. Optimal sizing of hybrid solar-wind distributed generation in an islanded microgrid using improved bacterial foraging algorithm. Proc. CSEE 31 (25), 17–25.

Srivastava, A.K., Kumar, A.A., Schulz, N.N., 2012. Impact of distributed generations with energy storage devices on the electric grid. IEEE Syst. J. 6 (1), 110–117.

Wang, L.F., Singh, C., 2006. Tradeoff between risk and cost in economic dispatch including wind power penetration using particle swarm optimization.International Conference on Power System Technology.

Yu, S.H., 2011. Calculation of Energy-Storage Capacity in Wind Farm and Research on Vitual Storage Technology. Shandong University, Shandong.

The optimal planning of wind power capacity and energy storage capacity based on the bilinear interpolation theory

18

Jinghua Li, Bo Chen, Jiasheng Zhou, Yuhong Mo
Department of Electrical Engineering, Guangxi University, Nanning, People's Republic of China

18.1 Introduction

The randomness and intermittency of wind power can cause negative influence on the power grid. Using energy storage system (ESS) for load shifting and peak smoothing can improve the peak-valley features of load, which is considered as an effective way of enhancing wind power accommodation. However, the existing ESS is costly and limited by capacity. Thus, it is urgent need to evaluate the wind power capacity and energy storage capacity properly, which is vital to the safe and economic operation of power systems.

In recent years, the assessment of wind power grid capacity is becoming a hot research issue. Wind power accommodation can be evaluated by setting an optimization model with constraints of power balance, safe operation and the line transmission capacity (Villumsen et al., 2013; Ahmadi and Ghasemi, 2012; Wu et al., 2012). However, the aforementioned models fail to consider the stochastic properties of wind power, so the simulation results may not meet the actual requirements of systems. Dietrich et al. (2012a,b) and Cervantes et al. (2013) have evaluated the wind power accommodation for daily scheduling operation. But the methods they proposed are not suitable for the system's long-term planning. Billinton et al. (2012), Meng and Zhuan (2012), and Muñoz et al. (2012) proposed to assess the wind power grid capacity for long-term planning. Nevertheless, the proposed method only considered the power indices, loss of load probability (LOLP) and loss of energy expectation (LOEE), which could not fully reflect the adequacy of the power system. With large-scale wind power integration, the peak-load regulation will become the main pressure for scheduling and operation. And ignoring the peak-load regulation adequacy could cause inaccuracy of evaluation. Therefore, it is necessary to propose an effective method for wind power capacity planning by fully considering the adequacy indices of power system.

With the rapid development of technology, nowadays ESS has broad application prospects in power system. The current researches mostly concentrate on the economic cost and operation control strategy. Lamont (2013) put forward a theory

Smart Power Distribution Systems. https://doi.org/10.1016/B978-0-12-812154-2.00018-3

framework to assess the economic value and optimal structure of large-scale ESS. Li et al. (2013) studied the allocation energy storage power for the wind power system by considering the double stochastic characteristics of the wind power and the load. However, a growing body of research tends to study the system control strategies to deal with the randomness and uncertainty of wind power. Zheng et al. (2014), Ghofrani et al. (2013), and Le et al. (2012) found different operation control strategies that affect the economy while ignored the influences of control strategies to system reliability. Sioshansi et al. (2014) and Xu and Singh (2012) have discussed the generation reliability of a system with wind power and ESS, which demonstrated that wind power can effectively improve the situation of insufficient power (Wei and Zhang, 2008). Wang et al. (2012a,b) proposed a set of reliability indices to measure the influence of the wind power capacity and ESS capacity. Yet there was no further reveal to the connections between the adequacy indices, ESS capacity and wind power capacity, and a solid reference for the coordination and configuration of the wind power and ESS cannot be provided accordingly.

To sum up, current researches mostly concentrate on guaranteeing the power system reliability, reducing the economic cost, enhancing the node voltage, and maintaining the stability of frequency. Based on peak-load regulation, this chapter proposes a method to improve the accommodation of wind power and regulate wind power reasonably by using ESS.

Therefore, this chapter analyzes in depth the relationship of adequacy indices, wind power capacity and ESS capacity, and proposes a method to reasonably optimize and allocate the wind power and energy storage. Four adequacy indices, which are peak-load regulation not enough probability (PRNEP), peak-load regulation not enough expectation (PRNEE), LOLP, and LOEE, are introduced to quantify the influence extent of wind-storage systems on adequacy. Based on the classification of peak-load regulation requirements and the comprehensive net load levels, the sequential models for wind power and the storage energy models can be defined. Combined with the results of kernel density estimate (Lu et al., 2013), this chapter puts forward a method of optimizing wind power and storage energy. At last, the proposed method is verified by being applied to a real case with wind power, and the obtained results possess practical significance of guiding the operation and planning for power systems with wind power integration.

18.2 Research status of wind power accommodation

18.2.1 Research on influencing factors of wind power accommodation

Accommodation of wind power refers to the maximum integrated wind power which meets certain constraints (Lei et al. 2002; Tang et al. 2015). Accommodation of wind power results from the combined action of many factors, to sum up, the influencing factors of wind power accommodation in power grid can be divided into technical factors (Jin, 2011; Liu et al., 2010; Keane et al., 2011), economy factors (Li et al. 2010;

Cui et al. 2014; Liu et al. 2014a,b), and politic factors (Wang and Li, 2011; Zhu et al., 2011) three categories.

18.2.1.1 Technical influencing factors

Technical influencing factors can be divided into two following categories.

The first one is the influencing factor of power grid, which describe the interaction between wind power and power grid, including the load level and load characteristics of power grid (Liu et al., 2010; Keane et al., 2011; Wei et al., 2010), power grid structure and network structure (Li et al. 2008a,b, 2012; Han and Yan, 2009; Zhang et al. 2011a,b), spinning reserve level of conventional units (Xue et al., 2014), peak-load regulation capability (Liu et al., 2014a,b; Xiao et al., 2010), optimization dispatch of conventional units of system (Cervantes et al., 2013; Liu et al., 2015; Zhang et al., 2014), power quality and stability of system (Rodriguez and Amaratunga, 2006; Zhang et al., 2008; Shen et al., 2002).

The above references are mostly based on one single factor to study the accommodation of wind power of power grid. However, the existing references pay more and more attention to the combined effects of multiple factors on the accommodation of wind power of power grid. Liu et al. (2012) established the simulation model of power system based on MATLAB/Simulink, assuming a wind power output, respectively, set voltage, load, and tie line quality coefficient (tie line quality coefficient is defined as the ratio of reactance to resistance) these three influencing factors. The simulation analysis was employed to update the output of wind power, so as to determine the accommodation of wind power. Wu et al. (2004) demonstrated the impacts of different factors on the accommodation of wind power through simulating the influence of different operation modes, compensation capacities, and the contact line (with changing resistance and reactance) such multiple factors on the maximum wind power injection. Based on prediction values of day-ahead wind power and load, Sortomme et al. (2010) and Zhang et al. (2015) took the maximum accommodation of wind power as the subject, considering the influence of peak-load regulation capability, network transmission capacity, reserve level, and load level to the accommodation of wind power, to establish estimation model of day-ahead accommodation of wind power. Liu et al. (2011) first considered the transmission capacity constraints, then the actual output of wind power is limited, until the wind power output meets the grid power limit requirements. The peak-load regulation capability was calculated based on the output, also whether the peak-load regulation can be met was judgd by correcting the output of wind power, and then the accommodation of wind power which meets the power transmission capacity, the peak-load regulation capacity can be acquired. According to the power structure and load characteristics of the system, Sun et al. (2011) calculated the reserve capacity demand of the system. Based on this reserve capacity, the accommodation of wind power of the grid is analyzed monthly according to the peak-load regulation capacity.

The second one is the influence factors of natural characteristics. For the characteristics of wind power, the factors are mainly reflected in the operation characteristics and wind power technology, including wind power prediction technology, low voltage ride through capability of wind unit, unit type, and dynamic reactive power compensation device.

Jia et al. (2012) took the prediction of day-ahead wind power and load as the basic information, the wind curtailment of system as the objective, obtained the day-ahead unit commitment of conventional units and the consumptive wind power output based on the unit commitment. Based on obtained result, with given unit commitment, calculate the consumptive wind power output as a reference for extension of positive and negative direction, so as to explore next-day limit of accommodation of wind power. Xu et al. (2011) first calculated the output of offshore wind farm by statistical analysis, then calculated the output distribution characteristic index of wind farm, compared the wind speed correlation of multiple scenarios, and then analyzed the output characteristics of offshore wind farm in detail. Finally, the output characteristics of offshore and onshore wind power were compared, which provided reference to the consumptive method. Shi et al. (2010) and Huang et al. (2013) have established a dynamic model of the power system with wind power, the maximum accommodation of wind power capacity of the isolated system was researched through simulation; and in this wind power capacity configuration, with the common point falling into contact fault, low voltage ride through capability of the isolated system was analyzed. Ma et al. (2008) analyzed the situation of the system in the case of reactive power to compensate the system using the capacitor, not only effectively improved the system node voltage, but also improved scale of the accommodation of wind power.

18.2.1.2 Economy influencing factors

The main economy influencing factors are the compensation cost of the wind power grid, the subsidy cost of the new energy generation in the wind farm, and the investment cost of the construction of the supporting facilities.

Based on the consideration of the economy of the dispatch scheme, the decision-making model of the wind power system is constructed based on the maximum accommodation of wind power as the goal (Li et al. 2010). Cui et al. (2014) established optimal accommodation of wind power from the economic point of view, considering the thermal units costs, wind power energy saving benefits, system spinning reserve costs and power grid security constraints, and other factors. Liu et al. (2014a,b) comprehensively considered the technical constraints (power balance constraints, power flow constraints, and the unit output constraints) and economic constraints while evaluating the accommodation of wind power to ensure the economy optimal with safe, stable, and reliable operation of the system.

18.2.1.3 National politic influencing factors

The national politic influencing factors mainly stem from the standpoint of wind farms, power grid, users, and so on. Publishing a series of favorable policies helps wind power integrate.

Wang et al. (2011a,b) put forward constructive comments to promote large-scale wind power integration from the national policy level: the development of wind power development planning reasonable, improve the relevant policies of wind power grid supporting system and the development of wind power and consumptive price policy, etc. Zhu et al. (2011) reviewed the accommodation of wind power limited by system regulation capability, power system transmission capability, wind power technical performance, and wind power dispatch operation level. According to the development characteristics of wind power in China, the countermeasures were put forward from the angle of technology and national policy, respectively.

From the analysis above, the accommodation of wind power is affected by various factors. To determine the optimal construction scale of wind farm and the accommodation of wind power, every influencing factor should be comprehensively analyzed. The influence of the factors should be thoroughly considered in the process of research to lay the foundation of determining the accommodation of wind power.

In summary, the classification of influencing factors of wind power integration is summarized in Fig. 18.1.

18.2.2 Evaluation methods for wind power accommodation

The common evaluation methods for accommodation of wind power are: numerical simulation method, steady-state power flow simulation method, optimization method, frequency constraint method, and hybrid method (method using several methods of combining). Through specific analysis and further comparison of the differences between various methods, the ideas of researching the accommodation of wind power are provided.

18.2.2.1 Numerical simulation method

Basic idea: first, assume a wind power output, and set the system parameters and operation mode to simulate the wind power integration, then test the safety and stability of the system. Through revising the output of wind power, the wind power accommodation can be determined (Jin, 2011).

This method belongs to exploratory class method, and its shortcoming lies in its large amount of calculation and poor practicability.

Huang et al. (2013) established detailed dynamic model of power system with wind power integration based on PSCAD (Power Systems Computer Aided Design)/ EMTDC (Electromagnetic Transients including DC). Based on this model, through numerical simulation method, the relationship between the accommodation of wind power and the low voltage ride through capability under the premise of ensuring the system to be stable can be obtained. Cui (2011) proposed a numerical simulation method to calculate the accommodation of wind power considering the characteristics of the wind power and the anti-disturbance capability of power grid. Yu (2013) analyzes the voltage stability of wind farms integration node by voltage-reactive power coordinated control.

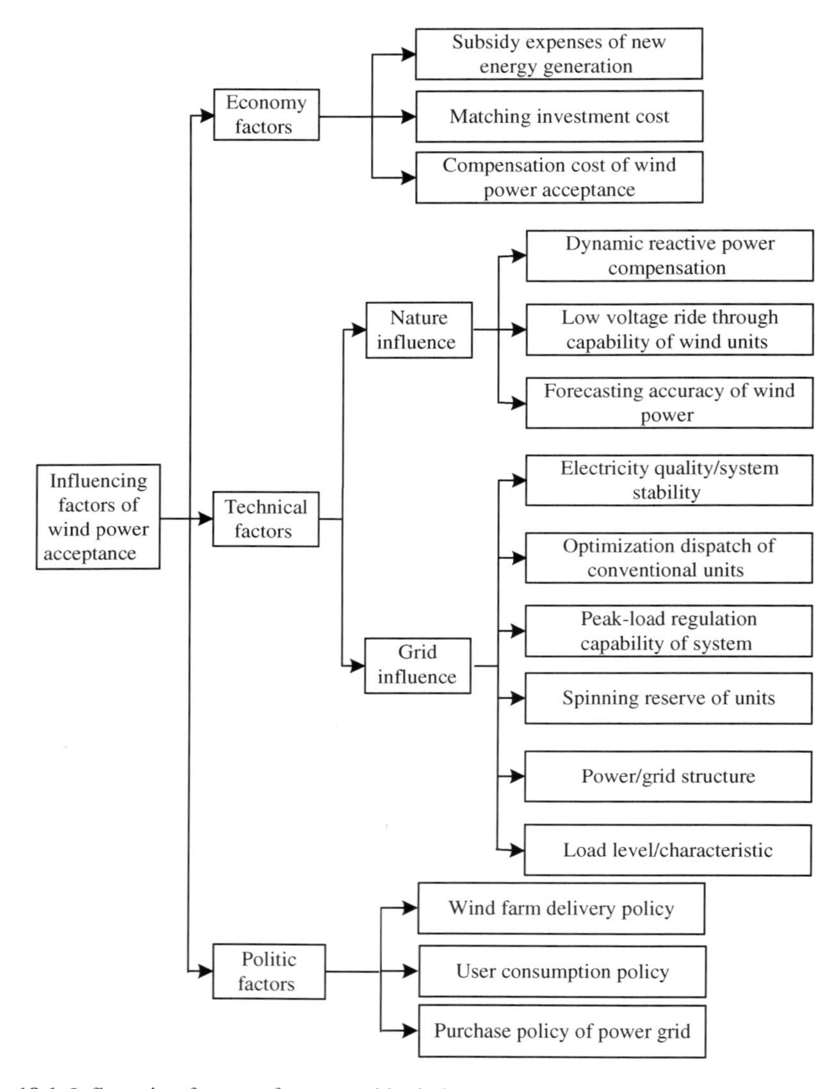

Fig. 18.1 Influencing factors of power grid wind power accommodation.

18.2.2.2 Steady-state power flow simulation method

Basic idea: establish a precise mathematical model of wind units, and solve the simultaneous equations of wind units model and system power, or solve the equations by alternating iteration (Wu et al., 2004).

This method is concerned about the change of power flow and node voltage in steady state after integration, while ignoring the impact of wind power on the system transient state. As a result, the accommodation of wind power has certain limitations.

Wu et al. (2005) established the steady-state analysis model of wind farm, considering the wake effect of wind farm and the influence factors of wind farm output. Based on this model, Keane et al. (2011) established a steady-state mathematical model of asynchronous generator, finding that the level of load and the maximum wind power accommodation of the system do not display simple linear relationship, the actual situation needs to be considered in the process of calculation. The steady-state mathematical model of the induction generator is established by using the power flow calculation (Wu et al., 2004), and the alternating iterative calculation of the power flow of the wind power system is realized. Combined with the analysis of actual system, it is proved that the accommodation of wind power is the result of multiple influence of various factors such as wind speed, grid structure, and so on.

18.2.2.3 Optimization method

The essential physical meaning of the accommodation of wind power is: under the existing control means and optimization operation strategy, the system accepts wind power (or capacity) to the maximum extent with the premise of meeting certain constraints. Therefore, the essence of transforming it into mathematics is the optimization problem.

The existing models usually use two methods to study this problem:

First, take wind power as a deterministic variable. The following mathematical models are established:
Objective function: maximum wind power accommodation of the system.
Equality constraint: the power balance equation of the system.
Inequality constraints: the system requirements for power quality and stability level.
The essence of accommodation of wind power is the optimization problem in mathematics.
That is:

$$
\begin{aligned}
&\max f(y)\\
s.t.&\\
&g(x, u, y) = 0\\
&h(x, u, y) \leq 0\\
&x, u : \text{common variable}
\end{aligned}
\tag{18.1}
$$

where u, x respectively represents control variables (physical variables that can be artificially changed and adjusted, such as the active and reactive power of system, unit commitment, phase shifter/adjustable transformer tap position, reactor/capacitor capacity) and state variables (physical variables that calculated by the control variables, such as the node voltage amplitude and phase angle, etc.); and y represents the accommodation of wind power.

Problems (Eq. 18.1) can be solved by mature algorithms such as nonlinear programming method, quadratic programming method, linear programming method, mixed programming method, modern interior point method, and artificial intelligence method.

The linear programming method was adopted to meet the system safety and stable operation, and the minimum cost of reactive power is taken as the objective to calculate the accommodation of wind power (Liew and Strbac, 2002). The problem of wind power accommodation was attributed to the maximization of the system under static security constraints, and the improved particle swarm optimization algorithm is used to solve the model (Yang et al., 2010b). Based on the daily dispatch operation, the accommodation of wind power is assessed (Kang et al., 2013), and the evaluation results can provide reference for the operation and dispatch of the system.

Second, the wind power is taken as an uncertain variable. Unlike other conventional units (such as hydropower units and thermal power units), wind power has strong randomness, the output of wind power is uncontrollable stochastic variables. If a deterministic variable is used to represent the wind power, the physical essence of wind power is veiled. Therefore, the power generated by the wind units can be considered as a nondeterministic variable. Besides the traditional control variables and state variables, the optimization variables also include stochastic variables, such as the wind power output, which makes the optimization problem of power system convert from the deterministic planning model into uncertain programming model.

The mathematical model of the optimization problem can be written as

$$
\begin{aligned}
&\max E(f(y)) \\
&s.t. \\
&\quad g(x, u, y, \xi) = 0 \\
&\quad h_1(x, u, y, \xi) \le 0 \\
&\quad \mathrm{pr}\{h_2(x, u, y, \xi)\} \ge \alpha \\
&\quad E\{h_2(x, u, y, \xi)\} \le \beta \\
&\quad x, u : \text{common variable} \\
&\quad \xi : \text{stochastic variable}
\end{aligned}
\tag{18.2}
$$

where u, x respectively represents control variables (physical variables that can be artificially changed and adjusted, such as the active and reactive power of system, unit commitment, phase shifter/adjustable transformer tap position, reactor/capacitor capacity) and state variables (physical variables that calculated by the control variables, such as the node voltage amplitude and phase angle, etc.); y represents the accommodation of wind power; $E(\cdot)$ represents the expectation function; and $\mathrm{pr}(\cdot)$ represents the probability function.

In mathematics, stochastic programming model (Eq. 18.2) can be solved by Monte Carlo, scenario method, analytic transformation of probability/mean function, and so on.

Lei et al. (2002), Tang et al. (2015), Wang et al. (2011a,b), Wu et al. (2007), Jiang et al. (2014), and Wu et al. (2006, 2008) took the stochastic characteristics of active power output of wind farm into account, and the problem of wind power accommodation was converted into a stochastic programming problem based on safe and stable operation constraints of the system.

18.2.2.4 Frequency constraint

The basic principle of frequency constraint method is: after the grid loses wind power, the frequency change of the power grid should be within the range of the regulation requirements after one or two times of frequency modulation for the conventional units (Xing, 2009).

This method considers only the single power index of frequency, and fails to take the influence of many factors on the accommodation of wind power into account.

Li et al. (2008a,b) gave the basic principle of frequency constraint method. Based on that, Xing (2009) calculated the maximum power injection of wind power with different wind units from the angle of local wind power stability and frequency level.

18.2.2.5 Hybrid method

The hybrid method is the combination of two or more than two methods.

The advantage of this method is that it is complementary in method and can overcome the limitation of adopting only one method. At the same time, the influencing factors of wind power accommodation can be considered systematically. The disadvantage of hybrid method is the complexity of modeling.

The method of combining steady-state analysis and numerical simulation verification was proposed (Li et al. 2008a,b). Use the steady flow method to determine the wind power capacity value, and then use the numerical simulation method to simulate of different disturbance the system bears, constantly update the wind power capacity, observe the change of dynamic system stability. As a result, the eventual accommodation of wind power can be determined with the system keeping the transient stability. The mathematical model of the unit was established (Wu, 2004), and then the mathematical model equation of the wind units and the power equation of the system were solved alternately and iteratively to obtain the accommodation capacity (initial value) of wind power. Then, the stability of the system under the wind power output is verified by numerical simulation method. Ning et al. (2013) used numerical simulation method combined with mathematical optimization, stochastic optimization model of wind power accommodation was first established, considering wind power and load double stochastic characteristics, substituted the optimization results into different operation modes to verify the stability of the test system, and then the accommodation of wind power can be determined.

All the above methods have their own characteristics, in practical applications, the specific circumstances of the problem should be considered, Table 18.1 summarizes the advantages and disadvantages of each method.

18.2.3 Research on the relationship between system adequacy and the wind power accommodation

At present, the research on the relationship between the system adequacy level and the accommodation of wind power is attracting more and more attention.

Table 18.1 The methods contrast table of the wind power integration capacity (Jin, 2011)

Method	Basic idea	Advantage	Disadvantage
Numerical simulation method	Assume a wind power output (to be revised), and set the system parameters and operation mode to simulate the wind power integration, then test the safety and stability of the system. Through revising the output of wind power, and then determine the accommodation of wind power	Clear physical meaning	The calculation is large and the practicability is poor
Steady-state power flow simulation method	Establish a precise mathematical model of wind units, and solve the simultaneous equations of wind units model and system power, or solve the equations by alternating iteration	Accurate steady-state mathematical model	Ignoring transient constraints
Optimization method	Under the safety constraints, establish the optimization model of wind power accommodation	Considering system static safety constraint	Ignoring dynamic constraints
Frequency constraint	After the grid loses wind power, the frequency change of the power grid should be within the range of the regulation requirements after one or two times of frequency modulation for the conventional units	Considering frequency regulation	The constraints are single
Hybrid method	The combination of two or more than two methods	Complementary in method	Complexity

From the angle of power generation, Tande (2007) and Wang et al. (2012a,b) considered the correlation among regions, and studied the generation adequacy assessment method for large-scale interconnected wind power systems. Wang and Li (2011), Chai (2012), and Hu et al. (2012) proposed a method of wind power consumptive mode to improve the level of adequacy in power system, which can significantly increase the accommodation of wind power. Jia et al. (2012) researched wind power-pumped storage optimization method combined with daily operation based on day-ahead load forecasting and wind power forecasting considering the adequacy of power system, and put forward a new model of wind power consumptive power. Atwa et al. (2010) and Madaeni et al. (2013) evaluate the adequacy of distributed power systems under different operating modes.

From the angle of transmission, Zhou et al. (2010) established the calculation model of the adequacy index of transmission system, and then used the nonsequential Monte Carlo simulation method to calculate the adequacy index of transmission system. Yan et al. (2009) considered the influence of the market on the transmission network, and put forward a transmission adequacy evaluation model considering market power index. Based on the actual situation of Jiuquan, Gansu Province, "wind-thermal-hydropower" combined delivery mode was proposed to improve the accommodation of wind power (Xu et al. 2010). The calculation model and method of adequacy index for multiple infeed high-voltage direct current (HVDC) transmission system were studied (Chen and Ren, 2005). Liu et al. (2011), Wang et al. (2011a,b), and Wu et al. (2013) studied the accommodation of wind power considering peak-load regulation constraints and transmission capability constraints, figured out that cross transmission is the most effective way to improve the wind power consumptive capability.

From the angle of accommodation, Zeng et al. (2014) comprehensively considered users accommodation method satisfaction and interaction benefit satisfaction, proposed real-time dispatch modeling method minimize the cost of interactive dispatch with load involved. Dietrich et al. (2012a,b) established day-ahead dispatch plan and operation simulation model considering the interaction of users' side.

Fan et al. (2012) and Toledo et al. (2013) comprehensively considered generation, transmission, and accommodation, and proposed to mobilize the wide and flexible adequacy resources involved in system regulation, drew the conclusion that the influence of peaking capacity resources on the wind power accommodation is the maximum one.

18.2.4 Existing problems

At present, the accommodation of wind power has made some research results. However, there are still some problems to be further studied.

(1) From the point of view of power generation, the evaluation index is too single. Wind power can be used as a load, but also can be used as the power supply, when the wind power used as a power supply can effectively improve power grid power shortage of the conventional unit; however, wind power will bring great pressure to the peak power regulation, restricting the accommodation of wind power.
(2) Aimed at the impact of system peak-load regulation capability on wind power integration, there is no uniform standard in current research.
(3) Focusing on the daily dispatch operation, the wind power accommodation of the grid is evaluated, but the evaluation results are difficult to provide a reference for the long-term planning of the system.
(4) Lacking of full use of the statistical characteristics of wind power output, only by experience or a certain extreme scenario to evaluate the wind power accommodation of the system, the results are difficult to accurately reflect the actual operation of the system.

18.3 Adequacy indices with wind power integration

After wind power integration, the pressure of peak-load regulation increased with the influence of the randomness and intermittency the wind power owes, conventional indices of power/electricity shortage expectation and power/electricity shortage probability cannot directly and accurately reflect the impact of wind power on power system. Therefore, more comprehensive indices should be introduced. Aimed at the influence wind power integration has on the peak-load regulation characteristics of system, it is necessary to research and analyze on the adequacy of operation system with wind power integration from multiple angles to evaluate the accommodation of wind power.

Based on this, in order to reflect the contradictory between wind power capacity and shortage of system peak-load regulation, this chapter introduces the index of peak-load regulation shortage expectation and probability to quantitatively analyze the influence of integration on system peak-load regulation adequacy. Therefore, in order to achieve the calculation of the adequacy index of the power system with wind power integration, this chapter mainly carries out the following two aspects of work:

(1) Use the kernel estimation theory to calculate the cumulative probability distribution function of the peak valley difference of the power system with or without wind power integration, respectively. Based on this, the peak valley difference characteristics of the system are analyzed.

(2) Use the kernel estimation theory to establish the classification level of the net load and the peak-load regulation demand, and the adequacy index of each classification level is calculated to obtain the overall adequacy index of the system.

18.3.1 Basic idea of calculating the adequacy indices of power system with wind power integration

The calculation steps calculating the adequacy indices of power system with wind power integration are shown in Fig. 18.2, and the process of calculation is:

Step 1: Use the load power and wind power data of power system, the net load curve of power system with wind power is obtained.

Step 2: Based on the above net load curve, the peak valley difference and peak-load demand curve of the power system with wind power are obtained, respectively.

Step 3: Use the kernel estimation theory, the cumulative probability distribution function of the system's net load, peak valley difference, and peak-load regulation demand are calculated, respectively.

Step 3.1: Based on the cumulative probability distribution function of peak valley difference, the peak valley difference characteristics of the system are analyzed.

Step 3.2: Based on the cumulative probability distribution function of net load and peak-load demand, the load and peak-load regulation demand are classified, respectively, and the level of each level and its corresponding probability are obtained.

Step 4: Based on the above classification level, the nonsequential Monte Carlo simulation method is used to calculate the adequacy indices of the system (including the peak-load regulation adequacy index and the generation adequacy index).

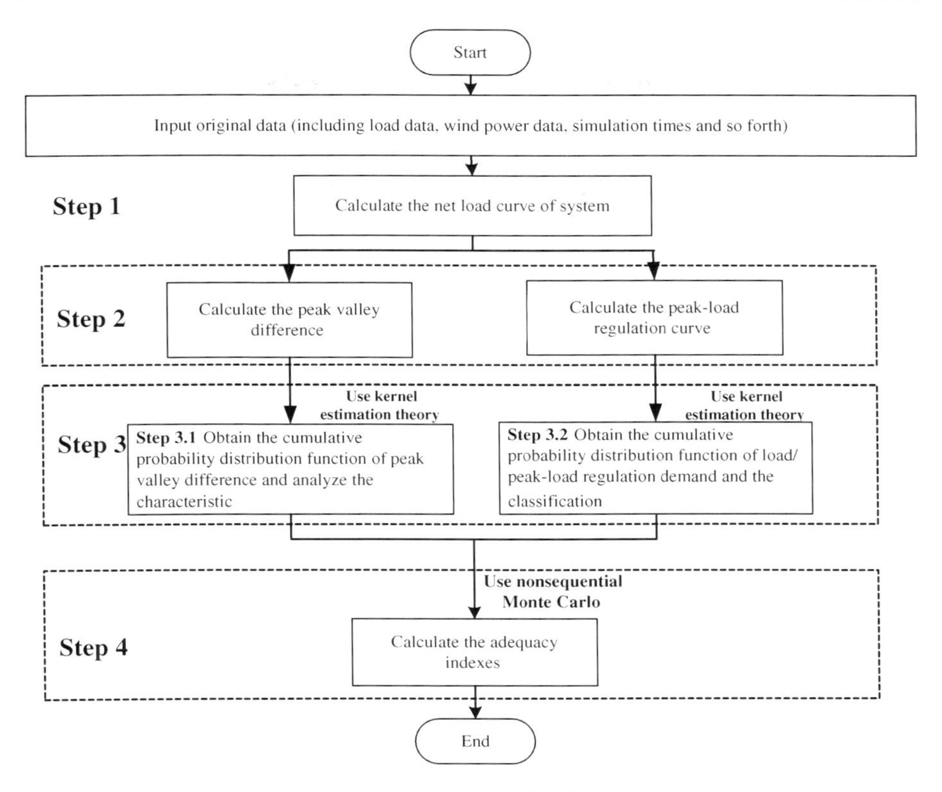

Fig. 18.2 The calculation idea of adequacy indices in wind power system.

From the idea of calculating indices of adequacy of power system with wind power integration, the net load curve, the peak valley difference, the peak-load regulation demand, and the classification model of load/peak-load regulation demand are the keys of current research, which will be introduced below.

18.3.2 The model involved in the calculation of the adequacy indices of power system with wind power integration

18.3.2.1 The calculation of net load curve

Generally, the net load is used to represent the load of the power system with wind power integration, that is, the system load minus the wind power output. The net load curve is shown in Fig. 18.3. The calculation process is shown in detail below.

Assume that $P_{wn} = l_1 \Delta P$, $E = l_2 \Delta E$ in the calculation scenario, when the system contains wind power only, $E = 0$; current wind power capacity of the system is P_{wn0}, historical wind power output is $P_{wind0,\ d,\ t}$, and historical load is $P_{load,\ d,\ t}$ (d represents the day, $d = 1, 2, \ldots, D$ and t represents the time period, $t = 1, 2, \ldots, T$), where $D = 365$, $T = 96$. The calculation process of net load is:

(1) Calculate the wind power $P_{wind,\ d,\ t}$ at time period t, day d using Eq. (18.3). And the sequence of wind power with different capacity P_{wn} can be approximately calculated:

$$P_{wind,d,t} = \frac{P_{wind0,d,t}}{P_{wn0}} l_1 \Delta P \tag{18.3}$$

where $P_{wind0,\ d,\ t}$ represents the historical wind power output at time period t, day d; P_{wn0} represents the current wind power capacity of the system; and $l_1 \Delta P$ represents the wind power capacity of current scenario.

(2) Calculate the net load $P_{netload,\ d,\ t}$ at time period t, day d using Eq. (18.4):

$$P_{netload,d,t} = P_{load,d,t} - P_{wind,d,t} \tag{18.4}$$

where $P_{load,\ d,\ t}$ represents the historical load at time period t, day d and $P_{wind,\ d,\ t}$ represents the wind power at time period t, day d.

18.3.2.2 The calculation of peak valley difference

The daily load curve of the power system depicts the change of the load within 24 h.

When the wind power is not considered, the maximum value of the load curve on day d of the system is called the maximum load of the day, namely peak-load, denoted as $P''_{max,\ d}$; the minimum value of the load curve on day d of the system is called the minimum load of the day, namely valley load, denoted as $P''_{min,\ d}$. The daily peak-load regulation demand represents the difference between peak-load and valley load, namely peak valley difference $P''_{dval,\ d}$, that is, $P''_{dval,\ d} = P''_{max,\ d} - P''_{min,\ d}$, as shown in Fig. 18.3.

When the wind power is considered, the net load curve model can be obtained, the maximum value of daily net load curve of the system is called peak net load $P_{max,\ d}$, the minimum value is called valley net load $P_{min,\ d}$, the peak valley difference is $P_{dval,\ d} = P_{max,\ d} - P_{min,\ d}$.

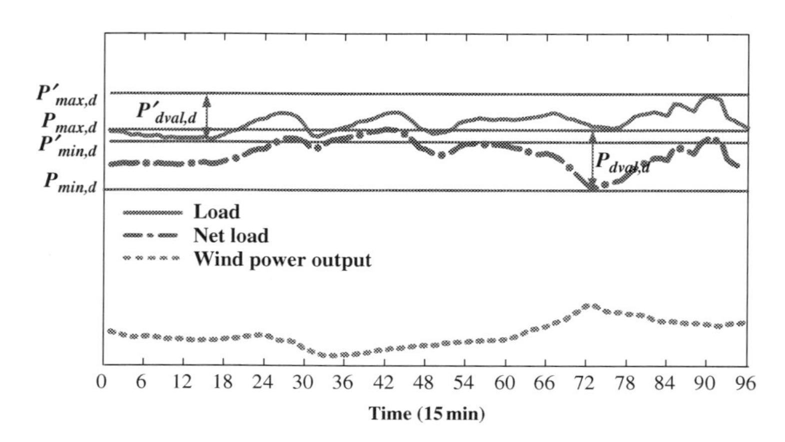

Fig. 18.3 Peak valley difference schemes.

18.3.2.3 The calculation of peak-load regulation demand

Based on the net load of the system, the peak-load regulation demand of the system can be calculated. The calculation method is: the peak-load regulation demand $L_{peakreq,\,d,\,t}$ at time period t, day d equals to the net load $P_{netload,\,d,\,t}$ minus the valley net load $P_{min,\,d}$, which is:

$$L_{peakreq,d,t} = P_{netload,d,t} - P_{min,d} \tag{18.5}$$

The peak-load regulation capacity $P_{reserve}$ can be calculated as:

$$P_{reserve} = \sum (P_{Gmax} - P_{Gmin}) \tag{18.6}$$

where P_{Gmax} represents the maximum technical output of system units and P_{Gmin} represents the minimum technical output of system units.

So far, the net load curve, peak valley difference, and peak-load regulation demand of each period of the system can be calculated. Based on this result, using kernel estimation theory, research will be carried out from the following two aspects:

(1) Calculate the cumulative probability distribution function of peak valley difference with and without wind power, respectively, based on this, analyze the peak valley difference characteristics.
(2) Establish the level of net load classification and the level peak-load regulation demand classification, calculate the adequacy index of each classification level, so as to calculate the overall adequacy index of the system.

Therefore, the kernel estimation theory and load estimation based on kernel estimation theory are introduced first.

18.3.2.4 The classification model of load/peak-load regulation demand based on kernel estimation theory

(1) Kernel density estimation (Wei and Zhang, 2008) is one of the most common methods in nonparametric density estimation. Let $X_1, X_2, \ldots, X_n$ be a sample from a unary continuous population (assume $X_1, X_2, \ldots, X_n$ represents the historical data of peak demand), and the kernel density estimator of the population density function $f(x)$ and the cumulative probability distribution function $F(x)$ at any point x is defined as:

$$\hat{f}(x) = \frac{1}{nh}\sum_{i=1}^{n} K\left(\frac{x - X_i}{h}\right) \tag{18.7}$$

$$\hat{F}(x) = \frac{1}{n}\sum_{i=1}^{n} K\left(\frac{x - X_i}{h}\right) \tag{18.8}$$

$$K\left(\frac{x - X_i}{h}\right) = \frac{1}{\sqrt{2\pi}} \exp\left[-\frac{1}{2} \times \left(\frac{x - X_i}{h}\right)^2\right] \tag{18.9}$$

$$h = 1.06 S n^{-0.2} \tag{18.10}$$

where $K(\cdot)$ is called kernel function; h is called window width; S represents the sample standard deviation; and n represents the number of the sample. In addition, the kernel function $K(\cdot)$ is required to satisfy:

$$K(x) \geq 0, \int_x^{\infty} K(x) = 1 \tag{18.11}$$

(2) Calculate the adequacy index using classification model, as shown in Fig. 18.4 (Li et al., 2014).

Averagely divide the peak-load regulation demand or the load curve into K levels, the curve represents the cumulative probability distribution function F of peak-load regulation demand or load, function F can be obtained by the historical load and peak-load regulation demand using kernel estimation method (Wei and Zhang, 2008). This chapter employs the data of a year at intervals of 15 min, 96 × 365 points in total, and the load level/peak-load regulation demand. L_1, L_2, …, L_k represent the peak-load regulation demand/load level, T_k represents the duration of level k, which equals to the duration of the historical data calculating curve F divides K, and P_k represents the probability of peak-load regulation demand/load level, which can be calculated by Eq. (18.10):

$$\begin{cases} P_1 = F^{-1}\left(\dfrac{L_1 + L_2}{2}\right) \\ P_k = F^{-1}\left(\dfrac{L_{k+1} + L_k}{2}\right) - F^{-1}\left(\dfrac{L_{k-1} + L_k}{2}\right), \quad k = 2, \ldots K-1 \\ P_K = 1 - F^{-1}\left(\dfrac{L_K + L_{K-1}}{2}\right) \end{cases} \tag{18.12}$$

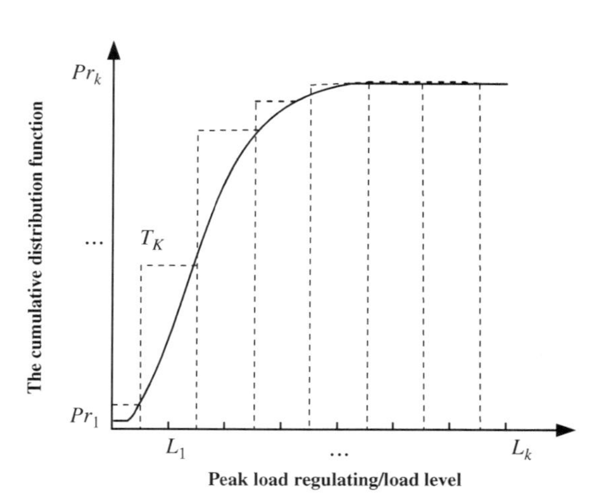

Fig. 18.4 Peak-load regulating requirement/load multi-level hierarchical model.

18.3.3 The calculation of adequacy indices of the power system with wind power integration

18.3.3.1 The adequacy index of the power system with wind power integration

The adequacy of power system represents that whether the power generation, transmission, and other equipment can meet the system load and operation requirements. Generally, two indices are employed (Sun et al., 2011; Billinton et al., 2012), which are the power shortage probability and the power shortage expectation. In order to reflect the contradictory between the wind power capacity and the peak-load regulation capacity shortage, this chapter introduced the influence of PRNEP and PRNEE, accordingly to guide the evaluation of wind power accommodation.

(1) PRNEP and PRNEE are denoted by P_{PRNEP} and E_{PRNEE}, respectively:

$$P_{\text{PRNEP}} = \sum_{k=1}^{K} P_{\text{PRNEP},k} \cdot P_k \tag{18.13}$$

$$E_{\text{PRNEE}} = \sum_{k=1}^{K} \left(\frac{T_k}{N_k} \sum_{i=1}^{N_k} P_{\text{RNE},k,i} \right) \tag{18.14}$$

where N_k represents the Monte Carlo sampling frequency calculating the adequacy index of level k; $P_{\text{PRNEP},\ k}$ represents the PRNEP of level k; P_k represents the probability of peak-load regulation demand of level k; and $P_{\text{RNE},\ k,\ i}$ represents the shortage of peak-load regulation at ith sampling of level k. The physical meaning of PRNEE is: to ensure the real-time balance of supply and demand of power system, the expectation to be cut off under a certain level of wind power capacity.

(2) LOLP and LOEE are denoted by P_{LOLP}, E_{LOEE}:

$$P_{\text{LOLP}} = \sum_{k=1}^{K} P_{\text{LOLP},\ k} \cdot P_k \tag{18.15}$$

$$E_{\text{LOEE}} = \sum_{k=1}^{K} \left(\frac{T_k}{N_j} \sum_{i=1}^{N_j} P_{\text{LNE},\ k,i} \right) \tag{18.16}$$

where N_j represents the Monte Carlo sampling frequency calculating the adequacy index of level k; $P_{\text{LOLP},\ k}$ represents the LOLP of level k; P_k represents the probability of load level of level k; and $P_{\text{LNE},\ k,\ i}$ represents the shortage of generation at ith sampling of level k. The physical meaning of LOEE is: the expectation of electricity shortage of the system under a certain level of wind power capacity.

18.3.3.2 The adequacy indices calculation based on nonsequential Monte Carlo

Under the condition of given wind power capacity and storage capacity of the system, the basic idea of calculating the system adequacy indices is calculated based on non-sequential Monte Carlo method is: first calculate the net load and peak-load regulation demand of the system, then use kernel density estimation theory to evaluate the cumulative probability distribution function of net load and peak-load regulation demand, and establish the classification level of net load and peak-load regulation demand. The adequacy index of each classification level is calculated, and finally the adequacy indices of all classification levels is calculated to obtain the adequacy indices of the system. The steps are as follows (Fig. 18.5).

18.4 Estimation of wind power accommodation

It is the bottleneck that whether the wind power grid can be integrated and how large is the wind power accommodation, also it is the focus of power system dispatch and planning department. The peak-load regulation capacity of power grid is the most fundamental factor that restricts the accommodation of wind power in power system. If the integrated wind power capacity exceeds the peak-load regulation limit of the grid, the power grid will be difficult to maintain power balance, resulting in the phenomenon of wind curtailment. If the integrated wind power capacity is too small, a large number of wind resources cannot be developed and utilized, resulting in a great waste of wind power resources. Therefore, the peak-load regulation adequacy is an important constraint to determine the capacity of wind power integration, which is of great significance for guiding the orderly development of wind power and ensuring the safe and reliable power supply of the power grid.

This chapter introduces PRNEP, PRNEE, LOLP, and LOEE these four indices to quantify the influence the accommodation of wind power on system adequacy, accordingly guide the evaluation of wind power accommodation, build the function between wind power capacity, storage capacity and adequacy index using the bilinear interpolation method, the accommodation of wind power meeting given adequacy levels can be evaluated. As a result, a practical evaluation method for wind power accommodation can be proposed, which combines wind power statistical characteristics, meets the reliability requirements and is more reasonable.

18.4.1 Evaluation idea of wind power accommodation

The evaluation steps for the adequacy of the system with wind power integration are shown in Fig. 18.6, the specific calculation process is as follows.

Step 1: Establish a combination of different wind power capacity and energy storage capacity of the system, call each combination as a scenario.

Step 2: Calculate the comprehensive net load curve under different scenarios, and calculate the peak- load demand of the system based on the comprehensive net load curve.

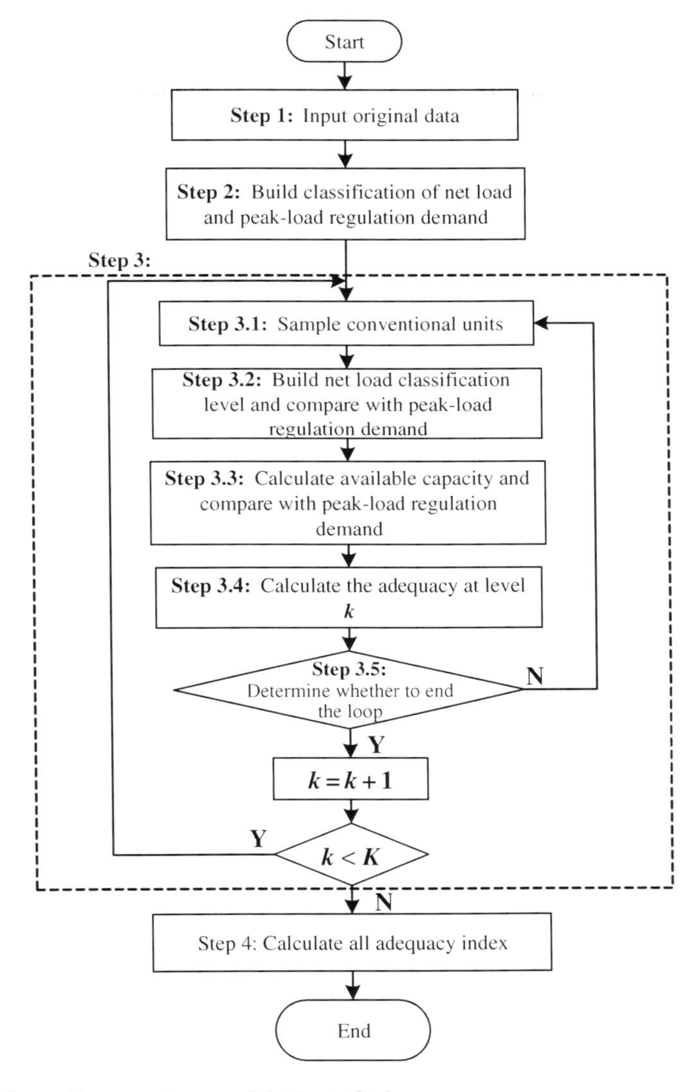

Fig. 18.5 Steps diagram of sequential Monte Carlo.

Step 3: Use the method of nonsequential Monte Carlo to calculate the adequacy index of the system under different scenarios, and obtain sample vectors the same number as scenarios (wind power capacity, storage capacity, the adequacy index).
Step 4: Based on the sample vectors obtained in step 3, a three-dimensional (3D) relation diagram between adequacy index, energy storage capacity, and wind capacity is established.
Step 5: The bilinear interpolation method is used to fit the sample vectors, and the function relationship of wind power capacity, energy storage capacity, and adequacy index is

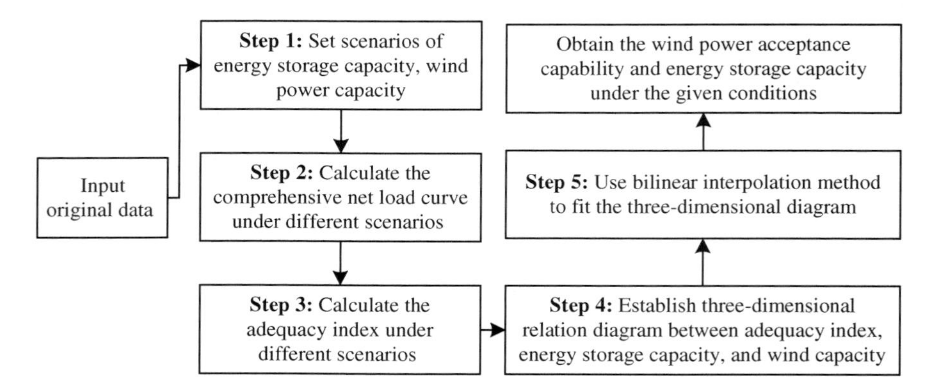

Fig. 18.6 The evaluation steps of wind power capacity.

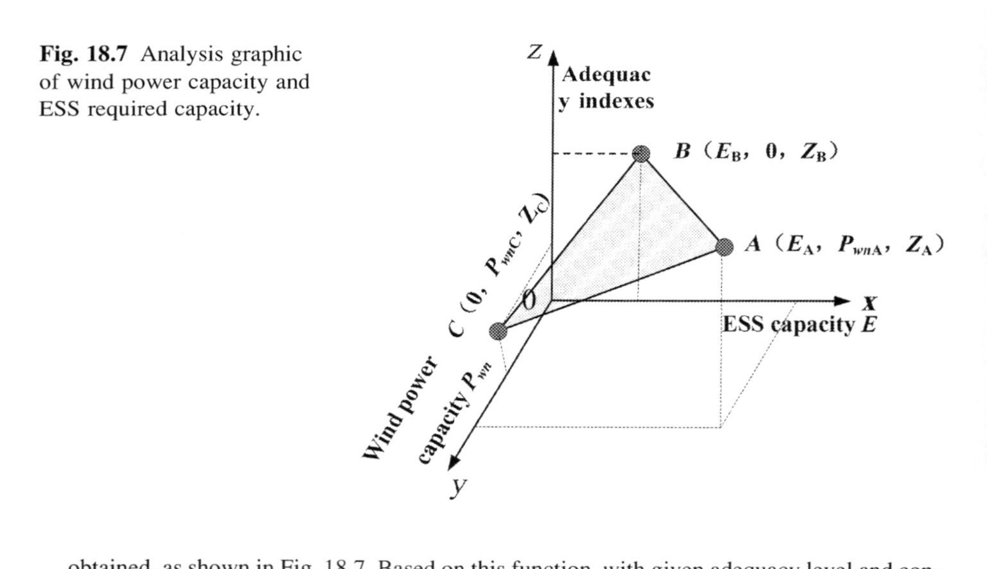

Fig. 18.7 Analysis graphic of wind power capacity and ESS required capacity.

obtained, as shown in Fig. 18.7. Based on this function, with given adequacy level and configurated storage capacity, the accommodation of wind power can be calculated; similarly, with given adequacy level and the accommodation of wind power, the storage capacity can be calculated.

From Fig. 18.7, the X-axis represents the energy storage capacity, the Y-axis represents the wind power capacity of the system, and the Z-axis indicates the adequacy index. Point A is an arbitrary point on the plane of wind power capacity, energy storage capacity, and adequacy index, which has two meanings:

(1) The energy storage capacity E_A to be configurated under the condition of adequacy index Z_A and wind power accommodation P_{wnA}.
(2) The accommodation P_{wnA} to be determined under the condition of adequacy index Z_A and energy storage capacity E_A.

In particular, when the point is located in x-o-z plane, such as point $B(E_B, 0, Z_B)$, this shows the relationship between the ESS capacity and adequacy indices without access to the wind power; when the point is located in y-o-z plane, such as point $C(0, P_{wnC}, Z_C)$, this indicates the relationship between the wind power capacity and adequacy indices without ESS.

Therefore, the energy storage capacity or the acceptable wind power capacity of the system can be easily obtained under the condition that the given adequacy index is satisfied.

Therefore, the setting of scenarios and the establishment of model are the key points, which will be introduced, respectively.

18.4.2 The setting of multiple scenarios

Denote ΔE and ΔP as the basic unit of energy storage capacity and wind power capacity, respectively. As a result, the possible values of energy storage capacity can be: $E = 0,\ \Delta E,\ 2\Delta E,\ 3\Delta E,\ \ldots,\ m\Delta E$; similarly, the possible values of wind power capacity can be: $P_{wn} = 0,\ \Delta P,\ 2\Delta P,\ 3\Delta P,\ \ldots,\ n\Delta P$. m and n limit the maximum value of energy storage capacity and wind power capacity, respectively. ΔE and ΔP can be selected according to the requirements of calculation accuracy, and the size of m and n is selected according to the actual system planning load capacity. Considering the calculation speed and accuracy, m and n are taken as 100.

Combine the energy storage capacity and the wind power capacity, four systems can be obtained as shown in Table 18.2.

System 1: $E = 0$, $P_{wn} = 0$ represents the conventional system, which does not consider the energy storage and the wind power.
System 2: $E = 0$, $P_{wn} \neq 0$ represents the system not containing energy storage but the wind power.
System 3: $E \neq 0$, $P_{wn} \neq 0$ represents the conventional system containing different energy storage capacity.
System 4: $E \neq 0$, $P_{wn} \neq 0$ represents the power system containing both energy storage and wind power.

Table 18.2 The combination of multiple scenarios setting

Energy storage capacity	**Wind power capacity**					
0	0	ΔP	$2\Delta P$	$3\Delta P$	...	$n\Delta P$
ΔE	0	ΔP	$2\Delta P$	$3\Delta P$	...	$n\Delta P$
$2\Delta E$	0	ΔP	$2\Delta P$	$3\Delta P$	...	$n\Delta P$
$3\Delta E$	0	ΔP	$2\Delta P$	$3\Delta P$	...	$n\Delta P$
...	0	...	...	...	...	...
$m\Delta E$	0	ΔP	$2\Delta P$	$3\Delta P$	...	$n\Delta P$

18.4.3 Calculation model for wind power accommodation

18.4.3.1 The calculation of comprehensive net load

In the wind power system, load minus wind power output is called net load. However, this traditional net load calculation failed to reflect the function of load shifting and peak smoothing of energy storage. So, this chapter puts forward comprehensive net load model, which includes the three aspects of load, wind power output, and ESS.

Assuming that in the current scene, $P_{wn} = l_1\Delta P$, $E = l_2\Delta E$, then set the current wind power capacity as P_{wn0}, the historical load curve as $P_{load,\ d,\ t}$ (d represents number of days, $d = 1, 2, \ldots, D$; t means periods of time, $t = 1, 2, \ldots, T$, the same as below), in this chapter, $D = 365$, $T = 96$. The specific process comes as follows:

1. According to Eq. (18.17), the wind power data $P_{wind,\ d,\ t}$ is derived from wind power capacity $l_1\Delta P$ at the period of t on day d. Thus, wind power sequence is obtained under the different capacity of P_{wn} approximately generated:

$$P_{wind,d,t} = \frac{P_{wind0,d,t}}{P_{wn0}} \times l_1 \Delta P \tag{18.17}$$

In this formula, $P_{wind0,\ d,\ t}$ is the wind power output date at period of t on day d; P_{wn0} is the initial wind power capacity; ΔP is the step length of wind power capacity; and l_1 is the total number of step length.

2. According to Eq. (18.18), the net load to $P_{netload,\ d,\ t}$ is calculated at period t on day d without inclusion of ESS as follows:

$$P_{netload,d,t} = P_{load,d,t} - P_{wind,d,t} \tag{18.18}$$

where $P_{netload,\ d,\ t}$ is the net load at the period of t on day d; $P_{load,\ d,\ t}$ is the historical load value of the system at period of t on day d; and $P_{wind,\ d,\ t}$ is the wind power data at the period of t on day d.

3. Using $E = l_2\Delta E$ to correct the net load curve to get the comprehensive net load at the period of t on day$d$.

Assuming the trough value of net load curve is $P_{min,\ d}$ in dth day, apply E to the r periods of time nearby the trough (r is empirical value, here $r = 5$), as shown in the shaded part of Fig. 18.8. Then, the trough value of net load increases from $P_{min,\ d}$ to $P'_{min,\ d}$, thus reducing the peak-valley difference of net load that day. The revised trough value of net load$P'_{min,\ d}$ is obtained from Eq. (18.19):

$$\sum_{t_1=1}^{r} \left(P'_{min,d} - P_{netload,d,t_1}\right)\Delta t = l_2 \Delta E \tag{18.19}$$

In this formula, Δt is the interval of 15 min; $P_{netload,\ d,\ t1}$ is the net load nearby trough at the r periods of time; ΔE is the step length of ESS capacity; and l_2 is the total number of step length of ESS capacity.

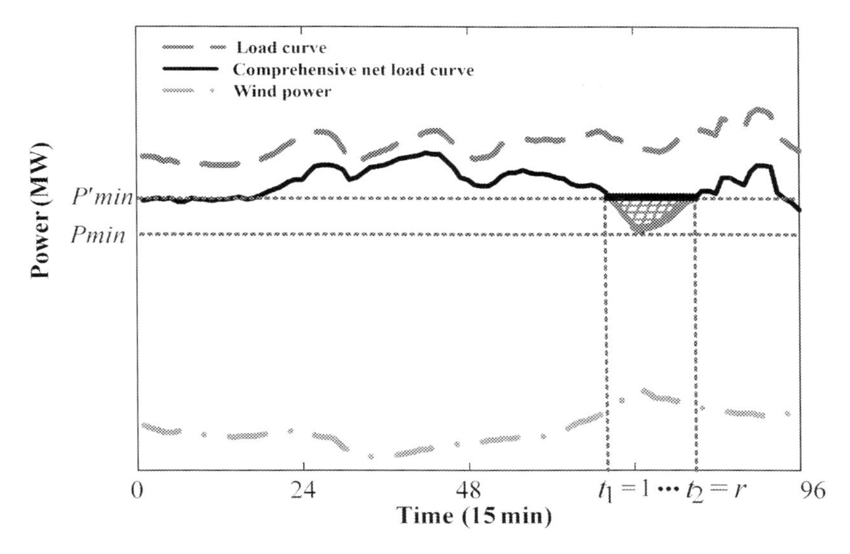

Fig. 18.8 The sketch map of comprehensive net load curve.

18.4.3.2 The calculation of peak-load regulation requirement

Based on the comprehensive net load data of the system with 96 periods of time and an interval of 15 min in 365 days, peak-load regulating requirement is calculated at every period every day. The peak-load regulating requirement of $L_{peakreq,\,d,\,t}$ at the period of t on day d equals the difference value between the comprehensive net load at this period and the trough value of the comprehensive net load of $P'_{min,\,d}$ that day, as indicated in Eq. (18.20):

$$L_{peakreq,d,t} = P_{comload,d,t} - P'_{min,d} \tag{18.20}$$

where $P_{comload,\,d,\,t}$ is the comprehensive net net load at the period of t on day d and $P'_{min,\,d}$ is the trough value of comprehensive net load on day d.

The formula for calculating peak-load regulating capacity of $P_{reserve}$ in the system comes as follows:

$$P_{reserve} = \sum (P_{Gmax} - P_{Gmin}) \tag{18.21}$$

where $P_{comload,\,d,\,t}$ is the comprehensive net load at the period of t on day d; P_{Gmax} is the maximum technical output power of the system's generator unit; and P_{Gmin} is the minimum technical output power of the system's generator unit.

The comprehensive net load and peak-load regulating requirement of the system at each period of each day are obtained through Table 18.2 set in all scenarios in Section 18.4.3. The cumulative probability distribution function of comprehensive net load and peak-load regulating requirement in different settings is got by estimating the data in 365 days via kernel density estimate (Xue et al., 2014). According to the computing results, by using hierarchical model of Section 18.2.2.4, this chapter calculates the adequacy indices of the system so as to obtain vector samples in the same number of scene (such as wind power capacity accepted, energy storage capacity, and adequacy indices). The function relation between wind power capacity accepted by the system and ESS capacity is arrived at by way of data sampling obtained and under the condition of satisfying the given adequacy got through the matching of bilinear interpolation.

18.5 The optimal allocation of the wind power capacity and ESS capacity based on bilinear interpolation

18.5.1 Fitting for the function relationship of adequacy indices, wind power capacity, and ESS capacity

This chapter fits the vector samples (wind power capacity, ESS capacity, and adequacy indices) obtained by using the bilinear interpolation (Liu et al., 2015). Set adequacy level and ESS capacity the known value and the point of wind power capacity can be obtained, or make adequacy level and wind power capacity integrated the known value and the point of ESS capacity can also be obtained. These points are called as the interpolating point. As shown in Fig. 18.9, point Q is interpolating point. A1–A4 whose values are A1(E_{A1}, P_{wnA1}, Z_{A1}), A2(E_{A2}, P_{wnA2}, Z_{A2}), A3(E_{A3}, P_{wnA3}, Z_{A3}), and A4(E_{A4}, P_{wnA4}, Z_{A4}) are closest to the interpolating point among the known samples.

The bilinear interpolation function of ESS capacity E, P_{wn}, and Z is mathematically expressed as

$$Z = (a \cdot E + b)(c \cdot P_{wn} + e) \tag{18.22}$$

where a, b, c, e are undetermined coefficients.

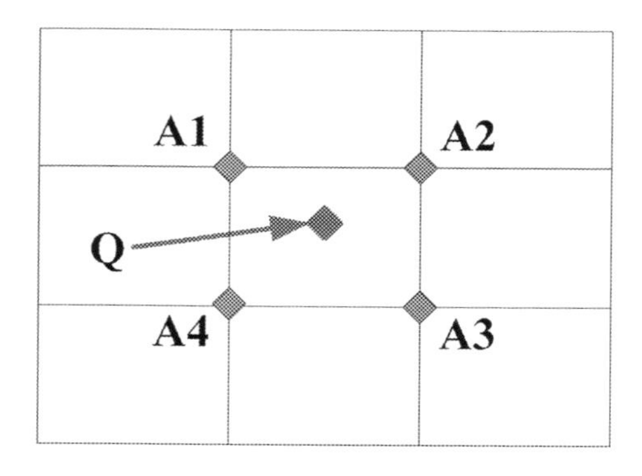

Fig. 18.9 The sketch map of bilinear interpolation grid.

Substitute coordinate values of points A1–A4, which are closest to the interpolation point into Eq. (18.22), and four equations are derived from. Then, connect and solve them, and four undetermined coefficients of a,b,c,e are obtained. Thus, the mathematical expression reflecting the relationship of the three interpolation points of wind power capacity, ESS capacity, and adequacy level are arrived at.

18.5.2 The optimal allocation of wind power capacity and ESS capacity

Based on Section 18.5.1, the function relation of ESS capacity E, wind power capacity P_{wn}, and adequacy indices are obtained. The threshold level of PRNEP, PRNEE, LOLP, and LOEE is set to α, β, γ, δ, respectively. The optimization solution of maximum wind power capacity and minimum ESS capacity will be obtained by using the four simultaneous equations.

1. When the adequacy indices and energy storage capacity are given, the calculation of the maximum wind power capacity accepted by the system is expressed as follows:

$$\min\left(P_{wn,1}, P_{wn,2}, P_{wn,3}, P_{wn,4}\right)$$
$$st.\begin{cases} P_{wn,1} \leq \left(\dfrac{\alpha}{(a_1 \cdot E + b_1)} - e_1\right)/c_1 \\ P_{wn,2} \leq \left(\dfrac{\beta}{(a_2 \cdot E + b_2)} - e_2\right)/c_2 \\ P_{wn,3} \leq \left(\dfrac{\gamma}{(a_3 \cdot E + b_3)} - e_3\right)/c_3 \\ P_{wn,4} \leq \left(\dfrac{\delta}{(a_4 \cdot E + b_4)} - e_4\right)/c_4 \end{cases} \tag{18.23}$$

2. ESS capacity required by the system is obtained when the adequacy indices and the wind power capacity are given, the formula is as follows:

$$\max\left(E_1, E_2, E_3, E_4\right)$$
$$st.\begin{cases} E_1 \leq \left(\dfrac{\alpha}{(c_1 \cdot P_{wn} + e_1)} - b_1\right)/a_1 \\ E_2 \leq \left(\dfrac{\beta}{(c_2 \cdot P_{wn} + e_2)} - b_2\right)/a_2 \\ E_3 \leq \left(\dfrac{\gamma}{(c_3 \cdot P_{wn} + e_3)} - b_3\right)/a_3 \\ E_4 \leq \left(\dfrac{\delta}{(c_3 \cdot P_{wn} + e_3)} - b_3\right)/a_3 \end{cases} \tag{18.24}$$

where E is the ESS capacity; P_{wn} is the wind power capacity; and the values of PRNEP, PRNEE, LOLP, and LOEE are set to α, β, γ, δ, respectively. Therefore, Eqs. (18.23), (18.24) are optimally solved to get the minimum energy storage capacity and the maximum wind power capacity accepted by the system.

18.6 Case study

18.6.1 Case description

This chapter adopts a certain actual power system. The total installed capacity of conventional power plants is 21,596 MW, of which the thermal power capacity is 20,906 MW, accounting for 96.8% of the total; and of which hydropower capacity is 690 MW, accounting for 3.2% of the total. Its maximum load is 22,303.11 MW. According to the adjustable capacity of the conventional unit, the calculation of the total peak-load regulating capacity of the system is 6151.5 MW. We adopt the actual data of the wind power in the corresponding year in a province where the maximum wind power output is 8635.19 MW that year, the capacity of ESS is 10 MWh. Assuming that forced outage rate of the generating unit is 0.05.

18.6.2 The relationship of the wind power capacity, ESS capacity, and adequacy indices

Under the condition of the different given threshold values, analyze the wind power acceptability and energy storage allocation in the system. And fit the vector samples of wind power capacity, ESS capacity, and adequacy indices obtained in various settings. Due to limited space, this chapter only illustrates how to obtain the function relationship between LOLP and wind power capacity and storage energy capacity.

First, substitute the points A1–A4 close to the interpolation point whose values are A1 (10,8000,0.002), A2(10,6000,0.0004), A3 (20,6000,0.000001), and A4(20,8000,0.0009) into Eq. (18.23) of the bilinear interpolation, get the undetermined coefficients of a_1,b_1,c_1,e_1 which reflect the bilinear interpolation function of LOLP. Based on the bilinear interpolation, fit the function relationship of LOEE, PRNEP, PRNEE, with wind power capacity and ESS capacity, respectively.

Setting the threshold levels of LOLP as 0.002% and 0.0015%, LOEE as 260 and 270 MWh/a, PRNEP as 2% and 1.5%, PRNEE as 3000 and 2500 MWh/a. So, the results of fitting are shown from Figs. 18.10 to 18.13.

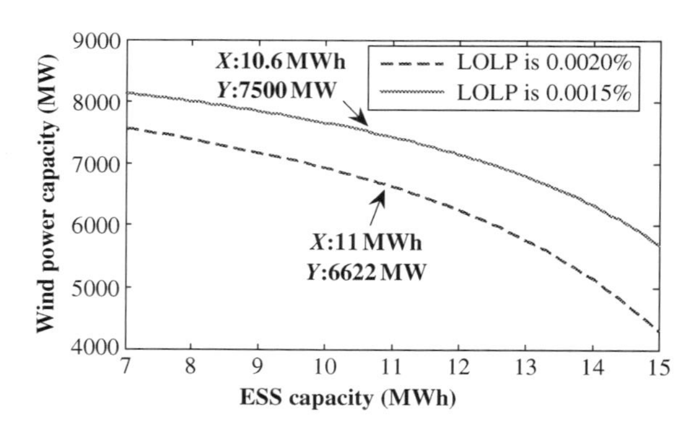

Fig. 18.10 α are 0.002% and 0.0015%, the fitting results of E and P_{wn}.

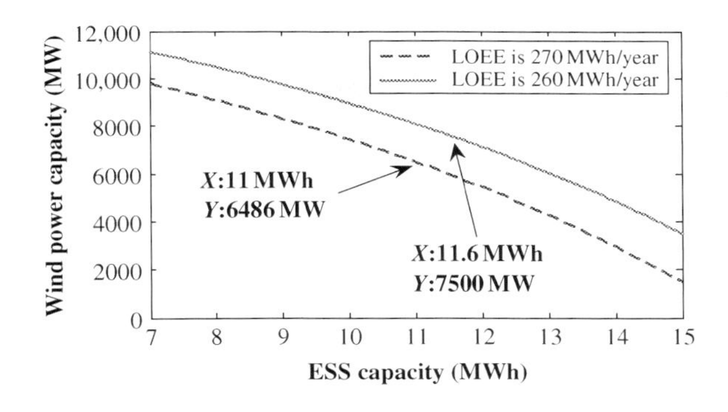

Fig. 18.11 β are 260 and 270 MWh/d, the fitting results of E and P_{wn}.

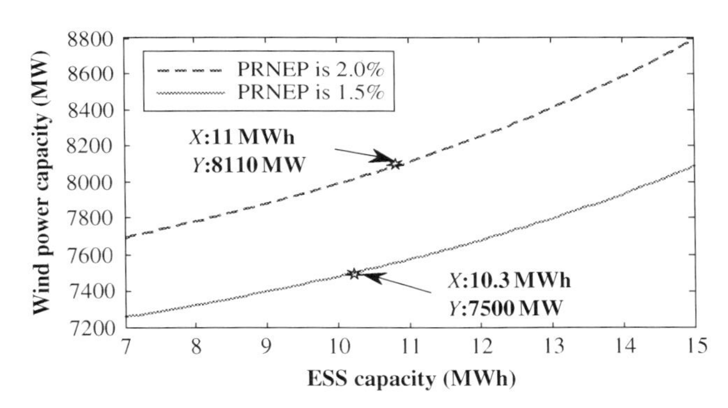

Fig. 18.12 γ are 2% and 1.5%, the fitting results of E and P_{wn}.

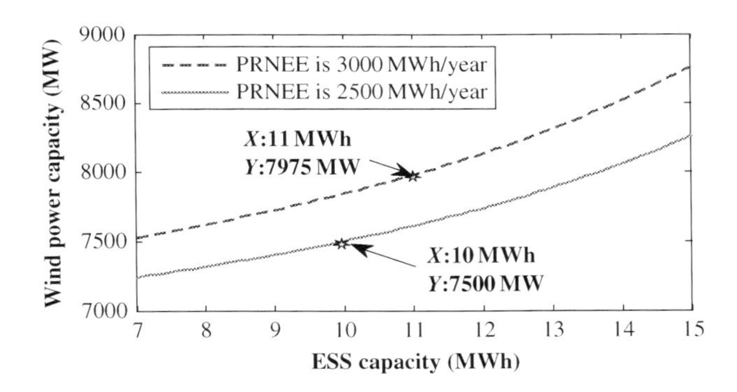

Fig. 18.13 δ are 3000 and 2500 MWh/d, the fitting results of E and P_{wn}.

As seen in Figs. 18.5 and 18.6, after wind power farm is connected into the electricity grid, and the ESS capacity holds constant, with the increase of wind power capacity in the system, the generating adequacy indices of the system (such as LOLP and LOEE) will gradually decrease. Under the condition of constant wind power capacity, as the ESS capacity increases, the generating adequacy indices of system will decline, too. Consequently, increase in the penetration level of wind energy or introduction of ESS system will, to some extent, improve the generating adequacy of the system.

From Figs. 18.12 and 18.13, it is easy to see that PRNEP and PRNEE in the system gradually increase with the enlargement of wind power integration, which shows that as the scale of wind power integration expands, the pressure of peak-load regulation will be increasingly growing. However, with the addition of ESS system, as ESS capacity increases, PRNEP and PRNEE in the system will be constantly decreasing, which indicates the addition of ESS will be helpful to better the situation of peak-load regulation of the system.

When the ESS capacity is 11 MWh, LOLP, LOEE, PRNEP, and PRNEE are respectively met as 0.002%, 270 MWh/year, 2% and 3000 MWh/year, the maximum integrated wind power capacity required by the system, respectively, are 6622, 6486, 8110, and 7975 MW. By optimizing Eq. (18.23), to meet the requirements of four adequacy indices at the same time, the required wind power capacity accepted by the system is 6486 MW.

When the wind power capacity is 7500 MW, and the values of LOLP, LOEE, PRNEP, and PRNEE are respectively met as 0.0015%, 260 MWh/year, 1.5%, and 2500 MWh/year, the ESS capacity allocation required by the system, respectively, are 10.6, 11.6, 10.3, and 10.0 MWh. By optimizing Eq. (18.24), to meet the requirements of four adequacy indices at the same time, the ESS capacity required the system is 11.6 MWh.

The function relationship of ESS capacity and wind power capacity under certain threshold levels of adequacy indices is shown in Figs. 18.10–18.13. Select upper limit values of the generation adequacy indices and adequacy indices of peak-load regulation, and obtain the optimization solution to the minimum ESS capacity and the maximum wind power capacity by the above simultaneous equations. Then, obtain wind power capacity which meets the given adequacy index under the condition of different ESS capacity, or obtain ESS capacity of configuration required by system with given wind power accommodation capacity. Thus, the optimization and coordination of wind power capacity and the ESS capacity accepted by the system are realized.

18.6.3 The analysis of the maximum wind power accommodation of the systems

According to the function relationship of wind power capacity, ESS capacity, and the four adequacy indices in Section 18.6.2, as well as with the different given threshold levels of the four adequacy indices and given ESS capacity, use Eq. (18.23) to obtain the system acceptability of wind power, and the results are shown in Table 18.2.

The optimal coordination results of energy storage and wind power system are listed in Table 18.3 by providing different adequacy indices. Set the threshold values of PRNEP,PRNEE,LOLP, and LOEE as 1.5%, 2500 MWh/a, 0.002%, and 270 MWh/a respectively in order to maximize the wind power acceptability, when the ESS capacity is 10 MWh, the maximum wind power capacity is 6926 MW, the wind power penetration limit is 31.1%. ESS increases the wind power penetration

Table 18.3 The accommodation of wind power with ESS

ESS capacity (MWh)	The value of the threshold level				The accommodation of wind power	
	LOLP (%)	**LOEE (%)**	**PRNEP (%)**	**PRNEE (MWh)**	**Wind power capacity (MW)**	**Percentage of load (%)**
10	0.2	270	1.5	2500	6926	31.3
10	0.15	260	2	3000	7662	34.4
50	0.2	270	1.5	2500	9340	41.9
50	0.15	260	2	3000	10,980	49.2
100	0.2	270	1.5	2500	13,587	60.9
100	0.15	260	2	3000	15,480	69.4

limit of the system. The utilization of ESS can raise the penetration rate of wind power. Table 18.3 shows that when the generating adequacy indices are lowered and meanwhile adequacy indices of the peak-load regulating are enhanced, the penetration rate of wind power to a certain degree can be improved. In addition, Table 18.3 indicates that after the increase of stored energy, the capability of system accepting wind power will be markedly improved.

18.6.4 Analysis of the minimum ESS capacity accepted by the system

According to Section 18.6.2 for the function relationship of wind power capacity, ESS capacity, and four adequacy indices, when the values of the four adequacy indices and wind power capacity were given in different threshold conditions, by using Eq. (18.24) and obtained the acceptability of wind power capacity, as the results are shown in Table 18.4.

Table 18.4 The allocation of ESS capacity under different scenarios

Wind power capacity (MW)	The value of the threshold level				ESS capacity (MWh)
	LOLP (%)	**LOEE (%)**	**PRNEP (%)**	**PRNEE (MWh)**	
7000	0.2	270	2	3000	35.7
7000	0.15	260	1.5	2500	39.1
8000	0.2	270	2	3000	51.1
8000	0.15	260	1.5	2500	55.9
9000	0.2	270	2	3000	68.4
9000	0.15	260	1.5	2500	78.1

Table 18.4 compares the different given values of the adequacy indices to obtain the energy storage capacity for system's optimal allocation. The table shows that when the wind power capacity is a constant, the energy storage capacity can be improved by reducing the adequacy indices. With the increasing of wind power capacity, to maintain the adequacy level of the system needs the increase of a certain amount of the ESS capacity, thus improving the effects caused by peak valley difference characteristics of wind power.

18.7 Conclusions

Pointing at the statistical characteristics of wind power, this chapter proposes a joint optimization and coordination method for wind power and ESS capacity. Based on the adequacy indices (e.g., PRNEP, PRNEE, LOLP, and LOEE), the influences wind power has on the power systems have been quantified. The relationship function of adequacy level, wind power accommodation, and ESS capacity is obtained, and then the optimization and coordination of wind power capacity and the energy storage capacity are realized. An analysis shows that with the integration of large-scale wind power, the randomness and intermittency of wind power sharply pose a greater difficulty on the system's peak-load regulation. The proposed method can not only satisfy the adequacy indices of the power system, but also can realize the wind power accommodation and required ESS capacity. Thus, it is enable to provide effective approaches for the planning and operation of power systems with large-scale wind power integration.

In addition, because of data collection is not completed, this chapter only takes the operation data of one year for analysis. The evaluation results will be better in reflecting the statistical characteristics if the data of more than five years are taken for analysis.

References

Ahmadi, H., Ghasemi, H., 2012. Maximum penetration level of wind generation considering power system security limits. IET Gener. Transm. Distrib. 6 (11), 1164–1170.

Atwa, Y.M., El-Saadany, E.F., Guise, A.C., 2010. Supply adequacy assessment of distribution system including wind-based DG during different modes of operation. IEEE Trans. Power Syst. 25 (1), 78–86.

Billinton, R., Karki, R., Gao, Y., et al., 2012. Adequacy assessment considerations in wind integrated power systems. IEEE Trans. Power Syst. 27 (4), 2297–2305.

Cervantes, J., Dai, T., Qiao, W., 2013. Optimal wind power penetration in the real-time energy market operation. Power and Energy Society General Meeting (PES), pp. 1–5.

Chai, P., 2012. Research on Optimization for Operation Way of Wind Pumped Storage Power System. Huazhong University of Science and Technology.

Chen, Y.J., Ren, Z., 2005. Adequacy evaluation model and method for multi-infeed DC power transmission systems. Power Syst. Technol. 29 (13), 15–19.

Cui, Y., 2011. Studies on the Coordination of Large Scale Wind Generation and the Incorporated Power Grid. North China Electric Power University.

Cui, Y., Zhang, C., Mu, G., et al., 2014. Analysis on power grid's optimal accommodation of wind power based on energy cost assessment. Power Syst. Technol. 38 (08), 2174–2179.
Dietrich, K., Latorre, J.M., Olmos, L., et al., 2012a. Demand response in an isolated system with high wind integration. IEEE Trans. Power Syst. 27 (1), 20–29.
Dietrich, K., Latorre, J.M., Olmos, L., et al., 2012b. Demand response in an isolated system with high wind integration. IEEE Trans. Power Syst. 27 (1), 20–29.
Fan, P.F., Zhang, L.Z., Xie, G.H., 2012. Analysis model for accommodation capability of wind power with adequacy resources involved in system regulation. Power Syst. Technol. 36 (05), 51–57.
Ghofrani, M., Arabali, A., Etezadi-Amoli, M., et al., 2013. Energy storage application for performance enhancement of wind integration. IEEE Trans. Power Syst. 28 (4), 4803–4811.
Han, X., Yan, Y., 2009. Wind power penetration limit calculation based on power system reliability. Nanjing, the 9th International Conference on SUPERGEN, pp. 1–4.
Hu, Z.C., Ding, H.J., Kong, T., 2012. A joint daily operation optimization model for wind power and pumped- storage plant. Autom. Electric Power Syst. 36 (02), 36–41.
Huang, H., Chung, C.Y., Chan, K.W., et al., 2013. Quasi-Monte Carlo based probabilistic small signal stability analysis for power systems with plug-in electric vehicle and wind power integration. IEEE Trans. Power Syst. 28 (3), 3335–3343.
Jia, W.Z., Kang, C.Q., Li, D., et al., 2012. Evaluation on capability of wind power accommodation based on its day-ahead forecasting. Power Syst. Technol. 36 (08), 69–75.
Jiang, X., Chen, H.K., Xiang, T.Y., et al., 2014. Wind power penetration capacity considering peak regulation characteristics. Electric Power Autom. Equip. 34 (12), 13–18.
Jin, H.L., 2011. Research on Impact of Wind Power Penetration on the Grid. North China Electric Power University.
Kang, C.Q., Jia, W.Z., Xu, Q.Y., et al., 2013. Capability evaluation of wind power accommodation considering security constraints of power grid in real-time dispatch. Proc. CSEE 33 (16), 23–29.
Keane, A., Milligan, M., Dent, C.J., et al., 2011. Capacity value of wind power. IEEE Trans. Power Syst. 26 (2), 564–572.
Lamont, A.D., 2013. Assessing the economic value and optimal structure of large-scale electricity storage. IEEE Trans. Power Syst. 28 (2), 911–921.
Le, H.T., Santos, S., Nguyen, T.Q., 2012. Augmenting wind power penetration and grid voltage stability limits using ESS: application design, sizing, and a case study. IEEE Trans. Power Syst. 27 (1), 161–171.
Lei, Y.Z., Wang, W.S., Yin, Y.H., et al., 2002. Wind power penetration limit calculation based on chance constrained programming. Proc. CSEE 22 (5), 32–35.
Li, F.T., Chao, Q., Dai, X.J., 2008a. The research of wind power optimized capacity configuration in hydraulic power system, electric utility deregulation and restructuring and power technologies.Nanjuing: The 3th International Conference on DRPT, pp. 6–9.
Li, Q., Zhang, Y., Song, X.K., 2008b. Researches on the capacity of wind power integration into power system. Power Syst. Protection Control 36 (16), 20–24.
Li, Z., Han, X.S., Yang, M., et al., 2010. Assessment of transient voltage stability of load bus considering mechanical torque of dynamic load. Autom. Electr. Power Syst. 34 (19), 15–19.
Li, Q.Y., Feng, L.M., Xu, Y.H., et al., 2012. Accommodation mode of wind power based on water Source Heat Pump Technology. Autom. Electr. Power Syst. 36 (17), 25–27, 61.
Li, J.H., Wen, J.Y., Cheng, S.J., et al., 2013. Method of minimum energy storage power allocation for electric power systems with large-scale wind power based on p-efficient point theory. Proc. CSEE 33 (13), 45–52.

Li, J., Long, Y., Wen, J., Luo, W., 2014. Assessment of wind power capacity in power systems to meet the adequacy indexes. Power Syst. Technol. 38, 3397–3404.

Liew, S.N., Strbac, G., 2002. Maximising penetration of wind generation in existing distribution networks. IEE Proc. Gener. Transm. Distrib. 149 (3), 256–262.

Liu, D.K., Yu, L., Zhou, X., 2010. Analysis and calculation of wind farms integration capacity. North China Electric Power 9, 5–10.

Liu, D.W., Huang, Y.H., Wang, W.S., et al., 2011. Analysis on provincial system available capability of accommodating wind power considering peak load dispatch and transmission constraints. Autom. Electr. Power Syst. 35 (22), 77–81.

Liu, X., Li, W.Y., Xu, L.J., et al., 2012. Influencing factors of the maximum power injection for grid-connected wind power. North China Electric Power 40 (03), 397–401.

Liu, W.X., Li, Y.Z., Li, H., et al., 2014a. Wind power accommodation capability considering economic constraints for western mountain areas. Electr. Power Autom. Equip. 34 (08), 19–24.

Liu, C., Wu, H., Gao, C.Z., et al., 2014b. Study on analysis method of accommodated capacity for wind power. Power Syst. Protection Control 42 (04), 61–66.

Liu, B., Liu, F., Wang, C., et al., 2015. Unit commitment considering flexibility and uncertainty of wind power generation. Power Syst. Technol. 39 (03), 730–736.

Lu, Q., Wang, W., Han, S., et al., 2013. A new evaluation method for wind power curtailment based on analysis of system regulation capability. Power Syst. Technol. 37 (7), 1887–1894.

Ma, Y.J., Zhang, J.D., Zhou, X.S., et al., 2008. Study on steady state voltage stability of power system containing wind farm based on bifurcation theory. Power Syst. Technol. 32 (09), 74–79.

Madaeni, S.H., Sioshansi, R., Denholm, P., 2013. Estimating the capacity value of concentrating solar power plants with thermal energy storage: a case study of the southwestern United States. IEEE Trans. Power Syst. 28 (2), 1205–1215.

Meng, X., Zhuan, X.T., 2012. Optimization of wind power capacity in an electricity grid. 2012 31st Chinese Control Conference (CCC), pp. 2453–2458.

Muñoz, C., Sauma, E., Contreras, J., et al., 2012. Impact of high wind power penetration on transmission network expansion planning. IET Gener. Transm. Distrib. 6 (12), 1281–1291.

Ning, H.M., Liu, T.Q., Li, X.Y., et al., 2013. A computational model of wind power penetration limit considering transient stability. Power Syst. Protection Control 41 (20), 47–52.

Rodriguez, C., Amaratunga, G.A.J., 2006. Dynamic maximum power injection control of AC photovoltaic modules using current-mode control. IEE Proc. Electr. Power Appl. 153 (1), 83–87.

Shen, H., Liang, J., Dai, H.Z., 2002. Calculation of wind farm penetration based on power system transient stability analysis. Power Syst. Technol. 26 (8), 8–11.

Shi, G., Cai, X., Cheng, M.Z., 2010. Simulation on wind power grid integration capacity and low voltage ride-through characteristics in an isolated power system. Autom. Electr. Power Syst. 34 (16), 87–91.

Sioshansi, R., Madaeni, S.H., Denholm, P., 2014. A dynamic programming approach to estimate the capacity value of energy storage. IEEE Trans. Power Syst. 29 (1), 395–403.

Sortomme, E., Al-Awami, A.T., El-Sharkawi, M., 2010. A multi-objective optimization for wind energy integration. New Orleans: Transmission and Distribution Conference and Exposition, pp. 19–22.

Sun, R.F., Zhang, T., Liang, J., 2011. Evaluation and application of wind power integration capacity in power grid. Autom. Electr. Power Syst. 35 (04), 70–76.

Tande, J.O.G., 2007. Impact of large scale wind power on system adequacy in a regional power system with weak interconnections. Tampa: Power Engineering Society General Meeting, pp. 24–28.

Tang, X.S., Yin, M.H., Zou, Y., 2015. An improved method to calculate wind power penetration limit considering wind speed correlation. Power Syst. Technol. 39 (2), 420–425.

Toledo, O.M., Oliveira Filho, D., Diniz, A.S.A.C., et al., 2013. Methodology for evaluation of grid-tie connection of distributed energy resources—case study with photovoltaic and energy storage. IEEE Trans. Power Syst. 28 (2), 1132–1139.

Villumsen, J.C., Brønmo, G., Philpott, A.B., 2013. Line capacity expansion and transmission switching in power systems with large-scale wind power. IEEE Trans. Power Syst. 28 (2), 731–739.

Wang, X.L., Li, Z.W., 2011. Multi-objective Optimization of System Combined Wind Power with Pumped Storage Power. Lanzhou University of Technology.

Wang, Y., Gu, W., Sun, R., et al., 2011a. Analysis on wind power penetration limit based on probabilistically optimal power flow. Power Syst. Technol. 35 (12), 214–220.

Wang, N.B., Wang, J.D., He, S.E., 2011b. Cross-border accommodation method and transmission scheme of Jiuquan wind power. Autom. Electr. Power Syst. 35 (22), 82–89.

Wang, P., Gao, Z., Bertling, L., 2012a. Operational adequacy studies of power systems with wind farms and energy storages. IEEE Trans. Power Syst. 27 (4), 2377–2384.

Wang, P., Gao, Z.Y., Bertling, L., 2012b. Operational adequacy studies of power systems with wind farms and energy storages. IEEE Trans. Power Syst. 27 (4), 2377–2384.

Wei, Y.H., Zhang, S.Y., 2008. Copula Theory and Its Application in Financial Analysis. China Environmental. Science Press, Beijing, pp. 10–23.

Wei, L., Jiang, N., Yu, G.L., et al., 2010. Research on Ningxia power Grid's ability of admitting new energy resources. Power Syst. Technol. 34 (11), 176–181.

Wu, J.L., 2004. Research on Technology Problems of Large Grid-Connected Wind Farm. Tsinghua University.

Wu, J.L., Zhou, S.X., Sun, J.F., et al., 2004. Analysis on maximum power injection of wind farm connected to power system. Power Syst. Technol. 28 (20), 28–32.

Wu, Y.C., Ding, M., Zhang, L.J., 2005. Power flow analysis in electrical power networks including wind farms. Proc. CSEE 25 (4), 36–39.

Wu, J., Li, G.J., Cheng, L., et al., 2006. Calculation of Maximum Injection Power of Large-scale Wind Farms Connected to Power Systems. Power System Technology, Chongqing, pp. 22–26.

Wu, J., Li, G.J., Sun, Y.Z., 2007. Maximum injection power calculation of wind farms connected to power systems based on stochastic programming. Power Syst. Technol. 31 (14), 15–19.

Wu, J., Li, G.J., Sun, Y.Z., 2008. Calculation of maximum injection power of wind farms based on chance-constrained programming.Chicago: Transmission and Distribution Conference and Exposition, pp. 21–24.

Wu Q., M. Zhou, Sun L. Y., et al. (2012) Wind Farm Penetration Limit Calculation Based on Evolutionary Programming Algorithm. Innovative Smart Grid Technologies—Asia, 1–5.

Wu, X., Wang, X.L., Li, J., et al., 2013. A model to analyze peak load regulation of provincial power system considering sending-out of wind power. Power Syst. Technol. 37 (06), 1578–1583.

Xiao, C.Y., Wang, N.B., Ding, K., et al., 2010. System power regulation scheme for Jiuquan wind Power Base. Proc. CSEE 30 (10), 1–7.

Xing, W.Q., 2009. Research on Maximum Power Injection of Wind Power in Xinjiang Regional Power Grid under Frequency Constraints. Xinjiang University.

Xu, Y.X., Singh, C., 2012. Adequacy and economy analysis of distribution systems integrated with electric energy storage and renewable energy resources. IEEE Trans. Power Syst. 27 (4), 2332–2341.

Xu, W., Yang, Y.L., Li, Z.G., et al., 2010. Participation mode of large-scale Jiuquan wind power farm in Gansu Province to electricity market and its utilization scheme. Power Syst. Technol. 34 (06), 71–77.

Xu, Q.Y., Kang, C.Q., Zhang, N., et al., 2011. A discussion on offshore wind power output characteristic and its accommodation. Autom. Electr. Power Syst. 35 (22), 54–59.

Xue, B.K., Zhang, X., Zheng, Y.X., et al., 2014. Influence of thermal unit spinning reserve on wind power accommodation. East China Electric Power 42 (08), 1550–1554.

Yan, Z., Luo, W., Yang, L.B., et al., 2009. A method to evaluate transmission adequacy based on market power index. Proc. CSEE 29 (16), 75–81.

Yang, Q., Zhang, H., Yang, J., 2010b. A new way of maximum injection power calculation of wind farms connected to power systems. Power and Energy Engineering Conference (APPEEC), pp. 1–4.

Yu, S., 2013. Research on Improving Wind Power Grid Integration Capacity Based on Voltage Control. Beijing Jiaotong University.

Zeng, D., Yao, J.G., Yang, S.C., et al., 2014. Optimization dispatch modeling for price-based demand response considering security constraints to accommodate the wind power. Proc. CSEE 34 (31), 5571–5578.

Zhang, F., Cao, Q., Liu, H., 2008. Steady-state analysis on switching-in regional power grid for wind farm adopting different control strategies. Power Syst. Technol. 32 (19), 89–92.

Zhang, J., Cao, Q., Duan, X.T., et al., 2011a. Research on maximum access capacity of wind farm based on dynamic constraints. Power Syst. Protection Control 39 (03), 62–66.

Zhang, Q., Wang, H.Q., Xie, Z.J., 2011b. Generation expansion planning with large-scale wind power integration into Jiangsu power grid. Autom. Electr. Power Syst. 35 (22), 60–65.

Zhang, T., Li, J.J., Zhang, Y.F., et al., 2014. Research of scheduling method for the wind power acceptance considering peak regulation. Power Syst. Protection Control 42 (21), 74–80.

Zhang, C.F., Liu, C., Wang, Y.F., et al., 2015. A fuzzy multi-objective optimization based evaluation model of wind power accommodation capability. Power Syst. Technol. 39 (02), 426–431.

Zheng, Y., Dong, Z.Y., Luo, F.J., et al., 2014. Optimal allocation of energy storage system for risk mitigation of DISCOs with high renewable penetrations. IEEE Trans. Power Syst. 29 (1), 212–220.

Zhou, X., Liu, M.B., Xie, M., 2010. Adequacy evaluation of transmission systems using Monte Carlo simulation and nonlinear programming technology. Power Syst. Protection Control 38 (20), 45–50, 56.

Zhu, N.Z., Chen, N., Han, H.L., 2011. Key problems and solutions of wind power accommodation. Autom. Electr. Power Syst. 35 (22), 29–34.

Further Reading

Cervantes J, Ting D, Wei Q. Optimal wind power penetration in the real-time energy market operation. Vancouver: Power and Energy Society General Meeting (PES) , 21–25.

Evans, J., Shawwash, Z., 2009. Assessing the Benefits of Wind Power Curtailment in a Hydro-Dominated Power System: Calgary: Integration of Wide-Scale Renewable Resources into the Power Delivery System. Power and Energy Society, pp. 29–31.

Wang, B.B., Liu, X.C., Li, Y., 2013. Day-ahead generation scheduling and operation simulation considering demand response in large-capacity wind power integrated systems. Proc. CSEE 33 (22), 35–44.

Yang, Q., Zhang, J.H., Yang, J., 2010a. A new way of maximum injection power calculation of wind farms connected to power systems. Power and Energy Engineering Conference (APPEEC), pp. 1–4.

Optimal energy dispatch in residential community with renewable DGs and storage in the presence of real-time pricing

Qiang Yang*, Ali Ehsan*, Le Jiang*, Xinli Fang[†,‡]
*College of Electrical Engineering, Zhejiang University, Hangzhou, People's Republic of China, [†]Powerchina Huadong Engineering Corporation Limited, Hangzhou, People's Republic of China, [‡]Hangzhou Huachen Electric Power Control Co. LTD., Hangzhou, People's Republic of China

Nomenclature

t	the time that appliance a may operate
x_a^t	appliance a's on or off condition at time t
d_a	length of operation time
P_a	rated power of appliance a
E_a	total energy needed for the operation
$[\alpha_a, \beta_a]$	allowable operational time range
t_a^{start}/t_a^{end}	operation start/end time slot
m	amount of uninterruptible loads
n	amount of interruptible loads
P_{must}^t	total baseline loads consumed at time t
$P_{flexiblet}$	total flexible loads consumed at time t
RTP_t	the real-time pricing at time t
c_a^t	energy to buy of appliance a at time t
SOC_t	battery's state of charge at time t
$SOC_{\min}$	lower bound limit of battery's SOC
$SOC_{\max}$	upper bound limit of battery's SOC
E_{charge}^h	charging or discharging energy of battery
$Popsize$	the population size of each household
pc/pm	probability of crossover/mutation
$store$	the former number of genomes
gen	the maximum times of iterations
X_a	appliance a's whole day working statues
X_{Chrom}^t	all flexible loads' working statues
P_{Chrom}^t	all loads' power consumption at time t
$[t_{up}^{begin}, t_{up}^{end}]$	RTP rising stages
$[t_{down}^{begin}, t_{down}^{end}]$	RTP falling stages
S_+	the area of $P_{DG}^t > P_{load}^t$

Smart Power Distribution Systems. https://doi.org/10.1016/B978-0-12-812154-2.00019-5

S_-	the area of $P_{DG}^t \leq P_{load}^t$
S	the difference value between S_+ and S_-
N	the number of all neighborhoods
E_{DG}	the excess DG
P_{DG}^t	DG power generation at a time slot t
P_{Load}	total loads consumed at a time slot
Q	battery capacity (kWh)
d_v	the amount of demand of a household
$D_{j \to i}$	the distance between j node and i node
Num	the serial number of households
c_v	the weight
x_i^t	household i's working statues at time t
$distri_i^t$	expected energy household $i$can distribute
$receive_i^t$	energy that household i can received
Buy_t	electricity to be bought from grid at time t

19.1 Introduction

The advancements in renewable distributed generation (DG) technologies have enabled high penetration of various small-scale renewable sources ranging in capacities from a few kilowatts (kW) to megawatts (MW), for example, PV panels, micro wind turbines, as well as battery-based storage units in the premise of households. In addition, real-time pricing (RTP) mechanism has been adopted by power utilities as a leveraging tool, enabling customers to adapt their energy use according to accessible price signals. Such recent changes enable the residents to promote the level of demand side response from different aspects, for example, flattening peak demand, as well as reducing electricity purchase cost by including RTP into decision-making of 1-day-ahead dispatch. However, this brings direct benefit through appropriate demand response actions, however, the intermittency of renewable DGs and dynamical pricing make it a nontrivial task (Liu and Ai, 2016). To fully use the installed DGs, their intermittent energy must be replaced elsewhere in the supply/demand loop (Bhandari et al., 2014; Finn and Fitzpatrick, 2014). This calls for a cost-effective demand response solution to efficiently manage the operations of appliances and allocate the domestic loads to appropriate time slots (e.g., sufficient DG generation or low RTP).

To address these challenges, a collective research effort has been made in the literature from different aspects. A number of domestic load management solutions have been presented considering different operational scenarios, for example, considering the operating patterns and constraints of appliances, and carrying out the load scheduling at different timescales as the response to the RTP (Barzin et al., 2016; EI-Baz and Tzscheutschler, 2015; Adika and Wang, 2014; Giorgio and Pimpinella, 2012; Munkhammar et al., 2015; Erdinc, 2014; Erdinc et al., 2015). In Barzin et al. (2016), the authors experimentally investigated the combination of weather forecasting and the price-based control for solar passive buildings, and showed that up to 90% electrical energy can be saved per day. In EI-Baz and Tzscheutschler (2015), a simplified and efficient day-ahead electrical load prediction approach for domestic EMS was proposed aiming to achieve supply-demand matching and minimizing the energy cost. This approach did not require being associated with prepared statistical or historical databases, or even measurement sensors, and its effectiveness was confirmed by the test over 25 households in Austria. In Adika and Wang (2014), the appliances were grouped into a set of clusters based on their operational preferences and a cluster-based scheduling approach was proposed to reduce the electricity cost. In Giorgio and Pimpinella (2012), the authors formulated the domestic electric energy management as an event-driven binary linear programming problem and presented an algorithmic solution. However, the aforementioned solutions have not fully considered the impact of RTP as well as the energy storage on the domestic energy management, and hence the potential benefit needs to be further investigated. To this end, novel control architectures (e.g., Munkhammar et al., 2015) considering domestic DGs and storage devices were studied to combine the long-term planning at the timescale of billing period (i.e., a month) and short-term corrective control to compensate the errors of DG and demand prediction. In Erdinc (2014), a more comprehensive evaluation based on the mixed-integer linear programming framework was carried out considering the impacts of various elements, for example, electric vehicles with bi-directional power flow capability via charging and V2H operating modes, energy storage systems (ESSs; peak clipping and valley filling), and DG unit enabling energy sell back to grid. The authors further investigated and presented interesting results regarding the sizing of additional (DG and ESSs to be applied in smart households in the presence of demand response activities (Erdinc et al., 2015).

These aforementioned solutions, however, were designed merely for the optimal energy management within a single household. In fact, due to the limited capacities of installed storage units and variability of household loads and DGs, the utilization efficiency of DGs can be undermined at individual households, for example, the surplus energy (when overall DG output exceeds demand) cannot be stored at certain times. To address this issue, some notable studies (Pedrasa et al. 2011; Giusti et al., 2014) have been carried out to explore the communication and trading between neighborhoods to improve the DG utilization and reduce the household electricity purchase cost at the residential community level. In Pedrasa et al. (2011), a methodology for making robust day-ahead operational schedules for controllable residential distributed energy resources (DERs) based on energy service decision-support tool was described using a stochastic programming approach formulated for the DER schedulers. In Giusti et al. (2014), the management of dispatchable loads in a residential microgrid was addressed by decentralized controllers deployed in each household to optimize conflicting objectives simultaneously: minimization of user energy costs and load flattening in an online fashion. However, the operation of such distributed control paradigm can be problematic in large-scale residential community due to inaccurate information update dissemination and prohibitive communication overheads induced from asynchronous communication and distributed coordination scheme. Our previous work (Wang et al., 2015) confirmed the potential benefits of energy dispatch across multiple households with preliminary results, but the DG resource utilization and performance assessment under uncertainties were not studied.

This chapter made the following technical contributions: it exploits a cost-effective optimal energy dispatch for large-scale residential community based on a centralized

1-day-ahead solution considering RTP. It is carried out at two levels: optimal dispatch within individual households and energy exchange among neighboring households, aiming to minimize the household electricity purchase cost, and optimize the global DG and storage utilization efficiency in the residential community. Unlike the distributed management approach (Giusti et al., 2014), all decision-making can be carried out in a centralized structure in an offline fashion, which can adopt advanced prediction and analysis algorithms. As the optimal dispatch in individual households is carried out locally, the communication overhead among neighboring households can be minimized (small amount of data transmission and no real-time requirements) and minimal hurdles of deployment in a massive number of households. The robustness of the solution under uncertainties (e.g., inaccurate prediction of RTP and DG output) is also assessed and confirmed.

The rest of this chapter is organized as follows: Section 19.2 outlines the system model and formulates the problem; the proposed energy dispatch algorithmic solution is presented in Section 19.3; Section 19.4 assesses its performance through simulations and presents the key numerical results; finally, some conclusive remarks are made in Section 19.5.

19.2 System model and problem formulation

The quantitative description of the stochastic characteristics of wind power, and the optimization operation theory and method of power system to deal with the randomness of wind power are two basic problems to realize the safe and stable operation of power system with large-scale wind power integration.

19.2.1 System model

Different forms of renewable DGs (PVs and micro wind turbines) and storage facilities (e.g., Li-ion battery) are considered in individual households which can provide energy to the local loads without purchase cost. A variety of appliances and smart meters are connected with the domestic controller via the home area network (HAN), as shown in Fig. 19.1.

In individual households, the load management controllers receive the broadcasted RTP prediction 1-day ahead from the remote energy dispatch center and derive the optimal energy dispatch plan locally for the next 24 h, that is, coordinating the DGs, loads, and storage. Note that the load control needs to meet the operational constraints, the patterns of the domestic appliances as well as the user preferences. In this work, the domestic appliances are categorized into the following three types based on their operational characteristics:

(1) Baseline load (nonschedulable): it is the must-run service that needs to be served immediately upon requests from the residents, for example, lighting, fridge, computer, and television. Hence it is the baseline load and is not schedulable.

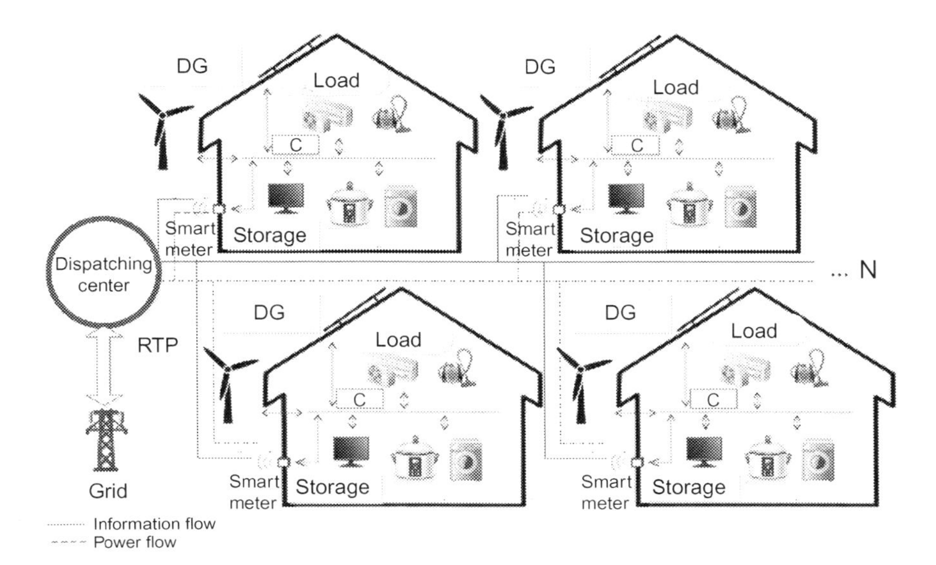

Fig. 19.1 The residential energy system (with micro wind turbines, solar panels, storage units, and controllers) under RTP.

(2) Uninterruptible load (schedulable): it refers to the domestic appliances that are schedulable but require to be operated continuously (starting and ending times can be flexibly set), for example, rice cooker, dish washer, and washing machine.

(3) Interruptible load (schedulable): it refers to the appliances that are schedulable and can start or stop operation at any given time interval, for example, air conditioner, clothes dryer, pool pump, floor cleaning robot, and electric radiator.

In the domestic energy dispatch, as the baseline load needs to be firstly met, the benefits of demand response can be exploited through optimally managing the schedulable appliances (i.e., uninterruptible and interruptible loads) in the presence of RTP. It is assumed that all appliances are operated at their rated power, and the detailed operational pattern and constraints are shown in Table 19.1. In individual households, the optimal energy dispatch needs to meet a collection of constraints given in Eq. (19.1). It should be noted that the operational period ($\beta_a - \alpha_a$) should be no less than the operating duration that required to complete the task of appliance a, and the total required power consumption needs to be met during the process.

$$\begin{cases} \alpha_a \leq t_a^{start} \leq t \leq t_a^{end} \leq \beta_a, \quad a = 1, \ldots, m+n \\ x_a^t = \{0, 1\}, \\ \beta_a - \alpha_a \geq d_a \\ E_a = P_a \times d_a \\ \sum_{t=\alpha_a}^{\beta_a} x_a^t \times P_a = E_a \end{cases} \tag{19.1}$$

Table 19.1 Parameters of schedulable loads in the simulation scenarios

Appliance	$\alpha_a \sim \beta_a$	Duration (h)	Power (KW)
Dish washer*	08:00–12:00	1.5	0.73
	20:00–23:00		
Rice cooker*	06:00–08:00	1	0.8
	18:00–20:00		
Washing machine*	06:00–09:00	2	0.38
	17:00–20:00		
Humidifier	00:00–09:00	4	0.15
	14:00–20:00		
Laundry drier	09:00–12:00	2	1.26
	20:00–23:00	1.5	
Floor cleaning robot	06:00–12:00	3	0.74
	08:00–18:00	2.5	0.7
	20:00—23:30	1.5	0.64
Water heater	04:30–08:30	3	1.64
	16:00–20:00	2	1.85
	20:00–24:00	3	1.64
Electric kettle	06:00–07:30	0.5	1.5
	16:00–19:00		
	20:00–23:00		
Air conditioner	00:00–08:00	3	1
	16:00–24:00	4	
	00:00–08:00	4	1.1
	18:00–24:00	4	
Electric radiator	12:00–17:00	2.5	2
Pool pump	07:00–18:00	4	1.8
Water pump	06:00–15:00	6	1.6
Oil press	09:00–18:00	3.5	0.35
Floor waxing	14:00–18:00	3	0.42
Electric oven	13:00–18:00	2	1.3
PHEV	00:00–08:00	3.5	2.4

Note: Appliances labeled with asterisk (*) indicate their operations are uninterruptible, otherwise they are interruptible; above appliances may operate with different time intervals, operation durations, or power ratings (e.g., different types of air conditioners may be installed in the same house).

Now we consider the scenario of a large-scale residential community consisting of N households (load profiles and capacities of installed DGs and storage units may differ) which are connected and supplied by the same distribution network feeder. The domestic energy management controllers in individual households are with the same configuration and functionality and are connected with the dispatch center via underlying communication network in a centralized structure, as illustrated in Fig. 19.1. In the dispatch center, RTP can be predicted 1-day ahead in an offline fashion by adopting sophisticated algorithms and tools based on the historical pricing data, and disseminated to individual households. Also, the energy consumption at all time slots of local dispatch plans in individual households will be made available to the dispatch center, which will further identify the energy trading plan among neighboring households for the next 24 h.

19.2.2 Problem formulation

The energy dispatch optimization in individual households needs to be firstly examined. The primary consideration in load control is to arrange the operation of the schedulable appliances into the appropriate time slots (30 min per slot and 48 time slots in all) so as to minimize the difference between the DG generation and load demand throughout the day. After fully utilizing the available DG generation, the excess loads, if any, need to be supplied by the power utility which can be further shifted to the times slots with lower electricity prices to reduce the purchase cost based on RTP. Hence, such load management can be considered as a multi-objective optimization problem, formulated as follows:

$$
\begin{aligned}
&\min \begin{cases} \sum_{t=1}^{48} \left| P_{DG}^{t} - P_{must}^{t} - \sum_{a=1}^{m+n} x_a^t \cdot P_a \right| \\ 0.5 \times c_a^{t1} \times RTP_{t1} + 0.5 \times c_a^{t2} \times RTP_{t2} \end{cases} \\
&s.t.\ \ t1 \in \left\{ h \mid P_{DG}^{h} < P_{must}^{h} + P_{flexible}^{h},\ x_a^h = 1,\ h = 1,\ \ldots,\ 48 \right\} \\
&t2 \in \left\{ h x_a^h = 0,\ \ h = [\alpha_a, \beta_a] \right\} \\
&c_a^{t1} = \begin{cases} 0,\ P_{DG}^{t1} \geq P_{must}^{t1} + P_{flexible}^{t1} - P_a \\ P_{must}^{t1} + P_{flexible}^{t1} - P_a - P_{DG}^{t1},\ P_{DG}^{t1} < P_{must}^{t1} + P_{flexible}^{t1} - P_a \end{cases} \\
&c_a^{t2} = \begin{cases} 0,\ P_{DG}^{t2} \geq P_{must}^{t2} + P_{flexible}^{t2} + P_a \\ P_{must}^{t2} + P_{flexible}^{t2} + P_a - P_{DG}^{t2},\ P_{DG}^{t2} < P_{must}^{t2} + P_{flexible}^{t2} + P_a \end{cases}
\end{aligned}
\tag{19.2}
$$

Here, the appliance a is chosen to be further shifted from time $t1$ to time $t2$ according to the RTP-based load scheduling, in order to reduce the overall electricity purchase cost.

Following the aforementioned load control, the storage units are installed in the domestic scope which can act as either the power demand or generator through flexible charging (when excess DG generation is available or RTP is low) and discharging (when DG generation is insufficient or RTP is high) process. Through incorporating the storage into the energy management, the global efficiency of DGs will be further improved and the electricity purchase cost is reduced. The basic idea behind this is to manage the charge/discharge pattern such that the DG energy can be scheduled to the time slots with insufficient generation, which can be implemented through minimizing the difference between the amount of energy of DG generation, power demand, and battery charge/discharge over the day. This can be formulated as the following form:

$$
\begin{aligned}
&\min\sum_{t=1}^{48}\left|\left(P_{DG}^{t}-P_{load}^{t}\right)\times 0.5-\left(SOC_{t}-SOC_{t-1}\right)\cdot Q\right| \\
&s.t.\ \ P_{load}^{t}=P_{must}^{t}+P_{flexible}^{t} \\
&E_{\text{charge}}^{t}=\begin{cases} if\ \ P_{DG}^{t}>P_{load}^{t}: \\ \min\left\{\left(P_{DG}^{t}-P_{load}^{t}\right)\times 0.5,\ \left(SOC_{\max}-SOC_{t-1}\right)\cdot Q\right\} \\ else\ if \\ \max\left\{\left(P_{DG}^{t}-P_{load}^{t}\right)\times 0.5,\ \left(SOC_{\min}-SOC_{t-1}\right)\cdot Q\right\} \end{cases} \\
&SOC_{t}=SOC_{t-1}+\frac{E_{charge}^{t}}{Q} \\
&SOC_{t}\in\left[SOC_{\min},SOC_{\max}\right]
\end{aligned}
\tag{19.3}
$$

Here, the state of charge (SOC) of the Li-ion battery needs to be constrained within the range of 20%–90% to protect its lifetime. In this study, with the observation of difference between DG supply and demand requirements, the capacity of Li-ion capacity can be determined based on the available capacity standards in the market, to guarantee that the SOC is within the range. The details of optimizing the operations of schedulable appliances for matching the DG generation and RTP are presented and discussed in Section 19.3.

Now, we look into the energy exchange across neighboring households in the residential community after local dispatch optimization. The key idea is to efficiently identify the appropriate mapping between the households with surplus energy and the neighboring households need extra supply with the minimum power loss. The optimal energy exchange among neighboring households can be formulated as follows:

$$
\begin{aligned}
&\min\sum_{t=1}^{48}\left(\left|E_{DG}-d_{v}\right|\times\left|D_{j\to i}\right|\right) \\
&E_{DG}=P_{DG}-P_{Load}-Q \\
&D_{j\to i}=Num_{j}-Num_{i}
\end{aligned}
\tag{19.4}
$$

Here, at any time slot, the following constraint applies:

$$
E_{DG}\leq\sum_{i=1}^{N}d_{v_i}
$$

19.3 Optimal energy dispatch approach

19.3.1 Energy dispatch in individual households

In individual households, the genetic algorithm (GA) is adopted to optimally allocate the loads and storage to match the DG output generation, respectively. The pseudo-codes of RTP-based load dispatch are given in Algorithm 19.1 and the operation schedule of the controllable loads can be well organized and determined such that the electricity purchase is significantly reduced.

Algorithm 19.1 DG and RTP based load control

Require: $P_{DG}^h, P_{must}^h, P_a, \alpha_a, \beta_a, d_a, RTP$

1: Initialize $Popsize = N$, $X_a = [x_a^1, x_a^2, \ \ldots, \ x_a^{48}]_{N\times 48}$ using (1), pc, pm, gen

2: Form $X_{Chrom}^k = [X_1^{1\times 48}; \ldots; \ X_m^{1\times 48}; \ldots; \ X_{m+n}^{1\times 48}]_{(m+n)\times 48}$, $k \in 1, \ \ldots, \ N$,
$P_{Chrom}^k = X_{Chrom(a)}^k \times P_a$

3: **FOR** $t \leftarrow 1$ to gen **DO**

4: Calculate fitness $Objvalue(t) = \sum_{h=1}^{48} |P_{DG}^h - P_{must}^h - \sum_{a=1}^{m+n} P_{Chrom}^{a\times h}|$ and select $X_{Chrom}(t)$ using the roulette method

5: **IF** $rand(0,1) < pc$ and appliance a is uninterruptible

6: Shift the time slots $[h_a^{start}, h_a^{end}]$ of $X_a^{1\times 48}$ in X_{Chrom}^i to the left or right side within $[\alpha_a, \beta_a]$

7: **ELSEIF** $rand(0,1) < pc$ and appliance a is interruptible

8: Cross the $X_a^{1\times 48}$ of X_{Chrom}^i and X_{Chrom}^{i+1} at a random position

9: **END IF**

10: **IF** $rand(0,1) < pm$

11: Choose the appliance a to do mutation and insure $sum(X_a^{1\times 48}) = d_a$

12: **END IF**

13: Choose the $X_{Chrom}(t)$ with the minimum $Objvalue$ (t), update X_{Chrom} and P_{Chrom}

14: **END FOR**

15: Calculate the purchasing power and cost by DG-based control, labeled as *BuyElec* and *BestCost*, respectively.

16: **FOR** $j \leftarrow 1$ to 48

17: **IF** $BuyElec\ (j) > 0$

18: Choose appliance $a \in \{a \mid x_a^j = 1, a \in 1, \cdots, m+n\}$ and calculate c_a^j // using (2)

19: **IF** appliance a is uninterruptible

20: **FOR** $k \leftarrow \alpha_a$ to $\beta_a - d_a + 1$

21: Calculate c_a^k and $Cost$ when a is shifted to $[k, k + d_a - 1]$, update $BestCost = \min\{Cost, BestCost\}$ and P_{Chrom}

22: **END FOR**

23: **ELSEIF** appliance a is interruptible

24: **FOR** $k \in \{k \mid x_a^j = 1, x_a^k = 0, \alpha_a \le k \le \beta_a\}$

25: Calculate c_a^k and $Cost$, update $BestCost$ and P_{Chrom}

26: **END FOR**

27: **END IF**

28: **END IF**

29: **END FOR**

In fact, after Algorithm 19.1, there still exists some room to further reduce the electricity cost by appropriately arranging the charging/discharging behaviors of the storage units based on the DG generation, demand and RTP through the storage-based energy dispatch, as given in Algorithm 19.2.

Through incorporating the storage units and optimally controlling rolling its charging/discharging under various conditions, the energy can be well managed to meet the demand while minimizing the electricity purchase cost and making the best use of the domestic DG generation.

Algorithm 19.2 Storage based dispatch in household

Require: P^t_{DG}, $P^t_{load} = \sum_{a=1}^{m+n} P^t_{Chrom}$, RTP

01: Initialize $SOC_{\min}$, $SOC_{\max}$, $SOC_{t=1}$, Q

02: Divide the working time into $[t^{begin}_{up}, t^{end}_{up}]$ and $[t^{begin}_{down}, t^{end}_{down}]$

03: **FOR**$t \leftarrow t^{begin}_{down}$ to t^{end}_{down}**DO**

04: Calculate E^t_{charge} and SOC_t using (3), update the power to buy for loads $Buy_t = \max\{(P^t_{DG} - P^t_{load}) \times 0.5 - E^t_{charge}, 0\}$

05: **END FOR**

06: **FOR** $t \leftarrow t^{begin}_{up}$ to t^{end}_{up} **DO**

07: Calculate S_+, S_-, $S = S_+ - S_-$, E^t_{charge}, SOC_t //using (3)

08: **IF** $P^t_{DG} > P^t_{load}$ and $t \in t^{begin}_{up}$

09: **IF**$S < 0$

10: Update $Buy_t = \min\{-S, \ (SOC_{\max} - SOC_t) \cdot Q - S_+\}$ for battery, $E^t_{charge} = (P^t_{DG} - -P^t_{load}) \times 0.5 + Buy_t$, and SOC_t at time $t \in t^{begin}_{up}$

11: **WHILE** $(SOC_{\max} - SOC_t) \cdot Q < S_- - (P^t_{load} - P^t_{DG}) \times 0.5$, $t \in P^t_{DG} < P^t_{load}$

12: Update $E^t_{charge} = 0$, SOC_t and $Buy_t = (P^t_{load} - P^t_{DG}) \times 0.5$

13: **END WHILE**

14: **END IF**

15: **ELSEIF**$P^t_{DG} < P^t_{load}$ and $t \in t^{begin}_{up}$

16: **IF** $(SOC_t - SOC_{\min}) \cdot Q > S_-$

17: Calculate E^t_{charge} and SOC_t //using (3)

18: **ELSE**

19: Update $Buy_t = (P^t_{load} - P^t_{DG}) \times 0.5 + E^t_{charge}$ for loads and battery, $E^t_{charge} = S_- - (-SOC_t - SOC_{\min}) \cdot Q - (P^t_{load} - P^t_{DG}) \times 0.5$ and SOC_t at time $t \in t^{begin}_{up}$, follow Step 11 to 13 afterwards

20: **END IF**

21: **END IF**

22: **END FOR**

19.3.2 Energy trading between neighborhoods

The energy dispatch across households can be carried out through solving the optimization problem formulated in Eq. (19.4) as the energy consumption states of individual households at any time slot are all available at the dispatch center. The searching algorithm firstly identifies the positive node (with surplus energy) with the largest amount of excess energy as the root node and supplies its neighboring negative nodes (i.e., households requires energy) with the searching radius of nodes (the nearest with the highest priority to be supplied), as illustrated in Fig. 19.2. In the case that the

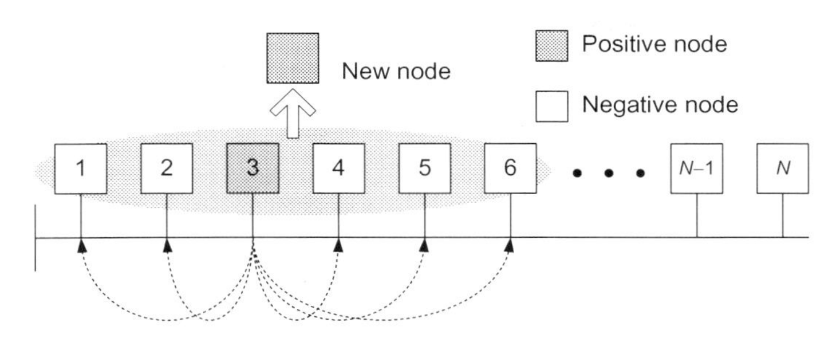

Fig. 19.2 Energy exchange across multiple neighboring households.

surplus energy is still available, a new positive node can be formed for further energy exchange, and the states of all nodes are updated. The algorithm iterates the above searching process until no positive or negative nodes are available. The pseudo-code of the searching algorithm is briefly given in Algorithm 19.3.

Algorithm 19.3 Energy exchange across households

Require: EDG, d_v, RTP

01:Initialized *pop size* $= N$, $X_i = [x_i^1, x_i^2, \ldots x_i^t, \ldots x_i^{48}]$, $i \in$[1,N], t$\in$[1,48] $i \in [1, N]$, $t \in [1.48]$

02: **IF** $x_i^t \geq 0$, **THEN** $x_i^t = E_{DG}$

03: **ELSE IF** $x_i^t = d_\nu$

04: **END IF**

05: **FOR** t: 1 to 48

06: **DO:** find the root node and other positive nodes with generation of zero, temp $E_{DG}^t = 0$, $x^t =$ temp E_{DG}^t; {assuming the number of neighboring household is n and the searching radius is 6 nodes;}

07: **IF** $E_{DG}^t(i) \leq \left(\sum_{j=i-3}^{i} d_\nu(j) + \sum_{j=i}^{i+3} d_\nu(j) \right), \quad i \in [1, n]$

08: **THEN** the weight $c_\nu(j) = \frac{|j-i|}{6}$, the expected exchanged energy is $distri_j^t = c_\nu \times E_{DG}^t(i)$

09: **FOR** j: $i - n/2$ to $i + n/2$ **DO:**

10: **IF** $d_\nu(j) - distri_j^t \geq 0$ and $j - i \leq 6$, this neighbouring household receives energy $receive_j^t = disri_j^t$

11: **ELSE IF** $receive_j^t = d_\nu(j)$

12: **END IF**

13: **END FOR**

14: **ELSE IF**

15: **FOR** j: $i - n/2$ to $i + n/2$

16: **DO:** $recieve_j^t = d_\nu(j)$.

Grouping nodes as a new positive node with the energy of $E_{DG}^t = E_{DG}^t(i) - \left(\sum_{j=i-3}^{i} d_\nu(j) + \sum_{j=i}^{i+3} d_v(j) \right)$;

17: **END FOR**

18: **END IF**

19: **END FOR**

19.4 Simulation experiment and numerical result

19.4.1 Simulation parameters

In this section, two optimization models for the ES are described. Model-I is a planning model that is used to minimize the ES for power system with large-scale integration of wind power. Model-II is used to verify the effectiveness of the results obtained with the Model-I by different scenarios.

For day-ahead dispatch of power system with high wind power penetration, the traditional deterministic unit commitment of conventional generators is firstly processed by a priority list method (Senjyu et al., 2003) based on the forecasted wind power and load demand on the following day. This can fully utilize the conventional power resources with relative low cost. At this stage, the forecast errors are not considered. As it is well known that precise prediction of the wind power is difficult to realize, the scheduled units including the spinning reserve (SR) may not be able to balance the fluctuation in actual wind power and load demand. ES is then scheduled to enhance the SR to cope with such unbalance. Model-I is used for this process.

For all simulations, we consider a residential community consisting of 200 households (with DGs and batteries installed). The RTP data are adopted from AEMO (Electricity Price and Demand, n.d.) (May 3, 2016) at the timescale of 0.5 h. In this work, we take the typical solar and wind generation from The California Energy Almanac (n.d.) and The Wind Power Almanac (n.d.), ranging from 0 to 5 KW, and consider that the DG generations and battery capacities (12V-600Ah Li-ion battery with the capacity of 7.2 kWh) are varied in individual households with the range of (−30% to +30%). The RTP, DG generation, and baseline loads profile are illustrated in Fig. 19.3. It is also assumed that the number of schedulable domestic appliances for a residence is up to 30 (including the intermittently operated appliances), and details can be seen in Table 19.1.

19.4.2 Performance evaluation

The performance of the proposed optimal dispatch solution is assessed through a comparative study of three scenarios: (I) without any optimal control; (II) household-level optimal dispatch (i.e. load and storage control under RTP); and (III) optimal dispatch within and across households (i.e., our solution). Two key metrics, that is, the generation-demand matching, and the electricity purchase cost under RTP, are adopted for the performance evaluation for both household and overall residential community, respectively. Also the DG utilization at any time slot over a day (48 time slots) is defined in the form of two conditions:

$$\text{Utilization}_t = \begin{cases} P^t_{Chrom}, & P^t_{DG} > P^t_{Chrom} \\ P^t_{DG}, & P^t_{DG} \le P^t_{Chrom} \end{cases}, \quad t \in 1, \ldots, 48. \tag{19.5}$$

Figs. 19.4 and 19.5 present the supply-demand matching performance of household (a representative household) and residential community, respectively. The result of

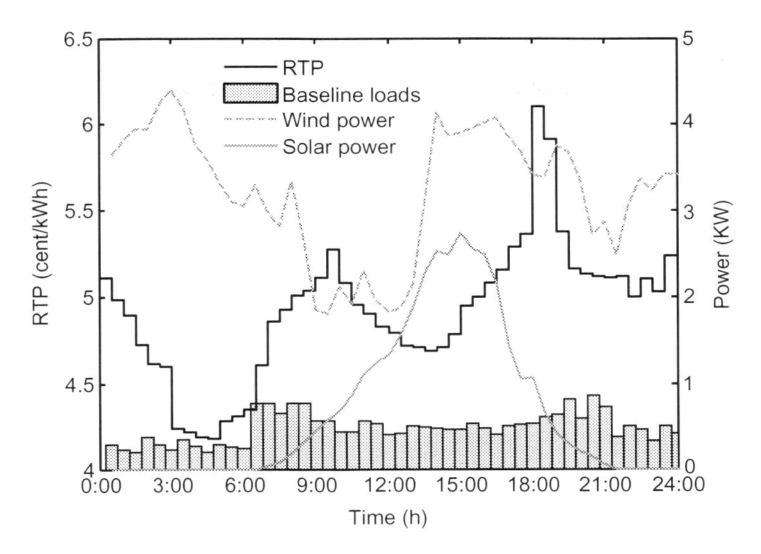

Fig. 19.3 The RTP, DG generation, and baseline loads profile.

power supply (shadow area) and demand (solid line) are plotted in a polar coordination over a day (0:00–24:00), respectively. Fig. 19.4A clearly shows the poor DG-demand matching (e.g., in the period of 6:00–12:00, the demand far outweighs the DG supply, while from 0:00 to 6:00, and 12:00 to 17:00, a large portion of DG generation is not utilized, and as a result, the resident has to purchase electricity of 99.3 cents from the power utility (calculated based on the RTP information over a day) to meet the demand requirement (i.e., the un-shadowed area between load demand and supply). In Fig. 19.4B, the match performance was improved after optimal dispatch within individual households. However, there still exists some energy that cannot be utilized, for example, 4:00–6:00, 15:00–18:00, and 22:00–24:00. In Fig. 19.4C, nearly perfect load and generation match is obtained after energy exchange with neighboring households, and only a small portion of electricity purchase is required (22:00–24:00 and 16:00–18:00), where the cost is reduced from 4.0 to 0.42 cents. Fig. 19.5A presents the result for the overall 200 households for two comparative scenarios (I and III), which obtain the similar results as in a single household and confirms our expectation.

The DG utilization efficiency is another important consideration for the cost-benefit analysis of residential DG facilities. Fig. 19.6 compares the results of DG generation utilization efficiency of scenario I and III over 200 households over a day. Without any control on energy dispatch, the overall utilization efficiency is around 80.76%, as given in Fig. 19.6A, and our solution can promote the efficiency up to 99.4% (Fig. 19.6B). This clearly highlights its effectiveness in making the best use of renewable energy through optimal energy dispatch in the household scope and residential scope, and in turn, the electricity purchase cost can be significantly reduced.

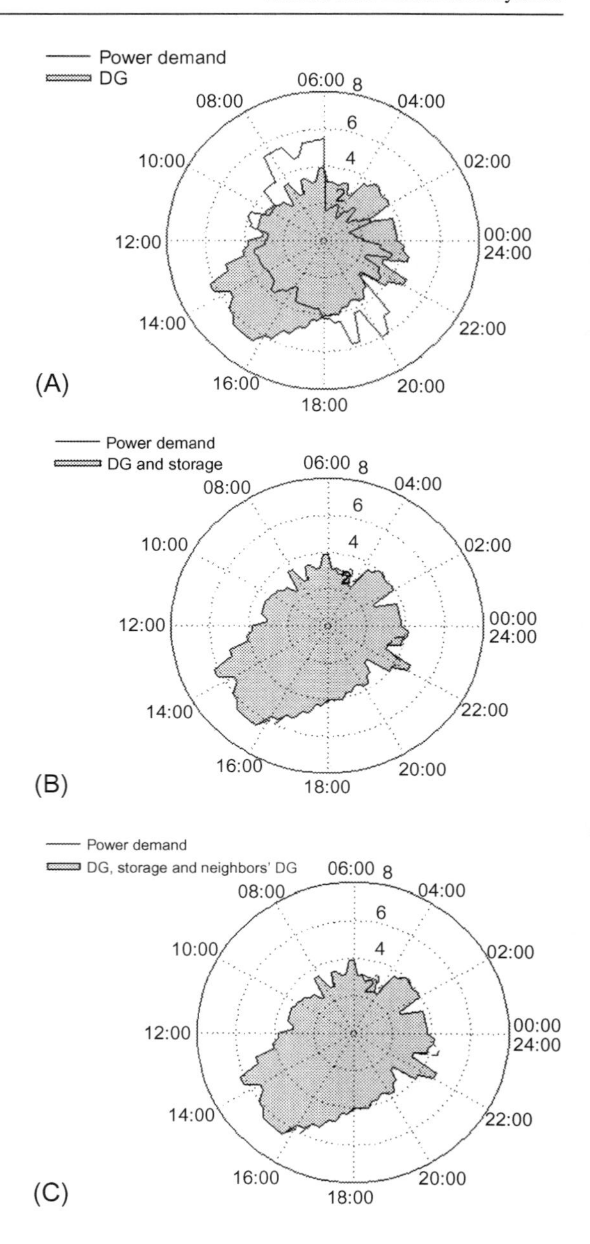

Fig. 19.4 Supply-demand match (single household): (A) scenario I: without control, (B) scenario II: household optimization, and (C) scenario III: our solution.

Fig. 19.7 presents the daily average electricity purchase over 3 months (50th and 95th percentile values are provided). In a quarter (3 months), the residential community (200 households) with the suggested solution (scenario III) can save 50402.17 cents in electricity purchase in comparison with scenario I.

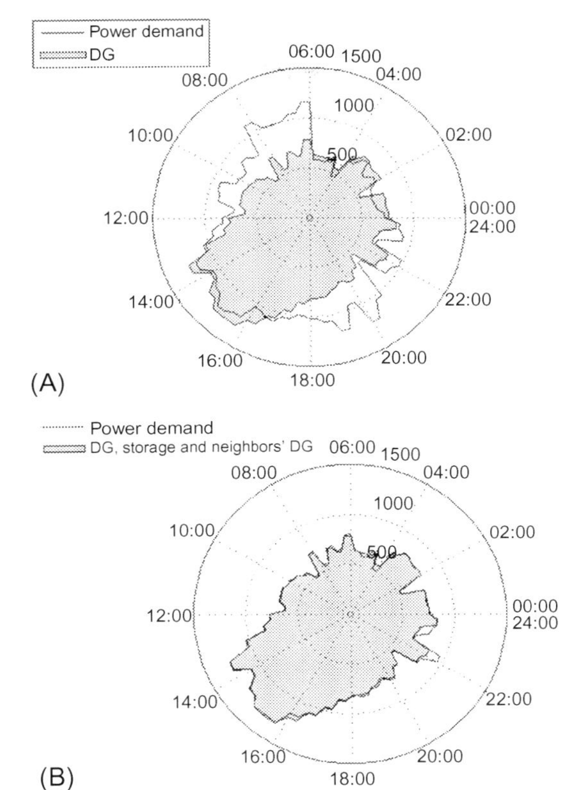

Fig. 19.5 Supply-demand matching performance (200 households): (A) scenario I (without any control) and (B) scenario III (our solution).

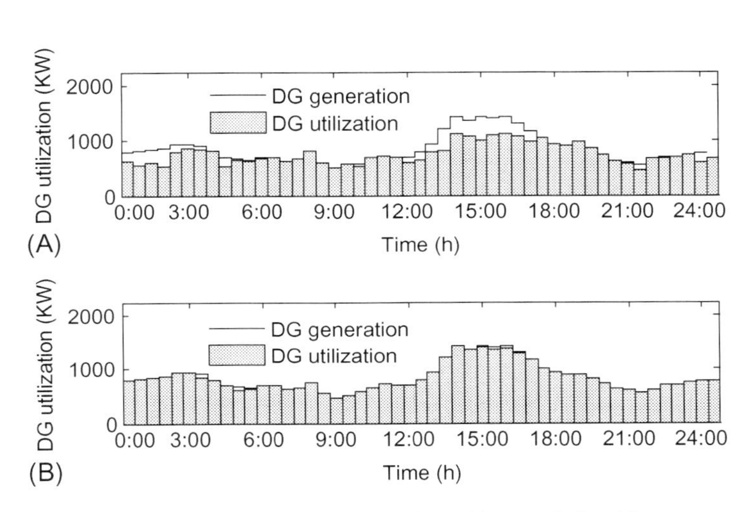

Fig. 19.6 DG utilization efficiency (200 households): (A) scenario I: without any control and (B) scenario III: our solution.

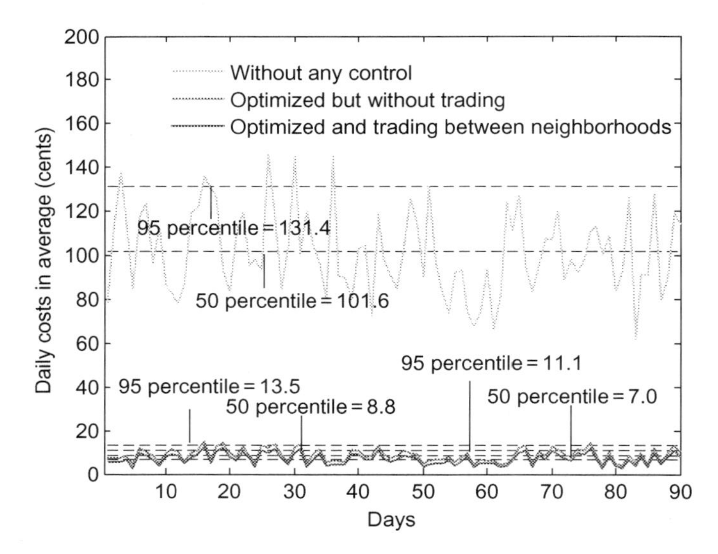

Fig. 19.7 Daily average electricity purchase cost (3 months, 200 households).

19.4.3 Performance under uncertainties

The findings obtained from the previous simulations are encouraging, however, it should be noted that the renewable DG generation and RTP information are with a stochastic nature which cannot be predicted accurately in practice, and data errors and unavailability can occur in the information exchanges (RTP) between the households and dispatching center due to congestion and temperate outage in underlying communication infrastructure. To this end, the robustness of the proposed solution is assessed assuming that the DG generation and RTP prediction are with random errors within the range of (−10% to +10%), as given in Fig. 19.8.

Fig. 19.9 presents a comparison between the energy supply (from domestic DGs, storage unit, and neighbors' DGs) and the domestic demand of a representative household in the presence of the given RTP and DG uncertainties. Compared with Fig. 19.4C, the performance is slightly degraded at certain time periods, for example, certain amount of DG generation is not utilized (e.g., 12:00–15:00) and renewable energy is not adequate (e.g., 04:00–09:00). In fact, such degradation under uncertainties can be improved through better sizing of battery capacity with sufficient margin and electricity purchase from the power utility, respectively. In addition, Fig. 19.10 gives the daily household electricity purchase costs over 90 days of the suggested solution for predicted and real RTP. It shows that the actual purchase cost in the presence of prediction errors is higher than the cost calculated based on the algorithm presented above, but it still outperforms the scenario without dispatch control. Overall, the results confirm that the proposed off-line dispatch solution can cope well with the uncertainties, with significant benefits of greatly reduced management and reinforcement complexity to implement demand-side management.

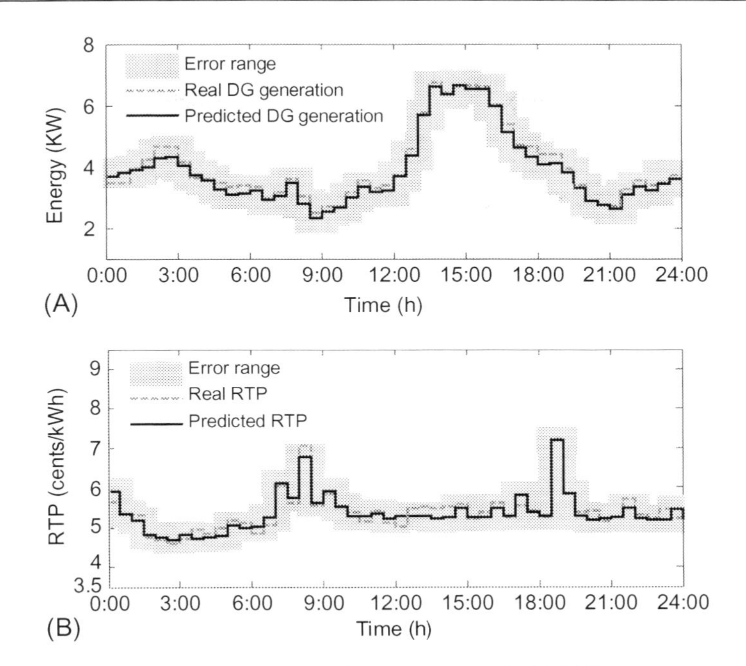

Fig. 19.8 Operational uncertainties: (A) DG error range and (B) RTP error range.

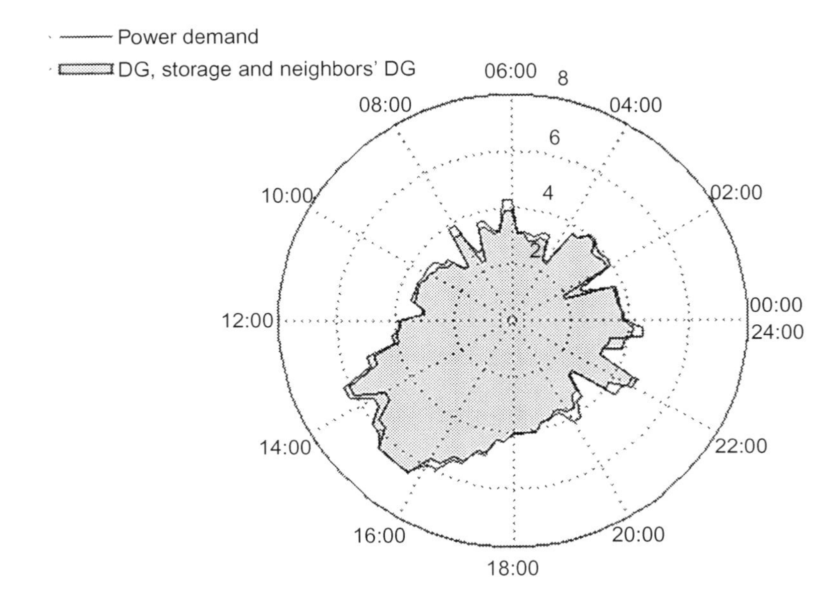

Fig. 19.9 Household supply-demand match under RTP and DG uncertainties.

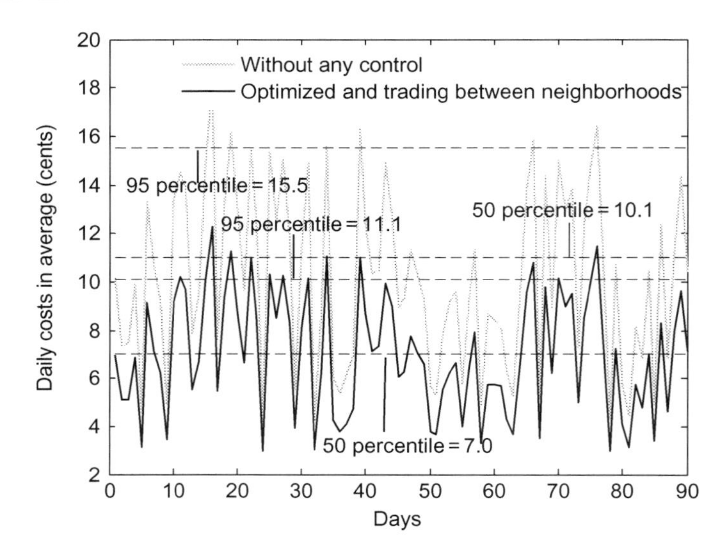

Fig. 19.10 Daily electricity purchase cost under uncertainties (3 months).

19.5 Conclusions and future work

This chapter presents a cost-effective energy dispatch solution for the residential community with renewable DGs and storage units under RTP. Through adopting such solution, both distribution network operators (DNOs) and the households can obtain direct benefit in terms of meeting residential demands in a more controlled manner and reduced electricity cost, respectively. The numerical results from a range of simulations confirm its effectiveness and robustness. In respect to the future work, two directions are considered worth further research effort. The performance of the proposed solution needs to be further assessed in realistic deployment of distribution network scenarios. It would be interesting to enclose the potential benefit from the coordination among DGs, storage, and loads into the domestic DG and storage capacity planning, so as the facility investment and energy purchase cost can be minimized from the resident's perspective. Also, the advanced prediction tools or error correction mechanisms need to be further exploited as the accurate DG and RTP prediction can promote the performance of the suggested solution.

Acknowledgment

This work is supported in part by the Nature Science Foundation of Zhejiang Province (Grant No. LZ15E070001).

References

Adika, C.O., Wang, L.F., 2014. Autonomous appliance scheduling for household energy management. IEEE Trans. Smart Grid 5, 673–682.

Barzin, R., Chen, J.J.J., Young, B.R., Farid, M.M., 2016. Application of weather forecast in conjunction with price-based method for PCM solar passive buildings – an experimental study. Appl. Energy 163, 9–18.

Bhandari, B., Lee, K., Lee, C.S., Song, C., Maskey, R.K., Ahn, S., 2014. A novel off-grid hybrid power system comprised of solar photovoltaic, wind, and hydro energy sources. Appl. Energy 133, 236–242.

El-Baz, W., Tzscheutschler, P., 2015. Short-term smart learning electrical load prediction algorithm for home energy management systems. Appl. Energy 147, 10–19.

Electricity Price and Demand. n.d. AEMO [Online]. Available from: http://www.aemo.com.au/Electricity/Data/Price-and-Demand

Erdinc, O., 2014. Economic impacts of small-scale own generating and storage units and electric vehicles under different demand response strategies for smart households. Appl. Energy 126, 142–150.

Erdinc, O., Paterakis, N.G., Pappi, I.N., Bakirtzis, A.G., Catalao, J.P.S., 2015. A new perspective for sizing of distributed generation and energy storage for smart households under demand response. Appl. Energy 143, 26–37.

Finn, P., Fitzpatrick, C., 2014. Demand side management of industrial electricity consumption: Promoting the use of renewable energy through real-time pricing. Appl. Energy 113, 11–21.

Giorgio, A.D., Pimpinella, L., 2012. An event driven smart home controller enabling consumer economic saving and automated demand side management. Appl. Energy 96, 92–103.

Giusti, A., Salani, M., Caro, G., Rizzoli, A., Gambardella, L., 2014. Restricted neighborhood communication improves decentralized demand-side load management. IEEE Trans. Smart Grid. 5(1).

Liu, T., Ai, Q., 2016. Interactive energy management of networked microgrids-based active distribution system considering large-scale integration of renewable energy resources. Appl. Energy 163, 408–422.

Munkhammar, J., Widén, J., Rydén, J., 2015. On a probability distribution model combining household power consumption, electric vehicle home-charging and photovoltaic power production. Appl. Energy 142, 135–143.

Pedrasa, M.A., Spooner, E.D., MacGill, I.F., 2011. Robust scheduling of residential distributed energy resources using a novel energy service decision-support tool.Proc. IEEE PES Eur. Conf. Innov. Smart Grid Technol. (ISGT)pp. 1–8.

Senjyu, T., Shimaukuro, K., Uezato, K., Funabashi, T., 2003. A fast technique for unit commitment problem by extended priority list. IEEE Trans. Power Syst. 18 (2), 882–888.

The California Energy Almanac n.d.[Online]. Available from: http://www.energyalmanac.ca.gov/renewables/solar/pv.html

The Wind Power Almanac n.d. [Online]. Available from: http://www.thewindpower.net/

Wang, Q., Yang, Q., Yan, W., 2015. Optimal dispatch in residential community with DGs and storage under real-time pricing.Proc. IEEE Conf. Information and Automation, Yunnanpp. 239–244.

Evaluation on the short-term power supply capacity of an active distribution system based on multiple scenarios considering uncertainties

20

Yajing Gao, Xiaojie Zhou*, Qitian Mu*, Jing Zhu*†
*School of Electrical and Electronic Engineering, North China Electric Power University, Baoding, China, †Maintenance Branch Company of State Grid Fujian Electric Power Co., Ltd, Xiamen, China

20.1 Introduction

To achieve the efficient use of energy resources and energy-saving emission reduction development goals, the proportion of wind power, photovoltaic (PV), and other renewable energy power generation system capacities of the total installed capacity in a power system is getting higher and higher (Meena et al., 2017). As the medium-, small-, and microdistributed power generation, energy storage (ES), electric vehicles, controllable load, and other distributed energy resources (DER) continue to access low-voltage and medium-voltage power grisd, the original only one-way power transmission network presents the features of "two-way trend" and "with the source" (Souza Machado et al., 2016; Gils et al., 2017). A two-way power flow will be generated between the future transmission and distribution networks. With the permeability of DER in the distribution network continuing to increase, the traditional distribution network presents the problems of renewable energy consumption difficulties, low level of interaction with electricity and scheduling, weak primary distribution frame, and other issues. Regardless of the mode of operation, the control mode or management model is difficult to adapt to new development needs. Therefore, it is necessary for the traditional distribution network to develop and transform into the active distribution system (ADS) with active control capability and load interaction capability.

20.1.1 Active distribution system

The ADS is the distribution system that has ability to manage the trend through a flexible network topology, in order to achieve active control and active management of the local DER (Zhigang et al., 2017). It can manage accessed DER actively through

Smart Power Distribution Systems. https://doi.org/10.1016/B978-0-12-812154-2.00020-1

advanced communication technology and achieve efficient DER consumption while making the distribution system operation more economical and stable.

The ADS is obviously different from the traditional passive distribution network. DER access not only changes the original one-way power flow and the formation of multiple between the two-way complex trend, but also makes the network structure difficult to form a stable balance because of the DER-based power output uncertainty. So, it needs a series of new technologies to meet the needs of the future power grid in planning, operation, management, services, etc. On the basis of analyzing the difference between the active distribution network and the traditional distribution network, some papers discussed the active power distribution system planning method with the aspects of constructing a distributed power and load uncertainty probability scenario, improving the utilization of the intermittent power supply, and achieving the minimum total cost through active management (Buhari and Kopsidas, 2014; Xing et al., 2016).

The CIGRE C6 Working Group presented the research framework and direction of ADS in 2008 (CIGRE Task Force C6.11. 2008). The existing ADS research results include: renewable and distributed energy access example system (C6.04), large intermittent energy grid (C6.08), demand-side integration (C6.09), low voltage (C6.11), ES system research (C6.15), and so on.

In view of the influence of distributed power supply and microgrid on ADS, foreign scholars mainly studied four aspects: the control and scheduling of power flow and ADS voltage (Tahir et al., 2016), ADS equipment thermal stability check (Fahmy, 2016), the impact on ADS short-circuit current levels (Hongjie et al., 2015), and the impact on ADS protection (Florin et al., 2014).

Regarding aspects of ADS operation and management, foreign scholars mainly studied ADS scheduling management technology through the regulation of distributed power supply capacity (PSC), distribution location, transformer on-load voltage regulator, voltage regulator, reactive power flow, and network reconstruction to maximize the utilization of the distribution network (Abdeltawab and Mohamed, 2017; Jianpeng et al., 2014; Tianjiao et al., 2016). As for the ADS feasibility technology implementation of ADS operation, foreign scholars have done a lot of research on demand-side management or demand-side response ES equipment, line dynamic measurement capacity, voltage and power flow control, short-circuit current limit, and advanced active distribution network protection measures (Yin et al., 2017).

20.1.2 PSC of distribution network

The PSC of the distribution network is the maximum load that the distribution network can supply under the conditions of the branch power constraint and the node voltage constraint (Hong et al., 2017). It is determined by the mode of operation of the distribution network and the growth pattern of the load. Due to the large number of DER access, the distribution network faces the challenges of network planning, operation, and management, while providing sufficient PSC; this is the fundamental task of the distribution network. Especially in the current rapid economic development scenario, energy saving, environmental protection, more stringent environment emission requirements, how to accurately evaluate the impact of DER on PSC, through DER's optimal access and operation to fully explore the PSC of the distribution network, and achieve the maximum benefit of the distribution network will be the problem in the distribution network's PSC aspect which need to be solved urgently.

The distribution network PSC of the study has gone through the following three stages:

- The stage of assessing the PSC of the distribution system based on the capacity of the transformer. This is performed via methods such as the capacity-load ratio method (Tan et al., 2016) and the trial and error method (Fu et al., 2016). This stage was based on the substation capacity of the substation, and the size of the PSC was evaluated from the macroscopic point of view. The calculation of this method is relatively simple, but it did not consider the influence of the substation's subordinate distribution network substation on PSC in detail.
- Network PSC calculation stage. Methods used included the peak load multiple method (Liu et al., 2015) and the network maximal flow method (Fu et al., 2016). This method used a feeder as the basis for calculating the PSC of the power grid and put forward the idea of calculating the PSC of network transfer, while calculated the substation capacity. But only using the feeder load to estimate the network transfer PSC is not accurate enough, as it ignores the impact of substation PSC.
- The calculation stage taking into account the $N-1$ safety criteria and combining the substation PSC with network transfer capacity. When carrying out the research on the power supply capability of the distribution network, this phase took into account the $N-1$ power supply safety guidelines on the city network planning that are mentioned in "Urban Power Network Planning and Design Guidelines," and put forward the idea of how to calculate the power supply ability under certain reliability criterion. At present, PSC research is mostly in this stage.

In existing PSC-related research, in order to meet the $N-1$ safe operation criteria, a study took the maximum load that the main transformer could provide, the main transformer overload, contact capacity constraints, and other factors into account, and established the analytical model of distribution network PSC (Jian et al., 2014a, b). Marzano et al. (2015) proposed a method of assessing the PSC of the distribution network taking into account the load ability constraints of the main transformer. This method included three stages: preevaluation, constraint analysis, and verification evaluation. The final results were obtained by recursive assessment based on the repetitive trend method. One study took the distributed generation's volatility and intermittent considerations into account, analyzed the PSC according to the fuzzy target theory, and improved PSC by the dynamic optimization method (Hashemi and Ostergaard, 2016). Another study considered the power limitation of the distributed power generation and the ES system, and established the PSC model of the autonomous control area and the active distribution network cable. Finally, it put forward the load transfer based on the distribution network's radial characteristics and the island's actual and stepwise optimization algorithm (Yu et al., 2015). Based on the consideration of the reliability of the operation environment of the grid, in order to solve the problem of how to evaluate the safety status of the grid on the basis of the $N-1$ criterion, domestic and foreign scholars have put forward a series of safety evaluation indices. These assessment indicators included the assessment indicators of the distribution network's largest PSC (Sun et al., 2017; Xiaolong et al., 2014), the indicators to assess the load transfer power supply recovery capacity (Liming and Xianjun, 2015), and the static safety analysis considered by the grid security rating system.

In the related research of uncertainty, for the uncertainty of DER output and load state, a study used Latin hypercube sampling to simulate the maximum output of wind power and the random state of node load power (Li et al., 2013). Wang et al. (2015a, b, 2016) established the probability assessment model considering the randomness of wind speed and the prediction error of PV and load. In view of the complexity of the grid structure caused by component failure, Wang et al. (2015a, b) and Vu and Turitsyn (2015) sorted the possible failures of the grid, and effectively simplified the model on the basis of taking into account the grid uncertainty. The existing studies have focused on the uncertainty of specific units, and have less consideration on assessing the impact of the uncertainty of the access elements on the PSC.

In order to analyze the influence of uncertainty factors on PSC, this chapter proposes a short-term PSC evaluation system based on multiscene technology. First, based on the $N-1$ failure that may occur in the distribution network, the probability output model is established for the uncertainties of each unit. Second, it based on multiscene technology to form a number of PSC assessment scenarios. Third, a series of probabilistic evaluation indices for evaluating the PSC are put forward from the idea of probability assessment. Finally, the PSC of each scene is evaluated and the evaluation index is obtained, which can provide reference for the optimization, operation, and scheduling of an active power distribution system.

20.2 Analysis of uncertainty factors in evaluating PSC

20.2.1 Considerations of uncertainty factors

Due to the complex topology of the distribution network, the high failure rate of the components, the large range of load changes, and the large amount of DER with strong uncertainty and volatility, the ADS faces many uncertainties. When conducting ADS scheduling and operation, we need to consider the impact of these uncertain factors to ensure the safe and economical operation of the distribution network (Kim and Cho, 2017).

20.2.1.1 Uncertainty of "element's" output

The "element" in the ADS means the load, the DER, and the ES unit. Among them, the load includes fixed load and flexible load. When discussing the impact of uncertainty factors, DER mainly refers to the PV power generation and wind power in this chapter.

The volatility and uncertainty of fixed load level and DER output are strong. The common prediction methods include Markov chain (Siltala and Granvik, 2017), neural network (Dolara et al., 2015), autoregressive and moving average (ARMA) (Abrahart and See, 2015), support vector machine (Shao et al., 2016), and so on. Most of these prediction mechanisms are supplemented by certain classification methods and rely on statistical methods to learn and deal with the historical output and the relevant influencing factors such as data generate predictive ability to reflect the effect.

The output of the flexible load and the ES components is determined primarily by the electricity price and its response to the price (Dan et al., 2014; Bo et al., 2014). The change of price is influenced by generation cost, power generation, demand of the electricity market, and policy guidance. Electricity price forecast is similar to DER output forecast, with uncertainty. Flexible load and ES components regarding the price of the specific response to the results determined by the user's operation, the user's decision-making results influenced by personal needs, power-related policies, and other factors, have both predictability and randomness.

Processing methods considering uncertainty include Monte Carlo (Qiang et al., 2013), estimation correction (Tsuyuki and Kanamitsu, 2015), error correction (Yajing et al., 2015a, b), probability modeling (Chalk and Corotis, 2017), etc. Among them, the Monte Carlo method estimates the probability and the value of the parameter by giving the frequency or expected value of a parameter. The estimated correction and error correction are mainly based on the historical error data, and superimpose the correction value of the prediction error on the basic prediction value. Probabilistic modeling, from the perspective of probability assessment, establishes a probability density model for the output or error of the prediction object, and has the ability to simulate various possibilities and probability of occurrence.

Existing studies have established probability models for load power, wind power maximum output (Yajing et al., 2015a, b), wind speed (Yaoqi, 2016), and PV and load prediction (Wang et al., 2015a, b) and have achieved good results. Commonly used probability density distributions include normal distribution (Chaichan and Kazem, 2016), Beta distribution (Weijia et al., 2015), location-scale distribution (Erto and Antonio, 2016), and nonparametric kernel density distribution (Nan et al., 2016). The normal distribution has a good distribution of simulation properties and is one of the most commonly used fitting methods. Beta distribution and location-scale distribution show some characteristics that are suitable for specific data groups. The nonparametric kernel density estimation method does not require any prior knowledge, studies the data distribution characteristics completely from the data samples, and has been successfully applied in load modeling, wind speed modeling, and reliability index calculation (Nagler and Claudia, 2016).

20.2.1.2 Uncertainty of network topology

In the ADS, the failure of components such as generators, lines, and transformers will lead to the change of distribution network topology and add to the complexity and uncertainty of the grid structure. In order to fully consider the impact of network topology on the PSC, it is necessary to analyze the possible faults in the ADS.

Power system failure is divided into horizontal fault and longitudinal fault. Horizontal faults are short-circuit faults, including single-phase short-circuit, two-phase short-circuit, and three-phase short-circuit fault. Longitudinal faults are all types of broken lines, including single-phase, two-phase, and three-phase disconnection failure. According to the actual operation data, it can be seen that the number of single-phase short circuits is about two-thirds of the total number of short-circuit faults, and the proportion of short-circuit faults in the power system is the largest. For longitudinal faults, although the rate of disconnection failures is relatively small, the consequences of the accident are quite severe (Xiaomin et al., 2015). Therefore, in the reliability assessment of the more complex power network, we are usually based on the $N-1$ safety criteria and only consider the single-line fault situation (Geng et al., 2015).

Based on the consideration of possible failures in the distribution system, Yeh (2015) used the occasional fault enumeration method combined with the Markov chain (Siltala and Granvik, 2017), through the enumeration of the normal or fault state of the system, combined with the Marco two-state model of the chain, to achieve the timing of the equipment in the system operating state of the forecast. In fact, the power distribution system is a complex, large scale, one-by-one enumeration system state, will spend a lot of running time, and is not conducive to ensuring the overall assessment efficiency. In addition, the affected degree of the distribution system varies with the specific fault situation. Only analyzing the consequences of a fault to a certain extent can not only guarantee the accuracy of the evaluation, but also save the running time and improve the calculation efficiency. Therefore, it is very important to diagnose and sort the fault consequence and its probability precisely for the complex distribution network.

The fault screening and sorting algorithm must take into account the evaluation of the extent of the impact of the fault, the speed, and accuracy of the guarantee (Dounas-Frazer et al., 2016). In this paper, the evaluation of PSC requires that under certain constraints, the load increases to the safe and stable operating limit at each time point, and the calculation speed is an important guarantee for the evaluation effectiveness. In a study, the $N-1$ faults that may occur in the distribution grid were sorted by defining and obtaining the accident consequence function. Considering the uncertainty of the grid, the model was simplified and the computation time was saved (Bartlett and Andrews, 2015). In the accident sorting method, state indicators or margin indicators were commonly used to reflect the severity of the failure (Farajollahi et al., 2016). On the basis of this, the consideration of failure probability and economic loss was added, and the fault order model with the expected loss as the sorting index was established (Huaidong et al., 2016).

20.2.2 Analysis of uncertainty of "element" output based on multiscene

The "element" in ADS, that is, load, DER, and ES unit, where the load, including fixed load and flexible load, in discussing the impact of uncertainty factors, DER mainly refers to PV power generation and wind power. According to the different modeling methods of "element" output, they can be divided into two categories. One is the output model based on the forecasting technique, including the fixed load and the DER. For this category, the output prediction is first carried out, and then the scene of the corresponding prediction error is simulated by the multiscene technique. Finally, the reduced error scene is combined with the predicted result to get the multiscene model. The other is the output model that is affected by the price, including the flexible load and the ES element. For this category, through establishing the probability model of the electricity price impact mechanism, we obtain the multiscene with its probability output.

20.2.2.1 Fixed load and DER output prediction model based on error correction

The basic method of fixed load and DER output prediction

The prediction method of fixed load and DER (PV and wind power) is forecasted by using the adaptive method of wavelet neural network (WNN) with additional adaptive dynamic programming (ADP) correction (Yajing et al., 2016). The idea of "estimating a correction" is introduced, and the ADP is added. The WNN parameters are updated with the actual measurement data to improve the prediction accuracy. The prediction principle is shown in Fig. 20.1. (See Boxes 20.1–20.5.)

In Fig. 20.1, $X_1, X_2, \ldots, X_n$ are the inputs of the prediction model; $Y_1, Y_2, \ldots, Y_m$ are the outputs of the prediction model; *wjk* is the weight between the hidden layer and the output layer of the WNN; *bj* is the expansion factor of the wavelet basis function; θj is the threshold; $P_{i1}(t), P_{i2}(t), \ldots, P_{in}(t)$ are the actual measurement data of the sampling

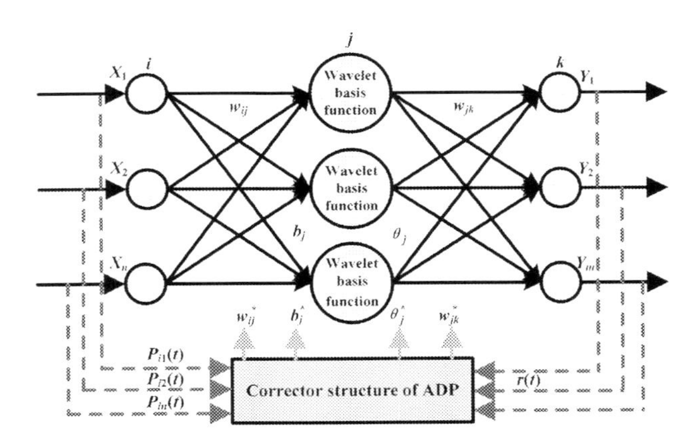

Fig. 20.1 Introducing WNN forecasting model of ADP correction.

> **Box 20.1 Research significance and research status**
>
> The first section is the introduction part; it introduces the background and the research significance. The research status of ADS, PSC of distribution network and uncertainty factor in the world are expounded.

Box 20.2 Analysis of uncertainty factors

In this section, the influence of the uncertain factors on the supply capacity is discussed, and the "element" in ADS is divided into two categories. For the fixed load and DER, we predict the output of DER and fixed load based on the idea of error correction, establish the corresponding prediction error probability model, and format contribution multiscenes based on multiscene technology. For the flexible load and energy storage component, the probability model is established based on its response mechanism to the price. Combined with the above two types of circumstances, the formation of ADS "element" output of the multiscene. The multiscene probability model fully considers the uncertainty of the output of "element," which is better for the simulation of the actual operation and provides the basis for the evaluation of the subsequent PSC.

Box 20.3 Evaluations of PSC

In order to quantify the probability assessment model, a series of evaluation indices of PSC are proposed from the aspects of numerical size, fluctuation situation, abundant situation, and the contribution degree of access DER.

Indicator	Formula
Expectation of PSC $E_{PSC}(t)$	$E_{PSC}(t)=\sum_{i=1}^{N_s} S_{PSC}(i,t)p(i)$
Variance of PSC $V_{PSC}(t)$	$V_{PSC}(t)=\sum_{i=1}^{N_s}[(S_{PSC}(i,t)-E_{PSC}(t))p(i)]^2$
Variation coefficient of PSC $\gamma_{PSC}(t)$	$\gamma_{PSC}(t)=\frac{\sqrt{V_{PSC}(t)}}{E_{PSC}(t)}$
Maximum value of PSC S max PSC(t)	$S_{PSC}^{max}(t)=\max\{S_{PSC}(i,t),\ i\in N_s\}$
Minimal value of PSC S min PSC(t)	$S_{PSC}^{max}(t)=\max\{S_{PSC}(i,t),\ i\in N_s\}$
Margin of PSC $M_{PSC}(t)$	$M_{PSC}(t)=E_{PSC}(t)-\sum_{i=1}^{N}p(i)\sum_{k=1}^{N_L}S_L(i,k,t)$
Dissatisfied probability of PSC $DP_{PSC}(t)$	$DP_{PSC}(t)=\sum p_{dp}(t)$
Dissatisfied amount of PSC $DA_{PSC}(t)$	$DA_{PSC}(t)=A-E_{PSC}(t)$
Contribution rate of DER to expectation of PSC $CR_{DTE}(t)$	$CR_{DTE}(t)=\frac{E_{PSC}(t)}{E_{PSC,N}(t)}-1$
Contributed amount of DER to expectation of PSC $CA_{DTE}(t)$	$CA_{DTE}(t)=E_{PSC}(t)-E_{PSC,\ N}(t)$
Contribution rate of DER to dissatisfied probability of PSC $CR_{DTDP}(t)$	$CR_{DTDP}(t)=1-\frac{DP_{PSC}(t)}{DP_{PSC,N}(t)}$

Box 20.4 PSC evaluation algorithm

On the basis of considering the uncertainty of "element" output in ADS, taking into account the possible failures in the distribution network, the multiscene model of network failure is formed based on the fault sorting method. Combining the multiscene of "element" with the multiscene of network fault, the multiscene probability evaluation model of power supply capability is obtained.

Box 20.5 Case study

Based on IEEE 14 nodes—example of the evaluation results shows that the probability of power supply capability more scenes containing evaluation index evaluation results for guiding the optimal operation of ADS, the excavation of the distribution network PSC, improves the distribution network plays an important role in guiding in DER given ability.

period; w_{ij}^*, w_{jk}^*, b_j^*, θ_j^* are the updated parameters after applying the ADP corrector; and $r(t)$ is the cost function.

Fixed load and DER prediction error correction

In order to control the deviation range of the predicted value, the effect and accuracy of the prediction are further improved, and the error correction is introduced on the basis of the above forecasting model; as a result the fixed load level is fluctuating and the DER output is random and uncertain. Error correction through the comprehensive analysis of the factors that produce the prediction error generates the prediction error of the fitting value, and uses it to compensate for the actual forecast error. In PV power generation, for example, to consider the error correction prediction research ideas shown in Fig. 20.2, the figure shows that the key steps of the forecast are the following points.

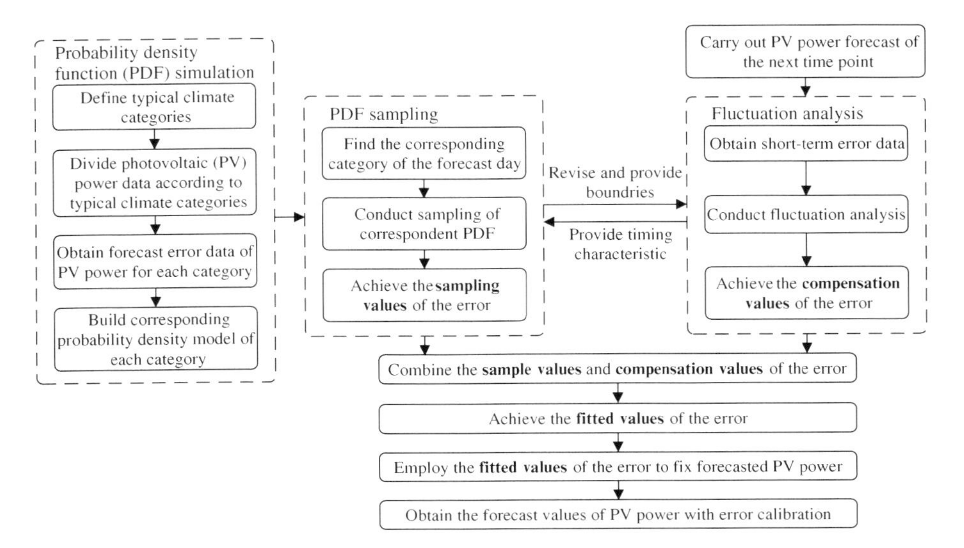

Fig. 20.2 Research on short-term photovoltaic (PV) power generation based on error correction under typical weather.

Definition of typical weather types PV power is closely related to climate conditions, which depend on a range of issues and can be described by many different approaches. To define the classification of climate categories using a harmonized standard, the China Meteorological Administration released "GB/T 22164—2008 Public Climate Service—Weather Graphic Symbols," on which 33 types of climate categories are defined. On that basis, the weather conditions can be classified into four categories that represent the most vital characteristics of these climates, which include sunny, cloudy, drizzle, and heavy rain (Trigo and Dacamara, 2015). However, the range of PV power for each category varies over the different seasons. For instance, the maximum PV output on sunny days in summer is apparently higher than it is in winter. Consequently, the classification of the typical climate categories is based on both the weather and the season, which generates a total of 16 categories, as shown in Fig. 20.3.

Establishment of probability density model According to the typical weather types, the historical prediction error data are classified, and the probabilistic density distribution fitting model is established according to the probability density function (PDF) of the prediction error probability of each classification, and the corresponding error sampling value is extracted.

The variances of the PDFs corresponding to the four climate types are expressed as σ_{A1}^2, σ_{A2}^2, σ_{A3}^2, and σ_{A4}^2. Fig. 20.4 shows the probability density distribution of WNN for historical predictive relative error data for a PV station under four typical climate categories. It is easy to see from the figure that the relative error distribution graphs of each type are approximate to the y-axis symmetry, and the PDF generally decreases with the increase of the absolute value of x and approaches the x-axis. The relative

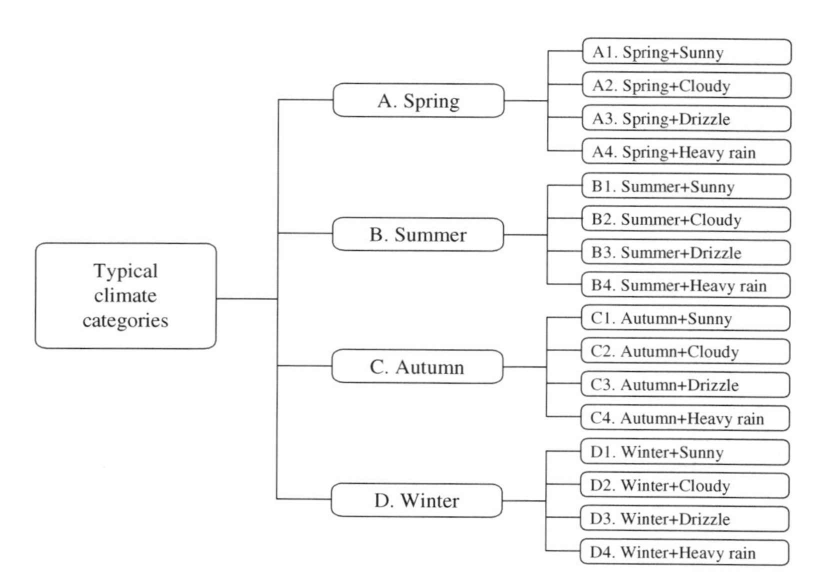

Fig. 20.3 Typical climate categories.

error (RE) values in sunny weather are the most concentrated, that is when the minimum variance σ_{A1}^2 of the four curves is achieved, and illustrates a "tall and thin" characteristic. In comparison, when it rains heavily, the PDF curve would appear to be "short and fat," which indicates a smaller variance σ_{A4}^2. Universally, the variances of the RE have a tendency to be larger when the weather types vary from sunny to cloudy to drizzle to heavy rain, successively, and therefore, $\sigma_{A1}^2 < \sigma_{A2}^2 < \sigma_{A3}^2 < \sigma_{A4}^2$.

In summary, the relative error of PV prediction can be classified into a more regular probability distribution according to the weather classification, and the nonparametric kernel density estimation method is used to fit the parameters of the relative error PDF of each classification.

Volatility analysis On the basis of this, the error estimation method and the compensation force under different fluctuation characteristics are determined by analyzing the fluctuation direction and amplitude regularity of short-term PV power prediction error. The prediction error of the PV power prediction error at the next time is obtained, and the error compensation value is obtained (Lin et al., 2016).

In volatility analysis, n long-term relative errors are selected as the measure of recent relative error analysis. Define the critical value $k1$ of the long-term variance level σ_l^2 and the absolute value of the fitted slope:

$$\sigma_l^2 = \frac{1}{n}\sum_{i=1}^{n}\left(\delta_i - \frac{1}{n}\sum_{i=1}^{n}\delta_i\right)^2 \tag{20.1}$$

$$k_l = |k_1 - k_2|/4 \tag{20.2}$$

where k_1 and k_2 are the up and down critical values of the one-sided probability interval determined by the simulated model and the probability level, respectively.

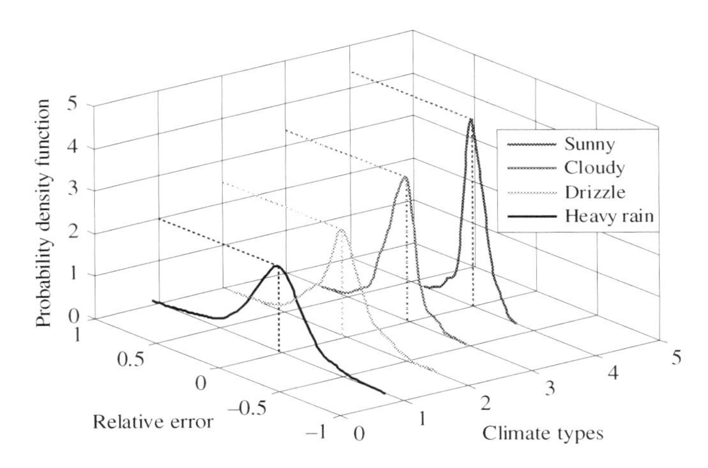

Fig. 20.4 Probability density functions of the relative error for different climate types.

Table 20.1 Investigation method of the RE fluctuation

Value judgment	Fluctuation	Investigation methods
$\sigma_s^2 < \sigma_l^2$ and $k_s < k_l$	Column 2	Moving average method
$\sigma_s^2 < \sigma_l^2$ and $k_s > k_l$	Low amplitude, not steady	Auto regressive moving average method
$\sigma_s^2 > \sigma_l^2$ and $k_s < k_l$	High amplitude, steady	Weighted moving average method
$\sigma_s^2 > \sigma_l^2$ and $k_s > k_l$	High amplitude, not steady	Liner method

The relative error of the a data points before the forecast point is taken as the sample value of the recent relative error, and the variance σ_s^2 and the slope absolute value k_s of the fitting straight line are calculated. Combined with the two can be similar to the recent prediction of the relative error of the volatility, the estimated point relative error can be estimated. Table 20.1 shows the specific analytical methods.

In order to avoid the introduction of new errors in volatility analysis, the principle of error compensation is established. On the basis of the relative error estimate, the relative error compensation method is determined based on the relative magnitude of the prediction point error δ_i and the previous time error δ_{i-1} and the threshold error δ_0 of dividing the large error and the small error as shown in Table 20.2.

Error correction Based on fluctuation analysis, the sampling results are sequenced using the compensation values. The sampling values closest to the compensation ones are elected to be the fitted values of the RE at the corresponding time points to correct the forecast values of the PV power, and moreover, to improve the forecast accuracy.

The multiscene generation and reduction of fixed load and DER output

For the fixed load, according to the theory of probability and statistics, the normal distribution has good properties and can be used to approximate the analysis of many probability distributions. Therefore, the error of fixed load forecasting is estimated by normal distribution.

Table 20.2 Compensation methods of estimated RE

Value judgment	Compensation methods
$\delta_i < \delta_0$ and $\delta_{i-1} < \delta_0$	None
$\delta_i < \delta_0$ and $\delta_{i-1} > \delta_0$	1 magnitude of the estimated values
$\delta_i > \delta_0$ and $\delta_{i-1} < \delta_0$	1 magnitude of the estimated values
$\delta_i > \delta_0$ and $\delta_{i-1} > \delta_0$	1/2 magnitude of the estimated values

For the wind power output prediction error, the actual wind power forecast error has a large peak and skewness; simply using the normal distribution description will have a greater error. Huang and Jiang (2017) pointed out that the wind speed prediction error is the main reason leading to the wind power prediction error. The wind power prediction error corresponding to the wind speed forecast is different, and the wind speed interval: the applicable range of the "0 error" distribution is [0, 2.1], [12.6, ∞]; the applicable interval for the exponential distribution is (2.1, 5.1]; the applicable interval for normal distribution is (5.1, 9.8), extreme value distribution for the interval [9.8, 12.6). In this paper, the error probability distribution of wind speed is established by using the error distribution of the corresponding wind speed to estimate the error of wind power output.

For the error of the PV output prediction, the use of the normal distribution description will also produce a large error. Weijia et al. (2015) proposed that the output level of PV power generation was closely related to the weather condition, and the prediction error distribution was different in different weather types. In this chapter, the idea of establishing different probability density distribution functions is used to estimate errors in PV output.

After many scenes are generated and reduced, N error samples are obtained. Where the probability of the nth sample corresponds to p_{1n} $(n \in N)$, the load forecast and the DER output value with random error can be expressed by:

$$P_{L,n} = P_{L,p}(1 + r_n\%) \tag{20.3}$$

$$P_{DER,n} = P_{DER,p}(1 + r_n\%) \tag{20.4}$$

where $P_{L,n}$ and $P_{DER,n}$ are the final load forecast value of the nth scene with simulation error and DER output prediction value; $P_{L,p}$ and $P_{DER,p}$ are load predictions and DER output predictions without simulation error; and r_n is the error generated by the simulation in the nth scene.

After the probability distribution function of the fixed load and the DER output in the ADS is obtained using the Latin hypercube sampling technique (Shields and Zhang, 2016), the corresponding output scene samples were generated and reduced for different distribution models (Yajing et al., 2015a, b), to ensure the effectiveness of sampling and representative of the "element " output multiscene samples and the corresponding probability of the scene, as shown in Fig. 20.5.

20.2.2.2 Modeling of flexible load and ES based on electricity price influence mechanism

Flexible load model

Modeling of interruptible load Interruptible load (IL) users can sign contracts with power supply companies, allowing them to break the power supply under different specific conditions. In this paper, the power failure threshold is used to characterize the power failure conditions. Before the power supply company interrupts the

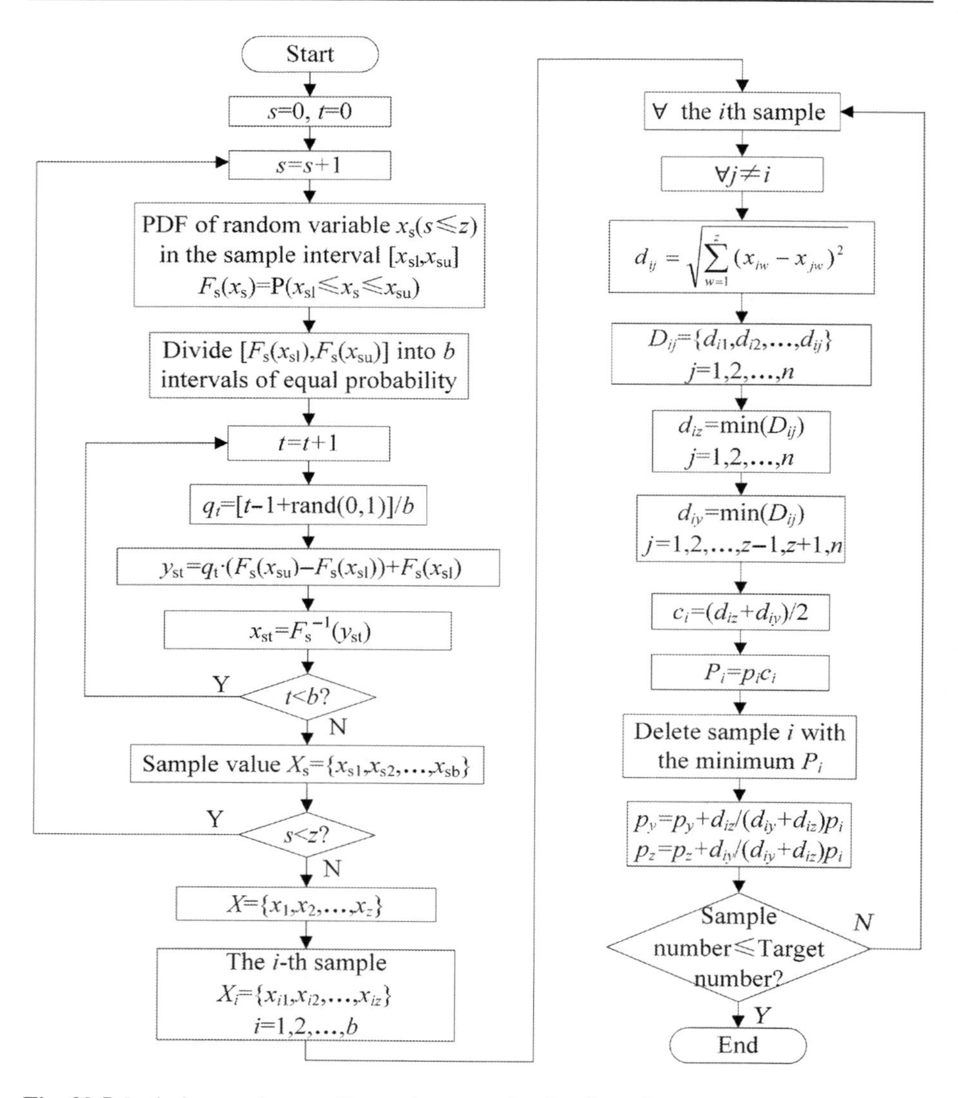

Fig. 20.5 Latin hypercube sampling and scene reduction flow chart.

supply of IL, its prediction method is the same as the fixed load, and the prediction error is fitted by the normal distribution. The S_{IL} is defined as a 0–1 state variable that characterizes the actual use of IL, as shown in the following equation:

$$S_{IL}(t)=\begin{cases}0, C(t)>C_{IL}\\1, C(t)<C_{IL}\end{cases} \tag{20.5}$$

where $S_{IL}(t)$ is the state variable at time t; $C(t)$ is the real-time electricity price at time t; and C_{IL} is the threshold value of the power supply.

Transferable load modeling The amount of transferable load (TL) is affected by the real-time electricity price, and its proportion fluctuates with the change of the price. The ratio of TL to electricity price for the load transfer rate α_{TL} can be defined.

The relationship between TL usage and electricity price $C(t)$ is shown as follows:

$$P_{TL}(t) = \alpha_{TL}(t)P'_{TL}(t)k + (1 - \alpha_{TL}(t))P'_{TL}(t) \tag{20.6}$$

$$k = \left(\frac{C(t)}{C_{TL}}\right)^{\beta} \tag{20.7}$$

where $P_{TL}(t)$ is the total amount of TL after the occurrence of t; $\alpha_{TL}(t)$ is the load transfer rate of TL at time t; $P'_{TL}(t)$ is the total amount of TL that does not occur at time t; k is the coefficient of influence of TL on TL; C_{TL} is the threshold price of TL transfer and transfer; and β is the elasticity coefficient of the demand price.

Assuming that the total amount of electricity demand for the user is constant within 1 day, the controlled demand for reduction will be delayed to other time periods; so the sum of the amount of change in TL usage in 1 day should be zero, as.

$$\begin{aligned}\sum_{t=1}^{24} \Delta P_{TL}(t) &= \sum_{t=1}^{24} (P_{TL}(t) - P'_{TL}(t)) \\ &= \sum_{t=1}^{24} \alpha_{TL}(t)P'_{TL}(t)(k-1) = 0\end{aligned} \tag{20.8}$$

In the case of TL access, $P'_{TL}(t) \neq 0$; so there is:

$$\sum_{t=1}^{24} \alpha_{TL}(t)(k-1) = 0 \tag{20.9}$$

β is usually less than -1; therefore, when $C(t) > C_{TL}$, TL is in the out of state, $k < 1$; when $C(t) < C_{TL}$, TL is in the transition state, $k > 1$. In view of this, if the TL transfer when the load transfer rate is subject to a different probability distribution, then type Eq. (20.9) can be derived as follows:

$$\sum_{t \in T_I} \alpha_{TL,I}(t)(k-1) = \sum_{t \in T_O} \alpha_{TL,O}(t)(1-k) \tag{20.10}$$

where T_I and T_O are the set of time for the transformation of the TL in a single day. $\alpha_{TL,I}$ and $\alpha_{TL,O}$ are the load transfer rates when the TL is transferred.

Since the α_{TL} is sampled according to its PDF, the sampling mean will be close to the expectation of its PDF when the number of samples is large. Thus, the formula (20.10) can be further simplified as follows:

$$\gamma = \frac{\mu_{TL,I}}{\mu_{TL,O}} \approx \frac{t_O \sum_{t \in T_O} (1-k)}{t_I \sum_{t \in T_I} (k-1)} \tag{20.11}$$

where γ is the ratio coefficient between the expected $\mu_{TL,I}$ and $\mu_{TL,O}$ of the distribution $\alpha_{TL,I}$ and $\alpha_{TL,O}$ and t_I and t_O are the number of times included in T_I and T_O, respectively.

When the next day electricity price is known, the threshold price set value, γ, is constant.

ES unit model

The basic model of ES components The ES unit can charge and discharge within its allowable range, and adjust the load state of the grid, in order to achieve the role of peak padding. In order to extend its service life, excessive charge and discharge is generally not allowed. The basic model of ES is as follows:

$$\begin{cases} \mathrm{SOC}(t) = \mathrm{SOC}(t-1) + \Delta \mathrm{SOC}_{ch}(t), & t \in T_{ch} \\ \mathrm{SOC}(t) = \mathrm{SOC}(t-1) - \Delta \mathrm{SOC}_{dis}(t), & t \in T_{dis} \end{cases} \tag{20.12}$$

$$\begin{cases} \Delta \mathrm{SOC}_{ch}(t) = P_{ch}(t)\eta_{ch}\Delta T S_{es}(t)/S_{ES}, & t \in T_{ch} \\ \Delta \mathrm{SOC}_{dis}(t) = P_{dis}(t)\eta_{dis}\Delta T S_{es}(t)/S_{ES}, & t \in T_{dis} \end{cases} \tag{20.13}$$

$$\begin{cases} S_{es} = 1, & \lambda_{es} < \lambda_0 \\ S_{es} = 0, & \lambda_{es} > \lambda_0 \end{cases} \tag{20.14}$$

where SOC(t) and SOC($t-1$) are the charge states of the ES element at t and $t-1$, respectively; $\Delta \mathrm{SOC}_{ch}(t)$ and $\Delta \mathrm{SOC}_{dis}(t)$ are the charge and discharge of ES at time t, respectively; $P_{ch}(t)$ and $P_{dis}(t)$ are the charge and discharge power of ES at time t, respectively; ΔT is the discharge time for the unit charge; η_{ch} and η_{dis} are the charge and discharge efficiency of ES, respectively; S_{ES} is the rated capacity of ES; $S_{ES}(t)$ is the charge or discharge state parameter of ES at time t; $\lambda_{es}(t)$ is the charge and discharge action parameters of ES at time t, which is the probability value of 0–1; λ_0 is the charge and discharge parameter threshold of ES; and T_{ch} and T_{dis} are the charge and discharge periods of ES, respectively, where they are equivalent to the tariff peak period T_f and the electricity price trough period T_g.

ES charge and discharge power and the relationship between electricity prices ES charge and discharge rules and electricity prices are closely related. The relationship between the charge and discharge power at time t and the price $C(t)$ is as follows:

$$\begin{cases} P_{\text{ch}}(t) = a_{\text{ch}}C(t) + b_{\text{ch}}, & t \in T_{\text{ch}} \\ P_{\text{dis}}(t) = a_{\text{dis}}C(t) + b_{\text{dis}}, & t \in T_{\text{dis}} \end{cases} \tag{20.15}$$

where a_{ch} and b_{ch} are the relationship between the charging time and the charging power and a_{dis} and b_{dis} are the relationship between the discharge time and the discharge power.

The initial charge state SOC(1) = 0 is taken for the day of the ES element, and the charge state returns to zero at the end of each day, that is, SOC(24) = 0. In view of this, at the time when the peak hourly price ends in a day, all ES will be fully discharged. If there is a power nonzero ES, when the discharge time from the end of time for its current required discharge time, it is forced to discharge to restore the initial charge state.

The output of "element" is generated in multiple scenarios

For the flexible load and ES elements, the probability force model is based on the response mechanism of the electricity price.

So, in combination with the above two categories of component models, ADS "element" multiscene generation is shown in Fig. 20.6.

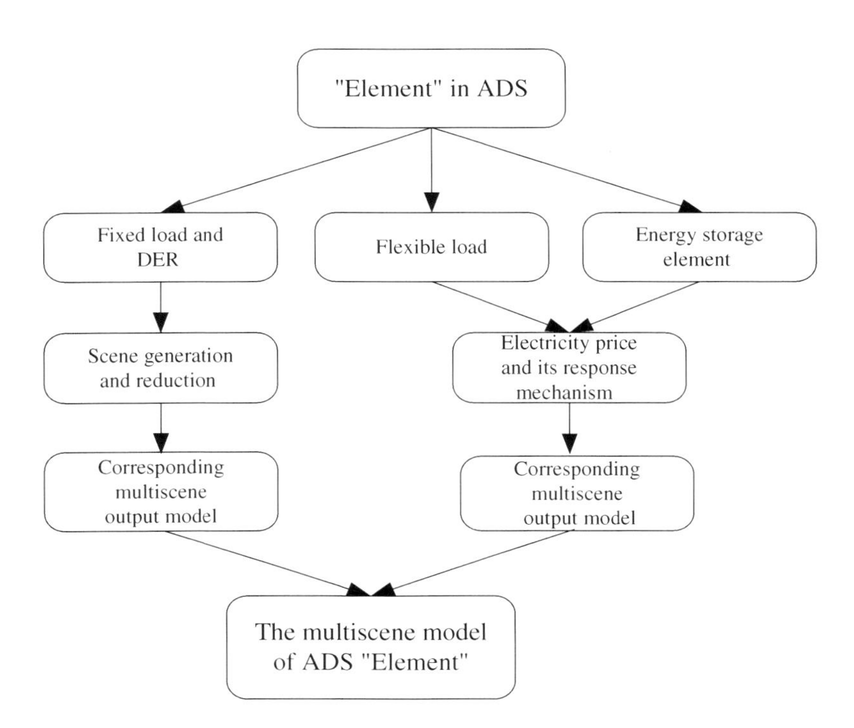

Fig. 20.6 The generation of multiscene model of active distribution system (ADS) "element."

20.2.3 Uncertainty analysis of distribution network topology

20.2.3.1 N − 1 test and expected accident assessment

At present, the more common network security operation requirement is to meet the $N-1$ test (Yin, 2016), that is, all the N lines in the arbitrary break a line, and the system of the operating indicators should meet the given requirements. In network planning at the early stage of formation of the network structure, the most important principle is to make the network free from overload; the network can meet the safety requirements of electricity; so we should take overload check one by one after breaking the line. When any one line is broken, this causes the other lines of the system to be overloaded or makes the system dismounted, indicating that the network does not meet the $N-1$ test.

A strict $N-1$ test requires all N-line disconnection analysis; the computational burden is heavy. In fact, some of the network after the break does not cause system overload; so we can accord to the possibility of overloading the system caused by the lines after breaking, and then in accordance with the order of the overload of the more likely to verify the line. When it is checked the line does not cause overload when broken, it is no longer necessary to verify the back of the line, which can significantly reduce the amount of calculation. This process is also called fault selection. At present, there are a lot of fault sorting methods; commonly used are the distribution coefficient method, fast decoupling load flow (FDLF)-1 method (Dounas-Frazer et al., 2016), load unloading method (Jian et al., 2014a, b), providing maximum load capacity method (Liang et al., 2016), and so on. Their criteria for judging system accidents are various.

Predictive accident assessment (also known as incidental analysis) is to assess the security of a system based on a subset (called anticipate the set of accidents) of all the possible perturbations in the system (Yin, 2016). In principle, the expected accident assessment should include both static security assessment and dynamic safety assessment. For the latter purpose, a number of articles have been published using the Lyapunov direct method (Wang et al., 2015a, b), coherence identification method (Gao et al., 2015), and image recognition method.

In the static safety assessment, the expected set of accidents should include line breaking and generator breaking, two disturbances at least. There are many kinds of analysis methods for these two accident assessments. These methods almost all use the linear system of overlapping principle, so we can directly solve the state after the accident variables. Although they have advantages like being simple, fast, easy to calculate in real time, and so on, the poor accuracy when dealing with too serious accidents, such as heavy-duty lines or large units break, cannot be ignored. In order to improve the precision, it should be involved in the iterative solution of some kind of AC power flow algorithm. The compensation method for line breaking and the generator breaking analysis combined with the fast decoupling tidal current algorithm, which takes into account the frequency characteristics of the governor, often achieves satisfactory results in accuracy, but has an increase in execution time. To this end, what kind of method to be used to carry out the expected accident analysis mainly depends on the research project in the progress and time requirements.

20.2.3.2 Predictive accident set generation based on fault sorting method

Studying the sorting of PSC needs to consider any branch of the accident. However, the probability of a single device failure is usually much greater than the probability of simultaneous failure of multiple devices. Therefore, let us only consider the system "$N-1$" failure situation here. In order to speed up the evaluation speed and improve the modeling efficiency, this chapter uses the fault sorting method to sort the severity of the system failure, and uses the performance index (PI) to quantify the consequences of the accident.

The troubleshooting process actually calculates the ΔPI value for all lines and sorts them from large to small. In the disconnection analysis, we need to break the line that has the greatest ΔPI value at first, then calculate and check the power flow of this breaking line, until we find the line no longer causes system overload after being broken. The possibility of system overload caused by the line with smaller ΔPI value is very small; so there is no need to do disconnection analysis. However, the use of such system behavior indicators may have a "shadowing" phenomenon, such as when there are individual lines overload and other line power flow is small, its ΔPI value may be less than the ΔPI value for lines that do not overload but the line power flow is relatively large.

So, according to this indicator for fault selection sorting there may be some error. Therefore, in the practical application, there should be a continuous check in a few broken lines and we should make sure all of them would not cause the system overload; then the disconnection analysis can be terminated.

Define the expected consequences of an $N-1$ break-off event in an ADS:

$$F=\Gamma\otimes PI \tag{20.16}$$

where F is the expected function of the impact of a failure when the system is in operation. Γ is the probability function of the system for a failure. PI is the performance index, that is, after the accident corresponding to Γ happened, the severity function of the impact of this accident on the system.

The concrete solution of PI and Γ is shown in Eqs. (20.17) and (20.18):

$$PI=\sum_{l=1}^{L}\alpha_l\omega_l\left(\frac{P_l}{S_l}\right)^2 \tag{20.17}$$

where P_l is the active power flow of line l. S_l is the transmission capacity in line l. α_l is the number of parallel branches in branch l. ω_l is the weight coefficient of line l, reflecting the impact of the fault line on the system. L is the number of network branches:

$$\begin{cases}\Gamma_k(t)=\prod\limits_{e=1,2,\ldots,L}^{e\neq k}(1-p_e(t))p_k(t)\\ \Gamma_0(t)=1-\sum\limits_{k=1}^{L}P_k\end{cases} \tag{20.18}$$

where $\Gamma_k(t)$ is the probability of occurrence of "$N-1$" fault on line k at time t. $\Gamma_0(t)$ is the probability that the system does not fail at time t.

20.3 Definition of PSC evaluation index

PSC is an important indicator for the safety and reliability of a power system; an accurate and effective evaluation system to evaluate the PSC can provide guidance for the optimal dispatching and economic operation of the distribution network; it can also provide the basis for the planning and construction of a distribution network. From the timescale division, distribution network PSC assessment can be divided into short-term assessment and long-term assessment. Among them, the medium and long-term PSC assessment in month and year as the assessment radius is mainly used to guide the planning of the distribution network, and short-term assessment refers to the assessment of PSC to guide distribution network operation and optimization.

The PSC of the distribution network is time varying; the adjustment of the network structure, the change of the position of the transformer tap, the switching of the reactive power compensation equipment, the power output and load fluctuation, and other uncertain factors will have an impact on it. When the traditional distribution network transformed to the ADS transition, the distribution network added the power components that are mainly composed by DER. So, there is a great deal of randomness and volatility, which further increases the impact on PSC's uncertainty factor.

In order to quantify the evaluation of PSC probability assessment results, the following PSC assessment indicators are defined:

- Expectation of PSC at time t $E_{\mathrm{PSC}}(t)$:

$$E_{\mathrm{PSC}}(t)=\sum_{i=1}^{N_s} S_{\mathrm{PSC}}(i,t)p(i) \tag{20.19}$$

where N_s is the total number of scenes constructed; $S_{\mathrm{PSC}}(i,t)$ is the power supply capability of the ith scene at time t; and $p(i)$ is the probability of occurrence of the ith scene.

- Variance of PSC at time t $V_{\mathrm{PSC}}(t)$:

$$V_{\mathrm{PSC}}(t)=\sum_{i=1}^{N_s}[(S_{\mathrm{PSC}}(i,t)-E_{\mathrm{PSC}}(t))p(i)]^2 \tag{20.20}$$

- Variation coefficient of PSC at time t $\gamma_{\mathrm{PSC}}(t)$:

$$\gamma_{\mathrm{PSC}}(t)=\frac{\sqrt{V_{\mathrm{PSC}}(t)}}{E_{\mathrm{PSC}}(t)} \tag{20.21}$$

- Maximum value of PSC at time t S max PSC(t):

$$S_{\mathrm{PSC}}^{\max}(t) = \max\{S_{\mathrm{PSC}}(i,t), i \in N_{\mathrm{s}}\} \tag{20.22}$$

- Minimal value of PSC at time t S min PSC(t):

$$S_{\mathrm{PSC}}^{\max}(t) = \max\{S_{\mathrm{PSC}}(i,t), i \in N_{\mathrm{s}}\} \tag{20.23}$$

- Margin of PSC at time t $M_{\mathrm{PSC}}(t)$:

$$M_{\mathrm{PSC}}(t) = E_{\mathrm{PSC}}(t) - \sum_{i=1}^{N} p(i) \sum_{k=1}^{N_L} S_L(i,k,t) \tag{20.24}$$

where N_{L} is the number of load nodes and N is the probability of the number of scenes. S_{L} (i, k, t) is the load size of the kth node of the ith scene at time t.

- Dissatisfied probability of PSC at time t, $DP_{\mathrm{PSC}}(t)$ characterizes the probability that the supply capacity is less than a given value:

$$DP_{\mathrm{PSC}}(t) = \sum p_{\mathrm{dp}}(t) \tag{20.25}$$

where $p_{\mathrm{dp}}(t)$ is the probability of the scene corresponding to the PSC whose value is less than the allowable value A at time t.

- Dissatisfied amount of PSC at time t, $DA_{\mathrm{PSC}}(t)$ represents the quantity of the power supply in the time of the t moment which is less than that of a certain allowable value:

$$DA_{\mathrm{PSC}}(t) = A - E_{\mathrm{PSC}}(t) \tag{20.26}$$

where A is the allowable value of PSC. And when $DA_{\mathrm{PSC}}(t) < 0$, the value is 0.

- Contribution rate of DER to expectation of PSC at time t, $CR_{\mathrm{DTE}}(t)$ characterizes the power supply capability which is expected to enhance the relative promotion of $E_{\mathrm{PSC}}(t)$, after characterizing the DER access at t moments:

$$CR_{\mathrm{DTE}}(t) = \frac{E_{\mathrm{PSC}}(t)}{E_{\mathrm{PSC,N}}(t)} - 1 \tag{20.27}$$

where $E_{\mathrm{PSC,N}}(t)$ is the expected value of the power supply capability of the system without DER at time t.

- Contributed amount of DER to expectation of PSC at time t, $CA_{\mathrm{DTE}}(t)$ represents the increment of $E_{\mathrm{PSC}}(t)$ after the DER access:

$$CA_{\mathrm{DTE}}(t) = E_{\mathrm{PSC}}(t) - E_{\mathrm{PSC,N}}(t) \tag{20.28}$$

- Contribution rate of DER to dissatisfied probability of PSC at time t, $CR_{\mathrm{DTDP}}(t)$ represents the relative decrease degree of $DP_{\mathrm{PSC}}(t)$ after the DER supply has been accessed at time t:

$$CR_{\mathrm{DTDP}}(t) = 1 - \frac{DP_{\mathrm{PSC}}(t)}{DP_{\mathrm{PSC,N}}(t)} \tag{20.29}$$

where $DP_{\mathrm{PSC,N}}(t)$ is the lower probability of the PSC without DER.

20.4 Short-term PSC evaluation algorithm based on multiscene technology

20.4.1 The calculation model of PSC evaluation in multiscene

20.4.1.1 Multiscene generation of PSC assessment

Combining the contribution of the "element" with the multiscene of the fault, the multiscene model of the power supply capability evaluation is generated. The total number of scenes is the product of the number of scenes in each part. The probability of each scene is the product of the probability of the corresponding subscene. Assuming the fixed load, PV, wind power, flexible load, ES components, the number of output scenarios were N_1, N_2, ..., N_5. The corresponding probabilities are $P_{1a}(a \in [1, N_1])$, $P_{2b}(b \in [1, N_2])$, ..., $P_{5e}(e \in [1, N_5])$. The probability of the number of scenes generated is N_6; the corresponding probability is $P_{6f}\,(f \in [1,\ N_6])$. The total number of power capacity assessment scenes $N_{\mathrm{se}} = \prod N_i\ (i = 1,\ 2,\ \ldots,\ 6)$, and the probability of occurrence of each assessment scene $P_{\mathrm{se}} = P_{1a}P_{2b}P_{3c}P_{4d}P_{5e}P_{6f}$. The probability modeling process for the uncertainty factors that affect the assessment of PSC is shown in Fig. 20.7.

20.4.1.2 Calculation model of single scene PSC

Objective function

$$\max P_{\mathrm{L}} = \sum_{i=1}^{N} P_{\mathrm{L}i} \tag{20.30}$$

where P_{L} is the maximum active load that the distribution network can supply. N is the load point. P_{Li} is the active load at the load point i.

Restrictions

- Trend calculation constraint

$$P_{\mathrm{G}i} + P_{\mathrm{DER}i} + P_{\mathrm{IL}i} + P_{\mathrm{TL}i} + P_{\mathrm{ES}i} - P_{\mathrm{L}i} = U_i \sum_{j=1}^{N} U_j \left(G_{ij}\cos\theta_{ij} + B_{ij}\sin\theta_{ij}\right) \tag{20.31}$$

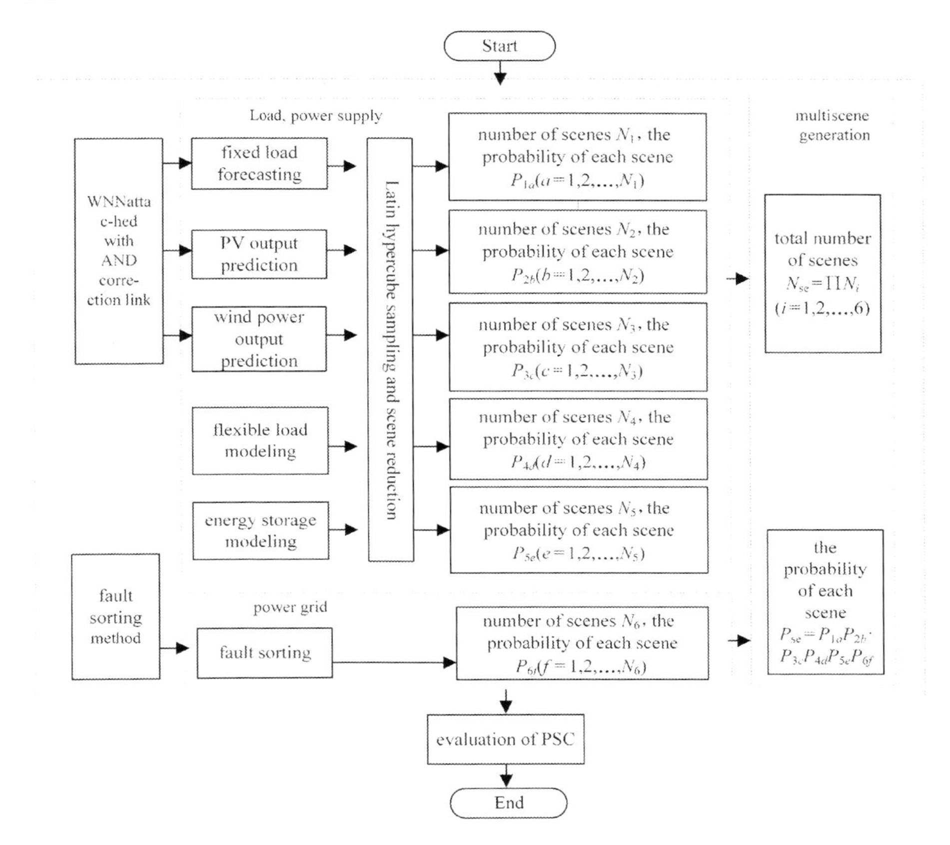

Fig. 20.7 Probabilistic modeling process for uncertain factors that affect the assessment of power supply capacity (PSC).

$$Q_{Gi}+Q_{DERi}+Q_{ILi}+Q_{TLi}+Q_{ESi}-Q_{Li} = U_i\sum_{j=1}^{N}U_j\left(G_{ij}\cos\theta_{ij}-B_{ij}\sin\theta_{ij}\right) \tag{20.32}$$

$$U_i^{\min}\le U_i\le U_i^{\max} \tag{20.33}$$

$$I_l\le I_l^{\max} \tag{20.34}$$

$$S_T\le S_T^{\max} \tag{20.35}$$

where P_{Gi}, P_{DERi}, P_{ILi}, P_{TLi}, P_{ESi}, and P_{Li} are the active power of the generator, DER, IL, TL, ES, and load at node i. Q_{Gi}, Q_{DERi}, Q_{ILi}, Q_{TLi}, Q_{ESi}, and Q_{Li} are the reactive power of the generator, DER, IL, TL, ES, and load at node i. G_{ij} and B_{ij} are the conductance and susceptance of branch i–j, respectively. θ_{ij} is the power angle between node i and node j. U_i, $U_{\max i}$, and $U_{\min i}$ are the voltage at node i and its upper and lower limits, respectively. I_l and $I_l^{\max}$ are the current of line l and its upper limit. S_T and $S_T^{\max}$ are the power of the transformer T and its upper limit.

- Flexible load constraint

$$\sum_{t \in T_I} \left(P_{TL}(t) - P'_{TL}(t)\right) \le P_{TL,I,\max} \tag{20.36}$$

$$\sum_{t \in T_O} \left(P'_{TL}(t) - P_{TL}(t)\right) \le P_{TL,O,\max} \tag{20.37}$$

where $P_{TL,I,max}$ is the maximum allowable load power of TL. $P_{TL,O,max}$ is the maximum allowable transfer load power of TL.

- ES component constraints

$$SOC_{\min} \le SOC(t) \le SOC_{\max} \tag{20.38}$$

$$0 \le P_{ch}(t) \le P_{ch,\max} \tag{20.39}$$

$$0 \le P_{dis}(t) \le P_{dis,\max} \tag{20.40}$$

where SOC_{min} and SOC_{max} are the minimum and maximum values of the charge state of the ES component, respectively. $P_{ch,min}$ and $P_{ch,max}$ are the upper power limit of the ES component.

Model solving

According to the definition, the PSC of an ADS is equal to the total load supplied by the distribution network when there is a critical state in the model that has an operative constraint. The most direct calculation method for this critical point is the repeated power flow method. The basic computing starts from an initial operating point, selecting the appropriate growth step, according to certain load mode. By increasing system load and repeated power flow calculation to determine the maximum load of the system can supply can determine the PSC of a system (Fahmy, 2016). The advantage of repeat power flow (RPF) is that it has a clear thinking, simple calculation, and short running time; it not only can get the PSC of the distribution network, but also can find the power supply bottleneck node with limited PSC (Yin, 2016).

In the traditional power flow calculation model, the load grows mostly according to the same proportion. A year-on-year growth model can simulate the rising trend of load to a certain extent, but it is difficult to distinguish the change trend of load at different load points. Its accuracy is largely influenced by the actual load distribution of the distribution network. Therefore, in order to increase the accuracy of modeling, this chapter uses load growth model based on relative growth rate to simulate the load growth of each node in ADS.

If the day is divided into 24 time periods, the relative growth rate at the i load point is as follows:

$$r_i = \frac{P_{\mathrm{L}i}^{\max}}{P_{\mathrm{L}i}^{\min}} \left(P_{\mathrm{L}i}^{\min} \neq 0\right) \tag{20.41}$$

where r_i is the relative growth rate of the load at the load point i. $P_{\mathrm{L}i}^{\max}$ and $P_{\mathrm{L}i}^{\min}$ are the maximum and minimum values of the load at the load point i in 24 periods, respectively.

Based on the above model, the load growth model based on the relative growth rate in the ADS is as follows:

$$P_{\mathrm{L}i}(m) = P_{\mathrm{L}i}(0)(1 + r_i\%)^m \tag{20.42}$$

$$P'_{\mathrm{L}i}(m) = P_{\mathrm{L}i}(m) - P_{\mathrm{DER}i} - P_{\mathrm{IL}i} - P_{\mathrm{TL}i} - P_{\mathrm{ES}i} \tag{20.43}$$

$$P_{\mathrm{L}}(m) = \sum_{i=1}^{N} P_{\mathrm{L}i}(m) \tag{20.44}$$

$$P'_{\mathrm{L}}(m) = \sum_{i=1}^{N} P'_{\mathrm{L}i}(m) \tag{20.45}$$

where $P_{\mathrm{L}i}(m)$ is the load at the load point i after the load growth m times. $P_{\mathrm{L}i}(0)$ is the initial value of the load at the load point i. $P'_{\mathrm{L}i}(m)$ is the payload value at the load point i after the load grows m times. $P_{\mathrm{L}}(m)$ is the load of the entire distribution network after the load is increased by m times. $P'_{\mathrm{L}}(m)$ is the payload size of the entire power distribution system after the load grows m times.

The model solution flow chart is shown in Fig. 20.8.

20.4.2 Short-term PSC evaluation steps based on multiscene technology

Specific steps are as follows:

- Step 1: using WNN based on error correcting additional ADP correction link to predict the fixed load and DER output in ADS. Establishing the corresponding probability distribution model by using the appropriate distribution, and generating the multiscene scenario of fixed load and DER output based on multiscene technology.
- Step 2: for the flexible load and ES component, establishing a multiscene probability output model based on the corresponding electricity price response mechanism by using the given electricity price curve.
- Step 3: using the fault ranking method to sort the $N-1$ disconnection fault that may occur on the line, recording the valid items of the fault order and the probability of each fault scene.
- Step 4: combining load, ES, power, and power grid architecture of the multiscene to generate PSC assessment of multiple scenes.

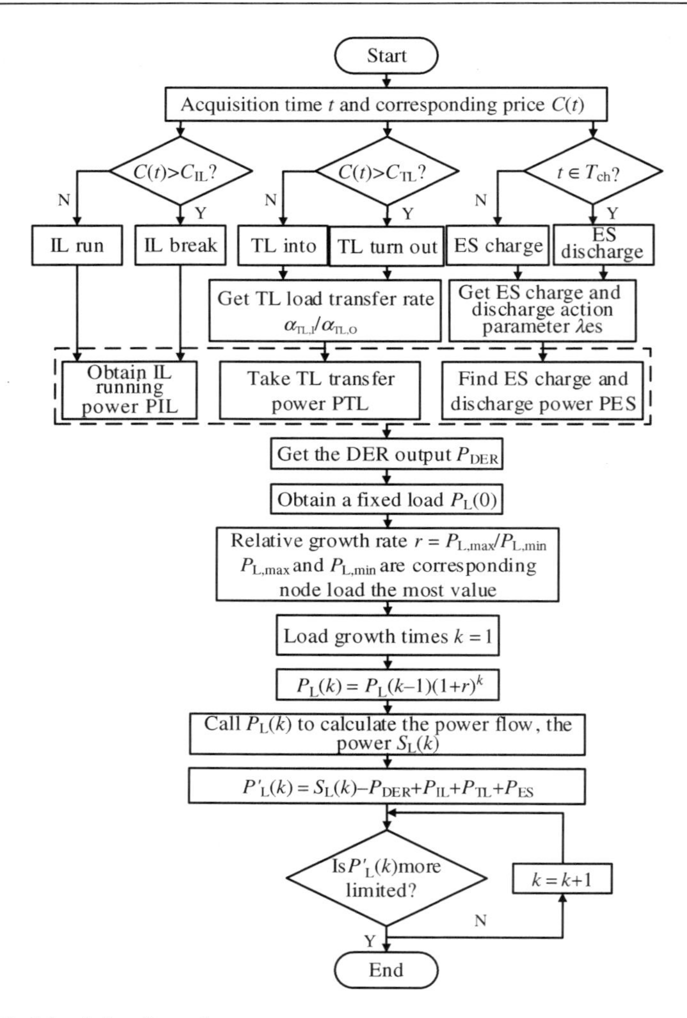

Fig. 20.8 Model solving flow chart.

- Step 5: dividing the day into 24 time periods. In each time period, substituting the probability calculation results of each scene into the system, and calculating the power flow capacity of each system by using the load growth model based on the relative growth rate combined with the RPF.
- Step 6: based on the calculation results of the PSC of each scene and the corresponding probability, obtaining the probability evaluation indices of PSC.
- Step 7: according to the evaluation index and the evaluation results of the PSC, synthetically analyzing the capacity of the power supply system and the influence of the "element" uncertainty factors in the ADS.

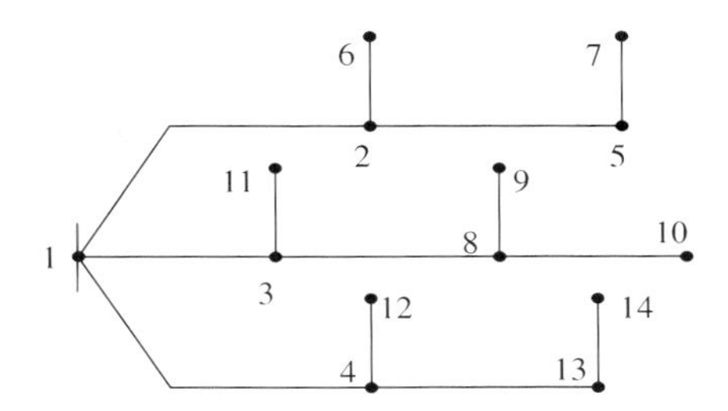

Fig. 20.9 IEEE14-node system structure diagram.

20.5 Case study

20.5.1 System overview

This chapter chooses the improved IEEE 14-node system to evaluate the PSC, and uses MATLAB to simulate the calculation. Considering that the distribution network is normally open-loop operation under normal circumstances, this paper removes 14th, 15th, and 16th three branches, making the system a 14-node 13-branch single-supply radiation network. The structure is shown in Fig. 20.9.

In Fig. 20.9, except that node 1 is the equilibrium node, the remaining nodes are all load nodes. The system base capacity is 100 MVA and the reference voltage is 23 kV. The upper limit of the line power is taken as the thermal stability limit capacity, and the allowable voltage variation range is $1\% \pm 5\%$.

In addition to the fixed load, consider access to ADS "element" including: wind generation (WG), PV, ES, IL, and TL. The access status is shown in Table 20.3.

Table 20.3 Transformer parameters table

Access node	Access capacity (kW)	Access type
2	300	PV
2	900	ES
3	300	IL
3	500	WG
4	200	PV
6	800	TL
8	300	WG
8	700	IL
9	500	IL
10	400	ES

20.5.2 Calculation of PSC

The modeling expected value μ_L of the normal distribution model of load forecasting error is 0.02, the PV prediction model expected value μ_{PV} is 0.07 (sunny, the error in line with the normal distribution), the modeling expected value μ_{WG} of the wind power prediction model is 0.05 (wind speed range 5.1–9.8, the error is normal distribution), the modeling expected values $\mu_{TL,I}$ and $\mu_{TL,O}$ of the TL load transfer rate are 0.5 and 0.7554, respectively, the modeling expected value μ_{ES} of the ES charge and discharge action parameters is 0.8, and the standard deviation σ of all models is 0.2. Latin super cube sampling times are taken as 2000. The reduced fixed load and IL number of uninterrupted time error samples $N1$ is 2. PV and wind power output prediction error sample number $N2$ and $N3$ are all 3. The TL load transfer rate, ES charge and discharge probability sample number $N4$ and $N5$ are all 2. Using Monte Carlo simulation to generate electricity price data, C_{IL} and C_{TL} take the electricity price mean. $\beta = -1.3$, $\eta_{ch} = \eta_{dis} = 0.85$, $\lambda_0 = 0.8$, $a_{ch} = -0.8327$, $b_{ch} = 0.6255$, $a_{dis} = 1.5895$, $b_{dis} = 0$. T_f 9:00–22:00 time period, T_g includes 1:00–8:00 and 23:00–24:00 time periods. Figs. 20.10–20.12, respectively, show the load and DER output prediction results, the results of flexible load output and ES components output results.

The results of the troubleshooting are shown in Table 20.4. Considering that there is a large transition between sort 4 and 5, the value of the desired function F is ignored. Therefore, the type of fault after sort 4 can be ignored, and take $N6$ as 4. In summary, a total of 288 scenes were built.

Fig. 20.13 shows the multiscene probability calculation results for PSC. Using a variety of colors to distinguish the PSC of the different numerical areas can clearly reflect the probability of occurrence of different PSC at different times.

Fig. 20.13A shows the three-dimensional (3D) graphs of multiscene PSC taking into account uncertainties. The probability of each scene is depicted on the X-axis, the time is on the Y-axis, and the PSC is the Z-axis. Let the X–Z plane be the front of the image and the Y–Z plane be to the right.

Fig. 20.13B is a top view of Fig. 20.13A, that is, an X–Y plan view. According to the color label on the right side, the size of the PSC at each time and its corresponding

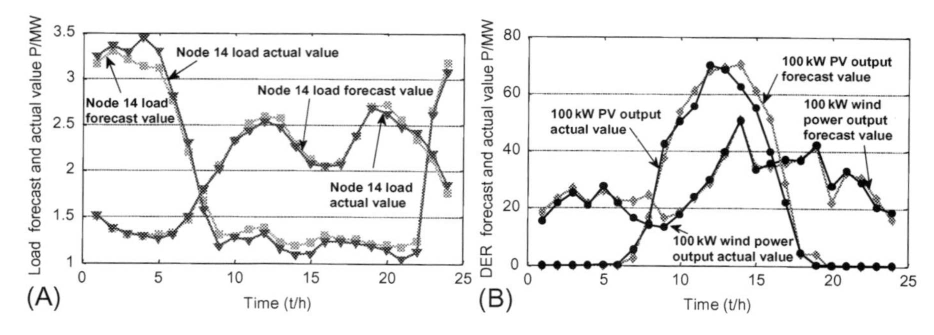

Fig. 20.10 Load and distributed energy resources (DER) output forecast results. (A) Partial node load forecasting results and (B) DER output forecast results.

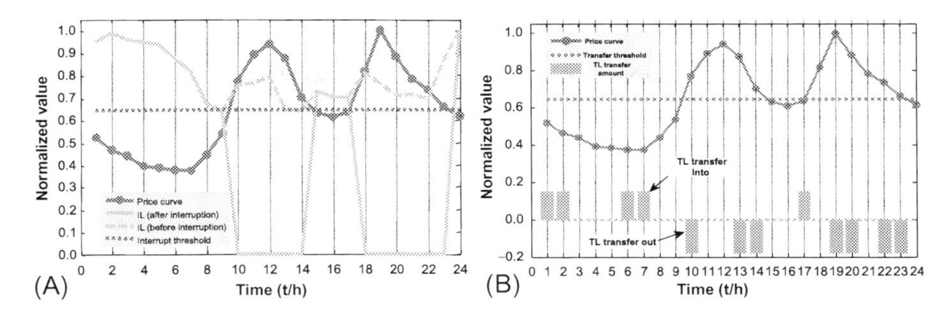

Fig. 20.11 Flexible load output results. (A) Results of interruptible load output and (B) results of transferable load output.

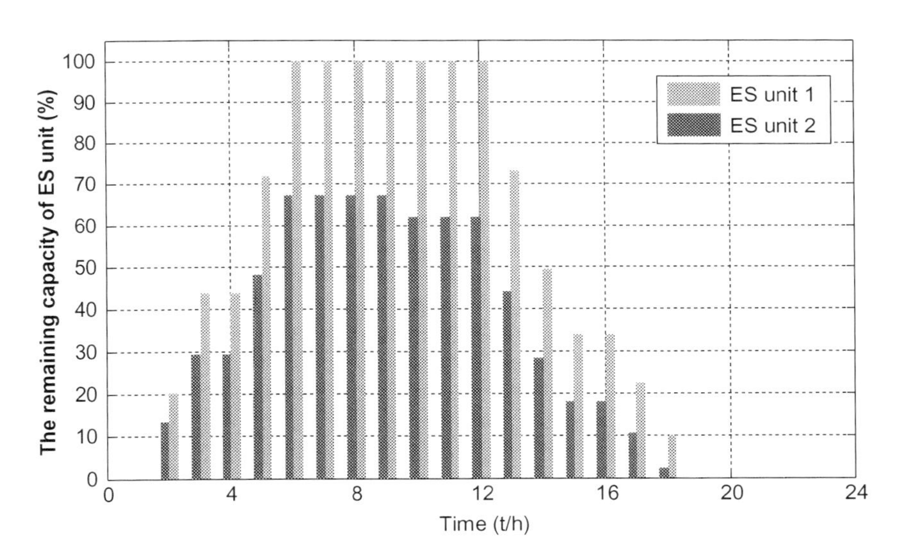

Fig. 20.12 Energy storage component output results.

Table 20.4 Fault sequencing results table

Fault sequencing	Line	F-value	Fault sequencing	Line	F-value
1	2	0.1214	8	5	0.0002
2	1	0.0219	9	13	0.0001
3	3	0.0094	10	6	0.0001
4	7	0.0081	11	10	0.0001
5	12	0.0015	12	11	0.0001
6	4	0.0011	13	9	0.0001
7	8	0.0004			

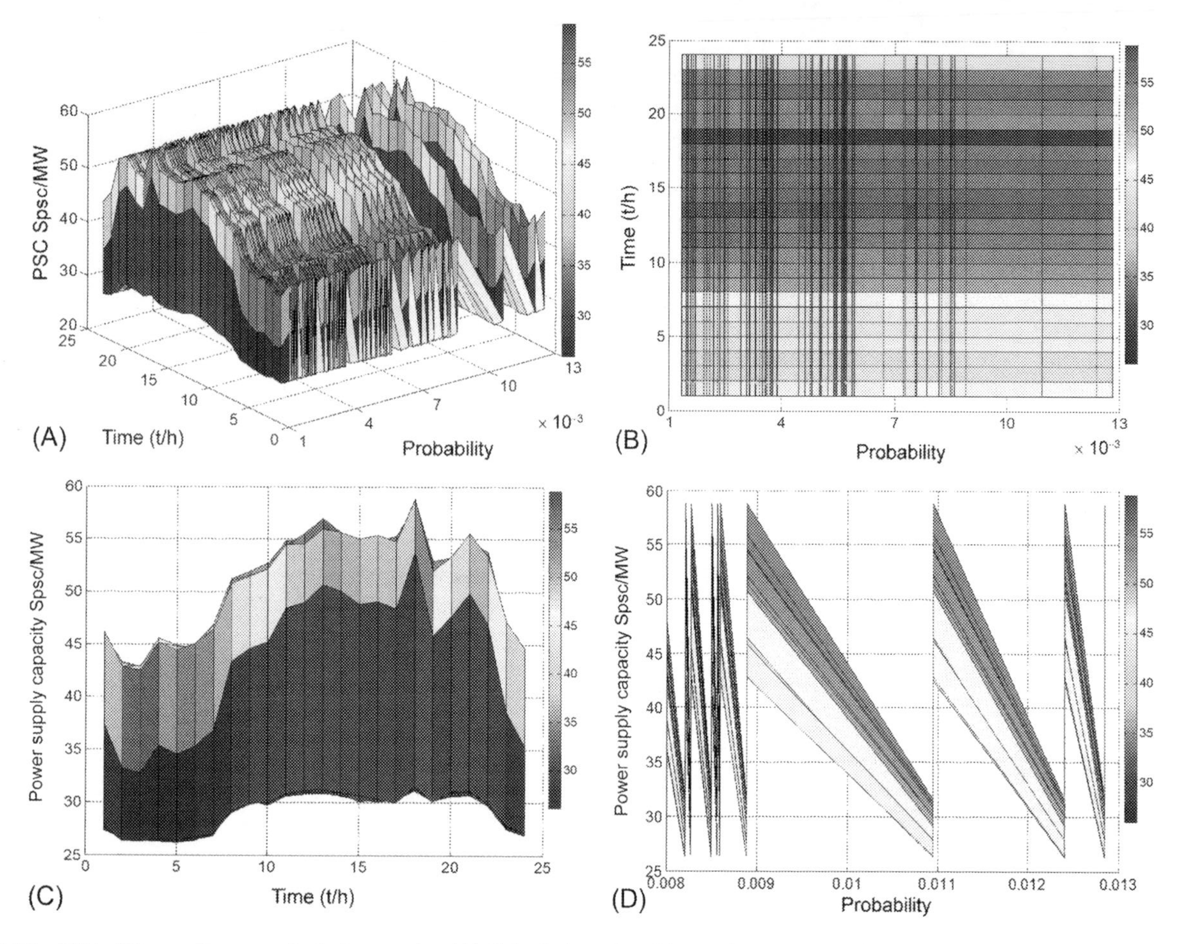

Fig. 20.13 The PSC of the multiscene calculation results. (A) Calculation of multiscenario power supply capacity taking into account uncertainties; (B) *X–Y* axis plan view of (A); (C) *Y–Z* axis plan view of (A); and (D) Partial view of the *X–Z* axis plan view of (A).

Table 20.5 The PSC evaluation indicators at $t = 3$ and $t = 12$

Evaluation indicators	$t = 3$	$t = 12$
$E_{PSC}(t)$/MW	35.4830	46.5856
$V_{PSC}(t)$	0.1127	0.2585
$\gamma_{PSC}(t)$	0.0090	0.0013
S max PSC(t)/MW	42.8979	55.5232
S min PSC(t)/MW	26.3124	31.8090
$M_{PSC}(t)$/MW	11.6187	11.9139
$DP_{PSC}(t)$	0.6843	0.2161
$DA_{PSC}(t)$/MW	9.4760	0.0000
$CR_{DTE}(t)$	0.0290	0.0220
$CA_{DTE}(t)$/MW	0.9590	1.1866
$CR_{DTDP}(t)$	0.1137	0.1598

probability can be clearly seen from Fig. 20.13B. It is easy to see that between 11 and 19 h, and between 21 and 23 h, there is a higher probability of achieving greater PSC.

Fig. 20.13A is the Y–Z plane view of Fig. 20.13A, that is, the Y–Z plane view, and also the Y–Z sectional view of Fig. 20.13A when the probability of taking the maximum value of 0.0128. The three S_{PSC}-t curves in Fig. 20.13C are the PSC curves for the three scenarios with a probability of 0.0128. At different probability values, the $S_{PSC} - t$ curve corresponding to each probability can be obtained by taking the Y–Z section plane in Fig. 20.13A.

Fig. 20.13D is a partial view of the main view of Fig. 20.13A in the probability interval [0.008, 0.013], that is, the local plot of the X–Z plane view. Each curve in Fig. 20.13D characterizes the probability of occurrence of the PSC at a given time.

20.5.3 Evaluation of PSC evaluation index

The evaluation of PSC evaluation indicators at times $t = 3$ and $t = 12$ is shown in Table 20.5. Fig. 20.14 has shown the comparison of the supply margin for the day with the total system load. Fig. 20.15 is a 24-h comparison chart of DER and DER-free power requirements.

The above results are analyzed as follows:

- The expected value of PSC and its maximum value $E_{PSC}(t)$, S max PSC(t), and S min PSC(t) (T) reflect the size of the PSC of each scene from the aspects of weighted average and the range of values, respectively. The variance $V_{PSC}(t)$ of the PSC characterizes the discrete degree of the PSC in each scene.
- $DP_{PSC}(t)$ and $DA_{PSC}(t)$ are indicators to evaluate whether the PSC is adequate or not. If the two indicator value are relatively large, it means that the system PSC cannot meet the set limits, and we need to increase the access capacity of DER, ES, and other components or consider line expansion and other measures.

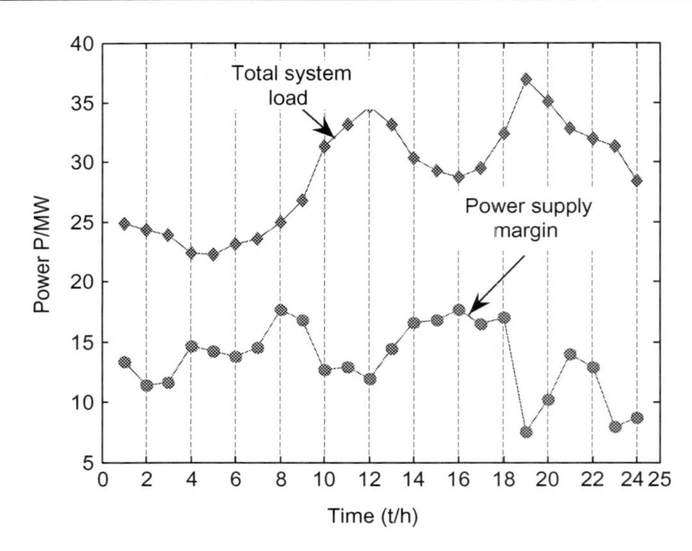

Fig. 20.14 Comparison of power supply margin and total system load in 24 h.

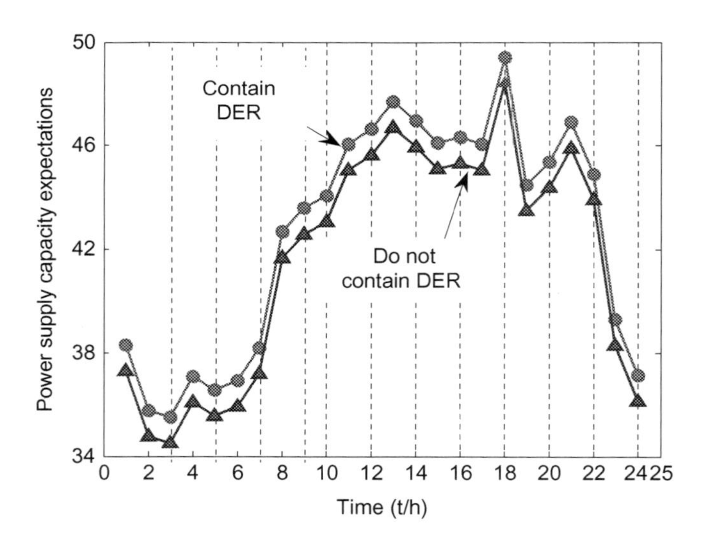

Fig. 20.15 Power supply capability expectation curve of 24 h containing DER and without DER.

- $CR_{\mathrm{DTE}}(t)$ and $CA_{\mathrm{DTE}}(t)$ characterize the impact of DER access capacity on the PSC expectations $E_{\mathrm{PSC}}(t)$. $CR_{\mathrm{DTDP}}(t)$ characterizes the influence of DER access capacity on the insufficient PSC probability $DP_{\mathrm{PSC}}(t)$. At $t=3$ and $t=12$, the three values mentioned above are all positive, reflecting the role of accessing DER to enhance the $E_{\mathrm{PSC}}(t)$ and reduce the $DP_{\mathrm{PSC}}(t)$.
- As can be seen from the comparison of power supply margin and total system load in 24 h, when the total load of the system is small, the value of the power supply margin value

$M_{PSC}(t)$ of the system is usually larger. On the contrary, when the total load value is large, the power supply margin value is generally smaller. Therefore, accurate and real-time assessment of PSC, especially in the peak load period, plays a particularly important role in ensuring reliable and safe operation of the grid.
- As can be seen from the power supply capability expectation curve of 24 h containing DER and without DER shows that the PSC expectations $E_{PSC}(t)$ of the system is improved after accessing DER, which again reflects the positive effect of accessing DER on improving the PSC of the system.

20.6 Conclusions

The PSC of the active power distribution system directly affects the load capacity of the power grid. Analysis the uncertain factors that affect the PSC is an important reference factor for the power system planning, operation, and dispatching to accurately evaluate the PSC of the distribution system. Based on the three aspects of power supply, grid structure, and load, this chapter systematically studies the short-term PSC evaluation of active power distribution system with uncertain factors and gets the following conclusions:

References

Abdeltawab, H.H., Mohamed, Y.A.R.I., 2017. Mobile energy storage scheduling and operation in active distribution systems. IEEE Trans. Ind. Electron. 64 (9), 6828–6840.

Abrahart, R.J., See, L., 2015. Comparing neural network and autoregressive moving average techniques for the provision of continuous river flow forecasts in two contrasting catchments. Hydrol. Process. 14 (11–12), 2157–2172.

Bartlett, L.M., Andrews, J.D., 2015. Efficient basic event ordering schemes for fault tree analysis. Qual. Reliab. Eng. Int. 15 (2), 95–101.

Bo, S., Qiangqian, L., Yudong, L., et al., 2014. Study on electricity price mechanism of sodium sulfur battery energy storage system. Power Syst. Technol. 38 (8), 2108–2113.

Buhari, M., Kopsidas, K., 2014. In: Probabilistic assessment of impacts of real-time line ratings on distribution networks.International Conference on Probabilistic Methods Applied To Power Systems, Durham, 7–10 July.

Chaichan, M.T., Kazem, H.A., 2016. Experimental analysis of solar intensity on photovoltaic in hot and humid weather conditions. Int. J. Sci. Eng. Res. 7 (3), 91–96.

Chalk, P.L., Corotis, R.B., 2017. Probability model for design live loads. J. Struct. Div. 106 (10), 2017–2033.

Dan, Z., Jianguo, Y., Shengchun, Y., et al., 2014. System dynamics simulation model of flexible load and power grid interaction. Proc. CSEE 34 (25), 4227–4233.

Dolara, A., Grimaccia, F., Leva, S., et al., 2015. A physical hybrid artificial neural network for short term forecasting of PV plant power output. Energies 8 (2), 1138–1153.

Dounas-Frazer, D.R., Bogart, K.L.V.D., Stetzer, M.K.R., et al., 2016. Investigating the role of model-based reasoning while troubleshooting an electric circuit. Phys. Rev. Phys. Educ. Res. 12 (1). 010137.

Erto, P., Lepore, A., 2016. Best unbiased graphical estimators of location-scale distribution parameters: application to the Pozzuoli's Brady seism earthquake data. Environ. Ecol. Stat. 23 (4), 605–621.

Fahmy, O.M., 2016. In: Analysis of existing power flow case for south operating area network. Smart Grid IEEE, Jeddah, 6–8 December.

Farajollahi, M., Fotuhi-Firuzabad, M., Safdarian, A., 2016. Deployment of fault indicator in distribution networks: a MIP-based approach. IEEE Trans. Smart Grid 99, 1.

Florin, C., Luis, F.O., Harag, M., et al., 2014. Assessing the potential of network reconfiguration to improve distributed generation hosting capacity in active distribution systems. IEEE Trans. Power Syst. 30 (1), 346–356.

Fu, Z., Xie, J., Zhang, R., et al., 2016. In: A summary of the maximum power supply capacity of the research of intelligent distribution network.International Conference on Intelligent Computation Technology and Automation, Nanchang, 19 May.

Gao, Q., Huang, Y., Zhang, H., et al., 2015. Discriminative sparsity preserving projections for image recognition. Pattern Recogn. 48 (8), 2543–2553.

Geng, J., Wang, B., Dong, X., 2015. A novel one-terminal single-line-to-ground fault location algorithm in transmission line using post-single-phase-trip data. Trans. China Electrotech. Soc. 30 (16), 184–193.

Gils, H.C., Scholz, Y., Pregger, T., et al., 2017. Integrated modelling of variable renewable energy-based power supply in Europe. Energy 123, 173–188.

Hashemi, S., Ostergaard, J., 2016. Efficient control of energy storage for increasing the PV hosting capacity of LV grids. IEEE Trans. Ind. Inform. 99, 1.

Hong, L., Jifeng, L., Jiaan, Z., et al., 2017. In: Consider the reliability of the medium voltage distribution system power supply assessment.Automation of Electric Power Systems, Milan, 6–9 June.

Hongjie, J., Wenjin, Q., Zhe, L., et al., 2015. Hierarchical risk assessment of transmission system considering the influence of active distribution network. IEEE Trans. Power Syst. 30 (2), 1084–1093.

Huaidong, L., Yu, C., Xiaojun, C., 2016. Power system fault screening and sorting based on expected loss index. Power Syst. Protect. Control 44 (14), 8–13.

Huang, G., Jiang, Y., 2017. Short-term wind speed prediction: hybrid of ensemble empirical mode decomposition, feature selection and error correction. Energ. Convers. Manage. 144, 340–350.

Jian, L., Qiang, Y., Zhihua, Z., 2014a. Evaluation and analysis of distribution power capacity of distribution network. Autom. Electr. Power Syst. 38 (5), 44–49.

Jian, L., Qiang, Y., Zhihua, Z., 2014b. Evaluation and analysis of layered power supply capacity of distribution network. Autom. Electr. Power Syst. 38 (5), 44–49.

Jianpeng, L., Yajing, G., Haifeng, L., et al., 2014. Calculation of power supply capacity of urban distribution network based on load forecasting. East China Electr. Power 42 (5), 873–877.

Kim, D.E., Cho, N., 2017. Fault analysis method for power distribution grid with PCS-based distributed energy resources. J. Electr. Eng. Technol. 12 (2), 522–532.

Li, X., Du, D.J., Pei, J.X., et al., 2013. Probabilistic load flow calculation with Latin hypercube sampling applied to grid-connected induction wind power system. Trans. Inst. Meas. Control 35 (1), 56–65.

Liang, T., Rong, J., Jianjian, C., 2016. Fault line-selection method in resonant grounded system based on CEEMD energy proportion. Electr. Eng. 17 (10), 36–45.

Liming, Z., Xianjun, Q., 2015. Load supplying capability for distribution network considering distributed generation randomness. Electr. Power Const. 36 (11), 38–44.

Lin, Y., Cheng, R., Yongning, Z., et al., 2016. Hierarchical analysis method for numerical characteristics of short-term wind power forecasting error. Proc. CSEE 36 (3), 692–700.

Liu, S.S., Qin, X.D., Wang, Z.C., et al., 2015. Research and application of divisional grid maximum power supply capacity based on optimal power flow. Power Energy 36, 470–475.

Marzano, L.G.B., Batista, F.R.S., Maceira, M.E.P., et al., 2015. In: A multi-area approach to evaluate the Brazilian power system capacity to supply the peak load demand using detailed simulation model of power plants operation.Power Systems Computation Conference, Wroclaw, 18–22 August.

Meena, N.K., Parashar, S., Swarnkar, A., et al., 2017. Improved elephant herding optimization for multiobjective DER accommodation in distribution systems. IEEE Trans. Ind. Inform. 99, 1.

Nagler, T., Czado, C., 2016. Evading the curse of dimensionality in nonparametric density estimation with simplified vine copulas. J. Multivariate Anal. 151, 69–89.

Nan, Y., Cui, J., Zheng, Z., et al., 2016. Research on nonparametric Kernel density estimation for modeling of wind power probability characteristics based on fuzzy ordinal optimization. Power Syst. Technol. 30, 335–340.

Qiang, C., Xiuli, W., Zuyong, L., 2013. Research on linkage price and benefit analysis of storage power station operation in market environment. Proc. CSEE 33 (13), 62–68.

Shao, H., Deng, X., Cui, F., 2016. Short-term wind speed forecasting using the wavelet decomposition and AdaBoost technique in wind farm of East China. IET Gener. Transm. Dis. 10 (11), 2585–2592.

Shields, M.D., Zhang, J., 2016. The generalization of Latin hypercube sampling. Reliab. Eng. Syst. Safety 148, 96–108.

Siltala, L., Granvik, M., 2017. Asteroid mass estimation using Markov-Chain Monte Carlo. Icarus 297, 149–159.

Souza Machado, I.B., Borba, S.M.C., Silva Maciel, R., 2016. Modeling distributed PV market and its impacts on distribution system: a Brazilian case study. IEEE Lat. Am. Trans. 14 (11), 4520–4526.

Sun, M., Dong, S., Xia, S., et al., 2017. Path description based calculation method for available capacity of feeder partition satisfied with N-1 security criterion. Autom. Electr. Power Syst. 41 (16), 123–129.

Tahir, M., Nassar, M.E., El-Shatshat, R., et al., 2016. In: A review of Volt/Var control techniques in passive and active power distribution networks.IEEE Smart Energy Grid Engineering, Oshawa, 21–24 August.

Tan, Y., Liu, D., Li, Q., et al., 2016. In: Confidential power supply and storage capacity calculation of the active distribution network planning.International Conference on Renewable Power Generation, Beijing, 17–18 October.

Tianjiao, P., Naishi, C., Xiaohui, W., et al., 2016. Application and architecture of multi-source coordinated optimal dispatch for active distribution network. Autom. Electr. Power Syst. 40 (1), 17–23.

Trigo, R.M., Dacamara, C.C., 2015. Circulation weather types and their influence on the precipitation regime in Portugal. Int. J. Climatol. 20 (13), 1559–1581.

Tsuyuki, T., Kanamitsu, M., 2015. One-month forecast experiments with a correction of systematic errors of the zonal mean temperature during the time integration. J. Meteorol. Soc. Jpn. 64, 805–815.

Vu, T.L., Turitsyn, K., 2015. A framework for robust assessment of power grid stability and resiliency. IEEE Trans. Autom. Control 62 (3), 1165–1177.

Wang, H., Lu, X., Xu, P., et al., 2015a. Short-term prediction of power consumption for large-scale public buildings based on regression algorithm. Procedia Eng. 121, 1318–1325.

Wang, L., Chen, Q., Gao, Z., et al., 2015b. Knowledge representation and general Petri net models for power grid fault diagnosis. IET Gener. Transm. Dist. 9 (9), 866–873.

Wang, J., Hu, J., Ma, K., 2016. Wind speed probability distribution estimation and wind energy assessment. Renew. Sustain. Energy Rev. 60, 881–899.

Weijia, Z., Ning, Z., Chongqing, K., et al., 2015. Probabilistic distribution estimation method for predictive error of photovoltaic power generation. Autom. Electr. Power Syst. 39 (16), 8–15.

Xiaolong, J., Yunfei, M., Hongjie, J., et al., 2014. The active reconfiguration strategy of the distribution network for the maximum power supply capacity is improved. Trans. China Electrotech. Soc. 29 (12), 137–147.

Xiaomin, G., Jun, H., Jianxiong, Y., 2015. Application of tidal current theory in risk assessment of transmission network disconnection. Power Syst. Technol. 39 (2), 486–493.

Xing, H., Cheng, H., Zhang, Y., et al., 2016. Active distribution network expansion planning integrating dispersed energy storage systems. IET Gener. Transm. Dis. 10 (3), 638–644.

Yajing, G., Dong, L., Xihua, C., et al., 2015a. Prediction model of short-term wind power output estimation based on data driven. Proc. CSEE 35 (11), 2645–2653.

Yajing, G., Ruihuan, L., Haifeng, L., et al., 2015b. Two step optimal dispatch based on multiple scenarios technique considering uncertainties of intermittent distributed generations and loads in the active distribution system. Proc. CSEE 35 (7), 1657–1665.

Yajing, G., Jing, Z., Xinhua, C., Haifeng, L., Peng, L., 2016. Evaluation of short-term power supply capacity of active power distribution system based on multi-scene. Proc. CSEE 22, 6076–6085.

Yaoqi, C., 2016. Study on short-term prediction of photovoltaic power generation based on typical weather types considering random prediction errors. Electr. Power 49 (5), 157–162.

Yeh, W.C., 2015. A novel node-based sequential implicit enumeration method for finding all d - MPs in a multistate flow network. Inform. Sci. 297 (C), 283–292.

Yin, J., 2016. Research on load friendly interactive technology for safe operation of UHV interconnected power grid. Proc. CSEE. 36(21).

Yin, C., Yang, D., Zhao, X., 2017. In: State estimation of active distribution system based on the factor graph analysis and belief propagation algorithm.IEEE International Conference on Environment and Electrical Engineering and IEEE Industrial and Commercial Power Systems Europe, Milan.

Yu, W., Liu, D., Huang, Y., 2015. Load transfer and islanding analysis of active distribution network. Int. Trans. Elect. Energy Syst. 25 (8), 1420–1435.

Zhigang, L., Guihong, Y., Liye, M., et al., 2017. Security level division of active distribution system based on power supply capability. Proc. CSEE 9, 2539–2551.

Further reading

Juxiang, H., 2016. Power supply reliability analysis of new distribution network with micro grid. Technol. Enterprise 01, 175–176.

Wang, P.K.C., 2015. Stability analysis of a simplified flexible vehicle via Lyapunov's direct method. AIAA J. 3 (9), 1764–1766.

Wang, J., Taaffe, M.R., 2015. Multivariate mixtures of normal distributions: properties, random vector generation, fitting, and as models of market daily changes. INFORMS J. Comput. 27 (2), 193–203.

Yeli, M., Bin, J., Zhigang, L., et al., 2014. Study on the security level of transmission grid based on static security and real-time power supply. Trans. China Electrotech. Soc. 29 (6), 229–237.

Multi-time-scale energy management of distributed energy resources in active distribution grids

21

Bishnu P. Bhattarai, Kurt S. Myers*, Robert J. Turk*, Birgitte Bak-Jensen†*
*Department of Power and Energy System, Idaho National Laboratory, Idaho Falls, ID, United States, †Department of Energy Technology, Aalborg University, Aalborg, Denmark

21.1 Introduction

Currently, the power system is undergoing two major transformations as shown in Fig. 21.1. On the one hand, increased environmental concern and favorable government policies are resulting in rapid growth of renewable energy sources (RESs), such as solar photovoltaic (PV) and wind turbine generators, which are continuously replacing the conventional generation sources. Limited dispatch capability of RESs on top of intermittent generation is creating several control and operational challenges (U.S. Department of Energy, 2010; The Danish Ministry of Climate, 2013). On the other hand, focus on using the available energy in a more efficient way leads to integrated energy systems or so-called multi-energy systems (MES) with increased electrification of the heating, transportation and gas sectors, and use of different kind of storage facilities (electrical or thermal) (Lund et al., 2014; Bhattarai et al., 2014a; Mancarella, 2014). This scenario is resulting in new electrical loads, such as electric vehicles (EVs), electric water heaters (EWHs), and heat pump (HP) to be integrated into the existing distribution network. Since most of the existing grids were not designed to withstand such increased penetrations of loads and RESs, such scenarios might threaten the power balancing and congest several distribution networks. As shown in Fig. 21.1, potential solutions to address these issues include:

- building large energy storages systems,
- expansion and upgrade of existing grids, and
- exploit flexibility from spatially distributed flexible resources.

Conventional solutions, such as building excess generation or grid-scale storages, are very capital intensive with the use of existing technologies (Bhattarai, 2015). One of the potential alternatives is to control intelligently spatially distributed small-scale flexible resources to make them follow the generation (Bhattarai, 2015; Mohanpurkar and Suryanarayan, 2013). This not only enables the end consumers to get reliable and cheap electricity but also enables the utility to prevent huge

Smart Power Distribution Systems. https://doi.org/10.1016/B978-0-12-812154-2.00021-3

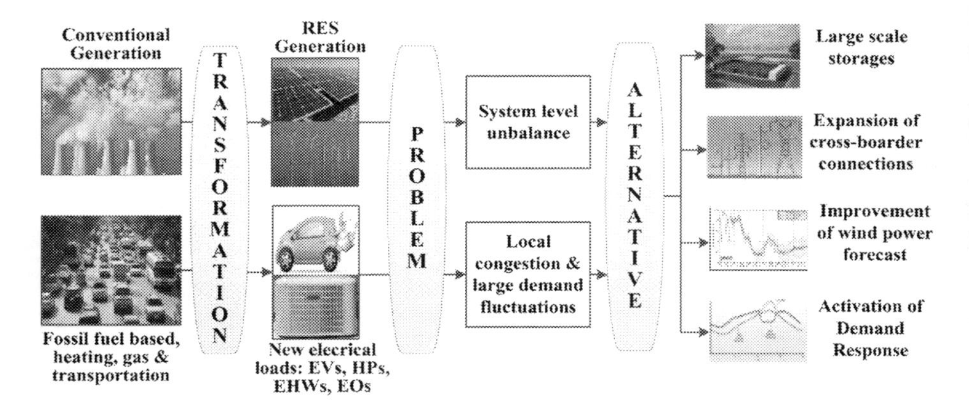

Fig. 21.1 Major transformation of electrical power system (Energinet.dk, 2010; Bhattarai, 2015).

investment in counterpart. Therefore, the theoretical foundation of this chapter is focused on better utilization of flexible resources for congestion (e.g., undervoltage (UV) violations) management and system balancing.

Recently, research communities and industries have been investing significant efforts to utilize flexible loads for system balancing or grid violations management. Xiu and Li (2012), Bhattarai et al. (2015a), Bhattarai et al. (2014b), Han and Han (2013), Masuta and Yokoyama (2012), Zhou et al. (2015), Kondoh (2013), and Gelazanskas and Gamage (2016) proposed various control algorithms and strategies for utilizing flexibility from EVs, EWHs, and HPs for system balancing purposes. In particular, frequency regulation using EV (Xiu and Li, 2012; Bhattarai et al., 2015a, 2014b), EV and HP (Han and Han, 2013; Masuta and Yokoyama, 2012), air-conditioners (Zhou et al., 2015), and residential EWH (Kondoh, 2013; Gelazanskas and Gamage, 2016), are developed to utilize EV flexibilities for system-level balancing. The effectiveness of the aforementioned methods greatly depends on how well the objectives of consumer, distribution system operator (DSO), and aggregator. The consumer tries to minimize its energy cost with simultaneous assurance of its comfort requirements, whereas the aggregator tries to maximize its profit by playing intermediate roles among consumers, DSOs, and electricity markets. Finally, the DSO tries to maximize grid performance (e.g., minimize power loss and voltage deviations). Wen et al. (2012) developed a decentralized control algorithm to maximize consumer comfort and Bhattarai et al. (2016a, 2017a) presented techniques to manage flexibility considering customer preferences. To ensure grid constraints, Bhattarai et al. (2013, 2017b) investigated various control algorithms for deploying demand flexibilities [e.g., energy shifting (Sundstrom and Binding, 2013), load profile smoothing (Bhattarai et al., 2017c), valley filling (Gan et al., 2013), peak shaving (Gerards and Hurink, 2016), and power loss minimization (Clement-Nyns et al., 2010)]. However, these literature decoupled the problem and analyzed the systems from a single actor's perspectives. This requires a more rigorous and generic approach to simultaneously assure consumer comfort, technical constraints, and consumer profit maximization, which will be dealt with in this chapter.

In addition to the loads, rooftop solar PV systems are also increasingly being deployed in distribution systems. Since PV systems often produce maximum power during the period when the load is lower, it might create overvoltage (OV) problems. Various control solutions have been developed in recent literatures for mitigating OVs resulting from increased PV penetration. Demirok (2012) and Bhattarai et al. (2015b) proposed reactive power controls to alleviate OV caused by distributed PVs. Despite, increased reactive power consumption to relieve OV normally creates additional power loss in the grid. Moreover, due to high R/X ratio of low-voltage (LV) distribution grids, the reactive power control is less effective (Demirok et al., 2011). To overcome these issues, a coordinated voltage control methodology has been developed by Tonkoski et al. (2011), where every distributed energy resources contribute to alleviate the OV. Moreover, Bhattarai et al. (2016b) proposed a combined predictive and real-time EV charging control to mitigate the OV. Nevertheless, utilizing EV might be ineffective to address the OV issues in residential feeders as the EVs are normally unavailable at home during the day when PV has maximum production. Therefore, integration of EWH and HP can serve as potential resources for addressing OV issues as well. Sizable ratings and energy storage capability of these resources provide significant demand flexibility that can potentially be utilized not only for solving grid violations, but also for supporting system balancing. In this study, a multi-time-scale energy management system considering coordinated demand response and onsite generation is presented. The key contributions of this chapter include:

- A hierarchical demand response (DR) control and coordination architecture consisting of scheduling, coordinative, and adaptive control layers for enabling flexible resources to contribute to grid balancing and local network congestion management.
- Coordinated control of flexible loads and onsite generation to maximize utilization of flexibility stemming from spatially distributed flexibility.
- A multi-time-scale control and energy management system to enable single DR resources to participate in multiple grid services and to enhance economic benefits to participating actors.
- The potential benefits of the proposed method to DSOs and consumers are studies using simulation performed in a co-simulation environment based on DIgSILENT Power Factory and MATLAB.

The rest of the chapter is structured as follows. First, in Section 21.2, details of the flexible resources modeling are presented. Next, the multi-time-scale energy management system is presented in details in Section 21.3. In Section 21.4, simulation configurations are presented. Detailed results and discussions are presented in Section 21.5 and the chapter concludes in Section 21.6.

21.2 System modeling

This section presents detailed mathematical models of EV, EWH, and PV which are set up in such a way that they can be used for MES modeling scenarios.

21.2.1 PV system

Rooftop solar PV is a common distributed energy resource integrated in a residential distribution feeder. In this study, we have used a simplified grid-tied PV system which is modeled as a dispatchable static generator capable of operating at the dispatched active and reactive power (PQ) set points. The capability curve of the rooftop PV system is designed in accordance with IEEE 1547 standard for distributed energy resources $\leq$30 kW as shown in Fig. 21.2B. Hence, the capability curve is designed to keep the power factor (PF) of the PV inverter to 0.85 lead/lag or higher at the point of connection (POC). The PV system, which is modeled in DIgSILENT PowerFactory is shown in Fig. 21.2A. It consists of four key parts: measurements, input model (i.e., solar irradiance), control strategy, and static generator. Note that reactive power injection/absorption is zero as long as the voltage at the POC is within acceptable limits and/or there is no dispatch request from the upstream control centers/grid operators. For the dispatched active/reactive power, real (i_d^{ref}) and imaginary (i_q^{ref}) components of currents are calculated as:

$$i^{pv} = \frac{P^{\mathrm{PV}}}{\sqrt{3} \cdot V_{\mathrm{POC}} \cdot \cos\varphi} \tag{21.1}$$

$$i_d^{ref} = i_{pv} \cdot \cos\varphi \& i_d^{ref} = i_{pv} \cdot \sin\varphi \tag{21.2}$$

where i^{pv} is the line current, P^{pv} is the power injected by the PV system, V_{poc} is the voltage at the POC of the generator, and $\cos\varphi$ is the PF of the PV system.

21.2.2 Electric water heater

An EWH is a sizable electrical load used in almost all residential households. It primarily consists of a water immersed electric rod(s) which uses electrical energy to heat the water inside a hot water storage tank (HWST). Therefore, energy consumption is

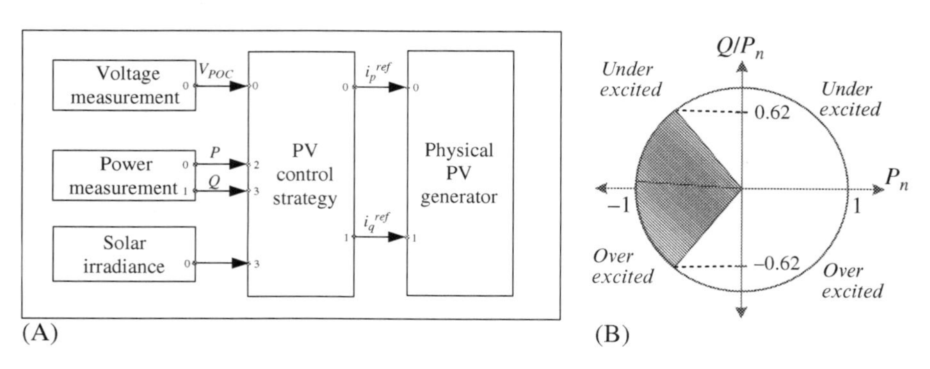

Fig. 21.2 Models of a photovoltaic (PV) system per IEEE 1547 standard for $\leq$30 kW. (A) DIgSILENT implementation of PV control. (B) PQ capability curve of PV system.

directly proportional to the thermal energy stored in the tank. As EWH is integrated with a thermostatic control, the EWH is turned on when the temperature of water inside the HWST goes below a predefined lower limit and is turned off when it goes above a predefined upper limit.

Fig. 21.3 illustrates a high-level diagram of the EWH and control model implemented in DIgSILENT Power Factory. It consists of measurements, thermal model, control strategy, and physical load. Assuming a single node, state of energy (SOE) of the HWST in per unit (p.u.) is determined using the following energy balance:

$$\mathrm{SOE}^{EWH} = \frac{E_{Ini} + E_{In} - E_{Dmd} - E_{Loss}}{E_{Cap}} \tag{21.3}$$

where E_{Ini}, E_{In}, E_{Dmd}, E_{Loss}, and E_{Cap} are the initial energy stored in the HWST, electrical input energy, hot water energy demand, thermal energy loss, and thermal capacity of the HWST, respectively. The energy loss, E_{loss}, is calculated as:

$$E_{Loss} = \frac{1}{k_t} \int \left[\mathrm{UA} \cdot \left(\frac{E_{Dmd} - E_{Cap}}{V_t \cdot Cp_w \cdot \rho \cdot k_n} - T_r \right) \right] dt \tag{21.4}$$

where k_t is the normalization parameter, UA is the heat transfer coefficient, V_t is the HWST volume in liters, Cp_w and ρ are the specific heat capacity and density of the water, and T_r is the temperature of the room where the appliance is installed. Similarly, thermal output energy is modeled as an integral of hot water demand (P_{Dmd}) as follows:

$$E_{Dmd} = \frac{1}{k_t} \int P_{Dmd} dt \tag{21.5}$$

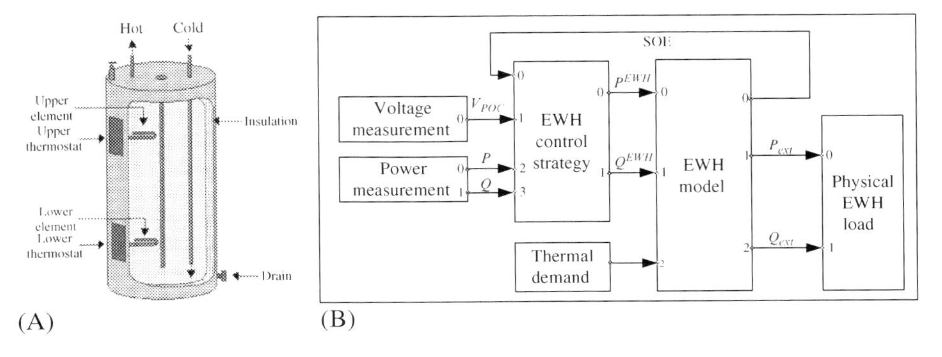

Fig. 21.3 Control diagram for electric water heater (EWH) implemented. (A) Electric water heater schematic. (B) Composite frame diagram implemented in DIgSILENT.

Finally, input energy is calculated by integrating the power consumed by the EWH (P^{EWH}) as follows:

$$E_{In} = \frac{1}{k_t} \int P_{In}{}^{*} dt \qquad 0 \le P_{In} \le P_R^{EWH} \tag{21.6}$$

where P_R^{EWH} is the rated power of the EWH. The control of EWH is designed such that it can adaptively consume/generate active and reactive power based on voltage at POC. It is worth mentioning that the advancement of power converter technology made it technically possible to regulate both active and reactive power of the EWH (Kondoh, 2013). This can provide additional flexibility and controllability of the EWH.

21.2.3 Electric vehicle

EVs are one of the highest power consuming electrical loads at residential consumers. From power system perspective, the EV is a storage device which is equivalent to an electrical load while charging and a static generator while discharging. So, the EV can be modeled as a three-state model as shown in Fig. 21.4.

During charging, EV is modeled as a constant power load such that its power is a continuous control variable ranging from zero to its maximum power rating ($P_{\max}^{\mathrm{PV}}$), while during discharging it is essentially a static generator that is capable of injecting power back to the grid. Mathematically, these two conditions can be represented as

$$\begin{aligned} P_{\mathrm{chg}}^{\mathrm{EV}} &\in \left[o,\ P_{\max}^{\mathrm{EV}}\right] \quad \forall SOC \le SOC_{\max} \\ P_{\mathrm{dsg}}^{\mathrm{EV}} &\in \left[o,\ -P_{\max}^{\mathrm{EV}}\right] \quad \forall SOC \ge SOC_{\min} \end{aligned} \tag{21.7}$$

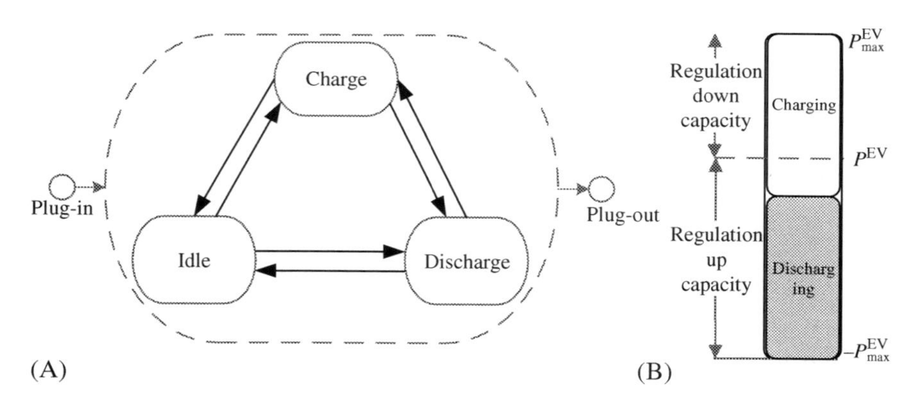

Fig. 21.4 Schematic diagram of electric vehicle (EV) modeling. (A) Three-state model of EV. (B) Regulation capability of EV.

where $P^{EV}_{chg/dsg}$ is the charging/discharging power and $SOC_{max/min}$ is the maximum/minimum allowable state-of-charge (SOC) of the EV. Please note that during idle mode, EV is connected to the grid, but it neither draws nor injects power. Despite, EV can still provide both upregulation and downregulation as shown in Fig. 21.4B), which is mathematically computed as follows:

$$\begin{aligned} P^{EV}_{Up} &= P^{EV} - \left(-P^{EV}_{max}\right) \\ P^{EV}_{Dn} &= P^{EV}_{max} - P^{EV} \end{aligned} \tag{21.8}$$

where P^{EV}_{Up} and P^{EV}_{Dn} are up and down-regulation capacities of the EV operating at P^{EV}. It is worth mentioning that the actual up/down-regulation amounts significantly vary depending upon the operating point. Fig. 21.5 illustrates the DIgSILENT implementation of the EV control framework. It consists of measurements, control strategy, EV model, and the actual loads. The measured V_{poc} is input to the actual EV model, where the EV model is combined with the control strategy to adjust the power consumption according to the observed V_{poc}.

21.3 Hierarchical multi-time-scale energy management system

Effective deployment of spatially distributed flexible resources requires proper coordination among key smart grid actors: consumer, aggregator, DSO, balancing and regulating markets (BRM), and energy market. As such, proper framework is crucial for coordinating response of different actors. It is worth mentioning that the analysis in this chapter is focused from consumer, DSO, and aggregator perspectives, while the

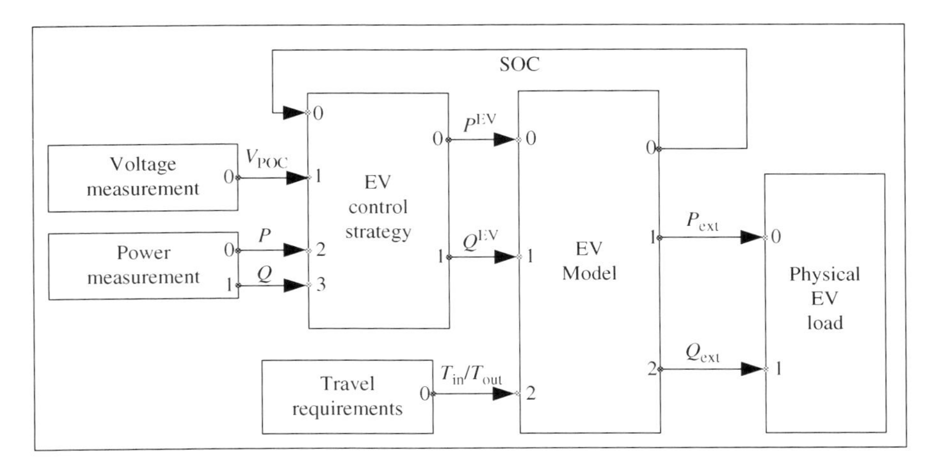

Fig. 21.5 Composite frame diagram of EV implemented in DIgSILENT power factory.

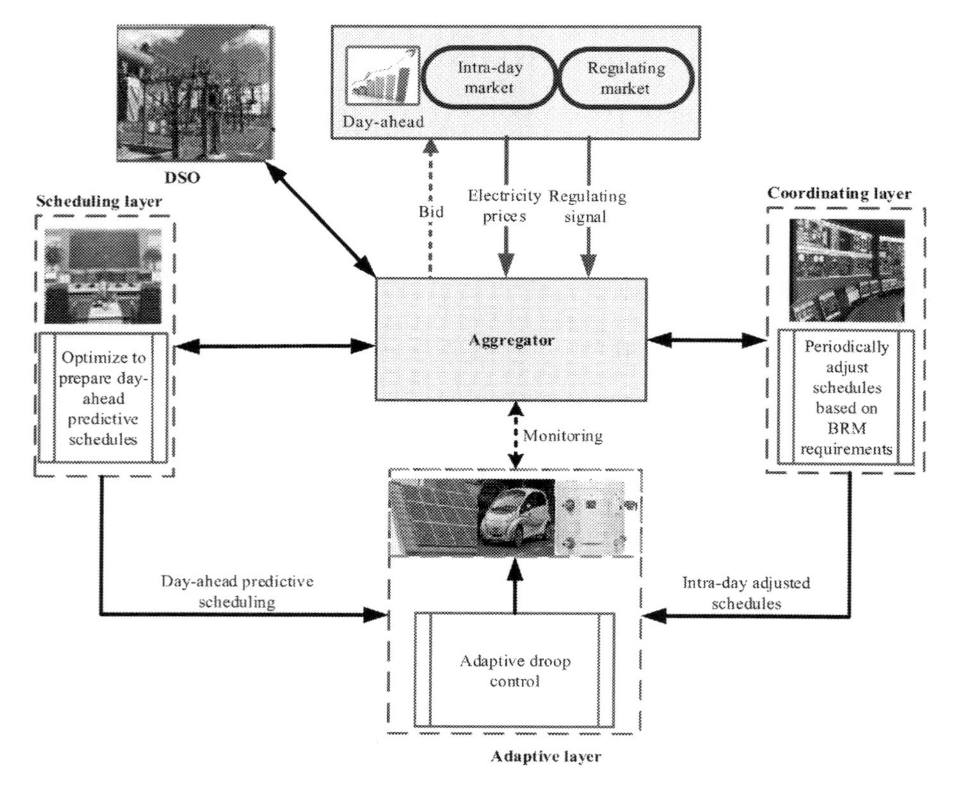

Fig. 21.6 Schematic diagram of hierarchical control architecture (HCA) and interactions among key actors for flexibility deployment.

impacts of BRM and energy markets are incorporated through the regulating signal and price signals, respectively. Fig. 21.6 illustrates a high-level overview of the proposed hierarchical control architecture (HCA) designed to realize multi-time-scale energy management.

The proposed multi-time-scale HCA framework works at three different stages starting from a day-ahead predictive scheduling, hourly balancing, and real-time adaptive adjustment. The HCA is designed with three control layers: scheduling, coordinative, and adaptive, such that it realizes the following functionalities:

- The HCA enables small and spatially distributed flexible resources to participate in multiple electricity markets.
- The HCA enables each inner layer to back up the immediate outer layer in smaller time resolution. For instance, any discrepancies on day-ahead scheduling are compensated in smaller time resolution by the balancing layer, and the discrepancies within the balancing layer are taken care of in near real time by real-time adaptive control.
- The HCA enables the same resources to provide multiple grid support functionalities. For instance, the resources can be used for day-ahead resource planning through the predictive

scheduling layer (SL), whereas it can also be used for grid constraints violation management through the adaptive control.

- The HCA provides a generic framework that can be utilized for exploiting flexibilities from load as well as onsite generation.

21.3.1 Day-ahead SL

The day-ahead SL is primarily responsible for preparing predictive operational schedules for the flexible resources. As accurate prediction of the characteristics of the individual flexible resources (e.g., plug-in/plug-out time of EVs, temperature settings of HPs/EWHs) is often difficult, a number of uncertainties may involve with the predictive scheduling. As depicted in Fig. 21.7, day-ahead electricity price, feeder loading,

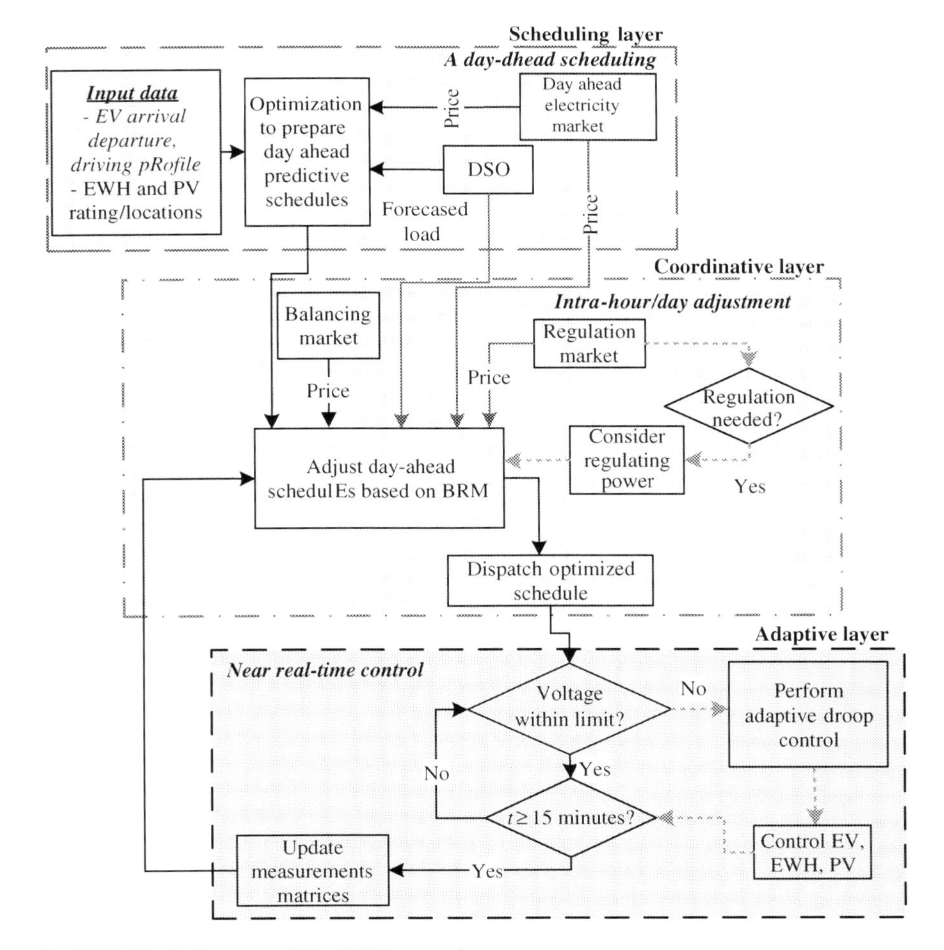

Fig. 21.7 Block diagram of the HCA operation.

and availability of flexible resources are used to compute a day-ahead predictive scheduling. Hourly electricity prices for the next day are obtained from the day-ahead energy market, forecasted load is obtained from the DSO, and availability of flexible loads (e.g., EV arrival/departure distribution) is obtained from consumers. Thereafter, the SL generates an optimum schedule for the available flexible resources. Mathematically, the optimization is formulated as

$$\text{Min.} \quad \sum_{k=1}^{K}\sum_{i=1}^{N} P_{i,k}^{Flx} * \Delta t^{k*} C^{k} \tag{21.9}$$

Subjected to the following set of constraints:

$$\begin{aligned} &\sum_{i=1}^{N} P_i^{\mathrm{Flx}} \leq \left(P_{\mathrm{Cap}}^{\mathrm{Fd}} - L_k^{\mathrm{Fd}}\right) \quad \forall k = 1:K \\ &V_{\min} < V_j < V_{\max} \qquad \forall j = 1:N \end{aligned} \tag{21.10}$$

where $P_{i,\,k}^{\mathrm{Flx}}$ is the predictive dispatch of ith flexible resource at kth time slot, C^k is the electricity price for the kth slot, and Δt_k is the time interval of kth time slot, respectively. Similarly, $P_{\mathrm{cap}}^{\mathrm{Fd}}$ and P_k^{Fd} are the maximum and actual feeder loadings at kth time slot, respectively; $V_{\mathrm{min/max}}$ is the minimum/maximum POC voltages of the ith resource, and N and K are the total number of flexible resources and total number of time slots, respectively.

Please note that Eq. (21.10) ensures thermal and voltage constraints in the network to be within the predefined limits. However, depending on the load types, individual consumer comfort may vary. For the case of EV, consumer comfort is constrained by keeping SOC within the given minimum/maximum limits, whereas for thermostatic loads (e.g., EWH), consumer comforts are ensured by keeping the thermostat temperature within predefined limits. After solving Eq. (21.9) subjected to constraints (21.10), the optimum predictive schedules are dispatched to respective resources.

21.3.2 Coordinative layer

The coordinative layer (CL) is designed to adjust the day-ahead schedules of the flexible resources based on observed uncertainties. Any discrepancies in the day-ahead schedules are optimally distributed among available flexible resources at that operating time. For instance, if EVs are the flexible resources, CL optimally adjusts the operational schedules of the EVs that are plugged in at that moment to cope with the intermittencies. In addition to compensating discrepancies on the day-ahead schedule, this layer also enables the flexible resources to provide balancing and regulating services. As BRM offers significantly higher prices, it is a great opportunity for all consumers, aggregators, and DSOs to maximize their benefits. Note that the actual regulating power each flexible resource can provide depends greatly on their operating point. For instance, EVs in a lower *SOC* regime can provide downregulation only, while the EVs in a higher *SOC* regime can provide only upregulation. In the CL, the predictive schedules dispatched from the SL will be rescheduled using the following formulation:

$$\text{Min.} \sum_{k=1}^{K}\sum_{i=1}^{N}\left(P_{i,k}^{\text{Fle}} * \Delta t^{k*} C^{k} - \left[P_{i,k}^{\text{Fle}} + P_{\max}^{\text{Fle}}\right] R_{k}^{\text{up}} - \left[P_{\max}^{\text{Fle}} - P_{i,k}^{\text{Fle}}\right] R_{k}^{\text{dn}}\right) \tag{21.11}$$

where $R_{k\text{Up}}$ and R_{k}^{Dn} are up and down-regulation prices at the kth slot, and $P_{i,\ k}^{\text{Up}}$ and $P_{i,\ k}^{\text{Up}}$ are the up and down-regulation power of the ith flexible resource at kth time slot. The first part of the objective function essentially minimizes the total EV charging cost based on the day-ahead electricity prices, while the second part of the objective function maximizes the benefit by participating in the BRM. Even though the CL provides a perfect platform for the flexible resources to participate in the BRM, it has no controllability to address any intra-slot discrepancies, especially real-time intermittencies. Therefore, an adaptive control layer that adapts the operation of the flexible resources based on real-time operating conditions of the local network is implemented.

21.3.3 Adaptive layer

The adaptive layer (AL) is designed to address any intra-slot discrepancies (e.-g., violations of network constraints in real time, consumer comfort violations) that cannot be captured by the SL and CL-based predictive scheduling. AL locally monitors and controls the operation of the flexible resources near real time according to the contemporary network conditions. Since this study is mainly focused on active distribution networks, voltage at the consumer POC is a very good indicator of grid conditions. Therefore, a voltage-based droop control which continuously monitors the voltage at POC and adjusts the operation of the flexible resources is implemented. The following subsections describe the adaptive control mechanisms for the EV, EWH, and PV.

21.3.3.1 Adaptive control of electric vehicle

Since EVs can be continuously controlled within its rated power (i.e., between $-P_{\max}^{\text{PV}}$ to $P_{\max}^{\text{PV}}$), a simple droop control can be implemented to adjust its charging/discharging power according to the measured POC voltage. The EV droop, illustrated in Fig. 21.8 can be formulated as follows:

$$P^{\text{EV}} = \begin{cases} -P_{\max}^{\text{EV}} & V > V_{th} \\ -P_{\max}^{\text{EV}} \dfrac{(V - V_{th})}{V_{\min} - V_{th}} & V_{\min} < V < V_{th} \\ P_{\max}^{\text{EV}} \dfrac{(V_{\text{TH}} - V)}{V_{\max} - V_{TH}} & V_{\text{TH}} < V < V_{\max} \\ P_{\max}^{\text{EV}} & V > V_{\max} \end{cases} \tag{21.12}$$

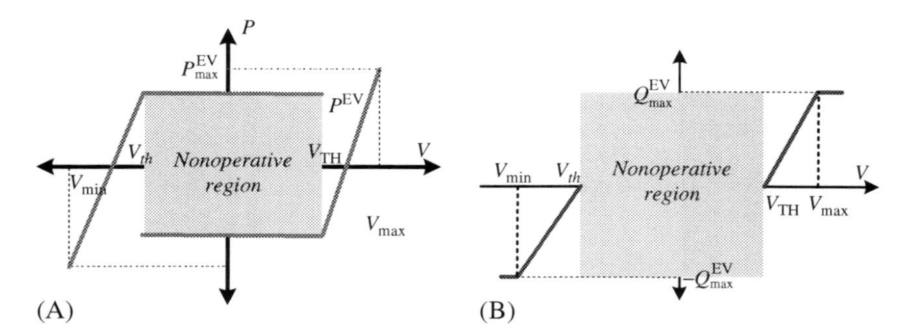

Fig. 21.8 Adaptive droop control for EV. (A) Active power droop control (*P–V*). (B) Reactive power droop control (*Q–V*).

where V, V_{th}/V_{TH}, and V_{min}/V_{max} are the measured voltage, UV/OV thresholds, and minimum/maximum cutoff voltage, respectively. Threshold voltages are essentially the voltages beyond which the EV droop starts functioning, while cut-off voltages are the ones beyond which EV discharges/charges at maximum power.

Similarly, the reactive power consumption or injection by the PV (Q^{EV}) is adjusted as follows:

$$Q^{EV} = \begin{cases} -Q_{max}^{EV} & V \leq V_{min} \\ -Q_{max}^{EV} \cdot (V - V_{th})/(V_{min} - V_{th}) & V_{min} < V \leq V_{th} \\ Q_{max}^{EV} (V_{TH} - V)/(V_{max} - V_{TH}) & V_{TH} \leq V < V_{max} \\ Q_{max}^{EV} & V \geq V_{max} \end{cases} \tag{21.13}$$

where Q_{max}^{EV} is the maximum allowable reactive power consumption or injection by the EV, which has operational limits that sets the upper minimum allowable PF as follows:

$$Q_{max} = P_{EV}{}^{*} \tan(\text{acos}(\phi_{Lim})) \tag{21.14}$$

where $\varnothing_{lim}$ is the maximum allowable PF. The *Q–V* droop is designed such that EV consumes reactive power in OV, whereas it injects in UV violations. However, for the normal operation ($V_{th} < V < V_{TH}$), both *P–V* and *Q–V* droops are nonoperational.

21.3.3.2 Electric water heater

The adaptive control of EWH is incorporated by using power-voltage droops for both active and reactive powers. As shown in Fig. 21.9, the droops are designed to adjust the *P* and *Q* consumption of the EWH to counteract the violation of the POC voltage. The *P–V* droop adjusts the consumption of the EWH (P^{EWH}) as

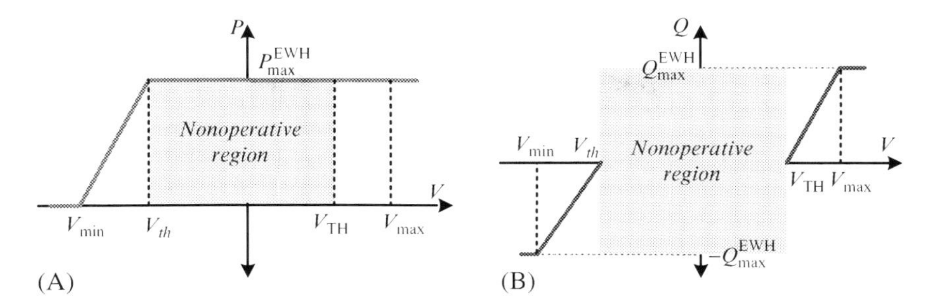

Fig. 21.9 Active and reactive power droop controllers for EWH. (A) Active power droop control (*P–V*). (B) Reactive power droop control (*Q–V*).

$$P^{\text{EWH}} = \begin{cases} 0 & V \le V_{\min} \\ P_{\max}^{\text{EWH}} * (V - V_{\min})/(V_{th} - V_{\min}) & V_{\min} < V \le V_{th} \\ P_{\max}^{\text{EWH}} & V \ge V_{th} \end{cases} \tag{21.15}$$

The *P–V* droop-based adaptive controller is designed to adjust P^{EWH} when V goes beyond V_{th}. Similarly, the reactive power consumption or injection by the EWH (Q^{EWH}) is adjusted as follows:

$$Q^{\text{EWH}} = \begin{cases} -Q_{\max}^{\text{EWH}} & V \le V_{\min} \\ -Q_{\max}^{\text{EWH}} . (V - V_{th})/(V_{\min} - V_{\text{th}}) & V_{\min} < V \le V_{th} \\ Q_{\max}^{\text{EWH}} (\text{V}_{\text{TH}} - V)/(V_{\max} - V_{\text{TH}}) & V_{\text{TH}} \le V < V_{\max} \\ Q_{\max}^{\text{EWH}} & V \ge V_{\max} \end{cases} \tag{21.16}$$

where $Q_{\max}^{\text{EWH}}$ is the maximum allowable reactive power consumption or injection by the EWH. $Q_{\max}^{\text{EWH}}$ has a limit which is often set based on minimum allowable PF and vary dynamically as

$$Q_{\max} = P_{\text{EWH}}{}^{*} \tan(\text{acos}(\phi_{\text{Lim}})) \tag{21.17}$$

where $\varnothing_{\text{lim}}$ is the maximum allowable PF. The *Q–V* droop is designed such that EWH consumes reactive power in OV whereas it injects in UV violations. However, for the normal operation ($V_{th} < V < V_{\text{TH}}$), both *P–V* and *Q–V* droops are nonoperational. Please note that the threshold and cut-off voltages for all droops (i.e., EV, EWH, and EV) are kept same in this study, but assigned with different random response time to avoid simultaneous switching of multiple resources.

To ensure consumer thermal comfort, SOE within HWST should stay within predefined limits. Therefore, EWH is subjected to an additional SOE-based ON-OFF control limits as

$$\begin{aligned} &P^{\text{EHW}} = P_{\max}^{\text{EWH}} * C_k \\ &\text{where}, C_k = \begin{cases} 1 & SOE \le SOE_{\min} \\ 0 & SOE \ge SOE_{\max} \end{cases} \end{aligned} \tag{21.18}$$

where C_k is the operating state of the EWH, and $SOE_{\min}/SOE_{\max}$ allowable minimum/maximum values of SOE in the tank. To exploit additional flexibility from EWH during voltage violations, an adaptive mechanism is developed to update the energy limits ($SOE_{\min}/SOE_{\max}$) according to the voltage as follows:

$$SOE_{\min/\max} = \begin{cases} SOE_{\min/\max} - 0.1 & V \leq V_{\min} \\ SOE_{\min/\max} + 0.1 & V \geq V_{\max} \end{cases} \tag{21.19}$$

As shown in Fig. 21.10, the adaptive controller is designed such that SOE limits are increased by 0.1 pu when the voltage at the POC exceeds $V_{\max}$, whereas it is decreased by 0.1 pu followed by the $V_{\min}$ violation. The key intent of shifting the energy band is to increase average power consumption during OV violations and vice versa during UV violations. In order to ensure consumer comfort, SOE-based control operates as a master control which has the capability to override the decisions made by the *P–V* and *Q–V* droops when the SOE limit is violated. However, as long as the SOE is within limit, the master control respects the decision made from the droops.

21.3.3.3 Adaptive control of solar PV

Similar to the EWH adaptive control, voltage-based droops are designed for adjusting PV power. As shown in Fig. 21.11, a *P–V* droop adjusts active power injection whereas a *Q–V* droop adjusts reactive power injection/consumption whenever the voltage deviates beyond predefined limits (V_{th} or V_{TH}). Mathematically, a piecewise linear droop is implemented to adjust injected active power (P^{PV}) as follows:

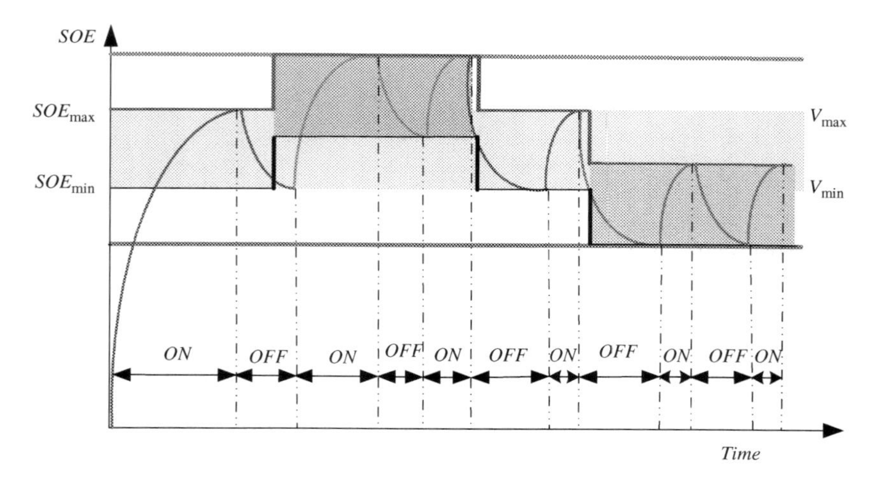

Fig. 21.10 Shifting of state of energy (SOE) bands per point of connection (POC) voltage.

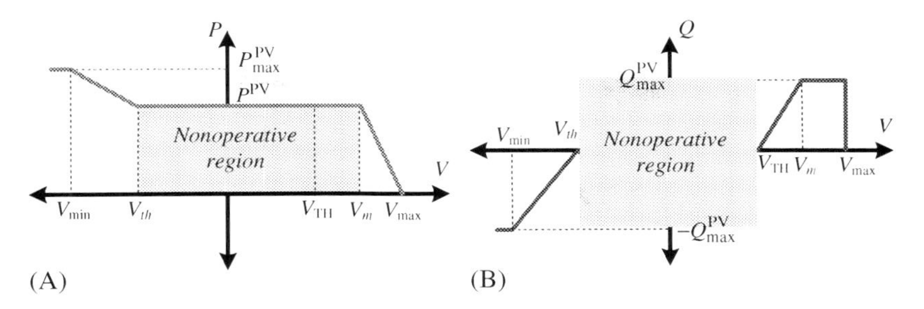

Fig. 21.11 Active and reactive power droops for PV. (A) Active power droop control (*P–V*). (B) Reactive power droop control (*Q–V*).

$$P^{PV} = \begin{cases} P^{PV}_{max} & V \leq V_{min} \\ P^{PV}_i + \left(P^{PV}_{max} - P_i\right)\dfrac{(V - V_{th})}{(V_{min} - V_{th})} & V_{min} < V \leq V_{th} \\ P^{PV}_i \dfrac{(V - V_{max})}{(V_{TH} - V_{max})} & V_m \leq V < V_{max} \\ 0 & V \geq V_{max} \end{cases} \tag{21.20}$$

where P^{PV}_{max} is the maximum power that the PV can generate per available solar irradiance and V_m is the mid-point voltage (average of V_{max} and V_{TH}) beyond which the *P–V* droop starts operation. The PV droop is designed such that it decreases the PV power injection in OV violation ($V > V_{TH}$) and increases the injected power (if applicable) in UV violation ($V < V_{th}$).

Similar to the active power droop, the reactive power droop adjust the consumption/injection as follows:

$$Q^{PV} = \begin{cases} -Q^{PV}_{max} & V \leq V_{min} \\ -Q^{PV}_{max}.(V - V_{th})/(V_{min} - V_{th}) & V_{min} < V \leq V_{th} \\ Q^{PV}_{max}(V_{TH} - V)/(V_{max} - V_{TH}) & VTH \leq V < V_m \\ Q^{PV}_{max} & V \geq V_m \end{cases} \tag{21.21}$$

where Q^{PV}_{max} is the maximum allowable reactive power injection/consumption by the PV inverter, which is limited by P^{PV} and allowable PF ($\emptyset_{max}$) as:

$$Q^{PV}_{max} = P^{PV}_i * \tan\left(\text{acos}(\phi_{max})\right) \tag{21.22}$$

It is worth mentioning that the *P–V* and *Q–V* droops are designed such that the *Q–V* droop fully exploit reactive power consumption before allowing the *P–V* droop to operate. Therefore, the threshold voltage for the *P–V* droop is equal to the maximum allowable voltage for the *Q–V* droop as shown in Fig. 21.11. Such scheme restricts active power curtailment as long as reactive power can resolve the voltage issues.

21.4 Simulation configuration and implementations

21.4.1 Test network description

The performance of the proposed methodology is demonstrated through a 24-h time sweep simulation. A typical winter and summer day load profile (i.e., maximum and minimum load profiles), as illustrated in Fig. 21.12A), are used to capture the worst-case operating scenarios from accommodating flexible loads and onsite generation perspectives. In addition, electricity prices as shown in Fig. 21.12B and a typical residential distribution network as illustrated in Fig. 21.13 are used for the simulation study.

21.4.2 Configuration of EV, EWH, and PV

Each consumer has base load, flexible loads (EVs, EWHs), and/or local generation (PV). In order to make the study interesting, penetrations of EV, PV, and EWH are chosen such that the penetrations create OV and UV violations. Configuration of EV data (e.g., arrival/departure time, availability) is done by using statistical information on traveling behaviors of light cars as shown in Fig. 21.14. Each EV is rated with three-phase 480 V supply (i.e., 11 kW, 25 kWh), rooftop PV with 6 kW, and EWH with 2.4 kW. Note that the location of EV, PV, and EWH are assigned randomly throughout the network.

21.4.3 Computational environment

The simulations are performed in a MATLAB-DIgSILENT co-simulation environment, where MATLAB is used for computations, including optimizations, forecasting, and data configuration, while DIgSILENT is used for time-sweep simulation. All the configuration parameters, such as consumer locations, feeder loadings, and electricity price are sent to the respective unit in DigSILENT for power simulations. Following the configuration, MATLAB runs the optimization procedure to compute

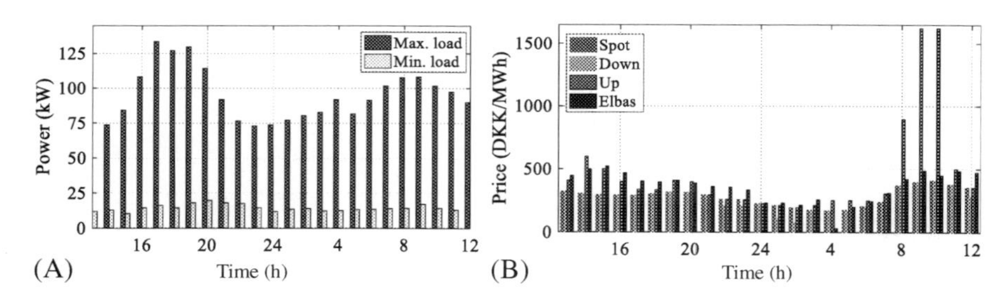

Fig. 21.12 Minimum and maximum load profile of the test network (Bhattarai et al., 2017d). (A) Maximum and minimum loading of test network. (B) Electricity cost at different time of day.

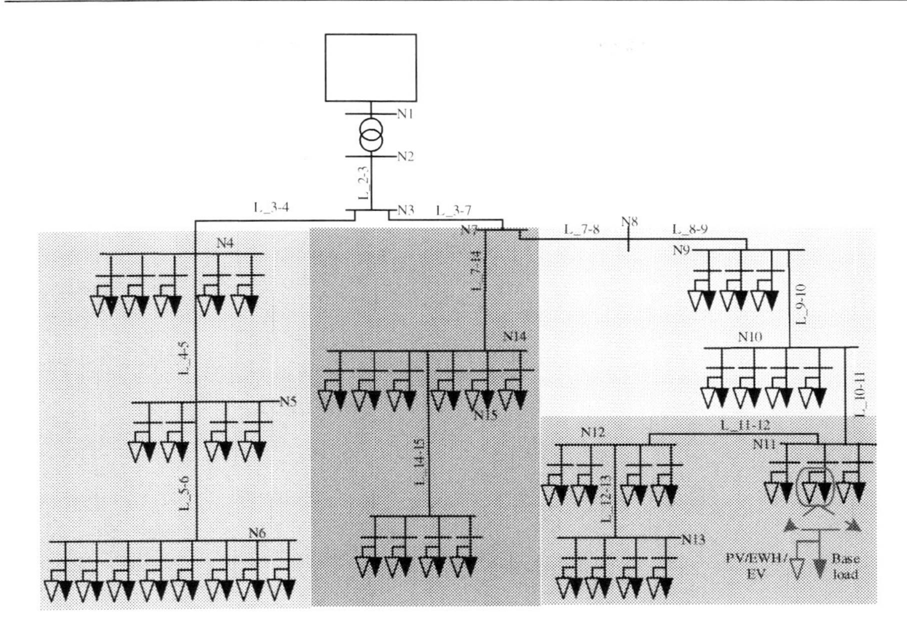

Fig. 21.13 A residential distribution network taken as test network.

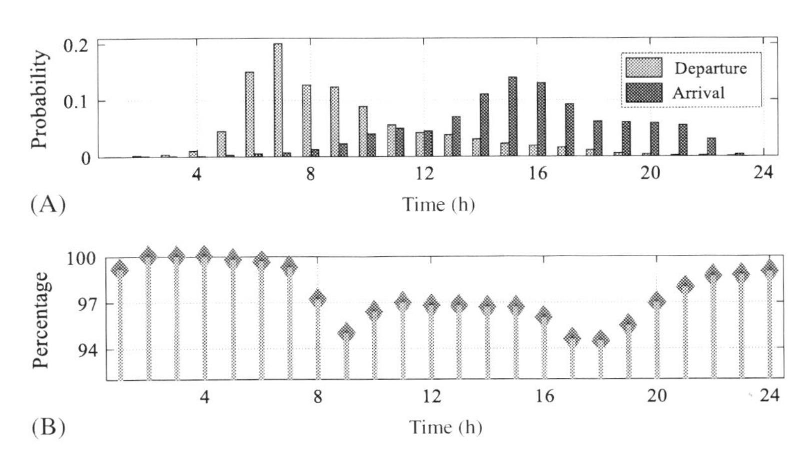

Fig. 21.14 Configuration of EV data for simulation studies. (A) Arrival and departure of EV over a day. (B) EV availability % over time.

predictive dispatch of flexible resources and send information to the DIgSILENT as a reference signal. Note that the adaptive control is performed in DIgSILENT only.

21.5 Results and discussion

This section presents results with multi-time-scale control algorithms considering EVs as flexible resources. In addition, different cases are presented for adaptive control considering EV only, EWH only, PV only, and EWH-PV.

21.5.1 Multi-time-scale energy management

21.5.1.1 Day-ahead predictive scheduling

The key idea of day-ahead predictive scheduling is to make optimum operational schedules for the flexible resources. It is worth mentioning that the operational schedules vary significantly depending upon consumer, DSO, and aggregator perspectives of scheduling. For instance, EV owners normally charge their EV during low price periods and discharge during high price periods irrespective of the network conditions. Therefore, day-ahead operational schedule from consumer perspective does not respect network limits. On the other hand, DSOs would have liked if the EVs were charged during low loading and discharged during high loading periods. Their key driver would be to avoid network congestion and improve its performance.

The day-ahead predictive scheduling of EVs looking from consumer perspective is illustrated in Fig. 21.15A. It can be observed that EVs are scheduled for charging at low price periods (2 am to 6 am) and for discharging during high price periods (8 pm to 10 pm). This leads to clear violations of feeder the capacity limit of 250 kVA. Moreover, the voltage at the farthest end node (as illustrated in Fig. 21.15C) is lower than the acceptable value (0.95 pu). More interestingly, the voltage and network limit violations are worse than the case with uncontrolled charging. Therefore, despite the consumer can significantly get economic benefit, such control schemes may not be technically feasible. Fig. 21.15C illustrates the EV charging profile while looking from DSO perspective. It can be observed that EVs are scheduled for charging during the lower demand period, which helps to improve the network utilization. DSO perspective respects both the voltage constraints as well as thermal constraints. Despite DSO perspective of scheduling improves the feeder utilization and respect network constraints, it allows EVs to be charged during higher price periods also. This leads to a higher EV charging cost compared with the EV charging case seen from consumer perspective. Such scenario demotivates the consumers to participate for network support only. Unlike consumer and DSO perspective, the aggregator perspective considers both charging costs and network limits. Fig. 21.16A illustrates the EV charging profiles for grid-to-vehicle (G2V) and vehicle-to-grid (V2G) cases. It is observed that the total feeder load is within the network limit of 250 kVA for both the cases. Further, it is observed from Fig. 21.16B that the voltage is also within

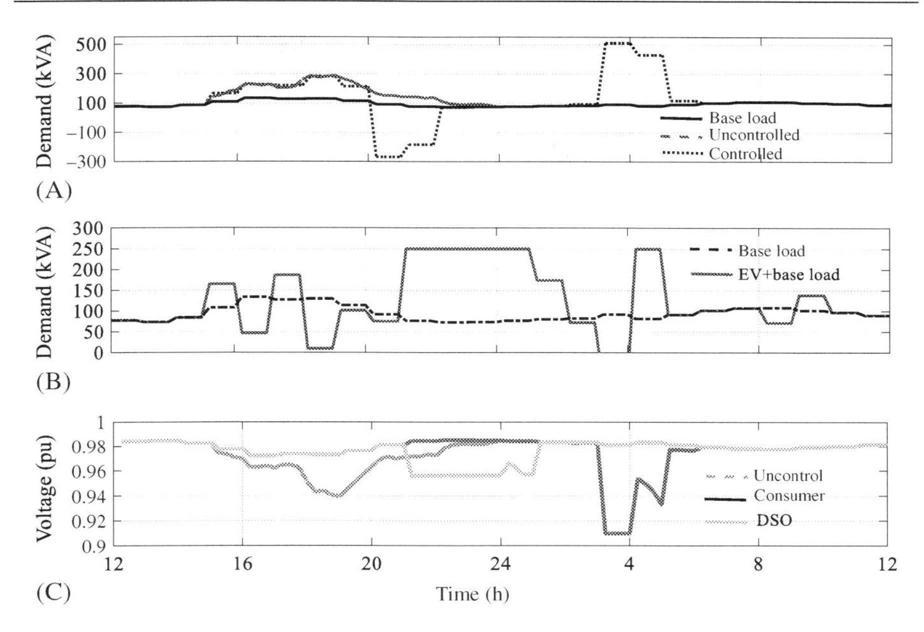

Fig. 21.15 EV charging/discharging from different actor perspective. (A) V2G/G2V from consumer perspective. (B) EVs charging from DSO perspective. (C) Voltage at farthest end node.

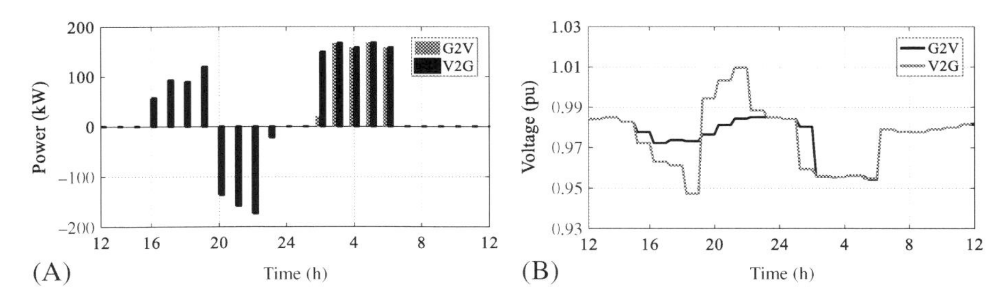

Fig. 21.16 EV charging/discharging from aggregator perspective. (A) EV charging/discharging profiles. (B) Voltage at farthest end node.

the limit throughout the charging horizon. This is done by rescheduling some of the EVs during the time periods when network limits are violated.

Even though technical and economical matrices can be improved while solving the problem from aggregator perspective, the monitory benefits resulting from the EV participation to only day-ahead market is often insufficient to motivate EV owners to provide flexibility for grid services. Therefore, an intraday compensating approach is presented, which allows the resources to provide the total cost by providing a framework to participate in multiple markets with simultaneous assurance of consumer comfort and network constraints. The HCA amplifies the economic advantage by allowing the EVs to participate in BRM which offers significantly higher prices for the demand flexibilities.

21.5.1.2 Intraday/hour optimal adjustment of predictive schedules

Unlike a day-ahead predictive scheduling perspective where EVs are scheduled based on their individual objectives, intraday/hour adjustment allows the flexible resources to participate in BRM. At this stage, HCA minimizes the total energy cost of the flexible resources by considering day-ahead, balancing, and regulating prices; network limits; and EV owner's requirements. Therefore, one of the notable attributes of the hierarchical multi-time-scale energy management is that it provides a framework for EVs to participate in BRM with the available up/down-regulation capability of the flexible resources. This not only maximizes utilization of availability flexibility, but also maximizes the economic benefits to the consumer. Fig. 21.17 depicts the regulating capacity of EVs during G2V and V2G scenarios. It is observed that up-regulation capacity is significantly smaller compared with the down-regulation capacity in the G2V case. This is due to the fact that the EV can provide upregulation during charging periods only. As the EVs stay idle for most of the time, it provides higher down-regulation capacity. However, during the V2G case, both up/down-regulations are significantly higher. This is due to increased regulation capacity resulting from V2G capability. One interesting observation from Fig. 21.17 is that the regulation capacity of EVs is greater than the feeder capacity limit especially during V2G case. Therefore, in reality the usable up/down-regulation powers are limited by the feeder capacity.

21.5.2 Adaptive control

In addition to the day-ahead predictive scheduling and intra-hour adjusting, there is always chance of discrepancies during actual operating conditions. For this reason, an adaptive real-time control is desired.

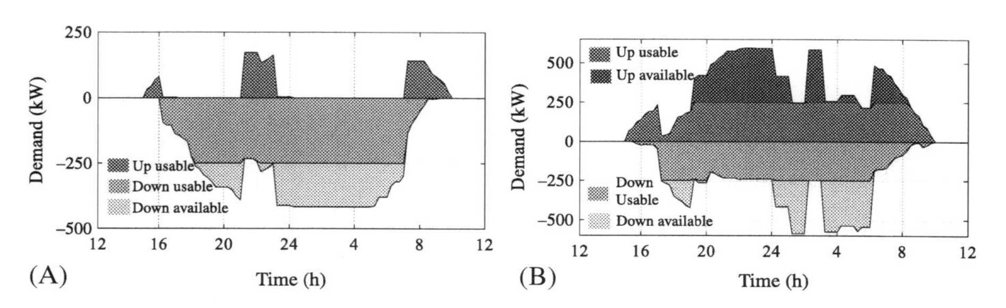

Fig. 21.17 Regulation capability from EV during (A) grid-to-vehicle (G2V) and (B) vehicle-to-grid (V2G) cases.

21.5.2.1 Adaptive control of EV

An adaptive control is implemented to adjust the charging/discharging power of the EVs when network limits are violated during actual operating conditions. Fig. 21.18 illustrates the voltage and power profile at the farthest end node in the network for a portion of a time period where voltage got violated beyond the predefined limits. It can be observed that total EV charging power profiles in the network after realization of the adaptive control is significantly better compared with the one before implementing the control. Particularly, the *P–V* droop controls the actual charging/discharging power of the EVs whose POC voltage is violated to bring the voltage back to the acceptable limits. As long as the voltage is between V_{th} and $V_{\min}$, the *P–V* droop decreases the EV power according to the observed POC voltage. Therefore, the droop locally improves voltage in real time which otherwise cannot be done by the SL and CL.

21.5.2.2 Adaptive control of EWH

In order to demonstrate the performance of the adaptive control with thermostatic loads (i.e., EWH), 24 h time-sweep simulation is performed with and without the adaptive control. Fig. 21.19 demonstrates adaptive control of EWH connected to the farthest end node (node 13) in the network. It can be seen that the adpative control adjusts the active and reactive power consumption of the EWH whenever the monitored voltage goes below V_{th} (Fig. 21.19C). First, reactive power is injected as shown in Fig. 21.19B for providing voltage support. Since the reactive power is not sufficient to alleviate the UV problem, the active power consumption is deceased according to the predefined *P–V* droop. Following the voltge violation below $V_{\min}$, the active power consumption of the EWH is decreased to zero. Followed by the $V_{\min}$ violation, SOE limits are decreased by 0.1 pu (Fig. 21.19A) so as to force the EWH to turnoff before reaching $SOE_{\max}$. Active and reactive power adjustment significantly improves the voltage compared with the case without control.

The aggregated power consumption of EWHs with and without adaptive control is illustrated in Fig. 21.20. When the adaptive control is implemented, the power consumed by EWHs is decreased significantly during the peak period to support the voltage. Both reactive power injection and active power reduction are realized to improve the voltage which was violated without control. This fact is reflected by decrement in total active power consumption and increment in reactive power injection as shown in Fig. 21.20B.

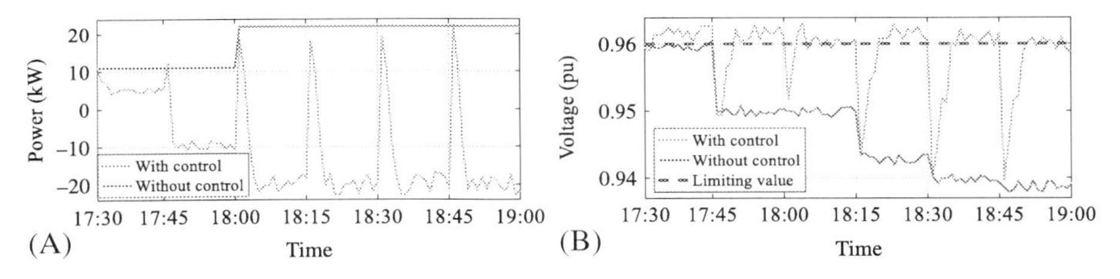

Fig. 21.18 Charging power and voltage from aggregator perspectives. (A) Flexible load at farthest node. (B) Voltage at farthest node.

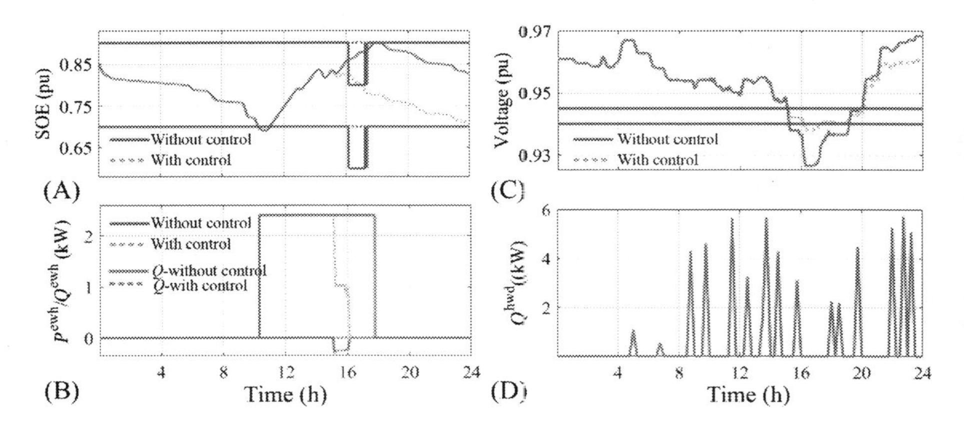

Fig. 21.19 Simulation results of EWH of a consumer connected at farthest end node (node 13). (A) SOE with and without control. (B) Power consumption. (C) Voltage at farthest node. (D) Thermal consumption.

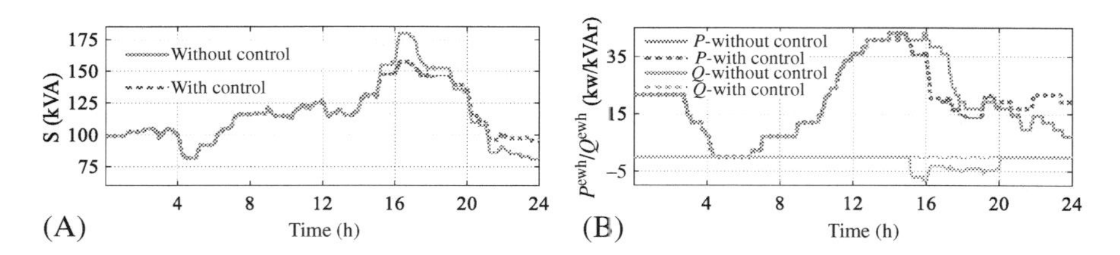

Fig. 21.20 Simulation results of EWH adaptive control for aggregated load. (A) Feeder apparent power. (B) Active and reactive power.

21.5.2.3 Adaptive control of PV

Adaptive control of PV is designed especially to address OV limit violations. This scenario is performed with minimum feeder loading so as to investigate the effectiveness of the proposed method during worst-case conditions. Fig. 21.21 depicts operations of the *P–V* and *Q–V* droops of the PV system connected to the farthest end node. From Fig. 21.21A and B, it can be observed that, whenever the POC voltage exceeds V_{TH}, *Q–V* adaptive control starts consuming reactive power. As long as the POC voltage is lower than V_m (i.e., midpoint between V_{max} and V_{TH}), the adaptive controller increase reactive power consumption. In this case, no active power curtailment occurs. However, when the voltage crosses V_m, the controller starts curtailing active power as well. As shown in Fig. 21.21, both active power curtailment as well as reactive power consumption are realized to keep the voltage within limit (i.e., below V_{max}). It can be seen that the proposed adaptive control significantly improves voltage compared with the case without the control.

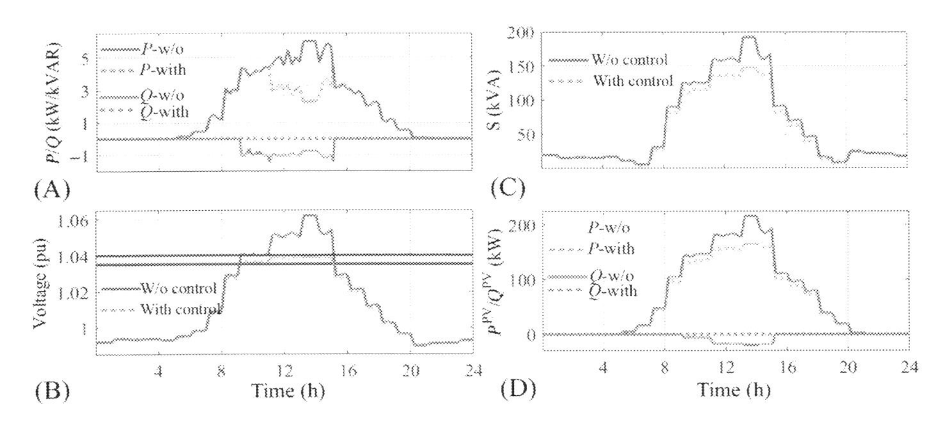

Fig. 21.21 Simulation results of PV for a single unit (A and B) and aggregated (C and D). (A) Adaptive control of PV at farthest node. (B) Voltage at farthest node. (C) Feeder loading. (D) PV active/reactive power.

Fig. 21.21C and D illustrates the aggregated adjustment of active and reactive power to support the network congestion. First, reactive power consumption is increased to provide voltage support. The active power is curtailed only when reactive power control is insufficient to alleviate the OV problem. It can be seen from Fig. 21.21 that total active and reactive power are significanlty regulated. To reduce the active power curtailment, an integrated control of solar PV and EWH is performed.

21.5.2.4 Adaptive control of EWH and PV

Since future distribution networks will have flexible loads as well as distributed generation, they are susceptible to both UV and OV violations. Fig. 21.22 illustrates a simulation with both EWH and PV providing real-time adaptive support. It can be

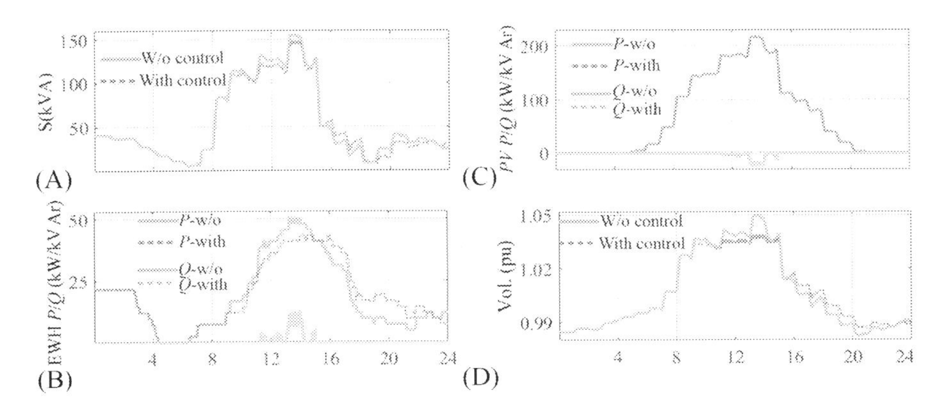

Fig. 21.22 Simulation results for EWH and PV with and without control. (A) Feeder power. (B) EWH power consumptions. (C) PV power consumption. (D) Voltage at farthest node.

seen that the active power consumption of the EWH is adjusted first to support OV violations. Then, EWH and PV reactive power consumptions are increased to support the voltage. However, as shown in Fig. 21.22C, there is no active power curtailment of the PV because the adjustment of active power consumption of EWH and reactive power consumptions of EWH and PV are sufficient to alleviate the OV problem. As such, the OV problem is effectively solved by the integrated control without incurring any revenue loss to the PV owner that could have resulted from active power curtailment of PVs.

21.6 Conclusion

This chapter provides a hierarchical framework for multi-time-scale energy management for active distribution networks. The HCA consisting of scheduling, coordinative, and ALs provides local network support as well as system-level balancing. The performance of the proposed multi-time-scale approach is demonstrated in a DIgSILENT-MATLAB co-simulation environment by using EV, EWH, and PV as distributed resources. First, multi-time-scale energy management predictively computes optimum operational schedules for the flexible resources, which are later adjusted based on intraday balancing requirements in intra-hour time resolution. Finally, an adaptive power management is performed to keep the network constraints within predefined limits. The simulation results demonstrate that the proposed method effectively deploys flexible resources in distribution networks and enable the single resources for multiple grid support functionalities.

References

Bhattarai, B., 2015. Intelligent control and operation of distribution system. PhD Thesis, Aalborg University.

Bhattarai, B., Bak-Jensen, B., Pillai, J.R., Gentle, J.P., Myers, K.S., 2015b. Overvoltage mitigation using coordinated control of demand response and grid-tied photovoltaics. IEEE. SusTech 1–6.

Bhattarai, B., Bak-Jensen, B., Pillai, J.R., Maier, M., 2014a. Demand flexibility from residential heat pump. In: IEEE PES General Meeting. pp. 1–5.

Bhattarai, B., Kouzelis, K., Mendaza, I., Bak-Jensen, B., Pillai, J., Myers, K.S., 2017b. Smart grid constraints violation management for balancing and regulating purposes. IEEE Trans. Ind. Inform. 3 (6), 2864–2875.

Bhattarai, B., Myers, K.S., Bak-Jensen, B., Mendaza, I.D., Turk, R.J., Gentle, J.P., 2017c. Optimum aggregation of geographically distributed flexible resources in strategic smart-grid/microgrid locations. Int. J. Electr. Power Energy Syst. 92, 193–201.

Bhattarai, B.P., Bak-Jensen, B., Mahat, P., Pillai, J.R., 2013. In: Voltage controlled dynamic demand response. Proceeding of IEEE PES Innovative Smart Grid Technologies – Europe. pp. 1–5.

Bhattarai, B.P., Bak-Jensen, B., Pillai, J.R., Mahat, P., 2014b. In: Two-stage electric vehicle charging coordination in low voltage distribution grids. Proceedings of IEEE PES Asia Pacific Power and Energy Engineering Conference. pp. 1–5.

Bhattarai, B.P., Levesque, M., Maier, M., Bak-Jensen, B., Pillai, J.R., 2015. Optimizing electric vehicle coordination over a hateregeneous mesh network in a scaled-down smart grid testbed. IEEE Trans. Smart Grid 6 (2), 784–794.

Bhattarai, B.P., Mendaza, I., Bak-Jensen, B., Pillai, J.R., 2016b. Local adaptive control of solar photovoltaics and electric water heaters for real-time grid support. CIGRE Paris Session.

Bhattarai, B.P., Mendaza, I., Myers, K., Baj-Jensen, B., Paudyal, S., 2017. Optimum aggregation and control of spatially distributed flexible resources in smart grid. IEEE Trans. Smart Grid. pp. 1–1. https://doi.org/10.1109/TSG.2017.2686873.

Bhattarai, B.P., Myers, K.S., Bak-Jensen, B., Paudyal, S., 2017d. Multi-time-scale control of demand flexibility in smart distribution networks. Energies 10, 1–18.

Bhattarai, B.P., et al., 2016a. Design and co-simulation of hierarchical architecture for demand response control and coordination. IEEE Trans. Ind. Inform. 13 (4), 1806–1816.

Clement-Nyns, K., Haesen, E., Driesen, J., 2010. The impact of charging plug-in hybrid electric vehicles on a residential distribution grid. IEEE Trans. Power Syst. 25, 371–380.

Demirok, E., 2012. Control of grid interactive PV inverters for high penetration in low voltage distribution networks. Ph.D. thesis, August.

Demirok, E., Gonzælez, P.C., Frederiksen, K.H.B., Sera, D., Rodriguez, P., Teodorescu, R., 2011. Local reactive power control methods for overvoltage prevention of distributed solar inverters in low-voltage grids. IEEE J. Photovoltaics 1, 174–182.

Energinet.dk, 2010. Strategy plan 2010. Technical report.

Gan, L., Topcu, U., Low, H., 2013. Optimal decentralized protocol for electric vehicle charging. IEEE Trans. Power Syst. 28, 940–951.

Gelazanskas, L., Gamage, K.A., 2016. Distributed energy storage using residential hot water heaters. Energies 9, 127.

Gerards, M.E., Hurink, J.L., 2016. Robust peak-shaving for a neighbourhood with electric vehicles. Energies 9, 594.

Han, S., Han, S., 2013. Economic feasibility of V2G frequency regulation in consideration of battery wear. Energies 6, 748–765.

Kondoh, J., 2013. Experiment of an electric water heater with autonomous frequency regulation. IEEJ Trans. Electr. Electron. Eng. 8, 223–228.

Lund, H., et al., 2014. 4th Generation district heating (4GDH) Integrating smart thermal grids into future sustainable energy systems. Energy 68, 1–11.

Mancarella, P., 2014. MES (multi-energy systems): an overview of concepts and evaluation models. Energy 65, 1–17.

Masuta, T., Yokoyama, A., 2012. Supplementary load frequency control by use of a number of both electric vehicles and heat pump water heaters. IEEE Trans. Smart Grid 3, 1253–1262.

Mohanpurkar, M., Suryanarayan, S., 2013. Accommodating unscheduled flows in electric grids using the analytical ridge regression. IEEE Trans. Power Syst. 28 (3), 3507–3508.

Sundstrom, O., Binding, C., 2013. Flexible charging optimization for electric vehicles considering distribution grid constraints. IEEE Trans. Smart Grid 3, 26–37.

The Danish Ministry of Climate, 2013. Energy, and Building "Smart Grid Strategy: The Intelligent Energy System of the Future". Technical report, May.

Tonkoski, R., Lopes, L.A.C., El-Fouly, T.H.M., 2011. Coordinated active power curtailment of grid connected PV inverters for overvoltage prevention. IEEE Trans. Sustain. Energy 2, 139–147.

U.S. Department of Energy, 2010. Smart Grid System Report. Technical Report.

Wen, C., Chen, J., Teng, J., Ting, P., 2012. Decentralized plug-in electric vehicle charging selection algorithm in power systems. IEEE Trans. Smart Grid 3, 1779–1789.

Xiu, X., Li, B., 2012. Study on energy storage system investment decision based on real option theory. Proceedings of IET Sustainable Power Generation and Supply, September. pp. 1–4.
Zhou, L., Li, Y., Wang, B., Wang, Z., Hu, X., 2015. Provision of supplementary load frequency control via aggregation of air conditioning loads. Energies 8, 14098–14117.

Further reading

Foster, J.M., Trevino, G., Kuss, M., Caramanis, M.C., 2013. Plug-in electric vehicle and voltage support for distributed solar: theory and application. IEEE Syst. J. 7 (4), 881–888.

Distribution Network planning considering the impact of Electric Vehicle charging station load

Sanchari Deb, Karuna Kalita†, Pinakeswar Mahanta†*
*Centre for Energy, Indian Institute of Technology, Guwahati, India, †Department of Mechanical Engineering, Indian Institute of Technology, Guwahati, India

22.1 Introduction

Driven by the objective of ameliorating environmental pollution, global warming accompanied by the latest expansion and progress of battery technology both developed as well as developing nations are accentuating on the electrification of transportation sector. The well to wheel studies has already established the superiority of Electric Vehicles (EVs) over conventional vehicles in terms of energy efficiency (Rousseau and Sharer, 2004). Thus, it can be speculated that in near future EVs will dominate the automobile industry. However, for large-scale deployment of EVs accessibility of charging infrastructure plays a significant role. Unplanned installation of EV charging stations will degrade the voltage stability, reliability, and other operating parameters of the distribution network. Thus, the shift of primary source of transportation energy from pipelines to power grid must be conducted in a strategic way. For proper planning of distribution network in the presence of EV charging station, loads it is essential to investigate the impact of EV charging station loads on different operating parameters of the distribution network. In this chapter, the impact of EV charging station loads on IEEE 69-bus distribution network is analyzed and an attempt has been made to optimally place the charging stations in the distribution network without disturbing the operating parameters of the network.

Literature (Dharmakeerthi et al., 2011; Zhang et al., 2016; Dharmakeerthi et al., 2014; March, Pazouki et al., 2014; Niitsoo et al., 2015; Dubey and Santoso, 2015; Qian et al., 2015) reports the impact of EV charging load on power grid. Dharmakeerthi et al. (2011) presented an overview of the impact of EV on power grid. Zhang et al. (2016) modeled the EV charger load as constant power load and investigated the impact of EV load on a two bus system with constant impedance, constant current, and constant power (ZIP) load. After rigorous analysis it is concluded by Zhang et al. (2016) that constant power loads are responsible for decrease in bus voltage. Dharmakeerthi et al. (2014) analyzed the impact of fast EV charging load on voltage stability of IEEE 43-bus typical industrial test system. They presented a

Smart Power Distribution Systems. https://doi.org/10.1016/B978-0-12-812154-2.00022-5

unique way of representation of EV charger load considering the impact of charger resistance on load modeling. Furthermore, they concluded that the bus voltages decrease with the introduction of EV charger load. Pea-da and Dechanupaprittha (2014) analyzed the impact of fast EV charging station load on radial distribution network of Thailand. The results of their analysis established the fact that the voltage of the node nearest to the substation is least affected after introduction of EV charging station load. Furthermore, the improvement in voltage profile after installing capacitor is also elaborated by Pea-da and Dechanupaprittha (2014). Pazouki et al. (2014) analyzed the impact of EV charging station load on reliability of IEEE 33-bus distribution network. The degradation of average sustained interruption duration index (ASIDI) and energy not served (ENS) index is analyzed in their work. Niitsoo et al. (2015) examined the impact of EV charging load on total harmonic distortion (THD) on a residential distribution network and made a remarkable conclusion that the network can withstand the EV penetration of at most 25%. Dubey and Santoso (2015) analyzed the impact of EV charging load on a real-world 13.8 kV radial distribution network and suggested that the negative impact of EV charging load can be mitigated to certain extent by coordinated charging in the charging station. Qian et al. (2015) analyzed the impact of EV charging load on thermal ageing of distribution transformers. The major contribution of Qian et al. (2015) is that they presented a methodology for determination of optimal charging time for which the deterioration of transformer is least. Pazouki et al. (2015) presented a scheme for optimal placement of EV charging station considering financial as well as technical terms. Sachan and Kishor (2016) presented a novel methodology for placement of EV charging station in the distribution network considering voltage sensitivity indices.

Although literature (Dharmakeerthi et al., 2011; Zhang et al., 2016; Dharmakeerthi et al., 2014; March), 2014; Pazouki et al., 2014; Niitsoo et al., 2015; Dubey and Santoso, 2015; Qian et al., 2015) reports various impacts of EV charging load on distribution network, however there are many factors not addressed in these works. The impact of EV charger load on the voltage stability indices formulated based on transferred active and reactive power equations of the branches which are crucial parameters of the power system are not analyzed in the aforementioned works. Also the impact of EV charging load on composite reliability indices developed taking into account frequency and duration of interruption which are regarded as the parameter of the network for determining customer satisfaction is not analyzed in all these works. Moreover, these works report the impact of EV charging load on much smaller networks. There are relatively less work which reports the impact of EV charging load on IEEE 69-bus network which is indeed a larger complex network with too many branches and nodes. All the aforementioned shortcomings of the existing literatures will be taken into account in this chapter and the objectives of this chapter are summarized as follows:

- to analyze the impact of EV charging load on a novel voltage stability index developed based on transferred active and reactive power equation of distribution line,
- to analyze the impact of EV charging load on Total Harmonic Distortion (THD) of the network,
- to analyze the impact of EV charging load on a new composite reliability index that takes into account both frequency and duration of interruption of the distribution network,

- to present a comparative analysis of impact of EV charging station load on all the aforementioned parameters in a quantitative way, and
- to present a methodology for optimal placement of EV charging stations in the distribution network

22.2 Different operating parameters of distribution network

For smooth operation of the distribution network its operating parameters must be within limit and should not deviate too much from the base case values even after being subjected to any sort of disturbance. Some of the operating parameters of the distribution network are voltage stability, reliability, THD, etc. A brief description of these parameters of the network is presented in this section.

22.2.1 *Voltage stability*

The increased load of the distribution network and other disturbances poses threat to voltage stability of the network. Voltage stability is defined as ability of power system to sustain fixed tolerable voltage at every single bus of the network under standard operating conditions as well as after being subjected to a disruption (Kundur, 1994). A heavily loaded system is very much prone to voltage collapse. Since the last two decades researchers have devised several methodologies like continuation power flow-based methods (Zhang, 2006), singularity of Jacobian-based methods (Ellithy et al., 2000), PV, QV-based methods (Sulaiman et al., 2015) to analyze voltage stability of power system. In this chapter, the voltage stability index developed by Eminoglu and Hocaoglu (2007) is utilized for determining the critical as well as strong nodes of the test network.

Fig. 22.1 illustrates the single-line diagram of a two-bus system where a and b are the two buses of the system. $V_a < \delta_a$ and $V_b < \delta_b$ are the voltage at bus a and bus b, respectively. I is the current flowing through the branch having resistance r and impedance x.

$$I = \frac{V_a - V_b}{r + jx} \tag{22.1}$$

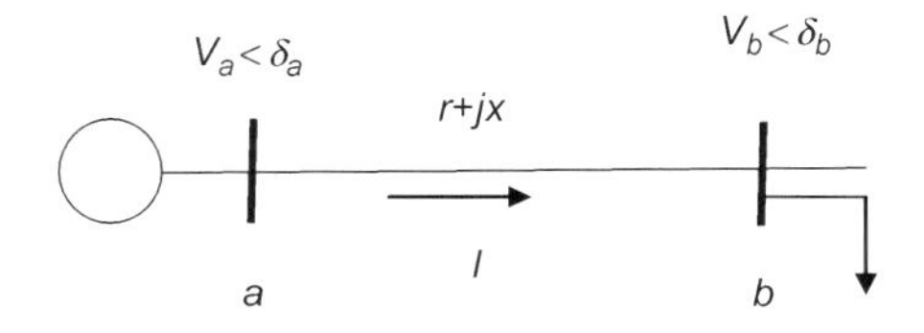

Fig. 22.1 Single-line diagram of a simple two-bus system.

and

$$P_b - jQ_b = V_b^* I \tag{22.2}$$

where P_b and Q_b are the active and reactive powers of bus b.

Substituting value of I in Eq. (22.2), equating real and imaginary parts and on further simplification, the following equation is obtained.

$$V_b^4 + 2V_b^2(P_b r + Q_b x) - V_a^2 V_b^2 + \left(P_b^2 + Q_b^2\right)|Z|^2 = 0 \tag{22.3}$$

From Eq. (22.3) the transferrable active and reactive power can be written as

$$P_b = \frac{A \pm \sqrt{B}}{|Z|} \tag{22.4}$$

where

$$A = -\cos\theta_z V_b^2$$

and

$$B = \cos^2\theta_z V_b^4 - V_b^4 - |Z|^2 Q_b^2 - 2V_b^2 Q_b x + V_a^2 V_b^2$$

$$Q_b = \frac{C \pm \sqrt{D}}{|Z|} \tag{22.5}$$

where

$$C = -\sin\theta_z V_b^2$$

and

$$D = \sin^2\theta_z V_b^4 - V_b^4 - |Z|^2 P_b^2 - 2V_b^2 P_b r + V_a^2 V_b^2$$

Thus, conditions of existence of transferrable active and reactive power are as

$$B \geq 0 \text{ and } D \geq 0 \tag{22.6}$$

Substituting the actual values of B, D and then after summing them, the following equation is obtained which gives the criterion of stability of the system.

$$2V_a^2 V_b^2 - 2V_b^2(P_b r + Q_b x) - |Z|^2\left(P_b^2 + Q_b^2\right) \geq 0 \tag{22.7}$$

The value of Eq. (22.7) which is considered as a criterion for determination of voltage stability will decrease with increase of active power. Increasing the active power beyond a certain limit will cause the system to become unstable. Thus for determination of voltage stability of a distribution network the voltage of all the nodes of the network needs to the calculated. For computation of bus voltages of radial distribution networks the typical methods of load flow analysis may not converge as there is possibility of the Jacobian matrix becoming singular owing to the fact that r/x ratio of these networks is relatively high. In this chapter forward backward sweep algorithm (Rupa and Ganesh, 2014; Eminoglu and Hocaoglu, 2008) will be used for solving the load flow problem.

22.2.2 Reliability

For satisfactory operation of power system its reliability is of utmost importance. In most of the works focusing on reliability analysis of distribution network either contingency load loss index (CLLI) (Kumar et al., 2015; Ganguly et al., 2010) or typical customer-based energy indices (Schneider et al., 1989; Chowdhury and Koval, 2011) like system average interruption frequency index (SAIFI), system average interruption duration index (SAIDI) are used. One of the inherent shortcomings of CLLI-based indices is their incapability of determining how events like change in load or increase in number of consumers affect reliability of the network. On the other hand SAIFI and SAIDI give the average frequency and duration of interruption per consumer, respectively. There are relatively fewer works which use a composite index of reliability (CIR) having amalgamation of both frequency and duration of interruption per consumer. In this chapter, a new reliability index having conglomeration of both frequency and duration of interruption per consumer is formulated. Mathematically, the CIR is as follows:

$$CIR = w_f I_f + w_d I_d \tag{22.8}$$

where w_f and w_d are the weights assigned to frequency index and duration index, respectively. I_f and I_d are the frequency and duration index of reliability, respectively, which are further elaborated as

$$I_f = \frac{\left(\frac{\Sigma\lambda_i N_i}{\Sigma N_i}\right) - \left(\frac{\Sigma\lambda_i N_i}{\Sigma N_i}\right)_{\text{base}}}{\left(\frac{\Sigma\lambda_i N_i}{\Sigma N_i}\right)_{\text{base}}} \tag{22.9}$$

$$I_d = \frac{\left(\frac{\Sigma u_i N_i}{\Sigma N_i}\right) - \left(\frac{\Sigma u_i N_i}{\Sigma N_i}\right)_{\text{base}}}{\left(\frac{\Sigma u_i N_i}{\Sigma N_i}\right)_{\text{base}}} \tag{22.10}$$

where λ_i denotes the failure rate of ith load point, N_i denotes the number of consumer connected at ith load point, and u_i denotes the interruption duration of ith load point.

22.2.3 Harmonics

In recent years, the problem of harmonic distortion has become more prominent due to presence of highly nonlinear electronic loads in the network. Power electronic loads like rectifiers, inverters, switched-mode power supplies are some of the most vital sources of harmonics (Pourarab et al., 2011). As the name indicates THD is the index which directly gives the extent of distortion of the waveform. The ratio of harmonic content to fundamental quantity of a certain waveform is termed as THD (Francisco, 2006).

Mathematically, THD is given by

$$\mathrm{THD} = \frac{\sqrt{\sum_{h=2}^{h_{\max}} N_h^2}}{N_1} \tag{22.11}$$

where N_h is the magnitude of hth harmonic component, N_1 is the magnitude of fundamental component of the waveform, and the THD in voltage is as

$$\mathrm{THD}_v = \frac{\sqrt{\sum_{h=2}^{h_{\max}} V_h^2}}{V_1} \tag{22.12}$$

where V_h is the hth harmonic component of voltage and V_1 is the fundamental component of voltage.

The harmonic component of voltage of a distribution network can be computed by harmonic load flow (Arunagiri and Venkatesh, 2011; Kumar and Reddy, 2013).

22.3 Impact of EV charging load on different operating parameters of distribution network

The nonlinear nature of the EV charging load affects all the operating parameters of the distribution network mentioned in Section 22.2. This section exemplifies the methodology of evaluating the impact of EV charging load on different operating parameters of the network.

22.3.1 Impact of EV charging station load on voltage stability index

For evaluating the impact of EV charging load on distribution network is modeling the EV charger load which is highly nonlinear in nature. In this chapter, the EV load model proposed by Dharmakeerthi et al. (2014) is used.

Mathematically, EV charger load is expressed as follows (Dharmakeerthi et al., 2014; Rahman et al., 2013):

$$\frac{P}{P_0} = 0.9061 + 0.0939\left(\frac{V}{V_0}\right)^{-3.715} \tag{22.13}$$

where P is the net input power, P_0 is the power consumption at reference voltage, V_0 is the reference voltage, and V is the input voltage

For elaborating the impact of EV charging load on distribution network a simple 10-bus radial distribution network is considered as shown in Fig. 22.2.

Fig. 22.2 represents a radial distribution network where bus 1 is the substation. The network has a single power source and all the feeders are coming out directly from the substation as shown in the figure. The bus and line data of this radial distribution network are as given in Tables 22.1 and 22.2, respectively.

Based on the methodology illustrated in Section 22.2 the voltage stability indices of the network are calculated for the base case without placement of any charging station. The bus having the highest value of voltage stability index is least prone to voltage collapse. So charging station is placed at that particular bus. After placement of charging station at the strongest bus the voltage stability index is again recalculated with the

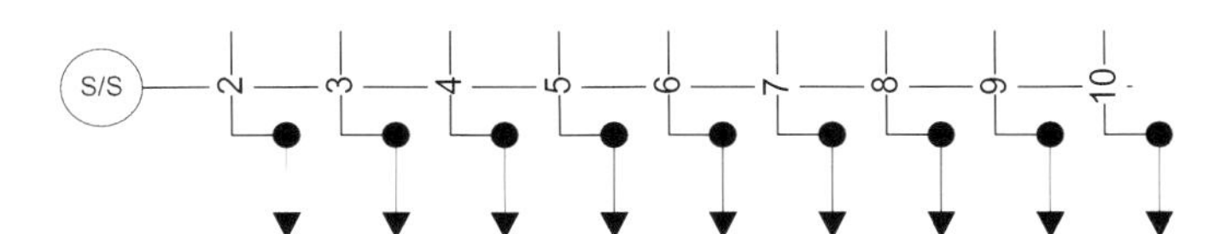

Fig. 22.2 10-bus radial distribution network.

Table 22.1 Bus data of the 10-bus distribution network (Babu and Samala, 2015)

Bus no	*P* (pu)	*Q* (pu)
1	0	0
2	0.0184	0.0046
3	0.0098	0.0034
4	0.0179	0.0045
5	0.0160	0.0184
6	0.0161	0.0060
7	0.0078	0.0011
8	0.0115	0.0006
9	0.0098	0.0013
10	0.0164	0.0002

Table 22.2 Line data of the 10-bus distribution network (Babu and Samala, 2015)

Branch no	R (pu)	X (pu)
1	0.1233	0.4127
2	0.0140	0.6050
3	0.7463	1.2050
4	0.6980	0.6084
5	1.9831	1.7276
6	0.9053	0.7886
7	2.0552	1.1640
8	4.7953	2.7160
9	5.3434	3.0264

Table 22.3 Voltage stability index before and after placement of charging station

Bus no	VSI before placement of charging station	VSI after placement of charging station at bus 2
2	0.9804	0.9801
3	0.9834	0.9487
4	0.9698	0.9353
5	0.9246	0.8909
6	0.8924	0.8593
7	0.8461	0.8139
8	0.8265	0.7947
9	0.7921	0.7610
10	0.7397	0.7097

new values of voltage of the buses obtained from the load flow solution. The change in voltage stability index is as

$$\%\text{change in voltage stability index} = \left| \frac{VSI_{\text{base}} - VSI}{VSI_{\text{base}}} \right| \tag{22.14}$$

where VSI represents voltage stability index after placement of charging station.

VSI_{base} represents the voltage stability index for base case.

Table 22.3 reports the voltage stability indices of all the buses before and after placement of charging station. As reported in Table 22.3 the voltage stability index of bus 2 is highest. As a consequence of which it is designated as the strongest bus of the network and is a candidate place for location of EV charging station. After placement of charging stations of 150 kW at bus 2 it is observed that the voltage stability indices of the buses have diminished thereby revealing the tendency of the network to become unstable. For a larger network with more number of charging stations the tendency of becoming unstable will be more prominent.

22.3.2 Impact of EV charging station load on reliability

Due to introduction of additional EV charging load, the failure rates as well as duration of disruption increase. The change in failure rate as well as duration of interruption due to increase in load follows a linear relationship as (Bhadra and Chattopadhyay, 2015)

$$\lambda_{new} = \frac{\lambda_{base}{}^{*}}{L_{base}}(L_{base} + \Delta L) \tag{22.15}$$

$$u_{new} = \frac{u_{base}{}^{*}}{L_{base}}(L_{base} + \Delta L) \tag{22.16}$$

where λ_{base} and u_{base} are the original values of failure rate and interruption duration for the load point before increase in load.

L_{base} is the load at that load point

ΔL is the increase in load at that particular load point.

For further exemplification of the methodology used for analyzing the impact of EV charging station on distribution network the example of 10-bus radial distribution network mentioned in the previous subsection is considered.

Table 22.4 gives the failure rate and duration of interruption data for the 10-bus radial network. The failure rate and interruption duration after placement of charging station are evaluated by Eqs. (22.15), (22.16), respectively. After computation of failure rate as well as duration of disruption for increase in load equivalent to placement of fast charging stations of 150 kW at bus 2 CIR is evaluated by the methodology presented in Section 22.2 which comes out to be 1.10.

22.3.3 Impact of EV charging station load on harmonic distortion

EV charger being a nonlinear load is a source of harmonics. EV charger is modeled as a constant source of current injection in harmonic power flow analysis (Aljanad and Mohamed, 2016). The line current harmonic content of a typical Nissan Leaf charger is as given in Table 22.5. For computation of THD at first the conventional power flow needs to converge to obtain the fundamental component of voltage and current. Then harmonic power flow is performed to get the harmonic components. For further elaboration the radial distribution network mentioned in the previous section is considered. Fast charging station having Nisan Leaf charger is placed at bus 2.

Table 22.4 Reliability data for 10-bus distribution network

Load point	Failure rate (failure/yr) for base case	Failure rate (failure/yr) after charging station placement	Interruption duration (h/yr) for base case	Interruption duration (h/yr) after charging station placement	Number of consumers
2	0.0500	0.2130	0.3000	1.2783	50
3	0.0400	0.0400	0.3000	0.3000	50
4	0.0600	0.0600	0.3000	0.3000	10
5	0.0300	0.0300	0.2000	0.2000	5
6	0.0300	0.0300	0.2000	0.2000	10
7	0.0900	0.0900	0.6000	0.6000	5
8	0.0300	0.0300	0.6000	0.6000	5
9	0.0300	0.0300	0.2000	0.2000	3
10	0.0200	0.0200	0.2000	0.2000	2

Table 22.5 Line current harmonic of a typical Nissan Leaf charger

Harmonic order	Magnitude (%)	Angle
3	100	−26
5	25	−94
7	17	−96
9	14.20	−72
11	9.69	−68
13	5.04	−49
15	1.80	−49
17	0.37	−46

Table 22.6 reports the fundamental as well as different harmonic components of voltage of the buses of the network after placement of charging station at bus 2. The dominance of seventh harmonics is very much prominent from the findings reported in Table 22.6. THD is computed by the equations elaborated in Section 22.3. Since the network and change of load is small the severity of THD is not so prominent in the network. Later on while dealing with larger and complex networks along with the presence of different variety of loads the severity of harmonics will be quite significant.

Table 22.6 Fundamental and harmonic component of voltage after placement of charging stations

Bus no	$V_{fundamental}$	V_3	V_5	V_7	V_9
2	0.9966	$0.1800*10^{-6}$	$0.1725*10^{-6}$	$0.2286*10^{-6}$	$0.1844*10^{-6}$
3	0.9941	$0.1800*10^{-6}$	$0.1725*10^{-6}$	$0.2286*10^{-6}$	$0.1844*10^{-6}$
4	0.9824	$0.1800*10^{-6}$	$0.1725*10^{-6}$	$0.2286*10^{-6}$	$0.1844*10^{-6}$
5	0.9749	$0.1800*10^{-6}$	$0.1725*10^{-6}$	$0.2286*10^{-6}$	$0.1844*10^{-6}$
6	0.9602	$0.1800*10^{-6}$	$0.1725*10^{-6}$	$0.2286*10^{-6}$	$0.1844*10^{-6}$
7	0.9555	$0.1800*10^{-6}$	$0.1725*10^{-6}$	$0.2286*10^{-6}$	$0.1844*10^{-6}$
8	0.9469	$0.1800*10^{-6}$	$0.1725*10^{-6}$	$0.2286*10^{-6}$	$0.1844*10^{-6}$
9	0.9329	$0.1800*10^{-6}$	$0.1725*10^{-6}$	$0.2286*10^{-6}$	$0.1844*10^{-6}$
10	0.9234	$0.1800*10^{-6}$	$0.1725*10^{-6}$	$0.2286*10^{-6}$	$0.1844*10^{-6}$

22.4 Optimal placement of charging stations in distribution network

Charging stations must be placed in the distribution network in such a way that its operating parameters are least affected. Thus the charging stations must be placed in such way that the reliability, voltage stability, THD, and other operating parameters are least affected. The different approaches of formulating the problem for optimal placement of charging stations in the distribution network are discussed in this section.

22.4.1 Problem formulation approach one

Voltage stability being one of the crucial parameters of the distribution network must be given utmost importance. This approach of problem formulation aims at optimally locating the charging stations in the distribution network with objective function being maximization of the voltage stability index elaborated in Section 22.1. The decision variables of this approach of dealing with this problem are buses in which charging stations will be placed and number of fast as well as slow charging stations placed at the buses. The detailed problem formulation is given as follows.

$$\text{Max} \sum_{i=2}^{N} \text{VSI}_i \tag{22.17}$$

Subject to the following constraints,

$$\left.\begin{array}{c} 0 \leq n_i \leq n_{\text{fastCS}} \\ 0 \leq n_i \leq n_{\text{slowCS}} \\ S_{\min} \leq S_i \leq S_{\max} \\ L \leq L_{\max} \end{array}\right\} \tag{22.18}$$

Here, N denotes the total number of buses of the distribution network, i is the index which denotes the bus, VSI denotes voltage stability index, n_i denotes number of charging stations located in ith bus, n_{fastCS} is the maximum number of charging stations that can be placed at each bus, n_{slowCS} is the maximum number of charging stations that can be placed at each bus. S denotes the reactive power, $S_{\min}$ and $S_{\max}$ denote the upper and lower bound of reactive power, respectively. L denotes the increase in load of the system after placement of charging station and $L_{\max}$ denotes the loading margin of the network.

22.4.2 Problem formulation approach two

The second approach of problem formulation is concerned with reliability of the distribution network. The optimization is aimed at minimization of the CIR. The CIR can be expressed as a function of failure rate, duration of interruption which is again dependent on change of load of the bus as elaborated by Eqs. (22.15),

(22.16). Thus the CIR is a function of bus in which charging station is placed and the number of fast as well as slow charging stations placed at that bus. The detailed problem formulation is as

Objective function is:

$$\text{Min CIR where } CIR = f(i, n_{\text{fast}}, n_{\text{slow}}) \tag{22.19}$$

Subject to

$$\left.\begin{array}{c} 0 \le n_i \le n_{\text{fastCS}} \\ 0 \le n_i \le n_{\text{slowCS}} \\ S_{\min} \le S_i \le S_{\max} \\ L \le L_{\max} \end{array}\right\} \tag{22.20}$$

22.4.3 Problem formulation approach three

The third approach of problem formulation is concerned with minimization of the overall cost. Overall cost includes installation cost, operational cost, penalty for voltage deviation, THD, as well as penalty for ENS. The detailed problem formulation is elaborated by the following equations.

Objective function is

$$\text{Min } \left(C_{\text{installation}} + C_{\text{operation}} + C_{\text{penalty}}\right) \tag{22.21}$$

where $C_{\text{installation}}$ is the installation cost of EV charging stations, $C_{\text{operation}}$ is the operating cost of charging stations, and C_{penalty} is the penalty for voltage deviation, THD, and ENS.

$$C_{\text{operation}} = \sum_{i=1}^{N_{CS}} CP_i{}^{*}8760^{*}\text{per unit cost of electricity} \tag{22.22}$$

N_{CS} is the total number of charging station and CP_i is the capacity of the charging station.

$$C_{\text{penalty}} = \text{Voltage deviation}_{\text{penalty}} + \text{AENS}_{\text{penalty}} + \text{THD}_{\text{penalty}} \tag{22.23}$$

$$\text{Voltage deviation}_{\text{penalty}} = P_{\text{VD}}{}^{*} \sum_{i=2}^{N} \text{Voltage Deviation}_i \tag{22.24}$$

$$\text{Voltage deviation}_i = V_i^{\text{base}} - V_i \tag{22.25}$$

$$\text{AENS}_{\text{penalty}} = P_{\text{AENS}}{}^{*}\text{AENS} \tag{22.26}$$

$$\text{AENS} = \frac{\Sigma L_i U_i}{\Sigma N_i} \tag{22.27}$$

$$\text{THD}_{\text{penalty}} = P_{\text{THD}}{}^{*} \sum_{i=2}^{N} \text{THD}_i \tag{22.28}$$

where Voltage deviation$_{\text{penalty}}$ is the penalty for voltage deviation. If after increase of load the voltage of buses drops to less than 0.9 pu then the utility has to pay the penalty for voltage deviation. P_{VD} is the penalty paid for per unit voltage deviation. V_i^{base} is the base voltage of ith bus. V_i is the bus voltage of ith bus. P_{AENS} is the penalty for per unit ENS. L_i is the load at ith bus. U_i is the duration of interruption at the ith bus or load point. N_i is the number of consumer at the ith load point. P_{THD} is the per unit penalty for THD.

Subject to

$$\left.\begin{array}{c} 0 \le n_i \le n_{\text{fastCS}} \\ 0 \le n_i \le n_{\text{slowCS}} \\ S_{\min} \le S_i \le S_{\max} \\ L \le L_{\max} \end{array}\right\} \tag{22.29}$$

22.4.4 Optimization algorithm

For solution of the optimization problem of the previous section, genetic algorithm (GA) is utilized. Despite the emergence of new optimization algorithms like particle swarm optimization (PSO), ant colony optimization (ACO) with excellent convergence criterion, GA continues to be popular among the researchers because of its computational simplicity and guarantee of convergence. The flowchart of GA applied to solve the optimization problem is as shown in Fig. 22.3. For reducing the computational complexity and time, first the strong nodes of the distribution network are evaluated based on VSI. Then the candidate locations of the charging stations are found out by GA among the already designated strong nodes.

22.5 Case study

In this section, the impact of EV charging station load on different operating parameters of IEEE 69-bus network is analyzed for different scenarios. IEEE 69-bus network is a radial distribution network with 69 nodes and 70 branches as shown in Fig. 22.4.

The bus, line, and reliability data of IEEE 69-bus network are taken from Attari et al. (2016).

Based on the methodology of computing VSI elaborated in Section 22.3.1, the VSI of the distribution network is computed for base case as well as for the cases reported in Table 22.8. Table 22.7 reports that the VSI of bus 36 for base case is 0.9968 which is the maximum compared to all the other buses. As a result of which it is specified as the

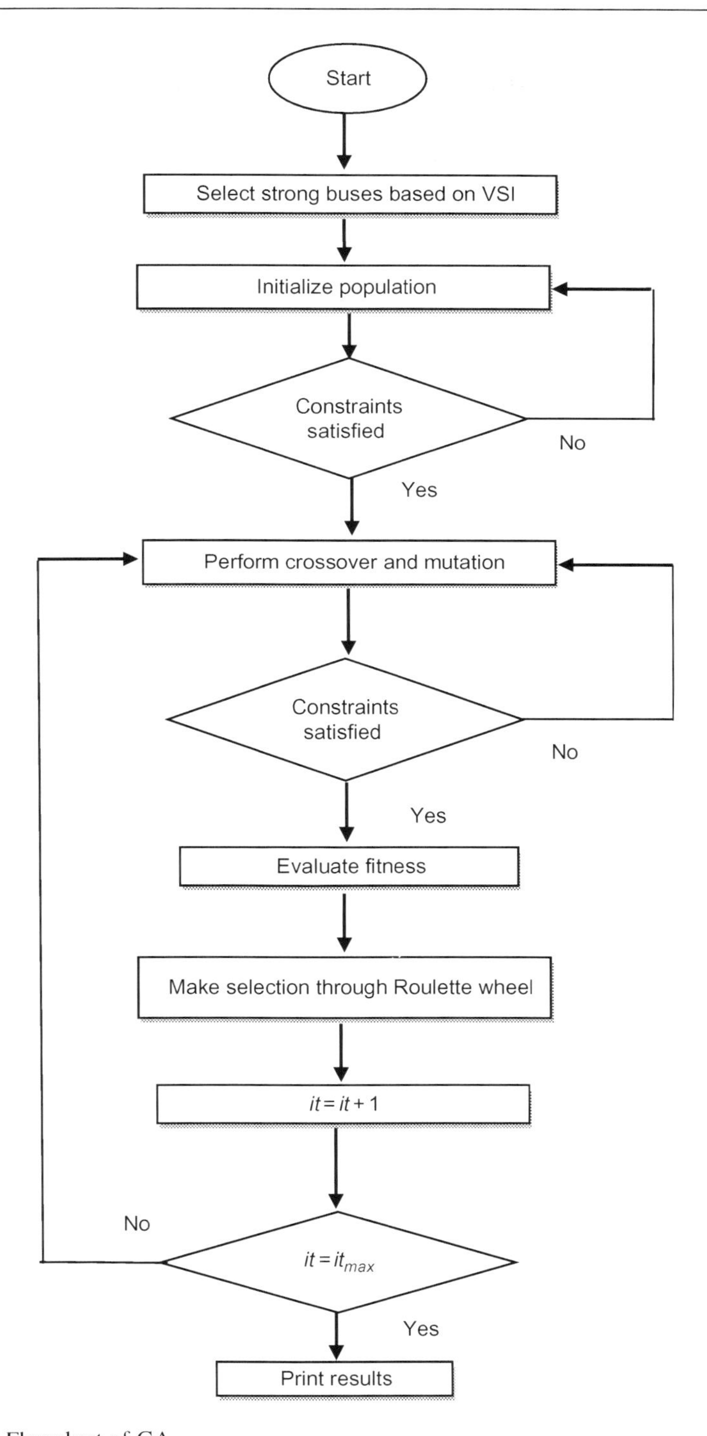

Fig. 22.3 Flowchart of GA.

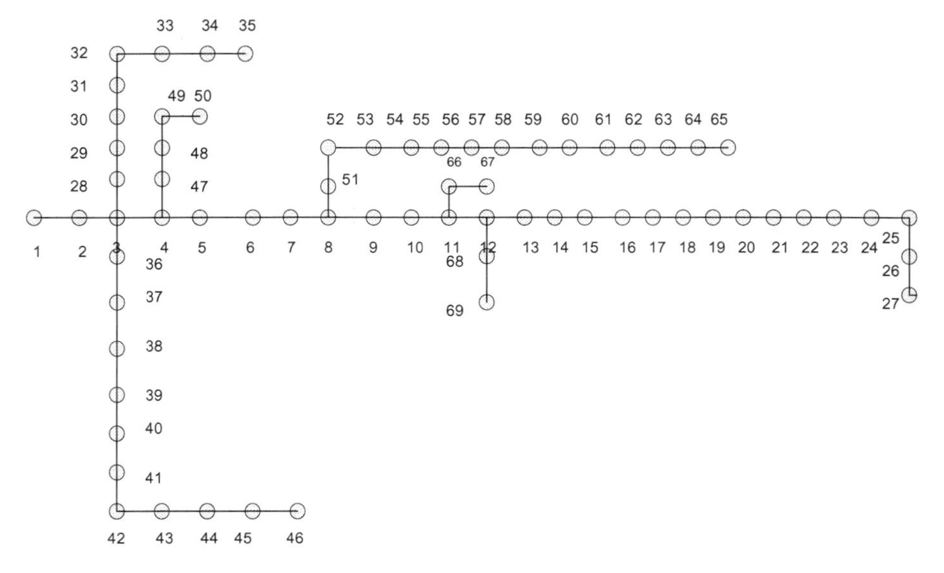

Fig. 22.4 69-bus radial distribution network

Table 22.7 Voltage stability index before and after placement of charging stations

Bus no	Base case VSI	Case 1 VSI	Case2 VSI	Case 3 VSI
2	0	1.0000	1.0000	1.0000
3	1.0000	0.9998	0.9992	0.9998
4	0.9999	0.9996	0.9984	0.9996
5	0.9996	0.9992	0.9980	0.9989
6	0.9992	0.9956	0.9944	0.9917
7	0.9953	0.9602	0.9590	0.9279
8	0.9584	0.9251	0.9239	0.8662
9	0.9281	0.9168	0.9157	0.8518
10	0.9248	0.9152	0.9140	0.8501
11	0.9233	0.8966	0.8955	0.8323
12	0.9046	0.8915	0.8903	0.8273
13	0.9014	0.8813	0.8801	0.8175
14	0.8965	0.8707	0.8696	0.8073
15	0.8891	0.8605	0.8593	0.7975
16	0.8819	0.8502	0.8490	0.7876
17	0.8739	0.8481	0.8470	0.7856
18	0.8740	0.8453	0.8442	0.7829
19	0.8726	0.8453	0.8441	0.7829
20	0.8734	0.8437	0.8425	0.7813
21	0.8718	0.8421	0.8409	0.7799
22	0.8690	0.8409	0.8398	0.7789
23	0.8699	0.8409	0.8398	0.7790
24	0.8704	0.8406	0.8394	0.7788

Table 22.7 Continued

Bus no	Base case VSI	Case 1 VSI	Case2 VSI	Case 3 VSI
25	0.8696	0.8402	0.8390	0.7786
26	0.8697	0.8395	0.8384	0.7781
27	0.8690	0.8393	0.8382	0.7781
28	0.8691	0.9996	0.9984	0.9996
29	0.9996	0.9995	0.9983	0.9995
30	0.9993	0.9993	0.9981	0.9993
31	0.9992	0.9989	0.9977	0.9989
32	0.9987	0.9988	0.9976	0.9988
33	0.9988	0.9983	0.9971	0.9983
34	0.9978	0.9974	0.9962	0.9974
35	0.9967	0.9966	0.9954	0.9966
36	0.9968	0.9993	0.9953	0.9996
37	0.9996	0.9989	0.9921	0.9994
38	0.9993	0.9982	0.9915	0.9988
39	0.9988	0.9976	0.9908	0.9982
40	0.9976	0.9974	0.9907	0.9980
41	0.9973	0.9974	0.9906	0.9980
42	0.9972	0.9946	0.9879	0.9952
43	0.9945	0.9935	0.9867	0.9940
44	0.9926	0.9933	0.9866	0.9939
45	0.9921	0.9932	0.9864	0.9938
46	0.9916	0.9929	0.9862	0.9935
47	0.9916	0.9992	0.9980	0.9989
48	0.9992	0.9988	0.9975	0.9985
49	0.9978	0.9902	0.9890	0.9899
50	0.9784	0.9778	0.9765	0.9775
51	0.9711	0.9168	0.9156	0.8517
52	0.9246	0.9111	0.9099	0.8394
53	0.9201	0.8931	0.8920	0.8023
54	0.9029	0.8828	0.8817	0.7811
55	0.8974	0.8710	0.8698	0.7568
56	0.8868	0.8551	0.8539	0.7243
57	0.8727	0.8379	0.8367	0.6854
58	0.8576	0.7630	0.7619	0.5496
59	0.7841	0.7274	0.7263	0.4894
60	0.7677	0.7143	0.7132	0.4672
61	0.7662	0.6898	0.6887	0.4242
62	0.7168	0.6763	0.6753	0.4044
63	0.7361	0.6755	0.6745	0.4038
64	0.7388	0.6721	0.6711	0.4012
65	0.7283	0.6679	0.6668	0.3980
66	0.7284	0.8929	0.8918	0.8287
67	0.9073	0.8928	0.8916	0.8286
68	0.9084	0.8811	0.8800	0.8174
69	0.8958	0.8802	0.8791	0.8165

Table 22.8 Different cases of placement of charging station

Case	Description	Increase in load	Number of EVs that can be charged simultaneously
1	Fast charging station is placed at bus 36	1500	30
2	Ten fast charging station is placed at bus 36	15,000	300
3	Fast charging station is placed at bus 61	1500	30

strongest bus of IEEE 69-bus network. On the other hand the VSI of bus 61 is 0.7662 which is the least as compared to all the other buses. So bus 61 is designated as the weakest bus of the system. Since bus 36 and bus 61 are the strongest and weakest bus of the system, respectively, the impact of EV charging load on the network is analyzed for the extreme cases of placement of charging stations at these two buses as mentioned in Table 22.8. The VSI of bus 61 for case 1, case 2, and case 3 are 0.6898, 0.6887, and 0.4242, respectively. The VSI for case 2 where ten charging stations are placed at bus 36 is less than case 1 where a single charging station is placed at bus 36. Also in case 3 when charging station is placed at the weakest bus the reduction in VSI of the weakest bus is even more prominent.

Fig. 22.5 represents the plot of voltage of bus 61 for base case as well as after placement of charging stations. The voltage drop of bus 61 after placement of charging station is highest for case 3. From the aforementioned results an inference can be drawn that placement of large number of charging stations at the weak buses is detrimental to the system and may also result voltage collapse.

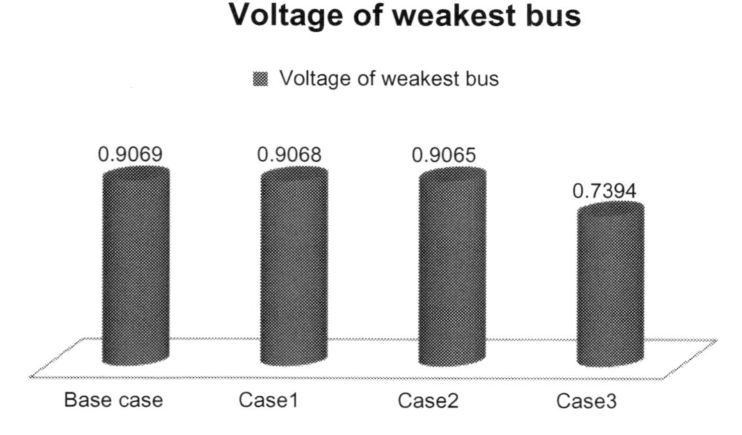

Fig. 22.5 Voltage of the weakest bus for different cases.

Table 22.9 Composite reliability index after placement of charging station

Case	CIR
1	0.0098
2	0.0976
3	0.0408

Table 22.9 reports the CIR of the IEEE 69-bus system for various cases of placement of charging stations. For case 3 CIR is 0.0976 which is more than case 1 as the number of charging stations place is more in case22. For case 3 where the same number of charging stations is placed at bus 61 the CIR is more than case 1 since bus 61 is the weakest bus of the network. Thus we can conclude that the severity of the impact of placement of charging stations in a distribution network depends on the bus in which charging stations are placed as well as the number of charging stations placed in that particular bus.

Table 22.10 reports the THD of IEEE 69-bus network after placement of charging stations. The THD of all the buses of the network increases on addition of EV charging station load. The increase in THD is even more prominent when the number of charging stations is increases in case 2. Furthermore, the value deteriorates even when charging station is placed at weak bus as in case 3.

Table 22.10 Total harmonic distortion after placement of charging station

Bus no	Case1 THD	Case2 THD	Case3 THD
1	0	0	0
2	$0.0903*10^{-8}$	$0.0796*10^{-7}$	$0.0004*10^{-5}$
3	$0.1807*10^{-8}$	$0.1591*10^{-7}$	$0.0008*10^{-5}$
4	$0.4517*10^{-8}$	$0.3978*10^{-7}$	$0.0019*10^{-5}$
5	$0.4517*10^{-8}$	$0.3978*10^{-7}$	$0.0111*10^{-5}$
6	$0.4517*10^{-8}$	$0.3978*10^{-7}$	$0.0714*10^{-5}$
7	$0.4517*10^{-8}$	$0.3978*10^{-7}$	$0.1344*10^{-5}$
8	$0.4517*10^{-8}$	$0.3978*10^{-7}$	$0.1496*10^{-5}$
9	$0.4517*10^{-8}$	$0.3978*10^{-7}$	$0.1496*10^{-5}$
10	$0.4517*10^{-8}$	$0.3978*10^{-7}$	$0.1496*10^{-5}$
11	$0.4517*10^{-8}$	$0.3978*10^{-7}$	$0.1496*10^{-5}$
12	$0.4517*10^{-8}$	$0.3978*10^{-7}$	$0.1496*10^{-5}$
13	$0.4517*10^{-8}$	$0.3978*10^{-7}$	$0.1496*10^{-5}$
14	$0.4517*10^{-8}$	$0.3978*10^{-7}$	$0.1496*10^{-5}$
15	$0.4517*10^{-8}$	$0.3978*10^{-7}$	$0.1496*10^{-5}$
16	$0.4517*10^{-8}$	$0.3978*10^{-7}$	$0.1496*10^{-5}$

Continued

Table 22.10 Continued

Bus no	Case1 THD	Case2 THD	Case3 THD
17	$0.4517*10^{-8}$	$0.3978*10^{-7}$	$0.1496*10^{-5}$
18	$0.4517*10^{-8}$	$0.3978*10^{-7}$	$0.1496*10^{-5}$
19	$0.4517*10^{-8}$	$0.3978*10^{-7}$	$0.1496*10^{-5}$
20	$0.4517*10^{-8}$	$0.3978*10^{-7}$	$0.1496*10^{-5}$
21	$0.4517*10^{-8}$	$0.3978*10^{-7}$	$0.1496*10^{-5}$
22	$0.4517*10^{-8}$	$0.3978*10^{-7}$	$0.1496*10^{-5}$
23	$0.4517*10^{-8}$	$0.3978*10^{-7}$	$0.1496*10^{-5}$
24	$0.4517*10^{-8}$	$0.3978*10^{-7}$	$0.1496*10^{-5}$
25	$0.4517*10^{-8}$	$0.3978*10^{-7}$	$0.1496*10^{-5}$
26	$0.4517*10^{-8}$	$0.3978*10^{-7}$	$0.1496*10^{-5}$
27	$0.4517*10^{-8}$	$0.3978*10^{-7}$	$0.1496*10^{-5}$
28	$0.1807*10^{-8}$	$0.1591*10^{-7}$	$0.0008*10^{-5}$
29	$0.1807*10^{-8}$	$0.1591*10^{-7}$	$0.0008*10^{-5}$
30	$0.1807*10^{-8}$	$0.1591*10^{-7}$	$0.0008*10^{-5}$
31	$0.1807*10^{-8}$	$0.1591*10^{-7}$	$0.0008*10^{-5}$
32	$0.1807*10^{-8}$	$0.1591*10^{-7}$	$0.0008*10^{-5}$
33	$0.1807*10^{-8}$	$0.1591*10^{-7}$	$0.0008*10^{-5}$
34	$0.1807*10^{-8}$	$0.1591*10^{-7}$	$0.0008*10^{-5}$
35	$0.1807*10^{-8}$	$0.1591*10^{-7}$	$0.0008*10^{-5}$
36	$0.1807*10^{-8}$	$0.1591*10^{-7}$	$0.0008*10^{-5}$
37	$0.1807*10^{-8}$	$0.1591*10^{-7}$	$0.0008*10^{-5}$
38	$0.1807*10^{-8}$	$0.1591*10^{-7}$	$0.0008*10^{-5}$
39	$0.1807*10^{-8}$	$0.1591*10^{-7}$	$0.0008*10^{-5}$
40	$0.1807*10^{-8}$	$0.1591*10^{-7}$	$0.0008*10^{-5}$
41	$0.1807*10^{-8}$	$0.1591*10^{-7}$	$0.0008*10^{-5}$
42	$0.1807*10^{-8}$	$0.1591*10^{-7}$	$0.0008*10^{-5}$
43	$0.1807*10^{-8}$	$0.1591*10^{-7}$	$0.0008*10^{-5}$
44	$0.1807*10^{-8}$	$0.1591*10^{-7}$	$0.0008*10^{-5}$
45	$0.1807*10^{-8}$	$0.1591*10^{-7}$	$0.0008*10^{-5}$
46	$0.1807*10^{-8}$	$0.1591*10^{-7}$	$0.0008*10^{-5}$
47	$0.4517*10^{-8}$	$0.3978*10^{-7}$	$0.0019*10^{-5}$
48	$0.4517*10^{-8}$	$0.3978*10^{-7}$	$0.0019*10^{-5}$
49	$0.4517*10^{-8}$	$0.3978*10^{-7}$	$0.0019*10^{-5}$
50	$0.4517*10^{-8}$	$0.3978*10^{-7}$	$0.0019*10^{-5}$
51	$0.4517*10^{-8}$	$0.3978*10^{-7}$	$0.1649*10^{-5}$
52	$0.4517*10^{-8}$	$0.3978*10^{-7}$	$0.2024*10^{-5}$
53	$0.4517*10^{-8}$	$0.3978*10^{-7}$	$0.2312*10^{-5}$
54	$0.4517*10^{-8}$	$0.3978*10^{-7}$	$0.2647*10^{-5}$
55	$0.4517*10^{-8}$	$0.3978*10^{-7}$	$0.3116*10^{-5}$
56	$0.4517*10^{-8}$	$0.3978*10^{-7}$	$0.3581*10^{-5}$
57	$0.4517*10^{-8}$	$0.3978*10^{-7}$	$0.5384*10^{-5}$
58	$0.4517*10^{-8}$	$0.3978*10^{-7}$	$0.6274*10^{-5}$
59	$0.4517*10^{-8}$	$0.3978*10^{-7}$	$0.6615*10^{-5}$
60	$0.4517*10^{-8}$	$0.3978*10^{-7}$	$0.7018*10^{-5}$
61	$0.4517*10^{-8}$	$0.3978*10^{-7}$	$0.7854*10^{-5}$

Table 22.10 Continued

Bus no	Case1 THD	Case2 THD	Case3 THD
62	$0.4517*10^{-8}$	$0.3978*10^{-7}$	$0.7854*10^{-5}$
63	$0.4517*10^{-8}$	$0.3978*10^{-7}$	$0.7854*10^{-5}$
64	$0.4517*10^{-8}$	$0.3978*10^{-7}$	$0.7854*10^{-5}$
65	$0.4517*10^{-8}$	$0.3978*10^{-7}$	$0.7854*10^{-5}$
66	$0.4517*10^{-8}$	$0.3978*10^{-7}$	$0.1496*10^{-5}$
67	$0.4517*10^{-8}$	$0.3978*10^{-7}$	$0.1496*10^{-5}$
68	$0.4517*10^{-8}$	$0.3978*10^{-7}$	$0.1496*10^{-5}$
69	$0.4517*10^{-8}$	$0.3978*10^{-7}$	$0.1496*10^{-5}$

Table 22.11 Specification of fast and slow charging station (Dubey and Santoso, 2015)

Parameter	Fast charging station	Slow charging station
Power consumed	50 kW	19.2 kW
Installation cost	3000$	2000$
Operational cost	65$/MWh	65$/MWh

Table 22.12 Specification of fast and slow charging station (Chowdhury and Koval, 2011)

Penalty for	Value
Unreliability (average energy not served)	0.18$/MWh
Voltage deviation	(Net voltage deviation)^2 *1000
THD	0.18$

Table 22.11 represents the different specifications of fast and slow charging stations. Table 22.12 reports the penalty imposed for violating limiting value of different operating parameters.

Tables 22.13–22.15 report the locations for optimal placement of charging station considering voltage stability, reliability, and cost as objective functions, respectively. The lower and upper limit of number of fast charging station is 2 and 1, respectively. The lower and upper limit of slow charging station is 1 and 3, respectively. Table 22.13 reports that the optimal locations of placement of EV charging stations giving priority to maximization of voltage stability index are bus 38, 2, 4, 36, and 29. The highest numbers of charging stations are placed at bus 36 out of which 3 are slow

Table 22.13 Optimal location for placement of charging stations considering voltage stability index as objective function (Approach 1)

Bus no	Number of fast charging stations	Number of slow charging stations
38	1	2
2	1	2
4	1	3
36	2	3
29	2	1

Table 22.14 Optimal location for placement of charging stations considering composite reliability index as objective function (Approach 2)

Bus no	No. of fast charging stations	Number of slow charging stations
47	2	2
2	2	2
28	2	2
31	2	2
5	2	1

Table 22.15 Optimal location for placement of charging stations considering cost as objective function (Approach 3)

Bus no	No. of fast charging stations	Number of slow charging stations
48	1	2
36	2	2
39	1	3
38	1	3
31	1	2

charging station and 2 are fast charging stations. Similarly Table 22.14 reports the optimal locations of EV charging stations considering the composite reliability index formulated in Section 22.3 as the objective function. The optimal locations are bus 47, 2, 28, 31, 5, which are quite different from the results of Table 22.13. The results reported in Table 22.15 are also quite different from 22.13 and 22.14. Thus on changing the objective functions the optimal locations change accordingly.

22.6 Conclusions

The growing concern related to environmental pollution as well as fossil fuel depletion has increased the popularity of EVs. With the increasing popularity of EVs, the development of charging infrastructure has become indispensable. This chapter analyzes the impact of EV charging station load on voltage stability, reliability, and harmonics of IEEE 69-bus network. All the aforementioned parameters must be within limit for smooth operation of power system. The degradation of these operating parameters on addition of EV charging load is very much prominent from the results presented in this chapter. Furthermore, the optimal locations of EV charging station placement with voltage stability, reliability, and cost as objective function are also presented in this chapter. Future work related to this area can include:

1. reconfiguration of the distribution network to reduce the negative impact of charging station load,
2. analysis of the positive impact of V2G scheme, and
3. multi-objective approach for planning of distribution network in the presence of EV charging station load

Thus, this chapter will be beneficial to power system engineers for planning of distribution network in the presence of EV loads.

References

Aljanad, A., Mohamed, A., 2016. Harmonic impact of plug-in hybrid electric vehicle on electric distribution system. Model. Simul. Eng. 2016.

Arunagiri, A., Venkatesh, B., 2011. Harmonic load flow for radial distribution systems. J. Eng. Sci. Technol. 6 (3), 300–310.

Attari, S.K.A., Bakhshipour, M., Shakarami, M., Namdari, F., 2016. A novel method based on teaching-learning-based optimization for recloser placement with load model consideration in distribution system. Indonesian J. Electr. Eng. Comput. Sci. 2 (1), 1–10.

Babu, G.K., Samala, R.K., 2015. Load flow analysis of 9 bus radial system using BFSLF algorithm. Int. J. Electron. Commun. Eng. 4 (6), 17–22.

Bhadra, J., Chattopadhyay, T.K., 2015. In: Analysis of distribution network by reliability indices. 2015 International Conference on Energy, Power and Environment: Towards Sustainable Growth (ICEPE), June. IEEE, pp. 1–5.

Chowdhury, A., Koval, D., 2011. Power Distribution System Reliability: Practical Methods and Applications. vol. 48. John Wiley & Sons, Hoboken, NJ, USA.

Dharmakeerthi, C.H., Mithulananthan, N., Saha, T.K., 2011. In: November), (Ed.), Overview of the impacts of plug-in electric vehicles on the power grid. Innovative Smart Grid Technologies Asia (ISGT), 2011 IEEE PES. IEEE, pp. 1–8.

Dharmakeerthi, C.H., Mithulananthan, N., Saha, T.K., 2014. Impact of electric vehicle fast charging on power system voltage stability. Int. J. Electr. Power Energy Syst. 57, 241–249.

Dubey, A., Santoso, S., 2015. Electric vehicle charging on residential distribution systems: impacts and mitigations. IEEE Access 3, 1871–1893.

Ellithy, K., Gastli, A., Al-Alawi, S., Al-Hinai, A., Al-Abri, Z., 2000. Voltage stability analysis of muscat power system during summer weather conditions. Sci. Technol. 5, 35–45.

Eminoglu, U., Hocaoglu, M.H., 2007. In: A voltage stability index for radial distribution networks. Universities Power Engineering Conference, 2007. UPEC 2007. 42nd International, September. IEEE, pp. 408–413.

Eminoglu, U., Hocaoglu, M.H., 2008. Distribution systems forward/backward sweep-based power flow algorithms: a review and comparison study. Electr. Power Compon. Syst. 37 (1), 91–110.

Francisco, C.D.L.R., 2006. Harmonics and Power Systems. CRC Press, Boca Raton, FL.

Ganguly, S., Sahoo, N.C., Das, D., 2010. A novel multi-objective PSO for electrical distribution system planning incorporating distributed generation. Energy Syst. 1 (3), 291–337.

Kumar, D., Samantaray, S.R., Kamwa, I., 2015. Multi-objective design of advanced power distribution networks using restricted-population-based multi-objective seeker-optimisation-algorithm and fuzzy-operator. IET Gener. Transm. Distrib. 9 (11), 1195–1215.

Kumar, S.A., Reddy, K.R., 2013. Computation of the power flow solution of a radial dis tribution system for harmonic components. Int. J. Energy Inf. Commun. 4.

Kundur, P., 1994. In: Balu, N.J., Lauby, M.G. (Eds.), Power System Stability and Control. vol. 7. McGraw-Hill, New York.

Niitsoo, J., Taklaja, P., Palu, I., Kiitam, I., 2015. In: Modelling EVs in residential distribution grid with other nonlinear loads. 2015 IEEE 15th International Conference on Environment and Electrical Engineering (EEEIC), June. IEEE, pp. 1543–1548.

Pazouki, S., Mohsenzadeh, A., Ardalan, S., Haghifam, M.R., 2015. Simultaneous planning of PEV charging stations and DGs considering financial, technical, and environmental effects. Can. J. Electr. Comput. Eng. 38 (3), 238–245.

Pazouki, S., Mohsenzadeh, A., Haghifam, M.R., 2014. In: December), (Ed.), The effect of aggregated plug-in electric vehicles penetrations in charging stations on electric distribution networks reliability. Smart Grid Conference (SGC), 2014. IEEE, pp. 1–5.

Pea-da, B., Dechanupaprittha, S., 2014. In: Impact of fast charging station to voltage profile in distribution system. 2014 International Electrical Engineering Congress (iEECON), March. IEEE, pp. 1–4.

Pourarab, M.H., Alishahi, S., Sadeghi, M.H., 2011. In: Analysis of harmonic distortion in distribution networks injected by nonlinear loads.Proceedings of the 21st International Conference and Exhibition on Electricity Distribution, CIRED, June.

Qian, K., Zhou, C., Yuan, Y., 2015. Impacts of high penetration level of fully electric vehicles charging loads on the thermal ageing of power transformers. Int. J. Electr. Power Energy Syst. 65, 102–112.

Rahman, M.M., Barua, S., Zohora, S.T., Hasan, K., Aziz, T., 2013. In: Voltage sensitivity based site selection for PHEV charging station in commercial distribution system. Power and Energy Engineering Conference (APPEEC), 2013 IEEE PES Asia-Pacific, December. IEEE, pp. 1–6.

Rousseau, A., Sharer, P., 2004. Comparing apples to apples: well-to-wheel analysis of current ICE and fuel cell vehicle technologies (No. 2004-01-1015). SAE Technical Paper.

Rupa, J.M., Ganesh, S., 2014. Power flow analysis for radial distribution system using backward/forward sweep method. Int. J. Electr. Comput. Electron. Commun. Eng. 8 (10), 1540–1544.

Sachan, S., Kishor, N., 2016. In: April), (Ed.), Optimal location for centralized charging of electric vehicle in distribution network. Electrotechnical Conference (MELECON), 2016 18th Mediterranean. IEEE, pp. 1–6.

Schneider, A.W., Raksany, J., Gunderson, R.O., Fong, C.C., Billington, R., O'Neill, P.M., Silverstein, B., 1989. Bulk system reliability-measurement and indices. IEEE Trans. Power Syst. 4 (3), 829–835.

Sulaiman, M., Nor, A.F.M., Bujal, N.R., 2015. voltage instability analysis on PV and QV curves for radial-type and mesh-type electrical power networks. Int. Rev. Electr. Eng. 10 (1), 109–115.

Zhang, X.P., 2006. In: Continuation power flow in distribution system analysis. Power Systems Conference and Exposition, 2006. PSCE'06. 2006 IEEE PES, October. IEEE, pp. 613–617.

Zhang, Y., Song, X., Gao, F., Li, J., 2016. In: October), (Ed.), Research of voltage stability analysis method in distribution power system with plug-in electric vehicle. Power and Energy Engineering Conference (APPEEC), 2016 IEEE PES Asia-Pacific. IEEE, pp. 1501–1507.

Distribution systems hosting capacity assessment: Relaxation and linearization

Mohammad Seydali Seyf Abad, Jin Ma*, Xiaoqing Han†*
*School of Electrical and Information Engineering, The University of Sydney, Sydney, NSW, Australia, †Shanxi Key Laboratory of Power System Operation and Control, Taiyuan University of Technology, Taiyuan, China

23.1 Introduction

Accommodating a high penetration of renewable energy sources (RESs) will be challenging in future power systems. Currently, solar and wind energy are the most popular and promising RESs. More than 50 GW photovoltaics (PV) was installed in 2015, and the worldwide cumulative capacity had reached 227 GW at the end of 2015 (Adib et al., 2016). The increasing penetration of RESs at distribution level does not come without technical challenges. In particular, this has led many electricity utilities to adopt newer methods to assess the amount of distributed energy resources (DERs) that can be accommodated in their grids (Tonkoski et al., 2011). Traditionally, distribution systems' planners had to deal with fewer interconnection requests and had the ability to analyze each one of them as they came in. However, the increasing penetration of DER can make the traditional method impractical due to the limited personnel, extensive required input data and the fact that a small DER has a negligible effect on some systems (this may be the case for the first few small DER applicants, but the cumulative capacity of them at some point affects the system). To address this issue, conservative simplified methodologies and practical rules of thumbs are often used to assess/control the interconnection of DERs to the main grid. These conservative limits are usually much less than the network's hosting capacity (HC)—the maximum capacity that can be connected to a given feeder without violating operational constraints. Moreover, these methods usually neglect the HC dependence on actual feeder characteristics. Therefore, industry needs a more accurate HC method that has the ability to distinguish among the aspects of feeder characteristics, as well as technology, size and location of DERs. The scope of this chapter is to model the HC as an optimization problem. For this purpose, we prove that convex conic relaxation of the original HC problem is not exact. Then, we reveal a condition under which one can linearize the original nonlinear HC model. Next, we propose a piecewise linear approximation based on the proven condition. Finally, the proposed method is compared with some of the most important linear models on the IEEE 33-bus system (Gharigh et al., 2015).

Smart Power Distribution Systems. https://doi.org/10.1016/B978-0-12-812154-2.00023-7

23.1.1 DERs' network impacts on the HC

High DER penetration level raises important technical issues especially in distribution systems. Fig. 23.1 demonstrates a comprehensive list of the technical criteria, based on which, distribution system operators (DSOs) have established evaluation methodologies for DERs integration.

In the following, technical issues related to the DER integration into the networks are described (Papathanassiou et al., 2014).

- *Thermal rating criteria:* every distribution infrastructure element such as lines, cables, and transformers is characterized by a current-carrying capacity, which is referred to as thermal rating. If this limit is exceeded a sufficient time, the element physical and/or electrical characteristics may be permanently damaged. Connecting DERs to a distribution network would change the current flows in the network. High DER penetration, especially in case of maximum generation and minimum load, could cause the current level to be higher than the thermal ratings in some parts of the system. In this case, the developer may opt to reinforce or uprate some elements. However, if the cost of this reinforcement is very high,

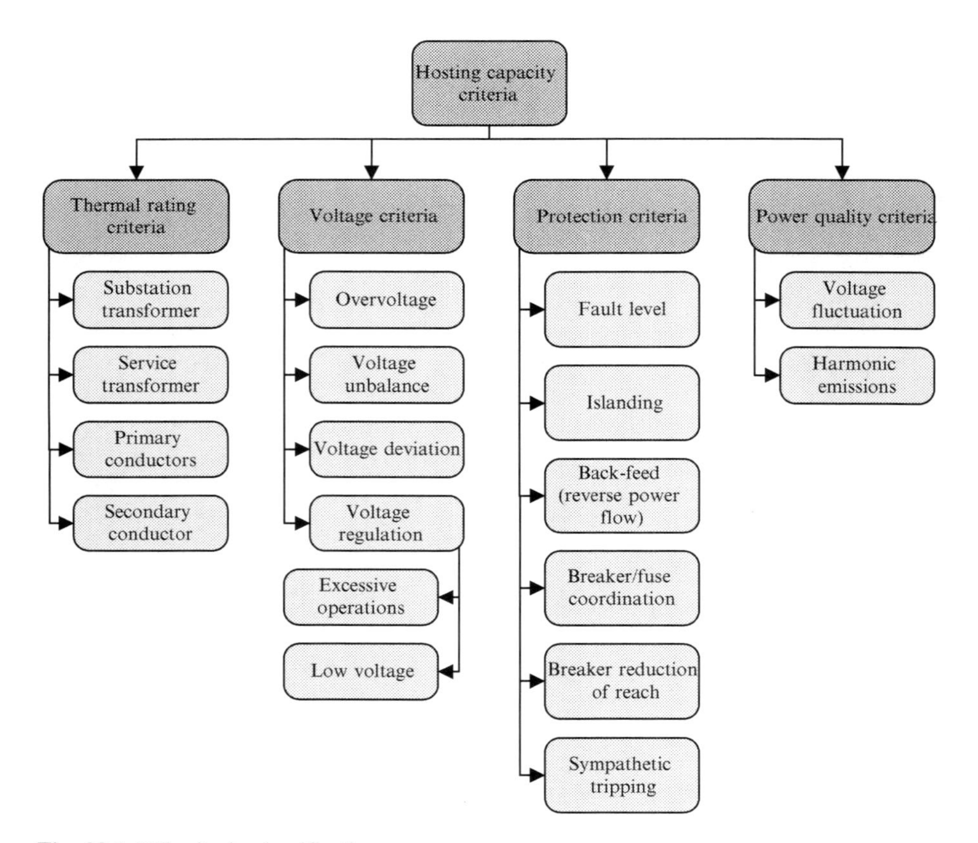

Fig. 23.1 HC criteria classification.

an alternative connection arrangement, possibly to a higher voltage level might be worth considering.

- *Voltage criteria*
  - *Overvoltage:* this criterion is the primary concern of DSOs. Traditionally, radial distribution systems have been operated based on this assumption that there is a voltage drop across the distribution transformer and the feeder's conductors. However, DERs can counter this voltage drop and if the DER penetration is high, under high generation and low-demand conditions, customers may experience voltages higher than normal service voltage.
  - *Voltage unbalance:* this is due to unbalance consumption and generation in distribution systems. In general, unbalance distribution of the single-phase DERs among the grids' phases leads to an unbalance flow of power, hence unbalanced voltage.
  - *Voltage deviation (rapid voltage change):* this may happen due to the fast variation in generation output and tripping or switching of DERs. The output power of PV systems usually varies more rapidly than the typical voltage regulation controls utilized in the system. Hence, the voltage controllers are not able to regulate properly the voltage of the system. Moreover, the rapid change of the output of rooftop PV systems can increase the operation of equipment such as tap changer and switching capacitors.
  - *Voltage regulations:* distribution networks' voltage regulation is usually achieved by on-load tap changers controlled by automatic voltage control (AVC) schemes. DERs could interfere in voltage regulation process and cause technical problems as follows (IEEE-SA, 2007):
    - *Excessive operations:* uncertainty in the output of RESs can disrupt the normal operation of voltage regulating devices and contribute to excessive tap changes of on-load tap changer or switching of capacitors.
    - *Low voltage:* voltage regulators are often equipped with line drop compensation (LDC) to control the voltage at a downstream point by raising the regulator output voltage to compensate for line voltage drop between the regulator and the load center. A DER may interfere with proper operation of regulator if the DER is located immediately downstream of a voltage regulator and its generation is a significant fraction of the heavy load seen by the regulator. In such cases, the feeder's voltage still drops from the DER to the load center, but the regulator output voltage is not increased due to the low loading seen by the regulator. Consequently, low voltage may occur at the load center.
- *Protection Criteria*
  - *Fault level:* fault level at a point in a distribution system is a measure of maximum fault current expected at that point. Fault currents need to be quickly detected and interrupted due to their extensive damage to cables, overhead lines, transformers, and other equipment. The circuit breakers' fault rating limits fault levels that can be permitted in the feeder. This limit is referred to as the design fault level (ABCSE, 2004). Design fault levels in distribution networks can sometimes be a limiting factor to the DERs. New DER connection can increase the fault level. The DER fault level contribution depends on several factors such as the DER type, the distance of the DER from the fault location, the configuration of the network between the DER and the fault, and method of coupling the DER to the network. It should be mentioned that DER with directly connected electrical generators would contribute significantly higher fault current than DER connected via power electronics interfaces such as PVs. Therefore, fault level is even more important for these DERs in comparison with the DERs connected via power electronics interfaces.

  - *Islanding:* this problem happens when some DERs continue to supply the load in a part of the network that is disconnected from the upstream grid. Using safety measures called anti-islanding such as IEEE 1547-2008 or making sure that the DER does not exceed the local load are common ways to prevent this problem.
  - *Back-feed (reverse power flows):* traditional distribution networks have been designed based on this assumption that power flows from a higher voltage network to a lower voltage network. However, increasing DER penetration can change the power flow direction when the total generation of DERs exceeds the network load. There are two main factors that limit the back-feed in a network. The first one is the reverse power rating of some elements, such as transformers, and the second one is the network's automatic control systems' ability to respond correctly under reverse power flow condition. For instance, reverse power flow in transformers can present a problem with the operation of the transformer's automatically controlled tap changer. Readjustment of control setting of control systems such as voltage regulators and replacement of protection relays can resolve the back-feed limit on the HC.
  - *Breaker/fuse coordination:* Fuses, reclosers, and over-current relays are the most common used protection devices in distribution systems. DER interconnection can change the fault current that protection devices sense. Fuse mis-coordination issues usually happen due to an increase in the fuse fault current relative to the breaker or recloser. In cases that the fault current is the same for both fuse and breaker/recloser, mis-coordination happens if the fault current is higher than the maximum coordination current.
  - *Breaker reduction of reach:* voltage support from DERs and their influence on the system Thevenin equivalent leads to a decrease in the fault current at the start of the feeder. The higher the distributed generation capacity is, the lower the current would be at the substation end of the feeder. When the fault current that flows through the breaker installed at the substation end of the feeder drops below the over-current setting, breaker may fail to operate.
  - *Sympathetic tripping:* this can happen due to circulating the zero-sequence current from DERs to a fault above the feeder circuit breaker. The increase in this current can lead to the breaker trip when it exceeds the setting of the ground current relay.
- *Power quality criteria:* a high DER penetration may raise power quality issues such as voltage fluctuations, flicker, and harmonics. In case of rooftop PV systems, power quality issues could be caused by voltage and power fluctuations due to irradiance uncertainty. The voltage fluctuation can cause excessive feeder's voltage regulator operation. Harmonics emissions may be another issue for the DERs connected via power electronic converters to the grid. Although advanced pulse width modulation (PWM) techniques and harmonic filters are usually used in converters, voltage distortion limits can be exceeded in high penetration levels. It should be mentioned that the evaluation of harmonics problems is somewhat complicated and is not in the routine investigations performed by the DSOs.

In many countries, DSOs are overwhelmed by numerous DER connection applications, which need to be evaluated in a fast and reliable manner. Due to aforementioned issues DSOs are reluctant to authorize new DERs' connections, unless detailed studies are performed. Such studies, however, significantly delay the DER integration and cause many complaints by the DER developers and investors. Hence, a need arises for simplified methodologies and practical rules that will allow DSOs to assess the network HC in a fast but reliable manner, without the need to resort to detailed studies.

23.1.2 HC determination

The value of the HC depends on the feeder physical characteristics, DER size, location, and technology of DER as well as evaluation criteria. Each criterion that has been mentioned in the previous section can be used to find a criteria-based HC. The lowest of the obtained capacities from the criteria-based HC analysis would be the HC of the feeder. Therefore, feeder's HC determination is not a straightforward process and HC is not a single value for any given feeder. Generally, three regions as shown in Fig. 23.2 can be defined for the HC of a feeder. Region (A) includes all the penetration levels that does not cause any criteria violation, regardless of location. Region (B) demonstrates the penetration levels that are acceptable in specific sites along the feeder. Region (C) includes all DER deployments that are not acceptable, regardless of location. Based on the definition of the regions, the border between (A) and (B) is the most conservative and the border between (B) and (C) is the least conservative hosting capacities (Smith and Rylander, 2012).

In this chapter, the feeder's HC considering the overvoltage and the thermal constraints is modeled as an optimization problem explained below in detail.

23.2 HC mathematical modeling

The aim of HC assessment is to identify the maximum DERs capacity that can be installed in a system without violating the technical criteria. If one can model the technical criteria mathematically, the HC assessment becomes a constraint optimization. Power flow equations, thermal rating criteria, and overvoltage constraints are the technical criteria that can be considered within mathematical models. So, the HC could be defined as the maximization of DERs capacity considering mathematical model of such criteria as the constraints. To solve this optimization problem, we require to discuss the convexity of the model. Generally, solving a convex optimization is easier than nonconvex problems. Thus, even if the model is not convex by itself, we would like to make it convex by using some techniques such as relaxing some constraints and

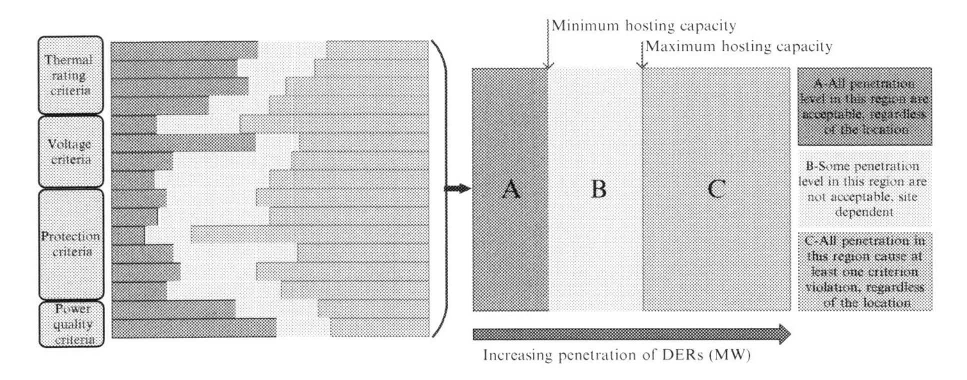

Fig. 23.2 Comprehensive method of the HC assessment.

convex approximation. In this section, the convexity of the HC will be assessed. To do so, the branch flow model, which was used for convexifying the optimal power flow (OPF) problem, will be discussed. The HC problem has the same constraints as OPF. In Farivar and Low (2013), it was proven that conic relaxation of branch flow model is exact, which means that solving the convex conic model results in the same solution as the original OPF problem. We prove here that the conic relaxation of branch flow model is not exact for the HC problem. Therefore, the only way to convexify the problem is a linear approximation, which is discussed in Section 23.3.

23.2.1 Branch flow model

Branch flow model is first proposed by Baran and Wu (1989a,b) for the optimal allocation of capacitors in distribution systems. A distribution system is a graph of buses and lines connecting these buses in a radial (tree) topology. Let $\mathcal{N}=\{0,\ldots,n\}$ denotes the set of buses and defines $\mathcal{N}^+:=\mathcal{N}\setminus\{0\}$. Let $\mathcal{B}$ denotes the set of all branches with cardinality $|\mathcal{B}|:=m$ and (i,j) or $i\rightarrow j$ a branch points from bus (i) to bus (j) in set $\mathcal{B}$. We use $i\sim j$ for a connection between i and j, that is, either $(i,j)\in\mathcal{B}$ or $(j,i)\in\mathcal{B}$ (but not both). For every bus $i\in\mathcal{N}$, let V_i denotes the complex voltage and defines $v_i := |V_i|^2$. Let $s_i=p_i+iq_i$ denotes the net complex power injection on bus i, $s_i^g=p_i^g+iq_i^g$ the generation complex power on bus i and $s_i^c=p_i^c+iq_i^c$ the load consumption on bus i. Let $z_i=r_i+ix_i$ denotes the shunt impedance from bus i to ground and $y_i:=1/z_i=:g_i-ib_i$. For every branch $(i,j)\in\mathcal{B}$, let $z_{ij}=r_{ij}+ix_{ij}$ denotes the complex impedance of the line, and $y_{ij}:=1/z_{ij}=:g_{ij}-ib_{ij}$ the corresponding admittance. Let $S_{ij}=P_{ij}+Q_{ij}$ denotes the sending-end complex power from bus i to bus j, and I_{ij} the complex current from bus i to bus j. It is also assumed that substation voltage V_0 is given. Fig. 23.3 demonstrates a summary of the notations.

Given $(z_{ij},(i,j)\in\mathcal{B})$, V_0 and $(s_i, i\in\mathcal{N}^+)$, the following equation can be written based on the Ohm's law:

$$V_i-V_j=z_{ij}I_{ij}\quad\forall(i,j)\in\mathcal{B} \tag{23.1}$$

Based on the complex power definition, the branch power flow is as follows:

$$S_{ij}=V_iI_{ij}^*\quad\forall(i,j)\in\mathcal{B} \tag{23.2}$$

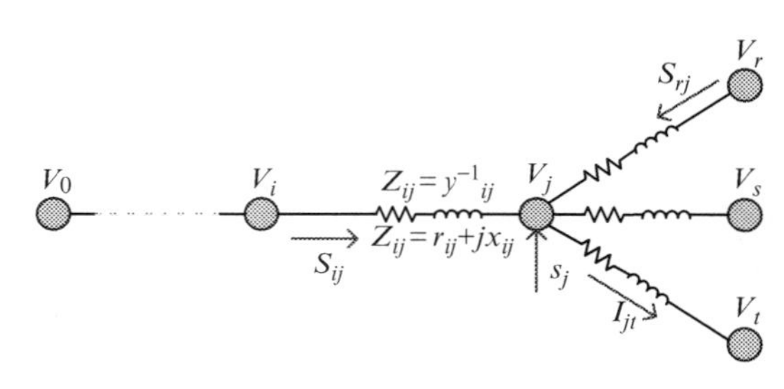

Fig. 23.3 Summary of the notations for branch flow model.

and power balance at each bus $j \in \mathcal{N}$ is as follows:

$$\int_j = \sum_{k:j \to k} S_{jk} - \sum_{i:i \to j} \left(S_{ij} - z_{ij} |I_{ij}|^2 \right) + y_j^* |V_j|^2 \ \forall j \in \mathcal{N} \tag{23.3}$$

Eqs. (23.1)–(23.3) are referred as branch flow model. This model consists of $2m + n + 1$ nonlinear equations in $2m + n + 1$ complex variables $\{ (S_{ij}\,(i,j) \in \mathcal{B}), (I_{ij}\,(i,j) \in \mathcal{B}), (V_i\, i \in \mathcal{N}^+), (\int_0) \}$.

23.2.2 Exact model of the HC

The mathematical structure of the constraints for the HC assessment is similar to the more familiar OPF problem. The main difference lies in the objective, which is the total generation capacity in the network as follows:

$$\max \sum_{i \in \mathcal{N}^+} p_i^g \tag{23.4}$$

Regarding the constraints, in addition to branch flow equations, the voltage and thermal capacity constraints should be considered. In other words, the voltage magnitude of buses must be maintained in the operational range and the current of branches should be less than their maximum currents. These constraints can be modeled as follows:

$$\underline{v}_i \leq |V_i|^2 \leq \overline{v}_i \ \forall i \in \mathcal{N} \tag{23.5}$$

$$|I_{ij}| \leq \overline{I}_{ij} \ \forall (i,j) \in \mathcal{B} \tag{23.6}$$

therefore, the HC problem is given as

HC:

$$\underbrace{\max}_{x, \int} \sum_{i \in \mathcal{N}^+} p_i^g \quad \text{subject to } (23.1) - (23.3), (23.5), (23.6) \tag{23.7}$$

The feasible set that is specified by constraints (23.1)–(23.3), (23.5), (23.6) is nonconvex in general, which implies that the HC problem is complex to solve. Considering that there is a convex formulation for branch flow model, which is the source of nonconvexity in constraints of the HC problem, the main question is whether using this convex branch flow model makes the HC problem convex or not. Based on the proposed method in Farivar and Low (2013), the main steps for convexification of the branch flow model are two relaxation, which are explained in the next subsection.

23.2.3 Relaxed branch flow model

In the first step of relaxation, the voltage and current angles will be eliminated from the branch flow equations. To do so, Eq. (23.2) is substituted in Eq. (23.1), which results in Eq. (23.8):

$$V_j = V_i - z_{ij}\frac{S_{ij}^*}{V_i^*} \tag{23.8}$$

taking the magnitude squared from Eq. (23.8) results in:

$$v_j = v_i + |z_{ij}|^2 \ell_{ij} - \left(z_{ij}S_{ij}^* + z_{ij}^* S_{ij}\right) \ \forall(i,j) \in \mathcal{B} \tag{23.9}$$

where:

$$\ell_{ij} = \frac{P_{ij}^2 + Q_{ij}^2}{v_i} \ \forall(i,j) \in \mathcal{B} \tag{23.10}$$

rewriting Eqs. (23.3), (23.9) in terms of real variables leads to following equations:

$$p_j = \sum_{k:j\to k} P_{jk} - \sum_{i:i\to j} \left(P_{ij} - r_{ij}\ell_{ij}\right) + g_j v_j \ \forall j \in \mathcal{N} \tag{23.11}$$

$$q_j = \sum_{k:j\to k} Q_{jk} - \sum_{i:i\to j} \left(Q_{ij} - x_{ij}\ell_{ij}\right) + b_j v_j \ \forall j \in \mathcal{N} \tag{23.12}$$

$$v_j = v_i - 2\left(r_{ij}P_{ij} + x_{ij}Q_{ij}\right) + \left(r_{ij}^2 + x_{ij}^2\right)\ell_{ij} \ \forall(i,j) \in \mathcal{B} \tag{23.13}$$

Authors in Farivar and Low (2013) referred to Eqs. (23.10)–(23.13) as the relaxed branch flow equations. In contrast to the original model, i.e., Eqs. (23.1)–(23.3), the relaxed model (23.10)–(23.13) consists of $2(m + n + 1)$ equations in $3m + n + 2$ real variables $\left\{\left(P_{ij}, Q_{ij}, \ell_{ij} \forall (i,j) \in \mathcal{B}\right), \left(v_i \ \forall i \in \mathcal{N}^+\right), p_0, q_0\right\}$. Structure of distribution systems is usually radial, which implies that $m = |\mathcal{B}| = |\mathcal{N}| - 1 = n$. Hence, the relaxed branch flow model in radial distribution systems consists of $4n + 2$ equations in $4n + 2$ real variables. The problem of the relaxed branch flow model above is still nonconvex due to the quadratic equality constraint in Eq. (23.10). Moreover, the solution of Eqs. (23.10)–(23.13) might be infeasible for the original branch flow equations. To explain this in more detail, suppose that there is a point in the complex plane. As it can be seen in Fig. 23.4, relaxing the angle of this point is similar to projecting this point to a circle with a radius equal to the distance of the point from the origin. However, a point on the circle may result in an angle different from the origin point.

To have a better understanding of the relation between relaxed branch flow equations and the original model, suppose that $\int := \left(\int_i \ \forall i \in \mathcal{N}^+\right)$ is fixed. To simplify the notation, let us define $S := (S_{ij}\ \forall(ij) \in \mathcal{B})$, $P := (P_{ij}\ \forall(ij) \in \mathcal{B})$, $Q := (Q_{ij}\ \forall(ij) \in \mathcal{B})$, $I := (I_{ij}\ \forall(ij) \in \mathcal{B})$, $\ell := (\ell_{ij}\ \forall(ij) \in \mathcal{B})$, $V := (V_i\ \forall i \in \mathcal{N}^+)$ and $\upsilon := (\upsilon_i\ \forall i \in \mathcal{N}^+)$. For a given $\int$, let $\mathbb{x}(\int)$ denotes the solution set of the original branch flow equations given in Eqs. (23.1)–(23.3) and $\hat{\mathbb{y}}(\int)$ the solution set of the relaxed model stated in Eqs. (23.10)–(23.13). So, $\mathbb{x}(\int) \subseteq \mathbb{C}^{2m+n+1}$ and $\mathbb{y}(\int) \subseteq R^{3m+n+2}$ can be defined as follows:

$$\mathbb{x}(\int) := \{x := (S, I, V, \int_0) | x \text{ solves } (23.1) - (23.3)\} \tag{23.14}$$

$$\hat{\mathbb{y}}(\int) := \{\hat{y} := (S, \ell, \upsilon, \int_0) | \hat{y} \text{ solves } (23.10) - (23.13)\} \tag{23.15}$$

Let $\hat{h}$ defines the projection of $(S, I, V, \int_0) \in \mathbb{C}^{2m+n+1}$ to $(P, Q, \ell, \upsilon, p_0, q_0) \in R^{3m+n+2}$ where:

$$P_{ij} = \text{Re}\{S_{ij}\},\ Q_{ij} = \text{Im}\{S_{ij}\},\ \ell_{ij} = |I_{ij}|^2\ \ \forall(i,j) \in \mathcal{B} \tag{23.16}$$

$$p_i = \text{Re}\{\int_i\},\ q_i = \text{Im}\{\int_i\},\ \upsilon_i = |V_i|^2 \forall i \in \mathcal{N} \tag{23.17}$$

Assume $\mathbb{y}(\int)$ denotes the set of all points whose projections are the solutions of the relaxed branch flow equations. So, $\mathbb{y}(\int) \subseteq \mathbb{C}^{2m+n+1}$ can be represented as follows:

$$\mathbb{y}(\int) := \{\hat{h}(y) := (S, \ell, \upsilon, \int_0) | \hat{h}(y) \text{ solves } (23.10) - (23.13)\} \tag{23.18}$$

Based on what was shown in Fig. 23.4, one can say that $\mathbb{x}$ is a subset of $\mathbb{y}$ and $\hat{h}(\mathbb{x})$ is a subset of $\hat{\mathbb{y}}$. This relation is illustrated in Fig. 23.5.

It was proven by Farivar and Low (2013) that for radial networks, $\hat{h}(\mathbb{x}) = \hat{\mathbb{y}}$, which implies that the angle relaxation is exact. In other words, there is always a unique inverse projection that maps any relaxed solution $\hat{y}$ to a solution of the original branch

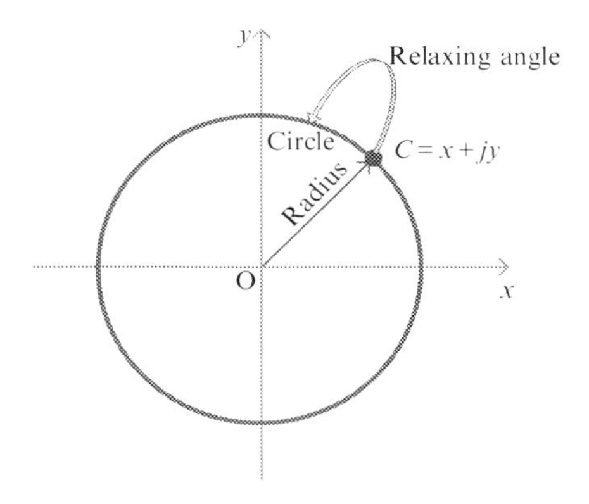

Fig. 23.4 Angle relaxation of a complex point.

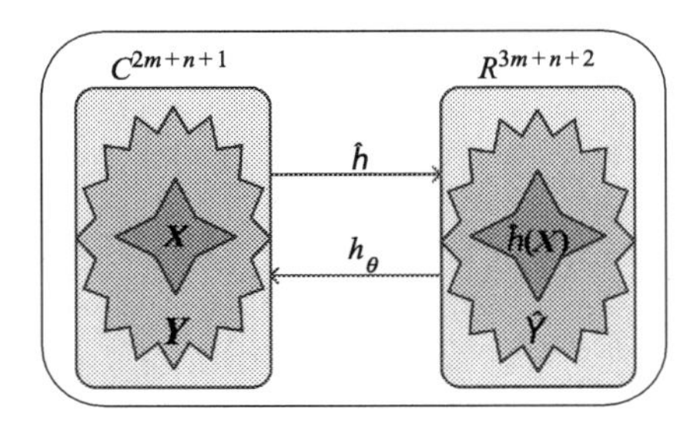

Fig. 23.5 Relation of the projection of the branch flow and relaxed model solutions.

flow model in $\mathbb{x}$. Relaxing the voltage and current angles in the constraints of the HC problem results in the relaxed problem as follows:

HC-ar

$$\underbrace{\max}_{x,\int} \sum_{i \in \mathcal{N}^+} p_i^g$$
$$\text{subject to } (23.5), (23.6), (23.10) - (23.13) \tag{23.19}$$

Since $\mathbb{x} \subseteq \mathbb{y}$, the HC-ar provides a higher bound on the HC. The problem is that HC-ar is still nonconvex. The source of nonconvexity is the quadratic Eq. (23.10). Eq. (23.10) can be rewritten as:

$$P_{ij}^2 + Q_{ij}^2 = \upsilon_i \ell_{ij} \ \forall (i,j) \in \mathcal{B} \tag{23.20}$$

Eq. (23.20) is similar to a rotating second-order cone. Relaxing (23.20) to an inequality leads to:

$$P_{ij}^2 + Q_{ij}^2 \le \upsilon_i \ell_{ij} \ \forall (i,j) \in \mathcal{B} \tag{23.21}$$

since υ_i and ℓ_{ij} are nonnegative, Eq. (23.21) is a rotating second-order cone. Therefore, the conic relaxation of HC can be given as follows:

HC-cr

$$\underbrace{\max}_{x,\int} \sum_{i \in \mathcal{N}^+} p_i^g$$
$$\text{subject to } (23.5), (23.6), (23.11) - (23.13), (23.21) \tag{23.22}$$

Let $\overline{\mathbb{y}}$ denotes the solution set of Eqs. (23.11)–(23.13), (23.21). So, $\overline{\mathbb{y}}(\int) \subseteq R^{3m+n+2}$ can be represented as follows:

$$\overline{\mathbb{y}}(\int) := \{\hat{y} := (S, \ell, \upsilon, \int_0) | \hat{y} \text{ solves } (23.11) - (23.13), (23.21)\} \tag{23.23}$$

Note that Eq. (23.10) is a specific case of Eq. (23.21), $\hat{\mathbb{y}}$ is a subset of $\overline{\mathbb{y}}$ $\left(\hat{\mathbb{y}} \subseteq \overline{\mathbb{y}}\right)$, which means HC-cr provides an upper bound to HC-ar. Nonetheless, the more important point is whether solving the HC-cr can lead to the exact solution of the HC or not. In order to assess this, the objective function of HC problem is reformulated as a minimization problem as follows:

HC-cr

$$\underbrace{\min}_{x,\int} \sum_{i \in \mathcal{N}^+} -p_i^g$$
$$\text{subject to } (23.5), (23.6), (23.11)-(23.13), (23.21) \tag{23.24}$$

To evaluate the exactness of the HC-cr, it is needed to check if any optimal solution of Eq. (23.24) attains equality in Eq. (23.21). Hence, assume that $(\hat{y}^*, \int^*) := (S^*, \ell^*, \upsilon^*, \int^{*g}, \int^{*c}, \int_0^{*g}, \int_0^{*c})$ is the optimal solution for HC-cr that attains equality in Eq. (23.21), i.e., $P_{ij}^{*2} + Q_{ij}^{*2} = \upsilon_i^* \ell_{ij}^* \forall (i,j) \in \mathcal{B}$, and inequality in Eq. (23.6), i.e., $\ell_{ij}^* < \bar{\ell}_{ij} \forall (i,j) \in \mathcal{B}$. For some $\mathcal{E} > 0$, consider another point $\left(\tilde{y}, \tilde{\int}\right) := \left(\tilde{S}, \tilde{\ell}, \tilde{\upsilon}, \tilde{\int}^g, \tilde{\int}^c, \tilde{\int}_0^g, \tilde{\int}_0^c\right)$ defined by

$$\begin{aligned}
&\tilde{\upsilon} = \upsilon^*, \tilde{\int}^c = \int^{*c} \\
&\tilde{\ell}_{ij} = \ell_{ij}^* + \mathcal{E}, \tilde{\ell}_{-ij} = \ell_{-ij}^* \\
&\tilde{S}_{ij} = S_{ij}^* + z_{ij}\frac{\mathcal{E}}{2}, \tilde{S}_{-ij} = S_{-ij}^* \\
&\tilde{\int}_i^g = \int_i^{*g} + z_{ij}\frac{\mathcal{E}}{2}, \tilde{\int}_j^g = \int_j^{*g} + z_{ij}\frac{\mathcal{E}}{2} \\
&\tilde{\int}_{-i,-j}^g = \int_{-i,-j}^{*g}
\end{aligned} \tag{23.25}$$

where a negative index means excluding the indexed elements from a vector. The incremental change in generation at buses (i) and (j) are set to counter the variation in (ℓ_{ij}^*) and (S_{ij}^*). As $\tilde{\int}_i^g = \int_i^{*g} + z_{ij}\mathcal{E}/2$, $\tilde{\int}_j^g = \int_j^{*g} + z_{ij}\mathcal{E}/2$, $\left(\tilde{y}, \tilde{\int}\right)$ has a strictly smaller objective value than $(\hat{y}^*, \int^*)$. If $\left(\tilde{y}, \tilde{\int}\right)$ is feasible, then the optimal solution of conic relaxation is different from HC-ar. To prove this, it suffices to demonstrate that there is a $\mathcal{E} > 0$ such that $\left(\tilde{y}, \tilde{\int}\right)$ holds for Eqs. (23.5), (23.6), (23.11)–(23.13), (23.21). Since $(\hat{y}^*, \int^*)$ is a feasible point, Eq. (23.5) holds for $\left(\tilde{y}, \tilde{\int}\right)$; and since $(\hat{y}^*, \int^*)$ attains inequality in Eq. (23.6), there is an $\mathcal{E} > 0$ that Eq. (23.6) holds for $\left(\tilde{y}, \tilde{\int}\right)$. Due to condition (23.25), $\left(\tilde{y}, \tilde{\int}\right)$ satisfies Eqs. (23.11), (23.12) for all buses except (i) and (j); in a similar manner, $\left(\tilde{y}, \tilde{\int}\right)$ satisfies Eqs. (23.13), (23.21) for all lines except $(i \rightarrow j)$. To prove the feasibility of $\left(\tilde{y}, \tilde{\int}\right)$, we only need to demonstrate that $\left(\tilde{y}, \tilde{\int}\right)$ satisfies Eqs. (23.11), (23.12) at buses (i), (j), and Eqs. (23.13), (23.21) over line $(i \rightarrow j)$. To do so, we reformulated Eqs. (23.11), (23.12) at bus (i) as follows:

$$
\begin{aligned}
\widetilde{\mathcal{f}}_i &= \widetilde{\mathcal{f}}_i^{g} - \widetilde{\mathcal{f}}_i^{c} = \mathcal{f}_i^{*g} + z_{ij}\frac{\mathcal{E}}{2} - \mathcal{f}_i^{*c} \\
&= \sum_{i \to j'} S_{ij'}^* - \sum_{k \to i} \left(S_{ki}^* - z_{ki}\ell_{ki}^*\right) + y_i^* v_i^* + z_{ij}\frac{\mathcal{E}}{2} \\
&= \sum_{\substack{i \to j' \\ j' \neq j}} \widetilde{S}_{ij'} - \sum_{k \to i} \left(\widetilde{S}_{ki} - z_{ki}\widetilde{\ell}_{ki}\right) + y_i^* \widetilde{v}_i + S_{ij}^* + z_{ij}\frac{\mathcal{E}}{2} \\
&= \sum_{i \to j'} \widetilde{S}_{ij'} - \sum_{k \to i} \left(\widetilde{S}_{ki} - z_{ki}\widetilde{\ell}_{ki}\right) + y_i^* \widetilde{v}_i .
\end{aligned}
\tag{23.26}
$$

Similarly, reformulated Eqs. (23.11), (23.12) at bus (j) can be given as:

$$
\begin{aligned}
\widetilde{\mathcal{f}}_j &= \widetilde{\mathcal{f}}_j^{g} - \widetilde{\mathcal{f}}_j^{c} = \mathcal{f}_j^{*g} + z_{ij}\frac{\mathcal{E}}{2} - \mathcal{f}_j^{*c} \\
&= \sum_{j \to k} S_{jk}^* - \sum_{i' \to j} \left(S_{i'j}^* - z_{i'j}\ell_{i'j}^*\right) + y_j^* v_j^* + z_{ij}\frac{\mathcal{E}}{2} \\
&= \sum_{j \to k} \widetilde{S}_{jk} - \sum_{\substack{i' \to j \\ i' \neq i}} \left(\widetilde{S}_{ki} - z_{ki}\widetilde{\ell}_{ki}\right) - S_{ij}^* + z_{ij}\ell_{ij}^* + y_j^* \widetilde{v}_j + z_{ij}\frac{\mathcal{E}}{2} \\
&= \sum_{j \to k} \widetilde{S}_{jk} - \sum_{\substack{i' \to j \\ i' \neq i}} \left(\widetilde{S}_{ki} - z_{ki}\widetilde{\ell}_{ki}\right) \\
&\quad - \left(\widetilde{S}_{ij} - z_{ij}\frac{\mathcal{E}}{2}\right) + z_{ij}\left(\widetilde{\ell}_{ij} - \mathcal{E}\right) + y_j^* \widetilde{v}_j + z_{ij}\frac{\mathcal{E}}{2} \\
&= \sum_{j \to k} \widetilde{S}_{jk} - \sum_{i' \to j} \left(\widetilde{S}_{ki} - z_{ki}\widetilde{\ell}_{ki}\right) + y_j^* \widetilde{v}_j
\end{aligned}
\tag{23.27}
$$

This implies that Eqs. (23.11), (23.12) hold at buses (i) and (j). To check if Eq. (23.13) holds over line ($i \to j$), we need to substitute (v_j^*, v_i^*, P_{ij}^*, Q_{ij}^*, ℓ_{ij}^*) in Eq. (23.13) with their equivalent value from Eq. (23.25) as follows:

$$
\begin{aligned}
v_j^* &= v_i^* - 2\left(r_{ij}P_{ij}^* + x_{ij}Q_{ij}^*\right) + \left(r_{ij}^2 + x_{ij}^2\right)\ell_{ij}^* \Rightarrow \\
\widetilde{v}_j &= \widetilde{v}_i - 2\left(r_{ij}\left(\widetilde{P}_{ij} - r_{ij}\frac{\mathcal{E}}{2}\right) + x_{ij}\left(\widetilde{Q}_{ij} - x_{ij}\frac{\mathcal{E}}{2}\right)\right) + \left(r_{ij}^2 + x_{ij}^2\right)\left(\widetilde{\ell}_{ij} - \mathcal{E}\right) \\
&= \widetilde{v}_i - 2\left(r_{ij}\widetilde{P}_{ij} + x_{ij}\widetilde{Q}_{ij}\right) + \left(r_{ij}^2 + x_{ij}^2\right)\widetilde{\ell}_{ij}
\end{aligned}
\tag{23.28}
$$

Hence, Eq. (23.13) holds for line ($i \to j$). Regarding Eq. (23.21) over line ($i \to j$), we need to substitute $\widetilde{v}_i, \widetilde{P}_{ij}, \widetilde{Q}_{ij}$, and $\widetilde{\ell}_{ij}$ in Eq. (23.21) by their equivalent from Eq. (23.25) as follows:

$$\begin{aligned}\widetilde{v}_i\widetilde{\ell}_{ij}-\widetilde{P}_{ij}^2-\widetilde{Q}_{ij}^2&=v_i^*\left(\ell_{ij}^*+\mathcal{E}\right)-\left(P_{ij}^*+r_{ij}\frac{\mathcal{E}}{2}\right)^2-\left(Q_{ij}^*+x_{ij}\frac{\mathcal{E}}{2}\right)^2\\&=\left(v_i^*\ell_{ij}^*-P_{ij}^*2-Q_{ij}^*2\right)+\mathcal{E}\left(v_i^*-r_{ij}P_{ij}^*-x_{ij}Q_{ij}^*-\mathcal{E}\left(\frac{r_{ij}^2+x_{ij}^2}{4}\right)\right)\\&=\mathcal{E}\left(v_i^*-r_{ij}P_{ij}^*-x_{ij}Q_{ij}^*-\mathcal{E}\left(\frac{r_{ij}^2+x_{ij}^2}{4}\right)\right)>0\end{aligned} \tag{23.29}$$

Since there is a $\mathcal{E}>0$ that $v_i^*-r_{ij}P_{ij}^*-x_{ij}Q_{ij}^*-\mathcal{E}\left(r_{ij}^2+x_{ij}^2\right)/4>0$, hence $\widetilde{v}_i\widetilde{\ell}_{ij}>\widetilde{P}_{ij}^2+\widetilde{Q}_{ij}^2$. We proved that there is a feasible solution for HC-cr with a lower objective value that attains inequality in Eq. (23.21), that is, $\widetilde{v}_i\widetilde{\ell}_{ij}>\widetilde{P}_{ij}^2+\widetilde{Q}_{ij}^2$. This indicates that the solution of convex conic relaxation model HC-cr is not necessarily the same as that of the original HC problem.

23.3 Linear model of HC

In the previous section, it was proven that conic relaxation is not exact for the HC problem. It means that linear approximation is the only way that allows us to develop a simple convex HC model. The need for this simple model comes from modeling the uncertainty that exist in the output power of DERs. Generally, robust and stochastic optimization are two approaches to deal with uncertain parameters in constraints. However, both approaches lead to computationally intractable counterparts. Thus, using approximations to reduce the complexity of robust and stochastic counterparts is inevitable. Generally, robust counterpart of linear programming (LP) is a second-order cone program (SOCP), and robust counterpart of a SOCP requires a semidefinite programming (SDP), and robust SDP becomes nondeterministic polynomial time (NP)-hard.

Linear approximation of HC problem leads to a LP model, which implies that the robust counterpart is SOCP. Therefore, linear approximation reduces the complexity of robust counterpart from SDP to SOCP. In this section, a proper linear model of the HC is presented. To do so, the equations with nonlinear terms are linearized as detailed below.

23.3.1 Linearizing the relaxed power flow model

Commonly, the linearization of the relaxed branch flow model is based on approximating the quadratic term in Eq. (23.10), which is the only source of nonlinearity and nonconvexity. A distribution system is a graph in which lines have no orientation. So, the relaxed branch flow Eqs. (23.11)–(23.13) holds for any graph orientation. Given an undirected graph $G(\mathcal{N},\mathcal{B})$, there are $2^{|\mathcal{B}|}$ orientations. In this section, we discuss

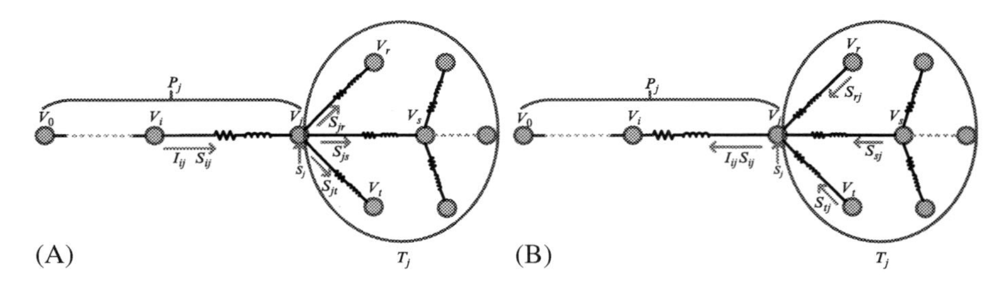

Fig. 23.6 Notation summary of the forward (A) and backward (B) orientations.

two orientations: (1) every bus points toward the substation (backward orientation); and (2) every bus points away the substation (forward orientation). Fig. 23.6 shows some notations for these two orientations.

I) Backward orientation

In this section, without loss of generality, it is supposed that $y_i = 0\ \forall i \in \mathcal{N}$. Considering this assumption, the relaxed power flow model reduces to Eqs. (23.30)–(23.33) in backward orientation:

$$\hat{\ell}_{ji} = \frac{\hat{P}_{ji}^2 + \hat{Q}_{ji}^2}{\hat{\upsilon}_j}\ \forall (i,j) \in \mathcal{B} \tag{23.30}$$

$$\hat{P}_{ji} = \sum_{k:k\rightarrow j} \left(\hat{P}_{kj} - r_{kj}\hat{\ell}_{kj}\right) + p_j\ \forall j \in \mathcal{N} \tag{23.31}$$

$$\hat{Q}_{ji} = \sum_{k:k\rightarrow j} \left(\hat{Q}_{kj} - x_{kj}\hat{\ell}_{kj}\right) + q_j\ \forall j \in \mathcal{N} \tag{23.32}$$

$$\hat{\upsilon}_j = \hat{\upsilon}_i + 2\left(r_{ji}\hat{P}_{ji} + x_{ji}\hat{Q}_{ji}\right) - \left(r_{ji}^2 + x_{ji}^2\right)\hat{\ell}_{ji}\ \forall (i,j) \in \mathcal{B} \tag{23.33}$$

if $j = 0$ then $\hat{P}_{ji}$:=0 and $\hat{Q}_{ji}$:=0. In addition, if j is a node that is connected to just one line, then all $\hat{P}_{kj}$:=0, $\hat{Q}_{kj}$:=0, and $\hat{\ell}_{kj}$:=0. As it was mentioned earlier, the key solution to convexify the problem lies in linearizing the relaxed model given in Eqs. (23.30)–(23.33) via approximating the quadratic equation $\hat{\ell}_{\mathrm{ji}}$ in Eq. (23.30). There are different methods to linearly approximate $\hat{\ell}_{ji}$. Let $\hat{\ell}_{ji}^{\mathrm{lin}}$ denotes the linear approximation of $\hat{\ell}_{ji}$. Then, the general linear model of backward orientation is as follows:

$$\hat{P}_{ji}^{\mathrm{lin}} = \sum_{k:k\rightarrow j} \left(\hat{P}_{kj}^{\mathrm{lin}} - r_{kj}\hat{\ell}_{kj}^{\mathrm{lin}}\right) + p_j\ \forall j \in \mathcal{N} \tag{23.34}$$

$$\hat{Q}_{ji}^{\mathrm{lin}} = \sum_{k:k\rightarrow j} \left(\hat{Q}_{kj}^{\mathrm{lin}} - x_{kj}\hat{\ell}_{kj}^{\mathrm{lin}}\right) + q_j\ \forall j \in \mathcal{N} \tag{23.35}$$

$$\hat{\upsilon}_j^{\text{lin}} = \hat{\upsilon}_i^{\text{lin}} + 2\left(r_{ji}\hat{P}_{ji}^{\text{lin}} + x_{ji}\hat{Q}_{ji}^{\text{lin}}\right) - \left(r_{ji}^2 + x_{ji}^2\right)\hat{\ell}_{ji}^{\text{lin}} \ \forall(i,j) \in \mathcal{B} \tag{23.36}$$

The important question is whether this approximation is valid for the HC problem. Linear HC model is valid as long as the constraints of original problem would not be violated with the solution of the linear model. Linearizing the relaxed model directly affects the voltages of the system. Therefore, voltages in linear model should be higher than the voltages of the original relaxed model. To identify the condition that $\hat{\upsilon}_j \leq \hat{\upsilon}_j^{\text{lin}} \forall j \in \mathcal{N}$, let $\mathbb{T}_j$ denotes the subtree rooted at bus j (including j). We use $k \in \mathbb{T}_j$ to refer to bus k in subtree $\mathbb{T}_j$ and $(k,l) \in \mathbb{T}_j$ to refer to line (k,l) in subtree $\mathbb{T}_j$. Let $\mathbb{P}_j$ denotes the set of lines on the path from substation to bus j. For a given υ_0 and $\int$, let $(\hat{P}^{\text{lin}}, \hat{Q}^{\text{lin}}, \hat{\ell}^{\text{lin}}, \hat{\upsilon}^{\text{lin}})$ and $(\hat{P}, \hat{Q}, \hat{\ell}, \hat{\upsilon})$ be solutions of the linear model (23.34)–(23.36) and the backward relaxed model (23.30)–(23.33), respectively. Then:

$$\hat{P}_{ji}^{\text{lin}} = \sum_{k \in \mathbb{T}_j} (p_k) - \sum_{(k,l) \in \mathbb{T}_j} \left(r_{kl}\hat{\ell}_{kl}^{\text{lin}}\right) \ \forall(i,j) \in \mathcal{B} \tag{23.37}$$

$$\hat{Q}_{ji}^{\text{lin}} = \sum_{k \in \mathbb{T}_j} (q_k) - \sum_{(k,l) \in \mathbb{T}_j} \left(x_{kl}\hat{\ell}_{kl}^{\text{lin}}\right) \ \forall(i,j) \in \mathcal{B} \tag{23.38}$$

$$\hat{P}_{ji} = \sum_{k \in \mathbb{T}_j} (p_k) - \sum_{(k,l) \in \mathbb{T}_j} \left(r_{kl}\hat{\ell}_{kl}\right) \ \forall(i,j) \in \mathcal{B} \tag{23.39}$$

$$\hat{Q}_{ji} = \sum_{k \in \mathbb{T}_j} (q_k) - \sum_{(k,l) \in \mathbb{T}_j} \left(x_{kl}\hat{\ell}_{kl}\right) \ \forall(i,j) \in \mathcal{B} \tag{23.40}$$

Considering the unique path from substation to each bus in radial distribution systems and Eqs. (23.36), (23.33), voltage of each bus can be calculated by

$$\hat{\upsilon}_j^{\text{lin}} = \upsilon_0 + \sum_{(k,l) \in \mathbb{P}_j} 2\left(r_{kl}\hat{P}_{kl}^{\text{lin}} + x_{kl}\hat{Q}_{kl}^{\text{lin}}\right) - \sum_{(k,l) \in \mathbb{P}_j} \left(r_{kl}^2 + x_{kl}^2\right)\hat{\ell}_{kl}^{\text{lin}} \ \forall j \in \mathcal{N} \tag{23.41}$$

$$\hat{\upsilon}_j = \upsilon_0 + \sum_{(k,l) \in \mathbb{P}_j} 2\left(r_{kl}\hat{P}_{kl} + x_{kl}\hat{Q}_{kl}\right) - \sum_{(k,l) \in \mathbb{P}_j} \left(r_{kl}^2 + x_{kl}^2\right)\hat{\ell}_{kl} \ \forall j \in \mathcal{N} \tag{23.42}$$

The linear approximation of $\hat{\ell}_{kl}$ is either larger or smaller than $\hat{\ell}_{kl}$. If $\hat{\ell}_{kl}^{\text{lin}} \leq \hat{\ell}_{kl}$, then:

$$\hat{P}_{kl}^{\text{lin}} \geq \hat{P}_{kl}, \hat{Q}_{kl}^{\text{lin}} \geq \hat{Q}_{kl} \ \forall(i,j) \in \mathcal{B} \tag{23.43}$$

So

$$2\left(r_{kl}\hat{P}_{kl}^{\text{lin}} + x_{kl}\hat{Q}_{kl}^{\text{lin}}\right) \geq 2\left(r_{kl}\hat{P}_{kl} + x_{kl}\hat{Q}_{kl}\right) \ \forall(i,j) \in \mathcal{B} \tag{23.44}$$

It means that following equations holds:

$$v_0 + \sum_{(k,l)\in\mathbb{P}_j} 2\left(r_{kl}\hat{P}_{kl}^{\text{lin}} + x_{kl}\hat{Q}_{kl}^{\text{lin}}\right) \geq v_0 + \sum_{(k,l)\in\mathbb{P}_j} 2\left(r_{kl}\hat{P}_{kl} + x_{kl}\hat{Q}_{kl}\right) \ \forall j\in\mathcal{N} \tag{23.45}$$

$$v_0 + \sum_{(k,l)\in\mathbb{P}_j} 2\left(r_{kl}\hat{P}_{kl}^{\text{lin}} + x_{kl}\hat{Q}_{kl}^{\text{lin}}\right) - \sum_{(k,l)\in\mathbb{P}_j} \left(r_{kl}^2 + x_{kl}^2\right)\hat{\ell}_{kl}^{\text{lin}}$$
$$\geq v_0 + \sum_{(k,l)\in\mathbb{P}_j} 2\left(r_{kl}\hat{P}_{kl} + x_{kl}\hat{Q}_{kl}\right) - \sum_{(k,l)\in\mathbb{P}_j} \left(r_{kl}^2 + x_{kl}^2\right)\hat{\ell}_{kl}^{\text{lin}} \ \forall j\in\mathcal{N} \tag{23.46}$$

$$\hat{v}_j^{\text{lin}} \geq v_0 + \sum_{(k,l)\in\mathbb{P}_j} 2\left(r_{kl}\hat{P}_{kl} + x_{kl}\hat{Q}_{kl}\right) - \sum_{(k,l)\in\mathbb{P}_j} \left(r_{kl}^2 + x_{kl}^2\right)\hat{\ell}_{kl}^{\text{lin}} \ \forall j\in\mathcal{N} \tag{23.47}$$

since we assumed that $\hat{\ell}_{kl}^{\text{lin}} \leq \hat{\ell}_{kl}$

$$v_0 + \sum_{(k,l)\in\mathbb{P}_j} 2\left(r_{kl}\hat{P}_{kl} + x_{kl}\hat{Q}_{kl}\right) - \sum_{(k,l)\in\mathbb{P}_j} \left(r_{kl}^2 + x_{kl}^2\right)\hat{\ell}_{kl}^{\text{lin}}$$
$$\geq \left[v_0 + \sum_{(k,l)\in\mathbb{P}_j} 2\left(r_{kl}\hat{P}_{kl} + x_{kl}\hat{Q}_{kl}\right) - \sum_{(k,l)\in\mathbb{P}_j} \left(r_{kl}^2 + x_{kl}^2\right)\hat{\ell}_{kl}\right] = \hat{v}_j \forall j\in\mathcal{N} \tag{23.48}$$

therefore,

$$\hat{v}_j^{\text{lin}} \geq \hat{v}_j \ \forall j\in\mathcal{N} \tag{23.49}$$

Following the same procedure, we can prove that if $\hat{\ell}_{kl}^{\text{lin}} \geq \hat{\ell}_{kl}$, voltage in original relaxed model is always higher than the voltage in the linear model ($\hat{v}_j^{\text{lin}} \leq \hat{v}_j \forall j\in\mathcal{N}$). Therefore, linearization is valid as long as the linear approximation of the quadratic term (23.10), is smaller than the original quadratic term.

II) Forward orientation

In this orientation, the relaxed power flow model reduces to Eqs. (23.50)–(23.53):

$$\acute{\ell}_{ij} = \frac{\acute{P}_{ij}^2 + \acute{Q}_{ij}^2}{\acute{v}_i} \ \forall (i,j)\in\mathcal{B} \tag{23.50}$$

$$\acute{P}_{ij} - r_{ij}\acute{\ell}_{ij} = \sum_{k:j\to k} \left(\acute{P}_{jk}\right) - p_j \ \forall j\in\mathcal{N} \tag{23.51}$$

$$\acute{Q}_{ij} - x_{jk}\acute{\ell}_{jk} = \sum_{k:j\to k} \left(\acute{Q}_{jk}\right) - q_j \ \forall j\in\mathcal{N} \tag{23.52}$$

$$\acute{\upsilon}_j = \acute{\upsilon}_i - 2\left(r_{ij}\acute{P}_{ij} + x_{ij}\acute{Q}_{ij}\right) + \left(r_{ij}^2 + x_{ij}^2\right)\acute{\ell}_{ij}\ \forall(i,j) \in \mathcal{B} \tag{23.53}$$

Note that if $j = 0$, then $\acute{P}_{ij}$:=0, $\acute{Q}_{ij}$:=0, and $\acute{\ell}_{ij}$:=0. In addition, if j is a node that is connected to just one line, then all $\acute{P}_{jk}$:=0 and $\acute{Q}_{jk}$:=0. Let $\acute{\ell}_{ij}^{\text{lin}}$ denotes the linear approximation of $\acute{\ell}_{ij}$. Then, the linear approximation of the forward orientation is as follows:

$$\acute{P}_{ij}^{\text{lin}} - r_{ij}\acute{\ell}_{ij}^{\text{lin}} = \sum_{k:j\to k}\left(\acute{P}_{jk}^{\text{lin}}\right) - p_j\ \forall j \in \mathcal{N} \tag{23.54}$$

$$\acute{Q}_{ij}^{\text{lin}} - x_{jk}\acute{\ell}_{jk}^{\text{lin}} = \sum_{k:j\to k}\left(\acute{Q}_{jk}^{\text{lin}}\right) - q_j\ \forall j \in \mathcal{N} \tag{23.55}$$

$$\acute{\upsilon}_j^{\text{lin}} = \acute{\upsilon}_i^{\text{lin}} - 2\left(r_{ij}\acute{P}_{ij}^{\text{lin}} + x_{ij}\acute{Q}_{ij}^{\text{lin}}\right) + \left(r_{ij}^2 + x_{ij}^2\right)\acute{\ell}_{ij}^{\text{lin}}\ \forall(i,j) \in \mathcal{B} \tag{23.56}$$

For a given υ_0 and $\int$, let $(\acute{P}^{\text{lin}}, \acute{Q}^{\text{lin}}, \acute{\ell}^{\text{lin}}, \acute{\upsilon}^{\text{lin}})$ and $(\acute{P}, \acute{Q}, \acute{\ell}, \acute{\upsilon})$ be solutions of the linear model represented in Eqs. (23.54)–(23.56) and the forward relaxed model given in Eqs. (23.50)–(23.53), respectively. Then:

$$\acute{P}_{ij}^{\text{lin}} = -\sum_{k\in\mathbb{T}_j}(p_k) + r_{ij}\acute{\ell}_{ij}^{\text{lin}} + \sum_{(k,l)\in\mathbb{T}_j}\left(r_{kl}\acute{\ell}_{kl}^{\text{lin}}\right)\ \forall(i,j) \in \mathcal{B} \tag{23.57}$$

$$\acute{Q}_{ij}^{\text{lin}} = -\sum_{k\in\mathbb{T}_j}(q_k) + x_{ij}\acute{\ell}_{ij}^{\text{lin}} + \sum_{(k,l)\in\mathbb{T}_j}\left(x_{kl}\acute{\ell}_{kl}^{\text{lin}}\right)\ \forall(i,j) \in \mathcal{B} \tag{23.58}$$

$$\acute{P}_{ij} = -\sum_{k\in\mathbb{T}_j}(p_k) + r_{ij}\acute{\ell}_{ij} + \sum_{(k,l)\in\mathbb{T}_j}\left(r_{kl}\acute{\ell}_{kl}\right)\ \forall(i,j) \in \mathcal{B} \tag{23.59}$$

$$\acute{Q}_{ij} = -\sum_{k\in\mathbb{T}_j}(q_k) + x_{ij}\acute{\ell}_{ij} + \sum_{(k,l)\in\mathbb{T}_j}\left(x_{kl}\acute{\ell}_{kl}\right)\ \forall(i,j) \in \mathcal{B} \tag{23.60}$$

Considering the unique path from substation to each bus in radial systems and Eqs. (23.56), (23.53), the voltage of each bus can be calculated by:

$$\acute{\upsilon}_j^{\text{lin}} = \upsilon_0 - \sum_{(k,l)\in\mathbb{P}_j} 2\left(r_{kl}\acute{P}_{kl}^{\text{lin}} + x_{kl}\acute{Q}_{kl}^{\text{lin}}\right) + \sum_{(k,l)\in\mathbb{P}_j}\left(r_{kl}^2 + x_{kl}^2\right)\acute{\ell}_{kl}^{\text{lin}}\ \forall j \in \mathcal{N} \tag{23.61}$$

$$\acute{\upsilon}_j = \upsilon_0 - \sum_{(k,l)\in\mathbb{P}_j} 2\left(r_{kl}\acute{P}_{kl} + x_{kl}\acute{Q}_{kl}\right) + \sum_{(k,l)\in\mathbb{P}_j}\left(r_{kl}^2 + x_{kl}^2\right)\acute{\ell}_{kl}\ \forall j \in \mathcal{N} \tag{23.62}$$

It can be proved for the forward orientation that linear approximation is valid if $\acute{\ell}_{kl}^{\text{lin}} \le \acute{\ell}_{kl}$. One of the most used linearization method is setting $\acute{\ell}_{kl}^{\text{lin}} = 0$ ($\hat{\ell}_{kl}^{\text{lin}} = 0$ for backward orientation). Based on what we proved in this section, since this linear approximation is always smaller than the original quadratic term, the voltage obtained by linear model is always higher than the voltage obtained by the original relaxed model.

In the following, we discuss how to define a linear model that approximate the original quadratic equation in the original HC problem. A new problem is obtained by substituting each nonlinear function by a piecewise linear function. To do so, let consider a continuous function $f(t)$ of the variable $t \in [a,b]$ aiming to approximate it with the piecewise linear function $\hat{f}(t)$.

First, the interval $[a,b]$ is divided into smaller intervals via the points $a = t_0 < t_1 < t_2 < \cdots < t_k = b$ as shown in Fig. 23.7. Function $f(t)$ can be approximated in interval $[t_v, t_{v+1}]$ as given in Eq. (23.65). Let $\lambda \in [0,1]$ to define $t \in [t_v, t_{v+1}]$ as a convex combination of t_v and t_{v+1}. Then:

$$t = \lambda t_v + (1-\lambda)t_{v+1} \tag{23.63}$$

$$\hat{f}(t) - \hat{f}(t_v) = \frac{\hat{f}(t_{v+1}) - \hat{f}(t_v)}{t_{v+1} - t_v}(t - t_v) \tag{23.64}$$

Substituting Eq. (23.63) in Eq. (23.64) results in:

$$\hat{f}(t) = \lambda\hat{f}(t_v) + (1-\lambda)\hat{f}(t_{v+1}) \tag{23.65}$$

More generally, function $f(t)$ can be approximated over interval $[a,b]$ by $\hat{f}(t)$ via points $a = t_0 < t_1 < t_2 < \cdots < t_k = b$ as follows (Stefanov, 2013):

$$t = \sum_{v=0}^{k} \lambda_v t_v, \ \sum_{v=0}^{k} \lambda_v = 1, \ \ \lambda_v \geq 0 \ \forall v \in \{0, 1, 2, \ldots, k\} \tag{23.66}$$

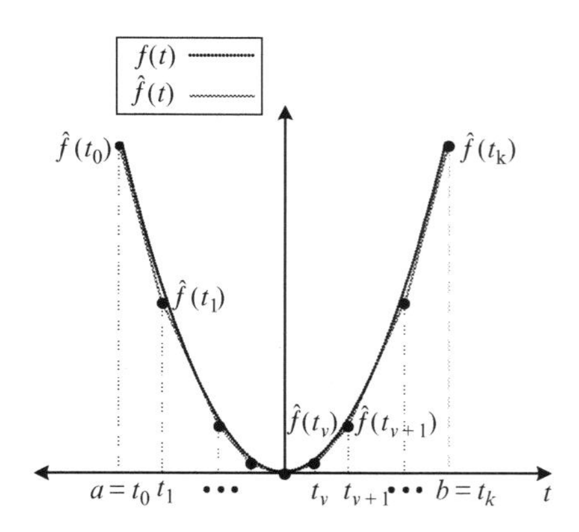

Fig. 23.7 Piecewise linear approximation of a function.

$$\hat{f}(t) = \sum_{v=0}^{k} \lambda_v f(t_v) \tag{23.67}$$

where, λ_v are special order set of type 2 (SOS2) variables, which means two of them are positive, and they must be adjacent. It is worth mentioning that increasing the number of points increases the approximation accuracy. The above piecewise linear approximation is valid for single variable functions. In the HC problem, the nonlinear Eq. (23.10) is a function of (P, Q, υ), which means that we still cannot use the suggested piecewise approximation. If it is supposed that the υ is fixed, then Eq. (23.10) becomes separable. Generally, a function $f(Var_1, Var_2, \ldots, Var_n)$ is said to be separable if it can be expressed as the sum of N single variable functions $f_1(Var_1)$, $f_2(Var_2)$, …, $f_N(Var_N)$. Therefore, each single variable function in Eq. (23.10) can be replaced by a piecewise linear approximation. For the sake of generality, let suppose that both objective and constraints are separable functions as follows:

$$\begin{gathered} SP: \text{minimize} \sum_{j=1}^{N} f_j(x_j) \\ \text{subject to} \\ \sum_{j=1}^{N} g_{ij}(x_j) \le d_i \quad \forall i \in \{0, 1, 2, \ldots, r\} \end{gathered} \tag{23.68}$$

Let define set $\mathcal{L}$ as follows:

$$\mathcal{L} = \{ j : f_j \text{ and } g_{ij} \text{ are linear } \forall i \in \{1, 2, \ldots, r\} \} \tag{23.69}$$

Considering the interval $[a_j, b_j]$ for each $j \notin \mathcal{L}$, the grid points of jth variable x_j are defined as follows:

$$a_j = x_{0j} < x_{1j} < x_{2j} < \cdots < x_{k_j j} = b_j \tag{23.70}$$

thus, the separable programming (SP) given in Eq. (23.68) can be approximated as follows:

$$\begin{gathered} \text{LASP}: \text{minimize} \sum_{j \in \mathcal{L}} f_j(x_j) + \sum_{j \notin \mathcal{L}} \hat{f}_j(x_j) \\ \text{subject to} \\ \sum_{j \in \mathcal{L}} g_{ij}(x_j) + \sum_{j \notin \mathcal{L}} \hat{g}_{ij}(x_j) \le d_i \ \forall i \in \{0, 1, 2, \ldots, r\} \end{gathered} \tag{23.71}$$

where

$$\begin{gathered} \hat{f}_j(x_j) = \sum_{v=0}^{k_j} \lambda_{vj} f_j(x_{vj}) \ j \notin \mathcal{L} \\ \hat{g}_{ij}(x_j) = \sum_{v=0}^{k_j} \lambda_{vj} \hat{g}_{ij}(x_{vj}), \ \forall i \in \{1, 2, \ldots, r\} \ j \notin \mathcal{L} \\ \sum_{v=0}^{k_j} \lambda_{vj} = 1, \lambda_{vj} \ge 0 \ \forall v \in \{0, 1, 2, \ldots, k\} \ j \notin \mathcal{L} \\ x_j = \sum_{v=0}^{k_j} \lambda_{vj} x_{vj} \ j \notin \mathcal{L} \end{gathered} \tag{23.72}$$

Note that two adjacent λ_{vj} are nonzero for $j \notin \mathcal{L}$, due to which, the linear approximation of separable programming (LASP) presented in Eq. (23.71) is not a linear program anymore. This restriction can be removed by using binary variables. So, considering this restriction result in a mixed integer program as follows:

$$\begin{gathered}
\text{MILASP: minimize} \sum_{j\in\mathcal{L}} f_j(x_j) + \sum_{j\notin\mathcal{L}} \sum_{v=0}^{k_j} \lambda_{vj} \hat{f}_j(x_{vj}) \\
\text{subject to} \\
\sum_{j\in\mathcal{L}} g_{ij}(x_j) + \sum_{j\notin\mathcal{L}} \sum_{v=0}^{k_j} \lambda_{vj} \hat{g}_{ij}(x_{vj}) \le d_i \ \forall i \in \{0, 1, 2, \ldots, r\} \\
0 \le \lambda_{0j} \le w_{0j}\, j \notin \mathcal{L} \\
0 \le \lambda_{vj} \le w_{v-1,j} + w_{vj},\ \forall v \in \{1, 2, \ldots, k_j - 1\}; j \notin \mathcal{L} \\
0 \le \lambda_{k_j j} \le w_{k_j-1,j}\, j \notin \mathcal{L} \\
\sum_{v=0}^{k_j-1} w_{vj} = 1\, j \notin \mathcal{L} \\
\sum_{v=0}^{k_j} \lambda_{vj} = 1\ j \notin \mathcal{L} \\
x_j = \sum_{v=0}^{k_j} \lambda_{vj} x_{vj}\ j \notin \mathcal{L} \\
w_{vj} = 0\, or\, 1,\ \forall v \in \{0, 1, 2, \ldots, k_j - 1\}; j \notin \mathcal{L}
\end{gathered} \tag{23.73}$$

It is worth mentioning that increasing the number of grid points to improve the accuracy of the approximation, increases the number of constraints and integer variables significantly. So, there is a compromise between the accuracy of the approximation and the number of grid points. The minimum number of points that are needed for approximating a nonlinear function is equal to two. It means that the approximation function is a linear function passing from those two points. In such situation, the approximation results in a LP. Fig. 23.8 demonstrates two linear approximations (by using two points) that can be used in linearizing (23.10). Since in both approximations $\hat{\ell}_{kl}^{\text{lin}} \le \hat{\ell}_{kl}$ ($\acute{\ell}_{kl}^{\text{lin}} \le \acute{\ell}_{kl}$ in forward orientation), the linear models are valid.

It is worth mentioning that increasing the number of points should not result in a piecewise approximation that violates the condition $\hat{\ell}_{kl}^{\text{lin}} \le \hat{\ell}_{kl}$ ($\acute{\ell}_{kl}^{\text{lin}} \le \acute{\ell}_{kl}$ in forward orientation). Therefore, identifying $\hat{f}(t_v)$ at interval points $a = t_0 < t_1 < t_2 < \cdots < t_v < t_{v+1} < \ldots < t_k = b$ for quadratic function is very vital. In order to find the proper $\hat{f}(t_v)$, consider two tangent lines to the quadratic function at points (P_v)

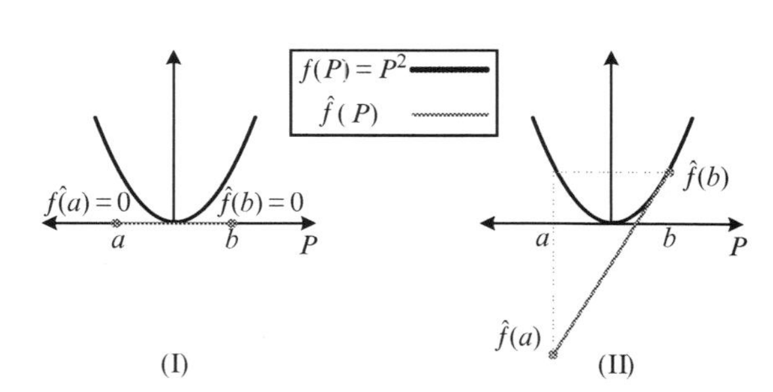

Fig. 23.8 Two linear approximation of the loss quadratic function (by using two points).

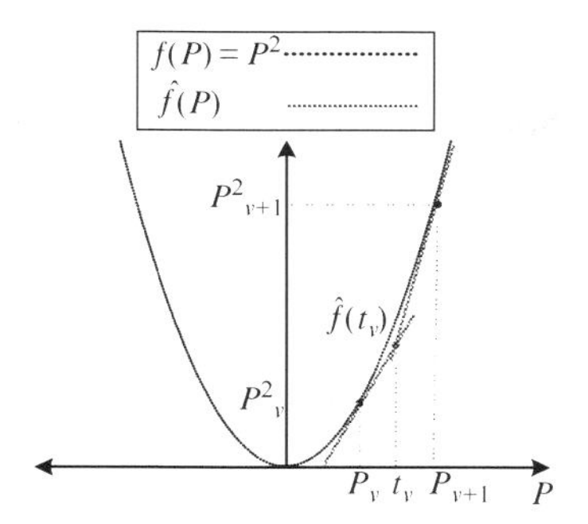

Fig. 23.9 Intersection point of two tangent line to the quadratic function.

and $(P_{\nu+1})$ as shown in Fig. 23.9. As can be seen, the tangent lines to the quadratic function are lower than it, which indicates their intersections are the proper points to be used in the piecewise linear model.

The intersection of the two tangent lines to the quadratic function at points (P_ν) and $(P_{\nu+1})$ results in the following point:

$$t_\nu = \frac{P_\nu + P_{\nu+1}}{2}, \ \hat{f}(t_\nu) = P_\nu \times P_{\nu+1} \tag{23.74}$$

which is the proper point to be used in the proposed piecewise linear approximation.

23.3.2 Traditional linear approximation in distribution systems

Neglecting the shunt impedance of the buses, the voltage deviation along the line $(i \rightarrow j)$ is as follows:

$$\Delta V_{ij} = z_{ij} I_{ij} \ \forall (i,j) \in \mathcal{B} \tag{23.75}$$

let $S^r_{ij} := S_{ij} - z_{ij}\left|I_{ij}\right|^2$ denotes the receiving complex power at bus (j) from bus (i). Considering bus (j) as the reference bus, we can determine I_{ij} as follows:

$$I_{ij} = \left(\frac{S_{ij}}{V_j}\right)^* = \left(\frac{P^r_{ij} - jQ^r_{ij}}{\left|V_j\right|}\right) \ \forall (i,j) \in \mathcal{B} \tag{23.76}$$

substituting I_{ij} in Eq. (23.75) results in Eq. (23.77):

$$\Delta V_{ij} = \frac{r_{ij}P^r_{ij} + x_{ij}Q^r_{ij}}{\left|V_j\right|} + j\frac{x_{ij}P^r_{ij} - r_{ij}Q^r_{ij}}{\left|V_j\right|} = \Delta V^{\text{real}}_{ij} + j\Delta V^{Im}_{ij} \ \forall (i,j) \in \mathcal{B} \tag{23.77}$$

Therefore, the voltage magnitude at bus (i) is as follows:

$$|V_i|^2 = \left(\left|V_j\right| + \Delta V_{ij}^{\text{real}}\right)^2 + \left(\Delta V_{ij}^{Im}\right)^2 \ \forall (i,j) \in \mathcal{B} \tag{23.78}$$

in which, the first right-hand side term is much bigger than the second right-hand side term. So, ΔV_{ij} can be approximated by Eq. (23.79):

$$\Delta V_{ij} = V_i - V_j \simeq \frac{r_{ij}P_{ij}^r + x_{ij}Q_{ij}^r}{\left|V_j\right|} \simeq \frac{r_{ij}P_{ij}^r + x_{ij}Q_{ij}^r}{|V_0|} \ \forall (i,j) \in \mathcal{B} \tag{23.79}$$

therefore,

$$\left|V_j\right|^{\text{lin}} = V_0 - \sum_{(k,\,l)\in \mathbb{P}_j} \Delta V_{kl} \ \forall j \in \mathcal{N}^+ \tag{23.80}$$

In the following, we aim to investigate if this approximation is valid for the HC problem or not. The linear model is valid if the $|V_j| \leq |V_j|^{\text{lin}}$. To check this condition, four cases as shown in Fig. 23.10 should be assessed.

For all four cases, actual and approximate voltage magnitudes are as follows:

$$\left|V_j\right| = |V_i| \times \cos(\delta) - \Delta V_{ij}^{\text{real}} \ \forall (i,j) \in \mathcal{B} \tag{23.81}$$

$$\left|V_j\right|^{\text{lin}} = |V_i| - \Delta V_{ij}^{\text{real}} \ \forall (i,j) \in \mathcal{B} \tag{23.82}$$

Condition $|V_j| \leq |V_j|^{\text{lin}}$ holds if $0 \leq \cos(\delta)$.

- ($P_{ij}^r \geq 0$, $Q_{ij}^r \geq 0$): in this case, $0 \leq \cos(\delta)$. So, $|V_j| \leq |V_j|^{\text{lin}}$.

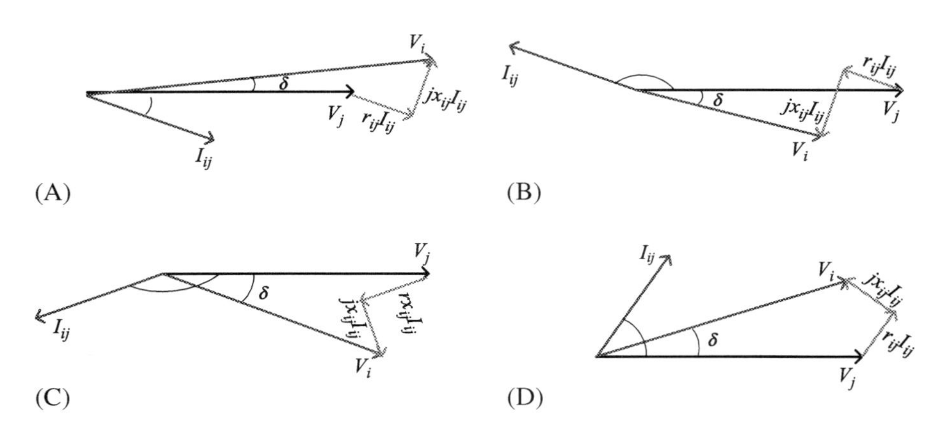

Fig. 23.10 Phasor diagram of a simple line based on the direction of the active and reactive power flows (A) $P_{ij}^r \geq 0$, $Q_{ij}^r \geq 0$, (B) $P_{ij}^r \leq 0$, $Q_{ij}^r \leq 0$, (C) $P_{ij}^r \leq 0$, $Q_{ij}^r \geq 0$, and (D) $P_{ij}^r \geq 0$, $Q_{ij}^r \leq 0$.

- ($P^{r}_{ij} \leq 0$, $Q^{r}_{ij} \leq 0$): in this case, if $|z_{ij}I_{ij}| \leq |V_j|$, then $0 \leq \cos(\delta)$. In other words, the condition $|V_j| \leq |V_j|^{\text{lin}}$ holds if the voltage drop ($|z_{ij}I_{ij}|$) on line ($i \rightarrow j$) is less than the voltage magnitude at bus (j).
- ($P^{r}_{ij} \leq 0$, $Q^{r}_{ij} \geq 0$): in this case, if $|r_{ij}I_{ij}| \leq |V_j|$, then $0 \leq \cos(\delta)$. In other words, the condition $|V_j| \leq |V_j|^{\text{lin}}$ holds if $|r_{ij}I_{ij}| \leq |V_j|$.
- ($P^{r}_{ij} \geq 0$, $Q^{r}_{ij} \leq 0$): in this case $0 \leq \cos(\delta)$.

In general, the linearization is valid if the voltage drop ($|z_{ij}I_{ij}|$) on line ($i \rightarrow j$) is less than the voltage magnitude at bus (j). This condition usually holds in a real distribution system.

23.3.3 Linearizing the thermal capacity constraint

It is important to consider the thermal capacity as one of the constraints in the HC problem. Eq. (23.6) represents this constraint in the original HC problem. Multiplying Eq. (23.6) with nominal voltage leads to:

$$|S_{ij}| \leq |\overline{S}_{ij}| \ \forall (i,j) \in \mathcal{B} \tag{23.83}$$

$$\sqrt{P_{ij}^2 + Q_{ij}^2} \leq |\overline{S}_{ij}| \ \forall (i,j) \in \mathcal{B} \tag{23.84}$$

Eq. (23.84) is a circle in (P, Q) coordination. In order to linearize this circle, we use a rotating line. The following line is tangent to the circle (Eq. 23.84) as shown in Fig. 23.11A:

$$Q_{ij} + P_{ij} \leq \sqrt{2}|\overline{S}_{ij}| \ \forall (i,j) \in \mathcal{B} \tag{23.85}$$

in two dimensions, we can rotate the point (P_{ij}, Q_{ij}) by using the counterclockwise rotation matrix as follow:

$$\begin{bmatrix} \grave{P}_{ij} \grave{Q}_{ij} \end{bmatrix} = \begin{bmatrix} \cos(\theta) & -\sin(\theta) \\ \sin(\theta) & \cos(\theta) \end{bmatrix} \begin{bmatrix} P_{ij} \\ Q_{ij} \end{bmatrix} \tag{23.86}$$

Rotating the line (Eq. 23.85) using the rotation matrix (23.86) leads to

$$(\cos(\theta) + \sin(\theta))\grave{Q}_{ij} + (\cos(\theta) - \sin(\theta))\grave{P}_{ij} \leq \sqrt{2}|\overline{S}_{ij}| \ \forall (i,j) \in \mathcal{B} \tag{23.87}$$

This equation implies that for each θ, there is a line. So, a set of lines can replace the circle. The smaller the step in θ is, the higher the number of the lines and the higher the accuracy of the linearization would be. Fig. 23.11B and C show the linear approximation of the thermal constraint with the proposed rotating line for step sizes $\theta = 45^{\circ}$ and $\theta = 10^{\circ}$, respectively. As it can be seen, the accuracy of approximation with θ=10° is higher than that of θ=45°. Note that the higher accuracy in (C) comes with its price, which is the higher number of constraint.

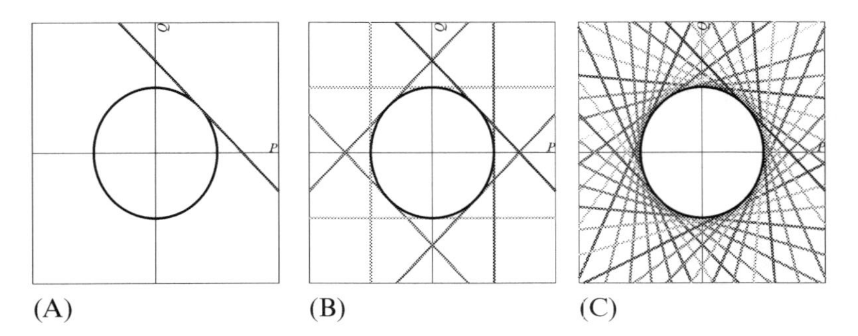

Fig. 23.11 Circular thermal capacity constraint linearization method for step sizes: (A) $\theta = 360°$, (B) $\theta = 45^{\circ}$, and (C) $\theta = 10^{\circ}$.

After linearizing the thermal capacity constraint, the HC problem can be formulated by a LP, which is proper to be used as the base model in stochastic or robust optimization. In the following section, results of different linear models are compared.

23.4 Simulations

In this section, the performance of different linear methods is compared. The test network is the IEEE 33 buses distribution system. This network is a radial as shown in Fig. 23.12, with a total load of 3.715 MW and 2.300 MVAr (Gharigh et al., 2015). It is supposed that rating current of all lines in the test system is 500 A. The HC nonlinear and linear models are implemented in A Mathematical Programming Language (AMPL) (Fourer et al., 2002) environment, and solved by nonlinear interior point trust region optimization (KNITRO) and CPLEX solvers, respectively.

First, the performance of the proposed piecewise linear model is assessed. Fig. 23.13 shows the voltage profile of the test grid for different number of points in Eq. (23.67) in

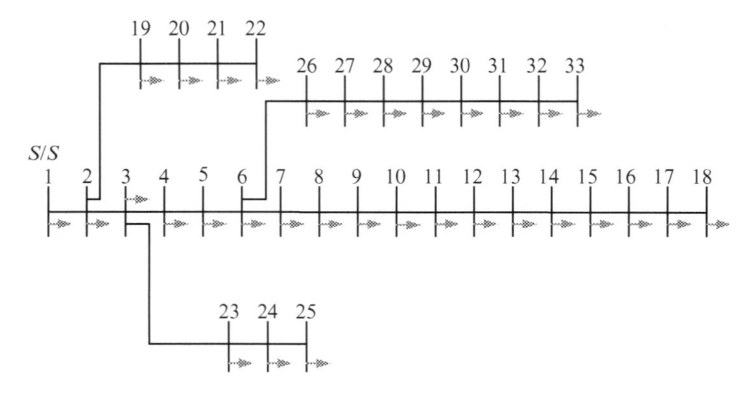

Fig. 23.12 IEEE 33-bus distribution system.

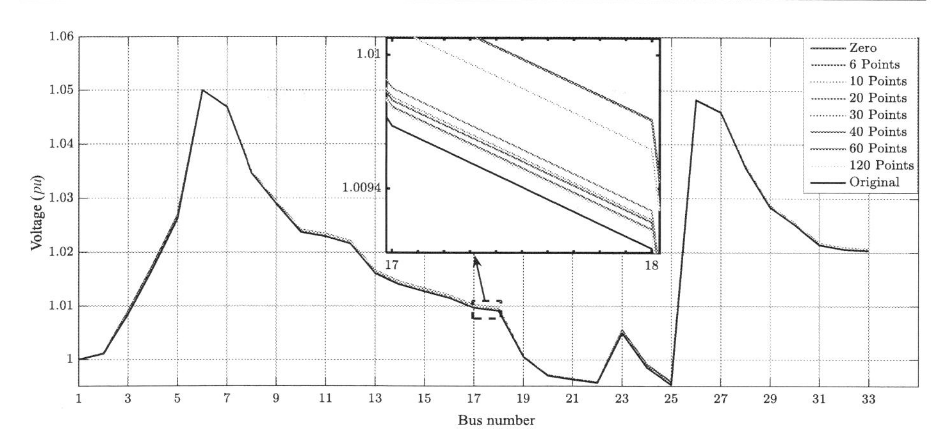

Fig. 23.13 Voltage profile of the test grid for different accuracy of the proposed piecewise linear method when the maximum distributed generation is installed at bus (6).

the proposed model when the maximum DER capacity is installed at bus (6). As it can be seen, increasing the number of points improves the accuracy of the proposed method. Moreover, as it was proven in Section 23.3.1, the voltage profile obtained by the proposed method is always higher than that of the original model. In addition, neglecting the loss quadratic term, which is a specific case of the proposed method, results in a higher voltage bound to the proposed model, so neglecting the loss term and the original model are two boundaries of the proposed method.

It appeared to us that neglecting the loss term (zero approximation) could have enough accuracy to approximate the voltage profile. So, what is the advantage of the proposed method? Fig. 23.14 demonstrates the advantage of our model compared with the zero approximation. As it can be seen, both the proposed method and the neglection of loss term result in the lower bounds to the locational HC. Nonetheless, the lower bound obtained by the proposed method is more accurate than the obtained bound by the zero approximation of the loss term. This is because the sensitivity of the HC is very high to the voltage and a small change in the voltage could result in a tangible change in the objective function, which is the HC.

HC of a system depends on quite a few technical constraints. As it was mentioned in Section 23.1.2, three range for HC can be defined based on each technical issue. The first range determines the capacities that would not cause any violation, regardless of the location of DER. The second range determines the capacities that are acceptable if DER is installed at specific locations, and third range determines the capacities that would result in the violation, regardless of the location of DER. Note that these three range for HC of a feeder can be found by considering the technical constraints altogether or by combining the stacked bar charts of all constraints. Fig. 23.15 shows the barchart diagram of the HC of the test grid for overvoltage and thermal capacity constraints. As it can be seen in this figure, the proposed linear models for branch flow equations and thermal capacity constraints approximate the HC bar charts with high

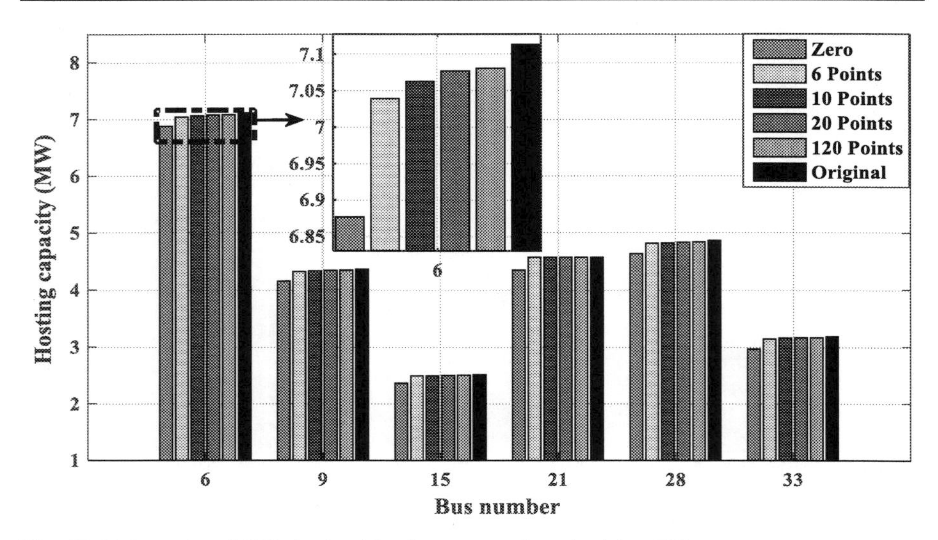

Fig. 23.14 Locational HC obtained by the proposed method for different accuracy.

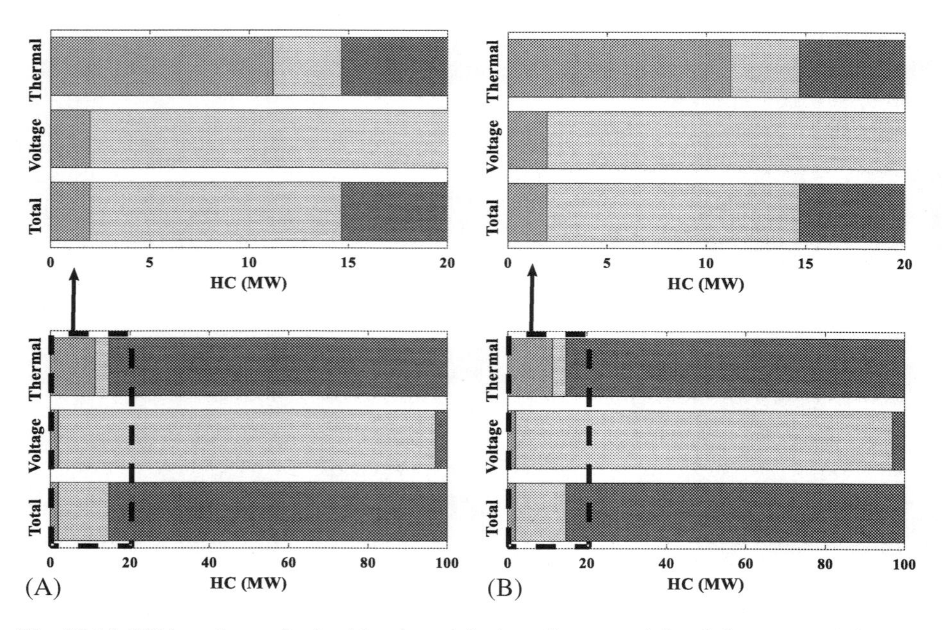

Fig. 23.15 HC bar charts obtained by the original nonlinear model and the proposed linear method: (A) Original nonlinear model and (B) Linear model.

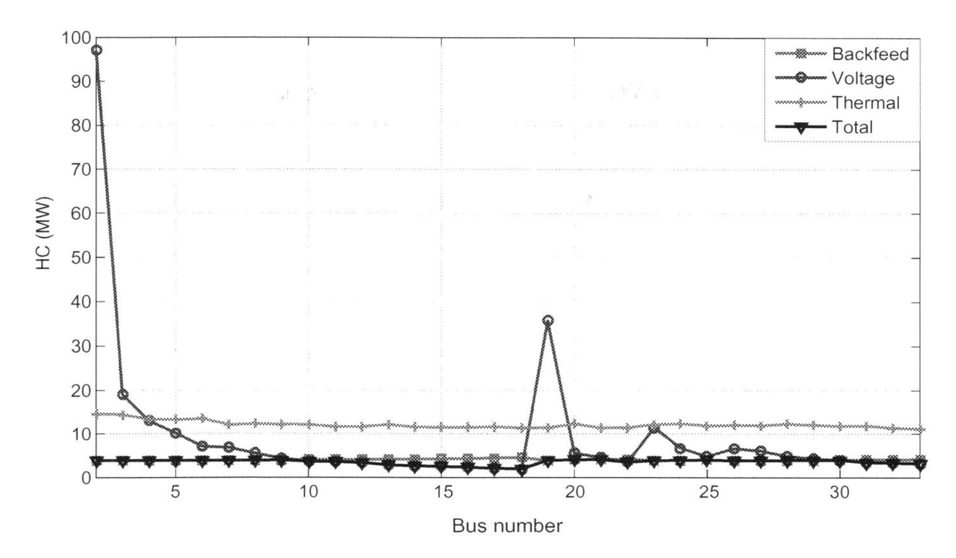

Fig. 23.16 Locational HC based on overvoltage, back-feed, and thermal capacity constraints obtained by the proposed linear method.

accuracy. Moreover, the combination of stacked bar charts of overvoltage and thermal capacity constraints give the same result as applying both of them at the same time.

Another important point to highlight is the importance of the constraints in the HC assessment. Fig. 23.16 demonstrates the HC obtained by the proposed method based on three constraints, namely, overvoltage, thermal, and back-feed constraints. As it can be seen, back-feed put a lower limit on the HC in comparison with the overvoltage and thermal constraints. Another observation is that the thermal constraint is more important than overvoltage constraint at the beginning of the feeder. Performing these evaluation analysis can assist DSOs to understand the available options to increase the HC effectively. For instance, Fig. 23.17 demonstrates the HC stacked bar charts with relaxation of the back-feed constraint. As it can be seen, this relaxation increases the

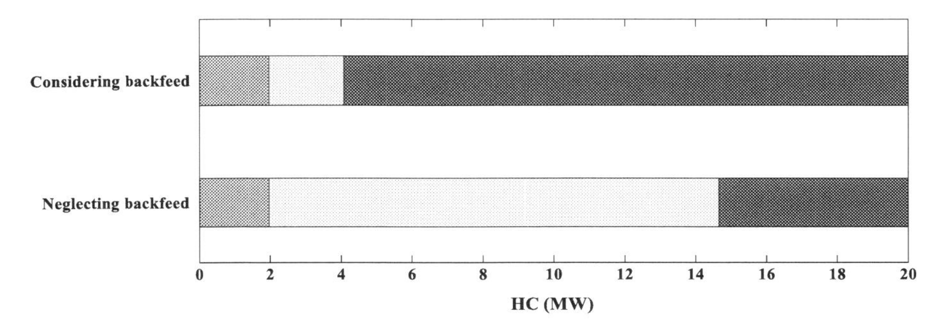

Fig. 23.17 Effect of the back-feed constraint relaxation on the HC.

second range of the HC. Therefore, one available option to increase effectively the HC could be modifying the network to allow bidirectional power flow. In this regard, DSOs may need to readjust the control setting of the voltage regulators or replace the protection relay of circuit breaker.

In the second part of the simulation results, we compare the performance of different linearization methods. The linear methods to be compared are as follows:

A. Linearizing the relaxed power flow with $\acute{\ell}_{ij}^{\text{lin}} = 0\,\forall(i,j)\in\mathcal{B}$ (Ma et al., 2017; Wang et al., 2016).

B. Linear relaxed power flow proposed in Robbins and Domínguez-García (2016): in this model, the $\acute{\ell}_{ij} \approx \acute{\ell}_{ij}^{\text{lin}} := \frac{1}{\upsilon_0}\left(2P_{ij0}P_{ij} + 2Q_{ij0}Q_{ij} - P_{ij0}^2 - Q_{ij0}^2\right)$, in which, P_{ij0} and Q_{ij0} are the line $(i \rightarrow j)$ initial active and reactive power flows.

C. Linearizing the relaxed power flow with $\acute{\ell}_{ij} \approx \acute{\ell}_{ij}^{\text{lin}} := \frac{1}{\upsilon_{j0}}\left(2P_{ij0}P_{ij} + 2Q_{ij0}Q_{ij} - P_{ij0}^2 - Q_{ij0}^2\right)$, in which, P_{ij0} and Q_{ij0} are the line $(i \rightarrow j)$ initial active and reactive power flows and υ_{j0} is the square of the initial voltage magnitude at bus (j).

D. Neglecting the $\acute{\ell}_{ij}$ in Eqs. (23.11), (23.12), and using the linear approximation $\acute{\ell}_{ij} \approx \acute{\ell}_{ij}^{\text{lin}} := \frac{1}{\upsilon_0}\left(2P_{ij0}P_{ij} + 2Q_{ij0}Q_{ij} - P_{ij0}^2 - Q_{ij0}^2\right)$ in Eq. (23.13).

E. Linearizing the relaxed power flow with $\acute{\ell}_{ij} \approx \acute{\ell}_{ij}^{\text{lin}} := \frac{1}{\upsilon_0}\left(P_{ij0}P_{ij} + Q_{ij0}Q_{ij}\right)$, in which, P_{ij0} and Q_{ij0} are the initial distribution line power flows.

F. The traditional linear power flow model: in this model, $\Delta V_{ij} \simeq \frac{r_{ij}P_{ij}^r + x_{ij}Q_{ij}^r}{|V_0|}$.

G. The traditional linear model proposed in Seydali Seyf Abad et al. (2017): in this model, $\Delta V_{ij} \simeq \frac{r_{ij}P_{ij}^r + x_{ij}Q_{ij}^r}{|V_{j0}|}$, in which $|V_{j0}|$ is the initial voltage magnitude of bus (j).

H. The proposed piecewise linear function with four points.

In the following, the performance of these methods in approximating the voltage of the system is assessed. Fig. 23.18 demonstrates the voltage profile of the test grid by using the above-mentioned linear power flow approximations. As it can be seen, the voltage profile obtained by the proposed piecewise linear method (H) and linear approximation (A) is always higher than the original profile. In addition, it can be observed that all these linear methods have reasonable performance in approximating the voltage. So, we need another index to compare the performance of these methods.

Fig. 23.19 shows the HC of the test grid obtained by different linear power flow model. As it can be seen, these linearization methods result in a lower approximation of the HC and the proposed piecewise linear method is even more accurate than other models. In addition, after the proposed piecewise linear approximation, the proposed linear model (D) has the best performance. It is also observed that the performance of other linear models (A, B, C, E, F, and G) depends on the characteristics of the system and each of them could have the highest performance at some locations.

Finally, it should be mentioned that the main reason of linearizing the HC model is building a proper model to be used for modeling the DERs uncertainties. Generally, robust and stochastic optimization are two approaches to deal with uncertain parameters in constraints. However, both approaches lead to computationally intractable

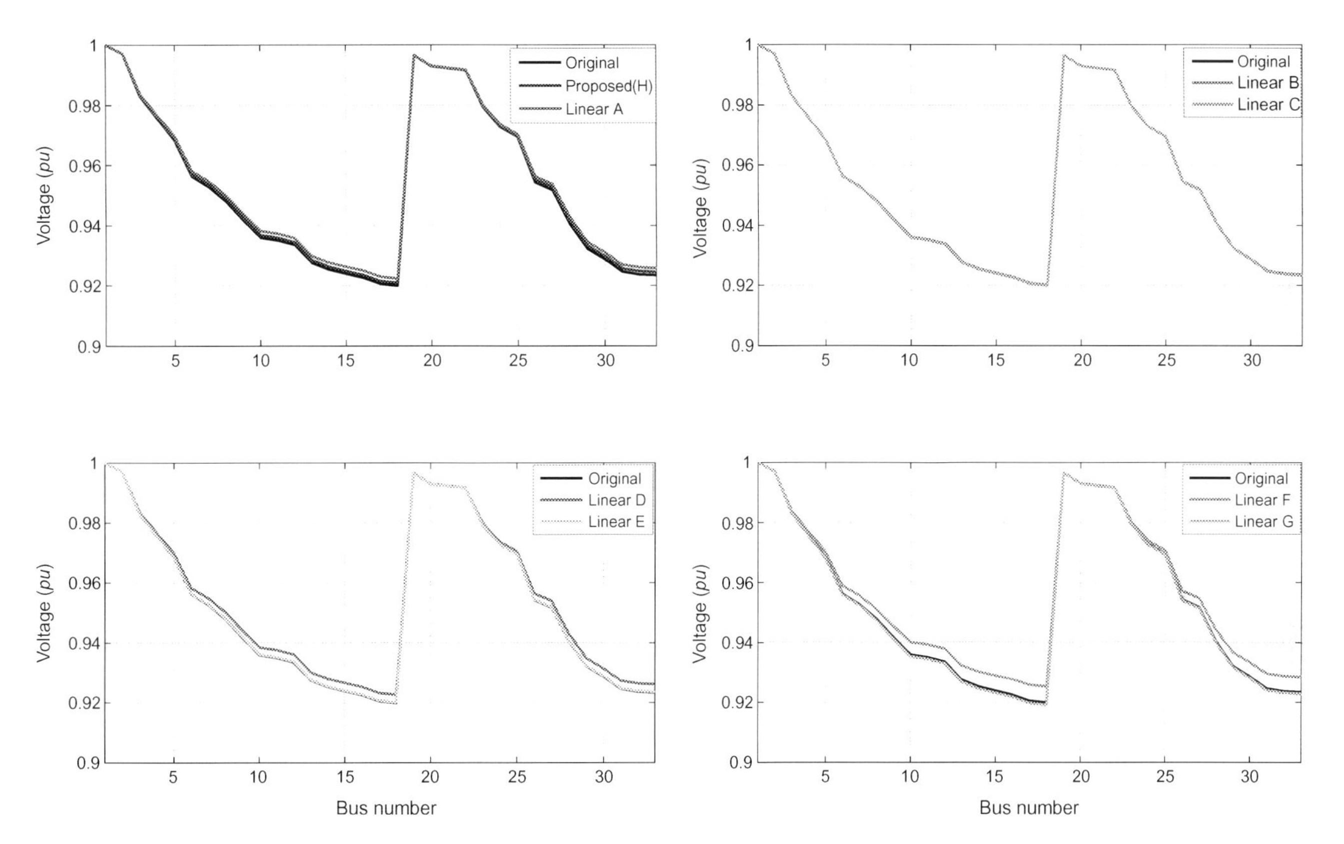

Fig. 23.18 Voltage profile of the test grid obtained by different linear models (A, B, C, D, E, F, G, and H).

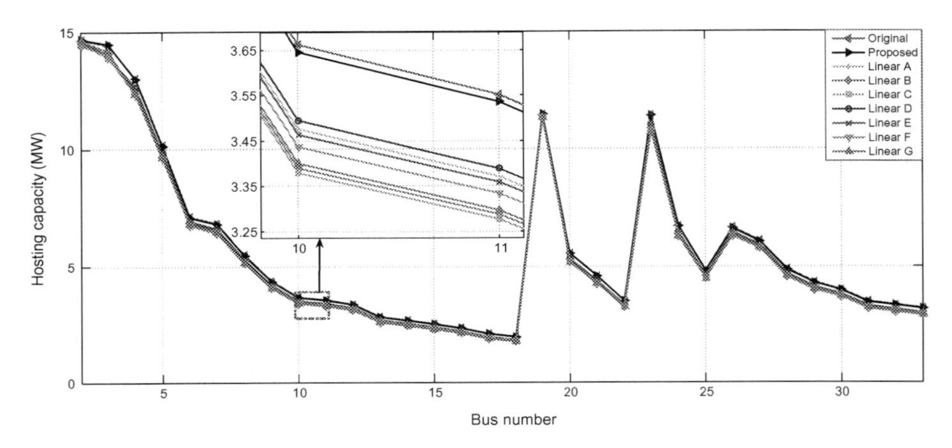

Fig. 23.19 Locational HC of the test grid obtained by different linear models (A, B, C, D, E, F, G, and H (proposed)).

counterparts if one use the original nonlinear HC model. Therefore, using approximations to reduce the complexity of robust and stochastic counterparts is necessary.

23.5 Conclusions

In this chapter, the HC modeling of radial distribution systems is presented. First, it is proved that conic relaxation for the HC problem is not exact. In other words, one cannot solve a convex conic program instead of the original nonconvex model. Then, the conditions under which the linearization of the relaxed branch flow model and the traditional voltage approximation are valid for the HC problem, are identified. It is proved that linearization of the relaxed power flow model is valid if the linear approximation of the branch losses is less than the actual quadratic term. It is also demonstrated that traditional voltage approximation in distribution systems is valid if the voltage at the receiving end of each branch is higher than the voltage drop over the corresponding branch. Further, a linear model for the thermal capacity constraint is proposed. Linearization of the power flow equations and thermal capacity constraints results in a linear HC model. Finally, different linear HC model were implemented on the IEEE 33-bus system to compare the effectiveness and suitability of each one of them. Based on the simulation results and using different linear HC models, the following conclusions can be made in order:

1. If the reverse power flow is considered as another HC constraint, the thermal capacity constraint would not limit the HC of the system. This is due to the fact that considering the reverse power flow constraint limits the HC to the total load of the system, which is tolerable for the conductors.
2. It was observed that if the conductor type is the same for all branches, the thermal capacity constraint is the main limiting factor of the HC for the location that are closer to the substation. However, for the locations that are far from the substation, the main limiting factor is

voltage constraint. In other words, increasing the distance from substation decreases the importance of the thermal capacity constraint and increases the importance of the voltage constraint.

3. The investigated linear models in Section 23.4 result in lower bounds to the HC of the feeder, which is close to actual HC. Moreover, proposed piecewise linear method and the proposed method (D) have the highest performance in the HC calculation.

Acknowledgment

This research is partly supported by the University of Sydney Bridging Funding and the Faculty of IT mid-career researcher support scheme.

References

ABCSE, 2004. Technical Guide for Connection of Renewable Generators to the Local Electricity Network. Australian Business Council for Sustainable Energy.

Adib, R., Murdock, H., Appavou, F., Brown, A., Epp, B., Leidreiter, A., LINS, C., Murdock, H., Musolino, E., Petrichenko, K., 2016. Renewables 2016 Global Status Report. Global Status Report Renewable Energy Policy Network for the 21st Century (REN21). .

Baran, M., Wu, F.F., 1989a. Optimal sizing of capacitors placed on a radial distribution system. IEEE Trans. Power Deliv. 4, 735–743.

Baran, M.E., Wu, F.F., 1989b. Optimal capacitor placement on radial distribution systems. IEEE Trans. Power Deliv. 4, 725–734.

Farivar, M., Low, S.H., 2013. Branch flow model: relaxations and convexification—part I. IEEE Trans. Power Syst. 28, 2554–2564.

Fourer, R., Gay, D., Kernighan, B., 2002. Ampl: A Modeling Language for Mathematical Programming. Duxbury Press.

Gharigh, M.K., Seydali Seyf Abad, M., Nokhbehzaeem, J., Safdarian, A., 2015. In: Optimal sizing of distributed energy storage in distribution systems. Smart Grid Conference (SGC), 2015. IEEE, Tehran, pp. 60–65.

IEEE-SA, 2007. IEEE 1547 Standard for Interconnecting Distributed Resources With Electric Power Systems. IEEE Standards Association, Piscataway.

Ma, Y., Seydali Seyf Abad, M., Azuatalam, D., Verbic, G., Chapman, A., 2017. Impacts of community and distributed energy storage systems on unbalanced low voltage networks. AUPEC 2017, Melbourne.

Papathanassiou, S., Hatziargyriou, N., Anagnostopoulos, P., Aleixo, L., Buchholz, B., Carter-Brown, C., Drossos, N., Enayati, B., Fan, M., Gabrion, V., 2014. Capacity of Distribution Feeders for Hosting DER. CIGRE Working Group C6.24. Technical Report. p. 149.

Robbins, B.A., Domínguez-García, A.D., 2016. Optimal reactive power dispatch for voltage regulation in unbalanced distribution systems. IEEE Trans. Power Syst. 31, 2903–2913.

Seydali Seyf Abad, M., Verbic, G., Chapman, A., Ma, J., 2017. In: A linear method for determining the hosting capacity of radial distribution systems.AUPEC 2017, Melbourne.

Smith, J., Rylander, M., 2012. Stochastic Analysis to Determine Feeder Hosting Capacity for Distributed Solar PV. Electric Power Research Institute, Palo Alto, CA. Technical Report 1026640.

Stefanov, S.M., 2013. Separable Programming: Theory and Methods. Springer Science & Business Media.

Tonkoski, R., Lopes, L.A., El-Fouly, T.H., 2011. Coordinated active power curtailment of grid connected PV inverters for overvoltage prevention. IEEE Trans. Sustain. Energy 2, 139–147.

Wang, S., Chen, S., Ge, L., Wu, L., 2016. Distributed generation hosting capacity evaluation for distribution systems considering the robust optimal operation of oltc and svc. IEEE Trans. Sustain. Energy 7, 1111–1123.

Index

Note: Page numbers followed by *f* indicate figures, *t* indicate tables, and *b* indicate boxes.

D

J

K

N

W

Y

Z

Printed in the United States
By Bookmasters